Cereal Grains

Related Titles

Steamed Breads: Ingredients, Processing and Quality
(978-0-08-100715-0)

Baking Problems Solved, 2e
(978-0-08-100765-5)

Cereal Grains for the Food and Beverage Industries
(978-0-85709-413-1)

Woodhead Publishing Series in Food Science, Technology and Nutrition

Cereal Grains

Assessing and Managing Quality

Second Edition

Editors

Colin Wrigley
QAAFI, University of Queensland, St. Lucia, QLD, Australia

Ian Batey
Sunset Cereal Services, Meadowbank, NSW, Australia

Diane Miskelly
Westcott Consultants, Goulburn, NSW, Australia

ELSEVIER

WP
WOODHEAD PUBLISHING
An imprint of Elsevier • elsevier.com

Woodhead Publishing is an imprint of Elsevier
The Officers' Mess Business Centre, Royston Road, Duxford, CB22 4QH, United Kingdom
50 Hampshire Street, 5th Floor, Cambridge, MA 02139, United States
The Boulevard, Langford Lane, Kidlington, OX5 1GB, United Kingdom

Library of Congress Cataloging-in-Publication Data
A catalog record for this book is available from the Library of Congress

British Library Cataloguing-in-Publication Data
A catalogue record for this book is available from the British Library

ISBN: 978-0-08-100719-8 (print)
ISBN: 978-0-08-100730-3 (online)

For information on all Woodhead publications
visit our website at https://www.elsevier.com/

Publisher: Nikki Levy
Acquisition Editor: Rob Sykes
Editorial Project Manager: Karen Miller
Production Project Manager: Julia Haynes
Designer: Greg Harris

Typeset by Thomson Digital

Contents

Appendix 1 Composition of Grains and Grain Products

Denise L. King, Bhaskarani Jasthi and Janet Pettit

Appendix 2 Equivalence Between Metric, US and Chinese Units for the Grain Industry

Colin Wrigley

List of Contributors

Anton J. Alldrick, Campden BRI, Chipping Campden, United Kingdom

Ian Batey, Sunset Cereal Services, Meadowbank, NSW, Australia

Zoltán Bedő, Agricultural Institute, Centre for Agricultural Research, Hungarian Academy of Sciences, Martonvásár, Hungary

Ferenc Békés, FBFD PTY LTD, North Parramatta, NSW, Australia

Walter Bushuk, Department of Food Science, University of Manitoba, Winnipeg, MB, Canada

Yang Cao, Academy of State Administration of Grain, Beijing, China

Pat J. Collins, Department of Agriculture and Fisheries, Agri-Science Queensland, Brisbane, QLD, Australia

Stephen Delwiche, USDA-ARS, Beltsville Agricultural Research Center, Food Quality Laboratory, Beltsville, MD, United States

Kwaku G. Duodu, Institute for Food, Nutrition and Well-being and Department of Food Science, University of Pretoria, Hatfield, Pretoria, South Africa

David R. Eagling, Plant Biosecurity Cooperative Research Centre, Canberra, ACT, Australia

Michael Edney, Canadian Grain Commission, Grain Research Laboratory, Winnipeg, MB, Canada

Melissa Fitzgerald, School of Agriculture and Food Science, University of Queensland, St. Lucia, QLD, Australia

Michael Gooding, Formerly University of Reading, Berkshire, United Kingdom

Zoltán Győri, Institute of Human Nutrition, University of Debrecen, Debrecen, Hungary

Ray Hare, Durum Wheat Improvement, Tamworth Agricultural Institute, New South Wales Department of Primary Industries, Tamworth, NSW, Australia

Crispin A. Howitt, CSIRO Agriculture and Food, Canberra, ACT, Australia

Marta S. Izydorczyk, Canadian Grain Commission, Grain Research Laboratory, Winnipeg, MB, Canada

Bhaskarani Jasthi, Nutrition Coordinating Center, Epidemiology and Community Health, University of Minnesota, Minneapolis, MN, United States

Denise L. King, Nutrition Coordinating Center, Epidemiology and Community Health, University of Minnesota, Minneapolis, MN, United States

László Láng, Agricultural Institute, Centre for Agricultural Research, Hungarian Academy of Sciences, Martonvásár, Hungary

Fujun Li, Academy of State Administration of Grain, Beijing, China

Bill Magee, Plant Biosecurity Cooperative Research Centre, Canberra, ACT, Australia

Diane Miskelly, Westcott Consultants, Goulburn, NSW, Australia

Janet Pettit, Nutrition Coordinating Center, Epidemiology and Community Health, University of Minnesota, Minneapolis, MN, United States

Mariann Rakszegi, Agricultural Institute, Centre for Agricultural Research, Hungarian Academy of Sciences, Martonvásár, Hungary

Yong Lin Ren, School of Biological Science and Biotechnology, Murdoch University, Perth, WA, Australia

Regine Schoenlechner, Institute of Food Technology, Department of Food Science and Technology, University of Natural Resources & Technology, Vienna, Austria

Dai Suter, Pymble, NSW, Australia

John R.N. Taylor, Institute for Food, Nutrition and Well-being and Department of Food Science, University of Pretoria, Hatfield, Pretoria, South Africa

Sándor Tömösközi, Department of Applied Biotechnology and Food Sciences, BUTE, Budapest, Hungary

Surjani Uthayakumaran, University of Sydney, Sydney, NSW, Australia

Colin Wrigley, QAAFI, University of Queensland, St. Lucia, QLD, Australia

Pamela Zwer, South Australian Research & Development Institute, SARDI, Adelaide, SA, Australia

Woodhead Publishing Series in Food Science, Technology and Nutrition

Chilled foods: A comprehensive guide
Edited by C. Dennis and M. Stringer

Yoghurt: Science and technology
A. Y. Tamime and R. K. Robinson

Food processing technology: Principles and practice
P. J. Fellows

Bender's dictionary of nutrition and food technology Sixth edition
D. A. Bender

Determination of veterinary residues in food
Edited by N. T. Crosby

Food contaminants: Sources and surveillance
Edited by C. Creaser and R. Purchase

Nitrates and nitrites in food and water
Edited by M. J. Hill

Pesticide chemistry and bioscience: The food-environment challenge
Edited by G. T. Brooks and T. Roberts

Pesticides: Developments, impacts and controls
Edited by G. A. Best and A. D. Ruthven

Dietary fibre: Chemical and biological aspects
Edited by D. A. T. Southgate, K. W. Waldron, I. T. Johnson and G. R. Fenwick

Vitamins and minerals in health and nutrition
M. Tolonen

Technology of biscuits, crackers and cookies Second edition
D. Manley

Instrumentation and sensors for the food industry
Edited by E. Kress-Rogers

Food and cancer prevention: Chemical and biological aspects
Edited by K. W. Waldron, I. T. Johnson and G. R. Fenwick

Food colloids: Proteins, lipids and polysaccharides
Edited by E. Dickinson and B. Bergenstahl

Food emulsions and foams
Edited by E. Dickinson

Maillard reactions in chemistry, food and health
Edited by T. P. Labuza, V. Monnier, J. Baynes and J. O'Brien

The Maillard reaction in foods and medicine
Edited by J. O'Brien, H. E. Nursten, M. J. Crabbe and J. M. Ames

Encapsulation and controlled release
Edited by D. R. Karsa and R. A. Stephenson

Flavours and fragrances
Edited by A. D. Swift

Feta and related cheeses
Edited by A. Y. Tamime and R. K. Robinson

Biochemistry of milk products
Edited by A. T. Andrews and J. R. Varley

Physical properties of foods and food processing systems
M. J. Lewis

Food irradiation: A reference guide
V. M. Wilkinson and G. Gould

Kent's technology of cereals: An introduction for students of food science and agriculture Fourth edition
N. L. Kent and A. D. Evers

Biosensors for food analysis
Edited by A. O. Scott

Separation processes in the food and biotechnology industries: Principles and applications
Edited by A. S. Grandison and M. J. Lewis

Handbook of indices of food quality and authenticity
R. S. Singhal, P. K. Kulkarni and D. V. Rege

Principles and practices for the safe processing of foods
D. A. Shapton and N. F. Shapton

Biscuit, cookie and cracker manufacturing manuals Volume 1: Ingredients
D. Manley

Biscuit, cookie and cracker manufacturing manuals Volume 2: Biscuit doughs
D. Manley

Biscuit, cookie and cracker manufacturing manuals Volume 3: Biscuit dough piece forming
D. Manley

Biscuit, cookie and cracker manufacturing manuals Volume 4: Baking and cooling of biscuits
D. Manley

Biscuit, cookie and cracker manufacturing manuals Volume 5: Secondary processing in biscuit manufacturing
D. Manley

Biscuit, cookie and cracker manufacturing manuals Volume 6: Biscuit packaging and storage
D. Manley

Practical dehydration Second edition
M. Greensmith

Lawrie's meat science Sixth edition
R. A. Lawrie

Yoghurt: Science and technology Second edition
A. Y. Tamime and R. K. Robinson

New ingredients in food processing: Biochemistry and agriculture
G. Linden and D. Lorient

Benders' dictionary of nutrition and food technology Seventh edition
D. A. Bender and A. E. Bender

Technology of biscuits, crackers and cookies Third edition
D. Manley

Food processing technology: Principles and practice Second edition
P. J. Fellows

Managing frozen foods
Edited by C. J. Kennedy

Handbook of hydrocolloids
Edited by G. O. Phillips and P. A. Williams

Food labelling
Edited by J. R. Blanchfield

Cereal biotechnology
Edited by P. C. Morris and J. H. Bryce

Food intolerance and the food industry
Edited by T. Dean

The stability and shelf-life of food
Edited by D. Kilcast and P. Subramaniam

Functional foods: Concept to product
Edited by G. R. Gibson and C. M. Williams

Chilled foods: A comprehensive guide Second edition
Edited by M. Stringer and C. Dennis

HACCP in the meat industry
Edited by M. Brown

Biscuit, cracker and cookie recipes for the food industry
D. Manley

Cereals processing technology
Edited by G. Owens

Baking problems solved
S. P. Cauvain and L. S. Young

Thermal technologies in food processing
Edited by P. Richardson

Frying: Improving quality
Edited by J. B. Rossell

Food chemical safety Volume 1: Contaminants
Edited by D. Watson

Making the most of HACCP: Learning from others' experience
Edited by T. Mayes and S. Mortimore

Food process modelling
Edited by L. M. M. Tijskens, M. L. A. T. M. Hertog and B. M. Nicolaï

EU food law: A practical guide
Edited by K. Goodburn

Extrusion cooking: Technologies and applications
Edited by R. Guy

Auditing in the food industry: From safety and quality to environmental and other audits
Edited by M. Dillon and C. Griffith

Handbook of herbs and spices Volume 1
Edited by K. V. Peter

Food product development: Maximising success
M. Earle, R. Earle and A. Anderson

Instrumentation and sensors for the food industry Second edition
Edited by E. Kress-Rogers and C. J. B. Brimelow

Food chemical safety Volume 2: Additives
Edited by D. Watson

Fruit and vegetable biotechnology
Edited by V. Valpuesta

Foodborne pathogens: Hazards, risk analysis and control
Edited by C. de W. Blackburn and P. J. McClure

Meat refrigeration
S. J. James and C. James

Lockhart and Wiseman's crop husbandry Eighth edition
H. J. S. Finch, A. M. Samuel and G. P. F. Lane

Safety and quality issues in fish processing
Edited by H. A. Bremner

Minimal processing technologies in the food industries
Edited by T. Ohlsson and N. Bengtsson

Fruit and vegetable processing: Improving quality
Edited by W. Jongen

The nutrition handbook for food processors
Edited by C. J. K. Henry and C. Chapman

Colour in food: Improving quality
Edited by D. MacDougall

Meat processing: Improving quality
Edited by J. P. Kerry, J. F. Kerry and D. A. Ledward

Microbiological risk assessment in food processing
Edited by M. Brown and M. Stringer

Performance functional foods
Edited by D. Watson

Functional dairy products Volume 1
Edited by T. Mattila-Sandholm and M. Saarela

Taints and off-flavours in foods
Edited by B. Baigrie

Yeasts in food
Edited by T. Boekhout and V. Robert

Phytochemical functional foods
Edited by I. T. Johnson and G. Williamson

Novel food packaging techniques
Edited by R. Ahvenainen

Detecting pathogens in food
Edited by T. A. McMeekin

Natural antimicrobials for the minimal processing of foods
Edited by S. Roller

Texture in food Volume 1: Semi-solid foods
Edited by B. M. McKenna

Dairy processing: Improving quality
Edited by G. Smit

Hygiene in food processing: Principles and practice
Edited by H. L. M. Lelieveld, M. A. Mostert, B. White and J. Holah

Rapid and on-line instrumentation for food quality assurance
Edited by I. Tothill

Sausage manufacture: Principles and practice
E. Essien

Environmentally-friendly food processing
Edited by B. Mattsson and U. Sonesson

Bread making: Improving quality
Edited by S. P. Cauvain

Food preservation techniques
Edited by P. Zeuthen and L. Bøgh-Sørensen

Food authenticity and traceability
Edited by M. Lees

Analytical methods for food additives
R. Wood, L. Foster, A. Damant and P. Key

Handbook of herbs and spices Volume 2
Edited by K. V. Peter

Texture in food Volume 2: Solid foods
Edited by D. Kilcast

Proteins in food processing
Edited by R. Yada

Detecting foreign bodies in food
Edited by M. Edwards

Understanding and measuring the shelf-life of food
Edited by R. Steele

Poultry meat processing and quality
Edited by G. Mead

Functional foods, ageing and degenerative disease
Edited by C. Remacle and B. Reusens

Mycotoxins in food: Detection and control
Edited by N. Magan and M. Olsen

Improving the thermal processing of foods
Edited by P. Richardson

Pesticide, veterinary and other residues in food
Edited by D. Watson

Starch in food: Structure, functions and applications
Edited by A.-C. Eliasson

Functional foods, cardiovascular disease and diabetes
Edited by A. Arnoldi

Brewing: Science and practice
D. E. Briggs, P. A. Brookes, R. Stevens and C. A. Boulton

Using cereal science and technology for the benefit of consumers: Proceedings of the 12P^{thP} International ICC Cereal and Bread Congress, 24 – 26P^{thP} May, 2004, Harrogate, UK
Edited by S. P. Cauvain, L. S. Young and S. Salmon

Improving the safety of fresh meat
Edited by J. Sofos

Understanding pathogen behaviour: Virulence, stress response and resistance
Edited by M. Griffiths

The microwave processing of foods
Edited by H. Schubert and M. Regier

Food safety control in the poultry industry
Edited by G. Mead

Improving the safety of fresh fruit and vegetables
Edited by W. Jongen

Food, diet and obesity
Edited by D. Mela

Handbook of hygiene control in the food industry
Edited by H. L. M. Lelieveld, M. A. Mostert and J. Holah

Detecting allergens in food
Edited by S. Koppelman and S. Hefle

Improving the fat content of foods
Edited by C. Williams and J. Buttriss

Improving traceability in food processing and distribution
Edited by I. Smith and A. Furness

Flavour in food
Edited by A. Voilley and P. Etievant

The Chorleywood bread process
S. P. Cauvain and L. S. Young

Food spoilage microorganisms
Edited by C. de W. Blackburn

Emerging foodborne pathogens
Edited by Y. Motarjemi and M. Adams

Benders' dictionary of nutrition and food technology Eighth edition
D. A. Bender

Optimising sweet taste in foods
Edited by W. J. Spillane

Brewing: New technologies
Edited by C. Bamforth

Handbook of herbs and spices Volume 3
Edited by K. V. Peter

Lawrie's meat science Seventh edition
R. A. Lawrie in collaboration with D. A. Ledward

Modifying lipids for use in food
Edited by F. Gunstone

Meat products handbook: Practical science and technology
G. Feiner

Food consumption and disease risk: Consumer–pathogen interactions
Edited by M. Potter

Acrylamide and other hazardous compounds in heat-treated foods
Edited by K. Skog and J. Alexander

Managing allergens in food
Edited by C. Mills, H. Wichers and K. Hoffman-Sommergruber

Microbiological analysis of red meat, poultry and eggs
Edited by G. Mead

Maximising the value of marine by-products
Edited by F. Shahidi

Chemical migration and food contact materials
Edited by K. Barnes, R. Sinclair and D. Watson

Understanding consumers of food products
Edited by L. Frewer and H. van Trijp

Reducing salt in foods: Practical strategies
Edited by D. Kilcast and F. Angus

Modelling microorganisms in food
Edited by S. Brul, S. Van Gerwen and M. Zwietering

Tamime and Robinson's Yoghurt: Science and technology Third edition
A. Y. Tamime and R. K. Robinson

Handbook of waste management and co-product recovery in food processing Volume 1
Edited by K. W. Waldron

Improving the flavour of cheese
Edited by B. Weimer

Novel food ingredients for weight control
Edited by C. J. K. Henry

Consumer-led food product development
Edited by H. MacFie

Functional dairy products Volume 2
Edited by M. Saarela

Modifying flavour in food
Edited by A. J. Taylor and J. Hort

Cheese problems solved
Edited by P. L. H. McSweeney

Handbook of organic food safety and quality
Edited by J. Cooper, C. Leifert and U. Niggli

Understanding and controlling the microstructure of complex foods
Edited by D. J. McClements

Novel enzyme technology for food applications
Edited by R. Rastall

Food preservation by pulsed electric fields: From research to application
Edited by H. L. M. Lelieveld and S. W. H. de Haan

Technology of functional cereal products
Edited by B. R. Hamaker

Case studies in food product development
Edited by M. Earle and R. Earle

Delivery and controlled release of bioactives in foods and nutraceuticals
Edited by N. Garti

Fruit and vegetable flavour: Recent advances and future prospects
Edited by B. Brückner and S. G. Wyllie

Food fortification and supplementation: Technological, safety and regulatory aspects
Edited by P. Berry Ottaway

Improving the health-promoting properties of fruit and vegetable products
Edited by F. A. Tomás-Barberán and M. I. Gil

Improving seafood products for the consumer
Edited by T. Børresen

In-pack processed foods: Improving quality
Edited by P. Richardson

Handbook of water and energy management in food processing
Edited by J. Klemeš, R.. Smith and J.-K. Kim

Environmentally compatible food packaging
Edited by E. Chiellini

Improving farmed fish quality and safety
Edited by Ø. Lie

Carbohydrate-active enzymes
Edited by K.-H. Park

Chilled foods: A comprehensive guide Third edition
Edited by M. Brown

Food for the ageing population
Edited by M. M. Raats, C. P. G. M. de Groot and W. A Van Staveren

Improving the sensory and nutritional quality of fresh meat
Edited by J. P. Kerry and D. A. Ledward

Shellfish safety and quality
Edited by S. E. Shumway and G. E. Rodrick

Functional and speciality beverage technology
Edited by P. Paquin

Functional foods: Principles and technology
M. Guo

Endocrine-disrupting chemicals in food
Edited by I. Shaw

Meals in science and practice: Interdisciplinary research and business applications
Edited by H. L. Meiselman

Food constituents and oral health: Current status and future prospects
Edited by M. Wilson

Handbook of hydrocolloids Second edition
Edited by G. O. Phillips and P. A. Williams

Food processing technology: Principles and practice Third edition
P. J. Fellows

Science and technology of enrobed and filled chocolate, confectionery and bakery products
Edited by G. Talbot

Foodborne pathogens: Hazards, risk analysis and control Second edition
Edited by C. de W. Blackburn and P. J. McClure

Designing functional foods: Measuring and controlling food structure breakdown and absorption
Edited by D. J. McClements and E. A. Decker

New technologies in aquaculture: Improving production efficiency, quality and environmental management
Edited by G. Burnell and G. Allan

More baking problems solved
S. P. Cauvain and L. S. Young

Soft drink and fruit juice problems solved
P. Ashurst and R. Hargitt

Biofilms in the food and beverage industries
Edited by P. M. Fratamico, B. A. Annous and N. W. Gunther

Dairy-derived ingredients: Food and neutraceutical uses
Edited by M. Corredig

Handbook of waste management and co-product recovery in food processing Volume 2
Edited by K. W. Waldron

Innovations in food labelling
Edited by J. Albert

Delivering performance in food supply chains
Edited by C. Mena and G. Stevens

Chemical deterioration and physical instability of food and beverages
Edited by L. H. Skibsted, J. Risbo and M. L. Andersen

Managing wine quality Volume 1: Viticulture and wine quality
Edited by A. G. Reynolds

Improving the safety and quality of milk Volume 1: Milk production and processing
Edited by M. Griffiths

Improving the safety and quality of milk Volume 2: Improving quality in milk products
Edited by M. Griffiths

Cereal grains: Assessing and managing quality
Edited by C. Wrigley and I. Batey

Sensory analysis for food and beverage quality control: A practical guide
Edited by D. Kilcast

Managing wine quality Volume 2: Oenology and wine quality
Edited by A. G. Reynolds

Winemaking problems solved
Edited by C. E. Butzke

Environmental assessment and management in the food industry
Edited by U. Sonesson, J. Berlin and F. Ziegler

Consumer-driven innovation in food and personal care products
Edited by S. R. Jaeger and H. MacFie

Tracing pathogens in the food chain
Edited by S. Brul, P. M. Fratamico and T. A. McMeekin

Case studies in novel food processing technologies: Innovations in processing, packaging, and predictive modelling
Edited by C. J. Doona, K. Kustin and F. E. Feeherry

Freeze-drying of pharmaceutical and food products
T.-C. Hua, B.-L. Liu and H. Zhang

Oxidation in foods and beverages and antioxidant applications Volume 1: Understanding mechanisms of oxidation and antioxidant activity
Edited by E. A. Decker, R. J. Elias and D. J. McClements

Oxidation in foods and beverages and antioxidant applications Volume 2: Management in different industry sectors
Edited by E. A. Decker, R. J. Elias and D. J. McClements

Protective cultures, antimicrobial metabolites and bacteriophages for food and beverage biopreservation
Edited by C. Lacroix

Separation, extraction and concentration processes in the food, beverage and nutraceutical industries
Edited by S. S. H. Rizvi

Determining mycotoxins and mycotoxigenic fungi in food and feed
Edited by S. De Saeger

Developing children's food products
Edited by D. Kilcast and F. Angus

Functional foods: Concept to product Second edition
Edited by M. Saarela

Postharvest biology and technology of tropical and subtropical fruits Volume 1: Fundamental issues
Edited by E. M. Yahia

Postharvest biology and technology of tropical and subtropical fruits Volume 2: Açai to citrus
Edited by E. M. Yahia

Postharvest biology and technology of tropical and subtropical fruits Volume 3: Cocona to mango
Edited by E. M. Yahia

Postharvest biology and technology of tropical and subtropical fruits Volume 4: Mangosteen to white sapote
Edited by E. M. Yahia

Food and beverage stability and shelf life
Edited by D. Kilcast and P. Subramaniam

Processed Meats: Improving safety, nutrition and quality
Edited by J. P. Kerry and J. F. Kerry

Food chain integrity: A holistic approach to food traceability, safety, quality and authenticity
Edited by J. Hoorfar, K. Jordan, F. Butler and R. Prugger

Improving the safety and quality of eggs and egg products Volume 1
Edited by Y. Nys, M. Bain and F. Van Immerseel

Improving the safety and quality of eggs and egg products Volume 2
Edited by F. Van Immerseel, Y. Nys and M. Bain

Animal feed contamination: Effects on livestock and food safety
Edited by J. Fink-Gremmels

Hygienic design of food factories
Edited by J. Holah and H. L. M. Lelieveld

Manley's technology of biscuits, crackers and cookies Fourth edition
Edited by D. Manley

Nanotechnology in the food, beverage and nutraceutical industries
Edited by Q. Huang

Rice quality: A guide to rice properties and analysis
K. R. Bhattacharya

Advances in meat, poultry and seafood packaging
Edited by J. P. Kerry

Reducing saturated fats in foods
Edited by G. Talbot

Handbook of food proteins
Edited by G. O. Phillips and P. A. Williams

Lifetime nutritional influences on cognition, behaviour and psychiatric illness
Edited by D. Benton

Food machinery for the production of cereal foods, snack foods and confectionery
L.-M. Cheng

Alcoholic beverages: Sensory evaluation and consumer research
Edited by J. Piggott

Extrusion problems solved: Food, pet food and feed
M. N. Riaz and G. J. Rokey

Handbook of herbs and spices Second edition Volume 1
Edited by K. V. Peter

Handbook of herbs and spices Second edition Volume 2
Edited by K. V. Peter

Breadmaking: Improving quality Second edition
Edited by S. P. Cauvain

Emerging food packaging technologies: Principles and practice
Edited by K. L. Yam and D. S. Lee

Infectious disease in aquaculture: Prevention and control
Edited by B. Austin

Diet, immunity and inflammation
Edited by P. C. Calder and P. Yaqoob

Natural food additives, ingredients and flavourings
Edited by D. Baines and R. Seal

Microbial decontamination in the food industry: Novel methods and applications
Edited by A. Demirci and M.O. Ngadi

Chemical contaminants and residues in foods
Edited by D. Schrenk

Robotics and automation in the food industry: Current and future technologies
Edited by D. G. Caldwell

Fibre-rich and wholegrain foods: Improving quality
Edited by J. A. Delcour and K. Poutanen

Computer vision technology in the food and beverage industries
Edited by D.-W. Sun

Encapsulation technologies and delivery systems for food ingredients and nutraceuticals
Edited by N. Garti and D. J. McClements

Case studies in food safety and authenticity
Edited by J. Hoorfar

Heat treatment for insect control: Developments and applications
D. Hammond

Advances in aquaculture hatchery technology
Edited by G. Allan and G. Burnell

Open innovation in the food and beverage industry
Edited by M. Garcia Martinez

Trends in packaging of food, beverages and other fast-moving consumer goods (FMCG)
Edited by N. Farmer

New analytical approaches for verifying the origin of food
Edited by P. Brereton

Microbial production of food ingredients, enzymes and nutraceuticals
Edited by B. McNeil, D. Archer, I. Giavasis and L. Harvey

Persistent organic pollutants and toxic metals in foods
Edited by M. Rose and A. Fernandes

Cereal grains for the food and beverage industries
E. Arendt and E. Zannini

Viruses in food and water: Risks, surveillance and control
Edited by N. Cook

Improving the safety and quality of nuts
Edited by L. J. Harris

Metabolomics in food and nutrition
Edited by B. C. Weimer and C. Slupsky

Food enrichment with omega-3 fatty acids
Edited by C. Jacobsen, N. S. Nielsen, A. F. Horn and A.-D. M. Sørensen

Instrumental assessment of food sensory quality: A practical guide
Edited by D. Kilcast

Food microstructures: Microscopy, measurement and modelling
Edited by V. J. Morris and K. Groves

Handbook of food powders: Processes and properties
Edited by B. R. Bhandari, N. Bansal, M. Zhang and P. Schuck

Functional ingredients from algae for foods and nutraceuticals
Edited by H. Domínguez

Satiation, satiety and the control of food intake: Theory and practice
Edited by J. E. Blundell and F. Bellisle

Hygiene in food processing: Principles and practice Second edition
Edited by H. L. M. Lelieveld, J. Holah and D. Napper

Advances in microbial food safety Volume 1
Edited by J. Sofos

Global safety of fresh produce: A handbook of best practice, innovative commercial solutions and case studies
Edited by J. Hoorfar

Human milk biochemistry and infant formula manufacturing technology
Edited by M. Guo

High throughput screening for food safety assessment: Biosensor technologies, hyperspectral imaging and practical applications
Edited by A. K. Bhunia, M. S. Kim and C. R. Taitt

Foods, nutrients and food ingredients with authorised EU health claims: Volume 1
Edited by M. J. Sadler

Handbook of food allergen detection and control
Edited by S. Flanagan

Advances in fermented foods and beverages: Improving quality, technologies and health benefits
Edited by W. Holzapfel

Metabolomics as a tool in nutrition research
Edited by J.-L. Sébédio and L. Brennan

Dietary supplements: Safety, efficacy and quality
Edited by K. Berginc and S. Kreft

Grapevine breeding programs for the wine industry
Edited by A. G. Reynolds

Handbook of antimicrobials for food safety and quality
Edited by T. M. Taylor

Managing and preventing obesity: Behavioural factors and dietary interventions
Edited by T. P. Gill

Electron beam pasteurization and complementary food processing technologies
Edited by S. D. Pillai and S. Shayanfar

Advances in food and beverage labelling: Information and regulations
Edited by P. Berryman

Flavour development, analysis and perception in food and beverages
Edited by J. K. Parker, S. Elmore and L. Methven

Rapid sensory profiling techniques and related methods: Applications in new product development and consumer research
Edited by J. Delarue, J. B. Lawlor and M. Rogeaux

Advances in microbial food safety: Volume 2
Edited by J. Sofos

Handbook of antioxidants for food preservation
Edited by F. Shahidi

Lockhart and Wiseman's crop husbandry including grassland: Ninth edition
H. J. S. Finch, A. M. Samuel and G. P. F. Lane

Global legislation for food contact materials
Edited by J. S. Baughan

Colour additives for food and beverages
Edited by M. Scotter

A complete course in canning and related processes 14th Edition: Volume 1
Revised by S. Featherstone

A complete course in canning and related processes 14th Edition: Volume 2
Revised by S. Featherstone

A complete course in canning and related processes 14th Edition: Volume 3
Revised by S. Featherstone

Modifying food texture: Volume 1: Novel ingredients and processing techniques
Edited by J. Chen and A. Rosenthal

Modifying food texture: Volume 2: Sensory analysis, consumer requirements and preferences
Edited by J. Chen and A. Rosenthal

Modeling food processing operations
Edited by S. Bakalis, K. Knoerzer and P. J. Fryer

Foods, nutrients and food ingredients with authorised EU health claims Volume 2
Edited by M. J. Sadler

Feed and feeding practices in aquaculture
Edited by D. Allen Davis

Foodborne parasites in the food supply web: Occurrence and control
Edited by A. Gajadhar

Brewing microbiology: design and technology applications for spoilage management, sensory quality and waste valorisation
Edited by A. E. Hill

Specialty oils and fats in food and nutrition: Properties, processing and applications
Edited by G. Talbot

Improving and tailoring enzymes for food quality and functionality
Edited by R. Yada

Emerging Technologies for Promoting Food Security: Overcoming the World Food Crisis
Edited by C. Madramootoo

Innovation and Future Trends in Food Manufacturing and Supply Chain Technologies
Edited by C. E. Leadley

Functional Dietary Lipids: Food Formulation, Consumer Issues and Innovation for Health
Edited by T. Sanders

Handbook on Natural Pigments in Food and Beverages: Industrial Applications for Improving Color
Edited by R. Carle and R. M. Schweiggert

Integrating the Packaging and Product Experience in Food and Beverages: A Road-Map to Consumer Satisfaction
Edited by P. Burgess

The Stability and Shelf Life of Food Second edition
Edited by P. Subramaniam and P. Wareing

Multisensory Flavor Perception: From Fundamental Neuroscience through to the Marketplace
Edited by B. Piqueras-Fiszman and C. Spence

Flavor: From food to behaviors, wellbeing and health
Edited by A. Voilley, C. Salles, E. Guichard and P. Etiévant

Developing Food Products for Consumers with Specific Dietary Needs
Edited by W. Morley and S. Osborn

Advances in Food Traceability Techniques and Technologies: Improving Quality Throughout the Food Chain
Edited by M. Espiñeira and J. F. Santaclara

Innovative Food Processing Technologies: Extraction, Separation, Component Modification and Process Intensification
Edited by K. Knoerzer, P. Juliano, and G. Smithers

Steamed Breads: Ingredients, Process and Quality
S. Huang and D. Miskelly

Handbook of Hygiene Control in the Food Industry, Second Edition
Edited by H. Lelieveld, D. Gabrić, and J. Holah

Handbook for Sensory and Consumer-Driven New Product Development: Innovative Technologies for the Food and Beverage Industry
M. O'Sullivan

Early Nutrition and Long-Term Health: Mechanisms, Consequences and Opportunities
Edited by J. M. Saavedra and A. Dattilo

Baking Problems Solved, Second Edition
S. P. Cauvain

Non-Equilibrium States and Glass Transitions in Food
Edited by B. Bhandari, Y. Roos

New Aspects of Meat Quality: From Genes to Ethics
Edited by P. Purslow

Advances in Food Authenticity Testing
Edited by G. Downey

Food for the Ageing Population, Second Edition
Edited by M. Raats, L. De Groot, D. van Asselt

Microbiological Quality of Food
Edited by A. Bevilacqua, M. R. Corbo and M. Sinigaglia

Swainson's Handbook of Technical and Quality Management for the Food Manufacturing Sector
Edited by M. Swainson

Food Protection and Security
Edited by S. Kennedy

Cereal Grains: Assessing and Managing Quality 2e
Edited by C. Wrigley, I. Batey, D. Miskelly

Preface to the Second Edition

Grain-based foods are essential to our nutrient requirements, supplying a major part of our diet. Thus the assessment and management of grain quality is critical for our health and well-being.

Grains have always have been a dietary essential, even way back into the early days of pre-history when grain-growing was the reason for people to settle down in one place because they could sow seed and harvest grain. The consequence was civilisation—the opportunity given by grain-based agriculture for people to think beyond that day's food supply.

In those days of subsistence farming, the harvested and stored grain was a familiar sight because milling and baking took place at home. In early English usage, the word 'corn' rather than 'grain' was used to refer to any of the grain species. For example, an early English measuring unit was the 'barley corn'. Imagine literally lining up 30 grains of barley, end-to-end, to measure about 25 cm (10 in.) (See Appendix 2).

This generic use of 'corn' continues even today in Britain, where the English word 'corn' may mean any type of 'grain' especially 'wheat'. But if we had entitled this book *Cereal Corns* there would be international confusion, because the word 'corn' crossed the Atlantic to be exclusively applied to the New World's novel cereal, *Zea mays*, the cereal species that is known in many English-speaking countries as 'corn'. So what title should we have given to Chapter 11? We called it 'Corn', but the specific word 'maize' often appears in this and other chapters.

Maize. The word 'maize' was originally Spanish, also coming from the botanical species name (*mays*), as well as from the word 'mahiz' in the Arawak language of Haiti. The settlers called it 'Indian corn'. The name in French-speaking regions is *maiz* or *blé d'Espagne* or *blé de Turquie* or *blé d'Inde*. The name in Italy is *granturco*, probably transliterating into 'Turkish grain'. From these origins, the grain species of maize (previously unrecognised by Europe) has grown in production volume to out-perform all other cereal species, with annual production exceeding a billion tonnes.

Wheat. Yet it is wheat (*Triticum* species) that attracts the greatest attention from the grain-quality viewpoint, because of its range of quality attributes and its unique food uses. Thus it is wheat that heads the list of species-specific chapters in this book, followed by the species most closely related to wheat. But wheat has experienced consumer backlash in recent decades due to the gluten-free trend. It is thus appropriate that some of the new chapters and text relate to this topic.

Rice. Rice (*Oryza sativa*) is a basic food source for about two-thirds of the world. Rice is thus one of the most important crops in the world. It has been grown in the tropics and sub-tropics probably for thousands of years.

All cereal grains. Production of the aforementioned 'big three' cereal species totals about 2.4 billion tonnes annually, accounting for most of the 2.8 billion tonnes for all cereal grains. This volume is so much greater than the 0.6 billion tonnes of the dicotyledonous grains (legumes and oilseeds). Yet each one of the range of grain species has a specific role to play in the utilisation avenues of food, feed, fuel and industrial applications. It is even some of the minor cereals that are attracting renewed interest, in some cases as replacements for the traditional species, such as wheat.

A SECOND EDITION

The need to describe these 'other' grain species was a major factor prompting Woodhead to request a second edition of *Cereal Grains*. Thus, this edition has added chapters on durum wheat (Chapter 6), on triticale (Chapter 8) and on several 'alternative' species—ancient grains and pseudo-cereals (Chapter 14). Developments in the analysis and management of grain quality for the full range of species have prompted the need for updates of information throughout this edition.

THE FUTURE

In the first edition, we noted the increasing power of consumer attitudes, preferences and eating habits to influence developments throughout the entire grain chain. Developing trends for ancient grains, whole grains, super-grains and gluten-free foods are expected to continue into the future. These developments must be assisted by application of new selection methods in breeding and management tools during production to ensure sustainability and the ability to adapt to future challenges.

ACKNOWLEDGEMENTS

In the production of this second edition, we have also appreciated the contributions of several experts who have acted as chapter reviewers, thereby ensuring authenticity. Essential contributions have been made by Woodhead editorial staff, especially Robert Sykes, Karen Miller, Josh Bennett and Julia Haynes. Thank you all.

Colin Wrigley
Ian Batey
Diane Miskelly

Preface to the First Edition

'The proof is in the eating.' So goes the old proverb, indicating that the customer is the final arbiter for assessing (getting the 'proof' of) the success of any venture, not least of all the production of a pudding, or of any other product resulting from the processing of grain—bread, beer or biofuel; cake, cookie or corn-on-the-cob; porridge, polenta or puff pastry; rice risotto or rye bread.

This proverb provides the underlying basis for this book. While getting proof or assessment of quality is inevitable for the final product from a process, the achievement of final success requires assessment throughout the process, coupled to suitable management of the process based on the results of the ongoing quality assessment.

Grain production and processing starts with the plant breeder producing genotypes (varieties) suited to the proposed end-use; the process continues through planting, on-farm management, harvesting, quality-based segregation, storage and transport, processing, packaging and retailing to the consumer who is the final arbiter, providing 'proof in the eating', either literally (for food) or figuratively (for the many non-food products from grains).

We trust that the advice provided in this book will find useful application with its readers, thereby assisting with the production and processing of cereal grains worldwide.

We gratefully acknowledge the contributions of many people to the production of this book. Dr. Ferenc Bekes was deeply involved in the initial stages of planning the contents, the chapters and potential authors, as well as making contact with authors. We were sorry when Ferenc had to withdraw from acting as an editor largely due to health reasons, but his input has been invaluable.

We have also appreciated the contributions of several experts who have acted as chapter reviewers, thereby adding their expertise to that of the chapter authors. Reviewers included Drs. Wal Anderson, Scott Bean, Ferenc Bekes, Bob Cracknell, Ray Hare, Ailsa Hocking, Pekka Lehtinen, Finlay MacRitchie, Laura Nystrom, Russell Reinke, Sophie Roumeliotis, Marc Savard, Siri Siriamornpun, Mike Sissons, Peter Stone, John Taylor, Surjani Uthayakumaran and Chuck Walker.

Several members of the editorial staff of Woodhead Publishers have made valuable contributions: Beatrice Bertram, Francis Dodds, Bonnie Drury, Mandy Kingsmill, Aidan Russell and Sarah Whitworth.

The old proverb about 'the proof of the eating' is also relevant to the production of a book: its success lies in the reading and application of its advice.

Colin Wrigley
Ian Batey

Part I

Cereal-Grain Quality

Chapter 1

Assessing and Managing Quality at all Stages of the Grain Chain

Colin Wrigley
QAAFI, University of Queensland, St. Lucia, QLD, Australia

1.1 INTRODUCTION

Where do these good things come from? Your breakfast cereal and toast (or did you have oatmeal porridge?); your lunch of bread roll, Danish pastry and beer; the sugar in your afternoon soft drink; the pasta and rice risotto that you had for dinner; the ethanol in your car's fuel; the smooth starch-based finish on that paper. Where did they come from? These and many more familiar items in our lives are all derived from the cereal grains (see Chapters 3 and 24). Add to these meat, milk and eggs, which largely come from grain-based feeds for animals and poultry.

Our everyday view of cereal-based products appears in the lower half of the sequence in Table 1.1, which summarises the grain chain—the chain of events from plant breeder to us, the final consumers. This sequence is similar for each of the cereal-grain species, extending from breeding, through on-farm production and harvesting, via various types of processing, to the multitude of products that reach us as consumers, both directly and indirectly, via both food and non-food products. The cereal grains are essential to the human diet, despite the distinctive variations from culture to culture and from region to region. Without the many contributions of the cereal grains to non-food products, our lives would be different.

The 'big three' cereal species are maize (corn), wheat and rice, accounting respectively for over 1,000, 700 and 700 million tonnes of grain each year (Table 1.2). Less in volume but still significant are other cereal species: barley, rye, triticale, oats, sorghum, the millets (only one example is given in Table 1.2) and the pseudocereals (classed as cereals by the Food and Agriculture Organisation). Average grain yields vary considerably from one grain species to another (Table 1.2). The grain characteristics of the range of cereal species are described in Chapters 5 to 14 (Part II).

Cereal Grains. http://dx.doi.org/10.1016/B978-0-08-100719-8.00001-2

TABLE 1.1 Products and Issues Associated with the Various Stages of the Grain Chain

Stages		Products	Issues for analysis and management
Breed	→	Improved varieties. Pure seed.	Yield factors, disease tolerance, marker-assisted segregation, GMO technology.
↓			
Sow	→		Soil analysis, time to sow, fertiliser, choose to target yield versus quality premium.
↓			
Grow	→	Forage and fodder for ruminants	On-farm management, climate variations, precision agriculture.
↓			
Harvest	→	Fibre for construction materials Fuel and fodder from straw	Grain moisture, protein content. Avoid pre-harvest sprouting. Respiratory problems for workers.
↓			
Deliver	→	Grain samples to test	Moisture, protein percent, contaminants, defects. Quality-based segregation and binning
↓			
Store	→	Seed to sow	Test and maintain grain quality, explosion risk.
↓			
Transport	→		Identity preservation.
↓			
Trade	→		Arrange terms of trade and transport costs.
↓			
Process		Food as whole grains Feed for animals, birds, fish Fuel as bioethanol Fats as germ oils Fluids as beverages	Sun cracking for rice. Amino-acid composition. A-type starch content. Oil content. Germination for malting.
↓			
Mill	→	Flour, starch, protein isolates	Moisture, milling yield, protein content.

TABLE 1.1 Products and Issues Associated with the Various Stages of the Grain Chain (*cont.*)

Stages		Products	Issues for analysis and management
↓			
Mix and bake	→	Food: Baked goods, noodles, pasta.	Dough quality, starch properties, food safety, Bakers' asthma risk.
↓			
Retail	→	Packaged foods	Composition analysis, food labelling.
↓			
Consumer satisfaction	→	Publicity; advertising	Appearance, taste, price. Pesticide residues? Organic? GMO?
↓			
Consumer nutrition	→	Fibre, protein, starch, vitamins, minerals.	Popular misconceptions.

TABLE 1.2 World-Wide Production and Yields in 2014 of Cereal Grains, Plus the Relevant Chapters in this Book for Each Species

Cereal (*Genus* and *species*)	Millions of tonnes	Yield (tonnes/ha)	Chapter number
Maize (*Zea mays*)	1,022	5.6	11
Rice, as paddy (*Oryza sativa*)	741	4.5	12
Wheat (*Triticum aestivum, T. durum*)	729	3.3	5, 6, 14
Barley (*Hordeum vulgare*)	144	2.9	9
Sorghum (*Sorghum bicolor*)	68	1.5	13
Millet (*Pennisetum glaucum*)	28	0.9	13
Oats (*Avena sativa*)	23	2.4	10
Triticale (×*Triticosecale* sp.)	17	4.1	8
Rye (*Secale cereale*)	15	2.9	7
Buckwheat[a] (*Fagopyrum esculentum*)	2.0	1.0	14
Quinoa[a] (*Chenopodium quinoa*)	0.2	0.8	14
Cereal grains, total	2,800	3.9	2

[a]*Pseudocereals, but included with the true cereals in FAO statistics*
Source: faostat3.fao.org.

The cereal grains are classed as grasses (monocotyledons, flowering plants) and are distinct from the dicotyledonous legumes and oilseeds, whose volume of production is rather smaller than that of the cereal family (see Chapter 2).

Production of all the cereal grains in total (2,800 million tonnes in 2014, Table 1.2) is equivalent to 1 kg of grain per day for each person on earth! However, significant parts of grain production are used as animal feed and for industrial purposes. In addition, the major regions of production are distant from many of the needy, who may lack the resources to obtain the needed grain.

1.2 THE CREATION OF PRODUCTS FROM THE CEREAL GRAINS

The contributions of cereal-based products are seen, in Table 1.1, to start with the young plants, which can be valuable forage for stock, the partly eaten plants then being allowed to mature and produce grain. After harvesting, the standing stubble and the straw also constitute fodder for ruminant animals.

In addition, the straw remaining after harvesting the grain can be used for fibre-based building materials, such as particle board. This approach is only a part of the grander scheme of whole-plant utilisation described by Munck (2016), whereby local factories make use of all parts of the mature plants, not only the grain. Furthermore, the lignocellulose straw residues show modest promise to become a raw material for the production of renewable fuel (Guragain et al., 2016).

The range of cereal-based products in Table 1.1 can be summarised as the following list of 'F-based' products, including food, feed, fuel, fats and fluids.

- *Food*. Grains directly supply over half of humanity's dietary energy and protein. In many cases, they are consumed as the whole grain. An obvious example is rice, which after milling to remove the bran layers is boiled or fried; other rice-grain uses include sake and extruded snack foods (Pallas, 2016). Barley is also consumed as the whole grain after pearling to remove the outer layers (Izydorczyk and Dexter, 2016).
- *Feed*. Animal feed includes the full range of cereal grains (Black, 2016), often after crushing the grain and pelletising, to permit the incorporation of other minor nutrients.
- *Forage and fodder*. Ruminant animals feed on the straw remaining after harvest, as well as foraging the early plant growth.
- *Fuel*. Maize and wheat grains are used for producing bioethanol after fermentation of the starch and de-watering of the resulting ethanol (Guragain et al., 2016).
- *Fats*. The germ tissue of the cereal grains is a rich source of oil.
- *Fermentation*. Virtually all species of cereals are used for fermentation into beer and for the production of distilled spirits. Other beverages include the use of roasted grain for *ersatz* coffee.

- *Flour.* Dry milling is the usual means of separating the endosperm from the whole grain for crushing into a fine powder, mainly for wheat (Sarkar and Dexter, 2016) and maize (Rausch and Eckhoff, 2016a). Wet milling is also used for these cereals, especially prior to separating the starch and protein fractions (Rausch and Eckhoff, 2016b; Velicogna and Shea Miller, 2016).
- *Flour, mixed and baked (or dried or boiled).* This is the basis of the wide range of foods such as bread, cookies, cakes, breakfast cereals, tortillas, porridge, pasta, noodles and many more.

1.3 GRAIN QUALITY

For each of the many uses of the cereal grains there are specific grain-quality requirements. The task of meeting these quality specifications is the theme of this book. This task requires the combined use of analytical methods to assess progress towards these specifications plus management to ensure that the specifications continue to be met throughout the grain chain to achieve consumer satisfaction.

What is meant by 'grain quality'? Just as 'Beauty is in the eye of the beholder', as the old saying goes, so also the assessment of grain quality depends on the needs of the grain user. Those needs are likely to involve different combinations of attributes, depending on the purpose for which the grain is being bought and used. Accordingly, the price that will be paid is determined by the quality—the suitability of the grain for the processor-customer's purposes.

In addition, the term 'quality' has different significance, depending on the stage being considered along the 'grain chain'—the continuum from breeding, through growing and management of the crop, to harvesting, segregation, storage and transport, to processing and final consumption (Table 1.1).

Effective management of grain quality along the consumption chain is thus essential to the satisfaction of all customers—the 'customer' being the next step along the grain chain, the ultimate customer being the retail consumer.

The degree of satisfaction is obviously related very closely to the price that will be paid at each stage of the value-added chain. Effective assessment of grain quality provides critical knowledge about the defects and advantages of the grain involved. The 'grain involved' may be thousands of tonnes in the hold of an ocean-going bulk-grain ship, or it may be a few grains being nurtured by a breeder as the basis for selecting a revolutionary new genotype. What happens to the grain after analysis depends largely on the analytical results. Thus, effective management of grain production and processing depends on effective analysis.

1.4 THE GRAIN CHAIN

The value-addition chain for grain has been dubbed 'field to fork' (Dixon, 2007). The many transactions involved are summarised in Table 1.1. This sequence of events is reflected in the order of the assessment-management-processing

FIGURE 1.1 An illustration of value addition as a result of the operation of the grain chain. This small amount of wheat grain (costing a few cents) is all that is needed to produce the flour for one of the pastries in Fig.1.2, selling for a few dollars.

chapters in this book (Parts III and IV, Chapters 15–24). The increase in value from harvest to retail can be seen, as depicted in Figs 1.1 and 1.2, by comparing the cost of a handful of wheat grain (say, 100 g, with a value of about two cents on a per-tonne basis) with the cost (two US dollars) of a glossy Danish pastry, containing the flour yielded by 100 g of grain, namely, 100-fold value increase.

1.4.1 Breeding

Despite the term 'chain', the linear sequences of Table 1.1 can be seen as a circle, starting and ending with the consumer, who must be the motivating force setting the grain-quality agenda for the breeder, who is otherwise at the top of the sequence of events that lead back ultimately to the consumer. That consumer may be represented by an export market in a country far from the breeder, who must become familiar with the grain-quality needs of processors in that market. The breeder is also responsible for the development of new varieties that suit the needs of the immediate customer—the grain grower who has specific agronomic

FIGURE 1.2 An illustration of value addition as a result of the operation of the grain chain. One of these pastries, selling for a few dollars, can be made from the small amount of wheat grain (costing a few cents) in Fig.1.1.

requirements that will optimise grain yields, plus grain-quality needs that will ensure premium payments for the harvested grain.

The breeder's opportunities to deliver these requirements are primarily limited to genetic potential. The ways in which that potential is expressed are dependent upon the vagaries of the growth conditions and farm-management practices. There is thus an interaction (G × E) between genotype (G, variety) and growth environment (E). Nevertheless, the breeder has significant opportunity to build-in genetic tolerance to various threats to quality loss, such as resistance to premature sprouting due to rain at harvest. Issues relevant to the breeding stage are described in Chapter 16.

1.4.2 Grain Growing

Beyond the breeder are the seed producers and grain growers. Initial assessment for them is the assurance that the seed being traded and sown is the correct variety and that the consignment is free of contaminating weed seeds and of seed-borne diseases. On-farm management of grain quality for the grower is interactive with grain yield, because of the need to maximise financial returns, represented by the combination of grain volume and market value. Thus management (M) of grain growing by the farmer adds to the interaction between genotype (G, variety) and growth environment (E) to provide the combination of three factors: (G × E × M). See Chapter 18 for the effects of growth conditions on grain quality. See Appendix 2 for information about the units used in the grains industry, especially for transferring between the units systems of metrics, USA and China.

1.4.3 Grain Receival

Beyond the farm gate is the critical stage of assessing the many aspects of grain quality that will determine the price to be paid to the grower. This assessment of quality (Chapters 4, 17 and 19) also determines the manner of segregation of the grain delivery, thereby ensuring that it is combined with grain of similar quality type. In many grain-growing regions, harvested grain is delivered directly from a regional storage structure, known as a 'country elevator' in northern America or a 'silo' in Australia (Figs 1.3 and 1.4).

FIGURE 1.3 Grain storage structures typical of Canada.

FIGURE 1.4 Grain storage structures typical of Australia.

Beyond this stage, there is the need to preserve the value of the aggregated grain that has been assessed to be of a specific quality, by storing and transporting it with its quality type preserved (Chapter 20).

1.4.4 Whole-Grain Processing

Suitability for processing is a basic factor determining quality-based segregation and management. Unless identity preservation can be assumed, quality assessment may again be a priority when the grain arrives at the flour mill, malt house, feed mill or other form of processing factory. The knowledge thus obtained will determine how a specific grain shipment will be stored, utilised and managed in the sequence of events leading to the product that is offered for final sale to the consumer.

If incoming grain is not of adequate quality, there is yet the possibility of adjusting processing methods to compensate for the inadequacies. That possibility introduces the further factor of Processing method (P) to add to the factors of Genotype (G), growth Environment (E) and on-farm management (M), to give the combination $G \times E \times M \times P$. However, the processor would rather not have to make adjustment to the process as that is likely to add to costs.

The grain-chain sequence to the point being considered is shorter than for ground grain because many grains are processed as the whole grain. Rice processing usually involves the milling and cooking of the whole grain. In the case of paddy rice grown by a subsistence farmer (e.g. Fig. 1.5), feedback is immediate with his family commenting on the eating quality. More elaborate

FIGURE 1.5 **'Paddy-to-Plate' value addition for rice starts with rice paddies in Indonesia.**

processing may involve the preparation of rice-based sushi displayed in an expensive restaurant (Fig. 1.6); this example of 'Paddy-to-Plate' also illustrates the value-addition resulting from grain processing.

There are many other examples of the processing of the whole grain without prior grinding, such as corn-on-the-cob, pearled barley, paella, risotto, bulgur and couscous (the last also from semolina). Fat extraction from germ is often a by-product of the whole-grain processing of grains such as rice and corn (Chapter 23) (Godber, 2016).

1.4.5 Milling and Baking

Further down the grain chain (Table 1.1) come the additional stages of flour milling (Chapter 22) and the baking of bread, cakes and many other foods, such as are shown in Fig. 1.2. While this stage most obviously applies to wheat, milling into flour is also performed for other grains, such as rice for rice noodles (Fig. 1.7), for rye bread and for corn-based tortillas.

Industrial processing generally proceeds after milling to permit the separation of starch granules from the protein matrix of the endosperm. The resulting starch preparations are further processed for the manufacture of food and non-food products (e.g. paper, adhesives, cosmetics and binding agents). Additionally, whole and milled grains are used as our foods via the less direct route of animal and poultry feeding to produce meat, milk, cheese and eggs.

FIGURE 1.6 'Paddy-to-Plate' value addition for rice may culminate with a display of sushi laid out in a high-class restaurant. In this case, the operation of the grain chain has proceeded only halfway down the sequence in Table 1.1, as whole-grain processing was involved.

Throughout all these stages there are the threats of contamination and other food-safety issues (Chapter 15) (Motarjemi and Lelieveld, 2013), requiring continuing vigilance for safeguarding quality from a wide variety of threat sources (Chapter 21).

At all stages there is also the matter of safety for workers involved. A significant example is the threat of dust explosions from grain or flour dust at the respective stages of grain handling and processing. Bakers' asthma is also an occupational risk for millers, bakers and pastry cooks who have daily exposure to flour dust.

1.4.6 The Consumer

The ultimate arbiters of quality are the consumers, who assess quality by voting with their buying power. Preference for the bread or cakes in Mr Jones' shop

FIGURE 1.7 'Paddy-to-Plate' value addition for rice may continue with the production of rice noodles as a product obtained after the grain has progressed further down Table 1.1 via grinding into flour.

means that the vote for quality may go to Mr Jones instead of the supermarket on some occasions, or possibly the reverse on other occasions. For the subsistence rice farmer, his family is the 'customer', assessing quality according to their individual preferences for cooked rice.

Eating quality is not the only basis for assessing quality at the consumer level. Various other factors come into play, one of which is consumer perception of nutritional quality (Table 1.1). Nutritional guidelines, developed for many countries, recommend the consumption of grain-based foods as a major part of food intake. This concept has been illustrated pictorially in the form of a pyramid, with grain-based foods forming the base, and thus the major contributor to the diet. Jones et al. (2015) have reviewed the various pictorial formats used in many countries to illustrate the need to emphasise some food sources (especially grains) in preference to others. This emphasis on grains in the diet provides a major boost for the grain-growing and grain-processing industries of the world.

Nevertheless, consumer preferences may stray from this central message to diverse concepts of nutritional quality involving the presence or lack of specific grain types or of additives. Gluten-free is one of these concepts that is fully justified for individuals with specific dietary restrictions, such as coeliac disease, but that may be less justified in other cases (see Chapter 14).

Appendix 1 offer readers the opportunity of obtaining appropriate information about the composition of grain-based foods. So, such questions may be answered as 'What nutrients are lost as a result of removal of the bran layers from the rice grain as a result of milling to produce white rice?' However, in an appendix such as this, it is possible to provide only one entry for each combination of nutrient and material, even though this single figure must represent a wider range of analyses given the factors of biological diversity and the effects on composition of growth conditions.

1.4.7 Consumer Feedback

The feedback loops from the consumer to others up the chain may be short or long. It may be as short as reactions within the family or village in a subsistence farming community. At the other extreme, feedback may involve great distances and long time periods, taking into account the extent of the world grain trade and the many years involved in bringing a new grain variety from an initial cross to the selection of an elite line ready for commercial sowing.

Another lag factor is the many months or even years that grain may be in storage and transport before it is processed. In addition, feedback may involve long distances, such as a flour miller in the Middle East or a legume trader in India reporting grain defects to grain traders or breeders in grain-producing countries such as the Americas, Australia or Europe.

1.5 THE ASSESSMENT OF GRAIN QUALITY

Such feedback obviously relies on effective and reproducible assessment of the many quality factors (Table 1.3) that determine the management of grain processing. The assessments may relate to a grain shipment that has just arrived: 'We checked for pesticide residues and we found that this grain shipment is contaminated at an unacceptably high level. It cannot be received. Send it back!' Assessment may relate to long-term experience that determines company policies: 'We continue to buy grain from a certain country because their grain is always clean and dry (both factors relate to good value) and our down-stream customers like the protein quantity and quality (suiting processing needs).'

TABLE 1.3 Major Aspects of Grain Quality Requiring Assessment at Various Stages of the Grain Chain

Stage of production	Major assessments needed	Analytical approaches
Breeding	Quality type matches target utilisation	Small-scale screening methods Gene and protein markers
Pure seed production	Authenticity of variety Germination viability	Variety identification based on protein or DNA composition Germination testing
At harvest—on farm and at grain delivery	Bulk density Moisture content Protein content (quality) Contaminants and defects Variety identification	Chondrometer test Sieve, weigh and inspect NIR spectroscopy
Storage and transport	Insect contamination Moisture content Mycotoxins	Visual inspection NIR spectroscopy Immuno-assays
Processing	Contaminants and defects Moisture and protein content	Visual inspection NIR
Food products	Food safety Appearance, taste	Immuno-assays Customer reaction
Animal-feed products	Mycotoxins, pesticides Energy value Essential nutrients	NIR spectroscopy Immuno-assays Chromatography

1.5.1 No Nasty Surprises

Bakers may complain that the quality of the flour provided by the flour mill varies in its dough-handling properties, and thus in its overall baking quality. Some bakers would even claim that they can cope with many types in flour quality, but that they want consistency in quality—'no nasty surprises'. The result may be for the blame game to flow up the value chain; for example, the baker blames the miller and the miller blames the quality of the grain supplied. Thus the grain buyer must make subjective decisions, in the context of commercial realities, about how far to go in assessing grain quality so as to fulfil the demands of millers and bakers for consistent and acceptable grain quality (Hajselova and Alldrick, 2003).

'No nasty surprises' is a generalisation that processors often quote to indicate their approach to quality assessment. Such a 'nasty surprise' may be a metal object hidden in the grain that would cause significant damage to processing equipment if undetected and if not removed.

Extensive contamination with a noxious agent such as ergot is another example, which in this case could cause illness in consuming humans or animals. Unacceptable contamination may involve a small amount of weed seeds that produce scent, thereby giving an off-flavour to the whole consignment. 'No nasty surprises' can overall be construed as the provision of a consistent and continuous supply of grain with uniform quality.

Another form of 'nasty surprise' is a large consignment of grain that is lacking in homogeneity, having wide variations in quality throughout the load. This problem accentuates the initial need (before even performing any quality assessment) of establishing adequate sampling procedures, so that quality is assessed on sub-samples that are representative of the whole consignment of grain (Hajselova and Alldrick, 2003).

1.5.2 Assessing Grain Quality Along the Grain Chain

Although general considerations of grain quality may be similar at all stages of the grain-supply chain (Table 1.1), there are specific concerns about grain quality at some stages (Table 1.3). As a result, the methods used to assess quality differ along the chain. In addition, methods of assessment have different constrictions according to speed and sample size. For example, the presence of pesticide residues is not relevant at the stage of breeding, but it is at grain receival and throughout processing. On the other hand, the breeder must be able to determine quality traits on very small samples, whereas having sufficient sample to test is not a problem at harvest.

Rapid methods are needed to check grain quality when the rush of the grain harvest may decree that a truck-load must not be left waiting for more than a few minutes before quality-grade has to be finalised to indicate where the load should be tipped (Table 1.3). On the other hand, the breeder requires analytical methods that can cope efficiently with large numbers of small samples, whereas methods may be acceptable for the breeder even if they require many hours or days to complete.

1.5.3 Technical Approaches to Assessing Grain Quality

Methods of analysis described in this book reflect the range of quality attributes that need to be assessed at the various stages of grain production and processing, together with the diversity of restrictions due to practical factors such as speed, labour requirement and accuracy.

The contrasts in such criteria are illustrated in Table 1.4 with respect to strategies for assessing the quality of grain delivered following harvest. Ideally, the major quality attributes should be determined during the few minutes that a truck-load is waiting prior to discharge of the grain. If this ideal condition cannot be met for some aspects of quality, it may be acceptable to have a grain sample sent to a regional laboratory for assessment, with the result being provided overnight.

TABLE 1.4 Criteria and Tolerances for Grain-Quality Test Methods Following Harvest

Criteria	Tests at grain receival	Tests at regional laboratory
Speed	Within a few minutes	Within an hour, but up to 24 h may be acceptable
Accuracy and precision of result	Requirements not so strict	Stricter requirement, for example ± 5%
Cost of equipment and consumables	Costs must suit cost-benefit analyses	Costs per sample must be justified, but more expensive equipment justified at regional laboratory
Safety concerns	Methods must satisfy safety requirements	Methods must satisfy safety requirements
Expertise needed for supervisor	Modest experience is expected	Laboratory technician with experience is acceptable
Expertise needed for general staff	No laboratory experience may be expected	Ideally, only modest laboratory experience

1.5.4 Standard Methods of Analysis

Most analyses involve conventional methodologies that require laboratory equipment and significant levels of expertise on the part of the operator. Conventional testing procedures have been developed and standardised to determine many of the grain-quality attributes required along the grain chain. The development of these standard methods is a laborious process, involving collaboration between various combinations of scientists from a range of laboratories involved in the type of analyses being evaluated (Bridges and Wrigley, 2016a,b) (Table 22.4).

Final agreement on procedural details often involves interactions via a relevant scientific society. Associations involved in developing and publishing standard methods of grain analysis include AOAC International (www.aoac.org), the International Seed Testing Association (www.seedtest.org), AACC International (www.aaccnet.org), the International Association for Cereal Science and Technology (www.icc.or.at), the European Brewing Convention, the International Organization for Standardization (www.iso.org) and the American Society of Brewing Chemists (www.asbcnet.org). Details of the various approved methods are available at the respective websites.

1.5.5 'Instant' Technologies of Quality Assessment

Although rapid analysis is especially needed at certain stages of the grain chain, swift provision of results is welcome at any stage. An ideal promise of rapid analysis is provided by near infrared spectroscopy (NIR) (see Chapter 19). Over recent decades, NIR methods have become widely adopted at mill or elevator receival for the determination of moisture and protein content in cereal grains generally, both for analysing single samples and for continuous monitoring online (Burns and Ciurczak, 2007). The results of this testing have been adopted as a primary means of grading grains, especially wheat and barley. Grain hardness for wheat is also routinely measured by NIR. Among further applications of this approach to grain analysis is monitoring the milling of rice (Saleh et al., 2008).

Originally NIR technology was applied to milled grain, but advances have brought the great advantage that this technology can be applied to whole grain. The consequence is that no sample preparation is required—a grain sample can be simply tipped into the top of the NIR machine and results are generated automatically. NIR works in a similar manner to any other spectroscopic technique; different chemical groups in the sample absorb radiation at characteristic wavelengths and the extent of absorption depends on the concentration of the analyte. Whereas NIR has been most effective for determining major grain components such as moisture, protein and oil contents; the analysis of minor components by NIR may not be so simple. Nevertheless, NIR technology has great potential to be extended to a wide range of analytical applications.

1.6 THE MANAGEMENT OF GRAIN QUALITY

Throughout the grain chain, efficiencies in management towards specified aims only become achievable if adequate analysis of quality trends is provided (Table 1.3).

1.6.1 Breeding to Grain-Quality Targets

The genetic improvement of grain genotypes has traditionally involved three major aims, namely: increasing grain yield, overcoming the impediments of biotic and abiotic stresses and improving the processing and end-use quality of the grain. The breeder's management approaches towards the last of these three aims (grain quality) are different in the early stages of the breeding process, compared to the final phase. Initially, when there are large numbers of lines to be screened, management's aim is to identify and exclude lines that are unlikely to meet quality targets. Their deletion at an early stage means that they will not be propagated unnecessarily, thus conserving resources. At this stage, tests that identify selected protein and DNA markers are especially effective, as they target aspects of genotype directly.

Later in the breeding process, there are fewer individual lines and larger amounts of each grain sample are available, permitting more extensive testing, such as the use of small-scale simulations of grain processing, namely, malting (for barley), actual milling and baking (for wheat) and cooking/tasting tests (for rice). Fig. 1.8 illustrates the propagation of 'elite' breeder's lines for which

FIGURE 1.8 **In the later stages of producing new varieties, plots are large enough to generate kilogram-sized amounts of grain, permitting evaluations such as test milling and baking (for wheat).**

larger amounts of grain become available for more detailed testing than is possible at earlier stages.

Results from these forms of quality assessment will be critical at the stage when specific lines are finally approved for release. At that stage, the management of registration will probably also involve 'DUS' testing, namely, assessing that the potential new variety is distinct, uniform and stable, thus meeting the guidelines of the International Union for the Protection of New Varieties of Plants (UPOV) (Mauria, 2000).

1.6.2 Sowing and Growing the Grain

The pure-seed supplier has a critical management responsibility to provide seed of the correct variety, free from contaminating seeds and free from seed-borne diseases. Detailed assessments of seed purity and identity should be provided to the farmer as assurance of the aspects of quality at the time of sowing, when management responsibility transfers to the farmer.

Grain quality in the harvested grain will depend on farm management and climatic factors, as well as on the quality of the seed. Many computer-based systems are now available to facilitate on-farm management towards specific quality goals, but none of these strategies can completely overcome the uncertainties of climatic factors. Nevertheless, the 'precision agriculture' approach to farm management offers possibilities for achieving quality as well as yield targets (Wrigley, 2005; Leonard, 2016). The increasing popularity of organically grown grain offers further possibilities for grain growers (Gelinas and David, 2016).

1.6.3 Grain Receival

Harvest time presents a major opportunity to manage grain quality, provided adequate analytical information is available promptly. As harvest time is busy for all concerned, the priority is more often to 'get the grain off' and into storage before wet weather threatens to damage quality, rather than to spend precious time checking that grain quality is adequately assessed to permit the separate binning of grain of different quality type.

Despite the time constraints of checking grain quality before deliveries are combined in the same storage cell, there is the possibility of retaining small samples of all the deliveries that contribute to each storage cell, so that this 'running sample' can later be analysed in detail, thereby providing information about the overall quality of the grain in each storage cell. This information, if available, is valuable in marketing the grain in specific lots, each of known quality, thus to satisfy the processing specifications of the food or feed manufacturer in the next step of the grain chain.

1.6.4 Grain Trading

The stage of grain trade is of critical importance to the grain grower. Before even sowing the crop, the grower has struggled with the dilemma of whether to target maximum grain yield, despite the lower quality and the lower price, versus targeting a higher price for the crop (but lower yield) based on a combination of appropriate premium variety, premium protein content and freedom from defects and contaminants (see Chapter 19).

The assessment of these latter aspects of grain quality are critical at grain delivery—if direct delivery to mill or silo is planned. An alternative management plan may be the provision of grain storage on the farm. This strategy allows the grower to delay the decision of trading the grain in anticipation that better sale conditions may be negotiated. But this strategy also involves the added costs and risks of storage and arranging quality testing, plus the cost of transport to the buyer's designated delivery site.

Analytical information is of critical importance to the grain buyer and trader, who is responsible to the processor receiving the grain. Poor or mixed grain quality will involve added costs to the processor. In the larger picture of international trade, the reputation of a specific country or region is an important part of trading management. In many such cases, international relationships between countries present confounding complications that even go beyond grain-quality considerations, because in such cases trading may be negotiated by government-associated agencies.

1.6.5 Processing

Whatever type of grain processing is involved, assessment of its quality is integral to efficient processing. Blending of grain streams from different storages or deliveries is the first avenue of management available to the processor. In the milling of wheat to white flour, it is estimated that about 75% of flour quality is determined by the quality of the incoming wheat; that is management factors (mill adjustment and environmental conditions) contribute only 25% (Posner and Hibbs, 2005). This example illustrates the importance of adequate assessment of grain quality as a basis for sound management with respect to safe-guarding the quality of the product—in this case, flour for the specific processing requirement at the next stage, possibly large-scale bread manufacture, pastry production or processing into cookies and confectionery.

The milling of wheat flour requires the miller to achieve tight specifications that are set by the flour customer. The blending of flour streams is an important management tool in meeting these specifications, but this approach is totally dependent on effective assessment of quality prior to and during milling (see Chapter 22).

Barley for malting is another grain for which detailed information about grain quality is critical to achieve effective management in processing. In this case, the maltster must have uniformity of grain size, of variety and of germination rate, thus to achieve uniform germination and maximise the extraction of soluble sugars for malting (see Chapter 9).

For feed mill operation, a primary concern is for the absence of disease or contaminants that may cause injury to stock. As a result, analyses for mycotoxins and contaminating grains are conducted on incoming grain.

1.7 FUTURE TRENDS

We can doubtless assume that the principle of the grain chain of events will continue into the future, because it is the essential route by which the consumer has access to all products derived from cereal grains. However, minor aspects of the sequence may vary or alter in importance. The matter of feedback up the chain may well evolve into different patterns.

For example, research into difficulties further down the chain often indicate a solution that is best applied at the stage of breeding. A classic example is the effect of rain at harvest causing premature sprouting and consequent loss of baking quality for wheat and loss of malting quality for barley. The breeder is thus requested to incorporate genetic tolerance via extra dormancy to solve these down-stream problems. As a result there are many varieties of wheat and barley that are largely tolerant to pre-harvest sprouting. In addition, management strategies are established to ensure the rejection at harvest of loads of partly sprouted grain.

The breeder will continue to be the agent for producing more predictable dough properties in wheat flour by manipulation of the alleles for the glutenin genes, which have been assigned with their various contributions in a list of thousands of wheat genotypes (Bekes and Wrigley, 2013). See http://www.aaccnet.org/initiatives/definitions/Pages/glutendatabase.aspx.

Ongoing research of this type is leading to the identification of marker proteins and genes that facilitate the use of marker-assisted selection (MAS) of desirable breeding lines in the early stages of breeding, long before sufficient grain is available at later generation for testing of the actual phenotype required (e.g. appropriate dough quality). Nevertheless, MAS technology does not substitute for the actual testing of phenotype in later generations, but MAS does reduce the likelihood of unnecessary propagation of poor-quality progeny (DePauw and O'Brien, 2016). In addition, the cereal grains are likely to gain further agronomic and quality improvements due to the application of novel genetic 'GMO' technologies, which have been spectacularly successful for species such as maize and soy.

However, the genetic improvement of varieties rarely obviates the need for having management strategies in place further down the grain chain to eliminate grain of sub-standard quality. Thus the principle of the grain chain is basic to understand the operation of grain production, processing and consumption.

A similar principle operates at the biochemical level as illustrated in Fig. 25.1. The genome (genomics) of a specific grain interacts with the growth conditions, so that protein synthesis occurs in the developing grain on the basis of the G × E interaction described in Section 1.4.1. The intermediate steps at this stage involve transcription (transcriptomics) from the active genes to produce the respective polypeptides (proteomics) which in turn associate to form active proteins (especially enzymes, hormones and storage proteins). These in turn are responsible for the synthesis of carbohydrates, lipids and other metabolites (metabolomics) to eventually create the mature grain.

As described in Section 1.4.1 and subsequently, the various stages of processing (in total, G × E × M × P) lead on to the great variety of grain-based products that we consume. This biochemical principle of the grain chain is complementary to the practical grain chain in Table 1.1, thus providing an underlying understanding at the molecular level of the operation of grain production, processing and consumption. This level of understanding will be a basis of future research and achievements.

SOURCES OF FURTHER INFORMATION

Useful websites:

- www.usda.gov—United States Department of Agriculture
- www.grainscanada.gc.ca—Canadian Grains Commission, Winnipeg, Canada

- www.niab.co.uk—National Institute of Agricultural Botany, United Kingdom
- www.campdenbri.co.uk—Campden BRI, Chipping Campden, Gloucestershire. UK. Also at other sites, including Hungary.
- www.aegic.org.au—The Australian Export Grains Innovation Centre, Perth and Sydney
- www.aaccnet.org—AACC International. Reference books are published by AACC International, describing the 'Chemistry and Technology' of each of the cereal grains, namely, wheat, rye, barley, oats, corn, rice, sorghum and the millets, as well as covering most stages of the grain chain. The AACCI website also provides free access to information about grains; go to sections such as 'Publications', 'Approved Methods Online', 'Definitions' and 'Check Samples'.
- www.icc.or.at—Similar services are available from the International Association for Cereal Science and Technology.

REFERENCES

Bekes, F., Wrigley, C.W., 2013. Gluten alleles and predicted dough-quality for wheats world-wide: A great resource, free on the AACCI website. Cereal Food World 58 (6), 325–328.

Black, J.L., 2016. Animal feed. Wrigley, C., Corke, H., Seetharaman, K., Faubion, J. (Eds.), Encyclopedia of Food Grains, vol. 3, Elsevier Ltd, Oxford, UK, pp. 215–223.

Bridges, A.R., Wrigley, C., 2016a. The standardization of methods for analyzing grains and grain-based products. Wrigley, C., Corke, H., Seetharaman, K., Faubion, J. (Eds.), Encyclopedia of Food Grains, vol. 2, Elsevier Ltd, Oxford, UK, pp. 300–308.

Bridges, A.R., Wrigley, C.W., 2016b. Lists of standardized testing methods of the analysis of grain and grain-based foods. Wrigley, C., Corke, H., Seetharaman, K., Faubion, J. (Eds.), Encyclopedia of Food Grains, vol. 4, Elsevier Ltd, Oxford, UK, pp. 450–459.

Burns, D.A., Ciurczak, E.W., 2007. Handbook of Near-Infrared Analysis. CRC Press, Baton Rouge.

DePauw, R., O'Brien, L., 2016. Wheat breeding: exploiting and fixing genetic variation by selection and evaluation. Wrigley, C., Corke, H., Seetharaman, K., Faubion, J. (Eds.), Encyclopedia of Food Grains, vol. 4, Elsevier Ltd, Oxford, UK, pp. 279–286.

Dixon, J., 2007. The economics of wheat; research challenges from field to fork. In: Buck, H.T., Nisi, J.E., Salomon, N. (Eds.), Wheat Production in Stressed Environments. Springer, Netherlands, pp. 9–22.

Gelinas, P., David, C., 2016. Organic grain production and food processing. Wrigley, C., Corke, H., Seetharaman, K., Faubion, J. (Eds.), Encyclopedia of Food Grains, vol. 4, Elsevier Ltd, Oxford, UK, pp. 154–161.

Godber, J.S., 2016. Oil from rice and maize. Wrigley, C., Corke, H., Seetharaman, K., Faubion, J. (Eds.), Encyclopedia of Food Grains, vol. 4, Elsevier Ltd, Oxford, UK, pp. 453–458.

Guragain, Y.N., Probst, K.N., Vadlani, P.V., 2016. Fuel alcohol production. Wrigley, C., Corke, H., Seetharaman, K., Faubion, J. (Eds.), Encyclopedia of Food Grains, vol. 3, Elsevier Ltd, Oxford, UK, pp. 235–244.

Hajselova, M., Alldrick, A.J., 2003. Analysing wheat and flour'. In: Cauvain, S.P. (Ed.), Bread Making; Improving Quality. Woodhead Publishing Limited, Cambridge, pp. 187–199.

Izydorczyk, M., Dexter, J.E., 2016. Barley, milling and processing. Wrigley, C., Corke, H., Seetharaman, K., Faubion, J. (Eds.), Encyclopedia of Food Grains, vol. 3, Elsevier Ltd, Oxford, UK, pp. 434–445.

Jones, J.M., Peña, R.J., Korczak, R., Braun, H.J., 2015. CIMMYT series on carbohydrates, wheat, grains, and health: carbohydrates, grains, and wheat in nutrition and health: an overview. Part I. Role of carbohydrates in health. Cereal Foods World 60 (5), 224–233.

Leonard, E.C., 2016. Precision agriculture. Wrigley, C., Corke, H., Seetharaman, K., Faubion, J. (Eds.), Encyclopedia of Food Grains, vol. 4, Elsevier Ltd, Oxford, UK, pp. 162–168.

Mauria, S., 2000. DUS testing of crop varieties—a synthesis on the subject of new PVP-opting countries'. Plant Var Seeds 13, 69–90.

Motarjemi, Y., Lelieveld, H., 2013. Food Safety Management, first ed. AACC International, St Paul, MN, USA.

Munck, L., 2016. Utilization of the whole cereal plant to maximize its value. Wrigley, C., Corke, H., Seetharaman, K., Faubion, J. (Eds.), Encyclopedia of Food Grains, vol. 4, Elsevier Ltd, Oxford, UK, pp. 228–234.

Pallas, L., 2016. Rice processing: beyond the farm gate. Wrigley, C., Corke, H., Seetharaman, K., Faubion, J. (Eds.), Encyclopedia of Food Grains, vol. 3, Elsevier Ltd, Oxford, UK, pp. 446–452.

Posner, E.S., Hibbs, A.N., 2005. Wheat Flour Milling, second ed. AACC International, St Paul, MN, USA.

Rausch, K., Eckhoff, S.R., 2016a. Maize: dry milling. Wrigley, C., Corke, H., Seetharaman, K., Faubion, J. (Eds.), Encyclopedia of Food Grains, vol. 3, Elsevier Ltd, Oxford, UK, pp. 458–467.

Rausch, K., Eckhoff, S.R., 2016b. Maize: wet milling. Wrigley, C., Corke, H., Seetharaman, K., Faubion, J. (Eds.), Encyclopedia of Food Grains, vol. 3, Elsevier Ltd, Oxford, UK, pp. 467–481.

Saleh, M.J., Meullenet, J.F., Siebenmorgen, T.J., 2008. Development and validation of prediction models for rice surface lipid content and color parameters using near-infrared spectroscopy: a basis for predicting rice degree of milling'. Cereal Chem. 85, 787–791.

Sarkar, A.K., Dexter, J., 2016. Wheat, dry milling. Wrigley, C., Corke, H., Seetharaman, K., Faubion, J. (Eds.), Encyclopedia of Food Grains, vol. 3, Elsevier Ltd, Oxford, UK, pp. 307–320.

Velicogna, R., Shea Miller, S., 2016. Wheat: wet milling. Wrigley, C., Corke, H., Seetharaman, K., Faubion, J. (Eds.), Encyclopedia of Food Grains, vol. 3, Elsevier Ltd, Oxford, UK, pp. 320–324.

Wrigley, C.W., 2005. Precision agriculture—a means of improving grain quality? Cereal Food World 50, 143–144.

Chapter 2

The Cereal Grains: Providing our Food, Feed and Fuel Needs

Colin Wrigley
QAAFI, University of Queensland, St. Lucia, QLD, Australia

2.1 INTRODUCTION

The cereal grains are a major source of our dietary energy and protein requirements. They are also needed for animal feeding and industrial processing. There are three major cereal species according to volume of production: wheat, maize and rice (Table 2.1; see also Table 1.2). These three represent nearly 90% of cereal-grain production; 700–1000 million tonnes of each are produced annually (Wood and Johnson, 2003; Champagne, 2004; Khan and Shewry, 2009). The other cereal grains of economic significance are triticale, rye, barley, oats, sorghum and the millets. Table 2.1 also includes the pseudocereals, which are classed by the Food and Agriculture Organization with the 'real' cereal grasses, although the pseudocereals are not in the grass family.

The range of cereal species is narrower than for the dicot grains, which are mainly oilseeds and legumes (Table 2.2). On the other hand, the volume of production of the cereals is much greater than the production of the dicot grains. For any of these many species, good grain quality is an important aim of production and a critical consideration for those industries which use the grain.

The term 'quality' has different significance, depending on the stage being considered along the 'grain chain' (see Chapter 1)—the continuum from breeding, through growing and management of the crop, to harvesting, segregation, storage and transport, to processing and final consumption.

The assessment of grain quality is essential for the effective management of grain production and processing. Analysis of grain quality provides critical knowledge about the defects and advantages of the grain involved, thereby assisting those concerned along the value-addition chain on how best to manage processing for the financial benefit of the specific stage of the overall process (Atwell, 2001; Dixon, 2007).

Cereal Grains. http://dx.doi.org/10.1016/B978-0-08-100719-8.00002-4

TABLE 2.1 The Cereal Species of Economic Significance

Common name	*Genus* and *species*
Cereals	
Barley	*Hordeum vulgare*
Maize (corn)	*Zea mays*
Millet	*Pennisetum glaucum*
Oats	*Avena sativa*
Rice	*Oryza sativa*
Rye	*Secale cereale*
Sorghum	*Sorghum bicolor*
Triticale	×*Triticosecale* sp.
Wheat (Common, bread)	*Triticum aestivum*
Wheat (Durum, pasta)	*Triticum durum*
Pseudocereals	
Buckwheat	*Fagopyrum esculentum*
Quinoa	*Chenopodium quinoa*

TABLE 2.2 World Production of the Major Groupings of Agricultural Grains in Millions of Tonnes for Specific Seasons

Grain group	2005	2006	2007	2013	2014
Monocots					
Maize	715	706	792	1018	1022
Paddy rice	632	642	660	738	741
Wheat	627	605	606	711	729
All cereal grains	2267	2239	2351	2759	2801
Coarse grains	1008	993	1086	1310	1331
Dicots					
Oilseeds	143	147	148	197	198
Soy beans	214	218	221	278	308
Legumes	60	60	56	77	78

Source: FAOSTAT, faostat3.fao.org.

2.2 CEREAL GRAINS AND PEOPLE

The cereal grains have been a main-stay for mankind's diet since pre-history. They thus occupy a central place in our diet, our culture, and even in our languages, customs and festivities.

2.2.1 Cereal Grains in our Diet

The cereal grains are produced by plants of the grass family (Poaceae or Gramineae) (Table 2.1). The species of agricultural significance are wheat, triticale, rye, barley, oats, maize, rice, sorghum and the millets. These are the grains described in this book. They are listed in approximate order of production volume in Table 1.2.

Most important are wheat, rice and maize (corn). The cereal group of plants has been very successful in contributing to mankind's food needs. The success is indicated by the cultivation of many trillions of cereal plants worldwide (probably over 10^{12}), resulting in an annual harvest totalling 2.8 billion tonnes of cereal grain each year. This level of production equates approximately to over a kilogram of grain per person per day for the world's population. However, world hunger does not reflect this statistic because of the vast distances between the major production regions and the areas of human need, and because of the considerable amounts of grain used for industrial processing and animal feeding.

An assortment of cereal grains is illustrated in Fig. 2.1. The word 'grain' relates to the use of this part of the grass plant by mankind. In contrast, the word 'seed' relates to the plant's means of producing another plant, thereby perpetuating the species. The words 'kernel' and 'caryopsis' are botanical terms for the grain. 'Kernel' is a relatively general term. The 'caryopsis' is defined as 'a one-seeded indehiscent fruit in which the pericarp is fused with a seed coat'. The grains of cereal grasses represent an entire fruit, and the term 'caryopsis' is generally limited to the cereal grains (as distinct from the grains of various dicotyledonous species).

The various grains share one central attribute that accounts for their success both in perpetuating themselves and in serving the needs of mankind; that characteristic is their ability to remain intact as a store of nutrients for long periods of time. For the seed, this period extends from the time that the mature plant sheds its dry seed until the 'rainy day', when the seed encounters moisture, triggering the processes of germination, sprouting and plant growth; in this way, the cycle of plant-seed reproduction goes on.

The waiting period of the apparently dead seed may be long or short, depending on the timing of the 'rainy day'. Safe storage of the seed's nutrients throughout this waiting period is essential to the plant's survival, irrespective of the conditions that may occur before the right combination of moisture and temperature arrives to initiate the germination response.

FIGURE 2.1 **Various grains of the cereal species.**

The ability of seed-bearing plants to preserve these stores of nutrients for long periods has been the major reason for our ancient ancestors' interest in the cereals as a food source. The harvested grain could be stored ready for their metaphorical 'rainy day', such as winter time, when other food sources were not available. In the same way, the storability of grains makes them an essential food source for us today.

The important role of carbohydrates in our diet is emphasised by Jones et al. (2015a,b) who cite the repeated emphasis on cereal grains and cereal carbohydrates as being the major food source in our diet. The nutritional guidelines for many countries have been assembled by Jones et al. (2015a), all being depicted diagrammatically in the form of pyramids and plates for several nations, a spinning top (Japan), a house (Hungary), a staircase (France) and a pagoda (China). All emphasise the basic role of cereal grains as the largest and foundational section of their recommended dietary combinations of food types.

2.2.2 Cereal grains in our Culture

Since the storability of grains helped ancient man to survive the 'rainy day' periods of his existence, the cereal grains were a major catalyst in transforming *Homo sapiens* from a hunter–gatherer into an agriculturalist, permitting ancient people to remain in one place, harvesting intentionally sown grain. As a result, fixed dwelling places were established near cereal crops, using the stored grain as a reliable source of food for family and feed for animals. This agricultural existence thus laid the foundations of a civilisation that could find spare time for the building of permanent dwellings, for cultural activities and for specialisation of occupation.

Grains have been recognised as an important food source since pre-history. The ongoing propagation and improvement of seed-bearing plants were important phases in the development of mankind. The selection of plant species that best suited our food needs commenced from those times. These activities of plant improvement were continued by early agriculturalists with the selection of genotypes that showed superior attributes, such as greater grain size, improved disease resistance and better processing quality.

Turkey is one of the regions that has been a significant part of the 'Fertile Crescent'—the region of domestication of ancient cereal plants from pre-history, and in recent years the region of discovery of many progenitors of modern cereal species. These sources of potentially valuable genes have been conserved in the Turkish Seed Gene Bank (Köksel and Cetiner, 2015). Turkey also represents a Middle East region where a diversity of cereal-based foods have been developed, especially variations on the theme of flat breads, as also described by Köksel and Cetiner (2015).

2.2.3 Cereal Grains in our Languages, Customs and Festivities

Our intimate and long-lasting relationships with the cereal grains, with wheat especially and with grain-based foods, are indicated by the many ways in which they enter in our culture and vocabulary, becoming symbols of social interaction. As examples, 'Give us this day our daily bread', 'Man shall not live by bread alone', and 'Cast thy bread upon the waters' are everyday expressions in English derived from the Bible. There are equivalent expressions in other languages, cultures and religions.

Centuries ago, celebration of the grain harvest in Western Europe was justification for celebration and thanksgiving by farmers and villagers. This tradition was carried across the Atlantic by the settlers in North America, leading to traditional Thanksgiving Day celebrations of the USA and Canada late in their fall season (autumn). In Eastern Europe, the completion of the grain harvest is still a great occasion for celebration, known as 'dozynki' in Poland. Traditionally, villagers donned colourful folk costumes, sang and played instruments for the Lord and Lady of the manor, presenting wreaths made of corn, wheat stalks and flowers, together with bread baked from the freshly harvested wheat.

Fig. 2.2 illustrates how these traditions are perpetuated even today. Year round, these cultures continue to show reverence to cereal-based foods, for example by the presentation of a loaf of bread to an honored guest. Likewise, in the Middle East, bread (Persian 'nan') may be taken as a gift when visiting friends. This special gift bread is commonly a sweet bread, covered with frosting. Unleavened bread is central to the Jewish Passover celebration, acting

FIGURE 2.2 Celebrations in present-day Poland of the end of the grain harvest (dozynki), involving ornamental structures made of wheat, barley and corn, augmented with flowers.

as a reminder of the Israelites' hurried departure from slavery in Egypt, when there was no time to leave the bread to rise overnight. In an Iranian wedding ceremony, bread often assumes a symbolic place on the table in front of the married couple.

2.2.4 Cereal Grains in Antiquity

We often use the word 'cereal' colloquially to describe ready-to-eat breakfast foods, such as corn flakes. This use is not incorrect, because breakfast cereals are mainly manufactured from various cereal grains, but the term 'cereal' botanically refers to a family of species in the grass family (Poaceae or Gramineae), whose seeds are used for human food and animal feed. The word 'cereal' comes to us originally from the Roman goddess of grain, Ceres (Figs 2.3 and 2.4). Her name meant 'wheat' or 'grain', coming from the root Latin word *creare*, referring to her ability to create or produce grain-based foods.

Demeter was the equivalent in Greek mythology, shown (in Fig. 2.5) with garlands of ripe stalks of wheat and barley as garlands in her hair and as sheaves in her arm and at her feet. As the goddesses of the harvest, Ceres and Demeter were credited with teaching humans how to grow, preserve and prepare grain and corn, thus being responsible for the fertility of the land.

FIGURE 2.3 **The Roman goddess Ceres, the source of our word 'cereal'.**

2.3 THE AGRICULTURAL GRAINS

Despite the countless number of species of seed-bearing plants, there is only a limited number of species that feature in agricultural production. The major agricultural species and grain groupings are listed in Table 2.2 with their annual production statistics. The entry for 'All cereal grains' includes maize, rice and wheat, plus the other significant cereal species listed in Table 2.1. 'Coarse grains' is the trade term for all cereal grains except wheat and rice. Soy beans are listed separately from oilseeds and legumes; soy can be considered as both an oilseed and a legume.

2.3.1 Relationships Between the Agricultural Grains

All the grains listed in Tables 2.1–2.3 contribute to our requirements of food, animal feed and fuel, but they do so in distinct ways due to their individual grain morphologies, their different growth habits, agronomic requirements and their general taxonomic relationships. Taxonomy is the scientific study of these differences and of the underlying uniformities (Morrison and Wrigley, 2016).

The Linnaean taxonomic system groups the grains into hierarchical families that reflect evolutionary relationships, dividing the grain genera and species between two major families, the monocotyledonous species (monocots) and the dicotyledonous species (dicots) (Table 2.2). This distinction is so named because of the number (1 or 2) of cotyledons (meaning 'seed leaves') produced as the first embryonic leaf or leaves that sprout from the germinating seed. The cereal grains and all other grasses are monocots (Table 2.1, except for the

FIGURE 2.4 **This statue of the Roman goddess Ceres is located in a small temple in the grounds of the Villa Cimbrone on the Amalfi Coast of Italy (see http://www.ravello.com/attractions/villa-cimbrone/).** *(Photo acknowledgement to www.pd4pic.com)*

pseudocereals, which are dicots). The many dicot families include the grains listed in Table 2.3.

2.3.2 Systematic Naming of the Agricultural Grains

Taxonomy is also the system that provides a universal naming system, consisting of genus and species names. The species (written in italics lower case) is a sub-set of the genus (written first, also in italics lower case, with the first letter capitalised). Thus common (hexaploid) wheat is *Triticum aestivum*, although some decades ago its designation was *Triticum vulgare*. (Taxonomy

FIGURE 2.5 The Greek goddess of cereal grains, Demeter.

is a dynamic discipline.) Durum wheat, used for making pasta is a distinct species, namely *Triticum durum*. In such a case, the full genus name would not be repeated providing only the initial letter, thus: *T. durum*.

This taxonomic naming system is needed because of the inadequacy of the common names that can be multiple for the same grain species, depending on variations from region to region. On the other hand, the reverse can occur, with different species having similar common names. The classical example of this type of duplication is common wheat (*T. aestivum*) and buckwheat (*Fagopyrum esculentum*). The genus name *Fagopyrum* translates as 'beech wheat', meaningful because its grain resembles the beechnut from beech trees (*Fagus* spp.) (Morrison and Wrigley, 2016). These two food grains are quite unrelated taxonomically, being respectively monocot and dicot.

A further source of potential confusion is Inca wheat, another dicot and pseudocereal, with the botanical name of *Amaranthus caudatus*. This grain is also

TABLE 2.3 The Major Dicotyledonous Plant Species (Legumes and Oilseeds) Producing Grains of Economic Significance

Grain species		
Common name	***Genus* and *species***	**Production**
Soybean	*Glycine max*	308
Cotton seed	*Gossypium* spp.	47
Canola, rape seed	*Brassica napus*	71
Groundnut (peanut, in shell)	*Arachis hypogaea*	42
Sunflower	*Helianthus annuus*	41
Beans (Navy and broad beans, dry)	Species include *Phaseolus vulgaris* and *Vicia faba*	25
Pea (dry)	*Pisum sativum*	11
Chickpea	*Cicer arietinum*	14
Cowpea	*Vigna unguiculata*	6
Lentil	*Lens culinaris*	5
Buckwheat	*Fagopyrum esculentum*	2
Linseed	*Linum usitatissimum*	3
Lupin (blue and white lupins)	*Lupinus angustifolius* and *Lupinus albus*	1
Safflower	*Carthamus tinctorius*	1

World production of each in 2014 is shown in millions of tonnes.
Source: FAOSTAT data, faostat3.fao.org.

known by multiple common (and colourful) names, including love-lies-bleeding, pendant amaranth, tassel flower, velvet flower, foxtail amaranth and quilete.

2.4 THE DICOT AGRICULTURAL GRAINS

World production of the non-cereal grains (Tables 2.2 and 2.3) is much less than for the cereals. The cereal grains (Table 2.1), with only one embryonic leaf (cotyledon), all belong to the hierarchical sub-class of monocotyledonous plants, which are thereby distinguished from the dicotyledonous oilseeds and legumes (Morrison and Wrigley, 2016).

Production of grains from dicot plants (Table 2.3) includes the major groupings of oilseeds and legumes (pulses), which together cover a much wider range of species and taxonomic families than the common cereals. Soy is the most significant of the dicot grains for volume of both production and processing. Due to its high oil content, the soybean may also be classed as an oilseed, but taxonomically it is a legume—a member of the large family Fabaceae

(Leguminosae) together with beans, peas, lentils, lupins and peanuts. The true oilseeds are classified separately. Canola (oilseed rape) is a member of the Brassicaceae (Cruciferae) family, cottonseed of the Malvaceae, linseed of the Linaceae and sunflower and safflower of the Compositae (Asteraceae).

The range of commercially grown dicot grain species extends beyond those listed in Table 2.3, which is restricted to species produced annually in excess of half a million tonnes. Other dicot grains of agricultural significance include amaranth, lentils, coix, sesame, quinoa and various mustards (Graybosch, 2016; Morrison and Wrigley, 2016).

In addition, there is the class of dicot grains termed 'pseudocereals' (see Chapter 14). These, like buckwheat, are mis-named, as they are not members of the cereal (grass) family; they are dicots—broadleaf plants. Their seed can be ground into flour and otherwise used as cereal grains. Examples of pseudocereals are amaranth, quinoa and buckwheat. The pseudocereals and the many other dicot grains have been reviewed by Belton and Taylor (2002), Abdel-Aal and Wood (2005), Tao et al. (2014) and Fletcher (2016).

2.5 THE MONOCOT AGRICULTURAL GRAINS

Despite the diversity of the grass family, only a relatively few species are used on a large scale for our food needs and for animal feed. The most common species are listed in Table 2.1. Of these, maize (corn), rice and wheat are grown to the greatest extent, each accounting for an annual production worldwide of over 700–1000 million tonnes. World production of the remaining six cereals totals less than half of the volume for any one of the 'big three'.

The statistics in Table 2.2 show several one-season snapshots of production; actual amounts of grain produced vary widely for specific grains in specific regions from year to year. Table 2.2 illustrates year-to-year variations for the 'big three' cereal species and for groupings of other major agricultural grains.

Wheat, triticale, rye, barley and oats are classified together in the broad family Pooids (subfamily Pooideae), distinct from rice (a Bambusoid), and separate from tef (Indian millet, a Chloridoid). Maize is classified as a Panicoid, together with sorghum, Japanese millet and pearl millet (Graybosch, 2016; Morrison and Wrigley, 2016). The descriptions of the specific cereals in this book (Chapters 5–14) are arranged roughly according to their taxonomic relationships, namely, common wheat, durum wheat, triticale, rye, barley, oats, maize (corn), rice, sorghum and the millets. Finally, in Part II, Chapter 14 describes less traditional grains (wheat progenitors and pseudocereals) that have attracted attention for specialty food uses.

The entry in Tables 2.2 and 1.2 for rice production is for the amount of 'paddy' rice, thus including the mass of the outer husk which is removed during initial processing and rice milling. The rice entry does not include the near relative of rice, known as 'wildrice' (not 'wild rice') (*Zizania aquatica*), whose production is much less than that for normal rice (Oelke and Porter, 2016).

The statistic for wheat in Table 1.2 includes both species [common (bread) wheat, and durum (pasta) wheat]; the production of durum is much less than for common wheat. The statistic for maize (known as 'corn' in the Americas) does not include a further 10 million tonnes of green plants that are harvested primarily for animal feed (Wood and Johnson, 2003).

Total cereal-grain production totals 2.8 billion tonnes per year. Table 2.2 gives a glimpse of how this total may vary from year to year for all cereal species. Cereals known as 'coarse grains' in world trade (and in FAO statistics) are a sub-set of the cereal species (plus a few non-cereals), generally referring to the common cereals other than rice and wheat, on the basis that these 'other cereals' are used to a lesser extent for direct food production, some being used for brewing (Chapter 9), industrial purposes (Chapter 23) and animal feed (Black, 2016).

However, this assumption is incorrect for many countries; for example most maize production in Mexico is for human food. On the other hand, much of the corn production of the USA goes to industrial processing, for starch and oil production. Furthermore, sorghum and the various millets are used as food sources in many developing countries, generally after de-husking and boiling, in the form of porridge, or as a source of starch for fermentation into beer-type drinks.

Statistics for world trade between countries differ considerably from those for production. International trade in wheat is highest, usually about 100 million tonnes annually, and much less for rice, since major regions of rice production and consumption are within the same country. This last generalisation applies even more to crops such as the millets and sorghum, which are both grown and favoured as food in the same region, often in developing countries.

The processing of cereal grains into their compositional fractions constitutes an important group of industries. For example starch and protein fractions are separated for wheat and corn in vast quantities (see Chapter 23). Furthermore, annual production of maize oil worldwide exceeds 2 million tonnes. Thus the grains may be used for food directly, or after a significant extent of preprocessing. For example cereal starch is a major source of glucose and fructose for use as sweetening agents in foods and drinks. The grains also contribute to many non-food uses, such as fuel alcohol for motor vehicles, adhesives, paper production and agents for mineral processing.

2.6 FUTURE TRENDS

Consumer dissatisfaction with conventional cereal grains is already creating interest in 'alternative' and 'specialty' grain species (Abdel-Aal and Hucl, 2014), such as einkorn (Abdel-Aal and Wood, 2005) and buckwheat (Tao et al., 2014). This interest and the positive consumer enthusiasm is likely to grow in the future. In many of these cases, there is unfounded fad reasoning that alternative grains will bring miraculous health benefits (Jones and Sheats, 2016). In other cases, the interest is stimulated by a genuine search for grain sources that could substitute for common wheat or durum, thus providing a basis for gluten-free foods. As explained in Chapter 14, that hope is fulfilled in the pseudocereals, but not in the ancient progenitors of modern wheats.

A logical extension of the assessment-management philosophy described in the book is the integration of analytical and management tools to provide 'expert systems' that can be used to assist in the overall process. Currently, decisions within the management function are largely taken by practised operators, acting upon analytical information, but also drawing upon the intangible characteristic that is summarised by the words 'experience' and 'wisdom'. Technological advances have been made in various management systems via the development of computer-based 'expert systems' that 'learn' to make decisions, or at least recommendations, based on previous experience. Maybe such approaches to grain utilisation will in the future serve to complement the accumulated experience of professional managers.

The future is certainly bringing us novel and improved analytical methods to provide more detailed and accurate information about grains in shorter time, probably permitting the move out of the laboratory literally into the 'field', thereby providing real-time analysis. These developments are likely to transform where, when and how many management decisions are made about grain quality and its processing.

Consumers in some countries have been slow to accept genetically modified grains, although food products containing GMO soy are often distributed widely and accepted. It is likely that remaining resistance will progressively dissipate in coming years, when the advantages to production, processing and nutrition become more evident, together with a fuller realisation of food safety.

Gene technology has the potential to transform breeding for specific quality traits and also to simplify the task of identifying specific aspects of processing quality at harvest. These changes will in turn facilitate segregation and identity preservation of grain according to quality type.

Future years will also see the fruition of current predictions about global warming and climate change. A significant aspect of these changes may be lower protein content for cereal grain, due to increasing starch synthesis under the influence of the 'carbon-fertilising' effect of higher levels of atmospheric carbon dioxide. These CO_2-related changes are more likely for C3 species such as wheat, but less likely for C4 species, for example corn. Other likely effects relate to altered rainfall and more frequent heat-stress episodes during grain filling. The latter are already known to alter grain quality in wheat by reducing dough strength and thus damaging baking quality. However, the expectation of less frequent frost episodes is expected to permit earlier sowing times, so that heat-stress episodes may be avoided.

REFERENCES

Abdel-Aal, E., Wood, P., 2005. Specialty Grains for Food and Feed. AACC International, St Paul, MN, USA.

Abdel-Aal, E.-S.M., Hucl, P., 2014. Einkorn: a functional wheat for developing high-lutein whole grain baked products. Cereal Foods World 59, 6.

Atwell, W.A., 2001. An overview of wheat development, cultivation, and production. Cereal Foods World 46, 59–62.

Belton, P.S., Taylor, J.R.N., 2002. Pseudocereals and Less Common Cereals: Grain Properties and Utilization. Springer, Berlin.

Black, J.L., 2016. Cereal grains as animal feed. Wrigley, C., Corke, H., Seetharaman, K., Faubion, J. (Eds.), Encyclopedia of Food Grains, vol. 3, Elsevier, Oxford, UK, pp. 215–222.

Champagne, E.T., 2004. Rice: Chemistry and Technology, third ed. AACC International, St Paul, MN, USA.

Dixon, J., 2007. The economics of wheat; Research challenges from field to fork. In: Buck, H.T., Nisi, J.E., Salomon, N. (Eds.), Wheat Production in Stressed Environments. Springer, Netherlands, pp. 9–22.

Fletcher, R.J., 2016. Pseudocereals: overview. Wrigley, C., Corke, H., Seetharaman, K., Faubion, J. (Eds.), Encyclopedia of Food Grains, vol. 1, Elsevier, Oxford, UK, pp. 274–279.

Graybosch, R.A., 2016. The grain crops: an overview. Wrigley, C., Corke, H., Seetharaman, K., Faubion, J. (Eds.), Encyclopedia of Food Grains, vol. 1, Elsevier, Oxford, UK, pp. 16–21.

Jones, J.M., Peña, R.J., Korczak, R., Braun, H.J., 2015a. Carbohydrates, grains, wheat, in nutrition and health: an overview. Part I. Role of carbohydrates in health. Cereal Foods World 60, 224–233.

Jones, J.M., Peña, R.J., Korczak, R., Braun, H.J., 2015b. CIMMYT series on carbohydrates, wheat, grains, and health: carbohydrates, grains, and wheat in nutrition and health: an overview. Part II. Grain terminology and nutritional contributions. Cereal Foods World 60 (6), 260–271.

Jones, J.M., Sheats, D., 2016. Grains and health: misinformation and misconceptions. Wrigley, C., Corke, H., Seetharaman, K., Faubion, J. (Eds.), Encyclopedia of Food Grains, vol. 2, Elsevier, Oxford, UK, pp. 23–28.

Khan, K., Shewry, P.R., 2009. Wheat: Chemistry and Technology, fourth ed. AACC International, St Paul, MN, USA.

Köksel, H., Cetiner, B., 2015. Grain science and industry in Turkey: past, present and future. Cereal Foods World 60 (2), 90–96.

Morrison, L.A., Wrigley, C.W., 2016. Taxonomic classification of grain species. Wrigley, C., Corke, H., Seetharaman, K., Faubion, J. (Eds.), Encyclopedia of Food Grains, vol. 1, Elsevier, Oxford, UK, pp. 31–40.

Oelke, E.A., Porter, R.A., 2016. Wildrice, *Zizania*: overview. Wrigley, C., Corke, H., Seetharaman, K., Faubion, J. (Eds.), Encyclopedia of Food Grains, vol. 1, Elsevier, Oxford, UK, pp. 130–139.

Tao, C.J., Yuan, P., Wang, X., Liao, B., 2014. Advances in the development of buckwheat products. Cereal Food Ind. 21, 52–54.

Wood, P.J., Johnson, L.A., 2003. Corn: Chemistry and Technology, second ed. AACC International, St Paul, MN, USA.

Chapter 3

The Diversity of Uses for Cereal Grains

Ian Batey
Sunset Cereal Services, Meadowbank, NSW, Australia

3.1 INTRODUCTION

For over a decade, cereal-grain production worldwide has exceeded two billion tonnes, with growth averaging around 2–3% per annum (Fig. 3.1). There have been some years when there has been a significant drop in worldwide production, for example in 2009 and 2010, but overall the trend for growth has continued. Production in 2015/16 is forecast to be lower than the record of 2014/15, but if the expected production is achieved, it will still be the second largest harvest ever. There has also been continued growth in utilisation with no real drop from one year to the next although there have been some years where the growth has been very small, but despite growing utilisation, world stocks of cereals have remained remarkably constant between 20% and 25% of production (Table 3.1). The concern expressed almost 10 years ago that utilisation would exceed production has so far been shown not to be justified.

On average, only 10–15% of cereal grain produced becomes a part of world trade although for wheat it is around 20% and for rice it is consistently below 10%. Almost half of cereal production is consumed as food, with another 35–40% used as feed and the remainder in other industrial uses. These other uses include production of starch and associated by-products and alcoholic beverages, and in recent years, an increasing amount is being converted to bio-ethanol as an automotive fuel. These statistics apply to total world production; for individual regions, and indeed for each cereal, the proportions vary considerably. This will be discussed under the heading for specific cereal groups later in the Sections 3.5.1–3.5.3.

End-use quality is one of the prime determinants of the value of a grain crop. In order to achieve a premium value for the crop, a farmer needs to deliver grain that meets the requirements of the next 'consumer' in the production–utilisation chain (Chapter 1). For most purposes, these requirements can be reasonably specific. Even for feed use, knowledge of the energy and nutrient composition

Cereal Grains. http://dx.doi.org/10.1016/B978-0-08-100719-8.00003-6

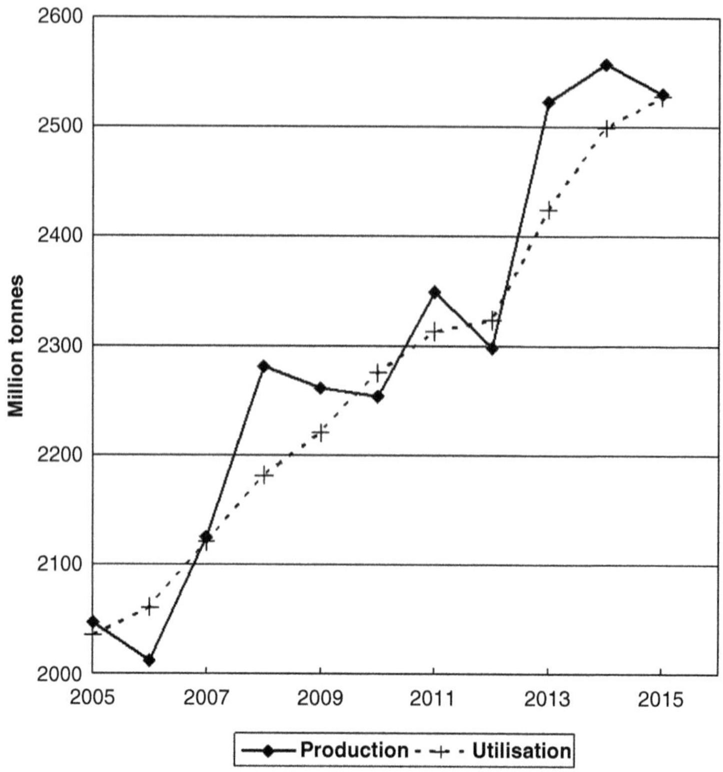

FIGURE 3.1 **Production and utilisation of cereals 2005–16.** *(FAO Cereal Supply and Demand Brief, www.fao.org)*

TABLE 3.1 World Cereal Production, Utilisation, Trade and Ending Stocks for the Period 2005–16

Year	Production (million tonnes)	Utilisation (million tonnes)	World trade as % of production	End stocks as % of production
2005/06	2047.4	2034.5	12.1	22.9
2006/07	2012.6	2060.9	12.8	21.2
2007/08	2124.9	2120.5	12.9	20.3
2008/09	2281.5	2181.4	12.6	22.6
2009/10	2261.7	2220.4	12.4	24.4
2010/11	2253.1	2275.1	12.8	23.3
2011/12	2348.7	2312.7	13.8	23.4
2012/13	2298.2	2323.8	13.8	23.1
2013/14	2522.7	2424.5	14.3	23.6
2014/15	2558.0	2498.6	14.7	25.1
2015/16	2529.6	2527.8	14.4	N/A

Source: FAO Cereal Supply and Demand Brief, www.fao.org

is essential, and for industrial use specific properties may be needed. Thus assessment of quality is an important part of maintaining the value of the grain.

3.2 FOOD

When agriculture began, food would have been the main, if not the only, purpose of growing grain. While it is not certain when grains became a raw material for fermentation and brewing in human pre-history, it is unlikely that the initial step to change from a nomadic to a settled society would have arisen from anything other than the ability for a tribe or group to cultivate and harvest sufficient food for sustaining itself in the immediate situation. In addition, the storability of grain, without the need for further processing, was a particularly attractive attribute of grain, compared to other food sources that were available to early humankind.

In many parts of the world, cereal grain is still grown principally for food (Table 3.2). In some places, a particular grain is grown for food, while in others the same species is used for feed or other purposes. For example the major use for barley in North Africa is for human food (see Chapter 9), while in most of the rest of the world, major purposes are feed and industrial uses. In the case of oats, the proportion used for food can be specific for a country, not just a region (see Chapter 10). In Britain and Brazil, the largest proportion goes to human food. In many other countries, including those close to Britain in Europe, the major use is for feed. It has even varied within a nation. For example Samuel Johnson (1709–84) defined oats as: 'A grain, which in England is generally given to horses, but in Scotland appears to support the people'. To which James Boswell is reported to have replied: 'And where do we find such fine horses as in England, or such fine men as in Scotland'.

Rice is the only cereal which is consumed principally as human food everywhere (Table 3.2) (see Chapter 12). Overall, rice for human consumption makes up over 90% of the total worldwide, but in a few countries, mostly in Europe, there is a significant use of rice for animal feed. This amount varies markedly, rising to as much as 50% in some years, and is clearly a result of price/availability issues of different grains for animal feed. Worldwide, the amount of rice going for purposes other than food or feed usually exceeds the amount consumed as feed. 'Other uses' (Table 3.2) include some industrial usages that result in human ingestion, such as beverage and snack-food manufacture.

Overall, wheat is also predominantly consumed as food, but in some countries feed or industrial usage are major outlets (Table 3.2). Oceania is the only region where food accounts for less than 50% of wheat use domestically, but this is because in Australia there is significant use of wheat for industrial processing as well as wheat is the main grain for feed purposes. In developing countries, the principal use of sorghum is food, but elsewhere it is feed (Kleih et al., 2007) (see Chapter 13).

The general analytical methods applicable to all grains for human use are those that are necessary to ensure safety (absence of chemical or biological

TABLE 3.2 Utilisation of Major Cereals Within Regions as Food, Feed and Other Uses

		Africa	Asia	Europe	North America	South America	Oceania
Wheat	Food	89.2	88.5	50.2	77.2	96.6	37.1
	Feed	8.9	9.1	45.7	21.7	3.0	51.7
	Other	1.9	2.4	4.1	1.1	0.4	11.2
Rice	Food	93.1	88.6	88.6	93.6	93.3	81.4
	Feed	3.0	8.2	8.8	0.0	1.9	0.0
	Other	3.8	3.1	2.5	6.4	4.8	18.5
Maize	Food	65.4	16.2	6.9	1.9	16.6	20.6
	Feed	33.3	70.9	88.3	62.0	74.9	74.1
	Other	1.3	12.9	4.8	36.1	8.5	5.2
Sorghum	Food	78.5	72.8	0.0	6.0	0.0	0.0
	Feed	14.3	27.0	99.6	94.0	99.6	100.0
	Other	7.2	0.3	0.4	0.0	0.4	0.0

Values are expressed as a percentage of total utilisation for the region.
Source: FAO Statistics website, www.fao.org

toxin residues) or end-product quality. Descriptions of these procedures will be found in the chapters relevant to the specific grain species and in Chapters 15, 17 and 24.

3.3 FEED

The 'coarse grains' (cereals other than rice and wheat) are mostly utilised for feed, with industrial use being greater than consumption as food in many countries. There are some exceptions; for example millets, which are almost exclusively used as food where they are grown as a major crop (Chapter 13). Maize has the largest annual production and accounts for around three quarters of the annual coarse-grain production. Most of its production is utilised as feed, with industrial uses accounting for more than human consumption in most countries (see Chapter 11).

Barley accounts for approximately 5–8% of world cereal production, with the principle use being feed. The development and production of high-lysine barley (Doll et al., 1973; Lange et al., 2007) illustrate this important aspect of its use. Sorghum is probably the next in the order of production, with an annual production fluctuating significantly around 50–70 million tonnes. The fluctuations are caused both by climate (mainly drought) and land-use changes by farmers. In developed countries, the main use of sorghum is feed.

For feed purposes, energy and nutritional value are the important considerations. Except where grain is used as emergency fodder, animal feed is formulated with careful consideration for the type and stage of development of the animal, the purpose for which the animal is being kept and the cost of the various grains that may comprise the feed material. Thus, analysis of the protein content and amino-acid composition, the digestible carbohydrate, lipid (oil) content and fibre content are all necessary to ensure that the final product meets the desired nutritional quality. As for human food, grains used for feed must meet standards for the presence of toxic or anti-nutritional components. For some animals, specific grains are considered less suitable and are avoided where possible.

3.4 INDUSTRIAL USES

The amount of grain utilised for neither food nor feed purposes has grown in recent years. In the period 1998–2003, grain for industrial use accounted for less than 6% of the amount consumed for food. Ten years later, it had grown to comprise 21% of food grain use (Fig. 3.2). It is this increase in industrial utilisation that has resulted in the trend growth for utilisation to rise above the trend for production. The main industrial use of grain is to isolate the starch component and then process it further. Starch comprises 60–80% of the dry matter of most cereal grains and it is isolated industrially from maize and wheat, and to a lesser extent from rice. The resulting starch may be utilised as it is or it may be processed further. Industrial processing of grains may provide products for human consumption, perhaps as alcoholic beverages or as starches added to foods to give desirable functional properties in the food.

Modified starches (anionic, cationic and cross-linked, to name a few) have a wide range of purposes in food and non-food applications. A significant amount

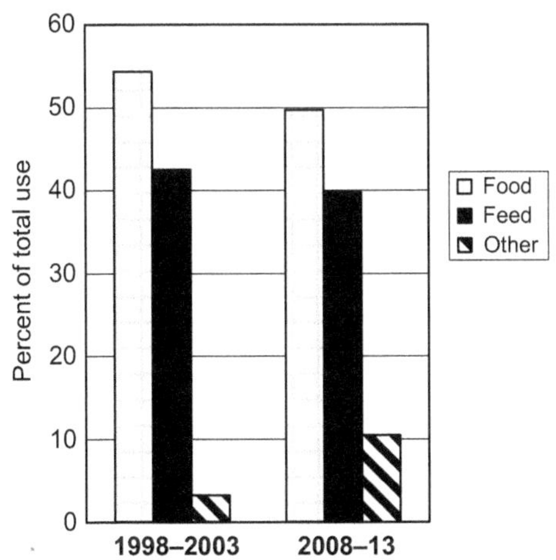

FIGURE 3.2 Utilisation of cereals as food, feed and other uses in the periods 1998–2003 and 2008–13. *(FAO Statistics website, www.fao.org)*

of starch is digested by enzymic or a combined acid/enzymic treatment to give glucose and some of this glucose is converted to high-fructose syrups for use as a sweetener. The starch may also be subjected to fermentation to ethanol. This use has traditionally been for the production of alcoholic beverages, but high oil prices have resulted in an increase in the amount of grain being converted to automotive fuel. Much of the increase in the amount of grain being used for purposes other than food and feed has been for bio-ethanol production. For the production of ethanol, whether for beverage or other use, grain is usually fermented without prior isolation of the starch component. There are some exceptions to this general principle. Some wheat-starch factories are associated with alcohol production, by utilising the B-starch (small and damaged granules) for fermentation. As small granules in wheat comprise around 20–30% of the total starch, this practice allows a major reduction in the water and energy required for purifying this fraction and permits production of the high value products (starch and gluten, as well as ethanol) at a lower cost.

In the United States, corn is still the major grain used for the production of ethanol, accounting for over 90% of ethanol production. In Europe and Australia, wheat is used more widely for ethanol, although corn is also used in Europe. In Australia, sorghum is also used and a purpose-built refinery has been constructed in one of the main sorghum-growing regions to take advantage of the local crop with minimal transport costs.

The paper industry is a major consumer of starch in its products, but it remains to be seen whether the trend towards digital format for books and newspapers will see a reduction in the requirement for starch in this industry.

Processing of wheat for industrial purposes is somewhat different from that for other grains. Although the starch is used as an essential industrial commodity, it is the protein of wheat, vital wheat gluten, which is the major product of value (about \$1500/tonne as against \$300/tonne for starch). The high value of this protein component makes starch production from wheat economically feasible. Most of the industrial applications of wheat involve the starch and/or gluten components.

3.5 UTILISATION OF INDIVIDUAL CEREALS

Most cereals are utilised in the country where they are grown and only 10–15% crosses borders as part of world trade (Table 3.1). Individual requirements for quality are set locally, and may vary from country to country depending on consumer preferences for food and the availability of methods and equipment to measure that quality. For international trade, specifications are often more rigid and analytical methods have been approved by organisations, such as AACC International (2000) and ICC (2011). Some aspects of quality are briefly discussed later for wheat and rice. For these grains, there are many aspects of quality used to determine suitability for particular purposes. See also Chapters 5, 12, and 15–17.

3.5.1 Wheat

Once physical issues, such as foreign matter have been considered by wheat traders, the important quality parameters are test weight, grade, protein content and dough strength (see Chapter 5). Cultivar (variety) is often important, but as one of the criteria used to assign wheat to a premium grade. Test weight will have an influence on milling yield, an important consideration for millers. Dough strength is affected by variety and by protein content, but for a specific protein content, the composition of the protein will also have an effect. Protein composition is not usually measured, as it is determined to a large extent by the genotype of the wheat. Knowledge of the variety will give an indication of the likely composition, although it can also be influenced by environmental conditions, such as heat stress during grain filling.

Protein content is often used as a primary way of selecting wheat for a particular purpose (Fig. 3.3). Hardness (or softness) is also used, although in commercial cultivars, soft wheat tends to have a lower protein content. Soft wheat is usually used for cakes and cookies where damaged starch (see Chapter 22) leads to undesirably increased water absorption. On the other hand, in wheat for

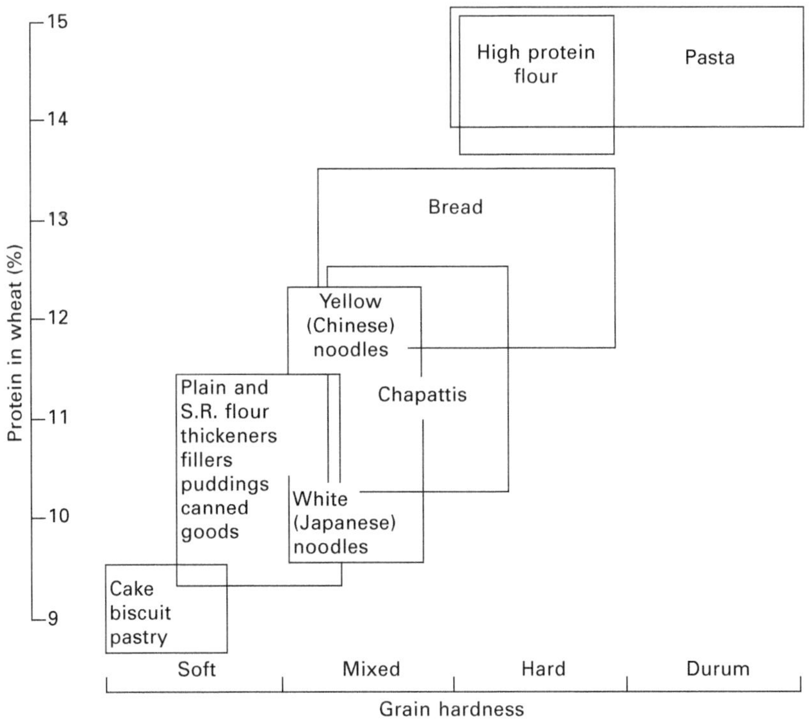

FIGURE 3.3 Range of wheat protein required for different products. *(Based on Moss, H.J., 1973. Quality standards for wheat varieties. J. Aust. Inst. Agric. Sci. 39, 109–115)*

bread making, increased water absorption is seen as desirable and so hard wheat is preferred for this purpose. Bread wheat also requires higher protein content than wheat for cakes and cookies. Between these protein contents, there are the requirements for the various types of noodle (excluding pasta), as indicated in Fig. 3.3.

Pasta is usually made from durum wheat with high protein content and the grain is very hard (see Chapter 6). Durum has a tetraploid genome, as distinct from the hexaploid genome of common or bread wheat. As well as hardness and protein content, endosperm colour is also a major determinant of durum quality. The presence of black point in durum grain is undesirable. Although there is no apparent loss in textural properties, the presence of black point may result in black specks occurring in the subsequent pasta. Durum grain should have highly vitreous endosperm, but may have non-vitreous regions (mottled) or even be almost completely non-vitreous (yellow berry). Durum with non-vitreous regions does not appear to yield lower quality pasta, but it tends to cause a reduction in yield of semolina from milling and so is avoided where possible. Grain containing significant amounts of mottled kernels will be downgraded. This property is usually assessed by visual inspection.

Wheat is usually tested for test weight, protein content, moisture, small grains, grain defects and foreign matter prior to receival. In those countries where on-farm storage is normally practiced, grain will be purchased based on the results of these and other tests. Testing is usually carried out before purchase and may include testing for variety (see Chapter 17). Where the grain goes into bulk storage, other tests as appropriate may be applied. These include testing for rain damage (premature sprouting) by Falling Number or Stirring Number methods, and the farmer delivering the grain is usually required to make a declaration of variety. Grain is then placed into storage with other grain of similar quality type and properties.

3.5.2 Rice

The quality of rice is often judged visually, as described in Chapter 12. As rice is usually purchased, cooked and eaten as milled (polished) rice, many aspects of its quality may be seen. Grain shape, size and the presence of broken or chalky kernels are key attributes which are important to the consumer. Increased chalky regions of the grain usually lead to more broken kernels, and in severe cases, may lead to the grain disintegrating into small fragments. This makes the rice unusable for many purposes, although much broken rice and rice flour is now used for extruded products. Broken grains also find a ready market in fermentation. Grain shape is often associated with particular varieties and types, sought by consumers for particular sensory properties, such as texture, flavour and odour. This may arise because consumer preferences lead breeders to emulate visual phenotypes in new varieties, not because the appearance actually affects the other sensory properties.

Starch provides many of the textural properties of rice after cooking, and aspects of starch quality need to be considered. Some varieties have a waxy allele, resulting in very low amylose content, and these rice types are sought for their textural properties. Amylose content, however, is not always a direct predictor of rice quality as rice samples of similar amylose content may have very different textures. Basmati rice has the characteristic of swelling longitudinally, and in some markets a premium is paid based on this property. Basmati rice also has a low glycaemic index and in Western countries where diabetes, obesity and heart disease are increasing, it is seen as a more desirable product than rice with higher glycaemic index.

3.5.3 Maize (Corn)

Corn is the third grain species of the 'big three', the three grains that are produced at over 600 million tonnes each year. Although a large proportion of corn production worldwide is used as feed and for industrial processing (see Chapter 11), it has long been an important human food source in some countries and regions (Table 3.2), and food uses are spreading to other cultures and countries. Africa is the only region where it is predominantly used as food with almost two-thirds being consumed for this purpose. In Mexico also, over 60% of maize is used as food. Nixtamalization is one of the traditional processes for corn treatment. The process involves treating corn grain with alkali to make the nutrients more available. Traditionally, the alkali used may have been either lime or sometimes potash from the cooking ashes (Berzok, 2005). In modern usage, lime is the preferred ingredient. Native populations that used alkaline treatment were able to have corn as their main food substance, supplemented by animal protein. Those that used unprocessed maize needed additional food sources. Although vitamins and minerals are more readily available after alkali treatment, it also results in the formation of lysinoalanine and other 'unnatural' amino acids, and a reduction in the amount of the essential amino acid, lysine, available for nutrition. The alkaline treatment may also have had benefits in reducing the amounts of mycotoxins resulting from fungal infection. Traditional tortillas are made solely from alkaline-treated maize, but modern commercial production often includes some wheat flour in the recipe.

3.6 GRAIN SUBSTITUTION

Chapters 5–14 that describe each of the major cereal species provide details of their various forms of utilisation, but it is relevant in this introductory chapter to consider the extent to which the traditional use of one cereal may be substituted by another. For example, no major cereal species can replace the role of maize to provide the eating experience of 'corn-on-the-cob'. On the other hand, the role of rice as a whole-grain cooked food can be substituted by a much cheaper alternative, namely sorghum, after milling by abrasion to remove the outer husk layers (as for rice). However, much longer boiling may be needed for sorghum before it is readily eaten (Welsch, 1972). Pearled barley is another example of eating a whole grain after removal of the bran layers.

The principle of making porridge by adding water to crushed grain is one of the most primitive of food preparation methods for the cereal grains, and one that is readily applicable to any of the starch-rich grains. However, the most popular grains for this purpose are still oats in Western civilisations and sorghum in developing countries, particularly in Africa (see Chapter 13).

Cross-substitution between cereals is evident with respect to fermentation, either for beverage production or for the manufacture of bio-ethanol–based fuel substitutes. For fuel, the raw material is chosen to maximize ethanol yield per unit input cost, and will take into account the starch content together with the availability of the starch for degradation and fermentation. Research has shown that millet could be an acceptable alternative to maize, as it gives a high yield of alcohol plus a dried distillers' grain of higher nutritional quality. Traditional beverages have usually used the grain most commonly grown, or at least available, locally. Barley is the traditional cereal used for beer production (see Chapter 9), but the incorporation of wheat flour as an 'adjunct' is common practice, and beer brewed from wheat alone is popular in some countries.

The ready availability of a particular grain as the preferred raw material is particularly evident for distilled spirits. In Scotland, malted barley is used for malt whiskies, but other grain whiskies utilize maize and wheat together with barley malt. Maize and rye are used in the USA and Canada with some malted grain and amylolytic enzymes. Elsewhere, other starch and sugar sources (potato, molasses, fruit juices) are used, often in association with some grain or malted grain. When fermenting starchy mixtures to ethanol, research has suggested that waxy grains (i.e. grains containing starch which consists, almost totally, of amylopectin) may give a higher yield of alcohol, in wheat at least (Zhao et al., 2009). Waxy wheats are a relatively recent development and grain yields are not yet as high as 'normal' wheat, so the overall production of alcohol would be lower per unit crop area.

The production of leavened bread from non-wheat sources is difficult, even though such a product is attractive to those who must avoid wheat gluten because of their dietary requirements, for conditions, such as coeliac disease or other forms of wheat intolerance. Mixed soy and wheat-flour breads have been shown to produce an acceptable alternative bread product when there are economic constraints on using 100% wheat flour (Maforimbo et al., 2006).

3.7 WHOLE-PLANT UTILISATION

The possibility of using the whole plant as a source of more than the traditional products from grain has been considered many times in recent history. As always, 'necessity has been the mother of invention' and in subsistence agriculture, all parts of the plant are used: as food, as fuel or for construction purposes.

In times of war, there has been renewed interest in converting agricultural waste material into products that can replace ones that are unavailable or in short supply. The energy crises of the 1970s and 1980s have given impetus to utilisation of the whole plant as a means of producing liquid fuels and chemicals normally prepared from oil (Talebnia et al., 2010).

The ready availability of cellulose, a β-linked polysaccharide of glucose, makes it an attractive target for the large-scale production of ethanol from the simple sugars that derive from cellulose breakdown. However, cellulose is much more resistant to enzymic digestion than starch, and ethanol production from cellulose is currently not an economic proposition. It is, however, the subject of increasing research interest and commercial processes may become feasible.

There are, however, other industries based on utilisation of waste parts of the plant. The first commercial production of furfural was from oat hulls (a residue from milling oats for food) in the 1920s. There are large amounts of hemicelluloses (pentosans) in many parts of grain plants, and these can be hydrolysed and dehydrated to form furfural (Adams and Voorhees, 1921). This compound is still mass produced from agricultural waste, such as corn cobs and sugar cane bagasse. Modern interest centres mainly on the production of alternative liquid fuels from plant residues.

3.8 FUTURE TRENDS

Changes in utilisation are already causing increased competition between the various basic uses of grains. Increases in ethanol production have caused the proportion of grain available for more traditional purposes to be reduced, and there has been a consequent increase in prices for feed and industrial grain. However, this has been offset to a certain extent by increased production. The premium paid for the higher quality grain consumed as human food has not, as yet, been affected by increased conversion of grain to ethanol. The price of quality grain for human consumption is determined largely by supply and affected by climatic variations. Demand has been increasing due to increasing affluence in some developing countries. With the trend towards higher prices for grain used for feed and industry, grain for human consumption may also be forced upward in price. This may be partially offset by improved technologies for conversion of grain to ethanol, and by new technologies for utilising other parts of the plant for producing liquid fuels. The choice of grain for ethanol production is determined by the price of the possible alternatives. In the United States, for example sorghum is sometimes being used with a concomitant reduction in the amount of maize. However, higher sorghum prices and increased maize production may see this trend reversed. Sweet sorghum is also being considered for the production of ethanol, but in this case it is the plant itself that provides the fermentable sugars. The stems are crushed in a similar manner to sugar cane. Lower oil prices worldwide have reduced the demand for cereals to be processed to fuel ethanol, although this reduction in demand may only be temporary.

Improvement in rapid testing should assist users in accepting grain for further processing. Although Falling Number and Stirring Number are the main ways for assessing sprouting damage, immunoassay-based tests have been developed (WheatRite) and may be applied rapidly with a minimum of equipment (Verity et al., 1999). This type of test is also being applied for detection of contaminants, such as pesticides and mycotoxins. Extension of this type of technology to other properties could revolutionise testing at grain receival.

SOURCES OF FURTHER INFORMATION

Campbell, G.M., Webb, C., McKee, S.L., 1997. Cereals: Novel Uses and Processes. Plenum Press, New York, NY, USA.

Carver, B.F., 2009. Wheat: Science and Trade. Wiley-Blackwell, Ames, IA, USA.

Champagne, E.T., 2004. Rice: Chemistry and Technology, third ed. AACC International, St Paul, MN, USA.

European Biofuels Technology Platform website: biofuelstp.eu/bioethanol.html

Kent, N.L., Evers, A.D., 1994. Technology of Cereals, fourth ed. Pergamon Press, Oxford, UK.

Khan, K., Shewry, P.R., 2009. Wheat: Chemistry and Technology, fourth ed. AACC International, St Paul, MN, USA.

The Food and Agriculture Organization web site: www.fao.org

The International Rice Research Institute web site: www.irri.org

White, P.J., Johnson, L.A., 2003. Corn: Chemistry and Technology, second ed. AACC International, St Paul, MN, USA.

REFERENCES

AACC International, 2000. Approved Methods of Analysis, eleventh ed. Method 44-19.01. Moisture—Air-oven method, drying at 135°. Approved April 13, 1961. AACC International, St Paul, MN.

Adams, R., Voorhees, V., 1921. Furfural Org. Synth. 1:49, Coll. vol. 1, 280. Available from: http://www.orgsyn.org/orgsyn/orgsyn/prepContent.asp?prep=cv1p0280

Berzok, L.M., 2005. American Indian Food. Greenwood Press, Westport, CT, USA.

Doll, H., Andersen, A.J., Koie, B., Ingversen, J., 1973. High lysine mutants in barley. Barley Genetics Newsletter 3, 12–13.

International Association for Cereal Science and Technology, 2011. ICC Standard Methods. ICC, Vienna, Austria.

Kleih, U., Ravi, S.B., Rao, B.D., Yoganand, B., 2007. Industrial utilization of sorghum in India ejournal.icrisat.org 3(1), 1–38.

Lange, M., Vincze, E., Wieser, H., Schjoerring, J.K., Holm, P.B., 2007. Suppression of C-hordein synthesis in barley by antisense constructs results in a more balanced amino acid composition. J. Agric. Food Chem. 55, 6074–6081.

Maforimbo, E., Skurray, G., Uthayakumaran, S., Wrigley, C.W., 2006. Improved functional properties for soy-wheat doughs due to modification of the size distribution of polymeric proteins. J. Cereal Sci. 43, 223–229.

Talebnia, F., Karakashev, D., Angelidaki, I., 2010. Production of bioethanol from wheat straw: an overview on pretreatment, hydrolysis and fermentation. Bioresour. Technol. 101, 4744–4753.

Verity, J.C.K., Hac, L., Skerritt, J.H., 1999. Development of a field enzyme-linked immunosorbent assay (ELISA) for detection of alpha-amylase in preharvest-sprouted wheat. Cereal Chem. 76, 673–681.

Welsch, D.E., 1972. Some speculations of the long run future of rice in Thailand, Staff Paper P72-2. Department of Agricultural and Applied Economics, University of Minnesota Institute of Agriculture, St Paul, MN, USA. Available from: http://purl.umn.edu/13262

Zhao, R., Wu, Z., Seabourn, B.W., Bean, S.R., Guan, L., Shi, Y.C., Wilson, J.D., Madl, R., Wang, D., 2009. Comparison of waxy vs. nonwaxy wheats in fuel ethanol fermentation. Cereal Chem. 86, 145–156.

Chapter 4

Cereal-Grain Morphology and Composition

Colin Wrigley
QAAFI, University of Queensland, St. Lucia, QLD, Australia

4.1 INTRODUCTION

The cereal grains share the uniformity of belonging to the family of grass plants, members of the Gramineae (or Poaceae) (Morrison and Wrigley, 2016). Within the bounds of that taxonomic uniformity, there is considerable diversity that distinguishes the various cereal grains with respect to their morphology and composition. These contrasts help to explain why some are better suited to certain types of utilisation than others. Diversity of grain shape and presentation on the plant at maturity is most evident when corn (maize) is contrasted with wheat. The corn cob is suited for direct consumption as cooked 'corn-on-the-cob' in a way that is out of the question for wheat, which is milled (ground to flour) for most wheat-based food products. On the other hand, rice is usually consumed as a whole grain, but only after a distinct form of 'milling' to remove the outer layers. The grains of wheat, rye and triticale, when harvested and threshed, are provided free of the outer glumes, but this is not so for most genotypes of barley, oats and rice. Morphology is thus an important consideration for utilisation and, of course, for distinguishing between species.

An aspect of morphology that indicates uniformity across the cereal-grain species is the consistent presence of a large amount of endosperm tissue, which acts as a food store for the developing embryo during germination and beyond, until the early leaves start to produce nutrients via photosynthesis. At the microscopic and molecular levels, diversity among the cereal species is provided by differences in the shapes of the starch granules of the endosperm and in the nature of the endosperm proteins.

The extent of taxonomic uniformity also applies to grain composition across the cereal grains, in that they all have a high starch content, moderate levels of protein and relatively minor amounts of lipid (fat). In all of the major cereal grains, most of the lipid is concentrated in the germ (embryo), with only small

Cereal Grains. http://dx.doi.org/10.1016/B978-0-08-100719-8.00004-8

amounts in the endosperm. See the tables of composition, provided in Appendix 1, which contrast the composition of the endosperm (white flour) with that of whole grain and of germ and bran for a few cereals. Within the taxonomic uniformity, genetic diversity is especially evident in protein composition, proteins being the first aspect of grain composition resulting from gene expression. A significant consequence of differences in protein composition is the unique dough-forming characteristic of the endosperm storage proteins of wheat, which alone is suitable for making leavened bread. These gluten-forming proteins are not present with the same rheological characteristics in the grains of any other cereal species, although there is some similarity for rye endosperm proteins.

This chapter complements the descriptions of specific cereals (Chapters 5 to 14) by contrasting the diversity of size and shape of the major cereal species, while also pointing out the uniformity that causes them to be grouped together taxonomically, as well as explaining how differences in composition and morphology serve useful purposes in the assessment and management of grain quality across all the cereal species.

4.2 THE STUDY OF PLANT, HEAD AND GRAIN MORPHOLOGY

The cereal species show considerable diversity in the size and shape (morphology) of their plants, of the grain-bearing structures and of the grains themselves. Obviously, the differences are greatest for the species that are most distant taxonomically, and least at the variety level (genotypes within the same species). Nevertheless, subtle morphological differences at the variety level can be usefully applied to the identification of varieties.

This opportunity is especially useful if it can be applied to grain samples, although this is probably the most difficult growth stage for identification, based only on visual examination. Morphological identification is also possible throughout the life of the plant. Examples of these opportunities are provided, with the accent on wheat, barley and oats as the examples. Illustrations are largely taken from the excellent handbook of Hervey-Murray (1980), now out of print and difficult to obtain. Reproduction in this chapter now offers an extension of the lifetime of some of his illustrations. Pre-dating Hervey-Murray is the classical book '*The Wheat Plant*' by Percival (1921).

Now, a few decades after Hervey-Murray, the need for morphological identification is still important, given the requirements for enforcing Plant Breeders' Rights via variety registration (by the International Union for the Protection of New Varieties of Plants, www.upov.int), based on the attributes of distinctness, uniformity and stability (DUS). For further information, see Jarman (1995), Mauria (2000), Cooke and Wrigley (2004) and the UPOV web site. Several books and review articles are available concerning the morphological characteristics that are useful for variety identification, some of them accompanied by identification descriptions and keys for the specific varieties in national variety collections; for example, Ferns et al. (1975) and subsequent editions (see 'Sources on further information') thereafter to the last by Fitzsimmons et al. (1986), Jarman (1995, 2004) and Agrawal (1997). See also the chapters in this book that describe the specific cereal species (Chapters 5 to 14).

4.3 THE CEREAL FLOWER AND POLLINATION

All cereal species are flowering plants, classified as angiosperms, reproducing sexually with the male and female organs being located in the flowers, although the flowers may be inconspicuous compared to those of other plants. The illustration of a wheat flower (floret), without the outer glumes (lemma and palea) (Fig. 4.1), shows the pollen-producing male anthers. At fertilisation (anthesis), the lodicules (at the base of the ovary) become swollen, forcing apart the lemma and palea. The anthers split open, releasing pollen grains which fall onto the stigma. Next, the anthers are pushed out of the floret and the remainder of the pollen is released beyond the floret.

Some of this excess pollen may fertilise neighbouring florets on the head. For wheat, however, there is only a small possibility that the pollen will be effective beyond the floret of origin because wheat is almost exclusively self-fertilising. Fertilisation of barley, oats and rice also occurs mainly within the

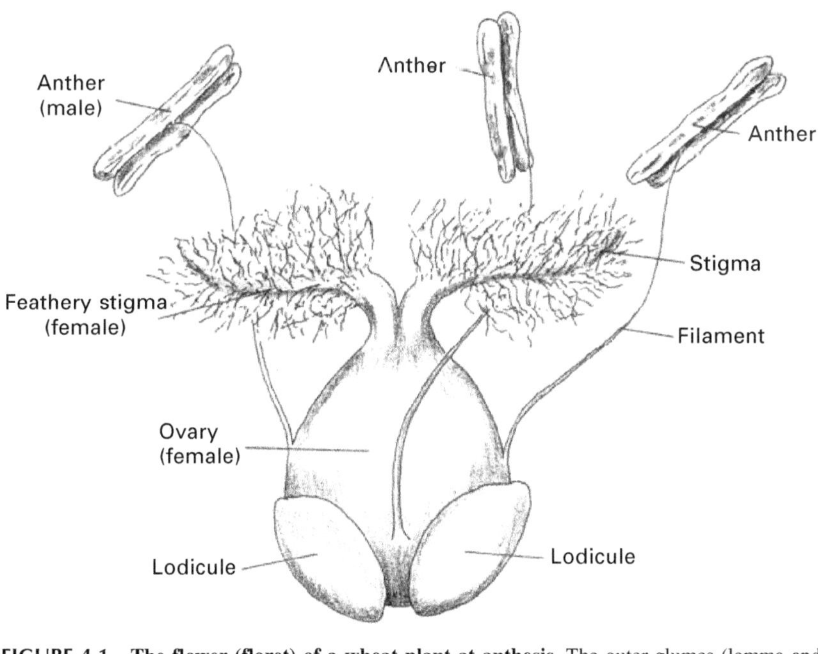

FIGURE 4.1 The flower (floret) of a wheat plant at anthesis. The outer glumes (lemma and palea) are not included in this illustration. *(Reproduced with permission from Hervey-Murray, C.G., 1980. The Identification of Cereal Varieties, RHM Arable Services Ltd., Cambridge, UK.)*

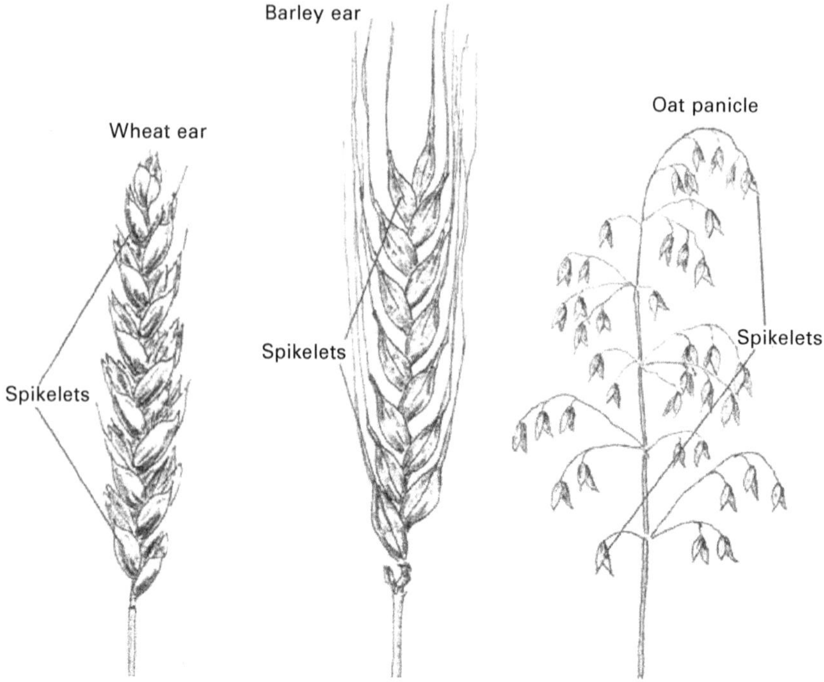

FIGURE 4.2 The contrasting morphology of the seed-bearing head (ear, panicle) of three major cereal species—wheat, barley and oats. *(Reproduced with permission from Hervey-Murray, C.G., 1980. The Identification of Cereal Varieties, RHM Arable Services Ltd., Cambridge, UK.)*

floret; rye is more likely to be out-crossing. The florets (spikelets) of wheat, rye and barley are attached close together on the rachis (the central backbone of the head), whereas they are well separated on the panicle of oats and rice (Fig. 4.2).

In contrast, the pollen of corn (maize) is released from the tassels on top of the plant, to fall onto the 'silks' that extend from the end of the cob's covering (Fig. 4.3). Corn is out-crossing, because the pollen is released from the tassels far above the target stigma, so that wind currents may carry the pollen well beyond the plant from which it originated. In both cases (corn and wheat), the next stage is similar; the pollen grain produces a tube-like structure which enters the style of the stigma, travelling down to fertilise the ovary and initiate a new grain, and thus (eventually) another plant.

Seeds in general are protected by a surrounding fruit, but in the case of the cereals, the 'fruit' is reduced to a thin outer tissue (bran) adhering to the seed. The cereals are distinguished taxonomically because they are monocotyledonous (monocots), distinct from the dicotyledonous grain species (dicots), such as legumes, pseudocereals and oilseeds. See lists in Chapter 2. The cereal-grain species are unified by having only one embryonic leaf (a cotyledon) in the germinating seed and young plant. The monocots are also characterised by having

FIGURE 4.3 The mature corn ear, showing the tassels and with the enfolding sheath partly opened to reveal the grains still fixed to the cob. *(Illustration provided by Prof Zoltán Győri and Dr Péter Sipos.)*

flower parts in threes, narrow parallel-veined leaves, scattered vascular bundles and a fibrous adventitious root system (Morrison and Wrigley, 2016). The cereal grains are all members of the grass family, known either as the Gramineae Family (old name) or the Poaceae Family (new name).

4.4 PLANT MORPHOLOGY

As cereal plants grow, the possibilities for distinction at the variety level increase, but even at the 3- to 6-leaf stage, some distinctions may be possible for wheat and barley varieties (Hervey-Murray, 1980; Ferns et al., 1975; Fitzsimmons and Wrigley, 1984). For example, Fig. 4.4 illustrates differences in growth habit, these being the extremes, whilst there may also be habits intermediate between these. Leaf dimensions (length and width) and

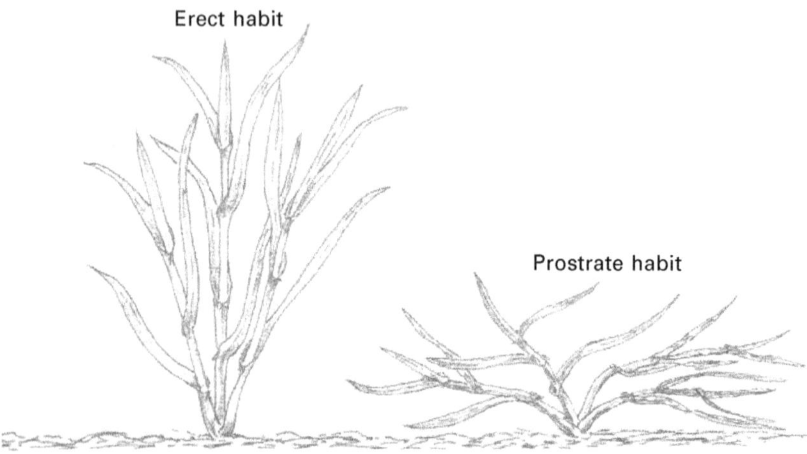

FIGURE 4.4 Young wheat plants, illustrating the two extremes of growth habit (erect and prostrate). *(Reproduced with permission from Hervey-Murray, C.G., 1980. The Identification of Cereal Varieties, RHM Arable Services Ltd., Cambridge, UK.)*

hairiness (pubescence) may also be useful for variety identification, but dimensions are generally subject to growth conditions as well as to genetic influences.

While a cereal grain is seen by us as a food source, it is really the means by which the respective plant perpetuates itself and by which it is multiplied. For cereal plants, the bulk of the grain is a repository of nutrients stored to benefit the new plant until its leaves are developed enough to provide its own energy needs via the processes of photosynthesis. The first stage (germination) is initiated by moisture, involving the swelling of the germ (embryo), followed by the splitting of the germ cover to permit the growth of the roots and coleoptile. Continued growth of the coleoptile leads to the appearance of the first green leaf, followed in turn by successive leaves unfolding from the leaf sheath (stem), each clasping to the stem by an auricle. A wheat plant is illustrated in Fig. 4.5 as a typical cereal, showing only a single stem for simplicity of illustration, but most wheat genotypes produce multiple stems (tillers), which arise from the base of the plant.

Cereals such as wheat show varietal differences in various aspects of plant morphology throughout the stages of growth, permitting limited opportunity for variety identification. As already illustrated in Fig. 4.4, in the early stages of growth when there are several leaves, growth habit may be classed as prostrate (lying almost flat against the ground) for some varieties, or as erect for others (the leaves standing reasonably upright).

At various growth stages, the abundance of short hairs on the leaf surface (leaf pubescence) is another attribute to use in variety identification, as are also leaf dimensions (length and width), but the latter are also dependent on growth

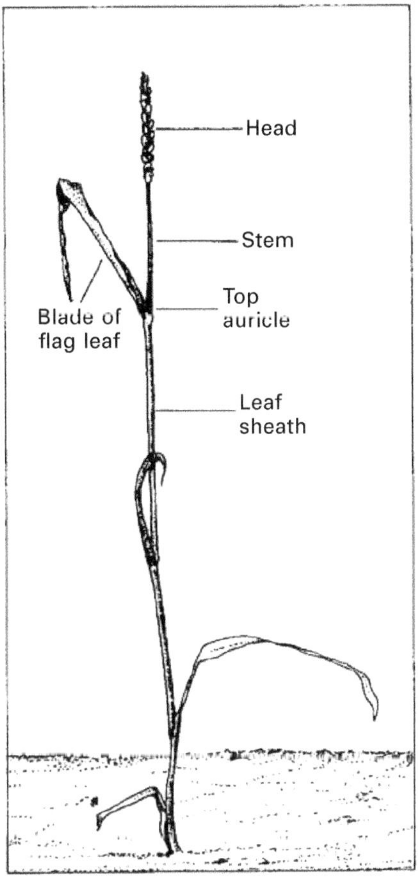

FIGURE 4.5 A wheat plant, showing only one stem, near to maturity with the stem fully elongated, carrying the head (ear) on its tip. *(Reproduced with permission from Ferns, G.K., Fitzsimmons, R.W., Martin, R.H., Simmonds, D.H., Wrigley, C.W., 1975. Australian Wheat Varieties: Identification According to Growth, Head and Grain Characteristics, CSIRO, Melbourne, Australia.)*

conditions. Fig. 4.6 shows how the position of the flag leaf (the last leaf to appear) of wheat may indicate varietal differences.

4.5 HEAD MORPHOLOGY

As the stem elongates, the grain-bearing head (ear, spike, panicle) (Fig. 4.2) develops inside the upper leaf sheath. Eventually the last leaf (the flag leaf) unfolds and the head appears, and flowering occurs. Wheat, triticale, oats and barley are largely self-pollinating, with little out-crossing so that pollen is not shared with neighbouring flowers. This means that the plant remains true to genotype during propagation once its genetic constitution has become stabilised. Following

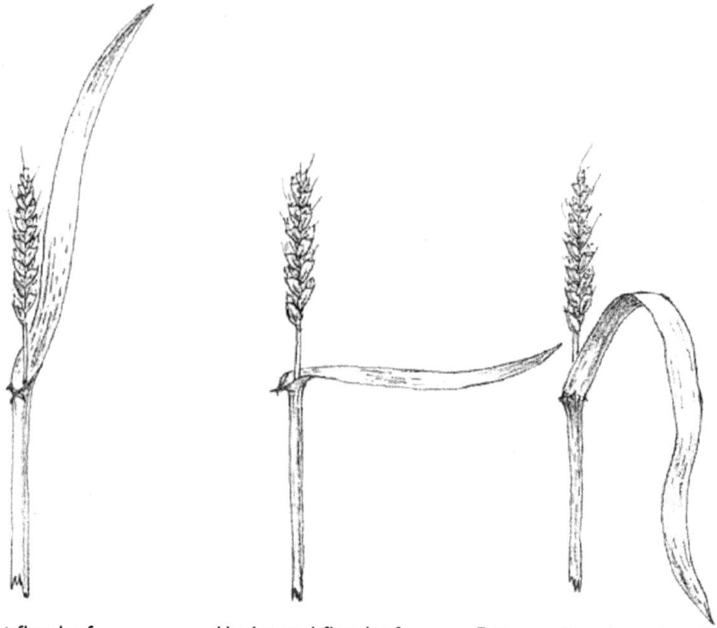

FIGURE 4.6 Varietal differences in the orientation of the flag leaf of a wheat plant approaching maturity. *(Reproduced with permission from Hervey-Murray, C.G., 1980. The Identification of Cereal Varieties, RHM Arable Services Ltd., Cambridge, UK.)*

flowering, the fertilised embryo develops inside the protecting glumes. For the purposes of distinguishing between varieties, it is useful to note plant height at maturity and straw strength (both dependent on growth conditions).

Time to maturity is an important genetic attribute as it is interactive with time of planting and expected time for harvest. In many plant-growing regions, the timing of these events is critical. The time for sowing must coincide with an unpredictable rain event to promote uniform germination, flowering must be timed to occur after the last frost, but the later stages of grain filling should not be so late as to occur when daily temperature maxima are excessive (e.g. over 35°C for wheat).

A major morphological difference for barley is provided by the number of spikelets attached side-by-side on the rachis—constituting either two-row or six-row types. The barley head in Fig. 4.2 is two-row. Fig. 4.7 shows that the two-row head has sterile spikelets, which are filled in the case of the six-row head. In some countries, the two-row/six-row distinction is used to differentiate malting (two-row) from feed barleys, respectively.

Several morphological characteristics of the wheat head are useful for variety identification. The most obvious characteristic is the presence or absence

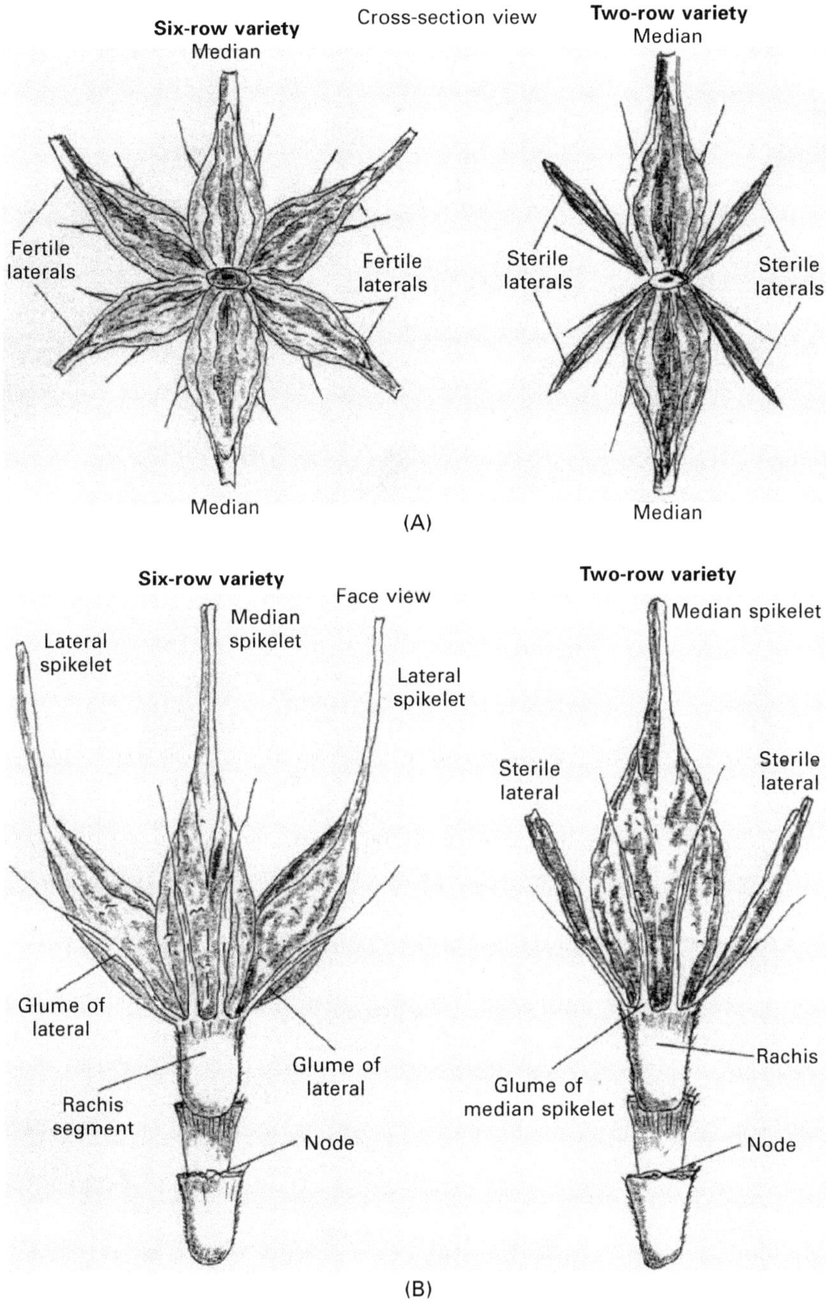

FIGURE 4.7 Morphological characters of the barley head that distinguish the two major types—six-row and two-row. (A) View from above. (B) Side view, with other florets removed. *(Reproduced with permission from Hervey-Murray, C.G., 1980. The Identification of Cereal Varieties, RHM Arable Services Ltd., Cambridge, UK.)*

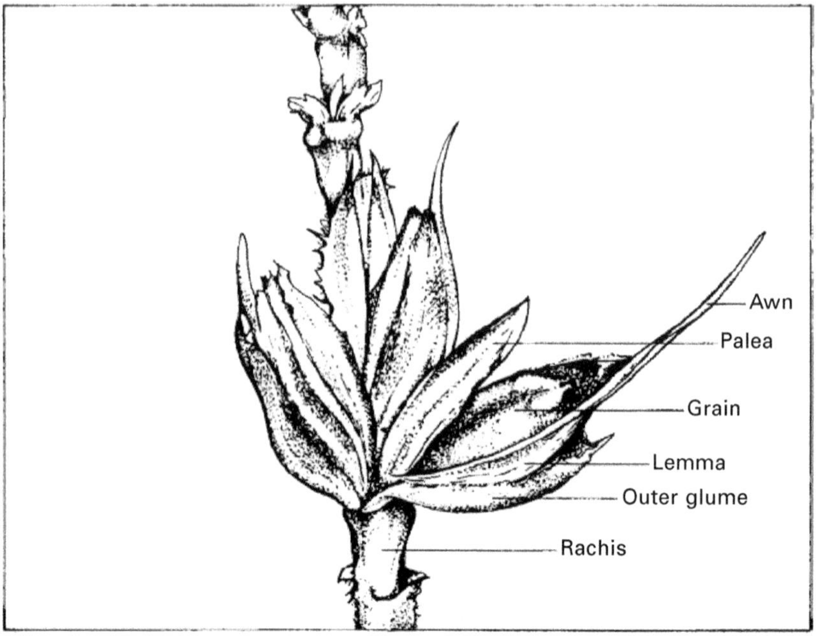

FIGURE 4.8 A single spikelet of wheat, with other florets removed to reveal the rachis, which is the 'backbone' of the head. The lemma and palea have been pulled apart to show the grain resting in the lemma. *(Reproduced with permission from Ferns, G.K., Fitzsimmons, R.W., Martin, R.H., Simmonds, D.H., Wrigley, C.W., 1975. Australian Wheat Varieties: Identification According to Growth, Head and Grain Characteristics, CSIRO, Melbourne, Australia.)*

of awns (beard), the long hair-like extension of the lemma glumes (Fig. 4.8), leading to three classes: unawned, half awned and fully awned (Fig. 4.9). This distinction is so clear that it can be seen in a wheat crop when driving past, provided it is sufficiently mature for the heads to be out of the flag leaf. On the other hand, awns are present on the heads generally for barley, rye and triticale.

After awnedness for wheat, head colour (due to glume colour) at maturity is another obvious distinguishing attribute—either white or brown. A third obvious characteristic is the club head, for which the florets are closely spaced on the rachis (Fig. 4.10). The club head is the extreme case (ten internodes occupying less than 4 cm of head) for the character known as head density. At the other extreme is the lax head, with ten internodes occupying more than 5 cm of head (Ferns et al., 1975). The clubbed wheat head in Fig. 4.10 is in striking contrast to the barley heads—two-row and six-row.

An additional distinguishing characteristic for wheats is head cross-section (either square or flattened when seen from above. The durum head is generally much wider when viewed from the side (looking between the spikelets) than in the face view (the dorso-ventral view, as seen with the spikelets in full face; the

FIGURE 4.9 Wheat heads, seen in dorso-ventral view, of three varieties that differ in awnedness. From left to right, these heads are classed as unawned, half-awned and fully awned. *(Reproduced with permission from Ferns, G.K., Fitzsimmons, R.W., Martin, R.H., Simmonds, D.H., Wrigley, C.W., 1975. Australian Wheat Varieties: Identification According to Growth, Head and Grain Characteristics, CSIRO, Melbourne, Australia.)*

view in Figs. 4.8 and 4.9). Head length may also be useful for identification, but it is partly dependent on growth conditions.

4.6 GLUME MORPHOLOGY

For wheat, useful distinctions can be made using the shapes of the husks (glumes) surrounding the grain in the head. At maturity, the glumes dry out completely while still enclosing the grain. The position of the grain in a floret can be seen in Fig. 4.8. In this expanded view, the grain is resting inside the lemma with the palea above it. The lemma continues up and out of the head as the awn (if awns are present). Each floret of the rachilla has a lemma and a palea, but only the lower florets have the outer glumes (Fig. 4.8). Only these outer glumes are useful for variety identification.

Distinctive glume shapes are shown in Fig. 4.11. Even when examining a wheat-grain sample for visual identification of variety, glume shape is often useful, as there may be glumes present that have not been winnowed away from the grain. Therefore, glumes are included with the illustrations of wheat-grain

FIGURE 4.10 The dramatic differences in the appearance of cereal heads. (A) Club heads of the Australian wheat variety Tincurrin. (B) Heads of the Australian two-row barley, Clipper. (C) Heads of the Australian six-row barley, Cape. *(Reproduced with permission from Ferns, G.K., Fitzsimmons, R.W., Martin, R.H. and Wrigley, C.W., 1978. Australian Wheat Varieties. Supplement No. 1. Eleven Recent Wheats, Grain Defects, and Appendixes on Foreign Seeds and Insects, CSIRO, Melbourne, Australia and from Fitzsimmons, R.W., Wrigley, C.W., 1984. Australian Barleys Identification of Varieties, Grain Defects and Foreign Seeds, second ed., CSIRO, Melbourne, Australia.)*

shape for three wheat varieties in Fig. 4.12. Differences in glume shape are very evident. Gatcher has a fully awned head, and like most awned varieties, there is a prominent and long 'beak' on its glumes. The other two varieties are not awned; accordingly they lack a prominent beak on the glume.

The glumes of the wheat head are the parts of head morphology that provide the distinguishing head colour (either white or brown) at maturity. Glume colour is thus another characteristic useful in grain samples that have not been thoroughly cleaned of husks.

4.7 GRAIN MORPHOLOGY

The ability to identify species and variety is of greatest value for grain samples. The first need for this ability is at the early stage of producing pure seed for sowing. Verification of identity at this stage can take the form of checking grain characteristics. At this stage, the grower also needs to see that the registered seed is free of contaminating weed seeds and defects.

It is after harvest that there is the main need for identification or verification of identity, yet at this stage it is difficult because many of the distinguishing characters of the plant and head have been lost. Visual examination of grain samples has the great advantage that it requires little time and no equipment (apart, perhaps, from a hand lens and a reference manual).

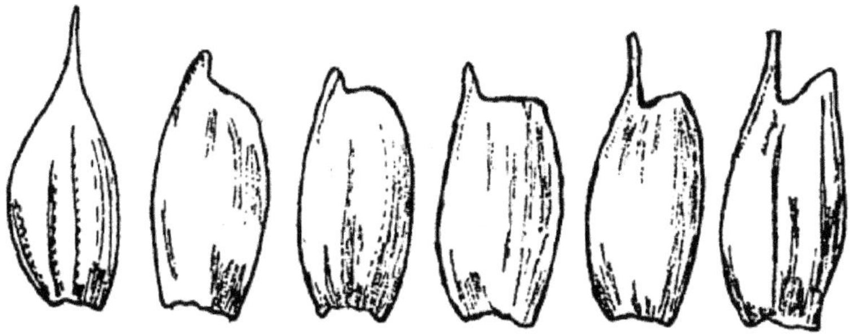

FIGURE 4.11 **Variations in glume shape are characteristic of variety.** *(Reproduced with permission from Agrawal, R.L., 1997. Identifying Characteristics of Crop Varieties, Science Publishers, Inc., Enfield, New Hampshire.)*

However, expertise and experience are required for variety identification based on grain samples, and distinction may never be unequivocal. Nevertheless, if verification of varietal identity can be performed at grain delivery (receival), immediately after harvest, then appropriate bulking of deliveries can be managed correctly, ensuring that deliveries of grain of different quality type are not mixed together. However, once unwanted mixing occurs, the process cannot be reversed.

The task is primarily verification of identity. For seed or delivered grain, varietal identity is already specified, and the visual examination is aimed at ensuring that the specification label is correct. Identification of variety with no indication of likely identity is very difficult, although keys for systematic identification are available in manuals covering regional or national sets of varieties.

Distinction between varieties based on grain morphology is difficult because many characters (size-related) that are potentially useful are also affected by growth conditions. Every grain in a sample may appear slightly different even though the sample is genetically homogeneous. The same comment may be made about the handwriting of an individual, but it is nevertheless possible to recognise the writer from the handwriting, if the observer has had sufficient experience. Although a word or letter is written slightly differently each time by a specific person, there is a consistent and characteristic similarity in that person's writing. A systematic approach to either task (handwriting or grain identification) is obviously needed, and assistance with the latter task is provided in the many relevant publications listed in this chapter and in the species-specific section of this book (Chapters 5 to 14).

4.7.1 Common Wheat Grain (See Chapter 5)

For wheat-grain identification, the hardness of the grain is valuable in providing distinction as well as being an important attribute for trade and processing quality (see Chapter 5). Grain hardness is judged by the ease of cutting a grain

FIGURE 4.12 Grain and glumes of the Australian wheat varieties (A) Gatcher, (B) Kite and (C) Olympic. *(Adapted from Ferns, G.K., Fitzsimmons, R.W., Martin, R.H., Simmonds, D.H., Wrigley, C.W., 1975. Australian Wheat Varieties: Identification According to Growth, Head and Grain Characteristics, CSIRO, Melbourne, Australia.)*

with a sharp blade or by biting it. Expertise in judging grain hardness by biting is an acquired skill, one that was essential to a breeder in the early days of wheat breeding. Many varieties are either hard or soft, but there may be some of intermediate hardness. Grains of durum (macaroni) wheat are very hard (see Section 4.7.2 and Chapter 6).

Grain hardness is related to the overall appearance of the grain, judged as grain texture (vitreosity); soft grains usually appear whitish due to having an opaque (white) endosperm (clearly visible by cutting the grain in half), whereas hard grains appear vitreous (horny or translucent endosperm). Both hardness and texture may be affected by seasonal variations, especially by rain at harvest.

Grain colour is also a major distinguishing character for wheat, especially for trade and for milling quality. The distinction between red and white wheats relates to the presence/absence of pigments in the outer bran layers, determined by the presence of up to three homoeologous genes, located on Group 5 chromosomes (Metzger and Silbaugh, 1970). Although the distinction is genetically determined, colour differences may be confused by the texture of the underlying endosperm. The red colour of the seed coat may be enhanced by immersing grains in aqueous 5% sodium hydroxide solution at room temperature for about 30 min (Hare, 1992).

Various aspects of grain size and shape are also useful in distinguishing between wheat varieties, although these attributes may be influenced by growth conditions and the position of an individual grain on the head; see Fig. 4.13 for relevant characteristics of the wheat grain. Additional features include the profile shape of the germ and brush ends and the length of the brush hairs (Ferns et al., 1975; Jarman, 1995, 2004).

From the dorsal (top, non-crease) side, the grains of different varieties may appear to be oval, ovate, elliptical, elongated or truncated (as illustrated in Ferns et al., 1975). As examples, see the shapes of the grains at the upper left for each of the varieties in Fig. 4.12. Kite and Olympic grains are classed

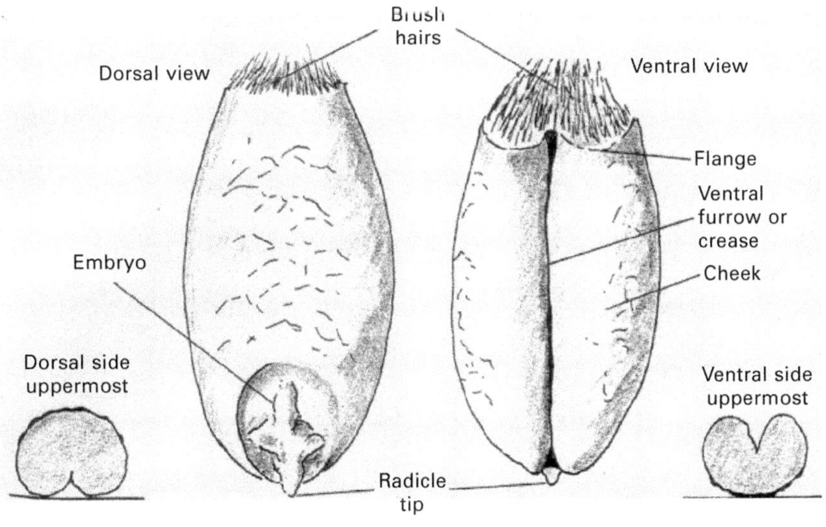

FIGURE 4.13 Distinguishing characters that assist in variety identification for wheat grains. *(Reproduced with permission from Hervey-Murray, C.G., 1980. The Identification of Cereal Varieties, RHM Arable Services Ltd., Cambridge, UK.)*

as oval, whereas Gatcher's brush end is pointed. Kite is classed by Ferns et al. (1975) as having a consistently longer grain than Gatcher or Olympic. Wheat kernels average 2.5–3.0 mm in thickness, 3.0–3.5 mm in width, 6.0–7.0 mm in length and 30–40 mg in weight. If grains are smaller than these dimensions, due to drought or frost, they may be unsuitable for characterisation by visual identification.

Wheat kernels are rounded on the dorsal side with a longitudinal crease running the full length of the ventral side. See the ventral view of the grain in the top right for each variety in Fig. 4.12. The embryo is located on the dorsal side at the end of the grain that is attached to the head (Fig. 4.13). At the opposite end of the grain (the apical end) is a tuft of hairs, known as the 'brush'.

Other distinguishing characteristics are illustrated by comparing grains of three varieties in Fig. 4.12. Multiple grains are included for each variety in Fig. 4.12 because there is inevitably some variation in size and shape from one grain of the same variety to another. The aim of classification methods is to use characteristics that provide the greatest distinction between varieties and that vary least due to changes in growth conditions. Such characteristics are evident in a comparison of the three varieties in Fig. 4.12. For example, grain of the variety Gatcher is pointed at the brush end when viewed from either the dorsal (back) view or from the side. In contrast, the grains of Kite and Olympic appear blunt in these views, Kite being classed as a large, oval grain. The long brush hairs of Gatcher are also evident. The germ face of Olympic is steeper than that of Gatcher.

4.7.2 Durum Wheat Grain (See Chapter 6)

The grains of durum wheat are usually larger than those of common 'bread' wheat. The durum grain is very hard and it is traditionally vitreous. In trade, its colour is given as 'amber'. The length of the brush hairs is usually short on durum grains (Ferns et al., 1975; Jarman, 1995 and 2004). Head and grain examples of a durum wheat are shown in Fig. 4.14. A feature distinguishing durum grains is the shape of the base of the grain (ventral side, seen in side view), which is often straight or even distinctively incurved. The glume beak for durum varieties is long, corresponding to the general status that durum heads are fully awned.

4.7.3 Triticale Grain (See Chapter 8)

Heads and grains of two triticale varieties are illustrated in Chapter 8. Most of the morphological characteristics that distinguish wheat varieties are also applicable to triticale. They are also likely to be useful for rye varieties, but because of the extent of rye's natural outcrossing, morphological characteristics are not generally used.

(A) (B)

FIGURE 4.14 **Heads (A), grains and glumes (B) of the Australian durum-wheat variety Durati.** *(Adapted from Ferns, G.K., Fitzsimmons, R.W., Martin, R.H. and Wrigley, C.W., 1978. Australian Wheat Varieties. Supplement No. 1. Eleven Recent Wheats, Grain Defects, and Appendixes on Foreign Seeds and Insects, CSIRO, Melbourne, Australia.)*

4.7.4 Barley Grain (See Chapter 9)

Verification of variety is important for barley grain due to the significant difference in value for malting varieties versus feed grain. Even for the higher value malting barley, single-variety consignments are often required as mixtures of malting varieties may germinate at different rates.

The many distinguishing characteristics for barley grains are illustrated in Fig. 4.15, as described in detail by Hervey-Murray (1980). Barley itself is distinct from wheat by having its lemma and palea fused to the grain after harvest (see Chapter 9). As a result, there are several other valuable distinctions available for identifying variety for barley grain, as illustrated in Fig. 4.15, namely, rachilla length and hairiness, the colour of the aleurone layer (just under the husk, blue or white), and the shape of the lemma husk at the point of attachment to the rachis (lemma base shape). The colour of the aleurone layer can be more clearly seen after abrasion of the outer husk layers, for example by the 'pearling' process. Even the shape of barley lodicules is useful, but to expose them the lemma husk must be peeled back from the base of the grain (Fitzsimmons and Wrigley, 1984).

The two-row or a six-row heads of barley appear dramatically different, as shown in Fig. 4.10. Even after threshing, it is possible to determine whether a barley grain came from a two-row or a six-row barley head. The distinguishing

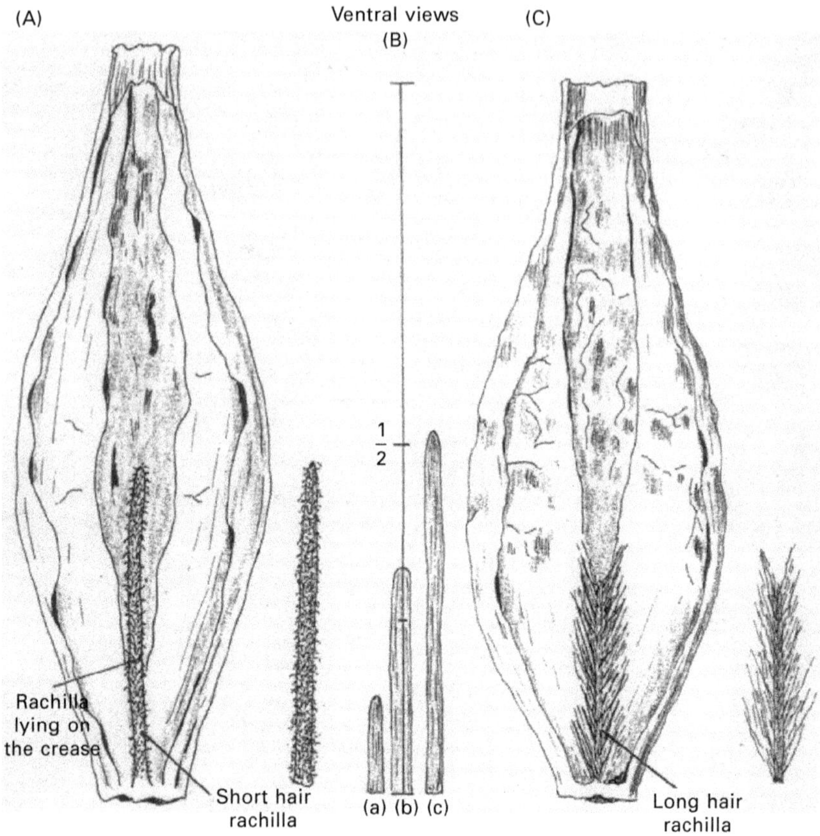

FIGURE 4.15 Distinguishing characters that assist in variety identification for barley grains. (A) Short hair rachilla (SHR) (B) Rachilla length scale (C) Long hair rachilla (LHR). *(Reproduced with permission from Hervey-Murray, C.G., 1980. The Identification of Cereal Varieties, RHM Arable Services Ltd., Cambridge, UK.)*

features are not completely lost as a result of threshing the grains from the head, because the lateral grains of the six-row head retain a twisted shape that is evident in two-thirds of the grains in a sample of six-row barley.

Some of these differences can be seen by examining grains of the two barley varieties in Fig. 4.16. On the one hand, it is evident that Cape is from a six-row head because of the twist in the grains at the top and bottom right and left positions, in contrast to the middle grains in the top and bottom rows. In contrast, none of the Clipper grains show this twist. The Cape grain has a blue aleurone colour, whereas it is white for Clipper. The rachilla hairs are short for Cape, but long for Clipper. [A condition known as 'Clipper itch' is due to these long rachilla hairs rubbing off the grains and irritating the skin of grain handlers.] See Fitzsimmons and Wrigley (1984) for further details of these and other distinguishing characters for barley grains.

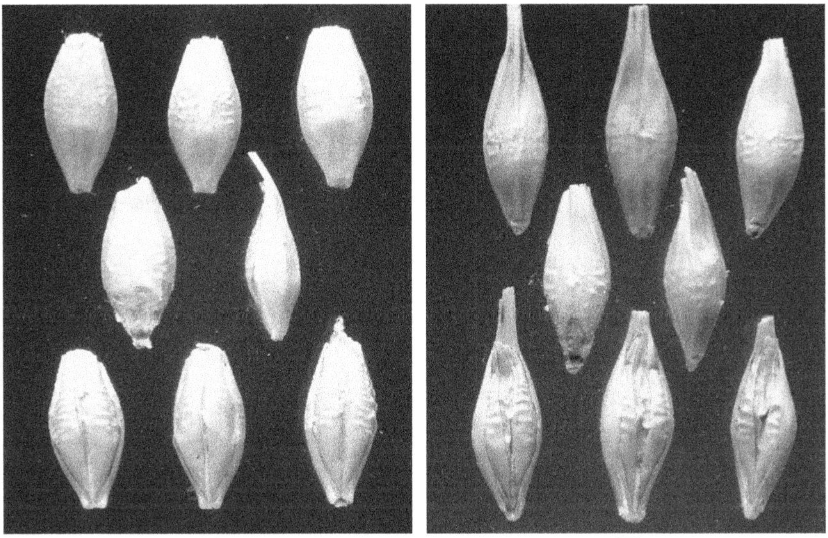

FIGURE 4.16 Grains of Clipper (A) (a two-row barley) and of Cape (B) (a six-row barley). The outside grains (top and bottom rows) for Cape show the twist at the base that is distinctive of 2/3 of the grains from a six-row head. This twist is not evident for any of the Clipper grains. *(Adapted from Fitzsimmons, R.W., Wrigley, C.W., 1984. Australian Barleys Identification of Varieties, Grain Defects and Foreign Seeds, second ed., CSIRO, Melbourne, Australia.)*

4.7.5 Oat Grain (See Chapter 10)

There are primarily two types of oat varieties—feed grain and grain suited for human food use, usually to be dehulled and processed into food products such as 'rolled oats' or for industrial processing (Zwer, 2016). Verification of variety at seeding and after harvest is thus important. A systematic approach to characterising oat varieties according to grain morphology is provided by Fitzsimmons et al. (1983a). See also Coffman (1961, 1977). Also important is distinction between cultivated oats (*Avena sativa*) and wild oats (variously classified as *A. fatua, A. ludoviciana* and *A. sterilis*). Advice on making these distinctions is also provided by Fitzsimmons et al. (1983a).

During harvesting and threshing, the structure of the oat panicle (Fig. 4.2) is disrupted, but the husks (lemma and palea) remain attached to the grain. For oats, the term 'groat' designates the inner part of the grain without the lemma and palea. Systematic identification of oat varieties by grain morphology requires distinction between the primary and secondary grains as they were originally in the floret. The primary grain is larger; it is positioned in the illustrations of Fig. 4.17 in the upper row; the secondary grains are in the lower row (outer positions). A groat is shown in the centre of the lower row.

Oat grains differ in colour, ranging from cream and yellow, through brown to black. This last possibility can obviously be confused with wild oats, also known

FIGURE 4.17 Grains of Australian oat varieties: Barmah (A) and Winjardie (B). *(Adapted from Fitzsimmons, R.W., Martin, R.H., Roberts, G.L., Wrigley, C.W., 1986. Australian Cereal Identification. Recent Varieties of Wheat, Barley, Oats and Triticale, CSIRO, Melbourne, Australia.)*

as black oats. To assess colour, concentrate on the back (dorsal surface) of the grain. Grain length and width can be distinctive, but are naturally affected by growth conditions. Evidence of the presence of awns is a valuable characteristic, but it is complicated by the likelihood that most of the awn has broken off during threshing. Examine the backs of primary grains for evidence of awns. The awn is still present on the grain of Winjardie in Fig. 4.17 (the top row), but not for the Barmah grains for which awns are classed as 'rare" by Fitzsimmons et al. (1986).

4.7.6 Rice Grain (See Chapter 12)

Appearance and kernel shape and important characteristics for rice. There is the famous saying that 'rice must look good to taste good'. That saying applies primarily to the polished grain, after removal of husk and bran layers; this is the form in which most rice is sold. In this form, it is immediately evident if there are chalky grains (with opaque patches) or is there is sun-cracking (partial fractures in the grain). The extent of broken grains is also evident.

Dimensions and length-width ratios vary considerably. The assessment of grain dimensions varies with the various markets; Basmati rice generally long and slender, in contrast to the shorter Japanese grain (Juliano, 2016).

4.7.7 Corn Grain (Maize) (See Chapter 11)

Corn is assessed on the basis of colour, shape and kernel hardness, these being the major characteristics that define categories. In addition, there are the soundness, plumpness, texture, kernel weight and moisture considerations.

Corn kernels come in many colours, with specific varieties ranging from white, yellow and red to purple (Scott and Emery, 2016).

4.8 THE USE OF MORPHOLOGY TO ASSESS AND MANAGE QUALITY

Considerable information about grain quality is provided by knowledge of the variety (genotype) involved. This knowledge about the suitability of the grain for processing has been 'built in' by the breeder, who has selected the specific genotype with relevant traits. For these reasons, it is common practice in grain-producing regions for a limited range of varieties to be specified for a certain grade or class of grain, probably thus making the grain qualify for a premium payment. Grain morphology can be the first, immediate step in verifying varietal identity, but there is a range of other test methods that are available for confirmation of identity (Miskelly and Wrigley, 2016).

Grain morphology is part of the management process for the registration of a new variety. This process is linked to Plant Breeders Rights (PBR) and the protection of new varieties so that royalties will be earnt. To qualify for this protection, a new variety must be shown to be distinct (D) from others, uniform (U) and stable (S) in its characteristics. Morphological attributes are primary considerations in the establishment of this 'DUS' requirement (Cooke and Wrigley, 2004).

Prior to harvest, it is common practice in grain-growing regions for those involved in harvesting and grain receival to drive around the district, examining the maturing crop; knowledge of plant and head characteristics provides an indication of what volumes of what varieties can be expected for receival. However, as already described, the main use of grain morphology is for verification (or otherwise) of the varietal identity declared by the grower/truck driver at the time of delivery. Many claims have been made that this process can be better handled, with greater objectivity, by machine vision. This involves the use of image analysis and on-the-spot analysis of the TV image. While such systems have proved practical for checking specific aspects of grain quality, such as contaminants, there has been no world-wide acceptance of a machine vision for the purposes of grain-quality management.

4.9 ULTRASTRUCTURAL MORPHOLOGY OF THE GRAIN

In addition to differences between cereal species and varieties in overall grain morphology, there are distinctions at the microscopic level with respect to the ultrastructure of the grain (Corke, 2016; Grundas and Wrigley, 2016). For example, grain species differ with respect to the nature and numbers of outer bran and aleurone layers (see chapters relevant to the cereal species). The starch granules differ with respect to their shape, as illustrated in the shapes of native starch granules isolated from wheat, rice and corn in Fig. 4.18. The starch

FIGURE 4.18 **Scanning electron micrographs of starch granules isolated from grains of (A) wheat, (B) rice and (C) corn (maize).**

TABLE 4.1 Starch Granule Properties for the Major Cereals

Cereal species	Size range (μm)	Size distribution	Gelatinisation onset (°C)	Amylose content (%)
Wheat	3–34	Bi-modal	61	26
Barley	2–35	Tri-modal	57	22
Rice	2–13	Normal	75	18
Maize	5–20	Normal	67	28
Waxy maize	4–18	Normal	68	0
Amylomaize	6–15	Normal	68	>70

Source: Adapted from Rahman, S., Li, Z., Batey, I., Cochrane, M.P., Appels, R., Morell, M., 2000. Genetic alteration of starch functionality in wheat. J. Cereal Sci., 31, 91–110.

granules of wheat and rye are characterised by having a bimodal distribution of starch granules (Table 4.1), consisting of large (A-type) and small (B-type) granules, as shown for wheat in Fig. 4.18(A) and in the broken surface of a sectioned wheat grain (Fig. 4.19). The size distribution for barley starch granules is trimodal (Table 4.1).

Many of the starch granules of rice and corn are irregularly shaped as illustrated in Figures 4.18 (B) and (C) (Regina et al., 2004). Consequently, the

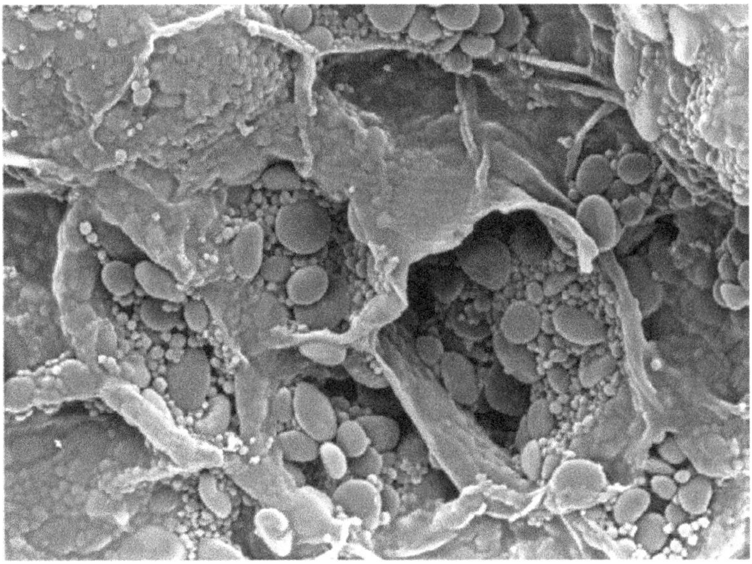

FIGURE 4.19 Scanning electron micrograph of the endosperm of a sectioned wheat grain, showing the contrasting large and small starch granules and the endosperm cell walls.

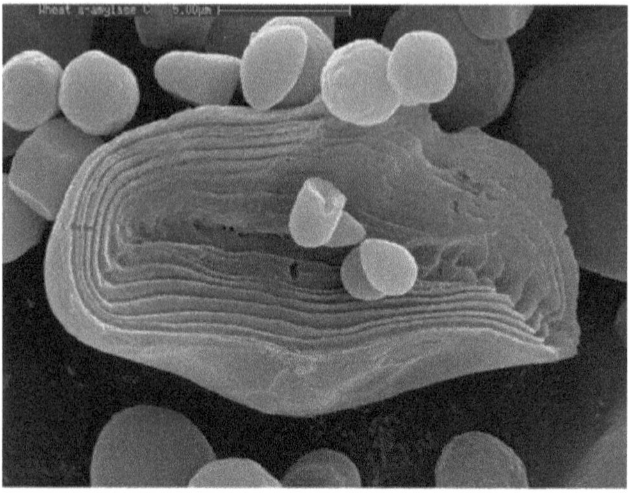

FIGURE 4.20 **Scanning electron micrograph of a wheat starch granule that has been enzymically eroded to show the concentric layers of starch deposition.** Small (B-type) starch granules are also present.

distinctive differences in starch morphology offer opportunities for an experienced microscopist to identify species origin (not variety) based on starch examination. Within the starch granule, the internal structure consists of concentric layers (Fig. 4.20), presumably due to the diurnal day-night cycle of starch synthesis. Native starch granules show birefringence in the form of a 'Maltese Cross' under crossed polariser microscopy. This characteristic is lost after gelatinisation.

An early attempt to distinguish between cereal varieties involved examining grain sections under the light microscope (Cobb, 1905). Cobb, an American phytopathologist working over a century ago in Sydney, Australia, claimed that he could distinguish between varieties of wheat according to the shapes of the aleurone and endosperm cells (Wrigley and Bekes, 2002). Needless to say, this approach has not been adopted generally.

4.10 THE USE OF GRAIN COMPOSITION TO ASSESS AND MANAGE QUALITY

Just as the various cereal species differ in appearance at the macro and micro levels, so also there are differences in the composition of their grains. On the other hand, there is a degree of uniformity of composition, most obviously evident when considering that they all have starch as the main source of stored energy. In addition, they all have a generous store of protein, providing the new plant with the essential resource of amino acids as building blocks for the synthetic mechanisms of the developing plant. These stored resources are the great attraction for the many predators (insects, animals, microorganisms and especially mankind) who wish to take advantage of the plant's mechanism of perpetuating its species. It is thus not surprising to find that the range of nutrients provided by cereal grains covers most of the requirements of mankind and the other predators who reply on grains for nutrition. However, there are a few exceptions, namely, the cereal grains are deficient in some vitamins, and their proteins do not supply a complete balance of amino acids, being slightly deficient in some essential amino acids, particularly lysine and to a lesser extent tryptophan.

4.10.1 Starch

The endosperm of the cereal grains consists mainly of starch, which is deposited in the form of the granules shown in Figures 4.18, 4.19 and 4.20. In most cereal species, the starch of normal genotypes is made up of about one quarter of amylose (smaller, straight-chain polymers of glucose) and three quarters of amylopectin (highly branched starch with large molecular weight). These proportions may vary considerably for rice, maize, barley and sorghum, which also have 'waxy' types (almost all amylopectin starch).

The starch type for wheat appeared to be fixed for wheat until only a few decades ago, when more recent knowledge about the biochemistry and genetics of starch synthesis (Regina et al., 2004) has led to the development of 'waxy' genotypes also in wheat (Zhao et al., 1998). The term 'waxy' relates to the translucent appearance of the endosperm of cereal grains. The starch from waxy grains has the distinct functional property of greater hot-paste viscosity when heated with water, and better freeze-thaw stability for the resulting gel, which is optically clear. In recent decades, there has been the recognition that wheats lacking one of the genes for starch synthesis (the 4A gene, *Wx-B1*) produce starch with properties well suited to noodle production because the starch has greater swelling power (Seib, 2000).

In addition, there are cereal genotypes that have high levels of amylose in their starch, but these have been more difficult to develop. Grain with low levels of amylopectin offer the functional advantages that they are associated with higher satiety, reduced glycaemic index, and reduced insulin release in comparison with normal-amylose starches. There is now a wide range of selected starch mutant types available throughout the cereal grain species, as is summarised by Regina et al. (2004).

4.10.2 Proteins

A history of methods for analysing grain composition would be a history of chemistry in general, because grain and flour samples have been readily available materials for developing and applying analytical methods as they have been developed (Békés and Wrigley, 2016). Indeed, wheat gluten was one of the first

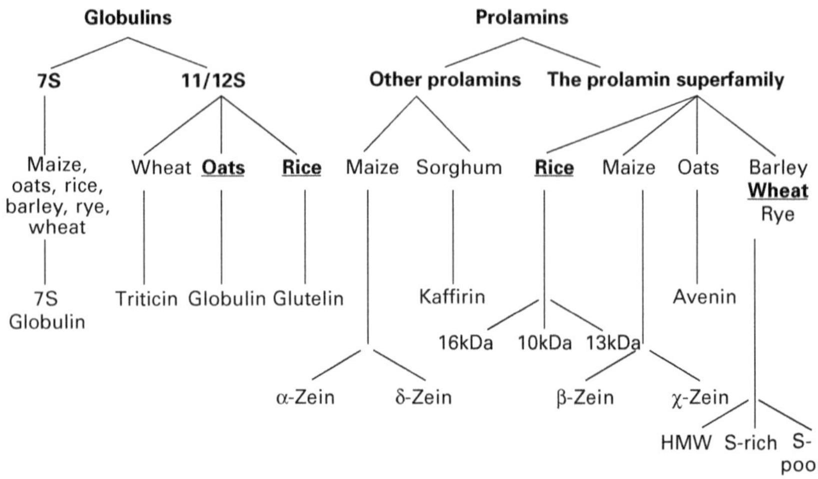

FIGURE 4.21 Classification of the cereals, according to the types of proteins in their grains. The major protein classes for specific cereals are indicated by bold and underline. *(Adapted from Shewry, P.R., 1996. Cereal grain proteins. In: Henry, R.J., Kettlewell, P.S. (Eds.), Cereal Grain Quality, Chapman and Hall, London, pp. 227–250.)*

proteins to be isolated in 1728 by the Italian scientist, Beccari. His experiments pre-dated the invention of the name 'protein' by a century and the development of adequate methods of protein analysis (Juhász et al., 2014, 2016).

The cereal-grain proteins were instrumental in the development of protein-fractionation methods by the American chemist, Thomas Burr Osborne, a century ago. He used solvent-fractionation procedures to distinguish four types of proteins—albumins, globulins, prolamins and glutelins. In wheat flour, he named the last two 'gliadin' and 'glutenin', respectively, these being the two components comprising gluten. The prolamins of other cereal species have specific names—'secalins' for rye, 'hordeins' for barley, 'avenins' for oats, 'zeins' for maize, 'kafarins' for sorghum (see Fig. 4.21 and the species-specific chapters). Because of an intervening century, Osborne's methods of fractionation and his nomenclature system have been largely replaced and updated, but many of his terms remain (see Fig. 4.21).

The classification system of Shewry (1996) in Fig. 4.21 shows that the grains of the various cereal genera have representatives of most of the globulins and prolamins, but that the amounts in specific cereals differ according to the names of some of the cereals in bold, underlined. The proteins listed in Fig. 4.21 are all found in the endosperm, except for the 7S globulins (extreme left) which are embryo and aleurone proteins. Also relevant to comparisons of the proteins of cereal species are the amino-acid compositions of the various cereal grains in Appendix 1. For example, oats and rice are unusual among the cereal grains in that their major protein class is the globulins, rather than the prolamins.

The amino-acid composition of rice is distinctive amongst the cereal grains because the globulins and prolamins are both present in significant levels. Whereas most other cereals accumulate prolamins as their primary nitrogen reserve, the major storage proteins in rice are the glutelins, which are homologous at the primary-sequence level to 11S globulin proteins, a class which is the dominant form of nitrogen deposition in legumes. Even the prolamin proteins of rice have a number of characteristics that are different from the prolamins of most other cereals. Although milled rice contains the lowest amount of protein among the major cereals, the overall amino-acid composition of rice protein is significantly better balanced, because of its relatively higher level of lysine.

The protein of oats also has a higher nutritional value than other cereal grains, based on essential amino acids, because oats has a lower content of the prolamin class of proteins, and relatively more of the globulin class of proteins (Fig. 4.21). Oat globulins account for about 75% of the seed protein content (Shotwell, 1999). Oat globulin is rich in glutamine and asparagine (amino acids that have two nitrogen atoms each), consistent with the role of storing nitrogen. However, oat-globulin protein is slightly deficient in the sulfur-containing amino acids, cysteine and methionine, being similar to the globulins of the legumes in this respect.

The grain proteins of sorghum are mainly prolamins (kafirins) but also glutelins. Like many cereals, sorghum is deficient in lysine, threonine and tryptophan, but high-lysine sorghum lines have been produced, providing the possibility of overcoming this nutritional deficiency. Similarly, genotypes of corn with high levels of lysine (opaque-2 maize) have been developed, but their acceptability has been hindered by their susceptibility to certain pests.

The gluten proteins of the wheat grain are the most studied of all cereal-grain proteins due to their importance in conferring the unique properties of dough formation (Juhász et al., 2014). The name 'gluten' has been applied indiscriminately to the grain proteins of various grains, including maize, and difficulties of nomenclature have resulted. The prolamins of rye, triticale and barley (sometimes also oats) are sufficiently similar in sequence to wheat gluten that people with coeliac disease must avoid them. However, only wheat gluten has the appropriate rheological properties to sustain the production of good quality leavened bread. This property results from the balance of the component protein classes of gliadin (single polypeptides) and glutenin, which contains very large disulfide-linked aggregations of polypeptides, whose overall molecular size is a major determinant of dough strength (see Chapter 5).

4.10.3 Fats

The cereal grains have generally low levels of lipids (fats) in their endosperm (white flour or pearled whole grain). Lipid levels are much higher in the germ (embryo), especially for rice and maize. Rice oil has the reputation of having a low content of linolenic acid, and being a rich source of tocopherols. Both rice and maize oils are good sources of the essential fatty acid linoleic acid. Commercial production of oil from these two major grains, a significant industry worldwide, makes use of the 'bran' fraction, containing both the germ and the outer layers that are removed in the respective milling processes (Godber, 2004). The difference between the endosperm and the outer layers of rice is shown by the contrast in fat levels for brown and white rice (about 2.9 and 0.7 g per 100 g of the respective materials) in Table 3 of Appendix 1. There is a 10-fold difference in fat levels between wheat germ and wheat flour (Table 1 of Appendix 1). Note also the differences in fat composition in these comparisons.

4.10.4 Fibre

Diets high in whole grains have long been advocated, largely because of the resulting elevated levels of fibre, as well as phytochemicals (Liu, 2007; Anderson et al., 2009; Tosh and Shea Miller, 2016). The grains of barley and oats, in particular, have reputations for being rich sources of beta-glucans, thus conferring a range of health benefits (Tables 3 and 5, respectively, of Appendix 1). Accordingly, these two grain species have been selected as suitable materials for the commercial preparation of beta-glucans (Bhatty, 1995). In addition, oat bran has attracted considerable attention as a special source of dietary fibre (Wood, 1993).

4.11 THE MANAGEMENT OF GRAIN PROCESSING BASED ON COMPOSITION

4.11.1 Milling

The main inner volume of the grain is occupied by the starchy endosperm which becomes the white flour when it is released and crushed to fine particles by the flour miller. The crease complicates the task of the miller because the bran of the crease, as well as the bran of the outer grain surface, must be separated from the endosperm during flour milling. The presence of the crease is undesirable because, if extreme, it can contribute to a low bulk density (test weight).

The main effect of milling is to separate the white endosperm from the germ (embryo) and bran layers, either for wheat and corn, for which the endosperm is subsequently crushed to flour, or for rice and barley, for which the milled grain is used as the whole of the remaining grain (endosperm).

The process of milling (polishing) rice provides a good example of the effects of a major form of processing (milling) on quality—both appearance and nutrition. During rice milling, some of the higher nutritional value of 'brown rice' is lost, because the (brown) aleurone layer carries the highest levels of protein and nutritionally important minor components. The sub-aleurone region of rice plays a very significant role nutritionally because it is several cell layers thick. These layers are also richer in the globulin class of proteins than the endosperm, so their lysine content is much higher than that of the proteins of white rice.

FIGURE 4.22 The fractions resulting for the milling of paddy rice: polished grain (left), bran and husks.

It is therefore desirable to mill as lightly as possible to retain most of the sub-aleurone layer on white rice, but on the other hand, consumer preferences demand a polished whiteness in their rice. Over the whole process of producing white polished rice, three major by-products are produced: husks (20%), bran (10%) and 'brokens' (10–22%), with protein contents of about 3, 17 and 8%, respectively (Fig. 4.22). Table 2 of Appendix 1 lists the full range of nutrients in these edible forms of rice, showing the progressive loss of minor nutrients.

4.11.2 Heat Treatment

Starch is the major component that is altered by heating in the presence of water. At room temperature, there may be modest swelling of wetted starch granules, but this process is reversible. With heat, however, irreversible swelling of the starch granules occurs (starting at about 60°C), as the granules swell, losing their crystalline structure and the viscosity of the starch suspension increases. These changes can readily be monitored with the light microscope using polarised light; as gelatinisation proceeds, the 'Maltese Cross' effect is lost. The gelatinisation process also makes the starch more readily digestible. Boiling in a great excess of water is a common practice for whole-grain processing, such as for rice, sorghum and pearled barley, but this process may lead to the removal of various minerals and vitamins into the water in which they were boiled.

4.11.3 Alkali Treatment

The alkaline treatment of maize (corn) leads to nutritional advantages. The production of 'masa' involves mixing maize with lime before cooking. The pericarp of the cooked maize (nixtamal) is partially degraded by the treatment so that vitamins and minerals become more readily available. Mycotoxin may be reduced if the sample has had fungal infection. However, the treatment may reduce the amount of lysine (an essential amino acid), due to its modification to form lysinoalanine.

4.12 FUTURE TRENDS

The practice of identifying grain and plants by their shape and size is very old and well established. Nevertheless, this approach to identification remains as a major part of the processes involved in variety registration. New technologies, such as image analysis, offer the likelihood that the traditionally subjective nature of this approach can be replaced by 'machine vision', but even this new technology must be applied correctly, or mis-information may result. In addition, chemical methods are also providing objective identification of variety, as described in Chapter 17.

Recent advances in breeding and genetics have provided novel grains, especially wheat, with novel starch composition and properties, mainly due to manipulation of the proportions of amylose and amylopectin. These advances open up new opportunities for novel food products and also for renewed interest in producing foods tailored to having desirable glycaemic index (GI) qualities. Intelligent manipulation of the GI of food products is still problematic, but novel starches and specific approaches to processing will assist in achieving these nutritional aims.

The many methods for determining grain composition have been thoroughly evaluated in collaborative testing between laboratories in many countries, resulting in the publication of a wide range of standard methods suited to the grain industry. This process and the resulting methods have been reviewed by Bridges and Wrigley (2016). Of the many analytical methods developed in recent decades for analysing grain composition, the most important for routine use are those based on spectroscopy in the near infrared (NIR) part of the spectrum. These are described in Chapter 19, following the sweeping statement: 'NIR spectroscopy has been the most important 'breakthrough' in the last 50 years for rapid assessment of grain quality'.

SOURCES OF FURTHER INFORMATION

For systematic information on grain, head and plant morphology, see the websites of the International Union for the Protection of New Varieties of Plants, www.upov.int., and the National Institute of Agricultural Botany, Cambridge, UK, www.niab.com

See also publications about plant and grain morphology, and national collections of cereal varieties, such as Coffman (1961, 1977), Jarman (1995, 2004), Agrawal (1997), Mauria (2000), Cooke and

Wrigley (2004), NIAB (2004), Ferns et al. (1975, 1978), Fitzsimmons and Wrigley (1979, 1984), Fitzsimmons et al. (1983a,b, 1985, 1986).
For further grain-composition tables, see:
Schakel et al. (2004), Kent and Evers (1994), Kulp and Ponte (2000).
USDA National Nutrient Database, www.nal.usda.gov/fnic/foodcomp.
The website of the Nutrition Coordinating Center, University of Minnesota, www.ncc.umn.edu.

REFERENCES

Agrawal, R.L., 1997. Identifying Characteristics of Crop Varieties. Science Publishers, Inc, Enfield, New Hampshire.

Anderson, J.W., Baird, P., Davis, Jr., R.H., Ferreri, S., Knudtson, M., Korayn, A., Waters, V., Williams, C.L., 2009. Health benefits of dietary fiber. Nutr. Rev. 67, 188–205.

Békés, F., Wrigley, C.W., 2016. The protein chemistry of cereal grains. Wrigley, C., Corke, H., Seetharaman, K., Faubion, J. (Eds.), Encyclopedia of Food Grains, vol. 2, Elsevier Ltd, Oxford, UK, p. 98.

Bhatty, R.S., 1995. Laboratory and pilot plant extraction and purification of beta-glucans from hull-less barley and oat brans. J. Cereal Sci. 22, 163–170.

Bridges, A.R., Wrigley, C.W., 2016. Lists of standardized testing methods of the analysis of grain and grain-based foods. Wrigley, C., Corke, H., Seetharaman, K., Faubion, J. (Eds.), Encyclopedia of Food Grains, vol. 2, Elsevier Ltd, Oxford, UK, p. 300.

Cobb, N.A., 1905. Universal Nomenclature for Wheat, Miscellaneous Publication No 539, Sydney, Australia, NSW Department of Agriculture.

Coffman, F.A., 1961. Oats and Oat Improvement. American Society of Agronomy, Madison, WI.

Coffman, F.A., 1977. Oat history, identification and classification. Technical Bulletin No. 1516, US Department of Agriculture, Washington, USA.

Cooke, R.J., Wrigley, C.W., 2004. Variety registration and breeders' rights. Wrigley, C., Corke, H., Walker, C. (Eds.), Encyclopedia of Grain Science, vol.3, Elsevier, Oxford, pp. 314–321.

Corke, H., 2016. Gra: Morphology of internal structure. Wrigley, C., Corke, H., Seetharaman, K., Faubion, J. (Eds.), Encyclopedia of Food Grains, vol. 1, Elsevier Ltd, Oxford, UK, pp. 41–50.

Ferns, G.K., Fitzsimmons, R.W., Martin, R.H., Simmonds, D.H., Wrigley, C.W., 1975. Australian Wheat Varieties: Identification according to Growth. Head and Grain Characteristics. CSIRO, Melbourne, Australia.

Ferns, G.K., Fitzsimmons, R.W., Martin, R.H., Wrigley, C.W., 1978. Australian Wheat Varieties. Supplement No. 1. Eleven recent wheats, grain defects, and appendixes on foreign seeds and insects. CSIRO, Melbourne, Australia.

Fitzsimmons, R.W., Wrigley, C.W., 1979. Australian Barleys: Identification of Varieties, Grain Defects and Foreign Seeds. CSIRO, Melbourne, Australia.

Fitzsimmons, R.W., Wrigley, C.W., 1984. Australian Barleys Identification of Varieties, Grain Defects and Foreign Seeds, second ed. CSIRO, Melbourne, Australia.

Fitzsimmons, R.W., Roberts, G.L., Wrigley, C.W., 1983a. Australian Oat Varieties. Identification of Plants, Panicles and Grains. CSIRO, Melbourne, Australia.

Fitzsimmons, R.W., Martin, R.H., Wrigley, C.W., 1983b. Australian Wheat Varieties: Identification According to Plant, Head and Grain Characteristics, second ed. CSIRO, Melbourne, Australia.

Fitzsimmons, R.W., Martin, R.H., Wrigley, C.W., 1985. Australian Wheat Varieties—Supplement No. 1 to Second Edition. Sixteen Recent Wheats and an Appendix on Australian Triticale Varieties. CSIRO, Melbourne, Australia.

Fitzsimmons, R.W., Martin, R.H., Roberts, G.L., Wrigley, C.W., 1986. Australian Cereal Identification. Recent Varieties of Wheat, Barley, Oats and Triticale. CSIRO, Melbourne, Australia.

Godber, J.S., 2004. Oil from rice and maize. Wrigley, C., Corke, H., Walker, C. (Eds.), Encyclopedia of Grain Science, vol.2, Elsevier, Oxford, pp. 375–380.

Grundas, S.T., Wrigley, C.W., 2016. Ultrastructure of the wheat grain, flour and dough. Wrigley, C., Corke, H., Seetharaman, K., Faubion, J. (Eds.), Encyclopedia of Food Grains, vol. 3, Elsevier Ltd, Oxford, UK, p. 384.

Hare, R.A., 1992. Anatomical location and inheritance of variegated red seed coat color in hexaploid wheat. Crop Sci. 32, 115–117.

Hervey-Murray, C.G., 1980. The Identification of Cereal Varieties. RHM Arable Services Ltd, Cambridge, UK.

Jarman, R.J., 1995. Visual identification by grain characteristics. In: Wrigley, C.W. (Ed.), Identification of Food-Grain Varieties. American Association of Cereal Chemists, St Paul, MN, pp. 35–54.

Jarman, R.J., 2004. Grains and plants; morphology. Wrigley, C., Corke, H., Walker, C. (Eds.), Encyclopedia of Grain Science, vol. 2, Elsevier, Oxford, pp. 38–46.

Juhász, A., Békés, F., Wrigley, C.W., 2014. Wheat proteins. In: Ustunol, Z. (Ed.), Applied Food Protein Chemistry. Wiley-Blackwell, John Wiley & Sons, West Sussex, UK, pp. 219–304.

Juhász, A., Haraszi, R., Békés, F., Skylas, D.J., Wrigley, C.W., 2016. Proteomic analysis relevant to cereal grains. Wrigley, C., Corke, H., Seetharaman, K., Faubion, J. (Eds.), Encyclopedia of Food Grains, vol. 2, Elsevier Ltd, Oxford, UK, p. 146.

Juliano, B.O., 2016. Rice: overview. Wrigley, C., Corke, H., Seetharaman, K., Faubion, J. (Eds.), Encyclopedia of Food Grains, vol. 1, Elsevier Ltd, Oxford, UK, p. 125.

Kent, N.L., Evers, A.D., 1994. Technology of Cereals: An Introduction for Students of Food Science and Agriculture, fourth ed. Woodhead Publishing, Cambridge, UK.

Kulp, K., Ponte, J.G., 2000. Handbook of Cereal Science and Technology, second ed. Marcel Deller Inc, New York, NY.

Liu, R.H., 2007. Whole grain phytochemicals and health. J. Cereal Sci. 46, 207–219.

Mauria, S., 2000. DUS testing of crop varieties. Plant Var. Seeds 13, 69–90.

Metzger, R.J., Silbaugh, B.A., 1970. Location of genes for seed color in hexaploid wheat, *Triticum aestivum* L. Crop Sci. 10, 495–496.

Miskelly, D.M., Wrigley, C.W., 2016. Identification of varieties of food grains. Wrigley, C., Corke, H., Seetharaman, K., Faubion, J. (Eds.), Encyclopedia of Food Grains, vol. 4, Elsevier Ltd, Oxford, UK, p. 29.

Morrison, L.A., Wrigley, C.W., 2016. Taxonomic classification of grain species. Wrigley, C., Corke, H., Seetharaman, K., Faubion, J. (Eds.), Encyclopedia of Food Grains, vol. 1, Elsevier Ltd, Oxford, UK, p. 31.

NIAB., 2004. NIAB Seed Identification Handbook. National Institute of Agricultural Botany, Cambridge, UK.

Percival, J., 1921. The Wheat Plant. Duckworth, London.

Regina, A., Rahman, S., Li, Z., Morell, M.K., 2004. Starch synthesis. Wrigley, C., Corke, H., Walker, C. (Eds.), Encyclopedia of Grain Science, vol. 3, Elsevier, Oxford, pp. 224–232.

Schakel, S.F., Van Heel, N., Harnack, J., 2004. Grain composition tables. Wrigley, C., Corke, H., Walker, C. (Eds.), Encyclopedia of Grain Science, vol. 3, Elsevier, Oxford, pp. 431–447.

Scott, M.P., Emery, M., 2016. Maize: overview. Wrigley, C., Corke, H., Seetharaman, K., Faubion, J. (Eds.), Encyclopedia of Food Grains, vol. 1, Elsevier Ltd, Oxford, UK, p. 99.

Seib, P.A., 2000. Reduced-amylose wheats and Asian noodles. Cereal Food World 45, 504–512.

Shewry, P.R., 1996. Cereal grain proteins. In: Henry, R.J., Kettlewell, P.S. (Eds.), Cereal Grain Quality. Chapman and Hall, London, pp. 227–250.

Shotwell, M.A., 1999. Oat globulins. In: Shewry, P.R., Casey, R. (Eds.), Seed Proteins. Kluwer Academic Publishers, Netherlands, pp. 389–400.

Tosh, S.M., Shea Miller, S., 2016. Health effects of β-glucans found in cereals. Wrigley, C., Corke, H., Seetharaman, K., Faubion, J. (Eds.), Encyclopedia of Food Grains, vol. 2, Elsevier Ltd, Oxford, UK, p. 236.

Wood, P.J. (Ed.), 1993. Oat Bran. American Association of Cereal Chemists, St Paul, MN.

Wrigley, C.W., Bekes, F., 2002. Grain-protein composition as a document of wheat-quality type; new approaches to varietal identification. In: Ng, P.K.W., Wrigley, C.W. (Eds.), Wheat Quality Elucidation: The Bushuk Legacy. American Association of Cereal Chemists, St Paul, MN, pp. 65–86.

Zhao, X.C., Batey, I.L., Sharp, P.J., Crosbie, G., Barclay, I., Wilson, R., Morell, M.K., Appels, R., 1998. A single locus associated with starch granule properties and noodle quality in wheat. J. Cereal Sci. 27, 7–13.

Zwer, P., 2016. Oats: Overview. Wrigley, C., Corke, H., Seetharaman, K., Faubion, J. (Eds.), Encyclopedia of Food Grains, vol. 1, Elsevier Ltd, Oxford, UK, p. 173.

Part II

Grain-Quality Characteristics for Specific Cereals

Chapter 5

Wheat: Grain-Quality Characteristics and Management of Quality Requirements

Surjani Uthayakumaran* and Colin Wrigley**
**University of Sydney, Sydney, NSW, Australia; **QAAFI, University of Queensland, St. Lucia, QLD, Australia*

5.1 INTRODUCTION

Wheat is one of the oldest food crops grown by man; it has achieved a central role as a staple food for all nations and cultures, each having its own distinctive type of wheat-based food. Wheat is a major member of the cereal-grain family which provides 'carbohydrate-staple foods that form the basis of most diets, both ancient and modern, around the world' (Jones et al., 2015a,b). These authors have also written to counter the 'many current claims that disparage and discourage the ingestion of carbohydrates [in] wheat and cereal grains, as well as to celebrate the versatility, nutritional and health benefits and contribution of these foods to the world food supply'.

In addition to its role as an important source of carbohydrates to the world, wheat adds the distinctive success as a food source that produces flour with the unique ability to form a cohesive dough (Fig. 5.1). Thus wheat flour can be made into leavened breads of many types and into an even wider range of other foods such as noodles, soups, biscuits and pasta (Morris, 2016a,b) (see Chapter 24).

In its various food forms, wheat provides a large proportion of the world's nutrition (Beta and Isaak, 2016). Compared to nutrition from all cereal grains, wheat provides about 44% of grain-based food worldwide. In addition, wheat contributes half of the protein and 40% of the fat that came from all cereal-grain sources (Uthayakumaran and Wrigley, 2009). These data contrast with the volume of wheat production (600–700 million tonnes annually), which is about 30% of the production of all cereal grains (2000–2500 million tonnes

Cereal Grains. http://dx.doi.org/10.1016/B978-0-08-100719-8.00005-X

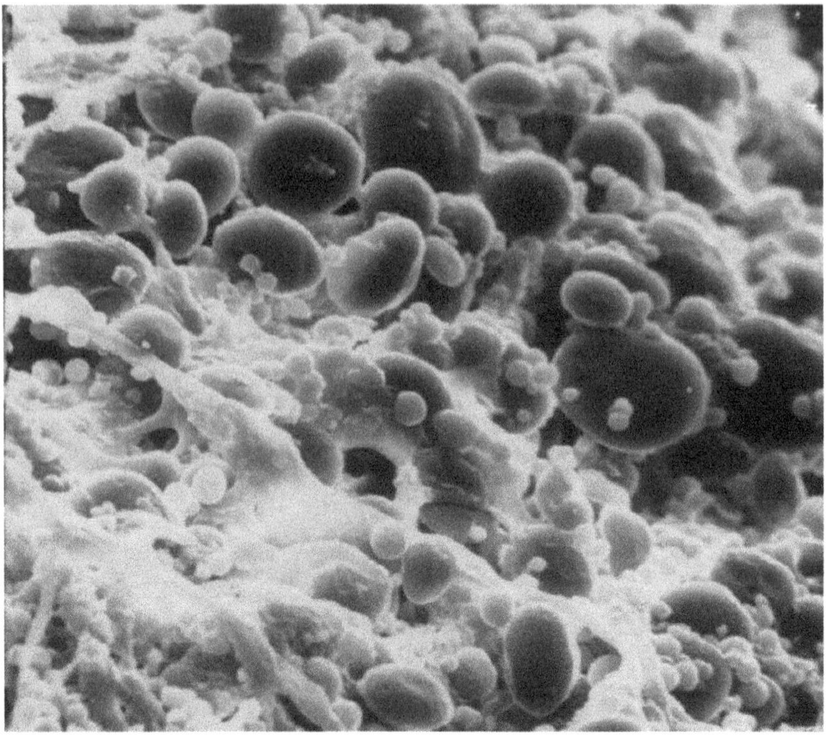

FIGURE 5.1 The surface of a dough piece, showing the continuous strands of gluten protein providing a matrix for the starch granules, both A-type large granules and B-type small ones. Scanning electron micrograph provided by W. Campbell.

annually) (Table 5.1). This contrast reflects the relative importance of wheat as a food grain, even though a significant amount is also used for animal feed and industrial purposes. The importance of wheat is also illustrated by the world-trade statistics for wheat (Table 5.1); its entry into world trade (about 44% of all cereal-grain trade) is much greater than the proportion (about 30%) of wheat in the production of all cereal grains. Table 5.1 also shows how the production and trade of wheat and all cereals have increased in the past decade, compared to the earlier parts of the 21st century.

In contrast to the importance of the wheat grain to humanity, the primary role of the grain for the plant is to provide a store of nutrients for the embryonic wheat plant. In anticipation of that role, the developing grain accumulates proteins, starch, lipids and sugars in the endosperm (Shewry and Tosi, 2016). We take advantage of this valuable package of nutrients—the mature wheat grain—when we mill it, releasing the inner endosperm from the bran and embryo to provide white flour for bread-making. The flour proteins are hydrated during the dough-mixing process, forming the protein mass referred to as 'gluten', which

TABLE 5.1 Annual Statistics for Production and Trade of Wheat Worldwide: Means and Ranges for Wheat Compared to all Cereal Grains in the Periods Specified

Statistics for the world	Harvest years	Wheat mean (*range*)	All cereals mean (*range*)
Production (million tonnes)	2001–07[a]	598 (560–632)	2191 (2028–2340)
	2007–16	675 (*601–736*)	2319 (*2012–2558*)
Trade[b] quantity (million tonnes)	2000–06[a]	118 (*110–126*)	281 (*268–301*)
	2007–16	138 (*114–157*)	313 (*258–375*)

[a]*Uthayakumaran and Wrigley (2009).*
[b]*Trade is expressed as export statistics.*
Source: Data for the recent decade came from www.fao.org.

permits gas bubbles to be retained during leavening. The baking step provides the light crumb that we associate with good bread.

An understanding of quality assessment and management for all stages of the grain chain (Table 5.2) requires a thorough knowledge of the grain and the relationship of its composition to the many forms of processing. This chapter describes the many ways in which wheat is used for food and non-food applications around the world. Despite the variety of uses, there are several basic means of assessing the processing value of wheat; in addition, there are more specialised approaches to elucidating the suitability of specific wheat consignments for particular uses. Knowledge about the quality specifications of wheat deliveries is basic to effective management of wheat utilisation at all the stages of producing items for the consumer—from breeding, through farm management to segregation, transport and on to processing and sales (Blakeney et al., 2009) (see Chapter 1).

5.2 WHEAT AS A GENUS

Wheat (*Triticum* species) is one of the oldest food crops. It has been known to exist since 10,000 BC as an agricultural species intentionally cultivated by man (Feldman, 2001; Gustafson et al., 2009; Morrison, 2016). Its popularity stems in part from the adaptability of wheat as a cultivated crop suited for many different soils and climatic conditions. Most significantly, wheat is unique because of the ability of wheat proteins to combine into the protein mass known as gluten (Békés and Wrigley, 2016). Wheat is thus the only effective source of flour for bread production in the world (Atwell, 2001; Moore, 2016). Its close relative, rye, shares some of the dough-forming ability of wheat, but most rye bread is made from a grist of wheat and rye flour (see Chapter 7).

TABLE 5.2 Wheat-Based Products and Issues Associated With the Various Stages of the Grain Chain

Stages		Products	Issues for analysis and management
Breed	→	Improved varieties. Pure seed	Grain quality. Yield factors. Disease tolerance. Marker-assisted segregation. GMO technology.
↓			
Sow	→		Soil analysis. Time to sow. Fertiliser.
↓			
Grow	→	Forage and Fodder for ruminants	On-farm management. Climate variations. Precision agriculture. Rising atmospheric CO_2 levels.
↓			
Harvest	→	Fibre for construction. Fuel from straw	Grain moisture. Protein content. Rain damage. Commercial fuel production from lignocellulose.
↓			
Deliver	→	Grain samples to test, thus segregating by grain quality	Moisture. Protein %. Contaminants, defects. Variety/quality-based segregation and binning.
↓			
Store	→		Monitor moisture and temperature; maintain quality.
↓			
Transport	→		Identity preservation to maintain quality segregation.
↓			
Process		Food as whole grains. Feed for animals, birds, fish. Fuel as bioethanol. Fats as germ oils. Fluids as beverages. Vital dry gluten. Wheat starch and hydrolysates.	Couscous. Bulgur. Breakfast cereals. Amino-acid composition, especially lysine. A-type starch content. Oil content and composition. Adjunct for brewing. Bread and food ingredient. A-type starch content. Low protein content.
↓			
Mill	→	Flour, starch, protein isolates	Moisture. Milling yield. Protein content.

TABLE 5.2 Wheat-Based Products and Issues Associated With the Various Stages of the Grain Chain (*cont.*)

Stages		Products	Issues for analysis and management
↓			
Mix and bake	→	Food: baked goods. Noodles.	Dough quality. Starch properties.
↓			
Retail	→	Packaged goods.	Composition analysis. Food labelling.
↓			
Consumer satisfaction	→	Feedback via media, via gossip, via internet.	Appearance. Taste. Allergic reactions? Gluten intolerance? Organic? GMO?
↓			
Consumer nutrition	→	Nutrients: protein, starch, fibre, vitamins, minerals.	Popular misconceptions. Gluten-free foods. FODMAP intolerance.

5.2.1 Taxonomy

The genus *Triticum* includes a wide range of species, but only two species are grown commercially to a large extent:

- *Triticum aestivum*—bread or common wheat, which is genetically hexaploid with genomes A, B and D;
- *Triticum durum*—macaroni or durum wheat, genetically tetraploid, with genomes A and B (see Chapter 6).

Over 90% of the wheat grown worldwide is *T. aestivum*. It is used for a wide range of applications, including bread, cakes, pastries, biscuits, puddings, thickeners and noodles. *T. durum* (durum wheat) is mainly used for pasta production. In addition, small amounts of a few specialty wheats are grown: spelt wheat (a hexaploid wheat—*T. aestivum var. spelta*), *Triticum monococcum* (small spelt wheat) and *T. timopheevii* (Georgian wheat) (see Chapter 14).

T. monococcum, *Aegilops speltoides*, *T. tauschii* and the wild *Aegilops* species, which is closely related to the modern *Ae. speltoides*, are ancestral diploid wheat species having seven pairs of chromosomes ($2n = 14$). The tetraploid *T. durum* is derived from the natural hybridisation of *T. monococcum* (A genome) and the ancestral *Ae. speltoides* (B genome) (Wrigley, 2009; Morrison and Wrigley, 2016). Hexaploid bread wheat ($2n = 42$) is a result of the natural hybridisation of *T. dicoccoides* (AABB) and *T. tauschii* (DD) (Mangelsdorf, 1953; Shewry et al., 2003; Gustafson et al., 2009).

The common wheats of today are divided into red and white wheats (based on the intensity of red pigmentation in the seed coat), hard and soft wheats (based on the resistance of the seed when it is crushed), and winter and spring types (differing in their requirement for a cold period to hasten or permit normal development towards reproductive development) (Gooding, 2009) (see Chapter 18).

5.2.2 Origins

Man assumed control over his own food production during the Neolithic Revolution. The pre-agricultural hunter–gatherer became familiar with nature's periodicity and with the life cycle of the dominant plants in his environment, finally succeeding to domesticate many of them. Wheat is one of the world's most important food crops to be domesticated. It is believed that wild relatives of wheat first grew in the Middle East, in the 'Fertile Crescent' (Feldman, 2001). Wheat was one of the first plants to be cultivated. It was grown about 11,000-years ago. Enormous changes in people's lives occurred because of wheat being grown. People began growing their own food, no longer needing to wander in search of food. Permanent settlements were established because wheat provided people with a stable food supply. Soon they grew enough wheat to feed others from surrounding regions.

Once there was extra wheat available, trade between various cultures developed. By 4000 BC, wheat farming had spread to Asia, Europe and North Africa. New species of wheat developed because early farmers selected kernels from their best wheat plants to use as seeds for planting the following year's crop. In that way, only the best wheat qualities were passed from one generation to the next. Soon wheat became an important worldwide crop.

Three main theories concerning the site of wheat domestication were proposed in the 19th century. According to the first theory, wheat was domesticated in the Near East. The evidence was from the Chaldean priest, Berosus, who mentioned that wild wheat occurred in Mesopotamia at about 2700 BP (Syncellus, Frag. Hist. Graec., vol. 2, p. 416). Another theory in 1899 proposed that the cultivation of wheat started in central Asia. The third theory proposed in 1908 suggested that wheat grew wild in Europe and was later domesticated there (Feldman, 2001; Morrison, 2016).

5.2.3 Genetic Constitution

Genetically, there are three main groups of cultivated wheat, namely, diploid, tetraploid and hexaploid wheats (McIntosh, 2016). The wild progenitor of cultivated diploid wheat einkorn was discovered in the Middle East, the progenitor for tetraploid wheat emmer and durum was discovered in the beginning of the

20th century. There are no equivalent progenitors for hexaploid wheat and it is clear that hexaploid wheats arose from hybridisations between cultivated tetraploid wheat and wild diploid species.

During the 20th century, breeders have developed many new wheat genotypes. These new types of wheat can produce good yields of grain, with resistance to cold, disease, insects and other crop threats. As a result, wheat production around the world has risen dramatically. During the last 30 years, there has been a tremendous increase in new knowledge of the biochemistry, genetics and functional properties of wheat. More recently, molecular biology has helped to elucidate the functions of the genes that determine the great diversity of wheat phenotypes (Jones et al., 2009; Akhonov, 2016; Henry, 2014, 2016).

Common or bread wheat (*T. aestivum*) is hexaploid with 21 pairs ($2n = 42$) of chromosomes comprising three homoeologous (similar) genomes (AA, BB and DD) each of 7 pairs. Durum and emmer wheats are tetraploids ($2n = 28$) possessing genomes, AA and BB, similar to these genomes in hexaploid wheat. Einkorn wheat is diploid (AA genome), offering a range of practical uses in the human diet (Abdel-Aal and Hucl, 2014) (see Chapter 14).

5.2.4 Plant and Grain Morphology

An understanding of plant, head and grain morphology (appearance, see Chapter 4) is useful for managing the stages of grain production, and also for the assessment of grain quality to the extent that morphology can be used for varietal identification—variety being a critical aspect of grain quality. Plant characteristics are an important part of the system for variety registration conducted by UPOV, the International Union for the Protection of New Varieties of Plants, for the purpose of enforcing Plant Breeders' Rights. Registration requires that a new variety should exhibit the attributes of distinctness, uniformity and stability (DUS), mainly with respect to morphology.

The procedures and attributes for DUS testing of wheat are provided by Cook and Wrigley (2004) and on the UPOV website (www.upov.int). Several books and review articles are available concerning the morphological characteristics that are useful for variety identification, some of them accompanied by identification descriptions and keys for the specific varieties in national variety collections as listed in Chapter 4, which provides a general description of morphology comparing the range of cereal species, accentuating wheat.

The wheat grain is botanically a single-seeded fruit, called a 'caryopsis' or 'kernel', developing within the glumes, which are modified leaves. The shape and dimensions of the grain have the potential to provide evidence for distinguishing among varieties, as described further in Chapter 4. It is the inner endosperm of the wheat grain that is the source of the flour produced by milling (see Chapter 22) to provide the multitude of wheat-based foods (see Chapter 24).

5.3 WHEAT PRODUCTION

The widespread popularity of bread as a staple food stems in part from the adaptability of wheat as a cultivated crop suited for many different soils and climatic conditions. Bread is also widely eaten in the 'rice bowl' countries of the world, where consumption of bakery foods has increased rapidly in the last few decades and continues to increase because bread is viewed as a more efficient 'convenience' food than rice (Nagao, 1998). Wheat is best adapted to being grown in temperate regions, with daily maxima of about 15°C and adequate moisture. However, wheat has long been grown in a wide range of climates around the world, including hot and dry regions ranging through to very cold parts of the world. In fact, wheat is cultivated from the Arctic Circle to the south of Argentina. It is also grown from sea level to 3000 m and in regions where the annual rainfall ranges from 250 to 1800 mm.

5.3.1 Grain Yields

The average wheat yield for the period 2001–07 in Table 5.3 (2.8 t/ha) is below that for the average yield relevant to all cereal grains in that period (3.2 t/ha). Wheat yields have increased considerably more recently (Tables 5.3 and 5.4). The ranges are narrow for yield data across the 7-year period in Table 5.3. This exercise in averaging smooths out the wide fluctuations in actual yield data from one region to another. However, there are great differences in wheat yields from one country to another (Table 5.4, for the years 2005 and 2014). For nearly every one of these countries, wheat yields have increased during the decade involved. At the upper range, yields of 7 t/ha are obtained in European countries (e.g. France, Germany and the United Kingdom), where intensive agriculture is practised with adequate moisture and modest temperatures. In these countries, many farmers pride themselves by belonging to 'Ten-Tonne Clubs', because their yields regularly exceed 10 t/ha.

In contrast, yields range down to below 2.5 t/ha for several countries where conditions are not ideal, being either hot and dry (e.g. India, Australia,

TABLE 5.3 Annual Statistics for Production and Yield of Wheat Worldwide Shows Increases in Both Parameters

	2001–07[a]	2013[b]	2014[b]
Statistics for the world	**Mean (*range*)**	**Actual**	
Production (million tonnes)	598 (*560–632*)	711	729
Yield (t/ha)	2.8 (*2.7–2.9*)	3.26	3.29

[a]*Uthayakumaran and Wrigley (2009).*
[b]*FAOSTAT, www.fao.org.*

TABLE 5.4 Production of Wheat in 1985, 2005 and 2014 for the Main Wheat-Producing Countries (in Millions of Tonnes) in Order of 1985 Volume of Production

1985[a]		2005[a]			2014		
Country	Production (million tonnes)	Country	Production (million tonnes)	Yield (t/ha)	Country	Production (million tonnes)	Yield (t/ha)
China[b]	85.8	China[b]	96.3	4.2	China[b]	126.2	5.0
USSR	73.2	India	72.0	2.7	India	94.5	3.0
USA	66.0	USA	57.1	2.8	USA	55.0	2.9
India	44.1	Russian Federation	47.6	Not available	Russian Federation	59.7	2.5
France	28.8	France	36.9	7.0	France	39.0	7.4
Canada	24.3	Canada	25.5	2.6	Canada	29.3	3.1
Turkey	17.0	Australia	24.1	2.6	Australia	25.3	2.0
Australia	16.0	Germany	23.6	7.4	Germany	27.8	8.6
Germany	13.8	Pakistan	21.6	2.6	Pakistan	26.0	2.8
Pakistan	11.7	Turkey	21.0	2.6	Turkey	19.0	2.4
		Ukraine	13.7	2.8	Ukraine	24.1	4.0
		UK	14.9	8.0	UK	16.6	8.6

Grain yield (t/ha) is provided for 2005 and 2014.
[a]*Uthayakumaran and Wrigley (2009).*
[b]*Mainland China.*
Source: Data for 2014 came from www.fao.org.

Pakistan, Turkey) or cold (e.g. Canada) or both (in different regions of the USA) (Table 5.4). These countries compensate for low yield potential by having very large areas under cultivation. Inevitably, wheat yield in any specific region or farm fluctuates greatly, depending on how closely growth conditions match ideal propagation conditions, and on how well the plants are fertilised or irrigated. In addition, plant breeders have concentrated successfully on breeding varieties that are suited to each of the diversity of conditions involved for the region addressed by a specific breeding program.

Global wheat production and yields increased dramatically through the period 1950–96 due to new varieties and technologies that produced tripling of yields especially in developed countries (Beta and Isaak, 2016). Even now, further increases in yield may be possible as similar improved technologies are used in developing countries.

5.3.2 Volume of Production

Worldwide, annual wheat production has averaged well over 700 million tonnes in the past few years, this being almost one-third of the grain production from all cereal species (Table 5.1); the proportion of wheat production has been increasing slightly. The range figures in Table 5.1 indicate the inevitable fluctuations in production for both wheat and cereal grains generally, although these worldwide figures for individual production years disguise the further fluctuations from one grain-growing region to another, as indicated in Table 5.4. All the data in Tables 5.1–5.4 indicate that yields and volume of production are continuing to increase for wheat and for cereals in general.

The major wheat-growing countries are listed in Table 5.4. The lists are topped by mainland China, but volumes are even greater if all the countries of the European Union (EU) are combined. In addition, the EU tops the yield score with an overall average of 5.5 t/ha (Wrigley, 2016b). Considerable increases in volume are seen for the near 31-year period covered across Table 5.3. For these top wheat-growing countries, there is little change in the countries represented, with only minor changes in the order of the countries with smaller volumes of production. The exception is the political change in the area involved in Russia/USSR. Over these many years, the production figures must be seen to represent single-year 'snap-shots', not fully reflecting inevitable year-to-year variations. Particularly susceptible to such fluctuations are the drought-prone countries, such as Australia, and others where the area of wheat production has only recently increased, for example Ukraine and Kazakhstan.

5.3.3 World Trade

Wheat grain represented about 44% of all cereal grains entering into international trade during the past decade (Table 5.1). About 20% of wheat production enters into international trade; this statistic is more than that for cereal-grain trade overall (13%) (Table 5.1). Many of the world's greatest wheat producers are also wheat exporters, but there are a few notable exceptions—the countries with populations that are large relative to production, namely, China, India and Pakistan. In some of these very large countries, wheat may be exported from one region and imported to another distant region because of transport economies, yet there is no net exportation. As a further example, in a recent year the USA imported about 2 million tonnes of wheat, whilst also exporting 23 million tonnes. Thus overall, the USA is included in the list of countries that export a large proportion of their wheat production, namely, Canada, Australia and Argentina.

5.4 UNIQUE PRODUCTS FROM WHEAT

Wheat flour is unique amongst grain species for its ability to be baked into good-quality leavened bread. When water is added to wheat flour with mixing, the matrix of storage protein becomes gradually hydrated, coming together to form the gluten strands visible in Fig. 5.1, while the individual starch granules (large and small) are trapped inside the gluten matrix. As gas bubbles develop in the dough due to yeast action, the consistency of the dough must be elastic enough to trap the bubbles (preventing them from merely rising to the dough surface) and it must also be viscous enough to allow the gas bubbles (mainly carbon dioxide from yeast action) to expand during baking, thus providing a well-expanded crumb and a large loaf volume. It is this ability of the gluten proteins to hold the gas bubbles during dough making and baking that makes wheat a unique cereal grain.

This ability is the major explanation for the wide range of uses of wheat worldwide. Wheat-based foods include leavened breads, cookies, cakes, biscuits, bagels, pizza, noodles, chapatti, breakfast cereals, macaroni, spaghetti, and other forms of pasta, meat substitutes and wheat beer. Some examples are shown in Fig. 5.2. Table 5.5 sets out the quality attributes needed in flour used for each of the many wheat-based products made worldwide; these attributes are thus also required correspondingly in the wheats used for flour milling. See Chapters 22 and 24 for fuller descriptions of the methods for assessing the grain-quality requirements for these and other wheat-based products (especially Tables 22.2–24.4).

The contribution of wheat as a major component of most diets of the world is due to its agronomic adaptability, its easy storage and nutritional benefits, and finally to its ability to produce a variety of interesting and tasty foods. Foods made with wheat are a major part of the diet for over a third of the world's people. In fact, wheat can be found in some form in almost every meal.

Bread is the most obvious food made from wheat flour (Fig. 5.2A–B). The unique viscoelastic properties of wheat dough are responsible for the production of good-quality yeast-leavened bread and also for a wide range of foods, some of which appear in Fig. 5.2. Wheat dough is better suited to these uses than the

FIGURE 5.2 Examples of products eaten or used by us as consumers. (A) Leavened bread loaf; (B) leavened bread rolls; (C) cookies; (D) cakes; (E) pancakes; (F) pastry spirals; (G) Danish pastries; (H) wafer cone; (I) liquorice and marshmallows; (J) soft drinks; (K) pet food.

TABLE 5.5 Quality Attributes Preferred in Wheat and Flour Used for Specific Products Worldwide

Product	Grain protein content	Grain hardness/softness	Water absorption of flour	Dough strength	Dough development time[a]	Starch peak viscosity[b]
Breads						
Pan bread (by method)						
Sponge-and-dough	>12%	Hard	High	Strong	High	High
Straight-dough	>12%	Hard	High	Medium	Medium	
Rapid-dough	>11%	Hard	High	Medium	Short	
Flat bread						
Middle Eastern	10.5–12%	Hard	High	Medium	Medium	Medium
Indian subcontinent	10–12%	Hard	High	Medium	Short	
Tortillas	10.5–12%	Hard	Medium	Medium	Medium	
Steamed bread						
Northern-China style	11–13%	Hard	High	Medium	Medium	
Southern-China style	10–12%	Medium-hard	Low-medium	Medium	Medium	

(Continued)

TABLE 5.5 Quality Attributes Preferred in Wheat and Flour Used for Specific Products Worldwide (*cont.*)

Product	Grain protein content	Grain hardness/ softness	Water absorption of flour	Dough strength	Dough development time[a]	Starch peak viscosity[b]
Noodles						
Yellow alkaline type	11–13%	Hard	High	High	Medium	Medium
White salted type	10–12%	Soft and hard	Medium	Medium	Medium	High
Instant type	10–12%	Hard	Medium	Medium	Medium	
Cookies and Cakes	8–9%	Very soft	Low	Weak	Low	Medium
Pasta (based on durum wheat semolina)	>13%	Extremely hard	Not high (semolina)	Strong	High	Medium
Starch-Gluten[c]	>13% preferred	Soft preferred	Low preferred	Medium	High	High preferred

Good milling quality is required for all wheat-grain types, giving a high yield of white flour.
[a]*In the Farinograph.*
[b]*In the Rapid ViscoAnalyzer.*
[c]*Genotypes preferred for starch-gluten manufacture should have soft grain, high protein content and high proportions of large (A-type) starch granules. However, economic factors may out-weigh these considerations.*
Source: Adapted from Wrigley, C., Asenstorfer, R., Batey, I., Cornish, G., Day, L., Mares, D., Mrva, K., 2009. The biochemical and molecular basis of wheat quality. In: Carver, B. (Ed.), Wheat: Science and Trade. Wiley-Blackwell, Ames, Iowa, pp. 495–520.

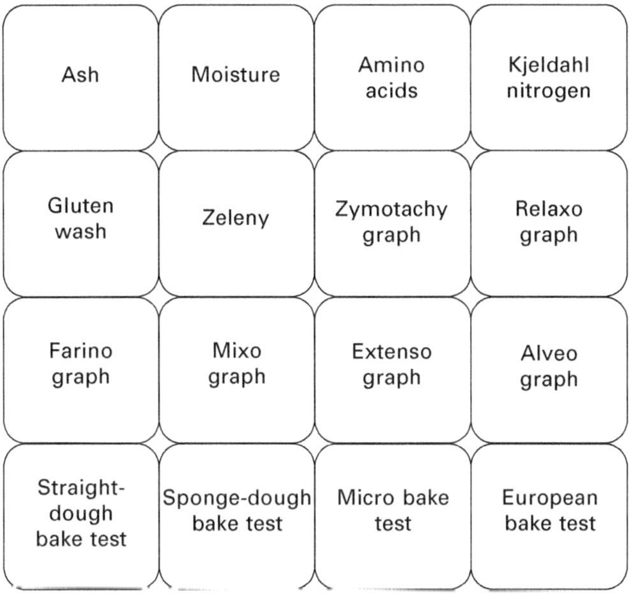

FIGURE 5.3 The various levels at which flour-quality may be analysed thus to provide information about the quality attributes needed, as listed in Table 5.5. The top row represents the analysis of basic composition; the second row lists methods to determine gluten function; the third row relates to dough properties; the lowest row specifies some of the approaches to determining baking quality, depending on the baking method and on the special requirements—the micro bake test suits the needs of breeders and researchers.

dough made from other cereals, with the possible exception of rye, but bread made from rye flour is usually augmented with wheat flour (see Chapter 7).

Flour for conventional baking would generally be milled from a hard-grained wheat with a protein content of over 12% (Table 5.5). Analysis of the suitability of such a flour would involve the tests listed at the top of Fig. 5.3, namely, determination of ash content (to determine the extent of bran and germ included in white flour during milling), moisture content (to be monitored before and during milling) and protein content (determined as nitrogen content by Kjeldahl analysis or by near infrared spectroscopy). The amino-acid composition is more likely to be relevant to the use of grain or grain by-products for pet foods (Fig. 5.2K) and for animal feeding (Black, 2016). See Chapter 22 for more details about wheat milling.

Strong-to-medium gluten properties would be appropriate for bread-making (Fig. 5.2A–B; Table 5.5), the gluten properties being tested by any of the methods in the second row of Fig. 5.3, or by testing the whole dough properties with the equipment in the third row (Morris, 2016b). The ultimate tests of baking quality are listed in the bottom row of Fig. 5.3, depending on the relevant baking method (see Table 24.1 for more details).

On the other hand, flour for making cookies and crackers, cakes, breakfast cereals, wafers, pancakes and pastries (represented in parts of Fig. 5.2) would be milled from a soft wheat of lower protein content and extensible gluten/dough properties (Table 5.5). The milling of a soft wheat gives flour of low starch damage, thus with minimal water absorption thereby reducing the baking time needed to dry out products such as cookies and crackers (HadiNezhad and Butler, 2009; Cauvain, 2016). In contrast, pasta production involves the milling of the very hard durum wheat (a distinct wheat species) to semolina giving strong dough properties (see Chapter 6).

Wheat flour is used widely in the food industry. Liquorice is an unsuspected example in Fig. 5.2I; the ingredient list for this sample includes wheat flour, glucose syrup made from wheat flour and wheat fibre. Marshmallows also contain wheat starch and sugar from wheat starch (Fig. 5.2I). The industrial separation of starch from gluten is a major use of wheat flour (Chapter 23). The resulting starch finds a wide range of uses including hydrolysis to make sweeteners (e.g. for candies and beverages—Fig. 5.2I–J) and fermentation to generate biofuel (Best, 2009); the other component is vital dry gluten used to bolster dough for bread manufacture and to augment the protein level and function in many foods (Table 5.5). Gluten is also modified physically and chemically for various food and non-food uses, such as for forming flexible films, as a foaming agent, extending its fibrous structure by extrusion technology to simulate meat fibres, and as a basis of adhesives (Day et al., 2006; Khosravi et al., 2014). In most of these cases, the gluten–protein complex is a key factor for processing. Indeed, protein content and quality are relevant to most of the quality-attribute columns in Table 5.5, including grain hardness as described later in the chapter.

5.5 WHEAT PROTEINS

The ability of wheat proteins to combine into the protein mass known as gluten has made wheat the most popular and effective source of flour for bread production in the world. Wheat gluten was first isolated by Beccari in 1728 (Bailey, 1941). The traditional method of preparation of gluten involves the mixing of flour and water to form dough, followed by gentle washing of the 'dough ball' in water or dilute salt solution to remove starch and soluble materials. Gluten is recovered as a rubbery mass containing approximately 80% of the total protein of the flour (Dill and Alsberg, 1924). This exercise is simple in principle and it has been offered as a student exercise (Wrigley, 2012). The production of gluten by water-washing of dough is a major industry worldwide (Bietz and Lookhart, 1996; Maningat et al., 1999; Day et al., 2006; Batey and Huang, 2016) (see Chapter 23).

Lipids, which make up 5–10% of the dried gluten, may not be naturally associated with the proteins, but are believed to be bound to them during dough formation (Day, 2016a). Occluded starch and moisture make up most of the remainder of gluten preparations (10–12%) and small amounts of non-starch carbohydrates may also be present.

5.5.1 Classification of Wheat Endosperm Proteins

Wheat proteins have been classified traditionally according to their solubility properties. The first comprehensive fractionation scheme for wheat proteins was developed by Osborne (1907). It was based on the different solubility of the proteins in various solvents:

- Albumin: soluble in water.
- Globulin: soluble in salt solution.
- Gliadin: soluble in 70% aqueous ethanol.
- Glutenin: soluble in dilute acid or alkali.

Gliadin and glutenin together represent about 80% of the total proteins in flour and are present in approximately equal amounts (Tatham and Shewry, 1985; Shewry et al., 2009; Békés, 2012; Békés and Wrigley, 2016). Gliadin and glutenin are the most important determinants of the functional properties of wheat flour. Ideal dough properties depend on an appropriate balance between gliadin (contributing to dough viscosity) and glutenin (contributing to the strength and elasticity of dough). It is the unique combination of these properties that comprises the functional properties of dough (Khatkar and Schofield, 1997; Wieser et al., 2006). The albumin and globulin proteins are less important to bread-making quality than the gluten proteins (Jones et al., 2006).

Prolamins are storage proteins soluble in alcohol–water mixtures; their amino-acid compositions are characterised by high contents of glutamine and proline, which together account for 30–70% of the total amino acids (Shewry et al., 1986; Juhasz et al., 2014). Comparison of the amino-acid sequences of individual prolamins has led to a classification based on primary-structure relationships (Shewry and Miflin, 1985). The three groups are:

- High-molecular-weight (HMW) prolamins: the HMW glutenin sub-units (HMW-GS).
- Sulfur-rich prolamins: the low-molecular-weight glutenin sub-units (LMW-GS), α-gliadins, β-gliadins and γ-gliadins.
- Sulfur-poor prolamins: ω-gliadins.

5.5.2 Genetics and Functional Properties of Endosperm Proteins

Many of the flour specifications required for wheat-based products are related to specific aspects of protein composition (and thus genes, see Fig. 5.4). As a result, it has been possible to specify marker proteins as a basis for predicting wheat or flour quality instead of performing quality tests such as dough testing, baking or starch pasting, for example those indicated in Fig. 5.4 (Cornish et al., 2006; Békés and Wrigley, 2013).

Protein markers may be non-functional polypeptides (indicated in the second row of Fig. 5.4), arising from gene action, via messenger RNA (the transcriptome). The amino-acid sequence of the polypeptides is specified by the

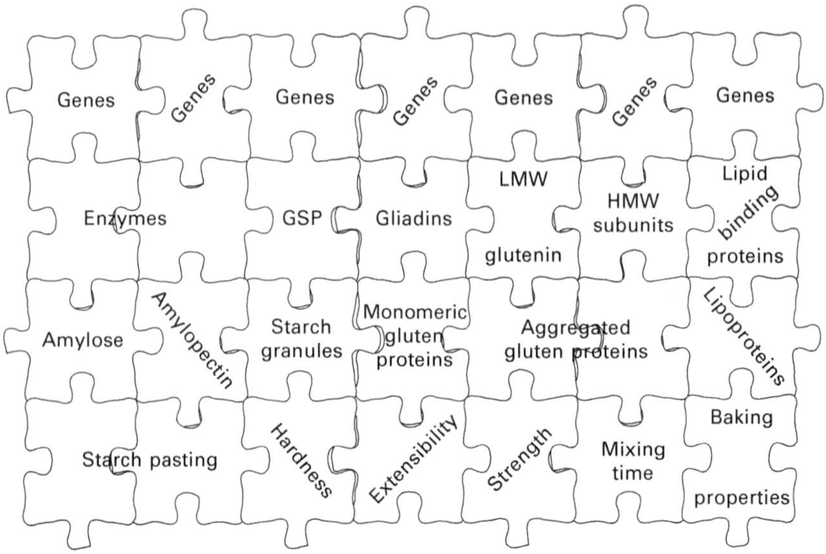

FIGURE 5.4 The interlocking nature of the functional properties is shown at the lowest line, interactive with the chemical components responsible (next line up), and with the proteins involved. The top line accentuates the primary influence of the genome in determining the synthesis of the proteins responsible for generating the necessary grain composition. *(Adapted from Wrigley, C.W., 1993. A molecular picture of wheat quality: finding and fitting the jigsaw pieces. Cereal Foods World 38, 68–74.)*

respective DNA and RNA sequences (Henry, 2014, 2016). The further stages of folding and disulfide bonding lead to the formation of functional proteins, such as enzymes, aggregated gluten proteins and lipoproteins in the second and third rows of Fig. 5.4.

The genes that code for the various gliadin and glutenin polypeptides have been identified. The *Glu-1* genes on the long arms of the homoeologous group-1 chromosomes code for the high-molecular-weight sub-units of glutenin (HMW-GS). The short arms of the same chromosomes carry the *Glu-3* genes, coding for the low-molecular-weight sub-units of glutenin (LMW-GS) and the *Gli-1* genes, coding for γ- and ω-gliadins. The α- and β-gliadins, however, are coded at the *Gli-2* loci on the short arms of group-6 chromosomes (Metakovsky et al., 1994, 2006). Each gliadin locus has been found to control the synthesis of several jointly inherited components (blocks) and 12–25 alleles have been assigned to each locus (Metakovsky et al., 2006). To indicate which locus is concerned, the genome symbol is appended to the general locus name, for example *Glu-D1* is the D genome HMW glutenin locus (Payne, 1987; Shewry et al., 1989, 2006). Each HMW-GS locus consists of two genes encoding an HMW *x*-type and an HMW *y*-type sub-unit (Shewry et al., 2006), the *y*-type generally having a greater mobility on SDS-polyacrylamide electrophoresis.

Therefore, hexaploid bread wheats could, in theory, contain six different HMW-GS. In fact, only three to five sub-units are present in most cultivars because of the silencing of some genes. A wheat cultivar with six HMW-GS has been identified (Johansson et al., 1993). All cultivars contain 1Bx, 1Dx and 1Dy sub-units, while some cultivars also contain 1By and/or 1Ax sub-units (Payne et al., 1984). The 1Ay sub-units are available in wild wheats. Payne and Lawrence (1983) surveyed alleles at the three glutenin loci in about 300 cultivars and found only three allelic forms at the *Glu-A1* (x sub-unit only) locus, one of which is a null form (i.e. no sub-unit expressed). Either 1 (the x) or 2 proteins (when the y sub-unit is also present) were found encoded by the *Glu-B1* locus, with 11 alleles involving 14 different proteins. Seven alleles were found in the *Glu-D1* locus coding for eight proteins.

Among a collection of 222 hexaploid wheats from 32 countries, 20 different electrophoretic band patterns were detected for LMW sub-units of glutenin, 6 for the *Glu-A3* locus, 9 for the *Glu-B3* locus and 5 for the *Glu-D3* locus (Gupta and Shepherd, 1990). There was also a null for the *Glu-A3* locus. The number of electrophoretic band combinations found for LMW-GS is much lower than expected based on random association of gene loci products, indirectly indicating that genes coding for these bands are closely linked. Most wheat varieties express more than 15 LMW-GS (Payne et al., 1984; Juhasz et al., 2014).

Recent decades have brought abundant evidence that the various genes for the glutenin sub-units contribute to grain quality, mainly with respect to dough properties (Cornish et al., 2006; Shewry et al., 2009; Juhasz et al., 2014). The HMW sub-units of glutenin (*Glu-1* alleles) make a greater contribution than the LMW sub-units (*Glu-3* alleles), and the *Glu-1* ranking system of Payne (1987) has been a valuable tool for assessing dough strength. However, we now know that the combined allelic contributions of both HMW and LMW sub-units should be considered when assessing potential parent lines and when selecting progeny (Cornish et al., 2006; Wrigley et al., 2009; Békés and Wrigley, 2013).

Nevertheless, the actual contribution to dough quality is made by the polymerised glutenin sub-units. The way in which these very large molecules are joined together by disulfide bonds is critical to dough quality (Wieser et al., 2006), even though some sub-units may make different contributions than others. This polymerisation process is under both environmental and genetic control, so growth conditions must be considered in addition to the genetic constitution of the glutenin alleles (Lafiandra et al., 1999).

About 50% of the variation in grain quality has been attributed to the allelic combination of the glutenin sub-units and the puroindolines—proteins that are associated with the starch granules (see Section 5.6) (Morris et al., 1994; Jones et al., 2006). In addition, another class of proteins, the serpins, has been implicated as contributing to grain quality (Cane et al., 2008). These proteins (serine protease inhibitors, classed as defence proteins) constitute about 4% of the grain mass. Their involvement is based on statistical analysis of data for a large population of wheat genotypes, as well as analysis of genotypes differing in the wild-type and null alleles of a serpin whose locus is located on chromosome 5B.

5.6 WHEAT CARBOHYDRATES

The wheat grain, at maturity, consists of 85% carbohydrate, most (80%) being the starch of the endosperm. The non-starch carbohydrate is made up of mono-, di- and oligo-saccharides and fructans (7%), plus cell-wall polysaccharides (12%) (Mares and Stone, 1973; Fincher, 2016). The main wheat carbohydrate (starch) is the predominant carbohydrate source for human diets, the substrate for producing alcoholic beverages and fuel ethanol by fermentation (Guragain et al., 2016), and the raw material for many other industries (Stone and Morell, 2009). Starch constitutes 60–70% of the mass of wheat flour. In general, soft wheats have less protein and more starch than hard wheats.

The importance of carbohydrates in our diet has been emphasised in articles by Jones et al. (2015a,b). They cite the repeated emphasis on cereal grains and cereal carbohydrates in all the nutritional guidelines for many countries. Visual representations are provided by these many countries, all with the same diagrammatic message: an emphasis on cereals and cereal carbohydrates being foundational in dietary recommendations (Jones et al., 2015a).

The contribution of starch to flour-processing quality has only recently been fully recognised (Rahman et al., 2000). Chemical properties (such as amylose content and granule-size distribution) and physical characteristics (such as temperature of gelatinisation, viscosity of starch gels and swelling power) are the main characteristics of starch that are important for processing quality. Starch is deposited in granules in the developing endosperm. At maturity, the granule sizes range from <1 mm to about 40 mm (see Chapter 4). The starch granules contain amylose and amylopectin, plus small amounts of protein, lipid and other polysaccharides (Rahman et al., 2000; Stoddard, 2016). The proteins may be present within the granules (granule bound) or on the surface of the granule (surface bound). The granule-bound proteins are enzymes which were involved in starch synthesis.

One of the surface-bound proteins known as 'friabilin' has been linked with the grain-softness protein (GSP in Fig. 5.4) and with differences in grain hardness (Greenwell and Schofield, 1986). This grain-softness protein has since been shown to be a mixture of several related proteins known as 'puroindolines' (Morris et al., 1994; Jones et al., 2006). The genetics of grain hardness have been demonstrated to be relatively complex, with a diversity of alleles being involved at the *Pina-D1* and *Pinb-D1* loci. New alleles are being discovered in various germplasm collections, offering breeders a range of opportunities to modify grain hardness and the starch properties that are related to noodle quality (Ma et al., 2009; Stoddard, 2016).

One of the granule-bound proteins, known as granule-bound starch synthase (GBSS), is associated with amylose synthesis (Regina et al., 2016). It is

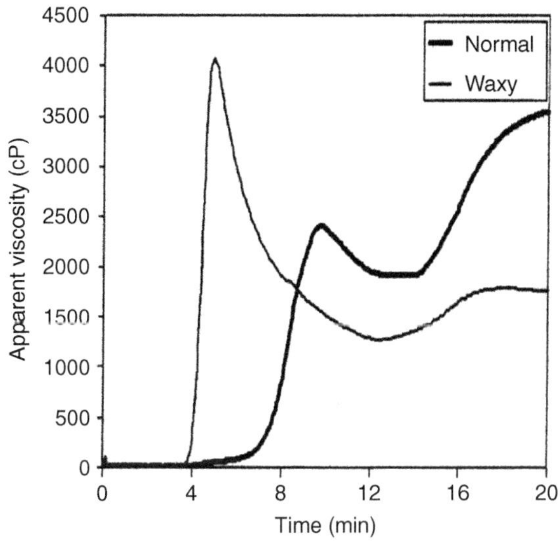

FIGURE 5.5 Pasting curves obtained using the Rapid ViscoAnalyzer for normal wheat starch and for waxy wheat starch.

encoded by loci on chromosomes 7A, 7D and 4A (Nakamura et al., 1995). The locus on chromosome 4A, known as '*Wx-B1*' was translocated naturally from chromosome 7B. The null allele at this locus is linked to superior noodle quality in Japanese Udon noodles (Zhao et al., 1998). The presence of all three GBSS enzymes, encoded at their respective loci, introduces a degree of redundancy in synthesis of amylose. Removal of one of the enzymes reduces the amylose content to 18–25%. Removal of the second enzyme further reduces the amylose content to 12–18%. When all forms of the enzymes are absent, the amylose content is reduced to almost zero (Nakamura et al., 1995). The resulting waxy wheat offers distinct starch pasting properties and novel applications of wheat (Purna et al., 2015) (Fig. 5.5). High-amylose content is also possible (Yamamori et al., 2000; Regina et al., 2006).

The size distribution of starch granules in wheat is important for processing quality. The distribution is bimodal. By weight, 70–80% of the starch is the large granules (A-type, greater than 10 mm), which contribute only 20% of the granules by number. Small starch granules (B-type) have a high surface-area-to-mass ratio and hence bind more water than the same mass of large granules (Larsson and Eliasson, 1997). A-type starch is the highly purified form comprising mainly large granules with very low protein content. B-type starch has higher protein content and comprises mainly small granules.

Starch-paste viscosity, an important aspect of starch quality, is measured using the Visco-amylograph or the Rapid ViscoAnalyzer (Fig. 5.5). The important parameters measured using this method are the peak viscosity, trough viscosity and final viscosity. The method involves a controlled temperature profile which heats and then cools the starch in the presence of excess of water, from an initial temperature of 50 to 95°C and then back to 50°C. During the heating phase, the starch gelatinises. However, the observed viscosity change does not occur until the gelatinised starch granule rapidly absorbs water and swells.

After reaching peak viscosity on heating, the viscosity reduces as the swollen granules rupture and degrade. The minimum viscosity reached is an indication of stability of starch gel at high temperature. When the gel is cooled, its viscosity increases. The increase in viscosity is contributed by amylose. The final viscosity reached is the indication of the gel strength of products at low temperature. Starch-paste viscosity is measured for two reasons (Stone and Morell, 2009): (1) starch-paste viscosity is an alternative to falling number measurements. This provides the impact of amylase on viscosity. (2) Starch-paste viscosity is also a predictor of end-use functionality in food production and for the non-food industry (e.g. paper manufacture).

The swelling power of starch measures the ability of starch to absorb water and increase in size under controlled conditions of starch–water ratio, temperature and agitation. Amylose content is the main factor which determines the ability of the granule to swell (Crosbie, 1991). This is a useful diagnostic tool to predict noodle quality. The consistency of the cooked product is an important attribute of Japanese Udon noodles and this is generated by the ability of the granule to swell and maintain integrity. The GBSS alleles determine the swelling power of granules. The triple-null genotype (waxy) has the greatest swelling power compared to the wild types having all three alleles (Fig. 5.5). Mutations in the homoeologous series of waxy protein genes (GBSS) on chromosomes 4AL (translocated from 7BS), 7AS or 7DS can, in various combinations, enhance the suitability of flour for noodle production (Zhao et al., 1998).

5.7 WHEAT LIPIDS

Wheat lipids make up 3–4% of the weight of the whole grain and 1–2.5% of straight-run milled flour. Even though lipids are a minor constituent in wheat, they play an important role in dough mixing, the baking process and consumer acceptance of the finished products. The lipids interact with the gluten proteins to form complexes, which contribute to the stabilisation of the gas-cell structure, thus having a significant effect on loaf volume and on the final texture of the baked product.

Grain lipids are a complex mixture of components which are distributed unevenly through the various parts of the wheat kernel (Chung et al., 2009; Day, 2016a). The wheat germ has the highest concentration of lipid. The wheat lipids can be divided into non-polar and polar lipids: 50% of the total non-polar lipids in wheat are the triglycerides, deposited in spherosomes bounded by monolayer membranes. Non-polar lipids are found in the outer layers of the kernel, or the bran and aleurone layers (Hargin and Morrison, 1980). The remaining non-polar lipids are mono- and diglycerides, fatty acids and sterol esters. Glycolipids and phospholipids are polar lipids found in all membranes.

In milled endosperm flour, the amount of glycolipids is often greater than phospholipids. These lipids are organised in a reversed hexagonal liquid crystalline phase located in a protein matrix closely bound to starch granules in the endosperm of wheat kernels (Al Saleh et al., 1986). Monogalactosyldiglyceride and digalactosyldiglycerides are the main glycolipids found in the wheat kernel. The major phospholipids are phosphatidylcholine, phosphatidylethanolamine and phosphatidylinositol. Wheat lipids are also grouped into saponifiable and non-saponifiable lipids. Fatty acids, acylglycerols, glycolipids, phospholipids and sterols are saponifiable lipids. Carotenoids and tocopherols are non-saponifiable lipids.

Although carotenoids are minor constituents, the colour contributed by carotenoids is an important factor in cereal foods especially in the use of durum wheat for pasta making. Wheat or flour lipids can also be grouped based on the location of lipids in the wheat or flour. Non-starch total lipids which constitute approximately two-thirds of the total flour lipids are present in wheat fractions other than starch granules and can be extracted at ambient temperature. It is the non-starch lipids that interact with flour proteins during mixing that have been proven to have technical value during the baking process. Starch lipids are predominantly phospholipids. They are tightly bound as an inclusion compound with amylose forming the amylose–lipid complex, which may affect the properties of starch.

Lipids play an important role in our diet. They are a source of high energy and are important structural parts of cell membranes. Lipids are great carriers of fat-soluble flavour and hence increase palatability (taste, aroma, texture) of food. The slow movement of the lipids in the gastrointestinal tract increases satiety value of foods. Free lipids in wheat flour play an important role in bread making. It has been shown that removal of free lipids from the baking formulation reduces loaf volume and crumb score (Chung et al., 2009) and also reduces the volume of steam bread and increases textural hardness (Pomeranz et al., 1991). It is also shown that removal of free lipids from cookie formulation decreases cookie diameter which is one of the major parameters in determining cookie quality (Clements and Donelson, 1981). In spaghetti, removal of free lipids increases surface stickiness and loss of yellow colouration (Lin et al., 1974). Lipids also play an important role in the quality of wheat extrudates (Faubion and Hoseney, 1982) and dry oriental noodles (Rho et al., 1989).

5.8 WHEAT MICRONUTRIENTS

The wheat grain is a source of a range of valuable bioactive compounds, in addition to the basics of carbohydrates, lipids and proteins. These many micronutrients have become the subject of intense research interest with respect to their roles in our health and nutrition as phytochemicals and antioxidants (Dey, 2014;

Sapirstein, 2016; Beta and Duodu, 2016). The presence and amounts of many of these compounds are listed in Appendix 1.

5.8.1 Vitamin B Complex

Major micronutrients of the wheat grain are vitamin E, several B vitamins and various minerals. These are unevenly distributed in the kernel and are mainly found in wheat germ and the bran; hence they are found in reduced amounts in refined flours. Wheat is known to be a good source of vitamin B especially thiamin, niacin, vitamin B6 and folate. Whether the wheat is consumed as whole-grain products or products from refined flour determines the contribution of wheat to recommended daily dietary allowance (Slavin et al., 2001; Sapirstein, 2016).

The contribution of vitamin B intake from 100 g of wholemeal flour to the recommended daily dietary allowance is provided in Appendix 1. Consumption of products made from whole meal contributes to 40% of the recommended daily allowance for thiamin, 10% for riboflavin, 22% niacin, 33% vitamin B6 and 13% folate. The bioavailability of vitamin B from cereal grains has been of concern. In cereals, niacin is found in a chemically bound form, nicotinic acid (van den Berg, 1997). Since bound niacin is less bioavailable than free niacin, around 70% of niacin in mature cereal grain may not be available after traditional cooking. Mild alkali treatment during cooking releases much of the bound nicotinic acid thus improving the bioavailability of niacin (van den Berg, 1997).

Folates in wheat generally exist as reduced derivatives of folic acid (pteroyl-L-glutamic acid, PGA) which are largely bound to polyglutamates. Polyglutamyl folates must undergo enzymatic deconjugation before absorption and hence their bioavailability is low. The study carried out by Fenech et al. (1999) showed that ingestion of cereal made from wheat aleurone flour contained natural folates which were readily bioavailable and significantly increased blood folate concentration.

The main vitamin B6 fraction in grain is glycosylated pyridoxine (PNG) (Ollilainnen, 1999; Sapirstein, 2016). Studies of Leklem et al. (1980) have shown that the vitamin B6 available from whole wheat bread was 5 to 10% less available than it is in vitamin-enriched white bread. Furthermore, Kies et al. (1984) reported that much of the naturally occurring B6 in wheat bread is biologically unavailable for the human diet. These reports presumably relate to the difficulty of access to such nutrients inside the bran cells, and the need for milling techniques to rupture these cells, thus providing access (Sapirstein, 2016) (see Chapter 22).

5.8.2 Vitamin E

The term vitamin E is used for four tocopherols and tocotrienols that exhibit the biological activity of α-tocopherol. Cereals are considered to be a moderate source of vitamin E, providing 6–20 mg of α-tocopherol equivalent per gram (Bramley et al., 2000; Beta and Duodu, 2016). The sum of all tocopherols and tocotrienols in wheat is in the range of 49–58 mg/g (Bock, 2000; Chung and Ohm, 2000; Chung et al., 2009). Wheat germ mainly consists of α- and β-tocopherols. Tocotrienols are concentrated in the pericarp, testa and the aleurone. In the human body, vitamin E is the most important lipid-soluble antioxidant, providing an effective network against oxidative stress together with other antioxidants such as vitamin C. Tocopherols and tocotrienols are absorbed together with dietary fat. The study by Kahlon et al. (1986) showed that natural vitamin E from wheat bran was poorly available in rats; the availability of vitamin E was reduced if coarse bran was fed.

Vitamins vary greatly in their stability and sensitivity to light, pH, temperature, oxygen and other process and storage conditions (Reddy and Love, 1999). The altering conditions of milling, processing and storage can have different effects on the stability and bioavailability of individual vitamins (Batifoulier et al., 2005). The most heat labile of the vitamins, thiamine, is affected by alkaline conditions during cooking. Vitamin B6 is unstable during prolonged heating and folate is significantly degraded during processing. Tocopherols and tocotrienols are easily oxidised by oxidising agents. Cooking of wheat grain results in a 55% loss of tocopherols (Wyatt et al., 1998).

5.8.3 Carotenoids

Carotenoids provide colour to wheat-based foods. The yellow pigment in wheat is mainly derived from xanthophylls. Carotenoid content in wheat grain is in the range of 1.8–5.8 mg/g (Kruger and Reed, 1988). Some wheat carotenoids such as β-carotene, α-carotene and β-cryptoxanthin have provitamin A activity. Carotenoids which do not have vitamin activity contribute to antioxidant activity. Lutein, β-cryptoxanthin, zeaxanthin and β-carotene are carotenoids found in wheat. Lutein and zeaxanthin, which are the only carotenoids in the macula of the eye, retard cataracts. Lutein also provides protection from ischemic stroke and together with other carotenoids protects skin from UV-induced damage. Not much data is available on the absorption and bioavailability of carotenoids other than β-carotene (Castenmiller and West, 1997; van den Berg et al., 2000). Carotenoids are easily oxidised, and thus they are readily lost during milling and heat processing.

5.8.4 Minerals and Trace Elements

Cereals contribute significantly to the dietary allowance of minerals and trace elements. According to Lewis and Buss (1988), in a typical Western diet, cereals and cereal products contribute to 50% of the dietary magnesium and iron, 30% of copper and magnesium and 20% of zinc and phosphorus. In wheat however, iron, zinc, copper and manganese contents are low and the minerals and trace elements are mostly situated in the outer part of the grain.

The calcium content of hexaploid wheat ranges from 0.1 to 0.8 mg/g (dry basis), magnesium ranges from 0.2 to 2.2 mg/g, phosphorus from 2.2 to 9.1 mg/g, potassium from 2.8 to7.3 mg/g, iron from 16 to 163 mg/g, manganese from 40 to 90 mg/g, copper from 1 to 14 mg/g and zinc from 15 to 102 mg/g (Piironen et al., 2009). These variations in mineral contents are due to differences between varieties and in growth conditions, the amounts of nutrients in the soil affecting the mineral and trace element content of the resulting grain. Single-figure data in the nutrition tables of Appendix 1 are within these ranges, but the aforementioned ranges emphasise the variability of content in a way that nutrition tables cannot do.

The intake of whole-grain products increases the intake of minerals and trace elements but it may decrease the absorption of these compared with refined wheat products. The reason for this apparent anomaly is that the bioavailability of minerals and trace elements is associated with the potent cation-exchanging capacity of dietary fibre which may therefore have a negative effect (Piironen et al., 2009).

5.9 ASSESSING GRAIN QUALITY

5.9.1 Physical Aspects

The point of grain delivery following harvest is the initial site for testing grain quality. This site may be the mill or grain elevator (silo). Testing may also be required at all other points along the grain chain (Table 5.2). Physical appearance (to the expert eye) is of primary importance, combined with testing for bulk density and sieving to determine how much grain is undersized and to examine for the presence of contaminants.

Visual inspection is the first approach to determining grain soundness, with particular attention to signs of sprouting, frost or heat damage and fungal or insect damage (Wrigley and Batey, 2003). These procedures, described fully in Chapter 19, are similar for cereal grains in general, but the exact specifications differ for particular grades and classes, and from one region to another (McFall and Fowler, 2009).

Grain sampling is a critical aspect at this initial stage of grain analysis. It cannot be assumed that a consignment of grain is homogeneous, so to obtain a representative sample, sub-sampling must be carried out at various points in the load, and the combined sub-samples must be thoroughly mixed before subsampling and analysis.

5.9.2 Variety

Knowledge about variety is a primary source of information about the quality type of a grain lot because the breeder has intentionally selected each variety to suit a specific range of processing needs. It is thus important to verify that the variety specified is correct. Accordingly, many wheat-producing countries have a system of variety declaration in place for delivered wheat, backed up by the capability to test that the variety declaration is correct. Visual identification offers limited possibilities for identification, as described in Chapter 4.

More definitive means of identification are provided by laboratory testing, most often by extracting the grain proteins and determining the protein composition. In recent decades, gel electrophoresis of grain proteins has been the routine method for doing so (Lookhart and Wrigley, 1995), although HPLC has also been used. Capillary electrophoresis in various forms has also proven effective (Uthayakumaran et al., 2004, 2005). See Chapter 17 for a description of methods, based on either protein or DNA composition.

5.9.3 Growth and Storage Conditions

In addition to the importance of genotype (variety) in determining grain quality, growth (Chapter 18) and storage conditions (Chapter 20) have the capacity to modify the built-in genetic potential for appropriate grain quality. All these conditions must be authenticated to justify claims for 'Organic' status for grain, thereby gaining recognition as a special grain and commanding premium value. The international regulations and restrictions relating to organic status are described by Gelinas and David (2016). Such restrictions include the use of 'chemicals' to prevent insect infestation but not the aeration of stored grain to reduce the temperature and moisture of the stored grain (Navarro et al., 2012; Annis, 2016), nor fumigation with carbon dioxide although the level of CO_2 must be maintained for over a week (Annis, 2016).

When grain is delivered at harvest, it will be known what the climatic conditions were at the growth site (farm). Useful prediction of quality can derive from this knowledge, depending on whether the harvested grain had been subject to extremes of heat or drought or whether there was rain at harvest at that site, prompting the testing for incipient sprouting (see Chapter 19). In the same way, knowledge of the storage history of a grain consignment may be a valuable indicator of possible loss of quality due to improper storage (see Chapter 20).

5.9.4 Chemical Composition

The moisture content of grain is a basic quality consideration that determines its safety during storage, although the maximum moisture content advisable varies depending on the grain species, regional variations and the expected storage temperature. Whereas dry weight after oven drying is the standard analytical method, near infrared spectroscopy is now the established routine procedure for determining grain moisture. NIR methods are also routine for determining protein content, which is an almost universal attribute for allocation to a quality class (see Chapter 19). Beyond the grain-receival site, laboratory testing may be needed to detect chemical residues at levels above regulations (Annis, 2016) and to determine the level of α-amylase in samples suspected of being sprout damaged (Morris, 2016b).

The full range of tests relevant to grain quality are provided on the websites of the scientific societies AACC International (formerly the American Association of Cereal Chemists; www.aaccnet.org) and the ICC (the International Association for Cereal Science and Technology; www.icc.or.at). These testing methods have been developed by collaborative research involving scientists from these associations, so that the procedures have the status of being standardised and recognised internationally. They are further explained and listed by Bridges and Wrigley (2016a,b) (see also Table 24.1).

5.10 MANAGING GRAIN QUALITY

5.10.1 Breeding and Genotype Selection

The suitability of the raw materials sourced makes a critical contribution to the success achieved in producing an end product that will satisfy consumer demands, as is portrayed in Table 5.2. Wheat quality means that the grain must be suitable to produce flour which in turn suits the processing needs for the specific wheat-based product (see Chapter 24). The initial step towards appropriate grain quality is selecting the right variety (or mix of varieties). The breeder needs to 'build-in' the necessary quality characters tailored to a narrow set of processing requirements when selecting a variety (genotype) for release (see Chapter 16).

The breeder not only aims to build in these quality attributes but also to add genetically-based resistance or tolerance to pests and genetically-based tolerance to environment variations that may reduce yield and quality. These are formidable goals for breeders, especially when global climate change is altering the growth conditions in many regions. In some wheat-growing areas, there has been the need to change to different types of agricultural production, with the transfer of wheat-growing to new regions. As a result, breeders have needed to re-focus their efforts with respect to targeted growth regions and conditions.

The primary effort of most of the breeders is building resistance to many fungi, bacteria, insects and nematodes in what is known as 'adapted material'. Traditionally, breeders have been primarily concerned with grain yield, selecting for it throughout the breeding process, with attention for grain quality being a secondary consideration, waiting until late generations before selecting for quality. This approach risks the unnecessary propagation of poor-quality lines that must be discarded later. Therefore, it is desirable that selection for quality characteristics be considered at an earlier stage to improve breeding efficiency (DePauw and O'Brien, 2016).

To select for grain quality early in the selection process, the breeder may have only a few grains and these are needed to grow on into another generation; hence, it is a great advantage to have non-destructive tests such as near infrared transmittance (NIT) with which to determine protein content and various quality traits on the whole grain. In addition, the breeder has the option of performing analysis on the brush end of the grain, leaving the germ end for propagation. A half-grain is sufficient for determining protein composition, thereby providing information about specific protein markers (Wrigley et al., 2009).

In addition, DNA from the grain or plant can be analysed for the presence of appropriate gene markers (Zhang et al., 2009; Patil et al., 2009). Specific traits to look for during early generation selection are grain colour, hardness, size and shape. The Single-Kernel Characterization System (Perten Instruments, Hägersten, Sweden) is designed to test the grain hardness of small grain samples. When there is about 50 g of sample, the near infrared reflectance spectroscopy (NIR) of the ground sample is used (Ross and Bettge, 2009).

To assess milling quality, a grain sample as small as 10 g or so can be milled in a micro mill such as described by Wrigley et al. (2011). The resulting flour is then available for small-scale dough testing, using mixing equipment such as the micro-doughLAB (Perten Instruments, Sweden) for testing mixing properties and Keiffer dough and gluten-extensibility rig (Stable MicroSystems, Surrey, UK) for testing dough extension.

Chemical tests such as the swelling index of glutenin test [indirect measure of dough strength, Wang and Kovacs (2002)], size exclusion HPLC [analysis of aggregated glutenin macro-polymer, Batey et al. (1991); Gupta et al. (1993)] are also used to predict end-product quality. The identification of molecular markers for quality traits has been used for early generation quality selection (Wrigley et al., 2009; Békés and Wrigley, 2013).

5.10.2 Farm Management

Following breeding and pure-seed propagation, the grower makes the critical decisions of what variety to sow and how best to manage the growth environment (see Chapter 18). Whereas the decision about genotype (variety) is a conscious one, the anticipation of climatic conditions is a factor that cannot be exact. Certain important aspects of final grain quality will be largely due to environment. However, some of the effects of growth conditions can be overcome to an extent by the breeder building in quality characters such as tolerance to damaging effects of the environment, for example tolerance to the effects of rain at harvest on pre-harvest sprouting.

Choice of variety for seeding plays a major role in quality and the subsequent use of harvested grain. Quality characteristics such as grain hardness are less likely to be affected by the influence of environment; however, quality traits such as kernel size, dockage and protein concentration are mainly determined by environment and agronomic factors.

Nutrients have obvious effects on quality. Grain quality is most responsive to nitrogen of all nutrients. Nitrogen is a major component of protein and hence nitrogen supply is the main factor influencing grain protein concentration. The supply of sulfur has considerable influence on protein quality. Managing weed infestation is important as weeds compete for soil nitrogen and may reduce grain protein. Disease and pests can affect grain quality; growing disease-resistant cultivars or applying fungicides and pesticides are important management practices in achieving grain-quality targets (see Chapter 18).

The timing of harvest will vary based on local conditions. Harvesting at low moisture can be detrimental to quality because brittleness of grain will lead to cracking and breaking. If wheat crops are left after they have dried to a suitable moisture content for harvest, the near-ripe grain may be exposed to rain or heavy dews that could lead to pre-harvest sprouting, depending on the genotype. White-grained wheat can tolerate little delay in harvest if wet weather occurs, whereas red-grained wheat can be left for longer periods after maturity.

5.10.3 Grain Receival, Segregation and Storage

The delivery of wheat of consistent quality is a key demand of flour millers. However, the quality of wheat can be variable not only between years but also between shipments that depend on the medium of transport. The variability is due to the interaction of environment with the inherent genetically-determined quality factors. Managing the grain receival and segregation is very important to ensure that the parcel of wheat that is created meets the needs of discriminating buyers, performs in a predictable and consistent way and exhibits quality attributes that are ideal for production of specific wheat-based foods (Cracknell and Williams, 2016) (see Chapter 19).

Grain receival is the first and most critical point at which food safety issues must be addressed, identified and excluded from general grain receival, thus to ensure that possible contaminants such as mycotoxins and other noxious contaminants are not admixed with the bulk of an otherwise sound grain consignment (Bianchini et al., 2015; McLean and Wrigley, 2016). Grain receival is also a stage at which biosecurity considerations come into play, with the opportunity to detect and remove contaminants and diseases that have the potential to cause wider damage in the agricultural sector. Biosecurity considerations also apply throughout the grain chain as described in Chapter 21.

In Australia, each truckload of a 20–25 million-tonne wheat harvest is sampled and tested for quality when it is delivered to the silo network around the country, using a range of tests such as those described in Chapter 19. Quick and accurate tests are needed at this stage to minimise delay and avoid the risk of weather damage due to delaying the harvest.

Knowledge about the variety of wheat being delivered is an important quality consideration. Currently, other than the variety declaration by the farmers, there are no routine on-the-spot tests used to test genetically determined quality traits such as milling performance, dough properties and suitability for processing into the diversity of food products. This range is broad (Fig. 5.2), including leavened breads, pasta, noodles, flat/pocket breads, steamed breads, biscuits, cakes, pastries and various food ingredients. However, each food product has specific and distinct dough specifications, depending on attributes such as strength, extensibility and tolerance to over-mixing.

The flour miller is thus required to provide flour of suitable dough properties to the respective food manufacturers, but it is a difficult task for the miller to produce flours with the required potential dough specifications. Examples of the range of dough properties for specific food applications are provided in Chapters 22 and 24, and also by Wrigley (2016a). The miller's role would be greatly simplified if grain of suitable dough-forming potential could be segregated and delivered for milling, thereby short-cutting the need for blending before and after milling in attempts to fulfil dough specifications.

The segregation of grain according to dough-quality potential requires test systems that could be applied prior to the bulking of grain loads in storage, or alternatively to 'running samples' that represent the contents of specific storage cells. It is clearly impractical for such a test system to involve milling samples to flour for conventional dough testing (Farinograph and Extensograph). Instead, predictive methodology is needed. This testing should be applicable to grain or wheatmeal, thus avoiding the need for milling to flour, and the methodology must be simple enough for use with large numbers of samples by staff with minimal experience. Possible test methods of this nature have been reviewed (Chung et al., 2003; Cauvain and Young, 2009).

Analysis of grain protein content is a valuable initial indication of dough properties, together with the knowledge of varietal composition, but predictive testing must at least include evaluation of protein quality, taking into account the extent to which growth conditions have modified the genetic potential of the variety. Studies carried out by Uthayakumaran et al. (2006) were aimed at identifying a test that would indicate harvest samples suitable for the respective specifications of wheat-flour customers, thereby moving the testing requirement as far back up the supply chain as possible. The site of such testing could be either the regional laboratory of the grain-handling company (based on samples representative of specific storage cells) or, if possible, the site of grain receipt and storage, where samples are taken by the grain handler.

The knowledge of grain behaviour during storage, safe guidelines for storage, grain and facility management and quality control procedures can be used to minimise quality loss in stored wheat (see Chapter 20). The storability of grain is influenced by the physical, chemical and biological characteristics of the grain along with the environmental conditions during growth, harvesting methods and handling practices prior to storage (Cenkowski and Zhang, 1995; Mills, 1992). Cleaning harvested material to remove high-risk debris, broken seeds, chaff and dust improves airflow and prevents hot spots. Drying the grain to optimum moisture content assists with pest control and loss of grain quality.

5.10.4 Processing

The assessment of the raw material quality is integral to efficient processing (Morris, 2016a,b). The first and foremost management for the processor is the blending of wheat grains from different storages or deliveries. When wheat is milled to white flour, it is estimated that about 75% of flour quality is determined by the quality of the incoming raw material, that is, wheat grain. The milling of wheat flour requires the miller to achieve tight specifications that are set by the flour customer (Sarkar and Dexter, 2016).

The blending of flour streams is an important management tool in meeting these specifications, but this approach is totally dependent on effective assessment and management of quality prior to and during milling. The other management factors such as mill adjustment and environmental conditions contribute to only about 25% of variation in flour quality (Posner and Hibbs, 2005). As mentioned in the section on grain receival, segregation and storage, a simple predictive test for dough-forming potential at segregation would make the miller's task simpler when attempting to meet dough-quality specifications. Another quality requirement is to get grain with absence of disease or contaminants, and hence analyses for mycotoxins and contaminating grains are essential for incoming grain, depending on its recent history (Waalwijk, 2003).

5.10.5 Satisfying the Consumer

Satisfying the consumer is the goal of all operators along the wheat value chain (Fig. 5.2). The ultimate person who determines quality is the consumer because it is he or she who buys the final end product (see Chapter 24). The consumer looks for subjective quality attributes when buying cereal products. This not only includes eating quality but also matching against recommended nutritional guidelines. Appropriate labelling of packaged foods is an important part of the interaction with consumers. This topic has been addressed by Jones et al. (2015a). The importance of grain-based products in a healthy diet has provided a major sales opportunity for the cereal-grain industry. Characteristics such as demographics, lifestyles, fads and personal attitudes influence trends in food consumption. These characteristics affect marketing, processing and the production of wheat-based products.

5.10.6 Nutrition and Intolerances

Wheat-based foods are basic to the diet of the whole world, despite the considerable variations in form depending on culture, region and preferences. These foods provide excellent nutrition, contributing energy, essential amino acids (even if slightly deficient in lysine), essential fatty acids, vitamins, minerals and fibre (reviewed by Dewettinck et al., 2008). The use of wholemeal foods is especially recommended, due to the enhanced nutritional and health advantages offered compared to foods made from white flour (Peressini and Sensidoni, 2009).

However, some of the non-endosperm components of the grain (bran and germ) detract from baking quality, posing management problems for the baker. Suggested solutions include management of particle size for the bran, the use of baking aids and enzymes and pre-treatment of the fibre (Katina, 2003). Glycaemic response has been identified as an additional health issue relevant to the consumption of wheat-based foods. The starch of most forms of bread is digested relatively quickly, but attention to the type of processing may permit changes to be made. For example sourdough processing has been reported to provide higher levels of resistant starch, thus reducing the glycaemic response (Scazzina et al., 2009).

Wheat-based foods are not suitable for people with some forms of wheat intolerance, especially coeliac disease (Sollid, 2000; Wieser and Koehler, 2008; Koehler et al., 2016). For these people, the remedy is life-long avoidance of gluten in the diet. If they are to include bread in their diet, it must be made without gluten, the protein complex that gives wheat flour the unique ability for bread making. Although it is difficult to make good-quality bread if gluten is to be excluded, various strategies have been employed to produce gluten-free breads (Gallagher et al., 2003; Arendt and Dal Bello, 2008; Casper and Atwell, 2014).

Negative aspects of wheat ingestion include the risk of toxic contaminants, such as mycotoxins. Due to the various measures used to identify wheat that is thus affected during grain receival and storage, the risk is small (Waalwijk, 2003). In addition, a process of dehulling has been developed to reduce or eliminate the presence of mycotoxins (Rios et al., 2009).

5.11 FUTURE PROSPECTS

In the western world's view, food is becoming less a source of energy for the body and more a source of health-giving nutrition. Consumers are therefore changing the basis of their food choices; for example, deciding to go gluten-free or salt-free; or attempting to maximise their intake of antioxidants or of bioactives or of some other specific components based on the latest infomercials, internet advice or gossip.

The wheat-processing industry is obliged to keep pace with genuine nutritional trends and to work to supply these needs. But in addition, the industry inevitably has to keep pace with food fads and customer-based pressures to supply specific requirements. One of these trends is for 'organic foods'. In the case of wheat-based products, the 'organic' label now means more than organically grown raw grains. It also means organically processed. The significance of this added aspect of the organic trend is described by Gelinas and David (2016). The wheat-production and processing industries are also obliged to keep pace with developments in food-safety requirements, including the ISO 22000 Food Safety Management System (Anon. 2015) (see Chapter 15).

There have been repeated recommendations that cereals must be foundational in our diet, based on government guidelines (Jones et al., 2015a). These recommendations should form an ongoing promotion to benefit the wheat-growing and wheat-processing industries. However, there have been movements advocating gluten-free eating, high-protein diets and low-carb diets. Although these fads may be based on poor scientific evidence (Wrigley et al., 2016), these movements affect the relevant industries, which must accept the resulting consumer trends or move their advertising and product formulations to pander to these trends.

For the wheat-growing and wheat-processing industry, going organic can best be seen as a financial opportunity to be accepted and capitalised upon. It is likely that the organic future will touch all parts of the grain chain (Table 5.2), including the breeding of varieties that are suited to the less-intensive growth situation of the organic farm—in many ways, the opposite of the conditions that produced the 'green revolution' of last century (Dixon et al., 2009).

Many of the research findings about grain quality are first delivered at the stage of breeding. As a result, an increasing load is placed on the breeder, who is expected to deliver a range of quality-specific attributes in addition to the many yield-related factors. However, increasing knowledge about the genes involved in these many traits will facilitate the more intelligent selection of parental lines, and the more efficient selection of desirable progeny. While genetically modified (GM) grains of corn and the dicot species have been accepted, this has not been generally so for wheat. As attitudes change, so will the combinations of attributes to be expected in GM wheat (Jones, 2005; Jones et al., 2009; Henry, 2014, 2016).

The management of actual wheat quality starts even before the grain is delivered after harvest. The grower has several important roles in growing grain to achieve quality targets, namely, variety selection and various on-farm management tools (see Chapter 18). A management tool that is becoming relevant to grain quality, as well as grain yield, is precision agriculture, the ability to know the performance of the crop on a metre-by-metre level (Srinivasan, 2006; Bramley, 2009; see Chapter 25). Sensors on the harvester linked to a global positioning system are used to construct a detailed map of the harvested area showing both yield and grain protein content, the latter being provided by continuous NIR monitoring (Whelan et al., 2009). This novel form of analysis will provide growers with a unique opportunity for quality management (see Chapter 18).

Climate-change predictions provide new challenges and opportunities for the grain grower (Wrigley, 2006). Increasing levels of carbon dioxide in the atmosphere will provide the effect of a carbon fertiliser, whereby biomass and grain yields will increase, whilst lowering the level of protein in the grain (Högy et al., 2009). In cooler growing regions, temperature increases during grain filling are expected to produce grains with stronger dough, whereas more frequent heat stress (a few days with maxima over 35°C) in other growing regions are expected to cause dough weakening. Breeding for tolerance to the effect of heat stress on dough may provide a partial remedy; in parallel, heat-stress episodes might be escaped by earlier sowing, especially if less-frequent frost can be expected.

The value of wheat as a specialist grain, rather than a commodity, will increase the justification for more detailed analysis of quality type at grain receival, leading to better management of storage and transport, providing identity preservation through to the mill or export terminal. As a result, the flour miller will have better opportunity to meet the specifications of food-processing customers, who in turn are expected to become more discerning with respect to sourcing flour that suits each of the separate requirements of the many wheat-based foods.

There will be better opportunities to assess specific aspects of grain quality at several stages as a result of research findings about marker proteins and genes to identify quality attributes (see Chapter 17). At the end of the grain chain, consumers will become more discerning with respect to food quality, price, value and importantly, nutrition. In view of this last factor, processors will need to improve dietary requirements, such as increased fibre, whilst maintaining perceptions of taste and mouth-feel.

The reduction of salt in wheat-based foods (mainly as sodium levels) will become an increasing health-related issue, and processors will need to find alternatives as processing aids, as well as for the taste-enhancing role of salt (Day, 2016b).

Whilst it is distantly possible that the unique dough-forming properties of wheat might be transferred to other cereal species via genetic transformation, it will be a long time before wheat loses its pride of place as the cereal grain that is uniquely suited for bread making.

SOURCES OF FURTHER INFORMATION

Literature

Anon, 2009. The ICC Handbook of Cereals, Flour, Dough & Product Testing: Methods and Applications. DEStech Publications, Inc, Lancaster, PA, USA.

Arendt, E.K., Zannini, E. (Eds.), 2013. Cereal Grains for the Food and Beverage Industries. Woodhead Publishing Ltd, Oxford, UK.

Blakeney, A.B., Cracknell, R.L., Crosbie, G.B., Jefferies, S.P., Miskelly, D.M., O'Brien, L., Panozzo, J.F., Suter, D.A.I., Solah, V., Watts, T., Westcott, T., Williams, R.M., 2009. Understanding Australian Wheat Quality. Grains Research and Development Corporation, Canberra, Australia.

Bonjean, A.P., Angus, W.J., 2001. The World Wheat Book. Lavoisier, Paris.

Carver, B.F. (Ed.), 2009. Wheat Science and Trade. Wiley-Blackwell, Ames, Iowa.

Cauvain, S.P. (Ed.), 2003. Bread Making: Improving Quality. Woodhead Publishing Ltd, Oxford, UK.

Cauvain, S.P. (Ed.), 2012. Bread Making: Improving Quality. second ed. Woodhead Publishing Ltd, Oxford, UK.

Faridi, H., Faubion, J., 1995. Wheat End Uses Around the World. AACCI, St Paul, MN, USA.

Khan, K., Shewry, P.R. (Eds.), 2009. Wheat Chemistry and Technology. fourth ed. AACC International, St Paul, MN.

Lookhart, G.L., Ng, P.K.W. (Eds.), 2007. Gluten Proteins 2006. AACC International, St. Paul, MN.

Moss, H.J., 1973. Standards for wheat varieties. J. Aust. Inst. Agric. Sci. 39, 109–115.

Posner, E.S., Hibbs, A.N., 2005. Wheat Flour Milling, second ed. AACC International, St Paul, MN.

Wrigley, C.W., Bekes, F., Bushuk, W. (Eds.), 2006. Gliadin and Glutenin. The Unique Balance of Wheat Quality. AACC International, St. Paul, MN.

Wrigley, C., Corke, H., Seetharaman, K., Faubion, J., 2016. Encyclopedia of Food Grains, vol. 1–4, Elsevier, Oxford, UK.

Websites

An extensive catalogue of alleles is provided for wheat genotypes worldwide, with respect to HMW and LMW sub-units of glutenin and gliadins, at the website of AACC International: http://www.aaccnet.org/initiatives/definitions/Pages/glutendatabase.aspx

The websites of the wheat-handling and marketing organisations, plus associated research laboratories, of various countries provide up-to-date information about quality requirements for grades and classes:

www.wheatquality.com.au
www.grainscanada.gc.ca
www.cgc.ca
www.cigi.ca
www.onic.fr
www.trigoargentino.com.ar
www.uswheat.org
www.usda.gov
www.fas.usda.gov
www.wheat.pw.usda.gov
www.crop.cri.nz
www.cimmyt.org
www.icarda.org

Scientific associations are good sources of information about wheat quality and methods of analysis:

www.aaccnet.org
www.icc.or.au
www.campden.co.uk
www.seedtest.org
www.aomillers.org
www.nabim.org.uk

REFERENCES

Abdel-Aal, E.-S.M., Hucl, P., 2014. Einkorn: a functional wheat for developing high-lutein whole grain baked products. Cereal Foods World 59, 6.

Akhonov, E., 2016. Wheat genetics and genomics. Wrigley, C., Corke, H., Seetharaman, K., Faubion, J. (Eds.), Encyclopedia of Food Grains, vol. 4, Elsevier, Oxford, UK, pp. 382–387.

Al Saleh, A., Marion, D., Gallant, D.J., 1986. Microstructure of mealy and vitreous wheat endosperms (*Triticum durum* L.) with special emphasis on location and polymorphic behaviour of lipids. Food Microstruct. 5, 131–140.

Annis, P.C., 2016. Chemicals for grain production and protection. Wrigley, C., Corke, H., Seetharaman, K., Faubion, J. (Eds.), Encyclopedia of Food Grains, vol. 4, Elsevier, Oxford, UK, p. 99.

Anon., 2015. User's Guide to ISO 22000 Food Safety Management System for the Grain Handling, Processing, Milling, and Baking Industries. AACC International, St Paul, MN, USA.

Arendt, E., Dal Bello, F., 2008. Gluten-Free Cereal Products and Beverages. AACC International, St Paul, MN, USA.

Atwell, W.A., 2001. Wheat Flour. Eagan Press, St. Paul, MN.

Bailey, C.H., 1941. A translation of Beccari's lecture 'concerning grain' (1728). Cereal Chem. 18, 555–561.

Batey, I.L., Gupta, R.B., MacRitchie, F., 1991. Use of size-exclusion high-performance liquid chromatography in the study of wheat flour proteins: an improved chromatographic procedure'. Cereal Chem. 68, 207–209.

Batey, I.L., Huang, W., 2016. Gluten and modified gluten. Wrigley, C., Corke, H., Seetharaman, K., Faubion, J. (Eds.), Encyclopedia of Food Grains, vol. 3, Elsevier, Oxford, UK, pp. 408–413.

Batifoulier, F., Verny, M.-A., Chanliaud, E., Remesy, C., Demigne, C., 2005. Effect of different breadmaking methods on thiamine, riboflavin and pyridoxine contents of wheat bread'. J. Cereal Sci. 42, 101–108.

Békés, F., 2012. New aspects in quality related wheat research: II. New methodologies for better quality wheat. Cereal Res. Commun. 40, 307.

Békés, F., Wrigley, C.W., 2013. Gluten alleles and predicted dough-quality for wheat varieties world-wide: a great resource—free on the AACC International website. Cereal Foods World 58 (6), 325–328.

Békés, F., Wrigley, C.W., 2016. The protein chemistry of cereal grains. Wrigley, C., Corke, H., Seetharaman, K., Faubion, J. (Eds.), Encyclopedia of Food Grains, vol. 2, Elsevier, Oxford, UK, pp. 98–108.

Best, D., 2009. Give biofuels a chance. Cereal Foods World 54, 88–89.

Beta, T., Duodu, K., 2016. Bioactives: antioxidants. Wrigley, C., Corke, H., Seetharaman, K., Faubion, J. (Eds.), Encyclopedia of Food Grains, vol. 2, Elsevier, Oxford, UK, pp. 277–282.

Beta, T., Isaak, C., 2016. Grain production and consumption: overview. Wrigley, C., Corke, H., Seetharaman, K., Faubion, J. (Eds.), Encyclopedia of Food Grains, vol. 1, Elsevier, Oxford, UK, pp. 349–358.

Bianchini, A., et al., 2015. DON occurrence in grains: a North American perspective. Cereal Foods World 60, 32–56.

Bietz, J.A., Lookhart, G.L., 1996. Properties and non-food potential of gluten. Cereal Foods World 41, 376–382.

Black, J.L., 2016. Cereal grains as animal feed. Wrigley, C., Corke, H., Seetharaman, K., Faubion, J. (Eds.), Encyclopedia of Food Grains, vol. 3, Elsevier, Oxford, UK, p. 215.

Blakeney, A.B., Cracknell, R.L., Crosbie, G.B., Jefferies, S.P., Miskelly, D.M., O'Brien, L., Panozzo, J.F., Suter, D.A.I., Solah, V., Watts, T., Westcott, T., Williams, R.M., 2009. Understanding Australian Wheat Quality. Grains Research and Development Corporation, Canberra, Australia.

Bock, M.A., 2000. Minor constituents of cereals. In: Kulp, K., Ponte, J.G.J. (Eds.), Handbook of Cereal Science and Technology. second ed. Marcel Dekker, New York, pp. 479–504.

Bramley, P.M., Elmadfa, I., Kafatos, A., Kelly, F.J., Manios, Y., Roxborough, H.E., Schuch, W., Sheehy, P.J.A., Wagner, K.-H., 2000. Vitamin E. J. Sci. Food Agric. 80, 913–938.

Bramley, R.G.V., 2009. Lessons from nearly 20 years of precision agriculture research, development, and adoption as a guide to its appropriate application. Crop Pasture Sci. 60, 197–217.

Bridges, A.R., Wrigley, C.W, 2016a. Standardized test methods for grains and grain-based products. Wrigley, C., Corke, H., Seetharaman, K., Faubion, J. (Eds.), Encyclopedia of Food Grains, vol. 2, Elsevier, Oxford, UK, pp. 300–307.

Bridges, A.R., Wrigley, C.W., 2016b. Appendix 2. Lists of standardized testing methods for the analysis of grain and grain-based products. Wrigley, C., Corke, H., Seetharaman, K., Faubion, J. (Eds.), Encyclopedia of Food Grains, vol. 4, Elsevier, Oxford, UK, pp. 45–459.

Cane, K., Sharp, P.J., Eagles, H.A., Eastwood, R.F., Hollamby, G.J., Kuchel, H., Lu, M., Martin, P.J., 2008. The effects on grain quality traits of a grain serpin protein and the VPM1 segment in southern Australian wheat breeding. Aust. J. Agric. Res. 59, 883–890.

Casper, J.L., Atwell, W.A., 2014. Gluten-Free Baked Products. AACC International, St Paul, MN, USA.

Castenmiller, J.J.M., West, C.E., 1997. Bioavailability of carotenoids. Pure Appl. Chem. 69, 2145–2150.

Cauvain, S.P., 2016. Cookies, biscuits and crackers: formulation, processing and characteristics. Wrigley, C., Corke, H., Seetharaman, K., Faubion, J. (Eds.), Encyclopedia of Food Grains, vol. 3, Elsevier, Oxford, UK, p. 37.

Cauvain, S.P., Young, L.S. (Eds.), 2009. The ICC Book of Cereals, Flour, Dough and Product Testing. DEStech Publications Inc, Pennsylvania, USA.

Cenkowski, S., Zhang, Q., 1995. Engineering properties of grains and oilseeds. In: Jayas, D.S., White, N.D.G., Muir, W.E. (Eds.), Stored Grain Ecosystems. Marcel Dekker, New York.

Chung, O.K., Ohm, J., 2000. Cereal lipids. In: Kulp, K., Ponte, J.G.J. (Eds.), Handbook of Cereal Science and Technology. second ed. Marcel Dekker, New York, pp. 417–477.

Chung, O.K., Ohm, J., Ram, M.S., Park, S.-H., Howitt, C., 2009. Wheat lipids. In: Khan, K., Shewry, P.R. (Eds.), Wheat Chemistry and Technology. fourth ed. AACC International, St. Paul, MN, pp. 363–399.

Chung, O.K., Park, S.-H., Tilley, M., Lookhart, G.L., 2003. Improving wheat quality. In: Cauvain, S.P. (Ed.), Bread Making: Improving Quality. Woodhead Publishing Ltd, Cambridge, UK, pp. 536–561.

Clements, R.L., Donelson, J.R., 1981. Functionality of specific flour lipids in cookies. Cereal Chem. 58, 204–206.

Cook, R.J., Wrigley, C.W., 2004. Variety registration and breeders' rights. Wrigley, C., Corke, H., Walker, C.E. (Eds.), Encyclopedia of Grain Science, vol. 3, Elsevier, Oxford, pp. 314–321.

Cornish, G.B., Bekes, F., Eagles, H.A., Payne, P.I., 2006. Prediction of dough properties for bread wheats. In: Wrigley, C.W., Bekes, F., Bushuk, W. (Eds.), Gliadin and Gluten: The Unique Balance of Wheat Quality. AACC International, St Paul, MN, pp. 243–280.

Cracknell, R.L., Williams, R.M., 2016. Wheat—grading and segregation. Wrigley, C., Corke, H., Seetharaman, K., Faubion, J. (Eds.), Encyclopedia of Food Grains, vol. 4, Elsevier, Oxford, UK, p. 1.

Crosbie, G.B., 1991. The relationship between starch swelling properties, paste viscosity and boiled noodle quality in wheat flours. J. Cereal Sci. 13, 145–150.

Day, L., 2016a. Lipid chemistry. Wrigley, C., Corke, H., Seetharaman, K., Faubion, J. (Eds.), Encyclopedia of Food Grains, vol. 2, Elsevier, Oxford, UK, p. 248.

Day, L., 2016b. Cereal food production with low salt. Wrigley, C., Corke, H., Seetharaman, K., Faubion, J. (Eds.), Encyclopedia of Food Grains, vol. 3, Elsevier, Oxford, UK, p. 396.

Day, L., Batey, I.L., Wrigley, C.W., Augustin, M.A., 2006. Gluten uses and food industry needs. Trends Food Sci. Technol. 17, 82–90.

DePauw, R., O'Brien, L., 2016. Wheat breeding: exploiting and fixing genetic variation by selection and evaluation. Wrigley, C., Corke, H., Seetharaman, K., Faubion, J. (Eds.), Encyclopedia of Food Grains, vol. 4, Elsevier, Oxford, UK, p. 279.

Dewettinck, K., Van Bockstaele, F., Kuhne, B., Van de Walle, D., Courtens, T.M., Gellynck, X., 2008. Nutritional value of bread: influence of processing, food interaction and consumer perception. J. Cereal Sci. 48, 243–257.

Dey, M., 2014. Dietary bioactives and health: the road ahead. Cereal Foods World 59, 60–62.

Dill, D.B., Alsberg, C.L., 1924. Some critical considerations of the gluten washing problem. Cereal Chem. 1, 222–246.

Dixon, J., Braun, H.-J., Kosina, P., Crouch, J., 2009. Wheat Facts and Futures 2009. AgEcon Search. Internet publication by CIMMYT. http://purl.umn.edu/56366.

Faubion, J.M., Hoseney, R.C., 1982. High temperature and short time cooking of wheat starch and flour. II. Effects of proteins and lipids on extrudate properties. Cereal Chem. 59, 533–537.

Feldman, M., 2001. Origin of cultivated wheat. In: Bonjean, A.P., Angus, W.J. (Eds.), The World Wheat Book. A History of Wheat Breeding. Lavoisier Publishing, Paris, pp. 3–53.
Fenech, M., Noakes, M., Clifton, P., Topping, D., 1999. Aleurone flour is a rich source of bioavailable folate in humans. J. Nutr. 129, 1114–1119.
Fincher, G.B., 2016. Cereals: chemistry and physicochemistry of non-starchy polysaccharides. Wrigley, C., Corke, H., Seetharaman, K., Faubion, J. (Eds.), Encyclopedia of Food Grains, vol. 2, Elsevier, Oxford, UK, p. 208.
Gallagher, E., Gormley, T.R., Arendt, E.K., 2003. Crust and crumb characteristics of gluten free breads. J. Food Eng. 56, 153–161.
Gelinas, P., David, C., 2016. Organic grain production and food processing. Wrigley, C., Corke, H., Seetharaman, K., Faubion, J. (Eds.), Encyclopedia of Food Grains, vol. 4, Elsevier, Oxford, UK, pp. 154–161.
Gooding, M.J., 2009. The wheat crop. In: Khan, K., Shewry, P.R. (Eds.), Wheat: Chemistry and Technology. fourth ed. AACC International, St Paul, MN, pp. 19–49.
Greenwell, P., Schofield, J.D., 1986. A starch granule protein associated with endosperm softness in wheat. Cereal Chem. 63, 379–380.
Gupta, R.B., Khan, K., MacRitchie, F., 1993. Biochemical basis of flour properties in bread wheats. I. Effects of variation in the quantity and size distribution of polymeric protein. J. Cereal Sci. 18, 23–41.
Gupta, R.B., Shepherd, K.W., 1990. Two-step one dimensional SDS-PAGE analysis of LMW subunits of glutenin. I. Variation and genetic control of the subunits in hexaploid wheats. Theor. Appl. Genet. 80, 65–74.
Guragain, Y.N., Probst, K.V., Vadlani, P.V., 2016. Fuel ethanol production. Wrigley, C., Corke, H., Seetharaman, K., Faubion, J. (Eds.), Encyclopedia of Food Grains, vol. 3, Elsevier, Oxford, UK, p. 235.
Gustafson, P., Raskina, O., Ma, X.-F., Nevo, E., 2009. Wheat evolution, domestication and improvement. In: Carver, B.F. (Ed.), Wheat: Science and Trade. Wiley-Blackwell, Iowa, pp. 5–30.
HadiNezhad, M., Butler, F., 2009. Association of glutenin subunit composition and dough rheological characteristics with cookie baking properties of soft wheat cultivars. Cereal Chem. 86, 339–349.
Hargin, K.D., Morrison, W.R., 1980. The distribution of acyl lipids in the germ, aleurone, starch and non-starch endosperm of four wheat varieties. J. Sci. Food. Agric. 31, 877–888.
Henry, R.J., 2014. Wheat genomics for grain quality improvement. Cereal Foods World 59, 22–24.
Henry, R.J., 2016. Genomics of food grains. Wrigley, C., Corke, H., Seetharaman, K., Faubion, J. (Eds.), Encyclopedia of Food Grains, vol. 4, Elsevier, Oxford, UK, pp. 360–364.
Högy, P., Zörb, C., Langenkämper, G., Betsche, T., Fangmeier, A., 2009. Atmospheric CO_2 enrichment changes the wheat grain proteome. J. Cereal Sci. 50, 248–254.
Johansson, E., Henriksson, P., Svensson, G., Heneen, W.K., 1993. Detection, chromosomal location and evaluation of the functional value of a novel high M_r glutenin subunit found in Swedish wheats. J. Cereal Sci. 17, 237–245.
Jones, B.L., Morris, C.F., Bekes, F., Wrigley, C.W., 2006. Proteins that complement the roles of gliadin and glutenin. In: Wrigley, C.W., Bekes, F., Bushuk, W. (Eds.), Gliadin and Glutenin. The Unique Balance of Wheat Quality. AACC International, St. Paul, MN, pp. 413–446.
Jones, H.D., 2005. Wheat transformation: current technology and applications to grain development and composition. J. Cereal Sci. 41, 137–147.
Jones, H.D., Sparks, C.A., Shewry, P.R., 2009. Transgenic manipulation of wheat quality. In: Khan, K., Shewry, P.R. (Eds.), Wheat: Chemistry and Technology. fourth ed. AACC International, St. Paul, MN, pp. 437–451.

Jones, J.M., Peña, R.J., Korczak, R., Braun, H.J., 2015a. Carbohydrates, grains, wheat, in nutrition and health: an overview. Part I. Role of carbohydrates in health. Cereal Foods World 60, 224–233.

Jones, J.M., Peña, R.J., Korczak, R., Braun, H.J., 2015b. CIMMYT series on carbohydrates, wheat, grains, and health: carbohydrates, grains, and wheat in nutrition and health: an overview. Part II. Grain terminology and nutritional contributions. Cereal Foods World 60 (6), 260–271.

Juhasz, A., Bekes, F., Wrigley, C.W., 2014. Wheat proteins. In: Zey Ustunol (Ed.), Applied Food Protein Chemistry. Wiley-Blackwell, John Wiley & Sons, West Sussex, UK, pp. 219–304, (Chapter 11).

Kahlon, T.S., Chow, F.I., Hoefer, J.L., Betschart, A.A., 1986. Bioavailability of vitamin A and E as influenced by wheat bran and bran particle size. Cereal Chem. 63, 490–493.

Katina, K., 2003. High-fibre baking. In: Cauvain, S.P. (Ed.), Bread Making: Improving Quality. Woodhead Publishing, Cambridge, UK, pp. 487–499.

Khatkar, B.S., Schofield, J.D., 1997. Molecular and physico-chemical basis of breadmaking—properties of wheat gluten proteins: a critical appraisal. J. Food Sci. Technol. 34, 85–102.

Khosravi, S., Nordqvist, P., Khabbaz, F., Öhman, C., Bjurhager, I., Johansson, M., 2014. Wetting and film formation of wheat gluten dispersions applied to wood substrates as particle board adhesives. Eur. Polym. J. 67, 476–482.

Kies, C., Kan, S., Fox, H.M., 1984. Vitamin B_6 availability from wheat, rice, and corn brans for humans. Nutr. Rep. Int. 30, 483–491.

Koehler, P., Wieser, H., Scherf, K.A., 2016. Celiac disease. Wrigley, C., Corke, H., Seetharaman, K., Faubion, J. (Eds.), Encyclopedia of Food Grains, vol. 2, Elsevier, Oxford, UK, p. 83.

Kruger, J.E., Reed, G., 1988. Enzymes and color. In: Pomeranz, Y. (Ed.), Wheat: Chemistry and Technology. third ed. AACC International, St. Paul. MN, pp. 441–500.

Lafiandra, D., Masci, S., Blumenthal, C., Wrigley, C.W., 1999. The formation of glutenin polymer in practice. Cereal Foods World 44, 572–578.

Larsson, H., Eliasson, A.-C., 1997. Face on the rheological behaviour of wheat flour dough. J. Texture Stud. 28, 487–501.

Leklem, J.E., Miller, L.T., Perera, A.D., Peffers, D.E., 1980. Bioavailability of vitamin B-6 from wheat bread in humans. J. Nutr. 110, 1819–1828.

Lewis, J., Buss, D.H., 1988. Trace nutrients. Minerals and vitamins in the British household food supply. Br. J. Nutr. 60, 413–424.

Lin, M.J.Y., D'Appolonia, B.L., Young, V.L., 1974. Hard red spring and durum wheat polar lipids. II Effect on quality of bread and pasta products. Cereal Chem. 51, 34–45.

Lookhart, G.L., Wrigley, C.W., 1995. Electrophoretic analysis. In: Wrigley, C.W. (Ed.), Identification of Food-Grain Varieties. AACC International, St. Paul, MN, pp. 55–71.

Ma, D., Zhang, Y., Xia, X., Morris, C.F., He, Z., 2009. Milling and Chinese raw white noodle qualities of common wheat near-isogenic lines differing in puroindoline b alleles. J. Cereal Sci. 50, 126–130.

Mangelsdorf, P.C., 1953. Wheat. Sci. Am. 189, 50–59.

Maningat, C.C., DeMeritt, G.K., Chinnaswamy, R., Bassi, S.D., 1999. Properties and applications of texturized wheat gluten. Cereal Foods World 44, 650–655.

Mares, D.J., Stone, B.A., 1973. Studies on wheat endosperm. I. Chemical composition and ultrastructure of the cell walls. Aust. J. Biol. Sci. 26, 793–812.

McFall, K.L., Fowler, M.E., 2009. Overview of wheat classification and trade. In: Carver, B.F. (Ed.), Wheat: Science and Trade. Wiley-Blackwell, Iowa, pp. 439–454.

McIntosh, R.A., 2016. Wheat genetics. Wrigley, C., Corke, H., Seetharaman, K., Faubion, J. (Eds.), Encyclopedia of Food Grains, vol. 4, Elsevier, Oxford, UK, pp. 376–381.

McLean, J., Wrigley, C.W., 2016. Contaminants of grain. Wrigley, C., Corke, H., Seetharaman, K., Faubion, J. (Eds.), Encyclopedia of Food Grains, vol. 4, second ed. Elsevier Ltd, Oxford, UK, pp. 83–98.

Metakovsky, E.V., Branlard, G.P., Graybosch, R.A., 2006. Gliadins of common wheat: polymorphism and genetics. In: Wrigley, C.W., Bekes, F., Bushuk, W. (Eds.), Gliadin and Glutenin. The Unique Balance of Wheat Quality. AACC International, St. Paul, MN, pp. 35–84.

Metakovsky, E.V., Pogna, N.E., Biancardi, A.M., Redaelli, R., 1994. Gliadin allele composition of common wheat cultivars grown in Italy. J. Genet. Breed. 48, 55–66.

Mills, J.T., 1992. Safe storage guidelines for grains and their products. Postharvest News and Information. 3. News and Information 3, 111N-5N.

Moore, T.R., 2016. Breads. Wrigley, C., Corke, H., Seetharaman, K., Faubion, J. (Eds.), Encyclopedia of Food Grains, vol. 2, Elsevier, Oxford, UK, p. 8.

Morris, C.F., 2016a. Cereals: overview of uses: accent on wheat grain. Wrigley, C., Corke, H., Seetharaman, K., Faubion, J. (Eds.), Encyclopedia of Food Grains, vol. 3, Elsevier, Oxford, UK, p. 1.

Morris, C.F., 2016b. Evaluation of wheat-grain quality attributes. Wrigley, C., Corke, H., Seetharaman, K., Faubion, J. (Eds.), Encyclopedia of Food Grains, vol. 3, Elsevier, Oxford, UK, p. 251.

Morris, C.F., Greenblatt, G.A., Bettege, A.P., Malkawi, H.I., 1994. Isolation and characterisation of multiple forms of friabilin. J. Cereal Sci. 20, 167–174.

Morrison, L.A., 2016. Cereals: domestication of the cereal grains. Wrigley, C., Corke, H., Seetharaman, K., Faubion, J. (Eds.), Encyclopedia of Food Grains, vol. 1, Elsevier, Oxford, UK, p. 86.

Morrison, L.A., Wrigley, C.W., 2016. Taxonomic classification of grain species. Wrigley, C., Corke, H., Seetharaman, K., Faubion, J. (Eds.), Encyclopedia of Food Grains, vol. 1, Elsevier, Oxford, UK, p. 31.

Nagao, S., 1998. Traditional and changing food consumption trends for wheat based foods. In: Blakeney, A.B., O'Brien, L. (Eds.), In: Proceedings of the 5th Pacific Rim Symposium, AACC International, St. Paul, pp. 13–23.

Nakamura, T., Yamamori, M., Hriano, H., Hidako, S., Nagamine, T., 1995. Production of waxy (amylose free) wheats. Mol. Gen. Genet. 248, 253–259.

Navarro, S., Noyes, R.T., Casada, M., Arthur, F.H., 2012. Grain aeration. In: Hagstrum, D.W., Phillips, T.W., Cuperus, G. (Eds.), Stored Product Protection. KSU Publisher, Manhattan, KS, USA, pp. 121–134.

Ollilainnen, V., 1999. HPLC analysis of vitamin B6 in foods. Agric. Food Sci. Finl. 85, 515–619.

Osborne, T.B., 1907. The Proteins of the Wheat Kernel. Carnegie Institution of Washington, Washington, DC, USA, Publ. No. 84.

Patil, R.M., Oak, M.D., Tamhankar, S.A., Rao, V.S., 2009. Molecular mapping of QTLs for gluten strength as measured by sedimentation volume and mixograph in durum wheat (*Triticum turgidum* L. ssp durum). J. Cereal Sci. 49, 378–386.

Payne, P.I., 1987. Genetics of wheat storage proteins and the effects of allelic variation on breadmaking quality. Annu. Rev. Plant Physiol. 38, 141–153.

Payne, P.I., Holt, L.M., Jackson, E.A., Law, C.N., 1984. Wheat storage proteins: their genetics and their potential for manipulation by plant breeding. Philos. Trans. R Soc. Lond. B (Biol. Sci.) 304, 359–371.

Payne, P.I., Lawrence, G.J., 1983. Catalogue of alleles for the complex gene loci, *Glu-A1*, *Glu-B1*, and *Glu-D1* which code for high-molecular-weight subunits of glutenin in hexaploid wheat. Cereal Res. Commun. 11, 29–35.

Peressini, D., Sensidoni, A., 2009. Effect of soluble dietary fibre addition on rheological and breadmaking properties of wheat doughs. J. Cereal Sci. 49, 190–201.

Piironen, V., Lampi, A.-M., Ekholm, P., Salmenkallio-Marttila, M., Liukkonen, K.-H., 2009. Micronutrients and phytochemicals in wheat grain. In: Khan, K., Shewry, P.R. (Eds.), Wheat Chemistry and Technology. fourth ed. AACC International, St. Paul, MN, pp. 179–222.

Pomeranz, Y., Huang, M., Rubenthaler, G.L., 1991. Steamed bread III. Role of lipids. Cereal Chem. 68, 353–356.

Posner, E.S., Hibbs, A.N. (Eds.), 2005. Wheat Flour Milling. second ed. AACC International, St. Paul, MN.

Purna, S.K.G., Shi, Y.-C., Guan, L., Wilson, J.D., Graybosch, R.A., 2015. Factors governing pasting properties of waxy wheat flours. Cereal Chem. 92, 529–535.

Rahman, S., Li, Z., Batey, I., Cochrane, M.P., Appels, R., Morell, M., 2000. Genetic alteration of starch functionality in wheat. J. Cereal Sci. 31, 91–110.

Reddy, M.B., Love, M., 1999. The impact of food processing on the nutritional quality of vitamins and minerals. Adv. Exp. Med. Biol. 459, 99–106.

Regina, A., Bird, A., Topping, D., Bowden, S., Freeman, J., Barsby, T., Kosar-Hashemi, B., Li, Z., Rahman, S., Morell, M., 2006. High-amylose wheat generated by RNA interference improves indices of large-bowel health in rats. Proc. Natl. Acad. Sci. USA 103, 3546–3551.

Regina, A., Rahman, S., Li, Z., Morell, M.K., 2016. Starch: synthesis. Wrigley, C., Corke, H., Seetharaman, K., Faubion, J. (Eds.), Encyclopedia of Food Grains, vol. 2, Elsevier, Oxford, UK, p. 181.

Rho, K.L., Chung, O.K., Seib, P.A., 1989. Noodles. VIII. The effect of wheat flour lipids, gluten and several starches and surfactants on the quality of oriental dry noodles. Cereal Chem. 66, 276–282.

Rios, G., Pinson-Gadais, L., Abecassis, J., Zakhia-Rozis, N., Lullien-Pellerin, V., 2009. Assessment of dehulling efficiency to reduce deoxynivalenol and Fusarium level in durum wheat grains. J. Cereal Sci. 49, 387–392.

Ross, A.S., Bettge, A.D., 2009. Passing the test on wheat end-use quality. In: Carver, B. (Ed.), Wheat: Science and Trade. Wiley-Blackwell, Ames, Iowa, pp. 455–493.

Sapirstein, H.D., 2016. Bioactives in wheat bran. Wrigley, C., Corke, H., Seetharaman, K., Faubion, J. (Eds.), Encyclopedia of Food Grains, vol. 2, Elsevier, Oxford, UK, pp. 268–276.

Sarkar, A.K., Dexter, J.E., 2016. Wheat: dry milling. Wrigley, C., Corke, H., Seetharaman, K., Faubion, J. (Eds.), Encyclopedia of Food Grains, vol. 3, Elsevier, Oxford, UK, p. 307.

Scazzina, F., Del Rio, D., Pellegrini, N., Brighenti, F., 2009. Sourdough bread: starch digestibility and postprandial glycemic response. J. Cereal Sci. 49, 209–421.

Shewry, P.R., D'Ovidio, R., Lafiandra, D., Jenkins, J.A., Mills, E.N.C., Bekes, F., 2009. Wheat grain proteins. In: Khan, K., Shewry, P.R. (Eds.), Wheat Chemistry and Technology. fourth ed. AACC International, St. Paul, MN, pp. 223–298.

Shewry, P.R., Halford, N.G., Lafiandra, D., 2003. Genetics of wheat gluten proteins. Adv. Genet. 49, 111–184.

Shewry, P.R., Halford, N.G., Lafiandra, D., 2006. The high-molecular-weight subunits of glutenin. In: Wrigley, C.W., Bekes, F., Bushuk, W. (Eds.), Gliadin and Glutenin. The unique balance of wheat quality. AACC International, St. Paul, MN, pp. 143–169.

Shewry, P.R., Halford, N.G., Tatham, A.S., 1989. The high molecular weight subunits of wheat, barley and rye: genetics, molecular biology, chemistry and role in wheat gluten structure and functionality. In: Miflin, B.J. (Ed.), Oxford Surveys of Plant Molecular and Cell Biology. Oxford University Press, Oxford, pp. 163–219.

Shewry, P.R., Miflin, B.J., 1985. Seed storage proteins of economically important cereals. Pomeranz, Y. (Ed.), Advances in Cereal Science and Technology, vol.7, AACC International, St Paul, MN, pp. 1–83.

Shewry, P.R., Tatham, A.S., Forde, J., Kreis, M., Miflin, B.J., 1986. The classification and nomenclature of wheat gluten proteins: a reassessment. J. Cereal Sci. 4, 97–106.

Shewry, P.R., Tosi, P., 2016. Protein synthesis and deposition. Wrigley, C., Corke, H., Seetharaman, K., Faubion, J. (Eds.), Encyclopedia of Food Grains, vol. 2, Elsevier, Oxford, UK, p. 115.

Slavin, J.L., Jacobs, D., Marquart, L., 2001. Grain processing and nutrition. Crit. Rev. Biotechnol. 21, 49–66.

Sollid, L.M., 2000. Molecular basis of celiac disease. Annu. Rev. Immunol. 18, 53–81.

Srinivasan, A. (Ed.), 2006. Handbook of Precision Agriculture. Principles and Applications. Haworth Press Inc, New York.

Stoddard, F.L., 2016. Starch: chemistry. Wrigley, C., Corke, H., Seetharaman, K., Faubion, J. (Eds.), Encyclopedia of Food Grains, vol. 2, Elsevier, Oxford, UK, p. 174.

Stone, B., Morell, M.K., 2009. Carbohydrates. In: Khan, K., Shewry, P.R. (Eds.), Wheat Chemistry and Technology. fourth ed. AACC International, St. Paul, MN, pp. 299–362.

Tatham, A.S., Shewry, P.R., 1985. The conformation of wheat gluten proteins. The secondary structures and thermal stabilities of α-, β-, γ- and ω-gliadins. J. Cereal Sci. 3, 103–113.

Uthayakumaran, S., Batey, I.L., Wrigley, C.W., 2004. Efficient identification of glutenin subunits in wheat by Lab-on-a-Chip capillary electrophoresis. In: Black, C.K., Panozzo, J.F., Rebetzke, G.J. (Eds.), Cereals 2004. Proc. 54th Aust. Cereal Chem. Conf. Melbourne, The Royal Australian Chemical Institute, pp. 411–414.

Uthayakumaran, S., Batey, I.L., Wrigley, C.W., 2005. On-the-spot identification of grain variety and wheat-quality type by Lab-on-a-chip capillary electrophoresis. J. Cereal Sci. 41, 371–374.

Uthayakumaran, S., Wrigley, C.W., 2009. Wheat: characteristics and quality requirements. In: Wrigley, C.W., Batey, I.L. (Eds.), Cereal Grains: Assessing and Managing Quality. first ed. Woodhead Publishing, Oxford, UK.

Uthayakumaran, S., Listiohadi, Y., Baratta, M., Batey, L.L., Wrigley, C.W., 2006. Rapid identification and quantitation of high-molecular-weight glutenin subunits. J. Cereal Sci. 44, 34–39.

van den Berg, H., 1997. Bioavailability of niacin. Eur. J. Clin. Nutr. 51, S64–S65.

van den Berg, H., Faulks, R., Granado, H.F., Hirschberg, J., Olmedilla, B., Sandmann, G., Southon, S., Stahl, W., 2000. The potential for the improvement of carotenoid levels in foods and the likely systemic effects. J. Sci. Food Agric. 80, 880–912.

Waalwijk, C., 2003. Detecting mycotoxin contamination of cereals. In: Cauvain, S.P. (Ed.), Bread Making: Improving Quality. Woodhead Publishing, Cambridge, UK, pp. 515–535.

Wang, C., Kovacs, M.I.P., 2002. Swelling index of glutenin test. I. Method and comparison with sedimentation, gel-protein, and insoluble glutenin tests. Cereal Chem. 79, 183–189.

Whelan, B.M., Taylor, J.A., Hassall, J.A., 2009. Site-specific variation in wheat grain protein concentration and wheat grain yield measure on an Australian farm using harvester-mounted on-the-go sensors. Crop Pasture Sci. 60, 808–817.

Wieser, H., Bushuk, W., MacRitchie, F., 2006. The polymeric glutenins. In: Wrigley, C.W., Bekes, F., Bushuk, W. (Eds.), Gliadin and Glutenin. The Unique Balance of Wheat Quality. AACC International, St. Paul, MN, pp. 213–240.

Wieser, H., Koehler, P., 2008. The biochemical basis of celiac disease. Cereal Chem. 85, 1–13.

Wrigley, C.W., 2006. Global warming and wheat quality. Cereal Foods World 51, 34–36.

Wrigley, C.W., 2009. Wheat: a unique grain for the world. In: Khan, K., Shewry, P.R. (Eds.), Wheat Chemistry and Technology. fourth ed. AACC International, St. Paul, MN, pp. 1–17.

Wrigley, C.W., 2012. Proteins—the basis of life. Teach. Sci. 58, 56–59.

Wrigley, C.W., 2016a. Cereals. In: Swainson, M. (Ed.), Swainson's Handbook of Technical and Quality Management for the Food Manufacturing Sector. Woodhead Publishing, Oxford, UK.

Wrigley, C.W., 2016b. Wheat: an overview of the grain that provides 'Our Daily Bread'. Wrigley, C., Corke, H., Seetharaman, K., Faubion, J. (Eds.), Encyclopedia of Food Grains, vol. 1, second ed. Elsevier Ltd, Oxford, UK, pp. 105–116.

Wrigley, C., Asenstorfer, R., Batey, I., Cornish, G., Day, L., Mares, D., Mrva, K., 2009. The biochemical and molecular basis of wheat quality. In: Carver, B. (Ed.), Wheat: Science and Trade. Wiley-Blackwell, Ames, Iowa, pp. 495–520.

Wrigley, C.W., Batey, I.L., 2003. Assessing grain quality. In: Cauvain, S.P. (Ed.), Bread Making: Improving Quality. Woodhead Publishing Ltd, Cambridge, pp. 71–96.

Wrigley, C.W., Corke, H., Faubion, J., 2016. The grains that feed the world. Wrigley, C., Corke, H., Seetharaman, K., Faubion, J. (Eds.), Encyclopedia of Food Grains, vol. 1, second ed. Elsevier Ltd, Oxford, UK, pp. 1–13.

Wrigley, C.W., Tömösközi, S., Békés, F., 2011. Hungarian-Australian collaborations in flour milling and test milling over 120 years. Cereal Res. Commun. 39 (2), 216–225.

Wyatt, C.J., Perez Carballido, S., Mendez, R.O., 1998. α- and γ- tocopherol content of selected foods in the Mexican diet: Effect of cooking losses. J. Agric. Food Chem. 46, 4657–4661.

Yamamori, M., Fujita, S., Hayakawa, K., Matsuki, J., Yasui, T., 2000. Genetic elimination of a starch granule protein, SGP-1, of wheat generates an altered starch and apparent high amylose. Theor. Appl. Genet. 101, 21–29.

Zhang, Y., Wu, Y., Xiao, Y., Yan, J., Zhang, Y., Zhang, Y., Ma, C., Xia, X., He H, 2009. QTL mapping for milling, gluten quality, and flour pasting properties in a recombinant inbred line population derived from a Chinese soft × hard wheat cross. Crop Pasture Sci. 60, 587–597.

Zhao, X.C., Sharp, P.J., Crosbie, G., Barclay, I., Wilson, R., Batey, I.L., Appels, R, 1998. A single genetic locus associated with starch granule and noodle quality in wheat. J. Cereal Sci. 27, 7–13.

Chapter 6

Durum Wheat: Grain-Quality Characteristics and Management of Quality Requirements

Ray Hare
Durum Wheat Improvement, Tamworth Agricultural Institute, New South Wales Department of Primary Industries, Tamworth, NSW, Australia

6.1 DURUM WHEAT: OVERVIEW

6.1.1 Origin of Durum

Evidence of humans consuming wild wheat goes back about 19,000 years to the shores of the Sea of Galilee, Israel (Feldman and Kislev, 2007). Extensive endemic stands of wild wheat could be harvested to provide a source of grain. This grain could be stored, milled and made into a range of nutritious foods. Where the land has not been disturbed, stands of wild wheat progenitors can still be observed in Syria, Lebanon and Israel.

Together with the wild diploid wheat species (*Triticum monococcum* subsp. *aegilopoides*, *Triticum urartu*), wild tetraploid species (*Triticum turgidum* subsp. *dicoccoides*, *Triticum timopheevii* subsp. *timopheevii*) were well distributed across the Middle East when humans commenced harvesting the wild wheat stands. The wild tetraploid species had advantages over the diploids, having a larger kernel (caryopsis) size, larger heads (spikes) and more kernels per head. The heads of *T. turgidum* subsp. *dicoccoides* shattered at maturity which proved a problem at harvest. The rachis of heads would disarticulate and fall to the ground resulting in harvest losses. Over time, humans selected a form, emmer wheat (*T. turgidum* subsp. *dicoccum*), which had a tough rachis not prone to shatter. The kernel was still hulled (non-free threshing).

In time, humans commenced cultivating seed derived from the wild types. Natural selection and human intervention, combined with cross pollination, random mutations and reproductive isolation resulted in the development of differentiated landraces adapted to various environmental and cultural circumstances.

Cereal Grains. http://dx.doi.org/10.1016/B978-0-08-100719-8.00006-1

At some time between 9000–7500 years ago, the first free-threshing kernels appeared. Spontaneous mutations allowed the kernels to separate from the glume, lemma and palea, leading to a naked kernel free of 'harvest trash'. Naked kernels were easier to process in food manufacturing.

The free-threshing form, durum wheat (*T. turgidum* subsp. *durum*), gradually replaced emmer wheat during the Greco-Roman period, when it became a major crop.

Many and varied durum wheat cultivars and landraces have been selected over the past 2000 years.

6.1.2 Genetic Relationship of Durum to Other Wheats

The wheats belong to the genus *Triticum*, within the Gramineae (Poaceae), grass family. The *Triticum* genus is a complex group of species related through shared genomes. Morris and Sears (1967), Kimber and Sears (1987) and MacKey (1966) have set out the relationship of the *Triticum* species with their genomic formulae. The most recent taxonomic classification of Van Slageren (1994) is adopted in this chapter. A "Triticum Comparative Classification Table" including the previously mentioned classifications can be found at www.ksu.edu/wgrc/Taxonomy/taxintro.html

Species and subspecies belong to each ploidy level from diploids ($2n = 14$) through tetraploids ($2n = 28$) to hexaploids ($2n = 42$). The tetraploids and hexaploids are allopolyploids of two or three related *Triticum* diploid species, respectively. Spontaneous or man-made hybrids can be produced between different species or subspecies. Interspecific hybrids, involving species containing a proportion of chromosomes in common, for example *Triticum aestivum* subsp. *aestivum* (genomic formula AABBDD) and *T. turgidum* subsp. *durum* (AABB), can result in partially fertile F_1 hybrids (AABBD) and fully (AABBDD or AABB) or partially fertile offspring in subsequent generations. However most interspecific hybrids are usually highly sterile as the homoeologous chromosomes of the two differentiated genomes do not pair precisely and result in unbalanced non-viable gametes.

Where partially fertile interspecific hybrids can be produced between the tetraploids and hexaploids or diploids, gene flow can occur between the ploidy levels. Chromosomes common to each parent can be resorted in fertile progeny at either ploidy level. For example a chromosome (or chromosomes) derived from a hexaploid parent could be included in the chromosome complement of a tetraploid progeny, and vice versa. Such gene flow across the ploidy levels has been used in breeding to enhance the genetic variability available for cultivar improvement. Recently, the sodium ion exclusion genes, *Nax1* and *Nax2*, were transferred from *T. monococcum*, through *T. turgidum* subsp. *durum* to *T. aestivum* subsp. *aestivum* (Huang et al., 2006). A number of stem rust resistance genes (e.g. *sr2*, *Sr12*, *Sr13*) have been transferred from various *T. turgidum* subspecies to *T. aestivum* subsp. *aestivum*. It is likely that the first Australian

Prime Hard bread wheat cultivar, Gabo, incorporated strong dough strength characteristics from the durum parent 'Gaza' when crossed to the hexaploid bread wheat 'Bobin Selection'. Genes for protein composition (i.e. molecular structure changes) are the most likely candidates for improved bread-making properties.

6.1.3 Production

Durum wheat is the only tetraploid wheat species of commercial importance that is widely cultivated today.

The main durum wheat growing nations include, European nations (France, Spain, Portugal, Italy and Greece), North America (Canada, United States, Mexico), Middle East (Turkey, Syria), North Africa (Morocco, Algeria, Libya, Tunisia) and Australia. Global production fluctuates around 35 million tonnes. In contrast, total global production of the various types of common wheat is over 700 million tonnes.

6.1.4 Processing and Products

Durum in Latin means "hard", and the species is the hardest of all wheats. This makes durum grain favourable for semolina production. The hard grain is more likely to shatter into coarse fragments and offer a high semolina yield on milling, when compared to the other wheat types. Industrial mills are specifically set up to mill durum grain into semolina and minimize the formation of flour. The semolina, high in strong gluten protein, offering a bright clean yellow colour is used to make premium pasta, bulgur, couscous and a range of other traditional products. Fines (flour) can be used in bread or as filler/binder in food products.

In the industrialised nations, high quality pasta is made in large extrusion plants. The semolina is hydrated with water, mixed to a stiff dough consistency under vacuum, extruded through dies of various shapes under considerable pressure and finally dehydrated carefully before the pasta is packaged.

The following discussion will focus principally on pasta quality, as most durum grain is used in the manufacture of pasta.

6.2 GRAIN, PROCESSING AND CONSUMER QUALITY

Pasta and the other products made from durum have relatively simple manufacturing processes and include few, if any, other ingredients. The quality attributes of the grain have a very significant influence on the finished product quality. Breeding for the expression of high quality attributes in varietal development is of paramount importance. Modern breeding programs devote considerable breeding effort and apply modern technologies to improve grain and processing quality.

Modern industrial plants apply sophisticated analytical tests and instrumentation to monitor and ensure end-product quality.

6.2.1 Grain Quality

6.2.1.1 Approved Cultivars

Most, if not all, nations that produce durum grain for pasta and other premium products provide growers with a list of approved cultivars. Approved cultivars have been tested extensively for a range of quality attributes and found to meet agreed market-quality standards. Grain buyers only purchase grain from approved cultivars. A number of quality attributes (e.g. protein quality) are under significant genetic control and not conveniently tested at grain delivery.

Acceptance of approved cultivars ensures that a bulk of such cultivars is within the agreed quality range for these attributes. Lists of approved cultivars are usually available to growers from the relevant state or marketing authorities.

6.2.1.2 Grain Size/Morphology—1000 Kernel Weight

In general, premium durum grain has a kernel size about 30% larger than the common wheats. Acceptable to desirable 1000 kernel weights vary from about 38 to 45 g/1000 kernels. Kernels are usually ovoid in shape (e.g. 8 mm long by 3.5 mm wide) (see Chapter 4). Large kernels have a relatively smaller surface area to mass ratio. Therefore, large kernels provide a greater proportion of endosperm, the grain component forming semolina, than the waste components (mill offal-bran, germ), an economic consideration in milling.

Large well-filled kernels are a breeding objective. Kernel size is under significant genetic control.

6.2.1.3 Grain Test Weight

Grain test weight (TW) is a measure of the grain packing density (kilogram per hectolitre), affected by both grain size and shape (AACC International Method 55-10.01, Test weight per bushel). TW is a widely used specification in durum grading as it is an internationally recognised measure of grain soundness and milling potential. TWs >76 kg/hL are desirable.

6.2.1.4 Grain Size Distribution

Grain going to the mill needs to have a reasonable uniform kernel size. In order to efficiently remove the bran layers (aleurone, pericarp) surrounding the endosperm during the first stages of milling, the grain is wetted or conditioned with water. This water penetrates the bran layers only, rendering them soft and therefore easier to remove. Small grains tend to condition faster than larger grains, with part of the endosperm taking up water. The soft endosperm breaks into small particles which are lost to the semolina mill stream, an economic loss.

Some modern high yielding cultivars tend to produce a proportion of small kernels greater than standards acceptable to the market. Small kernels arise in high yielding circumstances when grains in the 4th, 5th and 6th spikelet positions are competing with the earlier set kernels (1st to 3rd) for protein and starches. Out of specification small grain percentages reduce the market value of the grain and the benefit of the higher harvested grain yields. Breeders need to monitor kernel size distribution during breeding to avoid small kernel issues.

Any environmental factors (e.g. moisture stress, diseases, nutrient imbalances) that restrict grain yield can have an impact on grain size.

6.2.1.5 *Grain Hardness*

Modern durum cultivars have very hard kernels which makes durum wheat suitable for semolina production. Hard kernels shatter into coarse particles, a size that is ideal for semolina (a coarse grist of relatively uniform-sized particles). A number of factors can reduce grain hardness, including reduced protein content, which affects internal grain-structure components. The Perten Single Kernel Classification System instrument has a facility to provide a measure of grain hardness (Sissons et al., 2000). Individual grains are mechanically crushed between rollers and the force to break the kernel is recorded. An average hardness result is calculated for the grain sample.

6.2.1.6 *Mottling—Vitreous Kernel Count*

Sufficient protein is required to completely bond all starch granules together in a fully vitreous kernel. Insufficient kernel protein results in sectors where the starch granules are inadequately bound together. The lower-protein sectors contain void air spaces which appear as mottled, light-coloured patches (called “yellow berry” by some). The contents of these mottled sectors break into small particles, adding to the low-value fines (flour) mill fraction.

Factors, both genetic (e.g. low-protein achieving genes) and environmental (e.g. low-nitrogen nutrition, disease) that reduce protein content can influence the vitreous kernel count. Breeding for high-kernel protein-accumulating cultivars reduces the risk of mottled grain.

Grain buyers monitor vitreous kernel count by reference to photographic standards. An instrumental measurement method of vitreous count using image analysis is under development.

Minimum Hard Vitreous Kernel counts (HVK%) for Australian Durum No.1 (ADR1) is set at 80% while ADR2 is 70%, as listed on the Australian Classification system at www.graintrade.org.au

6.2.1.7 *Grain Protein Content*

Grain protein content is a key quality attribute in the manufacture of pasta. The protein content has a significant bearing on the end use of the grain and its monetary value. Grain exceeding 11.5% protein can be used for pasta, a high-value end product, while grain below 11.5% generally finds its way into the stockfeed market. The protein content influences grain hardness and hence semolina extraction. In addition, textural properties of the pasta are modified by the amount of protein. The integrity of the pasta during cooking and the feel of the pasta in the mouth, al dente, are in part determined by the protein content. Grain protein content is easily measured using near-infrared analysis allowing large numbers of experimental grain samples to be assessed in breeding programs, in commerce and in industrial situations (AACC International Method 39-25.01, Near infrared reflectance method for protein content in whole-grain wheat).

Significant advances have been made by breeders with new cultivars from Australia (EGA Bellaroi and Jandaroi) that are inherently high-protein achievers. These cultivars reduce the need to apply large amounts of expensive nitrogen (potentially polluting) fertilizer to achieve acceptable grain protein levels.

6.2.1.8 *Protein Quality*

Protein quality is determined by the gluten proteins, a complex combination of long-chain glutenin polymers (with molecular weights of many millions) and short-chain gliadin molecules (with molecular weights below about 100,000). These molecules have a vast array of structures, largely under genetic control. The complex chains of glutenin and the relatively compact gliadins combine to create a molecular mesh that is gluten. Disulfide bonds forming intra- and inter-chain links, together with non-covalent bonds (hydrogen bonds, ionic bonds, hydrophobic bonds) are important for the aggregation of gliadins and glutenins.

The mixing strength (viscosity) of semolina dough is largely determined by the strength of the developing gluten matrix. Strong dough, as opposed to weak dough, develops a greater number of cross-links between the protein chains, resisting the mechanical effort to change the shape of the dough. Strong dough properties are desired for pasta making. Such doughs retain their shape when extruded, the pasta retains its shape and integrity during cooking and the cooked product provides the desired textural resistance when eaten. The gluten matrix holds the starch granules within the pasta body during cooking and prevents dissolution of these starches into the cooking fluid (known as "pasta breakdown"). Pasta breakdown results in the formation of a rough sticky pasta surface leading to the adhesion of the individual pasta shapes during cooking and when served. Consumers prefer individual shapes well covered with flavoured sauces.

Protein quality is a critical quality attribute during the breeding of premium pasta-making cultivars, with considerable attention given to selection of this trait. A number of chemical (protein electrophoresis) and physical tests are used to evaluate breeding lines for protein quality (dough mixing with the Mixograph and Farinograph; SDS sedimentation using AACC International Method 56-61.02). New cultivars must produce strong stable dough during pasta manufacture.

6.2.1.9 *Pre-Harvest Sprouting Resistance*

Rain just before harvest can cause the grain to commence germination in the head. Enzymes are activated and mobilised resulting in the breakdown of starch and protein. The enzymatic cleavage of starch and protein chains results in the deterioration of grain quality suitable for pasta. Both dough strength and pasta integrity decline adversely.

Pre-harvest sprouting resistance is largely determined by seed dormancy. Durum has a useful range of genetic variation for seed dormancy. Breeders can select kernels under controlled laboratory conditions (temperature, moisture availability) for lines that take a number of days (4, 5 days) to germinate despite being fully imbibed (Hare et al., 1988). These resistant lines when re-dried appear to suffer little enzymatic deterioration.

The Falling Number test is used to measure the degree of starch and protein degradation. A Falling Number (FN) of 300 s for both ADR1 and ADR2 applies in Australia (AACC International Method 56-81.03, Determination of falling number). The activity of alpha-amylase (the main enzyme involved in starch breakdown) can be determined by measuring the changes in the viscosity of a starch paste using the Rapid Visco Analyser (AACC International Method 22-08.01, Measurement of alpha-amylase activity with the Rapid Visco Analyser).

6.2.1.10 Black Point

Black point refers to any black or dark discolouration of the grain. The cause of this discolouration is uncertain but believed to be due to phenolic compounds located in the kernel pericarp activated by humid conditions prevailing during grain fill. Blackened bran can be easily seen in semolina and pasta as small black specks. Such specks appear to the consumer as undesirable contaminants. Durum cultivars appear to vary in their expression of black point suggesting some genetic variability for sensitivity. Breeding for reduced sensitivity has proved difficult. In Australia, a maximum of 3% stained grain is allowed in ADR1 while 5% for ADR2.

6.2.1.11 Pink-Stained Grain

Pink-stained grain is caused by infection of the developing kernel by the fungi *Fusarium graminearum* or less often by *Epicoccum* spp. or *Drechslera* spp. The fungal species responsible for the development of pink staining may potentially produce toxic mycotoxins that are extremely harmful to both humans and animals. These harmful toxins are stable compounds and will remain active even after processing. Hence pasta made from contaminated semolina will contain mycotoxins (e.g. deoxynivalenol and zearalenone).

The fungi that cause pink staining are favoured by warm, humid conditions during anthesis. The symptoms develop when conditions are continuously warm and moist for a period of a few days. Prolonged favourable environmental conditions may increase the severity of symptoms.

There is no known genetic resistance to these fungi. Crop rotations involving non-host species (e.g. broadleaf species), while allowing for the complete breakdown of infected crop residues, will significantly reduce inoculum available to infect subsequent durum crops.

Grain buyers monitor pink-stained grain by reference to photographic standards (e.g. see Visual Recognition Standards Guide at www.graintrade.org.au).

A maximum limit of 2% applies to milling grades in Australia. Grain used for infant foods has a nil tolerance.

6.2.1.12 *Ash Content*

Ash content (% dry weight) is the inorganic residue remaining on complete combustion of all organic matter from a measured mass of grain (AACC International Method 08-12.01, Ash in farina and semolina).

Certain semolina millers use ash content as a measure of the extent of semolina extraction. As most of the inorganic compounds reside in the bran layers, high ash semolina could indicate high bran content. High-bran levels can cause pasta discolouration (browning), a defect. Cultivars with a high inherit ash content can yield semolina high in ash. To meet semolina ash specifications, extraction yields must be reduced when milling these high-ash cultivars to avoid bran inclusions and exceeding specifications. While not a high priority in breeding, the inherit ash levels are monitored to avoid the release of high-ash cultivars.

6.2.1.13 *Grain Moisture*

The moisture content of grain needs to be less than 12.5% w/w at harvest to allow the grain to be safely stored in bulk. The moisture in wet grain can migrate in the stack and allow fungal spoilage to develop. Where biological decay occurs, additional moisture is released as an end product of biological respiration. Respiration heat generated can lead to spontaneous combustion within the stack.

Wet grain has a high weight and where grain is purchased on weight, buyers do not want to pay for water at grain prices. Dry grain has an advantage in the market.

Grain moisture can be determined by using an air oven (AACC International Method 44-15.02, Moisture-Air-oven method). Alternatively, grain moisture can be conveniently and accurately measured using near-infrared analysis instruments.

6.2.1.14 *Artificially Dried Grain*

During wet harvests, growers can harvest grain above 12.5% moisture and dry the grain in heated silos to the required bulk storage standard. Care must be exercised during drying to ensure that the grain does not reach a temperature (>50°C) that will cause denaturation of the grain components (e.g. starches, proteins) and destroy the inherent quality for pasta making.

6.2.1.15 *Frost Damage*

Air temperatures at or below −3°C within the head canopy of the crop during grain filling can result in frost damage to the developing grain. Ice crystals form in the young grain tissues and disrupt these tissues resulting in cellular death. Affected grains can be distorted and shrivelled, if they survive the freezing. Frosted grain, lacking most of the endosperm, is regarded as a contaminant and of no value to the miller.

6.2.1.16 Other Crop Seed Contamination

Contamination of durum with seed from other cereals (e.g. bread wheat, barley) can lead to significant quality issues if they are not removed prior to milling. For example the softer bread-wheat kernels mill to fine white particles that can contaminate the semolina. Oilseed species have completely different technological properties. Growers need to ensure that their fields are free of non-durum crop volunteer plants, that the sowing seed is pure to type and that all grain-handling equipment (harvesters, bins, silos, transporters) are clean.

6.2.1.17 Weed Seeds

Certain weed seeds can cause significant contamination issues in semolina and pasta. Black seeds (e.g. *Tetragonia tetragonioides*, New Zealand spinach) shatter into small black particles resulting in high semolina- and pasta-speck counts. Some weed species have seeds that impart a strong undesirable flavour or odour to the semolina and pasta (e.g. *Melilotus indicus* Hexham scent). Within-crop control of weeds is essential to avoid grain contamination and downgrading. Published lists of weed-seed contaminants are available to growers (e.g. see www.graintrade.org.au).

6.2.1.18 Pesticide Residues

Grain buyers set low- or nil-pesticide residue limits. Only approved pesticides for durum can be applied to crops and to harvested grain using the approved application protocols. A variety of non-insecticide methods of controlling stored-grain insects can be used to avoid insecticide residues. Reduced storage temperatures (<15°C for e.g. *Rhyzopertha dominica*—lesser grain borer; *Sitophilus granarius*—granary weevil) and low humidity can prevent the reproduction of insects. Inert gases (carbon dioxide, nitrogen) have been used in sealed silos to asphyxiate insects.

6.2.1.19 Harvesting

6.2.1.19.1 Harvester Clean Down

Prior to commencing harvesting operations, a comprehensive clean down of the harvester will ensure contaminants (e.g. weeds seeds, crop seeds, trash, insects, vermin droppings) from previous harvesting operations are removed.

6.2.1.19.2 Harvester Adjustments

Adjustments to the grain-harvester threshing-drum concave settings are necessary to ensure the harvested grain is not cracked and is threshed free of 'white heads' (chaff). Correct selection of sieve sizes will facilitate the removal of small grains, weed seeds and soil particles, contaminants likely to lead to a reduced sale price of the grain.

6.2.1.19.3 Harvest Timing

Grain should be harvested as soon as it reaches dead ripeness. Buyers of durum grain consider grain appearance important and pay premiums for large, well-filled, hard, vitreous grain with a low percentage of mottled and bleached seeds. Rain on ripe grain can cause damage to the bloom and hardness of the grain, while prolonged wetting can lead to fungal staining.

6.2.1.20 Supply-Chain Cleanliness

A multitude of grain-handling facilities that store, move and transport the grain from farm to mill must be cleaned of contaminants (e.g. non-durum grain, grain insects, non-grain cargo residues, water).

Recently, millers have received grain of a specified description direct from the farm storage in sealed shipping containers. A full quality certificate precedes the container. Container shipments avoid in-transit contaminants and ensure that the grain received is not downgraded by co-mingling with grain of inferior grade. Transport costs may be higher using containers but the benefits can counter cost increases. Containers can be shipped through ports which do not have bulk-handling facilities and on container vessels with multiple cargo types.

6.2.2 Semolina Quality

6.2.2.1 Semolina Mill Yield

Maximising the yield of semolina from grain is a primary milling objective. Premium semolina is a grist of coarse relatively uniformly sized particles (200–425 μm) of the endosperm with minimal fines (flour) and bran content. Uniform particles hydrate evenly when the dough is developed. Fines tend to congregate together in pockets within the dough and form uneven and undesirable streaks or blotches in the extruded product.

Semolina milling yield is a difficult attribute to measure in the laboratory.

Each commercial semolina mill is set up differently and is constantly adjusted by a skilled miller as milling proceeds. Currently there are no small-scale (~100 g grain) predictive measures of semolina milling yield of acceptable accuracy. Where larger quantities of grain (~3 kg when using a Buhler laboratory mill) are available at advanced stages of cultivar breeding, milling performance can be assessed on experimental mills set up for semolina production. Given the variation in the configuration of commercial mills, experimental yields provide a guide to commercial performance and breeding-line rank. As larger quantities of grain are required from a number of trial sites per year, evaluation of mill yield can only be conducted at the later or advanced stages on a small number of elite lines. Thus selection pressure for inherited yield improvement is rather limited. Despite these difficulties, small yield increases have been achieved in Australia.

6.2.2.2 Semolina Colour

Good quality pasta is perceived by the consumer to have a bright clean yellow colour. It is a pivotal quality attribute. The yellow pigments are principally the carotenoids, lutein (~85%) and zeaxanthin (~8%) which reside in the endosperm. The concentration of these compounds is highly heritable and can be accurately quantified on small grain samples at an early stage in breeding. Significant genetic improvement has been achieved in recent times. Colour can be assessed accurately on whole or ground grain using reflectance spectroscopy by measuring the colour space parameters: brightness (L*), redness (a*) and yellowness (b*). A high concentration of endosperm pigments translates into a bright yellow semolina ($b^* > 30$) and the subsequent pasta.

6.2.2.3 Speck Counts

Consumers desire pasta that is a uniform translucent yellow product and free of specks. Specks are considered to be potential contaminants by consumers. Specks arise from the inclusion of bran fragments, non-durum crop seeds, weed-seed particles, mineral (stone/soil) and organic (plant/animal litter) fragments. Every effort needs to be taken to ensure non-durum and weed seeds, mineral and organic litter do not contaminate the grain from harvest to mill. Grain entering the mill stream is mechanically cleaned to remove these non-durum grain contaminants. Size and/or density differences are used to screen out most of these contaminants. Usually the main source of specks is from bran fragments not removed during semolina purification. These fragments are particularly evident if they arise from black- or dark brown-stained grain. Strict receival standards apply to stained grain for semolina production (see Section 6.2.1.10). Pasta makers have speck count specifications and conduct standardised speck counts on semolina purchased.

Lighter coloured streaks and blotches arise when fine particles accumulate together in the mixed dough and do not become incorporated into the uniformly hydrated dough matrix prior to extrusion (see Section 6.2.2.1). Removal of fines during milling should eliminate this issue.

6.2.3 Pasta Quality

6.2.3.1 Evaluation Tests

The assessment tests for pasta quality can be divided into two groups. The first group includes tests conducted on the grain and semolina. Many of the forgoing traits and associated tests provide good predictive data on the final pasta quality, given that pasta quality is closely related to the raw-material characteristics.

The second group involves tests conducted on the finished product. Mechanical strength and cooking quality tests must be conducted on pasta.

6.2.3.2 *Laboratory Pasta Making*

The AACC International Methods (66-41.01 and 66-42.01) describe agreed macro- and micro-scale pasta-processing methods. The macro-scale method simulates the commercial process using an extruder that makes from 1.5 to 10 kg of spaghetti. The micro-scale method is used when small grain samples from early-generation breeding lines are available.

6.2.3.3 *Dry Pasta Quality*

The visual appearance and mechanical strength of the dried product are important quality features that influence consumer acceptance at the point of purchase. Most pasta makers display the pasta through a transparent window in the packaging. Consumers can quickly recognize a bright clear yellow product free of specks, white blotches and broken strands.

6.2.3.4 *Mechanical Pasta Strength*

Broken pasta strands and shapes reduce the aesthetic appeal of the product.

White streaks and blotches represent one of the most common defects of dried pasta, due to the presence of un-hydrated particles during mixing, and is clearly noticeable in the finished product. This defect not only worsens the aesthetic appearance of the pasta, but is associated with the non-uniform structure of the protein network and renders the pasta very fragile in the dry and cooked states.

Well-developed and uniformly hydrated semolina dough, mixed under vacuum, will extrude into uniform translucent pasta free of air pockets and structural blemishes (white blotches).

Dry pasta has to have sufficient strength to resist breakage during cutting, packaging and transporting. Instrumental three-point bending tests applied to spaghetti strands is the most common mechanical test for resistance to breakage.

6.2.3.5 *Pasta Drying*

Pasta drying is a complex process involving strict regulation of temperature and humidity at various stages of the drying cycle. The aim is to remove the water from the dough in a manner that completely avoids the formation of mechanical stresses (shrinkage) and consequent fractures in the pasta mass. Water must be allowed to migrate from the internal pasta parts at a controlled steady pace. Product leaving the drier must be in moisture and temperature equilibrium with the surrounding factory atmosphere. Laboratory pasta driers must adhere to strict drying protocols to provide a test product suitable for ongoing evaluation.

6.2.3.6 Surface Properties

There is some debate about the surface appearance of pasta. A smooth surface usually has greater aesthetic appeal where the translucent product is better displayed. In contrast, a rough surface is usually less appealing in the dry state but has a greater retention of sauces and condiments when cooked. Due to the greater surface area provided by the rough surface, the amount of material released from the surface is higher, that is cooking losses are greater. There is evidence that the rough product is more prone to fracture.

Most pasta is extruded through Teflon-coated dies that impart a smooth surface to the product.

6.2.3.7 Pasta Colour

Pasta colour is important at two stages: following manufacture and after cooking, prior to eating. The consumer desires a bright clean yellow translucent product on purchase followed by a yellowish cooked product. The colour of the cooked product is usually masked by the adhering sauces and other ingredients in the serving.

Following extrusion and drying, pasta colour is the result of the desirable yellow component and the undesirable brown component. A reddening component can occur under certain drying conditions. Yellowness is determined by the carotenoid content in the kernel endosperm and subsequent semolina. On dough development prior to and during pasta extrusion, oxidative degradation of the pigments by lipoxygenase can occur. A high yellow-pigment levels in the semolina is no guarantee of acceptable colour in the pasta. Lipoxygenase activity is under environmental and genetic control. Breeders can select for low-lipoxygenase activity using enzyme or DNA-marker screens.

Undesirable colours have several causes. Elevated temperatures during dough mixing, extrusion and drying together with degraded starch (possibly due to pre-harvest weather damage) and alpha-amylase activity lead to the formation of reducing sugars which in turn react with free amino acids, producing dark highly coloured compounds through the Maillard reaction. Brown pasta colour also increases with higher semolina ash content, possibly due to elevated peroxidase and polyphenol oxidase activity. The undesirable colours mask the yellow colour and diminish the visual appeal of the pasta to the consumer.

Uncooked pasta colour is measured by reflectance spectrophotometric instruments, using the colour space parameters: brightness (white to black) L*, a* and b*. Yellow intensity is given directly by the b* value. Brown and grey colours are reflected in the L* values (AACC International Method 14-22.01, Color of pasta—reflectance colorimeter method).

6.2.3.8 Cooking Characteristics

The cooked characteristics of the pasta ultimately dictate consumer acceptance. Cooked pasta characteristics include features of mechanical texture such as firmness, elasticity, stickiness, chewiness together with aroma and taste. These can be evaluated by sensorial, instrumental and chemical tests. Sensory evaluation using trained panellists provides for an integrated analysis of the cooked quality. Instrumental and chemical tests only measure some appropriate physical and chemical components of cooked quality.

6.2.3.8.1 Cooking Protocols

Cooked pasta quality is complicated by the cooking protocol. The most common variables include cooking time, pasta/water ratio and water type (tap, mineral, distilled, deionised, that vary in concentration of dissolved solids and pH). The AACC International has set out a standard method (66-50.01; 2011).

6.2.3.8.2 Cooked Water Uptake

Cooking pasta in gently boiling water promotes the adsorption of a considerable amount of water (~1.5–2 times the dry mass of the pasta). When cooked, the pasta becomes palatable with a smooth textural consistency while retaining its physical shape. The proteins and starches have exceptional capacity to take up and hold water in a protein coagulant and gelatinised starch network. The degree of water adsorption is a quality attribute of pasta as measured by the increased mass of cooked over dried product.

6.2.3.8.3 Cooking Loss

Conversely while cooking proceeds, starches are transferred from the pasta surface to the cooking water. Cooking loss, a quality trait, measures the mass of solids lost to the water during cooking. As starch is the principle component, a standardised iodine–starch absorbance (650 nm) test can be used. The molecular structure of the glutenin proteins allows for the development of a network of cross-linked protein chains which hold the gelatinised starch in place within the pasta mass. Selection of proteins that form a strong stable network of cross-linked chains is a key breeding objective (see Section 6.2.1.8).

6.2.3.9 Cooked Pasta Texture

Cooked pasta texture is defined by five attributes which describe the performance of the product in the mouth. The six characteristics are: firmness—the force required to penetrate pasta with teeth; cohesiveness—the degree of deformation of the pasta mass before breakage; elasticity—the degree to which the pasta recovers its shape following the removal of the deforming force; stickiness—the force required to overcome the attraction between the pasta surface and surfaces of other items (e.g. teeth, tongue and fingers); bulkiness—as with stickiness, the force to separate pasta strands and chewiness—time or energy needed to masticate pasta in order to swallow.

6.2.3.9.1 Sensory Tests—Trained Assessors

Trained assessors, using standardised methods [The International Organization for Standardization (ISO) (ISO 7304) see Section 6.2.4] are considered the most reliable means of assigning firmness and stickiness for spaghetti. Such trained panels are expensive; thus instrumental methods of texture analysis have been devised.

6.2.3.9.2 Instrumental Bite or Cutting Tests

Various tool shapes are attached to the moving arm of the texture tester. A simulated tooth is brought to bear on strands of cooked spaghetti where the tooth cuts the strands. The maximum force recorded during the cutting cycle (peak force) and work, represented by the area under the curve, are measured (AACC International Approved Method 66-50.01; 2000).

6.2.3.9.3 Instrumental Compression Tests

Instead of a cutting tool attached to the moving arm, a flat plate probe compresses pasta strands between the moving probe and the fixed support plate.

Two traits are measured: the modulus of elasticity derived from stress and strain measurements, and the adhesiveness (stickiness), which is the work required to pull the compression probe from the sample.

Comparison of data between instruments and laboratories can be difficult due to differences in instrument settings and test rigs adopted by various laboratories. Despite these difficulties, valuable data can still be obtained.

Within a breeding or an industrial laboratory, the relative ranking of pasta samples can be achieved. New breeding lines can be referenced to well-established premium-quality cultivars and either accepted for release or rejected.

Pasta makers can determine if the production run is meeting their quality specifications. Cross-reference to other pasta brands can be undertaken.

6.2.3.9.4 Aroma and Flavour Tests

Currently there are no internationally standardised instrumental methods for assessing flavour and aroma.

6.2.3.10 Grain Safety Assurance

Present day consumers are well aware of food safety. The ability to meet established safety standards is now the most critical condition for sale of wheat and wheat products. Quality assurance of grain free of toxic substances must be provided by producers and traders through the observance of a strict quality management and a documentation trail.

Food safety hazards for grains may be classified as chemical, biological and physical. Toxic chemical contamination hazards can include organic and inorganic compounds (both natural and man-made), toxic elements and in certain regions, radioisotopes from nuclear accidents. For durum grain, pesticide residues (see Section 6.2.1.18), mycotoxins and heavy metals are the major concerns. Mycotoxins derived from several fungal species are of concern. The European Commission lists and sets maximum levels of the mycotoxins (deoxynivalenol, zearalenone derived from *F. graminearum*, *Fusarium culmorum*, *Fusarium sporotrichioides*), Ochratoxin A from *Penicillium verrucosum* and Aflatoxin B1 from *Aspergillus flavus*. The heavy metals, cadmium and lead must be less than maximum levels in grain, legislated by the European Commission (ec.europa.eu.legislation.index_en).

Modern analytical technology has given compliance laboratories tools to make accurate and highly sensitive measurements of each of the designated contaminants in grain and derived products. Based on approved cargo samples, independent compliance laboratories issue cargo certificates to the merchant and buyer showing the amounts of listed contaminants present and whether there is compliance with legal limits. A reserve sample is held in case a dispute arises which requires further analysis.

6.2.4 Standard Methods of Quality Analysis

Internationally agreed standardised analytical methods for quality determination are published by:

- The AACC International (St. Paul., Minnesota, USA), www.aaccnet.org/toc.aspx
- The International Association of Cereal Science and Technology (ICC) (Vienna, Austria), www.icc.or.at
- The International Organization for Standardization (ISO) (Geneva, Switzerland), www.iso.org

Examples of the comprehensive range of methods available have been referred to in this chapter.

6.2.5 In-depth Review of Durum Wheat Quality Science and Technology

An in-depth review of the latest knowledge and research findings on the chemistry and technology of durum wheat and derived products has been published by the AACC International as 'Durum Wheat—Chemistry and Technology', ed. M. Sissons, J. Abecassis, B. Marchylo, M. Carcea. Second Edition, 2012.

REFERENCES

AACC International Approved Methods of Analysis, eleventh ed. Method 08-01.01, Ash—Basic method; Method 08-12.01, Ash in farina and semolina; Method 14-22.01, Color of pasta—reflectance colorimeter method; Method 22-08.01, Measurement of alpha-amylase activity with the Rapid Visco Analyser; Method 39-25.01, Near infrared reflectance method for protein content in whole-grain wheat; Method 44-15.02, Moisture-Air-oven method; Method 55-10.01, Test weight per bushel; Method 56-61.02, Sedimentation test for wheat; Method 56-81.03, Determination of falling number; Method 66-41.01, Macro scale pasta processing; Method 66-42.01, Micro scale pasta processing; Method 66-50.01, Pasta and noodle cooking quality—Firmness. AACC International, St. Paul, MN.

Feldman, M., Kislev, M.E., 2007. Domestication of emmer wheat and evolution of free-threshing tetraploid wheat. In: Special Issue: A Century of Wheat Research—From Wild Emmer Discovery to Genome Analysis. Isr. J. Plant Sci. 55, 207–221.

Hare, R.A., Koth, B.E., Barnes, W.C., 1988. Pre-harvest sprouting resistance in *Triticum turgidum* ssp. *durum*. In: Proceedings of Seventh International Wheat Genetics Symposium, Cambridge, UK, pp. 951–956.

Huang, S., Spielmeyer, W., Lagudah, E.S., James, R.A., Platten, J.D., Dennis, E.S., Munns, R., 2006. A sodium transporter (HKT7) is a candidate for Nax1, a gene for salt tolerance in durum wheat. Plant Physiol. 142, 1718–1727.

Kimber, G., Sears, E., 1987. Wild Wheat: An Introduction (Special Report). College of Agriculture, University of Missouri, Columbia, MO.

MacKey, R., 1966. Species relationship in *Triticum*. In: Proceedings of the Second International Wheat Genetics Symposium. Hereditas Suppl. 2, 237–276.

Morris, R., Sears, E., 1967. The cytogenetics of wheat and its relatives. In: Quisenberry, K.S., Reitz, L.P. (Eds.), Wheat and Wheat Improvement. American Society of Agronomy, Madison, WI, pp. 19–87.

Sissons, M.J., Osborne, B.G., Hare, R.A., Sissons, S.A., Jackson, R., 2000. Application of the single kernel classification system to durum wheat testing and quality prediction. Cereal Chem. 77, 3–10.

The International Organization for Standardization (ISO) (TC 34 SC4 7304) (ISO 7304, 1985) International Organization for Standardization, Geneva, Switzerland.

Van Slageren, M.W., 1994. Wild Wheats: A Monograph of *Aegilops L.* and *Amblyopyrum* (Jaub. and Spach) Eig (Poaceae). Wageningen Agricultural University, Wageningen, The Netherlands, pp. 94–97.

Chapter 7

Rye: Grain-Quality Characteristics and Management of Quality Requirements

Colin Wrigley* and Walter Bushuk**

*QAAFI, University of Queensland, St. Lucia, QLD, Australia; **Department of Food Science, University of Manitoba, Winnipeg, MB, Canada

7.1 INTRODUCTION

Rye (*Secale cereale* L.) is a grass crop grown primarily for its grain, and also as a forage crop. Rye is a close 'cousin' of wheat (*Triticum* L. species)—close enough that it can be cross-bred with wheat to produce triticale (*x Triticosecale* Wittmack) (see Chapter 8). Like wheat, rye flour can be formed into a viscoelastic dough, but with poorer dough quality than wheat. As a result, rye bread is often made with a significant proportion of wheat flour because the dough-forming quality of rye flour alone is inadequate for normal baking procedures and bread-quality requirements.

The important uses of rye are for milling into flour or wholemeal for baking into various foods: 'black bread', rye bread, crisp bread, pumpernickel, and sour-dough breads (Fig. 7.1). To a lesser extent, rye is eaten as the boiled whole grain and is rolled and steamed into a product equivalent to rolled oats. It is also used for beverages: rye beer and some whiskies and vodkas. The major use is for livestock feed, accounting for about 50% usage worldwide (Sapirstein and Bushuk, 2016). Like other cereals, rye grain is also used for ethanol production. Rye straw is sought for animal bedding, for thatch and for building materials, as well as feed for ruminants.

To assess and manage the grain-quality of rye as a cereal grain (the accent of this book), it is necessary to understand the origins of rye, its genetic constitution, distinctive aspects of its composition, and the consequent ramifications of producing the end products via the grain chain. This sequence, as represented

Cereal Grains. http://dx.doi.org/10.1016/B978-0-08-100719-8.00007-3

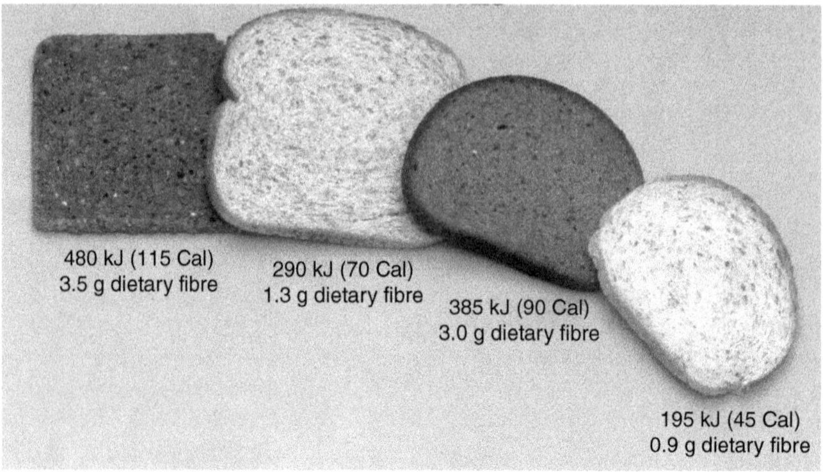

FIGURE 7.1 **Baked products made from rye flour.** From left to right: Pumpernickel (50 g portion), light rye bread (30 g, incorporating wheat flour), dark rye (40 g) and light rye (20 g, incorporating wheat flour). The fibre content and energy value is indicated for each portion.

in Table 7.1, involves all the steps from the breeding of improved varieties to growth, harvest, processing and consumption in many ways. Table 7.1 emphasises the aspects that are distinctive for rye, although it parallels the generic version in Chapter 1. Ergot is an example. Like most cereals, there is the possibility of defects in the harvested grain, but for rye there is the more common occurrence of ergot bodies taking the place of a grain in the head (Fig. 7.2).

7.2 WORLDWIDE PRODUCTION AND REGIONAL DISTRIBUTION

Despite similarities of rye to wheat, the country-by-country distribution of rye production (Table 7.2) is distinct from that of wheat. For example, several major wheat-producing countries have only minor rye production, due partly to rye being suited to cool growth temperatures and also due to large differences in regional preferences for rye-based breads. Rye production is most evident in Germany, Russia, Poland and Belarus, that is, in central, northern and northeastern countries of Europe. World trade in rye is small (about 13% of world production) in the context of international grain trade, because rye production is generally related to local food production and on-farm use.

Rye can grow better in poor soils and survive in cold conditions better than other cereals. Rye, due to its reputation for being better suited to marginal land, is often grown in adverse environments where other cereals may not provide an economic return. Nevertheless, grain yield statistics for rye vary greatly from one country to another (Table 7.2). In most situations, rye yields are less than

TABLE 7.1 Products Based on Rye; Distinctive Issues Associated with the Various Stages of the Grain Chain

Stages		Products	Issues for analysis and management
Breed	→	Improved varieties.	Pre-harvest sprouting. Winter or spring type.
↓			
Sow	→	F1 hybrid seed.	Suitability for marginal land.
↓			
Grow	→	Grazing for ruminants.	Winter hardiness. Drought tolerance.
↓			
Harvest	→	Straw for thatch and stock.	Pre-harvest sprouting. Ergot defect.
↓			
Store	→		Moisture content. Preserve identity.
↓			
Process		Food as whole grains. Stock feed. Fuel as bioethanol.	Anti-nutritionals. Pentosans. Ergot content. Alkyl resorcinols. A-type starch content (greater than for wheat).
↓			
Mill	→	Flour	Poor sieving. Bakers' asthma from dust.
↓			
Mix and bake	→	Black bread. Sourdough. Pumpernickel. *Knaeckebrot.*	Dough stickiness. Grain soundness. Sprout damage.
↓			
Consumer nutrition	→	Fibre. Bioactives. Vitamins. Minerals.	Types of fibre. Anti-nutritional factors. Phytochemicals.

those for wheat. However, agronomic advantages of rye, compared to wheat, include winter hardiness, drought resistance, compatibility with low-pH soils, and the ability to thrive in soils containing iron and aluminium. On the other hand, prices for rye grain are less than for wheat, since a significant proportion of rye production goes for animal feed. However, for either rye or wheat, the value of any consignment depends primarily on the quality of that grain.

Nevertheless, the main reason for the regional and worldwide distribution of rye (Table 7.2) relates to its value for food. The distribution of rye production

FIGURE 7.2 Ergot bodies have replaced grains in this wheat head. The occurrence of ergot is less common in wheat than in rye. *(Adapted from Ferns, G.K., Fitsimmons, R.W., Martin, R.H., Wrigley, C.W., 1978. Australian Wheat Varieties, Supplement No. 1: Eleven Recent Wheats, Grain Defects and Appendixes on Foreign Seeds and Insects, CSIRO, Melbourne.)*

is thus distinct from wheat (and barley) production. For example, Table 7.2 lists some major wheat-producing countries with negligible rye production—in particular India, Pakistan and Mexico. This distinction also reflects the reputation of rye for being better suited for moderate-to-cool growth temperatures. Distribution of rye production more obviously reflects regional preferences for food products from rye, versus wheat or other grains. In Scandinavian countries, such as Sweden, Denmark and Finland, rye production is modest, but considerable compared to the wheat production of these countries. While the statistics in Table 7.2 provide a generally valid comparison between countries, it should be pointed out that the data are for only 1 year, and significant year-to-year variations are inevitable.

TABLE 7.2 Production (in Thousands of Tonnes) and Yields (in Tonnes per Hectare) in 2014 of Rye Compared to Wheat Production for Countries Producing Significant Quantities of Rye, Plus a Few Major Wheat-Producing Countries

Country	Rye yield	Rye production	Wheat production
Germany	6.1	3,854	27,785
Russian Federation	1.8	3,280	59,710
Poland	3.2	2,793	11,630
Belarus	2.7	867	2,920
China, mainland	3.5	700	126,200
Denmark	6.4	678	4,940
Ukraine	2.6	480	24,110
Turkey	2.6	300	19,000
Austria	4.8	232	1,800
Spain	1.7	228	6,470
Canada	2.4	200	29,280
USA	1.7	180	55,400
Sweden	6.6	176	3,090
Czech Republic	5.1	130	5,440
France	4.9	128	38,970
Hungary	2.9	96	5,260
UK	7.0	70	16,620
Argentina	1.6	51	13,930
Australia	0.6	32	25,300
Bulgaria	2.0	28	5,350
Romania	2.4	24	7,590
Portugal	0.9	18	90
Netherlands	4.1	7	1,300
Chile	4.5	4	1,360
Brazil	1.4	4	6,260
Belgium	4.5	2	1,990
India	—	None recorded	94,480
Pakistan	—	None recorded	25,980
Mexico	—	None recorded	3,670

Source: FAO, via www.faostats.fao3.org.

Rye production is minor compared to the production of wheat. Worldwide, wheat production has been about 35 times more than rye production on a long-term basis. The disparity between rye and wheat is even greater when consumption for human food is considered; globally human consumption of rye is about 1 kg per person per year in contrast to about 70 kg for wheat (FAO statistics).

World trade in rye is minor when compared with trade in wheat and the other major cereal grains. This difference is due to rye production being often intended for local food production within the country of production and also because it is grown for on-farm use as stock feed. In the latter case, rye serves to feed animals in three ways, in the form of grazing of the young plants, later as the harvested grain, and lastly by feeding of the straw as hay (Table 7.1).

The main exporters of rye grain are Canada and European countries such as Germany and Poland. Major importers include Japan and Korea. Imports of rye to these latter countries reflect the use of rye for animal feed rather than a national preference for food products made from rye (Sapirstein and Bushuk, 2016).

7.3 THE ORIGINS AND TAXONOMY OF RYE

The English word 'rye' has Germanic and old Norse origins—appropriately, since the grain came into English use from those regions. The grain species is sometimes given the names 'ryecorn' or 'cereal rye' to distinguish it from the grass species perennial rye or ryegrass (*Lolium* species) (Fig. 7.3). The term 'ryecorn' is based on the old English use of 'corn' to mean any type of grain.

In botanical classification, rye is a member of the grass family Gramineae (also called Poaceae), the subfamily Pooideae, and the tribe Triticeae (Morrison and Wrigley, 2016). Almost all cultivated rye has the genus-species designation *S. cereale* (L.). A small amount of the species *Secale fragile* is grown in south western Asia. Modern rye has probably come from a perennial grass (*Secale montanum*) that still grows wild in southern Europe.

Rye shares common origins with other species of the tribe Triticeae (wheat and barley) in south western Asia (Ma et al., 2004), where it is believed to have evolved from *S. montanum*, a perennial weed in wheat and barley fields there (Morrison, 2016). Rye may have been domesticated later than wheat, since specific reference to it is not found in ancient writings or on Egyptian monuments. Nevertheless, its mention in early writings from northern Europe suggests that its actual cultivation may have started there in regions where today it is still very popular, namely, Poland, Germany, Russia and Scandinavian countries (Table 7.2).

The migration of rye into these regions is likely to have occurred during the first millennium BC, possibly northward via Russia (Seibel and Steller, 1988; Bushuk, 2001a; Ma et al., 2004). More recently, rye continued its spread, going westwards throughout Europe, and being carried to new regions with settlers in the Americas, Australia and South Africa. Rye is thus grown to a limited

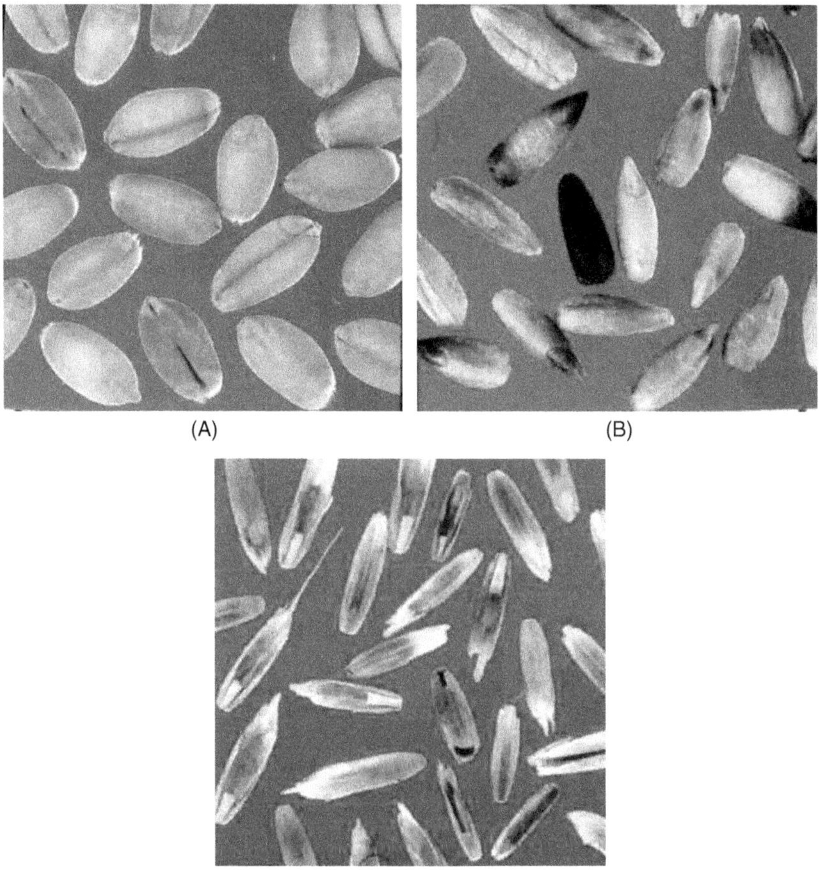

(A) (B) (C)

FIGURE 7.3 Grains of (A) common wheat, (B) rye, (C) annual ryegrass (*Lolium rigidum*). *(Adapted from Ferns, G.K., Fitsimmons, R.W., Martin, R.H., Wrigley, C.W., 1978. Australian Wheat Varieties, Supplement No. 1: Eleven Recent Wheats, Grain Defects and Appendixes on Foreign Seeds and Insects, CSIRO, Melbourne.)*

extent around the world, but the major production is in the countries where rye breads are most popular, such as Poland, Germany, western Russia, Belarus and Ukraine (Table 7.2).

7.4 THE GENETIC CONSTITUTION OF RYE

Most cultivated rye varieties are diploid, having seven pairs of chromosomes, numbered 1R to 7R (Schlegel, 2014; Sapirstein and Bushuk, 2016). These are homoeologous to the seven chromosome pairs of the three genomes of wheat (1A to 7A, 1B to 7B and 1D to 7D, respectively) (Naranjo and

Fernfindez-Rueda, 1991). It has been possible to develop a tetraploid rye by doubling the number of chromosomes using colchicine but the resulting plants have not been cultivated in commercial amounts, largely due to its added susceptibility to ergot and poorer cold tolerance (Sapirstein and Bushuk, 2016).

Rye genotypes may be distinguished as either 'winter rye' or 'spring rye', depending on their suitability, like wheat, for planting before or after (respectively) the severe winter of northern regions. Thus, winter rye genotypes have a vernalisation requirement and are late maturing. These distinctions relate primarily to countries with severe winter conditions, such as Canada and northern Europe. Rye has excellent winter-hardiness, so it can even be cultivated in regions where the winter is too severe for winter wheat (Sapirstein and Bushuk, 2016).

Ma et al. (2004) demonstrated that the maturity difference between spring and winter also has a genetically significant aspect. They used DNA techniques to examine genetic similarities among 20 spring and 22 winter accessions of agronomically different ryes from 14 countries. The 42 genotypes were readily classifiable into two main groups: spring and winter groups. Beyond the spring–winter relationships, groupings were evident along the lines of regional origins, with similarities evident, for example for rye genotypes from Germany, Finland and Estonia. On the other hand, there was only a distant relationship between 'Petkus' rye, from Germany, and the 'Mexico' variety, from Mexico (Ma et al., 2004).

7.5 F1 HYBRID RYE AND GENETIC IMPROVEMENTS

It was realised early in the 20th century that rye has the potential, similar to that of maize, for improved productivity based on hybrid vigour. However, genetic stocks did not become available for the practical implementation of this principle until the 1970s (Scoles and Evans, 1978). At present, about 70% of the rye grown in Germany is F1 hybrid rye, permitting considerable improvements in grain yield (Sapirstein and Bushuk, 2016). There are rye-breeding programs in many rye-producing countries.

Major breeding programs are centred in Germany and Poland, where a major accent is on the development of grain with quality suited to bread making. Promising approaches (Graybosch, 2001; Altpeter et al., 2004) include the incorporation of glutenin polypeptides (such as subunits 5 + 10) from wheat, in which these subunits have been demonstrated to contribute to gluten strength (Shewry et al., 2006; Békés et al., 2016), using conventional cross-breeding and genetic engineering. Breeding objectives also include improving the tolerance of rye to pre-harvest sprouting, and increasing the low Falling Number of harvested rye, which is one of its particular quality problems (Weipert, 1993, 1996; Schlegel, 2014). Disease-resistance objectives relate to snow mould and to *Fusarium* head blight (Czembor and Sowa, 2001).

Some aspects of the disease resistance of rye have made its genes attractive for improving disease resistance in wheat, due to the facility of inter-crossing between the two species. Translocation of chromosome portions of 1R into wheat chromosomes 1A or 1B have been especially successful (Dhaliwal et al., 1988; Bartos et al., 2008) (see Chapter 14). Many wheat varieties grown commercially around the world have 1A/1R or 1B/1R translocations, conferring useful disease resistance into wheat, mainly to rust (Singh et al., 1990). However, in most cases so far, these translocations into wheat have also introduced significant loss of dough quality and stickiness in the dough (Graybosch, 2001; Békés and Wrigley, 2016). As a result, many of these wheat varieties are unsuitable as strong-dough bread varieties, being better suited to weak biscuit/cookie grades.

7.6 DISTINCTIVE ASPECTS OF THE PLANT AND GRAIN MORPHOLOGY OF RYE

The rye plant and head (ear) are similar to those of wheat, but the rye plant is generally more slender. The rye head (Fig. 7.4) is also slender, as well as being long, lax and normally awned (bearded). The rye flower is more open than that of wheat, and as a result it tends to out-crossing, that is the rye flower is not always self-pollinating.

The rye grain is similar in size to wheat and is free threshing (Figs 7.3A and B). Thus the grain (caryopsis, kernel) is harvested free of the surrounding lemma, palea and outer glumes (chaff). The rye grain has a central crease (groove) like wheat, but its germ end is distinctively more pointed than that of a wheat grain (either durum or common wheat). The surface of the rye grain is often wrinkled. Unlike wheat, rye grains vary in colour from light to dark, even within an individual variety and even within the same sample (Fig. 7.3B). Sapirstein and Bushuk (2016) illustrate separate samples of rye kernels that are blue, red and white.

Ryegrass grains are illustrated (Fig. 7.3C) to emphasise its difference in morphology from the cereal-rye grain, despite the possible confusion due to their similar common names. There are a few species of ryegrass—all rather similar in appearance: annual ryegrass (*Lolium rigidum*) and perennial ryegrass (*Lolium perenne*). Similar morphologically, but slightly larger, is *Lolium temulentum* known as darnel, drake and tares (Ferns et al. 1978).

In texture, the endosperm of the rye grain is softer than that of bread or durum wheat. Thus the floury contents of the rye grain are readily released on milling. However, rye flour tends to clump, presenting the risk of clogging during milling, and a larger sifting area may be required for rye than for wheat milling. Rye grain is milled in Germany to produce a few levels of flour extraction: type 815 flour from 78% to 82% extraction with 0.90 ash content, type 997 flour from 82% to 86% extraction with 0.91–1.10% ash content, type 1150 at even higher extraction (1.11–1.30% ash), and wholemeal flour with 2.20% ash content (Seibel and Weipert, 2001a; Sapirstein and Bushuk, 2016).

FIGURE 7.4 **Illustration of the rye head.** *(Reproduced from Cooper, K.V., 1985. The Australian Triticale Cookery Book, Savvas Publishing, Adelaide, South Australia.)*

7.7 THE COMPOSITION OF THE RYE GRAIN

7.7.1 Polysaccharides

In the rye endosperm, and thus in rye flour, starch granule size is bimodal, that is there is a high proportion by weight of A-granules (15–35 μm in diameter), which have been initiated early in endosperm development, and a high

TABLE 7.3 Functional Qualities Compared for Rye, Compared to Wheat, Although it Must be Realised That Considerable Variations Occur Within Each Species

Attribute	Rye, compared to wheat
Grain	
1000 kernel weight	Similar
Hardness/softness	Much softer
Protein content	Similar range
Sprouting damage (α-amylase)	Much more susceptible
Flour	
Starch granule size	Larger
Starch damage during milling	Much less
Pentosan content	Greater
Water-binding capacity	Up to 8× more water bound
Water-soluble protein	Much more
Gluten quality	Much less effective
Gluten quantity	Less
Dough	
Rheology (viscosity/elasticity)	Poorly stretchable
Dough stickiness	Much more
Gas retention in baking	Much less
Bread	
Grain/texture	Coarser
Flavour	Stronger and characteristic
Shelf life	Poorer

Source: Adapted from Weipert, D., 1996. Rye and triticale. In: Henry, R.J., Kettlewell, P.S. (Eds.), Cereal Grain Quality, Chapman and Hall, London, UK, pp. 205–224 and Chen, A., Bushuk, W., 1970. Nature of proteins in triticale and its parental species. I. Solubility characteristics and amino acid composition of endosperm proteins, Can. J. Plant Sci., 50, 9–14.

proportion by number of much smaller B-granules (diameter less than 10 μm), initiated at a mid-stage of grain development (Simmonds and Campbell, 1976; Verwimp et al., 2004). The bimodal distribution of starch-granule size of rye is similar to that of wheat but overall, rye granules are slightly larger with more of the larger A granules (Table 7.3).

The amylose content (relative to amylopectin) of rye starch has been reported to be comparable to that of conventional wheat (Berry et al., 1971)—in the range of 20–30% amylose. The starch properties of rye are influenced by both genotype and growth conditions. Hansen et al. (2004) reported that starch properties and dietary fibre content were more influenced by growth environment and harvest year than by genotype in a study of 19 rye genotypes during three growth seasons.

The thick cell walls of rye aleurone and endosperm cells provide a rich source of fibre, averaging about 17% of the whole grain, of which about 4% is soluble fibre (Vinkx and Delcour, 1996; Shewry and Bechtel, 2001) (Table 7.3). Rye-related dietary fibre proved to be a significant component in diets surveyed in the European HEALTHGRAIN diversity screen (Nyström et al., 2008). Non-starch polysaccharides include pentosans, β-glucan, lignin and cellulose, plus a small amount of arabinoxylan (Shewry and Bechtel, 2001). The enhanced water-absorbing properties of rye flour (or wholemeal) are due largely to the enriched level of its non-starch polysaccharides.

Rye arabinoxylans, in particular, are capable of high water-binding capacity, of the order of nearly 5 g of water per gram of dry weight, forming a viscous solution. These factors contribute to the nutritional advantages of rye as a human food source and in food processing (Vinkx and Delcour, 1996), but make for difficulties in animal feeding. The high water-binding capability of rye may also contribute to the poor dough properties of rye and to dough stickiness via interactions with the gluten proteins of rye (Dhaliwal et al., 1988).

7.7.2 Protein and Amino-Acid Composition

The protein content of rye grain is similar to that of wheat, although in both cases, there are considerable variations (say, 8–15%) depending mainly on growth conditions (Békés and Wrigley, 2016). Variations in grain protein content have been reported to be more influenced by genotype than by growth conditions (Hansen et al., 2004). The storage protein of rye endosperm is classed as forming gluten on wetting, but the resistance to stretching of the resulting dough is less that that conferred by wheat gluten (from either common or durum wheat) (Table 7.3), even though rye flour has been reported to have a higher content of gluten than wheat flour (Chen and Bushuk, 1970). Such anomalies may result from different analytical methods.

Machine processing of rye dough is more difficult than for wheat dough due to the stickiness of rye dough. The resulting baked rye loaf is poorer in volume, with coarser crumb structure, but the distinctive flavour of rye makes it especially attractive to many customers. Rye flour is often mixed with wheat flour in baking applications thus overcoming the dough-strength inadequacies of rye, whilst retaining the unique flavour advantages of rye (Seibel and Weipert, 2001a).

The monomeric gluten proteins of rye are called secalins (equivalent to wheat gliadins) and the polymeric gluten proteins (polypeptides or subunits) take the generic name of glutelin (equivalent to wheat glutenin). The genes for the synthesis of the rye gluten polypeptides are on the R chromosomes that are homoeologous to those of the three wheat genomes, namely, the short arms of chromosomes 1R and 6R for the secalins (*Sec* loci), and on the short (low-molecular-weight subunits) and long arm (high-molecular-weight subunits) of chromosome 1R for the rye glutelin polypeptides (*Glu* loci) (Alonso-Blanco et al., 1993; Tatham and Shewry, 1995; Shewry and Bechtel, 2001).

The polymorphism of the rye and triticale grain proteins is well studied, providing the basis for distinguishing between varieties (Uthayakumaran et al., 2006) and for phylogenetic studies (Tatham and Shewry, 1995; Radzikowski et al., 2002). Such studies at the protein level are matched by analyses at the gene level (Ma et al., 2004; Bartos et al., 2008). Such gene markers have been used effectively to identify breeding lines carrying specific glutenin/glutelin subunits in rye and triticale crosses, especially for subunits that are difficult to distinguish by gel electrophoresis.

The rye genome has been a valuable source of stem-rust resistance genes. However, crosses made to provide these advantages have often involved the consequential incorporation of secalin genes, resulting in sticky-dough problems for the resulting wheat lines (Graybosch, 2001; Békés and Wrigley, 2016). A possible solution for these breeding problems is offered by the isolation of wheat-rye 1RS recombinants that break the linkage between the stem rust resistance gene SrR and secalin (Anugrahwati et al., 2008).

The gluten of rye is considered as being sufficiently similar to that of wheat to be toxic to people with coeliac disease (Wieser and Koehler, 2008; Anderson and Wieser, 2006). Thus triticale grain, having a combination of wheat and rye proteins, is also toxic for coeliacs. However, the grain protein of rye has the nutritional advantage of a lysine content (21.2 g/100 g total N; range 15.1–28.1) that is higher than that of wheat (17.9 g/100 g total N; range 13.1–24.9) (Chen and Bushuk, 1970; Seibel and Weipert, 2001b; Shewry and Bechtel, 2001), offering a minor advantage for rye in animal feeding. Rye has slightly less than wheat of the nitrogen-rich amino acids, such as glutamine and proline, but the conversion factor of 5.7 is still used for rye when converting total nitrogen results to protein content. The conversion factor of 5.7 contrasts with the factor of 6.2, used for this conversion with proteins having a more 'normal' amino-acid composition.

7.7.3 Anti-Nutritional Factors and Vitamins

The aleurone cells surrounding the endosperm of the rye grain are rich in phytic acid, present in phytin granules as potassium and magnesium salts (Shewry and Bechtel, 2001). Its anti-nutritional activity relates to its ability to form very insoluble salts with important mineral nutrients, especially calcium, iron, zinc and magnesium. A naturally occurring solution to the problem lies in the endogenous presence of phytase, which hydrolyses phytate to form inositol and phosphate, releasing the complexed cation.

A group of phenolics compounds, represented by 5-alkylresorcinols, occurs in rye bran at a level of about 0.5%. They can inhibit the growth of various animal species, especially pigs, also causing skin irritation (Seibel and Weipert, 2001b; Shewry and Bechtel, 2001). Their presence is one of several factors that limit the proportion of rye grain in animal feedstuff. On the other hand, the outer tissues of rye grain are a rich source of various vitamins and minerals. Their levels in whole rye grain and milling fractions from several literature sources are summarised by Shewry and Bechtel (2001). The presence of trypsin inhibitors in rye provides a further anti-nutritional factor (Shewry and Bechtel, 2001). Although these inhibitors of the digestive processes are present in rye at levels much lower than in many legumes, they constitute another factor to limit the use of rye and triticale in animal feeding.

7.8 GRAIN DEFECTS IN RYE

Rye is particularly susceptible to pre-harvest sprouting, with the risk that there will be excessive levels of α-amylase in the mature grain (Table 7.3). This deficiency is partly attributed to the open structure of the glumes of the rye head allowing easy access of water to the grain in the mature head just prior to harvest (Paulsen and Auld, 2004). The tendency to premature sprouting is accentuated by rye being grown in regions where moist climatic conditions at harvest are often conducive to sprout damage. This defect presents processing difficulties for the manufacture of rye breads and crisp breads. As a result, it is a priority at grain receival to test for sprout damage using the Falling Number test or equivalent procedures.

Contamination of harvested rye grain with ergot bodies (sclerotia) is a characteristic common to rye. Infection is likely to be more severe if ergot ascospores are present from the previous season and if there are windy conditions to spread the spores. Soon after the rye florets have flowered, infected florets in the rye head produce yellow droplets of liquid (honeydew) containing conidia of the ergot fungus (*Claviceps purpurea*) (Czembor and Sowa, 2001). Insects may transfer conidia to other florets where mycelia develop, eventually producing the blue-black ergot bodies (pinkish-white inside), ranging in length from 2 to 20 mm. When this occurs, the ergot body replaces the grain; in other cases, the ergot body may not develop but the floret is sterile, thereby reducing grain yield. Some of the ergot bodies may shatter and fall to the ground before harvest, but most are harvested with the mature grain. Ryegrass (*Lolium* species, Fig. 7.3C) is also susceptible to ergot infection. If there is significant contamination of the rye crop with ryegrass, ergots from the ryegrass may appear with the rye grain as contaminants.

Ergot is toxic if present in sufficient concentration in the diet due to the alkaloids contained in the ergot sclerotia, especially ergotamine, which produces hallucinations, agitation, and other symptoms associated with 'St Anthony's Fire' (De Costa, 2002). In recent times, ergotism is mainly associated with animals due to specifications being stricter for grain for human consumption. In past centuries, ergotism has been the cause of epidemics with millions of deaths in European countries where the poor had rye as a staple of their diet. For animals today, ergot poisoning may cause convulsions and, in chronic conditions, gangrene of the extremities. A full history, biology and management of ergot has recently been provided by Miedaner and Geiger (2015), who go so far as to state that 'the alkaloids produced by the fungus severely affect the health of humans and warm-blooded animals'.

Ergot is understood to be especially dangerous for breeding stock, affecting lactation and fertility. Therefore, there are strict tolerances for the presence of ergot bodies in rye grain on delivery (Lauber et al., 2008). In the USA, most rye deliveries have been considered 'ergoty' if they contain more than 0.3% ergot (visit www.gipsa.usda.gov/vri/ergot.aspx). Stricter specifications apply to #1 (0.05%) and #2 (0.20) grades of Canadian rye (Seibel and Weipert, 2001a). The mechanical removal of ergot bodies is possible, but of course not after milling.

Snow mould is a further disease that affects rye plants. As the name suggests, it is prevalent in cool climates where the presence of various fungi may greatly reduce grain yield (Czembor and Sowa, 2001). The main consequences for grain quality are a reduction in bulk-grain density and the presence of shrivelled grains.

7.9 UTILISATION AND NUTRITIONAL ADVANTAGES OF RYE

7.9.1 Rye in Food Products

The versatility of rye as a crop is indicated by its value at all stages of growth, thus fulfilling in practice the 'whole plant utilisation' concept of Munck (2016). As a green plant, rye provides livestock pasture and a green manure in crop rotations. At maturity, rye straw is harvested and valued as ruminant fodder and hay, and as animal-bedding material, especially for horses. The use of straw for biofuel production is being explored. Chemurgy of the straw offers opportunities for chemical industries to transform straw pentosans into xylose and furfural (Munck, 2016). Rye straw is sought for thatch and for building materials.

The rye grain has various avenues for utilisation. The highest value addition is milling into flour for baking into various forms of bread and many other foods (Seibel and Weipert, 2001a).

- Rye 'black bread', with a high content of rye flour, is favoured in Eastern Europe. The rye flour used has a high ash content (0.8–1.6%).
- Crisp bread (*Knaeckebrot*) is also made from rye flour, often whole-meal rye, possibly with some wheat flour added, with or without yeast leavening. Because of the processing method, the rye flour for this purpose must have a low level of α-amylase activity.
- Pumpernickel is also produced from 100% rye meal, using the sourdough process (Fig. 7.1). It is distinctive due to the very long baking

time involved—between 18 and 36 h. As a result, it is dark, with a very high dextrin content and a 'bitter-sweet' taste. The bread does not develop a normal crust; it has a long shelf life. Pumpernickel is distinctively German, originating in Westphalia, but there are variations in recipe and baking process from one country to another in (Eastern) Europe. For example, a light pumpernickel (with rye-wheat flour and possibly including a sour-dough process) is produced in North America as an alternative to the traditional product.

- Light-rye bread (Fig. 7.1). Gristing of rye and wheat flours provides lighter bread types (light-rye) and bread rolls for palates around the world. American rye bread is usually made with a very high proportion of wheat flour (60–80%), leaving the rye flour as a relatively minor ingredient.
- Sour-dough bread is also made from a wheat-rye grist, but in this case, 'old dough' is mixed with the grist to produce the characteristic 'sour' savour (Decock and Cappelle, 2005; Michalska et al., 2008).
- Biscuits. In addition to leavened breads, rye is popular for various biscuit and cracker types.
- Breakfast cereals. Rye has also been used in muesli as a breakfast cereal or as a muesli bar, with rye incorporated in the form of the cracked grain, as rye flakes or as extruded rye.
- Whisky. A distinct use of rye is fermentation and distillation into spirits as rye whisky (Seibel and Weipert, 2001b). [The spelling is 'whiskey', if originating in USA or Ireland, or 'whisky' if from Scotland, Canada or Japan.]
- Biofuels. Not unrelated, but distinct is the use of rye grain for the production of fuel ethanol and related biofuels (Wang et al., 1999; Sánchez and Cardona, 2008). For this application, the higher activity of α-amylase associated with sprouted rye grain is an unexpected advantage. In particular, winter rye has been identified as a bioenergy feedstock of special interest (Shao et al., 2015). Corn grain has been the mainstay for fuel ethanol production from the renewable resource of cereal grain, but the whole range of grain species has been examined for this purpose in recent years. In addition, cereal straw has been studied as a source of lignocellulosic material (Sun and Cheng, 2002). The main reason for considering rye relates to its regional availability and price compared to alternative grains and other uses, especially animal feed.
- Rye flour also finds a range of industrial uses, including the production of glue, adhesives, corrugated cardboard and papers. In some of these applications the distinct properties of rye starch offer advantages over wheat starch. An example is its use as a diluent in veneer and plywood manufacture (Seibel and Weipert, 2001b).
- Feed. Finally, the most obvious use of rye grain is for animal feed, generally for formulation with other grains. The offal from milling rye grain to flour also goes to animal feed.

7.9.2 Nutritional Advantages of Rye-Based Foods

The use of whole-grain flour for baking is advocated for all cereals, including rye, especially because whole grains are rich sources of fibre, vitamins, minerals and bioactive phytochemicals (Liu, 2007; Nyström et al., 2008). The phytochemicals of cereals are reported to complement those from other plant sources, such as fruits.

The high fibre content of rye-based foods is a distinct nutritional advantage for rye compared to wheat. Rye often has twice the pentosan content of the equivalent wheat-based food (Sapirstein and Bushuk, 2016). The progressive decrease in fibre content is shown in Fig. 7.1, going from pumpernickel to light rye bread. The increases in density are also evident by comparing the masses of the portions shown.

Some rye breads have a low Glycaemic Index, offering further nutritional advantages of these bioactives. It has even been claimed that in vivo antioxidant activity from cereal brans is greater than published values from in vitro studies (Fardet et al., 2008). The phenolic acids that are abundant in rye inhibit low-density lipoprotein oxidation. Michalska et al. (2008) reported specifically about antioxidant activity for rye breads, claiming that the baking process favours the formation of antioxidant compounds via the Maillard reaction. The nutritional and health advantages of phenolics in the human diet (phenolics acids, lignans and alkylresorcinols) have recently been reviewed by Bondia-Pons et al. (2009).

7.9.3 Rye as a Feed Grain

Rye has worldwide importance as a feed grain. This role, versus food use, is partly because it can be grown where other cereals may not provide good returns. As a result, rye production has exceeded the needs of local milling and baking industries, and the need has arisen to exploit the obvious alternative, namely, feed uses. Even in Germany, a classic producer of rye for bread manufacture, about two thirds of rye production goes for animal feed, generally mixed with other feed sources. At face value, the improved amino-acid balance of rye-grain proteins should make it preferable to other cereal grains for animal feeding, but in practice, this advantage is not necessarily a significant reason for feed manufacturers to pay a premium for rye grain. Nevertheless, prices and local availability make rye attractive for animal feed, generally mixed with other sources, including other grain species.

The presence of anti-nutritional factors (especially non-starch polysaccharides—mainly pentosans) in rye restricts its use in feed formulation; it must be combined with other feed sources. For chickens, a diet based largely on rye may lead to poorer weight gain and lower metabolisable energy (Ragaee et al., 2001; Mathlouthi et al., 2002). These problems have been attributed to increased viscosity properties due to non-starch polysaccharides, especially arabinoxylans. They create problems for poultry because of their high water-holding capacity, which leads to the formation of highly viscous substances in the gut of the bird or animal, there causing interference with the absorption of micronutrients (Choct, 2001). This undesirable effect can be mitigated by the supplementation of the rye diet with small quantities of pentosan-degrading enzymes, namely, xylanase and β-glucanase. These enzymes breakdown the pentosans, thus reducing viscosity and improving growth rate (Mathlouthi et al., 2002; Silva and Smithard, 2002). Commercial quantities of several enzyme preparations are now available to the mixed feed industry.

In addition, the 5-alkyl resorcinols are reputed to suppress growth in pigs, causing skin irritation and a condition colloquially known as 'thick blood condition'. These phenolic compounds are present in all cereal grains, but because their content is especially high in rye grain, the use of rye in animal feed rations is limited, especially for young pigs (Seibel and Weipert, 2001b). These compounds are concentrated in the bran layers, and their levels vary from one variety to another.

7.10 DISTINCTIVE ASPECTS OF ANALYSIS FOR RYE GRAIN

For most aspects of grain quality, assessment for rye is similar to that for other cereals, especially for wheat. Therefore, see Chapter 5 on wheat and the general chapters—Chapters 15–24. The following basic specifications apply to receivals of rye grain. Specific levels of these various parameters are provided by the respective grain-receiving organisations or companies; specifications may vary from one region to another (see 'Sources of further information').

Test weight (generally as kilograms per hectolitre) is an indication of milling quality based on grain size and the absence of shrivelled grains. Measurement involves weighing a specific volume of grain (details in Chapter 19). Test weight is a primary specification determining the grade level for rye-growing countries. A test-weight minimum of 72 kg/hL applies to the Canadian top grade of rye, falling to 69 and 63, respectively, for the two lower grades. These are similar to EU rye standards of 71 and 69 kg/hL. US rye grades are specified in pounds per bushel, namely, 56, 54, 52, down to 49 lb/bu. These convert into metric measures similar to the Canadian and EU specifications for rye. See Appendix 2 for US-metric conversions for weights and measures relevant to the grains industry, including conversion between kg/hL and lb/bu.

The many specifications for grades and classes are not static for the many aspects of grain quality, so the figures are likely to change with time. Up-to-date specifications are best obtained directly from the relevant grain-handling authority, for example from the respective website (see 'Sources of further information').

Variety or varietal mix determines quality attributes that have been 'built in' by the breeder. They are important in determining harvested grain quality, even though these genetically determined attributes may have been modified by growth conditions and farm management. Such environmental factors include climate, rainfall, fertiliser and harvest conditions. Methods of determining variety have traditionally involved visual examination by an expert (Jarman, 1995). However, laboratory methods are needed for definitive identification, such as the determination of grain-protein composition by gel electrophoresis or capillary electrophoresis (e.g. Uthayakumaran et al., 2006) (see Chapter 17).

Grain protein and moisture content are common specifications for admission of grain to a specific grade or class. Near infrared (NIR) spectroscopy of ground or whole grain is the usual procedure for determining both these factors (see Chapter 19).

Grain soundness (lack of sprouting) is especially important for rye, given its susceptibility to pre-harvest sprouting. Visual examination is effective in advanced cases of sprouting (the presence of roots and shoots, or split germ cover), but significant sprout damage may not be visually evident even when α-amylase activity is sufficient to cause processing problems. Inspection for visibly sprouted grains is the primary specification for the three main grades of Canadian rye covering the wide range of 0.5–10%. The latter high level indicates the difficulty of dealing with this intrinsic defect in rye, considerably more troublesome than in wheat.

Judging the extent of sprouting visually is difficult in borderline cases; a more objective evaluation of sprouting as a defect is provided by laboratory-based testing to determine α-amylase activity, using either the effect of the activity on endogenous starch as Falling Number or Stirring Number (AACC, 2002; ICC, 2002), or based on the determination of enzymic activity (Seibel and Weipert, 2001a; Wrigley, 2004) (see Chapter 19). The critical cut-off Falling Number value for rye is generally lower than is set for wheat or barley. Irrespective of testing, the likelihood of sprouted-grain deliveries may be anticipated by knowledge of the climate (rainfall and temperature) at harvest time for the growing area of a specific grain delivery.

Grain soundness also involves the absence of mouldy, frosted or damaged grains, as is evident by visual inspection. The US standards list the range of 2–15% damaged kernels for grades 1 to 4 of rye. These include grains that are diseased, insect-bored, heat-damaged and other possibilities termed 'otherwise materially damaged'.

Related to test weight is examination of a grain sample for 'unmillable material' by sieving using sieves with slot/hole sizes specified for the grade or class. Material larger than the test screen usually includes pieces of the head and stem, stones and other foreign material. Small material includes pinched grains, weed seeds and insects.

The presence of insects relevant to grain, or the presence of insect-damaged grain, may be sufficient to warrant the rejection of a grain load, or at least the requirement that the grain must be fumigated before it can be considered for receival. Absence of grain protectants below minimum residue levels is another requirement, but assessment of this specification may require laboratory testing (see Chapter 20).

The absence of contaminants is critical, depending on the nature of the contaminating material. Assessment involves visual examination, probably after sieving to concentrate the small material from a given volume of grain. Seeds of weeds and of other crops are common and levels admissible are specified for most grades of grain. Examples of these limits for a few grain-producing countries are given by Seibel and Weipert (2001a) and McLean and Wrigley (2016). Booklets illustrating likely seed contaminants are provided for specific regions, for example Ferns et al. (1978) for Australian cereal grains.

Ergot bodies are a particular concern for rye deliveries. The ergots are visually evident, unless they have been broken up during harvesting and transport. In this case, and for ground grain and milled flour, laboratory analysis is required, by methods such as described by Lauber et al. (2008). Local specifications vary from one region to another and from one grade to another, ranging up to 0.3% by weight. Specifications are obviously likely to be higher for feed grades than for grain destined for milling and food products.

Specifications for feed grade are less restrictive than for milling grades. Nevertheless, there is concern about the presence of ergot and of moulds and contaminants that could prove toxic. See the review by Wrigley (1999) for methodologies to evaluate grain quality for animal and bird feeding.

Whereas initial assessment of grain quality is performed by the above visual-examination procedures and by simple tests (e.g. test weight), more detailed assessment may require laboratory-based analysis, provided the costs involved are warranted. A wide range of such tests is available, many of them standardised by inter-laboratory collaboration, and endorsed by relevant grain-science associations, especially AACC International (AACC, 2002) and the ICC (2002). See 'Sources of further information', Chapters 19 and 22, and a review of the system for establishing standardised test protocols by Bridges and Wrigley (2016a,b).

7.11 DISTINCTIVE ASPECTS OF MANAGING RYE-GRAIN QUALITY

Opportunities for managing grain quality occur throughout the grain chain, as described in Chapter 1 (summarised in Table 7.1), and as amplified in Chapters 15–21. For all cereal grains, this opportunity begins in the breeding program, based on feed-back from the later stages of production, processing and consumer acceptance. For rye, the main aims of breeding have traditionally been aimed at increasing grain yield in its various aspects, with less emphasis on grain quality than for wheat or barley. Nevertheless, breeding offers opportunities for increasing tolerance to the characteristic disadvantages of pre-harvest sprouting and ergot contamination. At the stages of sowing and growing, farm-management strategies include the use of clean seed, freedom from weed seeds and from ergot spores, as well as reducing the presence of grasses on which ergot spores can over-winter between crops.

Grain receival is a critical stage at which the issues of grain quality come face-to-face with returns to the grower, depending on the assessment of quality and thus the grade or class to which a load of grain is allocated. At this stage, the grower receives the benefit for good farm management. At this stage also, the grain buyer and the processer commence quality management, based on the specifications for the grade of grain acquired, and based on the analyses conducted at grain receival. Given this information, the processor proceeds with cleaning (if necessary) and with milling, if proceeding on to food manufacture.

The miller has various strategies available, namely, producing whole-meal rye (by crushing the whole grain, with no sieving), 'gentle' milling to release inner endosperm (rich in starch, with less of the swelling materials—pentosans and proteins), or milling and sieving to provide various degrees of flour yield (extraction rate). These strategies determine the baking qualities expected; in general, heavier rye bread, with smaller loaf volume is expected as the extraction rate increases (Seibel and Weipert, 2001a).

Nevertheless, the use of high-extraction rye flour, or whole-meal flour, is reported to have favourable nutritional advantages—most obviously because of the added dietary fibre, but also because the baking process favoured the formation of antioxidant compounds. These advantages were diminished by using 'white' rye flour (Michalska et al., 2008). On the other hand, there was no significant decrease in the level of available lysine in the whole-meal rye bread.

Bakers' asthma appears to be a management issue that is distinctive for rye milling and baking. This occupational hazard is associated with the inhalation of rye flour, but skin irritation is also a risk for millers and bakers handling rye flour. Bakers' asthma has been reported to be worse with rye than with wheat flour (Letrán et al., 2008). Ramirez and Bahna (2009) reported that 'rye flour has gained increased attention as a causative inhaled allergen in baker's asthma, a phenomenon attributed to its increased cultivation and use.'

A case study of several bakers led to the identification of a baker with a history of more severe attacks of breathlessness following inhalation of rye flour compared with wheat flour (Baldo et al., 1980). This history was confirmed by formal bronchial challenge, but skin-prick testing was not a useful diagnostic. Bran particles were demonstrated to be at least as allergenic as flour for bakers' asthma. Appropriate management involves ensuring that workers exposed to these risks, take suitable precautions, such as the use of dust masks and protective clothing. The risk is reduced in modern, larger-scale mills and bakeries, where pneumatic conveyance of flour stocks limits inhalation and skin exposure.

7.12 FUTURE PROSPECTS

The distinctive agronomic advantages of rye make it an attractive prospect for genetic improvement with respect to its gluten-forming proteins. Traditional attempts to improve the bread-making quality of rye using conventional breeding methods have not brought the hoped-for baking quality. New hope for these aims has been brought by the many new breeding methodologies that are summarised by the title 'genetic engineering', creating genetically modified organisms (GMOs). Most obviously, these approaches have involved the use of transformation to introduce genes for the strength-conferring glutenin polypeptides of common wheat. The feasibility of this approach has been demonstrated in wheat, where the introgression of glutenin subunits resulted in added polymeric glutenin and increased baking quality (Vasil et al., 2001; Field et al., 2008).

For rye, similar studies have involved the introgression of glutenin subunits, especially subunits 1Dx5 and 1Dx10 from the D genome of bread wheat (Altpeter et al., 2004). This approach with rye has lagged behind similar studies on other grain species, due to rye being 'one of the most recalcitrant plant species for tissue culture and genetic transformation' (Altpeter, 2006). Nevertheless, these attempts are showing significant signs of success, with increased levels of the D-genome glutenin polypeptides and indications that these are incorporated into the glutenin polymer complex. Hopefully, this general approach will continue to bear the fruit of better dough strength, thus improving on this deficiency (weak dough) for rye. However, there are other characteristics of rye composition that may militate against the goals of turning rye into a grain with wheat-like bread-making quality; these difficulties include rye's high levels of non-starch polysaccharides. In addition, a transgenic rye would have to overcome perceived customer reticence against GMO material in the human food supply.

Miedaner and Geiger (2015) offer the following prospects for overcoming the ergot problems that have long been associated with rye: 'Molecular-based hybrid rye breeding could improve pollen fertility by introgressing effective restorer genes thus bringing down the ergot infection level to that of conventional population cultivars. A further reduction might be feasible in the future by selecting more resistant germplasm.'

Studies on the genetics of grain-quality factors for rye (Miedaner et al., 2012) have been successful in identifying major quantitative trait loci (QTLs) for several traits, namely, thousand-kernel weight, test weight, falling number and starch content. These had high frequencies of occurrence in cross validation. The expected outcome is that a large proportion of the genotypic variance can be exploited in marker-assisted selection programs, facilitating the genetic improvement of these traits in new rye varieties.

SOURCES OF FURTHER INFORMATION

Several books and reviews of the chemistry, genetics and uses of rye are available—Bushuk (1976), Seibel and Steller (1988), Weipert (1996), Bushuk (2001a,b), Altpeter (2006) and Sapirstein and Bushuk (2016).

For grain-quality specifications, see:

United States standards for rye, Grain Inspection, Packers, and Stockyards Administration, USDA: www.gipsa.usda.gov

Official grain grading guide, Canadian Grain Commission: www.grainscanada.gc.ca

For analytical procedures for processing quality at all stages of the grain chain:

AACC International website: www.aaccnet.org

International Association of Cereal Chemistry (ICC) website: www.icc.org.at

Other relevant websites:

Codex Alimentarius Commission (www.codexalimentarius.net)

Grain production and use statistics: FAOSTAT, accessed via the website at www.fao.org

REFERENCES

AACC, 2002. Approved Methods of the AACC, tenth ed. AACC International, St. Paul, MN, www.aaccnet.org.

Alonso-Blanco, C., Goicoechea, P.G., Roca, A., Giraldez, R., 1993. Genetic linkage between cytological markers and the seed storage protein loci *Sec2* [*Gli-R2*] and *Sec3* [*Gli-R1*]. Theor. Appl. Genet. 87, 321–327.

Altpeter, F., 2006. Rye (*Secale cereale* L.). Methods Mol. Biol. 343, 223–232.

Altpeter, F., Popelka, J.C., Wieser, H., 2004. Stable expression of 1Dx5 and 1Dy10 high-molecular-weight glutenin subunit genes in transgenic rye drastically increases the polymeric glutelin fraction in rye flour. Plant Mol. Biol. 54, 783–792.

Anderson, R.P., Wieser, H., 2006. Medical applications of gluten-composition knowledge. In: Wrigley, C., Bekes, F., Bushuk, W. (Eds.), Gliadin and Glutenin: The Unique Balance of Wheat Quality. AACC International, St Paul, MN, USA, pp. 387–409.

Anugrahwati, D., Shepherd, K., Verlin, D., Zhang, P., Mirzaghaderi, G., Walker, E., Francki, M., Dundas, I., 2008. Isolation of wheat-rye 1RS recombinants that break the linkage between the stem rust resistance gene *SrR* and secalin. Genome 51 (5), 341–349.

Baldo, B.A., Krilis, S., Wrigley, C.W., 1980. Hypersensitivity to inhaled flour allergens. Allergy 35, 45–56.

Bartos, J., Paux, E., Kofler, R., Havránková, M., Kopecký, D., Suchánková, P., Safár, J., Simková, H., Town, C.D., Lelley, T., Feuillet, C., Dolezel, J., 2008. A first survey of the rye (*Secale cereale*) genome composition through BAC end sequencing of the short arm of chromosome 1R. BMC Plant Biol. 8, 95.

Békés, F., Wrigley, C.W., 2016. The protein chemistry of cereal grains. Wrigley, C., Corke, H., Seetharaman, K., Faubion, J. (Eds.), Encyclopedia of Food Grains, vol. 2, Elsevier Ltd, Oxford, UK, p. 98.

Békés, F., Gianibelli, M.C., Wrigley, C.W., 2016. Proteins of the wheat grain in relation to flour quality. Wrigley, C., Corke, H., Seetharaman, K., Faubion, J. (Eds.), Encyclopedia of Food Grains, vol. 3, Elsevier Ltd, Oxford, UK, p. 375.

Berry, C.P., D'Appolonia, B.L., Gilles, K.A., 1971. The characterization of triticale starch and its comparison with starches of rye, durum, and HRS wheat. Cereal Chem. 48, 415–427.

Bondia-Pons, I., Aura, A.-M., Vuorela, S., Kolehmainen, M., Mykkanen, H., Poutenen, K., 2009. Rye phenolics in nutrition and health. J. Cereal Sci. 49, 323–336.

Bridges, A.R., Wrigley, C.W., 2016a. Standardized testing methods for the analysis of grain and grain-based foods. Wrigley, C., Corke, H., Seetharaman, K., Faubion, J. (Eds.), Encyclopedia of Food Grains, vol. 2, Elsevier Ltd, Oxford, UK, p. 300.

Bridges, A.R., Wrigley, C.W., 2016b. Appendix 2. Lists of standardized testing methods for the analysis of grain and grain-based foods. Wrigley, C., Corke, H., Seetharaman, K., Faubion, J. (Eds.), Encyclopedia of Food Grains, vol. 4, Elsevier Ltd, Oxford, UK, p. 450.

Bushuk, W. (Ed.), 1976. Rye: Production, Chemistry and Technology. first ed. AACC International, St. Paul, MN.

Bushuk, W. (Ed.), 2001. Rye: Production, Chemistry and Technology. second ed. AACC International, St. Paul, MN.

Bushuk, W., 2001b. Rye production and uses worldwide. Cereal Food World 46, 70–73.

Chen, A., Bushuk, W., 1970. Nature of proteins in triticale and its parental species. I. Solubility characteristics and amino acid composition of endosperm proteins. Can. J. Plant Sci. 50, 9–14.

Choct, M., 2001. Enzyme supplementation of poultry diets based on viscous cereals. In: Bedford, M.R., Partridge, G.C. (Eds.), Enzymes in Farm Animal Nutrition,. CAB International, Wallingford, UK, pp. 145–160.

Czembor, H.J., Sowa, W., 2001. Diseases and pests of rye. In: Bushuk, W. (Ed.), Rye: Production, Chemistry and Technology. second ed. AACC International, St. Paul, MN, pp. 37–52.

De Costa, C., 2002. St Anthony's fire and living ligatures: a short history of ergometrine. The Lancet 359, 1768–1770.

Decock, P., Cappelle, S., 2005. Bread technology and sourdough technology. Trends Food Sci. Technol. 16, 113–120.

Dhaliwal, A.S., Mares, D.J., Marshall, D.R., Skerritt, J.H., 1988. Protein composition and pentosan content in relation to dough stickiness of 1B/1R translocation wheats. Cereal Chem. 65, 143–149.

Fardet, A., Rock, E., Remesy, C., 2008. Is the in vitro antioxidant potential of whole-grain cereals and cereal products well reflected in vivo? J. Cereal Sci. 48, 258–276.

Ferns, G.K., Fitsimmons, R.W., Martin, R.H., Wrigley, C.W., 1978. Australian Wheat Varieties, Supplement No. 1: Eleven Recent Wheats, Grain Defects and Appendixes on Foreign Seeds and Insects. CSIRO, Melbourne.

Field, M.J., Bhandari, D., Bonet, A., Underwood, C., Darlington, H., Shewry, P., 2008. Introgression of transgenes into a commercial cultivar confirms differential effects of HMW subunits 1Ax1 and 1Dx5 on gluten properties. J. Cereal Sci. 48, 457–463.

Graybosch, R.A., 2001. Uneasy unions: quality effects of rye chromatin transfers to wheat. J. Cereal Sci. 33, 3–16.

Hansen, H.B., Moller, B., Andersen, S.B., Jorgensen, J.R., Hansen, A., 2004. Grain characteristics, chemical composition, and functional properties of rye (*Secale cereale* L.) as influenced by genotype and harvest year'. J. Agric. Food Chem. 52, 2282–2291.

ICC, 2002. Standard Methods. International Association for Cereal Science and Technology, Vienna, Austria, www.icc.org.at.

Jarman, R.J., 1995. Visual identification by grain characteristics. In: Wrigley, C.W. (Ed.), Identification of Food-Grain Varieties. AACC International, St. Paul, MN, pp. 35–54.

Lauber, U., Schnaufer, R., Gredziak, M., Kiesswetter, Y., 2008. Analysis of rye grains and rye meals for ergot alkaloids. Mycotoxin Res. 21, 258–262.

Letrán, A., Palacin, A., Barranco, P., Salcedo, G., Pascual, C., Quirce, S., 2008. Rye flour allergens: an emerging role in baker's asthma. Am. J. Ind. Med. 51, 324–428.

Liu, R.H., 2007. Whole grain phytochemicals and health. J. Cereal Sci. 46, 207–219.

Ma, R., Yli-Mattila, T., Pulli, S., 2004. Phylogenetic relationships among genotypes of worldwide collection of spring and winter ryes (*Secale cereals* L.) determined by RAPD-PCR markers. Hereditas 140, 210–221.

Mathlouthi, N., Lallès, J.P., Lepercq, P., Juste, C., Larbier, M., 2002. Xylanase and beta-glucanase supplementation improve conjugated bile acid fraction in intestinal contents and increase villus size of small intestine wall in boiler chickens fed a rye-based diet. J. Animal Sci. 80, 2773–2779.

McLean, J., Wrigley, C.W., 2016. Contaminants of grains. Wrigley, C., Corke, H., Seetharaman, K., Faubion, J. (Eds.), Encyclopedia of Food Grains, vol. 4, Elsevier Ltd, Oxford, UK, p. 83.

Michalska, A., Amigo-Benavent, M., Zielenski, H., del Castillo, M.D., 2008. Effect of bread making on formation of Maillard reaction products contributing to the overall antioxidant activity of rye bread. J. Cereal Sci. 48, 123–132.

Miedaner, T., Geiger, H.H., 2015. Biology, genetics, and management of ergot (*Claviceps* spp.) in rye, sorghum, and pearl millet. Toxins 7 (3), 659–678.

Miedaner, T., Hübner, M., Korzunm, V., Schmiedchen, B., Bauer, E., Haseneyer, G., Wilde, P., Reif, J.C., 2012. Genetic architecture of complex agronomic traits examined in two testcross populations of rye (*Secale cereale* L.). BMC Genomics 13, 706.

Morrison, L.A., 2016. Evolution of species. Wrigley, C., Corke, H., Seetharaman, K., Faubion, J. (Eds.), Encyclopedia of Food Grains, vol. 1, Elsevier Ltd, Oxford, UK, p. 86.

Morrison, L.A., Wrigley, C.W., 2016. Taxonomic classification of grain species. Wrigley, C., Corke, H., Seetharaman, K., Faubion, J. (Eds.), Encyclopedia of Food Grains, vol. 1, Elsevier Ltd, Oxford, UK, p. 31.

Munck, L., 2016. Whole plant utilization. Wrigley, C., Corke, H., Seetharaman, K., Faubion, J. (Eds.), Encyclopedia of Food Grains, vol. 3, Elsevier Ltd, Oxford, UK, p. 228.

Naranjo, T., Fernfindez-Rueda, P., 1991. Homoeology of rye chromosome arms to wheat. Theor. Appl. Genet. 82, 577–586.

Nyström, L., Lampi, A.-M., Andersson, A.A.M., Kamal-Eldin, A., Gebruers, K., Courtin, C.M., Delcour, J.A., Li, L., Ward, J.L., Fraś, A., Boros, D., Rakszegi, M., Bedö, Z., Shewry, P.R., Piironen, V., 2008. Phytochemicals and dietary fiber components in rye varieties in the HEALTHGRAIN diversity screen. J. Agric. Food Chem. 56, 9758–9766.

Paulsen, G.M., Auld, A.S., 2004. Preharvest sprouting of cereals. In: Benech-Arnold, R.L., Sánchez, R.A. (Eds.), Handbook of Seed Physiology: Applications to Agriculture. Haworth Press, Binghamton, NY, pp. 199–219.

Radzikowski, L., Nesic, L., Hansen, H.B., Jacobsen, S., Sondergaard, I., 2002. Comparisons of ethanol-soluble proteins from different rye (*Secale cereale*) varieties by two-dimensional electrophoresis. Electrophoresis 23, 4157–4166.

Ragaee, S.M., Campbell, G.L., Scoles, G.J., McLeod, J.G., Tyler, R.T., 2001. Studies on rye (*Secale cereale* L.) lines exhibiting a range of extract viscosities. 2. Rheological and baking characteristics rye and rye/wheat blends and feeding value for chicks of wholemeals and breads. J. Agric. Food Chem. 49, 2446–2453.

Ramirez, D.A., Bahna, S.L., 2009. Food hypersensitivity by inhalation. Clin. Mol. Allergy 7, 4.

Sánchez, Ó.J., Cardona, C.A., 2008. Trends in biotechnological production of fuel ethanol from different feedstocks. Bioresource Technol. 99, 5270–5295.

Sapirstein, H.D., Bushuk, W., 2016. Rye grain: its genetics, production, and utilization. Wrigley, C., Corke, H., Seetharaman, K., Faubion, J. (Eds.), Encyclopedia of Food Grains, vol. 1, Elsevier Ltd, Oxford, UK, p. 159.

Schlegel, R.H.J., 2014. Rye: Genetics, Breeding and Cultivation. CRC Press, Boca Raton, FL.

Scoles, G.J., Evans, L.E., 1978. Hybrid vigor obtained by crossing inbred lines of spring rye. Can. J. Plant Sci. 58, 745–751.

Seibel, W., Steller, W. (Eds.), 1988. Rye—Production, Processing, Trade (in German). Behrs Verlag, Hamburg.

Seibel, W., Weipert, D., 2001a. Breadmaking and other food uses around the world. In: Bushuk, W. (Ed.), Rye: Production, Chemistry and Technology. second ed. AACC International, St. Paul, MN, pp. 147–211.

Seibel, W., Weipert, D., 2001b. Animal feed and industrial uses. In: Bushuk, W. (Ed.), Rye: Production, Chemistry and Technology. second ed. AACC International, St. Paul, MN, pp. 213–233.

Shao, X., DiMarco, K., Richard, T.L., Lynd, L.R., 2015. Winter rye as a bioenergy feedstock: impact of crop maturity on composition, biological solubilization and potential revenue. Biotechnol. Biofuels 8, 35.

Shewry, P.R., Bechtel, D.B., 2001. Morphology and chemistry of the rye grain. In: Bushuk, W. (Ed.), Rye: Production, Chemistry and Technology. second ed. AACC International, St. Paul, MN, pp. 69–127.

Shewry, P.R., Halford, N.G., Lafiandra, D., 2006. The high-molecular-weight subunits of glutenin. In: Wrigley, C., Bekes, F., Bushuk, W. (Eds.), Gliadin and Glutenin: The Unique Balance of Wheat Quality. AACC International, St Paul, MN, USA, pp. 143–169.

Silva, S.S., Smithard, R.R., 2002. Effect of enzyme supplementation of a rye-based diet on xylanase activity in the small intestine of broilers, on intestinal crypt cell proliferation and on nutrient digestibility and growth performance of the birds. Brit. Poultry Sci. 43 (2), 274–282.

Simmonds, D.H., Campbell, W.P., 1976. Morphology and chemistry of the rye grain. In: Bushuk, W. (Ed.), Rye: Production, Chemistry and Technology. first ed. AACC International, St. Paul, MN, USA, pp. 63–110.

Singh, N.K., Shepherd, K.W., McIntosh, R.A., 1990. Linkage mapping of genes for resistance to leaf, stem and stripe rusts and ω-secalins on the short arm of rye chromosome 1R. Theor. Appl. Genet. 80, 609–616.

Sun, Y., Cheng, J., 2002. Hydrolysis of lignocellulosic materials for ethanol production: a review. Bioresource Technol. 83, 1–11.

Tatham, A.S., Shewry, P.R., 1995. The S-poor prolamins of wheat, barley and rye. J. Cereal Sci. 22, 1–16.

Uthayakumaran, S., Wrigley, C.W., Batey, I.L., Bushuk, W., Lookhart, G., 2006. Genotype identification. In: Wrigley, C.W., Bekes, F., Bushuk, W. (Eds.), Gliadin, Glutenin: The Unique Balance of Wheat Quality. AACC International, St Paul, MN, pp. 307–331.

Vasil, I.K., Bean, S., Zhao, J., McCluskey, P., Lookhart, G., Zhao, H.-P., Altpeter, F., Vasil, V., 2001. Evaluation of baking properties and gluten protein composition of field grown transgenic wheat lines expressing high molecular weight glutenin gene 1Ax1. J. Plant Physiol. 158, 521–528.

Verwimp, T., Vandeputte, G.E., Marrant, K., Delcour, J.A., 2004. Isolation and characterisation of rye starch. J. Cereal Sci. 39, 85–90.

Vinkx, C.J.A., Delcour, J.A., 1996. Rye (*Secale cereale* L.) arabinoxylans: a critical review. J. Cereal Sci. 24, 1–14.

Wang, S., Thomas, K.C., Sosulski, K., Ingledew, W.M., Sosulski, F.W., 1999. Grain pearling and very high gravity (VHG) fermentation technologies for fuel alcohol production from rye and triticale. Process Biochem. 34, 421–428.

Weipert, D., 1993. Sprouting damage and processing value of rye. Effects of alpha-amylase activity and pentosan content. In: Walker-Simmons, M.K., Reid, J.L. (Eds.), Pre-Harvest Sprouting in Cereals. AACC International, St. Paul, MN, pp. 448–454.

Weipert, D., 1996. Rye and triticale. In: Henry, R.J., Kettlewell, P.S. (Eds.), Cereal Grain Quality. Chapman and Hall, London, UK, pp. 205–224.

Wieser, H., Koehler, P., 2008. The biochemical basis of celiac disease. Cereal Chem. 85, 1–13.

Wrigley, C.W., 1999. Potential methodologies and strategies for the rapid assessment of feed-grain quality. Aust. J. Agric. Res. 50 (5), 789–805.

Wrigley, C.W., 2004. Sprouted grain: detection and aberrant results. Cereal Food World 49, 169–171.

Chapter 8

Triticale: Grain-Quality Characteristics and Management of Quality Requirements

Colin Wrigley* and Walter Bushuk**

**QAAFI, University of Queensland, St. Lucia, QLD, Australia; **Department of Food Science, University of Manitoba, Winnipeg, MB, Canada*

8.1 INTRODUCTION

Triticale was originally intended to be a new bread-making grain, for use in all the food-producing applications of wheat. However, it has not fulfilled these hopes, and its main usage is as a feed grain. Nevertheless, the flour from most triticale varieties has the potential for use as a replacement of rye in breads based on composite wheat-rye flours, in that situation providing the characteristic rye flavour and texture to the bread products. The wide range of these food products is illustrated in Fig. 8.1.

Triticale obviously reflects the properties of the two contributing parents—rye (the R genome) and wheat (mainly the A and B genomes, thus durum wheat). Nevertheless, the respective contributions may differ from one triticale genotype to another, depending on the proportions of wheat and rye chromatin incorporated. Many of the more modern triticale genotypes may have a greater contribution of genes from durum wheat than from rye, and they are more likely to include contributions from the D genome of common wheat. Accordingly, knowledge of bread wheat, durum wheat and rye (Chapters 5–7, respectively) is needed to assess and manage the grain-quality of triticale (the accent of this book).

8.2 THE HISTORY AND GENETICS OF TRITICALE

The genus name allocated to triticale (*x Triticosecale*) reflects the dual origins of this intergeneric hybrid. The inclusion of the letter '*x*' indicates the hybrid origin of triticale. The development of the wheat-rye hybrid triticale arose from

Cereal Grains. http://dx.doi.org/10.1016/B978-0-08-100719-8.00008-5

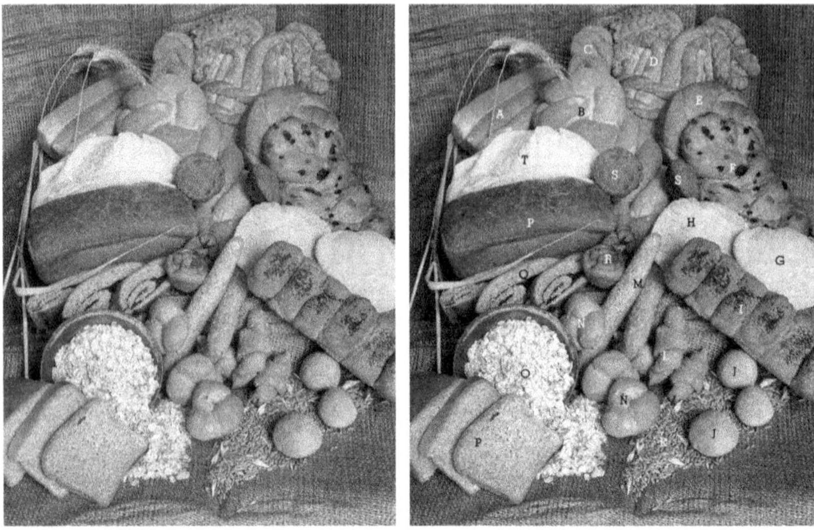

FIGURE 8.1 A variety of breads made from triticale flour, identified by letters. A, Triticale-wheat (hybrid) loaf; B and C, Hybrid plaits; D, Harvest loaf; E, Golden pumpkin loaf; F, Fruit plait loaf; G, Arabic bread, wholemeal; H, Arabic bread, white flour; I, Triticale loaf; J, Rolls; K, Triticale grain; L, Butter horns; M, Soft bread sticks; N, Cloverleaf rolls; O, Triticale flakes; P, Conventional high-top loaf; Q, Poppy seed loaf; R, Jam rolls; S, Triticale flake muffins; T, Triticale flour. *(Reproduced with permission from Cooper, K.V., 1985. The Australian Triticale Cookery Book, Savvas Publishing, Adelaide, South Australia.)*

the close genetic relationship between these two parent species, with the hope of combining the agronomic advantages for rye of winter hardiness and its ability to thrive in poorer soils (including acidic soils containing aluminium), with the hope that the baking quality of wheat would be retained (Tsen, 1974; Bushuk and Larter, 1980; Cooper, 1985; Darvey et al., 1991; Qualset, 2002; Oettler, 2005). This concept was explored in the very early days of cross breeding well over a century ago. The first triticale cross is attributed to the Scottish scientist A.S. Wilson in 1876, but the resulting plant was sterile (Furman, 2016).

In 1898, the Australian chemist F.B. Guthrie reported that his breeder colleague, W. Farrer, had made a rye-wheat hybrid by crossing 'a wheat known as Wilett's with Alabama Rye. The progeny from this cross was then crossed to the wheat Leak's R.R.' (Guthrie, 1898). This hybrid was claimed by Farrer to have 'five-sixteenth rye blood'. The colour of its flour was 'very bad' and its strength 'very low', but its gluten content was 'unusually high (16.5%)'. Guthrie (1896, 1898) included the 'rye-wheat hybrid' in a set of wheats for which he determined the gliadin–glutenin ratio as a function of 'dough strength' (Fig. 8.2).

The apparent gliadin content of the rye–wheat was much higher than that of the true wheats, and its glutenin content was correspondingly very low. The

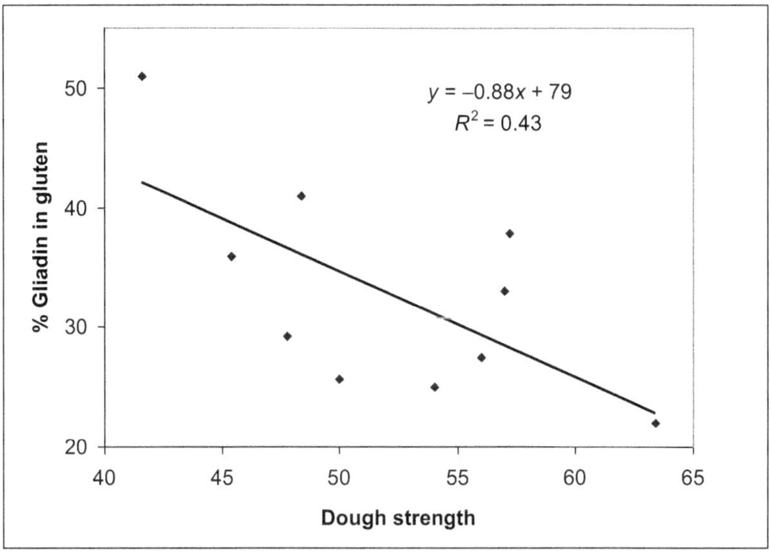

FIGURE 8.2 Dough strength and gluten composition of a 19th century rye-wheat line (the point at top left), compared to nine wheat varieties, as plotted from results reported by Guthrie (1896, 1898).

result for the wheat–rye hybrid reinforced Guthrie's attempt to show that dough strength decreased with higher gliadin content and increased with higher glutenin content. In the light of our more recent knowledge, we would consider that the high 'gliadin' level of the wheat–rye hybrid was the result of the partial substitution of wheat gliadin protein by rye secalin (the corresponding rye storage protein), which would be less effective in conferring gluten-like properties and more likely to cause weaker dough quality.

Successful development of viable triticale genotypes required the discovery and application of colchicine for chromosome doubling (Cooper, 1985). This achievement is attributed to Arne Müntzing from Sweden in 1938 (Furman, 2016), but a few more decades of improvement were needed to overcome agronomic problems. From the 1960s, breeding efforts were pursued in Mexico, Eastern Europe, Canada and the USA, leading to commercial releases in the late 1960s. Since that time, production volumes for triticale have gradually risen (Fig. 8.3) to the recent statistics shown in Table 8.1.

Originally, two triticale types became available for cultivation, namely, octoploid and hexaploid triticale (Bushuk and Larter, 1980). Octoploid triticale resulted from the hybridisation of hexaploid bread wheat (genomes AABBDD) with rye (genome RR) producing germplasm with a $2n$ number of 56 after chromosome doubling. Hexaploid triticale came from hybridising tetraploid durum wheat (genomes AABB) with rye, producing germplasm with a $2n$ number of 42 after chromosome doubling.

These two types are known as primary triticales, often being agronomically poor and genetically unstable. They have been improved by inter-crossing and by crossing back to wheat to produce substituted secondary triticales, which in turn have been further improved by substituting one or more rye chromosomes into the ABD background of wheat (Furman, 2016). Thus some triticale genotypes have a large proportion of wheat chromatin, despite the broad definition of triticale being a wheat–rye hybrid. Like rye and wheat, triticale is available in winter and spring types, depending on the time of sowing and the expected severity of conditions for the winter type. Given relevant growth conditions, triticale can out yield wheat. Or written another way, triticale can thrive in situations where wheat may not (McGoverin et al., 2011).

8.3 WORLDWIDE PRODUCTION AND REGIONAL DISTRIBUTION

Worldwide production of triticale is increasing, as shown by the graph of production from 1993 to 2014 (Fig. 8.3). This figure, with production starting at 5 Mt is not as dramatic as earlier versions which show the unique situation of production starting from zero in the early 1970s (Furman, 2016). Triticale production worldwide is considerably less than that of rye (compare data in Table 7.2) and greatly less than that of its other progenitor, wheat (Table 8.1). This table also shows that some major wheat-producing countries produce negligible amounts of triticale.

The grain-yield data in Table 8.1 show considerable variations in triticale yield, with the highest yields being obtained in European countries which also have high wheat yields. Comparisons with the rye yields in Table 7.2 show that in many countries, triticale yielded better than rye in 2014. However, this general comparison is based in these tables on only one season, and local fluctuations are hidden by the presentation of statistics for a whole country. Therefore, this generalisation may not apply for a specific combination of variety and growth site.

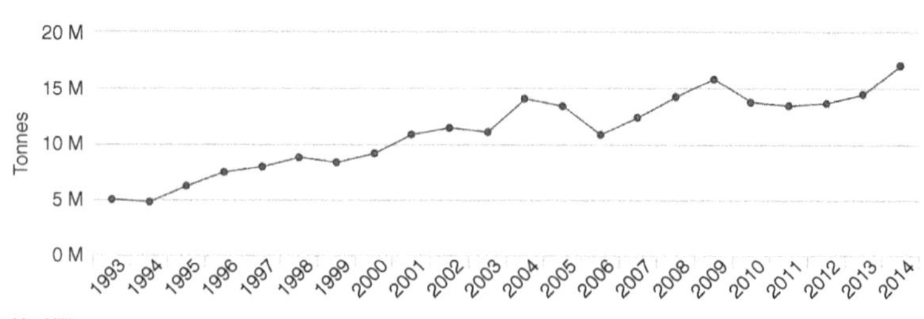

FIGURE 8.3 **World production of triticale (in millions of tonnes).** *(FAO, via www.faostats.fao3.org.)*

TABLE 8.1 Production (in Thousands of Tonnes in 2014) and Yields (in Tonnes per Hectare) of Triticale Compared to Wheat Production for Countries Producing Significant Quantities of Triticale, Plus a few Major Wheat-Producing Countries

Country	Triticale yield	Triticale production	Wheat production
Poland	4.0	5,250	11,630
Germany	7.1	2,970	27,785
Belarus	4.0	2,080	2,920
France	5.2	2,020	38,970
Russian Federation	2.6	650	59,710
China, Mainland	2.5	525	126,200
Hungary	3.9	486	5,260
Spain	2.3	450	6,470
Austria	5.9	300	1,800
Romania	3.6	280	7,590
Czech Republic	5.0	244	5,440
Sweden	5.9	226	3,090
Australia	1.6	126	25,300
Turkey	3.2	110	19,000
Denmark	6.2	100	4,940
Chile	5.1	100	1,360
Bulgaria	3.2	60	5,350
Brazil	2.2	52	6,260
UK	4.5	50	16,620
Belgium	9.0	50	1,990
Portugal	1.6	50	90
Canada	2.6	35	29,280
Tunisia	2.1	30	1,510
Netherlands	6.2	10	1,300
Argentina	—	None reported	13,930
India	—	None reported	94,480
Pakistan	—	None reported	25,980
Ukraine	—	None reported	24,110
USA	—	None reported	55,400

Source: FAO, via www.faostats.fao3.org.

Major producers of triticale include countries that are also major rye producers, particularly Poland, Germany and Belarus (Table 7.2). Exceptions include France, Brazil, Romania, Sweden, Denmark, Australia and Austria which tend to produce more triticale than rye. However, caution is needed in the interpretation of the country comparisons, as the production data are for a single season, and year-to-year variations are expected. World trade in triticale is minor, although it attracts a higher value internationally than rye. Statistics are not available for the consumption of triticale as food, as most utilization of triticale grain is for animal feed (Black, 2016).

8.4 HEAD AND GRAIN MORPHOLOGY

Comparisons of the head morphology of wheat and rye show the expected features, characterising triticale as intermediate between its parents (Figs. 8.4 and 8.5). Durum wheat (*Triticum durum*) is included in these comparisons in addition to bread wheat (*T. aestivum*) because the durum genomes (A and B) are the *Triticum* donors for most cultivated triticale genotypes. Heads of triticale, rye and durum are generally fully awned, whereas many varieties of common wheat are not awned or are only tip awned (Ferns et al., 1978).

Likewise, comparisons of the grain and glume morphology of triticale with those of wheat and rye also show the features for triticale as intermediate between its parents (Fig. 8.6). The durum wheat grain is more elongated that that of bread wheat, and this characteristic is evident in the triticale grain, which is often even longer than the durum grain. Like rye, the surface of the triticale grain is often wrinkled. The pointed germ end of triticale is more characteristic of rye than of either bread or durum wheat. On the other hand, triticale grains are uniformly coloured, not varying in colour as are rye grains.

As with other cereal grains, there are subtle differences between triticale varieties in their head and grain characteristics that can be used to provide visual identification at the variety level (Jarman, 1995). Such differences are evident for the two triticale varieties compared in Figs. 8.5 and 8.7. Systematic procedures for distinguishing among triticale varieties are described by Fitsimmons et al. (1986), based on head and grain morphology. Observation of the shapes of glumes (husks) is not as useful for triticale varieties as it is for wheats.

Whereas durum wheat is offered as the grain to compare with rye when considering triticale morphology, bread wheat is the basis of comparison in Table 8.2 for functional qualities because the great hope for triticale has been that it would offer attributes similar to bread wheat.

8.5 GRAIN COMPOSITION

As the comparison in Table 8.2 shows, triticale grain is generally softer than wheat, thus undergoing less starch damage on milling. However, grain hardness and milling yield equivalent to durum wheat has been observed in some triticale

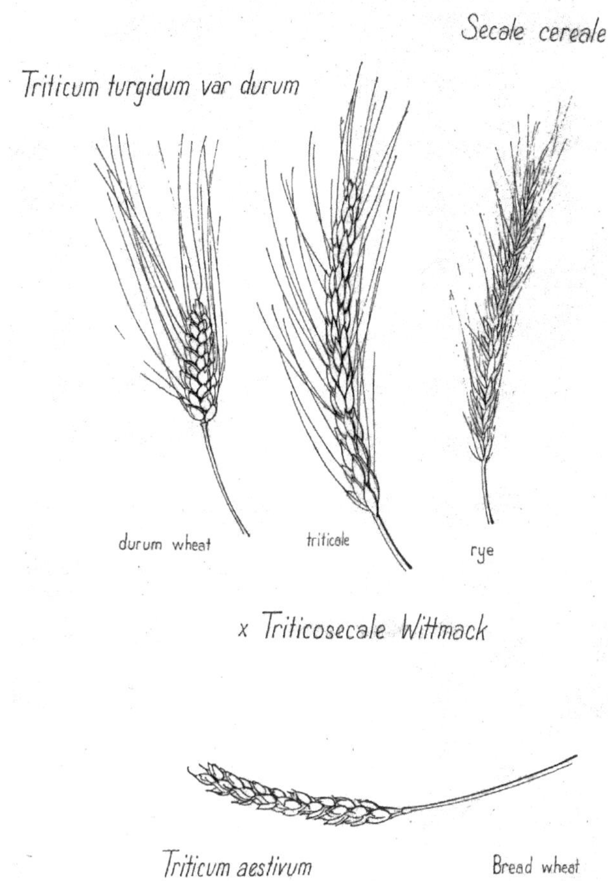

FIGURE 8.4 Head (ear) shapes of durum wheat (left), triticale, rye (right) and an awnless wheat variety (bottom of figure). *(Reproduced with permission from Cooper, K.V., 1985. The Australian Triticale Cookery Book, Savvas Publishing, Adelaide, South Australia.)*

genotypes, suggesting the occasional presence for these of a null allele at the rye softness-protein locus (Pattison and Trethowan, 2013).

The main constituent of triticale grain is obviously starch, the granule size being rather larger than for wheat (Table 8.2) (Berry et al., 1971). The amylose content of triticale starch is generally reported to be similar to that of 'normal' wheat (in the range 20–30%), but Dennett et al. (2009) surveyed a few hundred lines of triticale and found a much wider range of amylose content (13–35%), suggesting that diverse genotypes would be available for breeding purposes if warranted. Sharma et al. (2002) have also reported triticale genotypes that differ significantly in the 'waxy' genes that control amylose–amylopectin ratio.

FIGURE 8.5 **Heads of Australian triticale varieties.** (A) Toort (B) Samson. *(Adapted from Fitsimmons, R.W., Martin, R.H., Roberts, G.L., Wrigley, C.W., 1986. Australian Cereal Identification, Recent Varieties of Wheat, Triticale, Barley and Oats, CSIRO, Melbourne.)*

Cornejo-Ramírez et al. (2016) studied the synthesis of starch and related carbohydrates in developing kernels of two sets of triticales differing in genome composition and they were able to distinguish between the triticale genotypes according to their distinct patterns of starch (etc.) synthesis and polymerisation, leading to the possibility of selecting for factors that influence the chemical structure and physicochemical properties of specific triticales.

Triticale wholemeal has very good water-binding capacity, due to the presence of non-starch polysaccharides from the aleurone and endosperm cell walls. These include pentosans, arabinoxylans, β-glucan, lignin and cellulose (Fincher, 2016). The resulting high fibre content offers advantages for the human diet, but not for animal and poultry feeding. As rye is the main source of fibre enrichment, the extent of this attribute in any specific triticale genotype depends on the proportion of rye chromatin (Bushuk and Larter, 1980).

The dough made from triticale flour is weak, presenting handling problems of stickiness and poorer gas-retention ability compared to wheat flour. Consequently bread-making quality is poorer than that of wheat, with coarser crumb texture and poorer shelf life. These deficiencies are largely due to the composition of the grain storage proteins, triticale grain having comparatively less, weaker gluten-forming proteins and a higher content of water-soluble proteins.

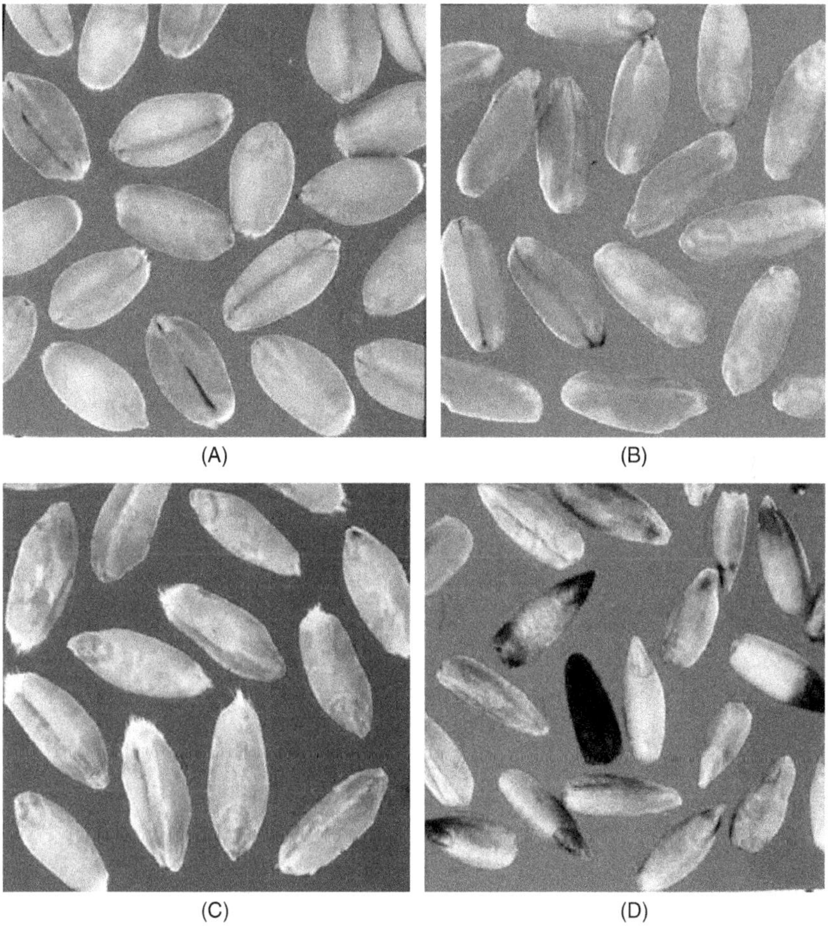

FIGURE 8.6 Grains of (A) common wheat, (B) durum wheat, (C) triticale and (D) cereal rye. *(Adapted from Ferns, G.K., Fitsimmons, R.W., Martin, R.H., Wrigley, C.W., 1978. Australian Wheat Varieties, Supplement No. 1: Eleven Recent Wheats, Grain Defects and Appendixes on Foreign Seeds and Insects, CSIRO, Melbourne.)*

In addition, the rye-related polysaccharides contribute to poor dough quality (see Chapter 7).

The lysine content of triticale is intermediate to that of rye and wheat. Rye and triticale have slightly less than wheat of the nitrogen-rich amino acids, such as glutamine and proline. Nevertheless, the conversion factor of 5.7 is still used for rye and triticale when converting total nitrogen results to protein content, contrasting with the conversion factor of 6.2 for proteins with a more 'normal' amino-acid composition.

The grain proteins of triticale can be readily fractionated by the wide range of protein-fractionation procedures applied to wheat-grain proteins (see

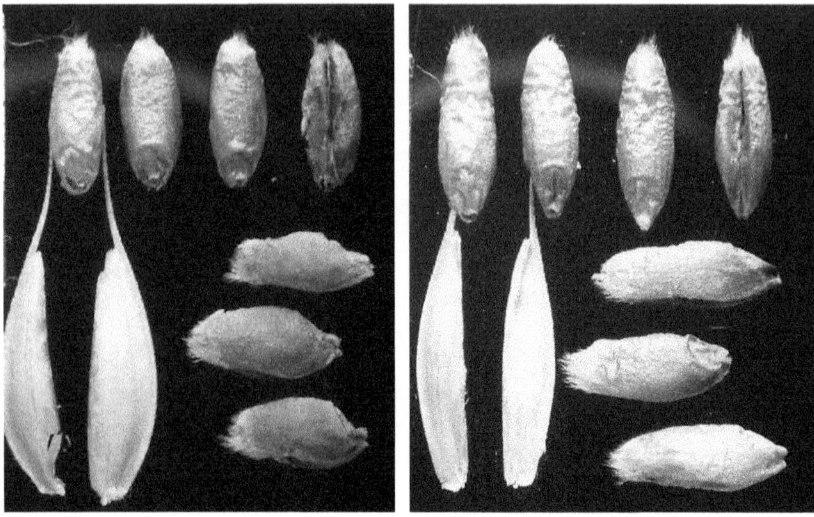

(A) (B)

FIGURE 8.7 Grains and glumes of Australian triticale varieties. (A) Toort (B) Samson. *(Adapted from Fitsimmons, R.W., Martin, R.H., Roberts, G.L., Wrigley, C.W., 1986. Australian Cereal Identification, Recent Varieties of Wheat, Triticale, Barley and Oats, CSIRO, Melbourne.)*

Chapters 5 and 17). In addition, Salmanowicz (2008) has described a procedure for capillary electrophoresis effective for fractionating and characterising the high-molecular-weight subunits of triticale glutenin. This approach assists in the identification of glutenin subunits that have been characterised in wheat and rye as conferring stronger or weaker dough properties (Bekes and Wrigley, 2013).

8.6 UTILISATION AND NUTRITIONAL (DIS)ADVANTAGES

Triticale may be substituted for rye in most of the uses of rye but in practice, rye is the preferred grain. Food uses of rye and triticale, in preference to wheat, reflect the distinctive rye attributes of flavour and the nutritional advantages offered in the human diet by its non-starch polysaccharides (dietary fibre). Triticale's higher lysine content, compared to wheat, is mainly seen as an advantage in its use in animal feed.

The range of uses of triticale is not as wide as for rye. Triticale is potentially suited to these uses, but it is generally used as animal feed, both as green feed and as grain, either alone or formulated with other materials. In addition, triticale is used for a range of baked goods, providing a distinctive 'nutty, aromatic and naturally sweet flavour' (Cooper, 1985). Baking with whole-grain triticale flour has the nutritional advantage of its high fibre content, but the addition of wheat flour to increase gluten strength is recommended for making

TABLE 8.2 Functional Qualities of Triticale, Compared to Wheat in General, but Considerable Variations Occur Within Each Species

Attribute	Compared to wheat, triticale is ...
Grain	
1000 kernel weight	Greater
Hardness/softness	Softer
Protein content	Similar range
Sprouting damage (α-amylase)	More susceptible
Flour	
Starch granule size	Slightly larger
Starch damage during milling	Slightly less
Pentosan content	Greater
Water-binding capacity	Very effective
Water-soluble protein	More
Gluten quality	Less effective
Gluten quantity	Less
Dough	
Rheology (viscosity/elasticity)	Moderately stretchable
Dough stickiness	More
Gas retention in baking	Less
Bread	
Grain/texture	Coarser
Flavour	Slightly stronger
Shelf life	Poorer

Source: Adapted from Weipert, D., 1996. Rye and triticale. In: Henry, R.J., Kettlewell, P.S. (Eds.), Cereal Grain Quality, Chapman and Hall, London, UK, pp. 205–224 and partly based on Chen, A., Bushuk, W., 1970. Nature of proteins in triticale and its parental species. I. Solubility characteristics and amino acid composition of endosperm proteins, Can. J. Plant Sci., 50, 9–14.

yeast-leavened goods (Cooper, 1985). Some of the attractive baked goods from triticale are illustrated in Fig. 8.1.

Although triticale was originally intended to be a new bread-making grain, for use in all the food-producing applications of wheat, it has not fulfilled these hopes, and its main usage is as a feed grain (Weipert, 1996). Nevertheless, the flour from most triticale varieties has the potential for use as a replacement of rye in breads based on composite wheat-rye flours, in that situation providing the characteristic rye flavour and texture to the bread products.

Malting and brewing is a potentially high-value application for triticale (Munoz-Insa et al., 2015). For this application, there are advantages for triticale in its susceptibilities to sprouting and to late-maturity α-amylase production, because of the need in malting for starch-hydrolysis. Accordingly, Munoz-Insa et al. (2015) have examined the metabolic changes in the malting of triticale, thus demonstrating the potential of triticale for the production of malt, beer and other cereal-based products. A related use is the production of bioethanol, the starch of triticale being as relevant to fermentation as various other cereal grains.

However, the fact remains that worldwide, triticale as mainly used as a feed grain. As such, the quality requirements are less stringent than for a food-quality grain. Like rye (see Chapter 7), feed use of triticale may be restricted by the presence of anti-nutritional factors (Madsen and Brinch-Pedersen, 2016), such as non-starch polysaccharides and alkyl resorcinols. For example, triticale's non-starch polysaccharides, especially arabinoxylans, create problems in some feeding uses, for example for poultry, because these polysaccharides have a high water-holding capacity, thus producing a high viscosity for the gut contents interfering with nutrient absorption (Choct, 2001).

The presence of alkyl resorcinols (as for rye) may suppress growth in pigs, causing skin irritation and malabsorption, especially for young pigs (Seibel and Weipert, 2001). Due to these problems, there may be restrictions on the proportion of triticale in feed formulation.

8.7 GRAIN-QUALITY ANALYSIS AND MANAGEMENT; POTENTIAL DEFECTS

To the extent that much of the triticale grown is destined for animal feed, quality considerations are not as stringent as for wheat or rye intended for milling and food use.

Some countries have several grades for triticale, specified primarily according to minimum test weight, ranging for example from 62 to 53 kg/hL for four US grades (Table 5.4 in Wrigley and Bushuk, 2010). These bulk densities are much less than for rye (63–72 kg/hL) or for wheat (73–79 kg/hL), partly at least due to the wrinkling of the outer bran surface of many triticale varieties. Up-to-date specifications are best obtained directly from the relevant grain-handling authority, for example from the respective website (see 'Sources of further information').

Determination of moisture and grain protein content are routine at receival in many triticale-growing regions, respectively being significant for grain storage and for utilisation. Methods of analysis have been described by Manley et al. (2013) and in the extensive descriptions of testing methods for all aspects of grain quality for triticale in the bulletin on Cereal Rye & Triticale Standards of Grain Trade Australia (2016).

Susceptibility to sprouting at harvest is a significant defect of triticale, reflecting its rye parentage rather than being due to the wheat parent. Thus testing for sprout damage by Falling Number, for example is a requirement for specific utilisation requirements. However, a degree of genetic tolerance is being conferred on specific triticale genotypes by suitable selection in breeding (Trethowan et al., 1994; Dennett et al., 2013).

Low Falling Number results may also be due to late-maturity α-amylase (LMA) (Wrigley, 2005; Mares and Mrva, 2008). (It is also known as prematurity α-amylase (PMAA) in Europe.) The genetic defect of LMA causes the development of α-amylase activity in the developing grain and the enzymic activity continues in the grain to harvest ripeness. A result of low Falling Number determined at grain receival may thus be due to either LMA or sprouting (or even both).

Like rye, triticale is susceptible to infection with ergot (Miedaner and Geiger, 2015). Ergot infection of the growing plant may reduce potential grain yields to a modest extent (5–10%), but the greater threat is the resulting contamination of harvested grain by toxic alkaloids. *Claviceps purpurea* produces three ergot alkaloids (clavine alkaloids, D-lysergic acid and its derivatives and ergopeptines) with the potential to cause severe health problems in both humans and animals. Thus the presence of ergot bodies is an important part of examination of grain at receival (see Chapter 7).

Weather-stained grains are considered defects; this condition may be caused by damp weather prior to harvest. They are distinguished as exhibiting various forms of staining, such as dark tipping, poor colour and signs of weather damage.

Grain quality analysis also involves examination for the range of contaminants that are specified in the standards for most cereal grains (see Chapter 19). The following list of contaminants of triticale is taken from the Cereal Rye & Triticale Trading Standards (2016/17 Season) of Grain Trade Australia (2016):

- cereal ergot
- chemicals not approved for triticale
- chemicals in excess of the MRL
- earth
- foreign seeds
- insects—large
- insects—small
- objectionable material
- other foreign material
- pickling compounds or artificial colouring
- ryegrass ergot
- sand
- snails
- stones
- stored-grain insects and pea weevil—live

8.8 FUTURE TRENDS

The opinions of many trade and scientific sources may continue to regard triticale as mainly a feed grain, with no prospects of further improvement or for a sudden increase in production worldwide. However, more positive prospects for genetic improvement are offered by Manley et al. (2013), by Dennett and Trethowan (2013a,b) and by Pattison and Trethowan (2013).

Triticale, in general, still has the reputation of thriving in situations that do not suit wheat. With the ongoing changes in world climate and 'global warming', there may be new opportunities for triticale to find otherwise unforeseen niche situations for its specific agronomic attributes (McGoverin et al., 2011).

SOURCES OF FURTHER INFORMATION

For historical review articles see Tsen, 1974; Bushuk and Larter, 1980; Cooper, 1985; Darvey et al., 1991; Weipert, 1996; Qualset, 2002; Oettler, 2005. For recent reviews, see Pattison (2013) and Furman (2016).

Relevant websites:

- Triticale production in many countries: http://www.fao.org/3/a-y5553e/y5553e02.pdf
- Triticale production in Europe: http://www.triticale-infos.eu/cultivation/market/index.html
- Triticale production in Canada: www.agric.gov.ab.ca; www.gov.mb.ca/agriculture/crops/cereals; www.agriculture.gov.sk.ca
- Triticale production in Australia: http://agriculture.vic.gov.au/agriculture/grains-and-other-crops/crop-production/growing-triticale
- Triticale production in the USA: https://edis.ifas.ufl.edu/an223
- World bank: 'Triticale: A reappraisal' G. Varughese, W.H. Pfeiffer, and R. J. Peña at: www.worldbank.org/html/cgiar/newsletter/april97/8tritic.html

For grain-quality specifications, see:

- United States standards for rye and triticale, Grain Inspection, Packers, and Stockyards Administration, USDA: www.gipsa.usda.gov
- Official grain grading guide, Canadian Grain Commission: www.grainscanada.gc.ca
- Grain Trade Australia (2016), Cereal Rye & Triticale Trading Standards (2016/17 Season). Published by Grain Trade Australia, Sydney. www.graintrade.org.au

For analytical procedures for processing quality at all stages of the grain chain:

- AACC International website: www.aaccnet.org
- International Association of Cereal Chemistry (ICC) website: www.icc.org.at

Other relevant websites:

- Codex Alimentarius Commission (www.codexalimentarius.net)
- Grain production and use statistics: FAOSTAT, accessed via the website at www.fao.org

REFERENCES

Bekes, F., Wrigley, C.W., 2013. Gluten alleles and predicted dough-quality for wheats world-wide: a great resource, free on the AACCI website. Cereal Food World 58 (6), 325–328.

Berry, C.P., D'Appolonia, B.L., Gilles, K.A., 1971. The characterization of triticale starch and its comparison with starches of rye, durum, and HRS wheat. Cereal Chem. 48, 415–427.

Black, J.L., 2016. Animal feed. Wrigley, C., Corke, H., Seetharaman, K., Faubion, J. (Eds.), Encyclopedia of Food Grains, vol. 3, Elsevier Ltd, Oxford, UK, p. 315.

Bushuk, W., Larter, E.N., 1980. Triticale: production, chemistry, and technology. Adv. Cereal Sci. Technol. 3, 115–157.

Choct, M., 2001. Enzyme supplementation of poultry diets based on viscous cereals. In: Bedford, M.R., Partridge, G.C. (Eds.), Enzymes in Farm Animal Nutrition. CAB International, Wallingford UK, pp. 145–160.

Cooper, K.V., 1985. The Australian Triticale Cookery Book. Savvas Publishing, Adelaide, South Australia.

Cornejo-Ramírez, Y.I., Ramírez-Reyes, F., Cinco-Moroyoqui, F.J., Rosas-Burgos, E.C., Martínez-Cruz, O., Carvajal-Millán, E., Cárdenas-López, J.L., Torres-Chavez, P.I., Osuna-Amarillas, P.S. Borboa-Flores, J., Wong-Corral, F.J., 2016. Starch debranching enzyme activity and its effects on some starch physicochemical characteristics in developing substituted and complete triticales (X *Triticosecale* Wittmack), Cereal Chem., 93, 64–70.

Darvey, N.L., Naeem, H., Gustafson, J.P., 1991. Triticale: production and utilization. In: Kulp, K., Ponte, Jr., J.G. (Eds.), Handbook of Cereal Science and Technology. Marcel Dekker, New York.

Dennett, A.L., Trethowan, R.M., 2013a. Milling efficiency of triticale grain for commercial flour production. J. Cereal Sci. 57, 527–530.

Dennett, A.L., Trethowan, R.M., 2013b. The influence of dual-purpose production on triticale grain quality. Cereal Res. Commun. 41 (3), 448–457.

Dennett, A.L., Schofield, P.R., Roake, J.E., Howes, N.K., Chin, J., 2009. Starch swelling power and amylose content of triticale and *Triticum timopheevii* germplasm. J. Cereal Sci. 49, 393–397.

Dennett, A.L., Wilkes, M.A., Trethowan, R.M., 2013. Characteristics of modern triticale quality: the relationship between carbohydrate properties, α-amylase activity, and Falling Number. Cereal Chem. 90 (6), 594–600.

Ferns, G.K., Fitsimmons, R.W., Martin, R.H., Wrigley, C.W., 1978. Australian Wheat Varieties, Supplement No. 1: Eleven Recent Wheats, Grain Defects and Appendixes on Foreign Seeds and Insects, CSIRO, Melbourne.

Fincher, G., 2016. Cereals: chemistry and physicochemistry of non-starchy polysaccharides. Wrigley, C., Corke, H., Seetharaman, K., Faubion, J. (Eds.), Encyclopedia of Food Grains, vol. 2, Elsevier Ltd, Oxford, UK, p. 208.

Fitsimmons, R.W., Martin, R.H., Roberts, G.L., Wrigley, C.W., 1986. Australian Cereal Identification, Recent Varieties of Wheat, Triticale, Barley and Oats, CSIRO, Melbourne.

Furman, B.J., 2016. Triticale: overview. Wrigley, C., Corke, H., Seetharaman, K., Faubion, J. (Eds.), Encyclopedia of Food Grains, vol. 1, Elsevier Ltd, Oxford, UK, p. 168.

Grain Trade Australia, 2016. Cereal Rye & Triticale Trading Standards (2016/17 Season), Grain Trade Australia, Sydney. www.graintrade.org.au.

Guthrie, F.B., 1896. On the absorption of water by the gluten of different wheats. Agric. Gaz. New South Wales 7, 583–589.

Guthrie, F.B., 1898. Notes on the constitution of wheat gluten. Agric. Gaz. New South Wales 9, 363–374.

Jarman, R.J., 1995. Visual identification by grain characteristics. In: Wrigley, C.W. (Ed.), Identification of Food-Grain Varieties. AACC International, St. Paul, MN, pp. 35–54.

Madsen, C.K., Brinch-Pedersen, H., 2016. The antinutritional components of grains. Wrigley, C., Corke, H., Seetharaman, K., Faubion, J. (Eds.), Encyclopedia of Food Grains, vol. 2, Elsevier Ltd, Oxford, UK, p. 283.

Manley, M., McGoverin, C.M., Snyders, F., Muller, N., Botes, W.C., Fox, G.P., 2013. Prediction of triticale grain quality properties, based on both chemical and indirectly measured reference methods, using near-infrared spectroscopy. Cereal Chem. 90, 540–545.

Mares, D., Mrva, K., 2008. Late-maturity α-amylase: low falling number in wheat in the absence of preharvest sprouting. Cereal Chem. 47, 6–17.

McGoverin, C.M., Snyders, F., Muller, N., Botes, W., Fox, G., Manley, M., 2011. A review of triticale uses and the effect of growth environment on grain quality. J. Sci. Food Agric. 91, 1155–1165.

Miedaner, T., Geiger, H.H., 2015. Biology, genetics, and management of ergot (*Claviceps* spp.) in rye, sorghum, and pearl millet. Toxins 7 (3), 659–678.

Munoz-Insa, A., Gastl, M., Becker, T., 2015. Influence of malting on the protein composition of triticale (× *Triticosecale* Wittmack) 'Trigold'. Cereal Chem. 93, 10–19.

Oettler, G., 2005. The fortune of a botanical curiosity—Triticale: past, present and future. J. Agric. Sci. 143, 329–346.

Pattison, A.L., 2013. Genetic improvement of grain quality for bread making in triticale. PhD thesis, University of Sydney, Sydney, Australia.

Pattison, A.L., Trethowan, R.M., 2013. Characteristics of modern triticale quality: commercially significant flour traits and cookie quality. Crop Pasture Sci. 64, 874–880.

Qualset, C.O., 2002. Triticale, ninth ed. McGraw-Hill, Inc, New York.

Salmanowicz, B.P., 2008. Detection of high molecular weight glutenin subunits in triticale (*x Triticosecale* Wittm.) cultivars by capillary zone electrophoresis. J. Agric. Food Chem. 56, 9355–9361.

Seibel, W., Weipert, D., 2001. Animal feed and industrial uses, second ed. AACC International, St. Paul, MN, pp. 213–233.

Sharma, R., Cooper, K.V., Jenner, C.F., 2002. Genetic variation for 'waxy' proteins and starch characteristics of triticale. In: Arseniuk E, Proceedings of the Fifth International Triticale Symposium. International Triticale Association, Radzikow, Poland, pp. 245–251.

Trethowan, R.M., Pena, R.J., Pfeiffer, W.H., 1994. Evaluation of pre-harvest sprouting in triticale compared with wheat and rye using a line source rain gradient. Aust. J. Agric. Res. 45, 65–74.

Tsen, C.C. (Ed.), 1974. Triticale: First Man-Made Cereal. AACC International, St. Paul, MN, USA.

Weipert, D., 1996. Rye and triticale. In: Henry, R.J., Kettlewell, P.S. (Eds.), Cereal Grain Quality. Chapman and Hall, London, UK, pp. 205–224.

Wrigley, C.W., 2005. Late-Maturity α-amylase—Apparent sprout damage without sprouting. Cereal Food World 51, 124–125.

Wrigley, C.W., Bushuk, W., 2010. Rye and triticale: characteristics and quality requirements. In: Wrigley, C.W., Batey, I.L. (Eds.), Cereal Grains: Assessing and Managing Quality. first ed. Woodhead Publishing Ltd, Oxford, UK.

Chapter 9

Barley: Grain-Quality Characteristics and Management of Quality Requirements

Marta S. Izydorczyk and Michael Edney
Canadian Grain Commission, Grain Research Laboratory, Winnipeg, MB, Canada

9.1 INTRODUCTION

Barley is one of the world's oldest and most versatile cereal crops. It has played a role in the human transition from a hunting and gathering to an agrarian lifestyle in the 'Fertile Crescent' of the Near East at least 10,000 years ago (Ullrich, 2011). Barley has been used through the centuries to feed livestock, poultry, people and produce alcoholic beverages (Newman and Newman, 2008). Barley matures under a short growing season allowing it to be grown under a diverse range of environments and geographical locations. These include sub-arctic Scandinavia through the equator to the Southern Hemisphere in Australia and South America, from below sea level near the Dead Sea to high altitudes in the Andes and Himalayas, and from regions with moderate precipitation and temperature in Europe to dry and hot areas in parts of North America, Africa and Australia. Barley also tolerates drought and saline growing conditions.

The long history and wide spread of barley cultivation are responsible for a broad range of barley utilisation and its changing prominence throughout centuries. In the ancient world, barley was used as human food, eaten raw or roasted, in bread, porridges or soups, and believed to provide strength and stamina. Barley lost its pre-eminent position as food staple in human nutrition at the beginning of the 20th century, when grains such as rice, wheat and maize, gained greater preference. The use of barley in the production of alcoholic beverages, especially beer, significantly increased after the Second World War. In recent times, the demand for barley for malting and brewing purposes has shown variable trends, with some drop in the West but significant growth in Asia. Cultivation of barley for feed is a more recent development, and the demand for feed barley has significantly surpassed its demand for other uses, with the majority of

Cereal Grains. http://dx.doi.org/10.1016/B978-0-08-100719-8.00009-7

barley production today for feeding animals. Most recently, however, there has been a resurgence of interest in barley for human nutrition due to many proven health benefits associated with its consumption.

As a result of this multiplicity of use, the desired attributes of barley can be quite variable highlighting the need for their proper definition and assessment taking into consideration the specific use of the grain. The preferred end use for barley is malting and brewing, where higher prices are always inviting; however, meeting the selection criteria for malting can be difficult due to stringent quality requirements. The majority of barley, therefore, is fed to animals because quality demands are less strict. The incorporation of barley in foods is presently limited to traditional uses but the future looks promising as unique barley properties are pursued for their benefits to human health.

The appeal for barley comes from its unique morphological traits such as endosperm cell walls consisting of water-soluble non-starch polysaccharides, a nutrient-dense aleurone layer capable of rapid production of high concentrations of enzymes and a hull that remains attached after harvest. These traits are key to meeting the quality requirements of barley end users such as the maltster's need for enzyme accumulation during germination, the ruminant feeder's need for hulls to limit bloat and the presence of the water-soluble non-starch polysaccharide, β-glucan, which has positive human health implications. The extreme climates under which barley grows offer challenges for grain storage in terms of germination maintenance and microbial infection. The release of new barley varieties is continually required in order to remain competitive in terms of agronomics and disease resistance while maintaining appropriate quality. The future for barley looks promising with barley's adaptability being an essential component for changing climates, increasing demand for beer and the positive health implications for food barley.

9.2 UNIQUE MORPHOLOGICAL AND BIOCHEMICAL CHARACTERISTICS OF BARLEY

Both inadvertent and targeted selection of barley over thousands of years has led to the development of unique morphological characteristics that have proven advantageous for a range of end uses. Breeders continue to select for these traits as well as investigate new ones that could be beneficial, particularly in terms of health benefits for humans.

9.2.1 Composition and Properties of Barley Cell Walls

The cell walls in barley exhibit a number of distinguishing features that are important during the processing of barley into value-added products like malt and beer and that have made major contributions to the emerging interest in cell wall non-starch polysaccharides as beneficial constituents in human health.

Non-starch polysaccharides, arabinoxylans and (1→3, 1→4)-β-D-glucans (β-glucans), make up a large proportion of the walls of the aleurone and starchy

TABLE 9.1 Approximate Proportion of Polysaccharides in the Cell Walls of Barley Grain Tissues

Component	Starchy endosperm (% by weight)	Aleurone layer	Pericarp
Arabinoxylans	8–20	60–75	50–60[a]
β-Glucans	65–75	20–25	4
Cellulose	1–2	1–2	20–30
Glucomannans	5	1	1–2
Arabinogalactans	5	0.5	

[a]*Mostly glucuronoarabinoxylans.*
Source: Adapted from Izydorczyk, M.S., 2010. Variations in content and molecular structure of barley nonstarch polysaccharides associated with genotypic and cellular origin. Cereal Chem. 87, 376–384; and Fincher, G.B., 2011. Biochemistry, physiology, and genetics of endosperm mobilization in germinated barley grain. In: Ullrich, S.E. (Ed.), Barley: Production, Improvement, and Uses. Wiley-Blackwell, Publishing Ltd., Chichester, UK, pp. 449–477.

endosperm of barley grain (Table 9.1) (Izydorczyk, 2010). The walls generally have low content of cellulose and no lignin; these features facilitate enzymic penetration and depolymerisation of cell walls during germination of the grain. While arabinoxylans (heteroxylans) occur commonly in walls of higher plants, β-glucans are restricted to only certain families of the order Poales. Barley endosperm cell walls consist of approximately 70% β-glucan compared to limited amounts in most other cereal grains (except oats) and barley is notable for its high level of β-glucans that may range from 3% to 12% depending on genotype, and on agronomic and climatic factors during grain development. The content of β-glucans in barley for malting and animal feed has been successfully lowered through appropriate selection to a target range of 3.5–5%; this is opposite to barley aimed for production of healthy foods, where a high β-glucan content (>6%) is considered a desirable trait (Fig. 9.1A–B).

β-Glucans consist of D-glucopyranosyl residues linked via a mixture of non-randomly distributed β-(1→3) and β-(1→4) glycosidic linkages. About 90% of the chain is made of cellotriosyl units [with 2 successive (1→4) linkages] and cellotetraosyl units [with 3 successive (1→4) linkages] joined by single (1→3) linkages. Longer cellulose-like oligomers (with 5–25 glucose units) are interspersed among the cellotriosyl and cellotetraosyl units and also connected by single (1→3) linkages (Lazaridou et al., 2006). The presence of (1→3) linkages, which breaks up the regularity of the β-(1→4) linkage sequences, and the random distribution of cello–oligosaccharides along the polymer chain contribute to the water solubility of β-glucans. The water solubility/extractability of β-glucans in barley grain is relatively high ranging from 40% to 60%. Solubilised β-glucan polymers occupy large hydrodynamic volumes in solutions due to their high molecular weight and random chain conformation with locally stiff

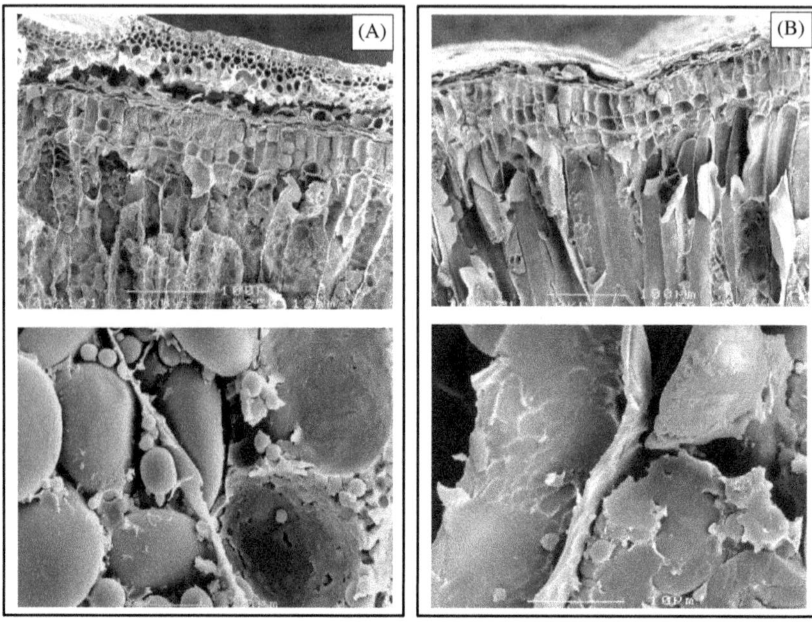

FIGURE 9.1 **Scanning electron micrographs of cross-section of barley grain showing differences in the abundance and thickness of the endosperm cell walls in malting variety AC Metcalfe (A) and high β-glucan food variety CDC Fibar (B).**

segments and, therefore, they impart high viscosity to aqueous solutions. The ability of β-glucans to impart high viscosity is a key attribute directly related to the beneficial effects of these polymers in lowering serum cholesterol and decreasing post-prandial glucose level in humans (Jenkins et al., 1978, 2002; Lund et al., 1989; Wolever et al., 2010). As the viscosity of gut contents increases in the presence of β-glucans, there is resistance to flow, which may affect the absorption of fat, cholesterol and bile acids.

The specific molecular features of barley β-glucans, described earlier, enable them to form a gel-like matrix in the walls and ensure that walls remain strong and control the transfer of water and other molecules across the wall. If not properly hydrolysed during germination, β-glucans will negatively affect biochemical reactions required for the production of good-quality malt, limit modification of the endosperm during mashing and contribute to the formation of hazes and precipitates in beer. The ability of β-glucans to produce viscous solutions is detrimental during brewing where high viscosities reduce flow rates and retard filtration. The rapid synthesis of high levels of enzymes that depolymerise β-glucans during germination is a desirable quality characteristic of malting grade barley (Fincher, 2011). The (1→3,1→4)-β-D-glucan endohydrolases are perhaps the most important enzymes for rapid solubilisation and complete degradation of β-glucans in endosperm walls of germinating barley. The high concentration of water-soluble β-glucan in barley likely results from selection over time for cell walls that quickly degrade during germination, thus, reducing processing times in the malt house. High concentrations of water-soluble β-glucans are problematic when barley is fed to poultry due to their negative effect on nutrient uptake and a need for β-glucanase supplements (Philip et al., 1995).

Arabinoxylans constitute another abundant group of barley cell wall polysaccharides (Izydorczyk, 2014). Unlike β-glucans, they are branched heteropolysaccharides consisting of a linear chain backbone of β-D-xylopyranosyl (Xyl*p*) residues linked through (1→4) glycosidic linkages with single α-L-arabinofuranosyl (Ara*f*) residues attached to some of the Xyl*p* residues at O-2, O-3 and/or at both O-2,3 positions. Depending on their genotypic and cellular origin, arabinoxylans exhibit wide variations in molecular features and properties (Izydorczyk, 2010). Recent studies also revealed positive effects of water-soluble cereal arabinoxylans on cecal fermentation, production of short-chain fatty acids, reduction of serum cholesterol and improved adsorption of calcium and magnesium (Lopez et al., 1999; Hopkins et al., 2003; Lu et al., 2004; Zunft et al., 2004; Garcia et al., 2007).

Compared to β-glucans, arabinoxylans are only marginally solubilised and hydrolysed during malting. Nevertheless, they have been shown to have a negative effect on the viscosity and filterability of wort (Egi et al., 2004; Lee et al., 1998; Sadosky et al., 2002; Li et al., 2005). Although the importance of arabinoxylans in the malting and brewing processes has been recognised, barley arabinoxylans have not yet been sufficiently investigated to define specific quality requirements regarding their content or physicochemical properties.

9.2.2 Potential for Synthesis of Excess Enzymes by Aleurone Cells

The barley aleurone layer is located between the endosperm and the seed coat. It consists of three layers of cells which contrast with the single layer in most other cereals. The aleurone layer of the mature barley grain is alive and consists of thick-walled cells rich in storage reserves: lipids, proteins and minerals. Once germination is initiated, the gibberellic acid (GA) that is synthesised in the embryo diffuses through the grain and progressively activates gene expression in the aleurone layer. Aleurone cells are responsible for the synthesis and secretion of a range of hydrolytic enzymes that break down endosperm compounds providing nutrients for the growing barley embryo. Selection pressures during breeding have resulted in the aleurone that produces excess levels of protein-, cell wall- and starch-degrading enzymes normally required by the germinating barley grain. The malting and brewing industries, whose selection pressures over the years are at least partly the cause of the efficiency of barley's aleurone layer, rely on the excess enzymes to efficiently destroy the cell walls and protein matrix of the endosperm during malting and completely convert starch into fermentable sugars during mashing. Starch granules are released and made available for gelatinisation and hydrolysis in the brewery. Proteases release amino acids for use in enzyme synthesis by aleurone cells but they also produce the free amino acids required for yeast nutrition in the brewery.

9.2.3 Hull Retention

The majority of the barley presently grown in the world retains its hull after harvest, in contrast to wheat. The hulls are advantageous in the malt industry where they protect the growing acrospire as the barley moves through the malt house. The hull also prevents exposure of the seed coat to wet conditions during steeping which otherwise would render the seed coat sticky, restricting movement of steeped barley. During the kilning phase of malting, the hull insulates the aleurone and endosperm helping to preserve enzymatic activity for use in the brewery. The hull is also essential in the brewery where lauter tuns are used to separate wort from spent grain. Lauter tuns rely on the hull to add structure to filter beds improving porosity and wort runoff.

Barley hull is also useful for ruminants. Bloat often results from excess fines produced from processing wheat and hull-less barley for ruminants. The fines, which cause rapid starch digestion leading to the bloat, are less of a problem in covered barley where the hull reduces their production. However, in contrast, the hull reduces the feeding value of barley for monogastrics due to indigestibility and the dilution effect on overall energy content.

Hull-less barley, which loses its hull during harvest, continues to be pursued for possible advantages as a feed for monogastrics but commercial uptake has been limited. The food industry is also interested in using hull-less barley for easier processing as they develop new products based on the health benefits of β-glucan.

9.2.4 Proteins Unique to Barley

Barley has similar levels of protein to other cereal grains with averages in commercial shipments between 9% and 15% dry matter depending on end use and barley source. Barley proteins are traditionally classified on the basis of their extraction with various solvents according to the concept developed by Osborne (1907). Albumins and globulins, the water- and salt-soluble proteins, respectively, function mostly as enzymes, metabolic regulators or inhibitors of particular enzymes. Prolamins, called hordeins in barley, are the major storage proteins located in the starchy endosperm; they can be extracted from grain with warm aqueous alcohol. The residual fourth group of barley proteins, glutelins, can be extracted from grain with either dilute acid or alkali solutions or in the presence of detergents. Glutelins serve mainly as structural proteins. Gubatz and Shewry (2011) provided a summary of different groups of barley proteins and their functions as related to the Osborne classification (Fig. 9.2)

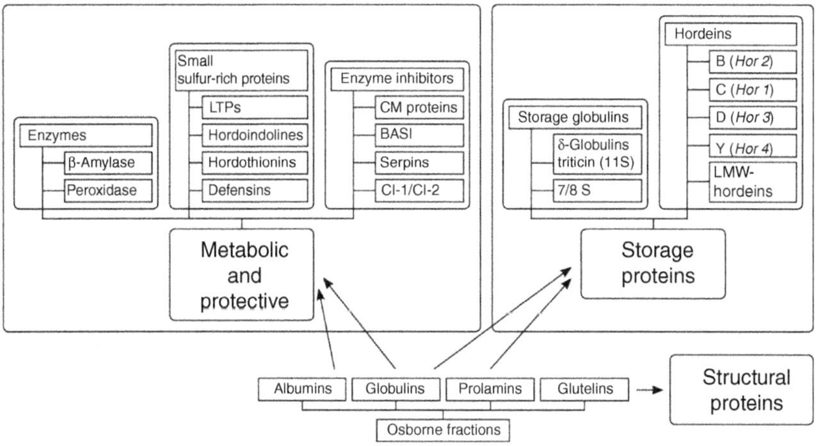

FIGURE 9.2 Summary of various groups of proteins in ripe barley grain. *(Re-printed with permission from Gubatz, S., Shewry, P.R., 2011. The development, structure, and composition of the barley grain. In: Ullrich, S.E. (Ed.), Barley: Production, Improvement, and Uses. Wiley-Blackwell, Publishing Ltd., Chichester, UK, pp. 391–477.)*

Hordeins account for 35–50% of the total grain nitrogen and their relative amount is related to the total protein content in the grain and availability of nitrogen. Hordeins are diverse in structure and sequence and have been classified into B-, C-, D- and γ-hordeins according to their electrophoretic mobility (Finnie and Svensson, 2014). Extraction of grain with propanol (50%) and dithiothreitol (DTT) affords isolation of free monomeric C-hordeins and free and bound polymeric B- and D-hordeins. The polypeptides of different hordein groups can be separated by reversed-phase chromatography, and since the patterns of hordein components differ among different genotypes it can used to confirm varietal identity and homogeneity (Fig. 9.3). Izydorczyk et al. (2004) used matrix-assisted laser desorption/ionisation time of flight mass spectrometry (MALDI-TOF MS) to obtain detailed mass patterns of polypeptides constituting

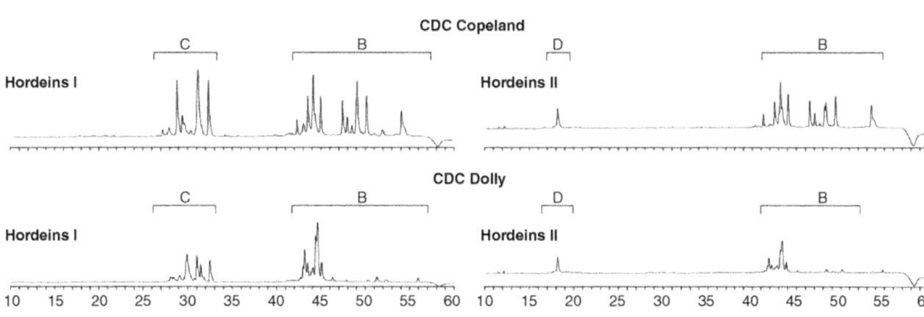

FIGURE 9.3 Reversed-phase chromatography (RP-HPLC) profiles of reduced hordeins I (left) and hordeins II (right) in CDC Copeland (malting barley) and CDC Dolly (feed barley). Polypeptides constituting the C, B and D hordeins are indicated above the peaks.

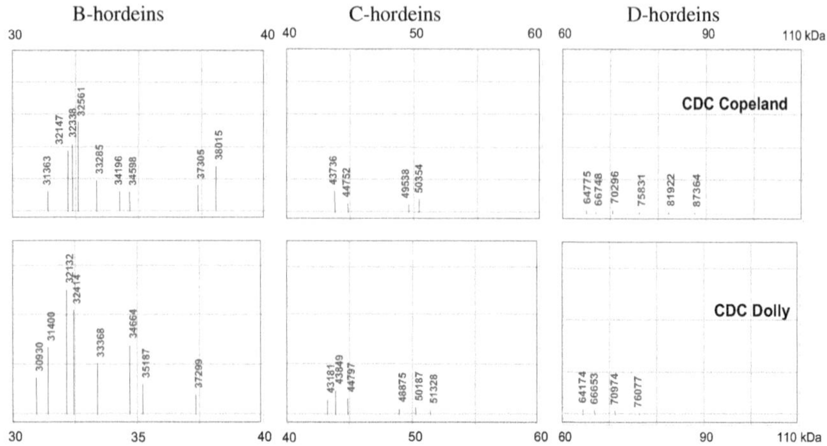

FIGURE 9.4 Mass spectra of polypeptides (as shown in Fig. 9.3) constituting the B- (left column), C- (middle column) and D- (right column) hordeins in CDC Copeland (malting barley) and CDC Dolly (feed barley).

the C-, B- and D-hordeins (Fig. 9.4). It was shown that the B-hordeins consisted of polypeptides with molar masses ranging from 30 to 38 kDa, C-hordeins contained mostly polypeptides with molar masses of 43 and 44 kDa with a few polypeptides with m/z of 48–51 kDa, and the D-hordeins consisted of a main polypeptide with m/z of 71 kDa and a few minor polypeptides with m/z of 64–87 kDa. The C-hordeins lack cysteine residues and hence are present in grain only as monomers, whereas B-, D- and γ-hordeins contain a number of cysteine residues allowing them to form disulfide bridges and polymeric complexes. The polymeric 'gel proteins' containing all D-hordeins and some B-hordeins have been implicated in the quality of malting barley (Smith and Lister, 1983), however, Brennan et al. (1998) showed that D-hordein null mutants did not differ in malting quality.

The influence of specific types of barley protein on malt and beer quality is complex and not fully elucidated because of the large number of proteins present in the barley grain and the diversity of functional roles they play during the malting and brewing processes as well as in the final product (Shewry and Darlington, 2002). In addition to hordeins, barley shows some distinct proteins such as β-amylase and serpins (called protein Z) present in relatively high concentrations. Both β-amylase and protein Z contribute to the desirability of barley for brewing as they enhance starch-degrading capacity and improve beer foam, respectively. β-Amylase is synthesised during grain development and stored in the mature endosperm, in free and bound forms, in readiness for digestion of starch during germination. The enzyme behaves like a storage protein as it increases under conditions of high nitrogen availability. The serpins are a family of serine proteinase inhibitors with a defensive role of inhibiting chymotrypsin-like enzymes of insects and pathogens (Roberts et al., 2003).

Barley serpins were initially identified as allergens in beer (Hejgaard, 1984) and they were shown to be present in greater amounts in the early generation high-lysine lines (Hejgaard and Boisen, 1980). Other proteins known to stabilise beer foam are non-specific lipid-transfer proteins (LTPs). Named initially after their apparent ability to transfer phospholipids in vitro, both classes of these proteins (LTP1 and LTP2) are restricted to the cells of the aleurone layer, where likely they contribute to defense against fungi infections (Gorjanovic et al., 2005). Jegou et al. (2000) showed that the reduced and glycated form of LTP1 is more effective in foam stabilisation than other forms. In addition to the well-characterised surface-active proteins, such as LTPs and protein Z, several others were also identified in beer foams including barwin protein, α-amylase inhibitors (CMd sub-unit) trypsin inhibitors, hordoindolines and protein zx (Izydorczyk et al., 2006). The family of α-amylase and trypsin inhibitors, classified as CM proteins based on their solubility in chloroform/methanol mixtures, is now known as the 'prolamin superfamily' with seven major sub-units in barley active towards exogenous enzymes (Jenkins et al., 2005). Barley hordoindolines appear to play a role in grain texture, but no relationship has yet been established between grain texture and allelic differences in hordoindoline sequences (Darlington et al., 2000, 2001).

The nutritional quality of barley proteins is compromised somewhat by relatively low contents of the essential amino acids, especially lysine, followed by threonine, methionine and valine. An additional problem is represented by the high contents of the non-essential amino acids proline and glutamine originating from the hordein fractions. These amino acids are not utilised by livestock with the surplus nitrogen being excreted (Finnie and Svensson, 2014). Much effort has been put into generation of barley with increased lysine content; most recently high-lysine lines with uncompromised yield and grain quality were developed using antisense technology, making it a promising tool in improving nutritional quality of barley protein (Lange et al., 2007; Hansen et al., 2007).

9.2.5 Elevated Levels of Tocopherol and Tocotrienol

Vitamin E, the lipid-soluble vitamin found naturally in plant tissues, is made up of a family of eight structurally related compounds referred to as vitamers: α-, β-, γ- and δ-tocopherols (α-, β-, γ-, δ-T) and α-, β-, γ- and δ-tocotrienols (α, β-, γ-, δ-T3). Barley grain is unique as a source of vitamin E because, unlike other cereals and oilseeds, it contains all eight vitamers (Fig. 9.5) (Falk et al., 2004). The structural differences among the vitamers are small but have a significant impact on their metabolism and their specific health benefits. α-Tocopherol is the best-known form of vitamin E for its anti-oxidant properties; however, more recently scientific attention has shifted to other forms, specifically to γ-tocopherol and tocotrienols. γ-Tocopherol has been implicated in anti-inflammatory and cancer prevention properties, whereas tocotrienols have been shown to possess powerful hypocholesterolemic, anti-cancer and

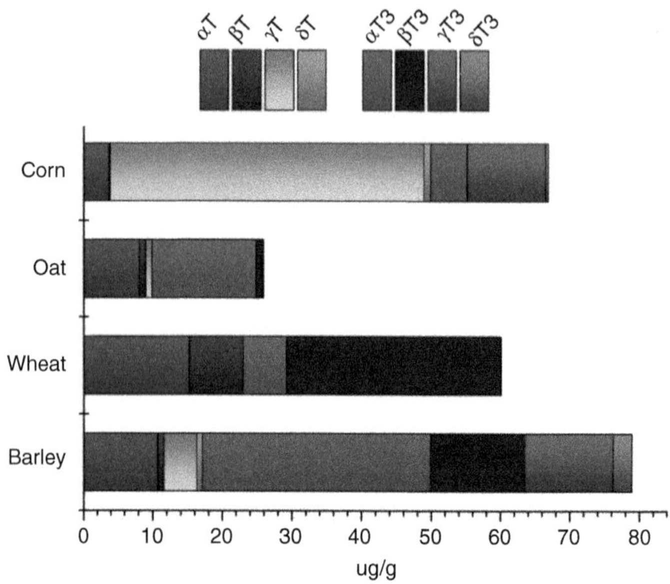

FIGURE 9.5 Content and composition of tocochromanols (different forms of vitamin E) in cereal; grains. αT, βT, γT, δT indicate α-, β-, γ- and δ-tocopherols, respectively; αT3, βT3, γT3, δT3 indicate α-, β-, γ- and δ-tocotrienols, respectively.

neuroprotective properties not exhibited by tocopherols (Jiang et al., 2004; Sen et al., 2007; Aggarwal et al., 2010; Cardenas and Ghosh, 2013). The ratio of T3/T has been proposed as a criterion of the bioactivity of grains as sources of vitamin E. The distribution of vitamin E in barley kernel is not uniform throughout the kernel and the levels of total tocochromanols in barley germ or in the outer tissues being several-fold higher than in the whole grain (Fig. 9.6). As shown in Fig. 9.6, tocopherols are preferentially localised in the germ, whereas tocotrienols are concentrated in the seed coat and aleurone layers. Fractionation of whole grain barley by physical means (pearling, milling and sieving) has been investigated as means of concentrating tocopherol and tocotrienol levels from barley for use as supplements or as a source of oil extraction.

9.3 PRODUCTION AND MAJOR USES OF BARLEY

The annual world production of barley in 2013 was estimated at 144 million tonnes slightly higher than the last 10-year average of 140 million tonnes. Production of barley peaked in the 1980s, but in general the past 3 decades have seen significant swings in barley production (Table 9.2) due to a range of factors. Political incentives in Europe during the 1970s and 1980s were responsible for large increases in barley acreages on that continent. Increased beer consumption contributed to greater barley production in Asia and South America. The increased beer consumption in south-eastern Asia also increased

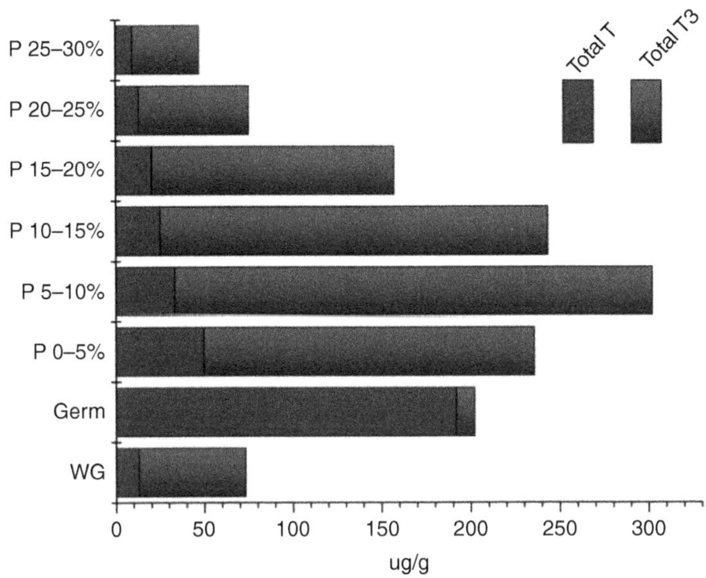

FIGURE 9.6 Content of total tocopherols (T) and total tocotrienols (T3) in whole grain (WG), germ and various pearling fractions (P) of barley grain. The percentage values indicate the weight of pearling fractions successively removed by scouring.

demand for barley from Australia, although, Australian production figures have remained inconsistent due to frequent droughts. North American production has fallen off in recent years, predominately in the United States, because of increased incidence of *Fusarium graminearum*. Reduced world production in recent years has mainly reflected reduced production of feed barley in Europe.

Barley has three distinct uses: malting, feed and food (Table 9.3). The use for malt and feed increased steadily into the 1980s but then levelled off with some drop in feed use (Fig. 9.7). In contrast, the use of barley for food decreased into the 1980s but has now levelled off at a low, but still significant, 7 million tonnes per year.

9.3.1 Malting Barley

The best-known end use for barley is as an ingredient for alcoholic beverages, predominately beer and Scotch whisky. However, raw barley must be malted prior to use in the beverage industry. Yeasts, which ferment simple sugars to ethanol in the brewery and distillery, do not produce the enzymes necessary for degrading barley starch. It was discovered thousands of years ago that barley produces its own starch-degrading enzymes during germination. The malting process, therefore, was developed using controlled germination to produce the enzymes that degrade starch to simple sugars in the brewery. The β-glucan in endosperm cell walls is also degraded during germination releasing starch from

TABLE 9.2 Average Annual Barley Production Across the Continents (Kilo Tonnes)

Years	Europe[a]	Asia	North and Central America	Oceania[b]	Africa	South America	World
2013	85,838	22,514	15,516	7,888	7,387	5,612	144,755
2012	79,108	19,871	13,841	8,660	6,023	6,004	133,507
2011	81,225	20,174	11,636	8,363	6,431	5,121	132,951
2010	43,440	20,165	12,204	7,602	6,449	3.822	123,683
2009	95,541	21,290	14,987	8,344	9,225	2,394	151,782
2008	105,378	16,580	17,793	8,405	3,919	2,712	154,788
2007	82,843	20,800	16,213	7,516	4,413	2,391	134,175
2006	88,850	23,198	14,366	4,534	6,238	2,337	139,522
2005	83,092	22,326	17,053	9,785	4,654	1,769	138,680
2004	96,512	21,445	19,580	8,043	6,290	2,035	153,905
2003	83,203	21,445	19,309	10,760	5,803	2,045	142,565
1993–2002	90,902	23,488	20,180	6,212	4,056	1,328	146,165
1983–92	115,947	19,525	23,065	4,787	5,174	0,971	169,468
1973–82	110,958	15,905	20,238	3,183	3,902	0,999	155,184
1963–72	70,888	15,787	16,510	1,757	3,591	1,190	109,722

[a]*Includes Russia.*
[b]*Oceania consists of Australia, New Zealand and South Pacific Islands.*
Source: Food and Agriculture Organization of the United Nations, 2015. FAOSTAT data, http://faostat.fao.org/site/567/default.aspx.

TABLE 9.3 Major End Uses of Barley Around the World

Malt barley	Feed barley	Food barley
Malt for brewing beer	Large ruminants (cattle)	Flour/fibre-rich fractions
Malt for distilling whiskey	Small ruminants (sheep)	Couscous
Food malt	Pigs	Rice extender
	Laying hens	Pearled (soup)
	Broiler chickens	Barley tea
		Shochu

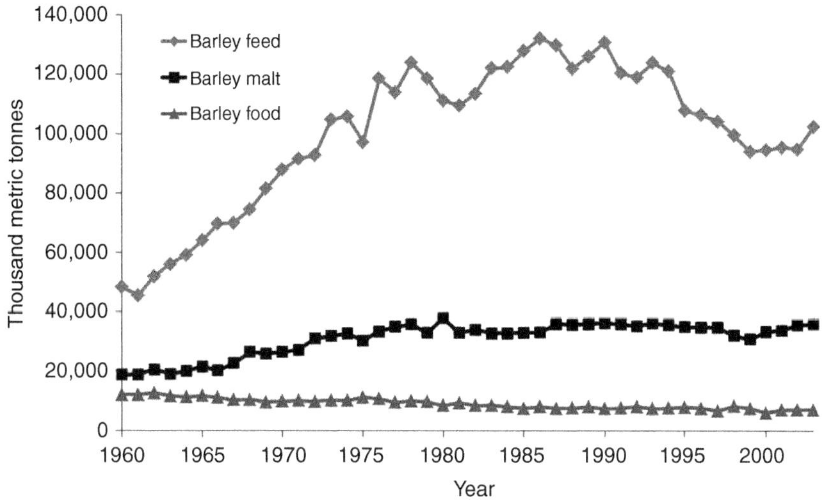

FIGURE 9.7 Amounts of barley used for malt, feed and food around the world for the past 50 years. *(Source: USDA Foreign Agricultural Service (2009).)*

within cells and reducing potential problems associated with β-glucan, particularly poor filtration, in the brewery and distillery. Kilning, the final drying phase of malting, contributes to malt flavour. This typical flavour is not only desirable for beer and whisky but it is also used for flavouring and as a colour enhancer by food processors. The baking industry also uses malt flour in bread dough formulations because of its high content of α-amylase.

The amount of barley used for malting has increased slowly but steadily over the past 45 years. Europe (including Russia) doubled its use of malt barley during this period. Malt use in Europe has remained relatively flat since the late 1970s but at 25 million tonnes still accounts for more than 50% of world use. Asia has recently become the next biggest consumer of malting barley having quadrupled its use over the past 45 years with half the increase coming since 1990. Use in 2008 was still only a third of Europe but 40% greater than North America where demand increased steadily through the 1960s, 1970s and into the 1980s but has remained steady since. South America shows the steepest increase in malt barley use but still used less than 2 million tonnes in 2008. Oceania (Australia, New Zealand and south Pacific islands) has shown erratic domestic production of malt but has averaged 1 million tonnes in recent years, 3 times its production of the 1960s.

Recent changes in malt demand are accounted for by reduced beer consumption in the more traditional western markets while consumption has increased in Asia, South America and Eastern Europe including Russia. Greater use of adjuncts, non-malt sources of starch for brewing, has contributed to the reduction in demand for malt. The Scotch whisky industry has expanded significantly in recent years but a limited requirement for malt reduced the effect of this demand on world supply of malting barley.

9.3.2 Feed Barley

Feed is the major end use for barley around the world. Barley is mainly fed to ruminants but also pigs, especially in Europe, and a limited amount is fed to layer and broiler chickens. The hull on covered barley severely restricts energy content, the major nutrient supplied by barley, for monogastrics. In contrast the hull improves the feeding value of barley for ruminants, compared to wheat, as it reduces the incidence of acidosis and bloat. Novel uses have been proposed for feed barley such as inclusion in trout diets but commercial use has yet to develop (Sealey et al., 2008).

Feed barley use has decreased over the past 10 years, largely due to low prices of other feed stuffs, but only after steady increases during the previous 40 years (Fig. 9.7). Even today over 75% of total world barley production is consumed as feed. The major decrease in feed barley use has been in Europe where barley has been replaced by cheaper and more energy-rich grains, for pigs, such as wheat and corn. Asia is the next biggest user of feed barley at *20 million tonnes in 2008* with over half going to the Middle East, an area where feed barley use has significantly increased over the past 20 years. North America remains a consistent user of feed barley (10 million tonnes annually), mainly for cattle, but demand has fluctuated with supply and price competitiveness. Australia and Africa are both significant users of feed barley (3 million tonnes annual average each) with both showing significant increases since the 1980s. Feed barley use remains insignificant in South America.

9.3.3 Food Barley

Food barley use worldwide has stabilised in recent years (Fig. 9.7) after a steady decrease over the past 45 years. A large market for food barley continues to exist in northern Africa, including Morocco, Algeria and Ethiopia, where on an average over 2 million tonnes of barley are incorporated annually into foods such as flour and couscous. Asia remains a major food barley user (2 million tonnes annually) but use has dropped to two-thirds the levels seen in the 1960s. The major loss has been in Eastern Asia, Korea and Japan in particular, where barley had been a staple food product often used as a rice extender to reduce costs. Rice extenders have been revived to some extent in Japan where the health implications of barley, such as increased levels of protein, calcium and dietary fibre compared to white rice, have been promoted. Barley tea remains an important use of barley in Japan with health benefits of β-glucan once again being promoted. Shochu, a barley liquor, is also a significant market for raw barley in Japan.

Barley is a food staple in some parts of South America such as the high Andes but annual use is restricted to less than 0.5 million tonnes of barley. In Europe pearled barley for soup remains a small but significant market for food barley. Health implications have led to significant increases in food barley use over the past 10 years and now accounts for over 1 million tonnes of barley annually. North America continues to use small quantities of pearled barley in soups at a quarter of a million tonnes annually.

The North American barley community has anticipated increased human consumption of barley after a USA government ruling (Food and Drug Administration, 2005) allowing a health claim, 'reduced incidence of cardiovascular disease', for β-glucan soluble fibre from barley products (Baik and Ullrich, 2008). In response new products such as high-fibre fraction flours and concentrated dietary supplements have been developed for inclusion in human diets. Demand remains limited but commercial success is expected as incorporation of barley products into processed foods gains consumer acceptance.

9.4 SPECIFIC QUALITY REQUIREMENTS FOR BARLEY

9.4.1 Malt Barley

Assessing the quality of malt barley is unlike that of any other cereal because the major use of malt barley involves two biological processes, barley germination and yeast fermentation. These processes can be unforgiving, offering limited opportunity to adjust processing conditions when dealing with barley of varying quality. The automated nature of modern malt houses and breweries further emphasises the importance of barley and malt quality. The food-nature of the final products, malt and beer, necessitates concern with food safety of both the barley and malt. As a result strict criteria must be met before barley and malt are deemed acceptable for malting and brewing, respectively. Assessment is divided into two components, barley quality prior to malting and the quality of malt (Tables 9.4–9.7).

TABLE 9.4 Typical Analyses Used to Determine Suitability of a Barley Sample for Malting

Barley analysis	Rationale
Varietal purity	Each malt variety processes differently necessitating segregation of varieties to produce good malt quality.
Germination	Good germination is imperative for the production of quality malt.
Kernel appearance	Indication of microbial infestation.
Kernel peeling	Loose hulls reduce uniformity of germination and increase handling problems.
Protein	Barley with too much protein (>13% dry matter) modifies slowly and produces less malt extract. Barley with too little protein (<9.5%) can compromise enzyme levels.
Kernel sizing	Plumper kernels result in greater malt extract potential. Processing is simplified with more uniform kernel size.

TABLE 9.5 Typical Tests for Assessing Germination of Malt Barley With Rationale

Germination test	Acceptable values	Rationale
Germinative energy: standard 3-day test (4 mL)	>95%	Indicates the degree to which a sample will germinate in the malt house.
Water sensitivity: standard 3-day test (8 mL)	<10% difference between 4 and 8 mL tests	Larger differences mean greater water sensitivity and more need to alter steeping or reject sample.
Detecting pre-germination: Rapid Visco Analysis (RVA)	<50 RVU unacceptable 50–120 RVU caution >120 RVU acceptable	Indication of pre-harvest sprouting. See text for storability of a sample.
Germinative capacity: 1. Peroxide method—breaking dormancy 2. Tetrazolium—staining live embryos	>98%	Germinative capacity and energy tests need to be compared. A sample with good capacity (>98%) shows potential for breaking dormancy and eventually improving germinative energy.
Germinative index (GI)[a]	Closer the GI is to 10 the better	Indicates the speed at which a sample will germinate in the malt house. The higher the GI the less time required to achieve adequate processing.

[a]*Calculated according to Eq. (9.1).*

TABLE 9.6 Examples of Germination Results for Malt Barley Samples and Their Suitability for Malting

Germinative energy (4 mL test) (%)	Kernels germinated after 24 h	Kernels germinated after 48 h	Kernels germinated after 72 h	Germinative index[a]	Comments
98	95	3	0	9.7	Good
98	0	9	89	3.4	Good, slow processing
95	88	5	2	9.1	Acceptable
96	0	56	40	4.1	Acceptable, slow processing
80	72	8		9.1	Reject poor germination

[a]*Calculated according to Eq. (9.1).*

TABLE 9.7 Conventional Malt Analyses and Target Specifications for Different End Uses

Malt analysis	Typical targets for adjunct brewing	Desirable targets for all malt (craft) brewing	Meaning
Malt extract	>81%	>80%	Higher values mean more beer.
Wort-soluble protein	>5%	<5%	Required for good beer foam but too much can produce beer hazes.
KI, soluble/total protein ratio	42–47	38–45	
FAN	>200 ppm—high gravity brewing with maltose syrups (adjunct)	<200	Essential for good yeast nutrition.
Wort β-glucan	<100 ppm	Variable, 135 ppm max	Less β-glucan gives lower viscosities and fewer problems with wort and beer filtration.
Diastatic power	Medium/high: 120–140/>140°L	Low/medium: 100–120/120–140°L	Higher levels required with high starch adjuncts and for distilling.
α-Amylase	45–65 DU		Often overlooked for the ability to produce β-amylase substrate.

FAN, Free amino nitrogen; °L, degrees Lintner.

Assessment of malt barley quality begins shortly after harvest and can involve a number of subsequent verifications depending on how direct delivery is from field to malt house. The initial assessment will determine the suitability of a sample for malting. Subsequent assessments, which can occur at each transition (producer to grain company to merchant to final end user—the malt house), are used to verify quality specifications. Analyses cover a range of physical, biochemical and metabolic characteristics, which are assessed following methods standardised by the industry. These include the Methods of Analysis of the American Society of Brewing Chemists (2004) widely used in the Americas, the Analytica-EBC (European Brewery Convention, 1998) of the European Brewery Convention, which are used in much of Europe and other countries around the world, and the Mitteleuropäische Brautechnische Analysenkommision (MEBAK) methods (MEBAK, 2006) in the German speaking countries in Europe and in the Czech Republic. The Institute of Brewing (IOB) Approved

Methods of Analysis (1997) were used in the United Kingdom, but recently harmonised with those of the European Brewery Convention (1998). The types of analyses will vary at each point becoming more complex and stricter as the barley approaches processing.

9.4.1.1 Varietal Purity

The first assessment in determining the suitability of a sample for malting is verifying the variety and varietal purity of a barley sample. Each variety differs in its malting potential and processing characteristics. Maltsters require specific varieties to meet a customer's malt specification. Varietal purity is an important assessment as varietal mixtures generally process poorly with each variety requiring specific processing conditions to realise the variety's full potential and to produce the best-quality malt possible without excessive malt losses.

Producer samples seldom contain mixtures and variety is generally declared verbally on delivery; or in some countries a written declaration of variety may be required. However, as barley moves through the supply chain, mixing opportunities arise increasing the need for assessing varietal purity. In the past variety purity has been assessed by visual inspection but the proliferation of malting varieties has reduced the effectiveness of a visual inspection. Assortments of protein-based methods, including polyacrylamide gel electrophoresis (Marchylo and LaBerge, 1981), high-performance liquid chromatography (Marchylo and Kruger, 1984) and mass spectrometry (Bloch et al., 2001) have been developed. However, due to rather limited variability of seed storage proteins among elite malting barley varieties, the level of resolution and accuracy with these approaches has become insufficient. More recently, attention has shifted to DNA-based methods, which with essentially limitless number of markers offer much greater resolving power (Terzi et al., 2005; Perry et al., 2014). Most laboratory methods presently determine purity by assessing kernels individually; this requirement complicates the analysis as large numbers of kernels must be analysed individually to provide reliable results. Recently, however, Perry et al. (2014) reported simple, rapid individual-kernel DNA preparation methods for use in conjunction with the multiplexed markers to provide a convenient and relatively inexpensive tool for barley variety identification (see Chapter 17).

9.4.1.2 Germination

Germination is the most important quality requirement for malting barley. Barley must germinate if quality malt is to be produced in a trouble-free manner. The importance of assessing barley germination is unique among cereal grains and a number of specialised tests have been developed (Table 9.5) and standardised (American Society of Brewing Chemists, 2004; European Brewery Convention, 1998). The most common test is germinative energy (3-day test), a test where 4 mL of water is added to a Petri dish containing 100 seeds and 2 pieces of filter paper. Germinated kernels are removed every 24 h. A sample with acceptable quality should have at least 95% germination after 72 h. A similar 3-day test uses 8 mL of water to test for water sensitivity, a phenomena where healthy, non-dormant barley kernels do not germinate in excess water. Water sensitivity is a calculated difference between the 4 and 8 mL results. Some degree of water sensitivity can be tolerated in malt houses, through modifications to the steeping process, but barley with a difference in excess of 10% is routinely rejected or held in storage for possible reduction in sensitivity. Water sensitivity can be a form of dormancy.

Dormancy helps prevent barley from sprouting prior to harvest which can occur when dry barley is exposed to rain or heavy dew prior to harvest. Sprouted barley can be persuaded to germinate in the malt house but this ability is reduced with time, especially under poor storage conditions (hot and humid). Detection can be difficult and had been restricted to visual assessment until the Rapid Visco Analyser was implemented as a more objective method for determining pre-harvest sprouting (Bason et al., 1993). The test is based on indirect measurement of α-amylase synthesised upon the onset of germination. Even a small amount of α-amylase can reduce the viscosity of barley slurry resulting in a lower RVA value. RVA values are used to detect and measure the degree of pre-germination as well as predict the risk of losing the germination energy upon storage. Samples with final viscosity values above 120 (rapid visco units, RVU) are considered sound and the probability that they will retain germination energy (GE) after storage is very high. Samples with RVA values between 50 and 120 RVU are moderately pre-germinated, whereas samples with RVA values below 50 RVU are substantially pre-germinated and the probability that they will lose GE during storage is high. They should be malted as soon as possible. To predict safe storage time more accurately, not only the RVA values, but also the storage conditions (T and RH) and the initial moisture content of the grain have to be taken into account (Izydorczyk, 2005).

Dormant barley is unacceptable in the malt house because dormant kernels will not germinate, resulting in incomplete modification and poor malt quality. Dormancy will break and selectors are unwilling to reject a good sample that may soon break dormancy. Two tests are used to determine dormancy, or germinative capacity as it is sometimes called. Peroxide may be used to break dormancy and the standard method is then somewhat similar to the germinative energy test. The tetrazolium technique involves the staining of individual kernels followed by visual assessment of embryos. The method is quicker but is subjective and requires an experienced operator.

A quick start to germination can reduce the processing time required to achieve properly modified malt. Germinative Index (GI) is a newer method (Riis and Bang-Olsen, 1991) of assessing the speed and uniformity at which germination commences. GI is calculated from the standard 3-day germination test (4 mL) where germinated seeds are counted on removal every 24 h and the results are substituted into the formula indicated in Eq. (9.1). The closer GI is to 10, the maximum GI which results from all kernels germinating the first day, the quicker barley will germinate in the malt house, thus reducing the required processing time (Table 9.6). The GI has yet to become a routine assessment tool.

$$GI = (A \times 10) \div ((B \times 1) + (C \times 2) + (D \times 3)) \quad (9.1)$$

A, Total kernels germinated after 72 h; *B*, kernels germinated after 24 h; *C*, kernels germinated between 24 and 48 h; and *D*, kernels germinated between 48 and 72 h.

9.4.1.3 Protein Level

The level of protein in barley grain is a key predictor of barley and malt quality and it is taken into consideration when purchase agreements are prepared. Protein levels in malting barley must be within a narrow range because elevated levels (>12.5%) cause poor endosperm modification, reduce levels of malt extract, and produce problems during beer-making, while barley with insufficient protein (<9.5%) may fail to produce adequate levels of enzymes. The required protein levels may vary among countries; for example 2-rowed European varieties are expected to have <11.5% proteins, whereas the desired range of proteins for 2-rowed North American varieties is 11–12.5% (12–13.5% for 6-rowed North American barley).

Barley kernels with high protein content may absorb water more slowly during steeping and this can lead to modification problems and non-uniformity especially if high and low protein barley kernels are mixed (Kramer, 2006). There is an inverse relationship between protein and starch content, and generally higher protein will yield lower levels of fermentable extracts (Briggs, 1998; Kunze, 1999). Higher protein grain will also produce higher levels of soluble nitrogen compounds in wort which can lead to several problems. Higher amount of soluble proteins in wort and beer may cause filtration problems due to formation of insoluble complexes with other wort or beer constituents. The formation of chill- or permanent hazes in beer has been linked to cross-linking of proline-rich soluble proteins with polyphenols (Siebert and Lynn, 1998). Soluble amino acids and low molecular nitrogenous compounds participate in the Maillard reactions and their excessive levels may lead to dark colour development in malt during kilning and undesired colour of wort and beer (Bathgate, 1973). Excessive level of free amino acids may also alter the growth and fermentation pattern of yeasts.

Even though more problems result from excessive protein levels in barley than from inadequate proteins levels, the grain protein levels below 9.5% may lead to insufficient production of hydrolytic enzymes, inadequate formation of amino acids to support the growth of yeast cells and poor foam formation in beer.

9.4.1.4 Physical Grain Characteristics

Loose or missing hulls, as a result of peeling, significantly reduce the quality of a barley sample. The hull offers protection to the barley kernel during harvesting, transportation and malting. Peeled kernels show reduced uniformity in germination, increased susceptibility to microbial infection and they present handling problems such as additional cleaning needs. Varieties can be resistant to peeling, especially when they have thicker hull-less but hot, dry growing conditions, especially at ripening, can exacerbate the problem. Peeling increases with more handling. Peeling is assessed using standard methods of visual assessment (Canadian Grain Commission, 2009).

Endosperm structure affects speed of water uptake and endosperm modification during malting. Steely, also referred to as vitreous kernels, result from the starch granules being tightly packed in the protein matrix. Such kernels take up water and modify more slowly than mealy/floury kernels. In the past steeliness was assessed visually by examining a number of transverse sections of kernels but the development of a specialised LTm machine, based on laser light, has automated and increased the objectivity of the assessment (Chandra et al., 2001). The method is good for assessing the malting potential of commercial barley as results are dependent on growing conditions and not barley variety. In contrast, a milling energy method, which assesses kernel hardness by quantifying the energy required to mill a barley sample, is more dependent on barley variety and it has found favour with barley breeders in their culling of inferior barley lines (Allison et al., 1979).

Variability of physical characteristics among kernels has received increased attention with the development of the 'Single Kernel Characterization System' (SKCS). The SKCS was developed for assessing wheat hardness but it has the capability of measuring kernel hardness, diameter, weight or moisture in any cereal grain (Osborne and Anderssen, 2003). The system analyses 300 individual kernels for the four parameters and calculates averages and standard deviations. Maltsters desire samples with limited variability as kernels with different sizes take up water and germinate at different rates. Limited variability means malt processing conditions can be set to maximised endosperm modification while limiting malt losses. When samples have variable kernel size, the maltster must compromise by limiting over-modification in smaller kernels to limit malt losses, while attempting to achieve the necessary modification in the large kernels. In addition to all the specialised testing outlined earlier, malting barley is assessed on receival for more traditional cereal quality factors, such as kernel plumpness and test weight, which are described in more detail in Chapter 19.

9.4.1.5 Microbial Load

Microbial contamination is a concern for all cereals but it may be particularly troublesome during processing of malting barley where moderate temperatures and high humidity during the steeping and germination steps of the malting process result in conditions that are favourable for the multiplication of the contaminants (Noots et al., 1999; Bokulich and Bamforth, 2013). Barley, like other cereals, is exposed to microflora from several sources during cultivation in the field (water, composted manure, soil, etc.), harvest, storage and transport.

Barley grain, covered by a fibrous husk, is normally colonised by a wide variety of bacteria, yeasts and filamentous fungi [for a comprehensive list see references by Flannigan (2003) and Noots et al. (1999)], but only a few plant-pathogenic fungi relevant to beer quality have been studied intensively. There is also a general consensus that fungi have a greater impact on malt quality than bacteria (Kaur et al., 2015). Grain-associated microbes are usually associated with negative effects on grain quality; however, some microbes may also have beneficial effects during malting. For example the fungal community may augment the malting process by production of hormones and enzymes to stimulate germination, by production of cell wall degrading enzymes, such as β-glucanases and xylanases, and proteolytic enzymes to improve malt modification, or by contribution to vitamin production or enhancing the bioavailability of components such as minerals (Laitila, 2007; Laitila et al., 2011).

Negative effects of fungal contamination of barley include safety concerns, problems with processing and poor quality of the final beer (Van Nierop et al., 2006). Fusarium head blight (FHB) of barley, caused principally by *F. graminearum* and *Fusarium culmorum*, results in the accumulation of mycotoxins in the grain: deoxynivalenol (DON), its acetylated forms 3-ADON or 15-ADON as well as the related mycotoxins T-2 and HT-2. Other *Fusarium* species may produce other toxins, including nivalenol (*F. poae* and *F. sporotrichioides*), moniliformin (*F. avenaceum*) and zearalenone (*F. oxysporum, F. moniliforme and F. equiseti*). The toxicogenic effects of DON and other toxins are well established for animals and humans, leading to adoption of strict standards for DON levels in barley (Pestka, 2010). Barley for malting or feed is normally rejected with levels in excess of 2 ppm of DON.

Favourable moisture and temperature conditions were shown to proliferate *Fusarium* species during air rests (steeping stage), germination and even the first step of kilning. *Fusarium* mould colonises barley kernels and besides potential production of mycotoxins, depletes grain nutrients, such as starch and protein, resulting in significant malting losses. *Fusarium*-infected grains were reported to produce worts with reduced β-glucans and viscosity, and greater levels of free amino nitrogen (FAN), soluble proteins and wort colour as compared to wort produced from control malt (Schwarz, 2003; Schwarz et al., 2002; Oliveira et al., 2012, 2014; Nielsen et al., 2014). These changes in wort characteristics are indicative of the action of fungal enzymes.

Fungal infection of barley has also been implicated as a cause of gushing in beer, a phenomenon of spontaneous release of gas, over-foaming and beer gushing out of the bottle in seconds after removal of the crown (Sarlin et al., 2005). Fungi have been shown to produce surface active, hydrophobic polypeptide molecules (hydrophobins) that act as nucleation sites for formation and growth of CO_2 bubbles (Shokribousjein et al., 2011). The link between excess fungal growth and gushing in well known, and suppression of *Fusarium* growth by application of lactic acid bacteria (LAB) has been shown to control and diminish gushing (Reininkainen et al., 1999). Gushing has also been linked to non-specific LTPs elicited by a stress response to fungal infections of barley (Hippeli and Elstner, 2002; Gorjanovic, 2010).

Beer fermentation can also be affected by microbes or their by-products. An example is premature yeast flocculation (PYF), a condition whereby yeast cells flocculate prior to completion of fermentation of sugars. Yeast flocculation occurs when cell wall mannoproteins bind lectin-like glycoproteins on other cells, resulting in aggregation and settling of yeast cells (Stratford and Carter, 1993). This phenomenon normally should happen at the end of fermentation when the sugar levels are very low and no longer prevent the interactions between the yeast cells. PYF is responsible for incomplete fermentation, off-flavours in beer and inefficiency within breweries (Panteloglou et al., 2012). The phenomenon of PYF has been associated with fungal infection of barley, although the exact mechanism has not been entirely elucidated. MacIntosh et al. (2014) demonstrated that PYF behaviour was induced by infection of malting barley in the field specifically with *F. graminearum* and *Cochliobolus sativus*. It has been proposed that PYF is caused by a range of polysaccharides occurring in the barley husk and released either in response to microbial infection or by degradation of the husk by microbial enzymes (Koizumi, 2008, 2009; Van Nierop et al., 2004).

The microbial load on barley grains used to be based on visual assessment of colour or general appearance of grain samples. Highly stained kernels are indicative of contamination with yeast and certain genera of filamentous fungi; for example red or black discolouration indicates the presence of *Fusarium* and *Alternaria*, respectively. Subsequently, more detailed identification of microbial communities was done using conventional culture-dependent methods. These conventional methods have been deemed biased towards the selective enrichment of fast-growing microorganisms adapted to specific nutrients present in the growing medium that potentially represented only a fraction of the entire microbial community. Nowadays, molecular methods relying on DNA-sequencing technologies have replaced traditional, culture-based detection methods and revolutionised the study of microbial community profiling in grain and other systems. The availability of several high-throughput methods for accurate identification of microbial communities (e.g. fluorescence in situ hybridisation, quantitative PCR, denaturing gradient gel electrophoresis, terminal restriction length polymorphism or next generation sequencing) and the possibility of their integration (Bokulich and Mills, 2012; Bokulich et al., 2015) may provide better insight into the relationships between microbes present on the barley grain and technological, sensorial and safety aspects of malt and beer, as well as to better management of microflora.

9.4.1.6 Malt Quality

Malt quality must be accurately assessed to ensure it meets the tight specifications demanded by the brewer. The maltster routinely assesses malt quality immediately on completion of processing. This provides quality control for processing and aids decisions on blending. The quality is reassessed after blending, which usually occurs just prior to shipment, to ensure the final product meets the brewer's specifications. The brewer often performs a final malt analysis on delivery depending on relationships between the maltster and brewer.

Malt quality is generally analysed following standard methods of analysis (American Society of Brewing Chemists, 2004; European Brewery Convention, 1998; Institute of Brewing, 1997). Typical malt analyses (Table 9.7) include:

1. Malt extract (fine grind), the proportion of malt that will solubilise during mashing indicating a malt's beer production potential.
2. Soluble protein, a requirement for adequate foam stability in beer, although, too much soluble protein can result in beer hazes and darker coloured beers.
3. FAN, an indication of available nitrogenous compounds, predominately free amino acids, for yeast nutrition.
4. β-Glucan, which indicates the extent of endosperm modification.
5. Diastatic power and α-amylase, methods for quantifying activity of starch-degrading enzymes.

Specifications for standard malt analyses can vary considerably among breweries and beer brands depending on recipes and in particular, the use or not of adjuncts. All-malt breweries, ones that do not use adjuncts, require well-modified malts with high extract levels and low levels of starch-degrading enzymes and FAN. In contrast, breweries using starchy adjuncts require higher levels of enzymes. Levels of enzymes can be too high for beer brands that require greater mouth feel, a property requiring higher levels of unfermented dextrins and, thus, less enzymes. Breweries that use high maltose syrups as adjuncts have reduced need for starch-degrading enzymes but increased need for FAN, especially when brewing at high gravities. Distilleries require high enzyme malt as their goal is to maximise conversion of starch to fermentable sugars.

Standard malt analyses are often considered poor predictors of brew house performance (Axcell, 1998) and alternate methods of analysis are always being pursued. In particular, methods for better prediction of fermentability have gone in several directions. Brewers have traditionally relied heavily on diastatic power as an indicator of fermentability, however, analysis of individual starch-degrading enzymes: α-amylase, β-amylase and limit dextrinase, have proven to be a better predictor (Evans et al., 2008). The brewing industry has often provided specifications for α-amylase in the past but not yet for the other two enzymes, continuing to rely on diastatic power. Duke and Henson (2008) found that measuring osmolyte concentration, in extracts of barley and green malt, gave a better indication of fermentable sugars than standard malt extract. The method is simple and could also be used to predict pre-harvest sprouting and endosperm modification. Barley breeders have shown interest in the method because of the reduced need for expensive malt analysis.

PYF is a common problem in breweries around the world but predicting a potential problem with malt remains difficult. Lake et al. (2008) recently reported a micro test for testing fermentability which should predict PYF but the practical significance of the test has yet to be demonstrated.

9.4.2 Feed Barley

Quality assessment of feed barley is limited as barley price is the major factor in determining acceptability. Only a few simple tests are used routinely to assess feed barley quality. Test weight (see Chapter 19) remains the most consistently measured feed barley parameter. It gives limited information on feeding value, unless test weight levels are extremely low, and is generally used as a pricing mechanism. Plumpness is of more practical significance as the majority of barley fed to cattle is rolled. Consistent plumpness is essential where rollers can be adjusted to prevent production of fines from over-rolling plump kernels, or whole kernels from under-rolling thin kernels. Problems arise when samples contain a mixture of plump and thin kernels but mixtures are hard to quantify, a possible opportunity for SKCS (see Section 9.4.1.4).

Barley moisture can also significantly affect rolling properties. The factor of greatest consequence for feeding is microbial contamination and the presence of toxins. The major concerns with barley are deoxynivalenol from *F. graminearum* and ochratoxin A mainly from *Penicillium verrucosum*. The other major concern is ergot which can have dire consequences for most animals if ingested, including death. Very few other toxins are found in barley due to the temperate regions where barley is produced and stored. Toxins are avoided, like with malting barley, through visual assessment and assaying for deoxynivalenol (see Chapter 15).

The feeding value of barley is known to vary significantly among samples but assessment is difficult. The ideal method is through animal-feeding trials, an unreasonable expectation for barley travelling through the supply chain. Prediction of metabolisable energy is possible with NIR (Edney et al., 1994) but commercial use has been limited except for proprietary use within feed companies and in barley-breeding programs. Feeding value has traditionally been predicted from proximate analysis as well as more specific fibre analyses, such as neutral detergent fibre and acid detergent fibre, as described by the AACC International (2008).

9.4.3 Food Barley

Quality requirements for food barley have been ill defined compared to those for malting barley. In modern times, the food industry and consumers have largely ignored barley as a food grain, resulting in a shortage of good-quality barley ingredients and lack of readily available attractive and tasty barley food products. This situation has partly resulted from the reputation of barley as sustenance for the masses throughout history, especially in times of scarcity of other cereals. There is also the general belief that barley grain, due to its versatility, does not require any specific quality traits for food production. For a long time, a synchronised research effort to define the quality traits of barley for food use and to develop barley varieties uniquely fit for food use was mostly absent.

In recent years only a few specific barley products, such as shochu and barley tea, have had clearly defined barley quality characteristics because of their importance in subsequent processes and resulting products. One of the requirements for barley for shochu production is satisfactory pearling performance. Pearling, which is a process of abrasive scouring, must gradually remove the hull, pericarp and aleurone layer without production of excess broken kernels. The Japanese have been the most progressive in developing methods for predicting suitability of barley for pearling. They rely on sound kernel ratios for which there is no standard method. Generally the test calculates the percentage of sound kernels remaining after pearling a 180-g sample, under standard conditions, to 65% (117-g remaining). Sound kernels exclude broken and thin kernels (Iwami et al., 2003; Edney et al., 2002). Sound kernel ratios are calculated based on Eq. (9.2). Small representative samples from the pearled barley can be used to count broken and thin kernels and to calculate the ratio (see e.g. in Table 9.8).

$$\text{Sound kernel ratio} = (A) \div (A + B + C) \times 100 \tag{9.2}$$

A, Sound kernels after pearling; *B*, broken kernels after pearling; and *C*, thin kernels after pearling (through 2.0 × 1.6 mm rectangle-punched sieve).

The hardness index from the SKCS also predicts suitability of barley for pearling and provides additional information on suitability for shochu processing. The high SKCS hardness indices correlate positively with pearling performance but negativity with fermentation, one of the subsequent processes of the shochu production (Iwami et al., 2003). As a result, the SKCS specifications have been used for commercial transactions of barley for shochu.

Barley tea, a cold summer drink popular in eastern Asia, is prepared by roasting whole kernel barley, boiling the roasted product and then cooling. The

TABLE 9.8 Sound Kernel Ratios of Barley Pearled to 65% and Suitability for Pearling

Pearling suitability	Sound kernel ratio (%)[a]	Sound kernels (g)	Broken kernels (g)	Thin kernels (g)
Very good	82.2	29.6	4.8	1.6
Good (thin too high)	77.2	24.1	3.8	3.3
Acceptable (broken too high)	70.1	22.3	8	1.5

[a]*Calculated according to Eq. (9.2).*

barley must have good roasting capability and ability to produce consistent flavours. The major assessment for quality is identification of appropriate varieties. Individual samples then must be of uniform size with moderate levels of protein (11–13% dry matter) to help maintain consistent roasting quality and flavour.

9.4.3.1 High β-Glucan Food Barley for Health Benefits

Consumer interest in health-promoting foods and functional grain ingredients is increasing as the link between diet and health is substantiated by positive clinical trials. Barley grain has a potential to prosper in the functional and natural food product market due to β-glucans, the major dietary fibre constituents in barley with proven ability for reducing serum cholesterol and the risk of cardiovascular disease (Food and Drug Administration, 2006; European Food Safety Authority, 2011; Health Canada, 2012) and implications in the ability to reduce the glycaemic index. Barley is also an excellent source of other dietary fibre constituents, mainly arabinoxylans, and bioactive constituents such as tocopherols and tocotrienols, vitamin B complexes, minerals and phenolic compounds. However, the use of barley for health benefits is a relatively new avenue and the desirable quality traits have not yet been fully specified. The health benefits of barley β-glucans are dose-dependent, and the approved health claims stipulate consumption of 3 g of barley β-glucans per day and at least 0.75–1 g of β-glucans per serving. Although there are several advantages of consumption of whole grain barley, achieving the recommended daily intake of β-glucans from whole grain may require consumption of large quantities of grain, making it unappealing to consumers concerned with high starch and/or caloric intake. In recent years several new high β-glucan barley varieties have been developed specifically for use in healthy food products. In addition to having a hull-less trait and altered starch characteristics (waxy or high amylose) these varieties contain 1.5–2 times more β-glucans than malting and feed barley cultivars. Several dry fractionation techniques, including roller-, pin- or stone-milling, were shown to further increase the concentration of β-glucans in the resultant fibre-rich fractions (Izydorczyk et al., 2014). The concentration of β-glucans in the fibre-rich fractions was shown to be proportional to the content of β-glucans in the grain. It is prudent to recommend, therefore, that barley varieties for human health food must contain a high level of β-glucans with a minimum of 6% (dwb). The measurement of β-glucan in food barley destined for health benefits becomes an important test for proper evaluation of both the monetary and functional value of barley grain. The use of hull-less barley for food is attractive because it eliminates the need for pearling or de-hulling to make edible products. However, hull-less barley has a propensity to retain its hull after harvest. Barley with greater than 5% adhering hulls, as assessed visually and by weight, is degraded and/or rejected as food barley (Canadian Grain Commission, 2009). Barley varieties with altered starches, waxy or high amylose, tend to have higher levels of β-glucan but also provide interesting functional properties. The starch type of delivered barley is generally identified through variety declaration as the tests for starch-type are complex and time consuming (Yanagisawa et al., 2006).

9.5 MAINTAINING BARLEY QUALITY

Maintaining barley quality begins with the availability of varieties with the genetic potential to produce the quality in demand. The goal of achieving and maintaining that genetic potential requires favourable growing conditions and adequate storage, including preparation for storage, such as drying.

9.5.1 Breeding for Quality

The genotype (variety name) defines the potential quality of malting barley and that potential is the result of breeding. Names of food barley varieties are beginning to be associated with a defined quality, especially with respect to specialised starch properties and increased levels of functional compounds, such as β-glucans. In contrast, although feed barley breeders continue to pursue varieties with specific feed traits, commercial interest has remained low.

Barley is bred for improved agronomic properties and appropriate disease resistance to help maximise a producer's return, but when breeding malting barley the greatest concern is quality. As a result malt barley breeding lines are probably more thoroughly assessed for quality than any other commodity. Breeders must begin the monitoring of quality at the earliest generation possible. Sample size is often the limiting factor and specialised equipment, techniques and methods have been developed to mimic malt houses and breweries and to produce meaningful information on a breeding line's malting potential (Sage, 1992). Results do not need to be as accurate as industrial analyses since malt breeders are answering a simple question: 'Does the line have the appropriate quality or not?' Techniques include NIR prediction of malt quality as well as the use of molecular markers. Numerous quantitative trait loci (QTL) have been reported for malting barley quality but their complexity has severely restricted their value in breeding programs (Fox et al., 2003) and markers for specific traits, such as heat stability of β-amylase, have proven to be of greater value (Eglinton et al., 1998).

Malt breeders must eventually provide industrial-like quality data on potential varieties in order to convince the malting and brewing industries to commercially test the lines. Only after successful plant-scale testing (malt house and brewery) are malt varieties commercially grown.

Considerable resources are dedicated, around the world, to the breeding of feed barley but quality concerns are limited. Breeding targets tend to concentrate on agronomic traits and disease resistance. Test weight and kernel plumpness are often monitored and specialised traits are sometimes bred for, such as increased metabolisable energy, low phytate barley (Thacker et al., 2003) and low ruminal digestibility (Bowman and Blake, 1996).

The food industry provides limited direction for quality breeding targets in barley. Breeders concentrate on agronomic traits and disease resistance with some consideration for constant kernel size which improves pearling suitability. Breeding efforts have been directed at developing hull-less barley with unique traits such as specialised starch-types, for example waxy and high amylose, and elevated levels of β-glucan.

9.5.2 Effects of Growing Conditions

The genetic potential of barley can only be realised when barley is grown under favourable conditions. Weather obviously has the greatest effect, and is uncontrollable; however, several factors are under the producer's control starting with selection of an appropriate variety. Local conditions, such as maturity needs, lodging resistance, drought-resistance and disease pressures should be considered in selecting an appropriate variety. Seed selected for the chosen variety should be relatively pure with respect to variety. The seed must have good germination and not be excessively contaminated with other cereals or weed seeds. The use of certified seeds (specifically selected and inspected for seeding purposes), while not essential does guarantee varietal purity and germination.

In addition to environmental factors, certain agronomic practices, such as seeding rate, seeding date and nitrogen fertilisation rates may also have a significant impact on the production and quality of barley. Studies conducted in the Northern Great Plains of North America reported that seeding barley relatively early often leads to positive outcomes including higher yields and improved malting quality with lower protein concentration (Lauer and Partridge, 1990; Weston et al., 1993; Juskiw and Helm, 2003; McKenzie et al., 2005). Recently, O'Donovan et al. (2012) conducted 3-year (2006–08) field trials in eight locations in western Canada to evaluate the effects of seeding date (relatively early and late) and seeding rates (100, 200, 300, 400 and 500 seeds/m^2) on the yield and quality of AC Metcalfe, the most commonly grown Canadian malting barley. Delayed seeding often resulted in negative results including increased protein content and decreased kernel plumpness and yield. Seeding rate at 300 seeds/m^2 was determined optimal as it maintained or improved yield, decreased protein concentration, increased kernel uniformity and decreased tillering (O'Donovan et al., 2012). Edney et al. (2012) reported that seeding rates of 400 compared to 200 seeds/m^2 produced smaller, but more uniform kernels with lower protein concentration; this leads to malt with higher extract and lower wort β-glucan (Table 9.9).

Application of high nitrogen (N) fertilisation rates, essential for maximising barley yields, usually result in higher grain protein levels, thinner kernels and greater variability in kernel size (O'Donovan et al., 2011). Excessive grain protein concentration ($>12.5\%$), which usually depresses endosperm modification and reduces malt extract levels (Edney et al., 2012), is often the reason why barley is rejected for malting purposes and sold at a lower price in the feed market (Brewing and Malting Barley Research Institute, 2015). However, acceptable

TABLE 9.9 Effect of Seeding Rate on Grain and Malt Quality Variables

Parameter	200 seeds/m^2	400 seeds/m^2	LSD ($P < 0.05$)
Barley protein (g/kg DM)	**114**[a]	**111**	**1.2**
Average kernel diameter (mm)	**2.56**	**2.51**	**0.009**
Standard deviation kernel diameter	**0.315**	**0.309**	**0.004**
Average grain hardness (SKCS units)	51.5	51.1	0.79
Standard deviation grain hardness	14.14	14.27	0.17
Germinative energy (%)	97.9	98.3	0.30
Water sensitivity (%)[b]	93.1	92.8	1.05
Germination index	**6.9**	**7.3**	**0.09**
Steep-out moisture (g/kg)	**456**	**459**	**0.9**
Malt yield (g/kg)	**921**	**920**	**0.84**
Fine grind malt extract (g/kg DM)	**805**	**806**	**1.0**
Soluble protein (g/kg DM)	48.2	48.3	0.4
Kolbach index (%)	**42.0**	**42.9**	**0.36**
Wort β-glucan (mg/L)	**143**	**112**	**8.07**
Diastatic power (°L)	**133**	**129**	**2.01**
α-Amylase (DU)	**57.4**	**58.2**	**0.56**
Friability modification (%)	**78.7**	**81.6**	**1.00**
Calcofluor modification (%)	**92.1**	**93.9**	**0.62**
Calcofluor homogeneity (%)	**83.5**	**85.4**	**0.78**

DM, Dry matter; DU, dextrinizing units; °L, degrees Lintner.
[a] *Pair means in bold differ significantly (P < 0.05).*
[b] *Presented as % germination with 8 mL of water.*
Source: Adapted from O'Donovan, J.T., Turkington, T.K., Edney, M.J., Clayton, G.W., McKenzie, R.H., Juskiw, P.E., Lafond, G.P., Grant, C.A., Brandt, S., Harker, K.N., Johnson, E.N., May, W.E., 2011. Seeding rate, nitrogen rate, and cultivar effects on malting barley production. Agron. J. 103 (3), 709–716.

levels of proteins might be achievable at high N-fertilisation rates because individual barley varieties may differ in their response to uptake of nitrogen and synthesis of proteins. O'Donovan et al. (2015) investigated the responses of relatively new Canadian barley cultivars to increasing N rates compared to the response of the most commonly grown malting cultivar. The experiments, conducted for 3 consecutive years at five N rates (0–120 kg/ha), indeed showed that some varieties produce significantly lower protein concentration, possibly

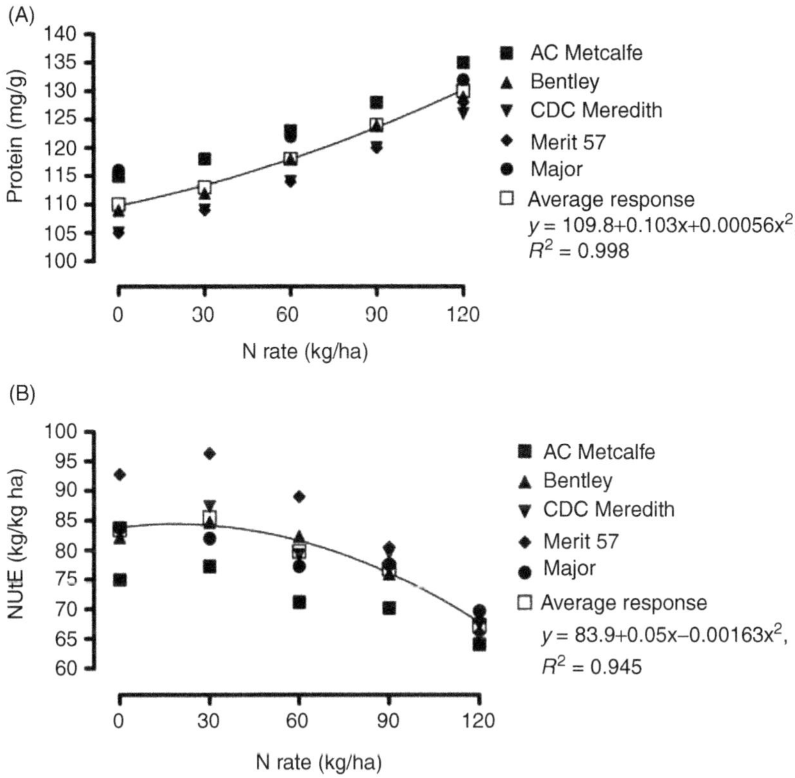

FIGURE 9.8 Effects of increasing N rates on (A) barley protein concentration in several malting barley varieties and (B) barley N utilisation efficiency (NUtE) *(Reprinted with permission O'Donovan, J.T., Anbessa, Y., Grant, C.A., Macleod, A.L., Edney, M.J., Izydorczyk, M.S., Turkington, T.K., Juskiw, P.E., Lafond, G.P., May, W.E., Harker, K.N., Johnson, E.N., Beres, B.L., McAllister, T.A., Smith, E.G., Chapman, W., 2015. Relative responses of new malting barley cultivars to increasing nitrogen rates in western Canada. Can. J. Plant Sci. 95, 831–839.)*

due to partitioning proportionally more N into leaf chlorophyll development (Fig. 9.8). The higher leaf chlorophyll content may increase photosynthesis which results in more efficient utilisation of N for grain formation.

Elevated protein levels are not concerns for feed barley where maximum yield, and thus maximum fertilisation, is desired. The timing of barley harvest will vary considerably based on local conditions but the aim should be minimum grain moisture. Extra time in the field should be avoided to reduce exposure of near-ripe grain to rain or heavy dew that could lead to pre-harvest sprouting (see Chapter 18 for more on growing conditions).

9.5.3 Storage Conditions

Rarely is there sufficient demand at harvest for all the barley produced that year. Barley, therefore like other cereal grains, must be stored with significant implications for germination loss as well as potential insect and microbial infection. Concerns are best avoided by storing grain with low moisture, by diligent use of aeration with grain turning as required and by proper use of insecticides when necessary (see Chapter 20 for more on storage). Concerns with germination loss are unique for malting barley where germination is an essential component of end-use processing. Similar to other storage concerns, low-moisture grain is essential for preserving germination. Storing grain at low moisture is advantageous even when barley has been sprouted prior to harvest. RVA testing can be used to predict optimal or maximum storage times for sprouted grain.

9.5.4 Processing

The most common process for maintaining the quality of all barley types is drying prior to storage, which if done properly can also break dormancy. Drying is done both in bulk and with on-farm grain dryers, depending on anticipated storage type, bulk or on-farm. Proper bulk drying, especially for breaking dormancy, starts with medium-length storage at warm temperatures followed by cooling and storage at ambient temperature. Drying with the small grain dryers is quicker but must be done with care to ensure grain temperatures do not exceed 43°C otherwise losses in germination, malting potential and nutritive value can be expected. Dry barley is less susceptible to microbial infestation; however, not all microbes are detrimental and barley has even been inoculated, prior to malting, with specific microbes known to be beneficial in terms of competition with detrimental microbes that are often present on the barley (Laitila et al., 2006).

9.6 FUTURE

Barley will continue to have three major end uses, feed, malting and food. Quality concerns will remain top priority across the whole value chain as processors continue to expect stable and predictable quality, whereas modern consumers are prepared to pay extra for premium quality, safe, healthy and tasty foods and beverages with distinct attributes. Even though the majority of physical and compositional characteristics of barley grain can be measured with the current methods and tests, there is a continuous search for alternate, fast and accurate methods of quality assessment, measurement and prediction. There is also a need for better understanding the relationships between the physical and compositional traits and processing characteristics and product quality. Climate change is expected to produce extreme weather with a resulting need for producing more climate tolerant varieties and increased monitoring of barley quality. There has been a renewed focus on agronomic research and the investigation of production systems that generate better-quality barley and lead to greater acceptance rates for malting barley producers. Globalisation has contributed to spreading of diseases and there will be continuing need not only to develop specific resistance in the barley plant to protect the quality of barley grain but also to monitor for levels of toxins in barley. There is major potential for increased food use of barley as processors develop products based on the functionality of specialised barley starches and on the health benefits of β-glucan. Beer consumption is expected to continue to rise but at a slower rate. Changes in technology and trends, for example the use of adjuncts and high gravity brewing or liquid sugar brewing, could affect malt demand and malt-quality requirements. The growth of craft brewing sector with its all-malt beers has already increased malt demand and changed the targets for the desirable malting barley traits compared to characteristics required by the mainstream adjunct brewing. Feed use is expected to remain relatively constant or decline; however, feed quality is uniquely suited for improvement. Alternative feed barley types, such as low phytate, high starch and high metabolisable energy, including hull-less, could become of interest to feed mills, especially if measurement of quality can be simplified at the mill driveway. Barley homogeneity, for all uses, could become an important quality parameter as methods based on SKCS and NIR are further developed. Other methods, such as RVA analysis of pre-harvest sprouting, could become essential for predicting storability of malt barley as pre-harvest sprouting continues to be a problem with extreme weather at harvest. The development of single-kernel NIR provides some of the most interesting possibilities for predicting kernel variability of several different parameters like protein, kernel size and metabolisable energy.

9.7 FURTHER INFORMATION

The development and release of barley varieties with distinct traits, for example with particular starch properties, low phytate, low gluten or lipoxygenase (LOX)-free barley, will continue with developments reported in journals like *Crop Science, Journal of Functional Foods* and *Journal of Plant Registrations*. The use of new types of barley will require development of innovative processes by researchers and industry. Annual meetings of organisations such as AACC International, the Canadian Institute of Food Science And Technology and the Institute of Food Technologists; as well as publications such as *Cereal Chemistry, Cereal Foods World, Food Chemistry, Journal of Cereal Science* and *Journal of the Science of Food and Agriculture* provide for a fair presentation and discussion of new processing technologies.

Alternative methods for assessing barley, including improvements to existing methods, are continually being addressed for all end uses of barley. Successful developments are presented in many of the aforementioned journals as well as specialised journals such as *Biosystems Engineering and the Journal of Near Infrared Spectroscopy*. Improved methods for assessing barley quality are also published in journals associated with specific end uses such as *Animal Feed Science and Technology*, the *Journal of Animal Science*, the *Journal of the American Society of Brewing Chemists* and the *Journal of the Institute of Brewing*. Standard methods of societies are constantly being reviewed and updated with improved methods being listed in their official methods as they become standardised and accepted (MEBAK, 2006; American Society of Brewing Chemists, 2004; Association of Analytical Communities, 2005; European Brewery Convention, 1998).

REFERENCES

AACC International, 2008. Approved Methods of the American Association of Cereal Chemists, tenth ed. AACC International, St. Paul, MN.

Aggarwal, B.B., Sundaram, C., Prasad, S., Kannappan, R., 2010. Tocotrienols, the vitamin E of the 21st century: its potential against cancer and other chronic diseases. Biochem. Pharmacol. 80 (11), 1613–1631.

Allison, M.J., Cowe, I.A., Borzucki, R., Bruce, F.M., McHale, R., 1979. Milling energy of barley. J. Inst. Brew. 85, 262–264.

American Society of Brewing Chemists, 2004. Methods of Analysis, ninth ed. The Society, St. Paul, MN.

Association of Analytical Communities, 2005. Official Methods of Analysis of AOAC International (OMA), eighteenth ed. AOAC International, Gaithersburg, MD.

Axcell, B., 1998. Malt analysis: prediction or predicament. MBAA Tech. Quart. 35, 28–30.

Baik, B.-K., Ullrich, S.E., 2008. Barley for food: characteristics, improvement, and renewed interest. J. Cereal Sci. 48, 233–242.

Bason, M.L., Ronalds, J.A., Wrigley, C.W., 1993. Prediction of safe storage life for sound and weather-damaged malting barley. Cereal Foods World 38 (5), 361–363.

Bathgate, G.N., 1973. Biochemistry of malt kilning. Brewers Dig. 48, 60–65.

Bloch, H.A., Petersen, M., Sperotto, M.M., Kes¸mir, C., Radzikowski, L., Jacobsen, S., Søndergaard, I., 2001. Identification of barley and rye varieties using matrix-assisted laser desorption/ionization time-of-flight mass spectrometry with neural networks. Rapid Commun. Mass Spectrom. 15, 440–445.

Bokulich, N.A., Bamforth, C.W., 2013. The microbiology of malting and brewing. Microbiol. Mol. Biol. Rev. 77 (2), 157–172.

Bokulich, N.A., Bergsveinson, J., Ziola, B., Mills, D.A., 2015. Mapping microbial ecosystems and spoilage-gene flow in breweries highlights patterns of contamination and resistance. eLife 4, 1–21, e04634.

Bokulich, N.A., Mills, D.A., 2012. Next-generation approaches to the microbial ecology of food fermentations. BMB Rep. 45 (7), 377–389.

Bowman, J., Blake, T., 1996. Barley feed quality for beef cattle. Proceedings Seventh International Barley Genetics Symposium, Saskatoon, University Extension Press, University of Saskatchewan, vol. 1, pp. 82–90.

Brennan, C.S., Smith, D.B., Harris, N., Shewry, P.R., 1998. The production and characterization of *Hor 3* null lines of barley provides new information on the relationship of D hordein to malting performance. J. Cereal Sci. 28, 291–299.

Brewing and Malting Barley Research Institute, 2015. http://www.bmbri.ca.

Briggs, D.E., 1998. Malts and Malting. Blackie Academic and Professional, New York.

Canadian Grain Commission, 2009. Barley. Canadian Grain Commission Official Grain Grading Guide (Chapter 6). http://grainscanada.gc.ca/oggg-gocg/06/oggg-gocg-6-eng.htm

Cardenas, E., Ghosh, R., 2013. Vitamin E: A dark horse at the crossroad of cancer management. Biochem. Pharmacol. 86, 845–852.

Chandra, S., Wheaton, L., Schumacher, K., Muller, R., 2001. Assessment of barley quality by light transmission—the Rapid LTm meter. J. Inst. Brew. 107 (1), 39–47.

Darlington, H.F., Rouster, J., Hoffmann, L., Halford, N.G., Shewry, P.R., Simpson, D.J., 2001. Identification and molecular characterization of hordoindolines from barley grain. Plant Mol. Biol. 47, 785–794.

Darlington, H.F., Tecsi, L., Harris, N., Griggs, D.L., Cantrell, I.C., Shewry, P.R., 2000. Starch granule associated proteins in barley and wheat. J. Cereal Sci. 32, 21–29.

Duke, S.H., Henson, C.A., 2008. A comparison of barley malt quality measurements and malt sugar concentrations. J. Am. Soc. Brew. Chem. 66 (3), 151–161.

Edney, M.J., Morgan, J.E., Williams, P.C., Campbell, L.D., 1994. Analysis of feed barley by near–infrared reflectance technology. J. Near Infrared Spectrosc. 2, 33–41.

Edney, M.J., O'Donovan, J.T., Turkington, T.K., Clayton, G.W., McKenzie, R., Juskiw, P., Lafond, G.P., Brandt, S., Grant, C.A., Harker, K.N., Johnson, E., May, W., 2012. Effects of seeding rate, nitrogen rate and cultivar on barley malt quality. J. Sci. Food Agric. 92, 2672–2678.

Edney, M.J., Rossnagel, B.G., Endo, Y., Ozawa, S., Brophy, M., 2002. Pearling quality of Canadian barley varieties and their potential use as rice extenders. J. Cereal Sci. 36, 295–304.

Egi, A., Speers, R.A., Schwarz, P.B., 2004. Arabinoxylans and their behavior during malting and brewing. Mas. Brew. Am. Assoc. Tech. Quart. 41, 248–267.

Eglinton, J.K., Langridge, P., Evans, D.E., 1998. Thermostability variation in alleles of barley beta-amylase. J. Cereal Sci. 28, 301–309.

EBC (European Brewery Convention), 1998. Analytica-EBC, Nürnberg, Germany, Verlag Hans Carl Getränke-Fachverlag.

European Food Safety Authority, 2011 [Online]. http://www.efsa.europa.eu/en/efsajournal/pub/2471.htm

Evans, D.E., Li, C., Eglinton, J.K., 2008. Improved prediction of malt fermentability by measurement of the diastatic power enzymes beta-amylase, alpha-amylase, and limit dextrinase: I. Survey of the levels of diastatic power enzymes in commercial malts. J. Am. Soc. Brew. Chem. 66 (4), 223–232.

Falk, J., Krahnstver, A., van der Kooij, T.A.W., Schlensog, M., Krupinska, K., 2004. Tocopherol and tocotrienol accumulation during development of caryopses from barley (*Hordeum vulgare* L.). Phytochemistry 65, 2977–2985.

Fincher, G.B., 2011. Biochemistry, physiology, and genetics of endosperm mobilization in germinated barley grain. In: Ullrich, S.E. (Ed.), Barley: Production, Improvement, and Uses. Wiley-Blackwell, Publishing Ltd, Chichester, UK, pp. 449–477.

Finnie, C., Svensson, B., 2014. Barley grain proteins. In: Shewry, P.R., Ullrich, S.E. (Eds.), Barley Chemistry and Technology. AACC International Inc., St. Paul, Minnesota, pp. 123–168.

Flannigan, B., 2003. The microbiota of barley and malt. In: Priest, F.G., Campell, I. (Eds.), Brewing Microbiology. third ed. Kluwer Academic/Plenum Publishers, New York, pp. 113–180.

Food and Drug Administration, 2006. Food labelling: health claims; soluble dietary fiber from certain foods and coronary heart disease. Federal Register 70 (246), 76150–76162, http://edocket.access.gpo.gov/2005/pdf/05-24387.pdf.

Fox, G.P., Panozzo, J.F., Li, C.D., Lance, R.C.M., Inkerman, P.A., Henry, R.J., 2003. Molecular basis of barley quality. Aust. J. Agric. Res. 54, 1081–1101.

Garcia, A.L., Otto, B., Reich, S.C., Weickert, M.O., Steiniger, J., Machowetz, A., Rudivich, N.N., Mohlig, M., Katz, N., Speth, M., Meuser, F., Doerfer, J., Zunft, H.J.F., Pfeiffer, A.H.F., Koebnick, C., 2007. Arabinoxylan consumption decreases postprandial serum glucose, serum insulin and plasma total ghrelin response in subjects with impaired glucose tolerance. Eur. J. Clin. Nutr. 61, 334–341.

Gorjanovic, S., 2010. A review: the role of barley seed pathogenesis-related proteins (PRs) in beer production. J. Inst. Brew. 116, 111–124.

Gorjanovic, S., Spillner, E., Beljanski, M.V., Gorganovic, R., Pavlovic, M., Gojgic-Cvijanovic, G., 2005. Malting barley grain non-specific lipid-transfer protein (ns-LTP): importance for grain protection. J. Inst. Brew. 11, 99–104.

Gubatz, S., Shewry, P.R., 2011. The development, structure, and composition of the barley grain. In: Ullrich, S.E. (Ed.), Barley: Production, Improvement, and Uses. Wiley-Blackwell, Publishing Ltd, Chichester, UK, pp. 391–477.

Hansen, M., Lange, M., Friis, C., Dionisio, G., Holm, P.B., Vincze, E., 2007. Antisense-mediated suppression of C-hordein biosynthesis in the barley grain results in correlated changes in the transcriptome, protein profile, and amino acid composition. J. Exp. Bot. 58, 3987–3995.

Health Canada, 2012. Summary of health Canada's assessment of a health claim about barley products and blood cholesterol lowering [Online]. http://www.hc-sc.gc.ca/fn-an/label-etiquet/claims-reclam/assess-evalu/barley-orge-eng.php

Hejgaard, J., 1984. Gene products of barley chromosome 4 and 7 are precursors of the major antigenic beer proteins. J. Inst. Brew. 90, 75–87.

Hejgaard, J., Boisen, S., 1980. High lysine proteins in Hiproly barley breeding: identification, nutritional significance and new screening methods. Hereditas 93, 311–320.

Hippeli, S., Elstner, E.F., 2002. Are hydrophobins and/or non-specific lipid transfer proteins responsible for gushing in beer? New hypotheses on the chemical nature of gushing inducing factors. Z. Naturforsch. 57, 1–8.

Hopkins, M.J., Englyst, H.N., Macfarlane, S., Furrie, E., Macfarlane, G.T., McBain, A.J., 2003. Degradation of cross-linked and non-cross-linked arabinoxylans by the intestinal microbiota in children. Appl. Environ. Microbiol. 69, 6354–6360.

Institute of Brewing, 1997. Methods of Analysis. The Institute of Brewing, London, England.

Iwami, A., Kajiwara, Y., Omori, T., 2003. Estimating barley character for Shochu using a single-kernel characterization system (SKCS). J. Inst. Brew. 109 (2), 129–134.

Izydorczyk, M.S., 2005. Prediction of germination energy of malting barley during long-time storage. Canadian Grain Commission Scientific and Technical Reports. http://grainscanada.gc.ca/technology-technologie/germination/pgemb-pegob-eng.htm

Izydorczyk, M.S., 2010. Variations in content and molecular structure of barley nonstarch polysaccharides associated with genotypic and cellular origin. Cereal Chem. 87, 376–384.

Izydorczyk, M.S., 2014. Barley arabinoxylans: molecular, physicochemical, and functional properties. In: Shewry, P.R., Ullrich, S.E. (Eds.), Barley: Chemistry and Technology. second ed. AACC International, Inc., St. Paul, MN, pp. 97–122.

Izydorczyk, M.S., Bazin, S., Liu, H., Preston, K., Qian, K., Ens, W., Lefol, E., Harvey, B. 2004. Investigation of the molecular characteristics of barley hordeins by HPLC and MALDI-TOF MS techniques. In : Spunar, J., Janikova, J. (Eds.), Proceedings of the Ninth International Barley Genetics Symposium, 449–454. Brno June 2004.

Izydorczyk, M.S., Ens, W., Qian, Y., Bazin, S., Chornick, T., 2006. Identification of foam- and haze-forming barley proteins using chromatographic and MALDI-TOF mass spectrometry techniques. American Association of Brewing Chemist, Annual Meeting.

Izydorczyk, M.S., Miller, S.S., Beattie, A.D., 2014. Milling food barley: functional fractions enriched with β-glucans and other dietary fiber components. Cereal Food World 59, 277–285.

Jegou, S., Douliez, J.-P., Molle, D., Boivin, P., Marion, D., 2000. Purification and structural characterization of LTP1 polypeptides from beer. J. Agric. Food Chem. 48, 5023–5029.

Jenkins, J.A., Griffiths-Jones, S., Shewry, P.R., Breiteneder, H., Mills, E.N.C., 2005. Structural relatedness of plant food allergens with specific reference to cross-reactive allergens: and in *silico* analysis. J. Allergy Clin. Immunol. 115, 163–170.

Jenkins, A.L., Jenkins, D.J.A., Zdravkovic, U., Wursch, P., Vuksam, V., 2002. Depression of the glycemic index by high levels of β-glucan fiber in two functional foods tested in type 2 diabetes. Eur. J. Clin. Nutr. 56, 622–628.

Jenkins, D.J.A., Wolever, T.M.S., Leeds, A.R., et al., 1978. Dietary fibres, fibre analogues, and glucose tolerance: importance of viscosity. Br. Med. J. 1, 1392–1394.

Jiang, Q., Wong, J., Fyrst, H., Saba, J.D., Ames, B.N., 2004. γ-Tocopherol or combinations of vitamin E forms induce cell death in human prostate cancer cells by interrupting sphingolipid synthesis. Proc. Natl. Acad. Sci. USA 101 (51), 17825–17830.

Juskiw, P.E., Helm, J.H., 2003. Barley response to seeding date in central Alberta. Can. J. Plant Sci. 83, 275–281.

Kaur, M., Bowman, J.P., Stewart, D.C., Evans, D.E., 2015. The fungal community structure of barley malts from diverse geographical regions correlates with malt quality parameters. Int. J. Food Microbiol. 215, 71–78.

Koizumi, H., 2008. Barley malt polysaccharides inducing premature yeast flocculation and their possible mechanisms. J. Am. Soc. Brew. Chem. 66, 137–142.

Koizumi, H., 2009. Structural features of barley malt polysaccharides inducing premature yeast flocculation. J. Am. Soc. Brew. Chem. 67, 129–134.

Kramer, P., 2006. Barley, malt, and malting. Ockert, K. (Ed.), MBAA Practical Handbook for the Specialty Brewer. Raw Materials and Brewhouse Operations, vol. 1, Master Brewers Association of the Americas, St. Paul, MN.

Kunze, W., 1999. Technology of Brewing and Malting, second ed. VLB, Berlin.

Laitila, A., 2007. Microbes in the Tailoring of Barley Malt Properties. Academic Dissertation in Microbiology. VTT Publications, 645, Helsinki.

Laitila, A., Sweins, H., Vilpola, A., Kotaviita, E., Olkku, J., Home, S., Haikara, A., 2006. *Lactobacillus plantarum* and *Pediococcus pentosaceus* starter cultures as a tool for microflora management in malting and for enhancement of malt processability. J. Agric. Food Chem. 54, 3840–3851.

Laitila, A., Sarlin, T., Raulio, M., Wilhemson, A., Kotaviita, E., Huttunen, T., Juvonen, R., 2011. Yeasts in malting, with special emphasis on *Wickerhamomyces anomalus* (synonym *Pichia anomala*). Antonie Van Leeuwenhoek 99, 75–84.

Lake, J.C., Speers, R.A., Porter, A.V., Gill, T.A., 2008. Miniaturizing the fermentation assay: effects of fermentor size and fermentation kinetics on detection of premature yeast flocculation. J. Am. Soc. Brew. Chem. 66 (2), 94–102.

Lange, M., Vincze, E., Wieser, H., Schjoerring, J.K., Holm, P.B., 2007. Suppression of C-hordein synthesis in barley by antisense constructs results in a more balanced amino acid composition. J. Agric. Food Chem. 55, 6074–6081.

Lauer, J.G., Partridge, J.R., 1990. Planting date and nitrogen rate effects on spring malting barley. Agron. J. 82, 1083–1088.

Lazaridou, A., Biliaderis, C.G., Izydorczyk, M.S., 2006. Cereal β-glucans: structure, physical properties, and physiological functions. In: Biliaderis, C.G., Izydorczyk, M.S. (Eds.), Functional Food Carbohydrates. CRC Press, Boca Raton, pp. 1–72.

Lee, R., Allan, G., Greig, R., Dunn, C., Stuart, M., 1998. The effect of (1-3)(1-4)-β-glucan and arabinoxylans on malt quality and brewhouse performance. Proc. Conv. Inst. Guild Brew. Asia Pacific Sect. 25th, Perth, Institute & Guild of Brewing, London, UK, pp. 33–37.

Li, Y., Lu, J., Gu, G., Shi, Z., Mao, Z., 2005. Studies on water—extractable arabinoxylans during malting and brewing. Food Chem. 93, 33–38.

Lopez, H.W., Levrat, M.A., Guy, C., Messager, A., Demigné, C., Rémésy, C., 1999. Effects of soluble corn bran arabinoxylans on cecal digestion, lipid metabolism and mineral balance (Ca, Mg) in rats. J. Nutr. Biochem. 10, 500–509.

Lu, Z.X., Walker, K.Z., Muir, J.G., O'Dea, K., 2004. Arabinoxylan fibre improves metabolic control in people with Type II diabetes. Eur. J. Clin. Nutr. 58, 621–628.

Lund, E.K., Gee, J.M., Brown, J.C., Wood, P.J., Johnson, I.T., 1989. Effect of oat gum on the physical properties of the gastrointestinal contents and on the uptake of D-galactose and cholesterol by rat small intestine in vitro. Br. J. Nutr. 62, 91–101.

MacIntosh, A.J., MacLeod, A., Beattie, A.D., Eck, E., Edney, M., Rossnagel, B., Speers, R.A., 2014. Assessing the effect of fungal infection of barley and malt on premature yeast flocculation. J. Am. Soc. Brew. Chem. 72 (1), 66–72.

Marchylo, B.A., Kruger, J.E., 1984. Identification of Canadian barley cultivars by reversed-phase high-performance liquid chromatography. Cereal Chem. 61, 295–301.

Marchylo, B.A., LaBerge, D.E., 1981. Barley cultivar identification by electrophoretic analysis of hordein proteins II. Catalogue of electrophoregram formulae for Canadian-grown barley cultivars. Can. J. Plant Sci. 61, 859–870.

McKenzie, R.H., Middleton, A.B., Bremer, E., 2005. Fertilization, seeding date, and seeding rate for malting barley yield and quality in southern Alberta. Can. J. Plant Sci. 85, 603–614.

MEBAK (Mitteleuropäische Brautechnische Analysenkommision), 2006. Rohstoffe, Freising, Germany.

Newman, R.K., Newman, C.W., 2008. Barley for Food and Health. John Wiley and Sons Inc, New Jersey.

Nielsen, L.K., Cook, D.J., Edwards, S.G., Ray, R.V., 2014. The prevalence and impact of Fusarium head blight pathogens and mycotoxins on malting barley quality in UK. Int. J. Food Microbiol. 179, 38–49.

Noots, I., Delcour, J.A., Michiels, C.W., 1999. From field barley to malt: detection and specification of microbial activity for quality aspects. Crit. Rev. Microbiol. 25 (2), 121–153.

O'Donovan, J.T., Anbessa, Y., Grant, C.A., Macleod, A.L., Edney, M.J., Izydorczyk, M.S., Turkington, T.K., Juskiw, P.E., Lafond, G.P., May, W.E., Harker, K.N., Johnson, E.N., Beres, B.L., McAllister, T.A., Smith, E.G., Chapman, W., 2015. Relative responses of new malting barley cultivars to increasing nitrogen rates in western Canada. Can. J. Plant Sci. 95, 831–839.

O'Donovan, J.T., Turkington, T.K., Edney, M.J., Clayton, G.W., McKenzie, R.H., Juskiw, P.E., Lafond, G.P., Grant, C.A., Brandt, S., Harker, K.N., Johnson, E.N., May, W.E., 2011. Seeding rate, nitrogen rate, and cultivar effects on malting barley production. Agron. J. 103 (3), 709–716.

O'Donovan, J.T., Turkington, T.K., Edney, M.J., Juskiw, P.E., McKenzie, R.H., Harker, K.N., Clayton, G.W., Lafond, G.P., Grant, C.A., Brandt, S., Johnson, E.N., May, W.E., Smith, E., 2012. Effect of seeding date and seeding rate on malting barley production in western Canada. Can. J. Plant Sci. 92, 321–330.

Oliveira, P.M., Mauch, A., Jacob, F., Waters, D.M., Arendt, E.K., 2012. Fundamental study on the influence of Fusarium infection on quality and ultrastructure of barley malt. Int. J. Food Microbiol. 156, 32–43.

Oliveira, P.M., Zannini, E., Arendt, E.K., 2014. Cereal fungal infection, mycotoxins, and lactic acid bacteria mediated bioprotection: from crop farming to cereal products. Food Microbiol. 37, 78–95.

Osborne, T.B., 1907. The Proteins of the Wheat Kernel. Carnegie Institute of Washington, Washington, DC.

Osborne, B.G., Anderssen, R.S., 2003. Single-kernel characterization principles and applications. Cereal Chem. 80 (5), 613–622.

Panteloglou, A.G., Smart, K.A., Cook, D.J., 2012. Malt-induced premature yeast flocculation: current perspectives. J. Ind. Microbiol. Biotechnol. 39, 813–822.

Perry, D.J., Fernando, U., Lee, S.-J., 2014. Simple sequence repeat-based identification of Canadian malting barley varieties. Can. J. Plant Sci. 94, 485–496.

Pestka, J.J., 2010. Deoxynivalenol: mechanisms of action, human exposure, and toxicological relevance. Arch. Toxicol. 84, 663–679.

Philip, J.S., Gilbert, H.J., Smithard, R.R., 1995. Growth, viscosity and β-glucanase activity of intestinal fluid in broiler chickens fed on barley-based diets with or without exogenous β-glucanase. Br. Poult. Sci. 36, 599–603.

Reininkainen, P., Peltola, P., Lampinen, R., Haikara, A., Olkku, J., 1999. Improving the quality of malting barley by employing microbial starter cultures in the field. Proceedings of the Twenty-Seventh Congress of the European Brewery Convention. European Brewery Convention, Brussels, Belgium, pp. 551–558.

Riis, P., Bang-Olsen, K.B., 1991. Germination profile—a new term in barley analyses. Proceedings of the European Brewery Convention Congress, Lisbon, Oxford, IRL Press, 101–108.

Roberts, T.H., Marttila, S., Rasmussen, S.K., Hejgaard, J., 2003. Differential gene expression for suicide-substrate serine proteinase inhibitors (serpins) in vegetative and grain tissues of barley. J. Exp. Bot. 54, 2251–2263.

Sadosky, P., Schwarz, P.B., Horsley, R.D., 2002. Effect of arabinoxylans, β-glucan, and dextrins on the viscosity and membrane filterability of a beer model solution. J. Am. Soc. Brew. Chem. 60, 153–162.

Sage, G.C.M., 1992. Current and future breeding of barley for malting and brewing (Cass Lecture Autumn 1991). Brewers' Guardian 121 (2), 27–29.

Sarlin, T., Laitila, A., Pekkarinen, A., Haikara, A., 2005. Effects of three *Fusarium* species on the quality of barley and malt. J. Am. Soc. Brew. Chem. 63 (2), 43–49.

Schwarz, P.B., 2003. Impact of head blight on the malting and brewing quality of barley. In: Leonard, K.J., Bushnell, W.R. (Eds.), *Fusarium* Head Blight of Wheat and Barley. American Phytopathological Society Press, St. Paul, MN, pp. 395–419.

Schwarz, P.B., Jones, B.L., Steffenson, B.J., 2002. Enzymes associated with *Fusarium* infection of barley. J. Am. Soc. Brew. Chem. 60, 130–134.

Sealey, W.M., Barrows, F.T., Hang, A., Johansen, K.A., Overturf, K., LaPatra, S.E., Hardy, R.W., 2008. Evaluation of the ability of barley genotypes containing different amounts of beta-glucan to alter growth and disease resistance of rainbow trout *Oncorhynchus mykiss*. Anim. Feed Sci. Technol. 141 (1–2), 115–128.

Sen, C.K., Khanna, S., Rink, C., Roy, S., 2007. Tocotrienols: the emerging face of natural vitamin E. Vitam. Horm. 76, 203–261.

Shewry, P.R., Darlington, H., 2002. The proteins of the mature barley grain and their role in determining malting performance. In: Slafer, G.A., Molina-Cano, J.L., Savin, R., Araus, J.L., Romagosa, I. (Eds.), Barley Science: Recent Advances from Molecular Biology to Agronomy of Yield and Quality. Food Product Press, New York, pp. 503–521.

Shokribousjein, Z., Deckers, S.M., Gebruers, K., Lorgouilloux, Y., Baggerman, G., Verachtert, H., Delcour, J.A., Etienne, P., Rock, J.M., Michiels, C., 2011. Hydrophobins, beer foaming and gushing. Cerevisia 35, 85–101.

Siebert, K.J., Lynn, P.Y., 1998. Comparison of polyphenol interactions with polyvinylpolypyrrolidone and haze active protein. J. Am. Soc. Brew. Chem. 56, 24–31.

Smith, D.B., Lister, P.R., 1983. Gel-forming proteins in barley grain and their relationship with malting quality. J. Cereal Sci. 1, 229–239.

Stratford, M., Carter, A.T., 1993. Yeast flocculation—lectin synthesis and activation. Yeast 9, 371–378.

Terzi, V., Morcia, C., Gorrini, A., Stanca, A.M., Shewry, P.R., Faccioli, P., 2005. DNA based methods for identification and quantification of small grain cereal mixtures and fingerprinting of varieties. J. Cereal Sci. 41, 213–220.

Thacker, P.A., Rossnagel, B.G., Raboy, V., 2003. Phosphorus digestibly in low-phytate barley fed to finishing pigs. Can. J. Anim. Sci. 83, 101–104.

Ullrich, S.E., 2011. Significance, adaptation, production and trade of barley. In: Ullrich, S.E. (Ed.), Barley: Production, Improvement, and Uses. Wiley-Blackwell, Publishing Ltd., Chichester, UK, pp. 3–13.

USDA Foreign Agricultural Service, 2009. Production, Supply and Distribution. http://www.fas.usda.gov/psdonline/psdHome.aspx

Van Nierop, S.N.E., Cameron-Clarke, A., Axcell, B.C., 2004. Enzymatic generation of factors from malt responsible for premature yeast flocculation. J. Am. Soc. Brew. Chem. 62, 108–116.

Van Nierop, S.N.E., Rautenbach, M., Axcell, B., Cantrell, I.C., 2006. The impact of microorganisms on barley and malt quality—a review. J. Am. Soc. Brew. Chem. 64 (2), 69–78.

Weston, D.T., Horsley, R.D., Schwarz, P.B., Goos, R.J., 1993. Nitrogen and planting date effects on low-protein spring barley. Agron. J. 858, 1170–1174.

Wolever, T.M., Tosh, S.M., Gibbs, A.L., et al., 2010. Physicochemical properties of oat β-glucan influence its ability reduce serum LDL cholesterol in humans: a randomized clinical trial. Am. J. Clin. Nutr. 92, 723–732.

Yanagisawa, T., Domon, E., Fujita, M., Kiribuchi-Otobe, C., Takayama, T., 2006. Starch pasting properties and amylose content from 17 waxy barley lines. Cereal Chem. 83 (4), 354–357.

Zunft, H-J.F., Lueder, W., Koebnick, C., Imhof, D., Meuser, F., 2004. Reduction of the postprandial glucose and insulin response in serum of healthy subjects by an arabinoxylans concentrate isolated from wheat starch plant process water. Asia Pac. J. Clin. Nutr. 13, S147.

Chapter 10

Oats: Grain-Quality Characteristics and Management of Quality Requirements

Pamela Zwer
South Australian Research & Development Institute, SARDI, Adelaide, SA, Australia

10.1 INTRODUCTION

Oats have played a major role in farming systems due to the versatile uses of the grain and plant. Oats are used for human consumption, feed for animals and non-food products. Grain quality characters differ for each of these end uses.

Food products from oats are differentiated from other cereal grains by using the whole kernel or groat after removal of the hull, the plant tissue that tightly encloses the groat. Fig. 10.1 shows the grain with the hull and the groat after hull removal. The use of the whole kernel in porridge and other oat products is responsible for a number of health benefits beyond basic nutrition. Oats have been called 'nature's functional food' (Kapica, 2001). Oats were clinically shown to reduce cholesterol levels, reduce the risk of heart disease and enhance the feeling of satiety for weight control (Kapica, 2001).

The highest proportion of the oat crop is utilised as animal feed for horses, cattle and sheep. The fibre in the hull along with the nutritive composition of the kernel is a good combination for animal nutrition and protection against acidosis. Naked oats are especially suited for poultry and pigs, which do not have the enzyme systems to cope with the hull, and for race horses which require high energy rations.

Non-food uses and industrial oat products vary from cosmetics, lotions and shampoos to the production of cardboard products and the manufacture of furfural and furan compounds used for solvents, adhesives, filtering aids and the construction of board material and cellulose pulp.

Although oat production is declining in many countries, there is a significant increase in the consumption of oats for food products and the opportunity to expand into non-food products to add value to the oat crop. Moreover, specialisation

Cereal Grains. http://dx.doi.org/10.1016/B978-0-08-100719-8.00010-3

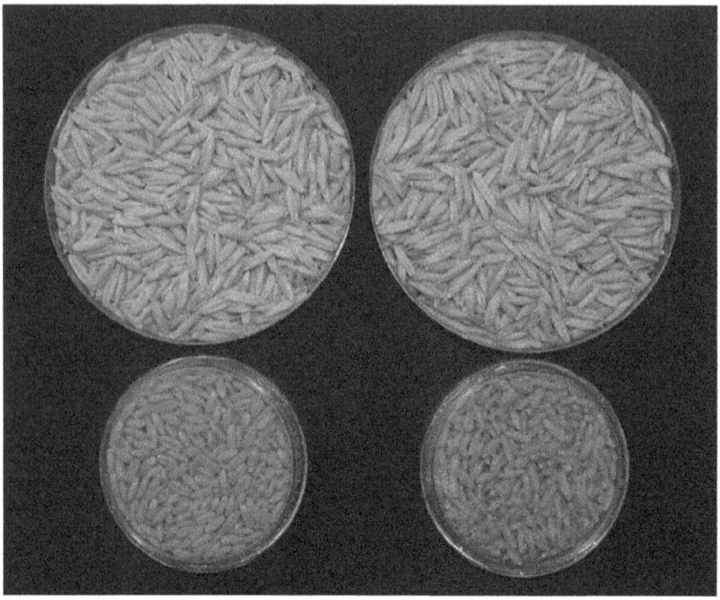

FIGURE 10.1 The grain with the hull and the groat, after hull removal, of two Australian varieties, Yallara (left) and Mitika (right).

of oat varieties for specific end uses, such as food, animal feed, health and beauty, industrial and pharmaceutical products will be dependent on the components of the oat grain; these topics are discussed in the following sections.

10.2 CLASSIFICATION, ORIGIN AND HISTORY

The oat genus, *Avena* L. (Poaceae), belongs to the tribe Aveneae of the family Gramineae. The primary species cultivated is *Avena sativa*. However, *Avena byzantina* and *Avena strigosa* are also grown in some regions for animal feed and fodder.

The species described in *Avena* form a polyploidy series varying from one to three chromosome sets with the basic chromosome number seven. Three distinct diploid genomes A, C and D hybridised in the development of cultivated hexaploid oats. Although it is not certain where the centre of origin for oats is located, the greatest genetic diversity is found in the Mediterranean, the Middle East, and the Himalayan regions.

Farming and trading flourished in Europe by 2000 BC. Wheat and barley seed were traded bringing oat and rye with them as weed contaminants. Although the historical record is limited, cultivated oats were identified in northern regions of Western Europe between 4500 and 400 BC where climatic conditions were wet and cold. Oats were favoured in these conditions compared to wheat and barley. Oats were ranked fourth in importance after wheat, barley and rye by the Renaissance. Oats could flourish in areas where wheat and barley produced marginal yields. Although the primary use of oats was for animal feed, by 1500 it was the principal grain crop for human consumption in Scotland, Wales, Ireland and Britain.

Immigrants and explorers from Great Britain and Spain introduced oats to North America. The English introduced oats into Canada, New England and eastern USA between 1500 and 1600 as animal feed. Scottish immigrants continued to use oats for porridge and other foods. The Spanish brought oats to the Pacific Coast in the early 1800s to feed their horses. Because of the medicinal properties, the general public used oats for the sick, purchasing the product at chemists. Oat production moved west to the Upper Mississippi Valley and into Canada by the 1880s. Oatmeal as a breakfast food began to flourish by 1900 because local mills began producing breakfast cereals, which were sold in grocery stores rather than pharmacies.

Governor Phillip introduced oats into New South Wales, Australia, in 1791 to be used as a green fodder and hay for horses, dairy cattle and pigs. Immigrants from Great Britain were primarily responsible for introducing rolled oats as a breakfast food. Leonard and George Parsons emigrated from England to Australia in 1861 and by the late 1880s began manufacturing rolled oats.

10.3 UNIQUE MORPHOLOGICAL AND BIOCHEMICAL CHARACTERISTICS

Although wheat, barley, rice, maize and oats have similar tissues and physiological processes in the grain, oats are unique chemically and structurally compared to the other cereal grains. See Chapter 4 for more information about head and grain morphology. Oat protein in the groat ranges from 6% to 18% and has a superior balance of amino acids, particularly lysine, compared to other cereals (Lockhart and Hurt, 1986). Wheat, barley and maize have simple starch granules compared to oat and rice starch, which have compound granules. Oat lipids have a fatty acid composition possessing high oleic and linoleic acids, resulting in a favourable polyunsaturated to saturated fatty acid ratio. Oat lipid content exceeds wheat and barley 2–5 times in oat varieties. Oats contain β-glucan, which possesses viscosities in solution that exceed other biological hydrocolloids, resulting in some of the healthful benefits. Oats are also rich in the antioxidants; tocopherols, tocotrienols, phenolic acids, flavonoids, sterols, phytic acid and avenanthramides, unique to oats.

The hulls, tightly enclosing the groat, are dry and brittle at physiological maturity and contribute 25–35% of the total grain weight. The hull is high in dietary fibre composed of cellulose, hemicellulose and lignin and is not easily digested by non-ruminants. Once the hulls are removed from the grain for food processing, the groat is long and narrow possessing abundant hairs or trichomes. The commercial fractions of a groat are separated into the bran, germ or embryo and starchy endosperm (Fig. 10.2). The most important component of bran is the aleurone layer. The germ or embryo is a small proportion of the groat composed of rudimentary leaves and roots, coleoptile and coleorhizae. The starchy

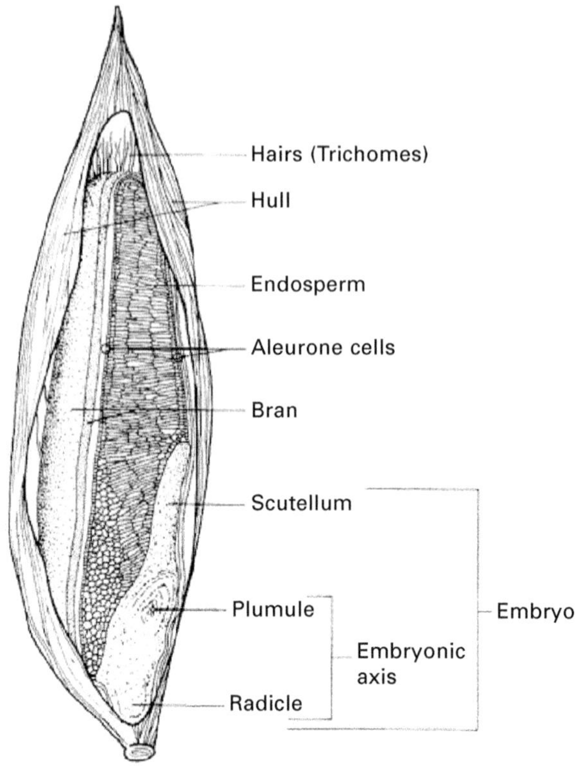

FIGURE 10.2 Composition of an oat kernel. *(Reproduced with permission from Youngs V.L., 1986. Oat lipids and lipid-related enzymes. In: Webster, F. H. (Ed.), Oats: Chemistry and Technology. The American Association of Cereal Chemists, St. Paul, MN.)*

endosperm is 80% of groat dry weight and contains the major reserves of starch, protein, lipids and β-glucan.

10.3.1 Starch

Starch is the major carbohydrate in oats and comprises from 65% to 85% (by weight) of the groat depending on the groat protein content (Paton, 1986). The primary components of starch are amylose and amylopectin. Early literature reported significant quantities of starch intermediates called anomalous amylose and anomalous amylopectin in oats (Banks and Greenwood, 1967). However, Tester and Karkalas (1996) provide evidence suggesting the starch intermediates are most likely contamination of the amylose fraction due to coprecipitation. The main variation in oat starch composition is due to the relative proportions of amylose and amylopectin along with the characteristics associated with the amylopectin molecule and this has an effect on the properties of the starch.

Oat starch exists in clusters formed from irregularly shaped granules compared to other cereals that have distinct A- and B-type granules. Oat starch granules are similar in size to rice starch, but smaller than wheat, maize and potato starch granules. However, oat starch is highly aggregated with composite or compound granules. Isolated oat starch was found to contain from 1% to 3% lipid content compared to 0.8% in wheat and 0.6% in maize (Paton, 1986; Ellis et al., 1998). The protein content of isolated oat starch varies from 0.44% to 0.60% with the enzyme α-amylase the most significant protein affecting pasting properties. The starch of oats is not easily separated from protein and lipids in contrast to wheat and maize.

Viscosity or pasting is the behaviour of starch in water treated with varying heat regimes. This is the most important functional characteristic of starch. Three instruments have been used to determine pasting properties: the Amylograph, the Ottawa Starch Viscometer and the Rapid Visco Analyser. The pasting properties determine the quality of the food product. Paton (1977) demonstrated that oat starch has unique pasting characteristics, requiring a short time to peak viscosity and low pasting temperatures. This is an advantage during processing as sufficient gelatinisation would occur during the short processing times used commercially.

10.3.2 Storage Proteins

Compared to other cereals, oats have a higher protein content varying from 15% to 20% in the groat (Peterson and Brinegar, 1986). Furthermore, the protein quality surpasses other cereals due to the amino acid balance.

Seed storage proteins occur in three tissues, the embryo or germ, the aleurone layer in the bran, and the starchy endosperm (Fig. 10.2). Although groat protein concentration is higher in the bran, the total contribution to groat storage protein is equal in the bran and starchy endosperm. There are four storage proteins in oats, globulins, albumins, prolamins and glutelins. Globulins are the main storage protein in oats and comprise from 52% to 75% of the total protein (Peterson and Brinegar, 1986). In contrast, the major storage protein is prolamin in wheat, barley and rye (Shewry and Halford, 2002). Glutelins are the second most prevalent storage protein, comprising 21–27% of the total groat protein. The oat groat possesses 11–19% albumins and 7–13% prolamins. Each storage protein has a distinctive amino acid composition affecting nutritional quality.

Lysine is recognised as the most limiting amino acid in cereals (Lockhart and Hurt, 1986). The high content of the globulin storage protein in oats is responsible for higher nutritional values compared to wheat and barley. This is because the globulin storage proteins have the highest concentration of lysine and represent the major storage protein in oats. Oats may contain up to double the concentration of lysine compared to wheat and barley, and lysine represents about 3.2–5.2% of the total protein in oat groats (Fulcher, 1986; Lockhart and Hurt, 1986). In contrast, the major storage protein in wheat and barley is prolamin, which is low in lysine and high in glutamine-glutamic acid. Results from human metabolic studies show that oat protein has higher nutritive value than other cereals (Lockhart and Hurt, 1986).

10.3.3 Lipids

Although oats contain the highest lipid or oil concentration of the cereals, it does not compete as a source of edible oil when compared to the concentrations in oilseed crops. Lipid content and composition are important for nutritional quality, energy content, flavour/off-flavour attributes and pasting properties. Oat lipids are highly unsaturated. Lipid content varies from 2% to 11% and is dependent on variety (Zhou et al., 1999). The highest lipid concentration is in the embryo compared to bran and endosperm, but the embryo is small in proportion to the endosperm and bran. Hence the endosperm and bran account for most of the total lipids.

The fatty acid composition of oat lipids is significant to the nutritional and physical qualities of oats. The three primary fatty acids in oat varieties are linoleic, oleic and palmitic, comprising about 95% of the total fatty acids (Zhou et al., 1998). The two major fatty acids, linoleic and oleic acid, both unsaturated, are important for human and animal nutrition. Palmitic acid, the major saturated fatty acid, increases oil stability against peroxidation.

Generally lipids are associated with negative food flavours. Lipids in sound groats, stored at 20°C and at 12–14% moisture, remain stable. However, when the groat is damaged or processed, the enzyme lipase is released, producing free fatty acids that cause rancidity and a bitter taste. Lipase is inactivated with steam treatment, heating to 90–100°C at moisture levels above 12% (Zhou et al., 1999).

10.3.4 β-glucan

β-glucan is a mixed linkage polysaccharide, (1→3)(1→4)-β-D-glucan with major concentrations, 75–80%, found in the endosperm cell walls (Miller et al., 1995) and minor amounts located in aleurone cell walls. The β-glucan found in aleurone cell walls may improve the water-binding capacity of oat bran, resulting in improved effectiveness as dietary fibre. There are varietal differences in β-glucan content, ranging in general from 3% to 6%, but there are exceptions where varieties have been developed with higher β-glucan levels.

Soluble fibre primarily in the form of β-glucan is approximately 55% of oat dietary fibre with the remaining 45% fibre insoluble. A primary health benefit is the effectiveness of lowering blood cholesterol reported in more than 40 human studies over the last 30 years (Kapica, 2001). Oat consumption was shown to reduce total cholesterol; low-density lipoprotein (LDL) blood cholesterol levels were reduced an average of 3–16% depending on baseline levels in men and women aged 20–70 years (Kapica, 2001). The 'good' cholesterol, high-density lipoprotein (HDL), was unchanged and in some cases increased. A daily serve of 1½ cups cooked porridge, containing about 3 g of β-glucan results in lowering cholesterol levels (Davidson et al., 1991). Kapica (2001) suggested about 75% of the cholesterol reduction is due to the soluble fibre, β-glucan. Kapica (2001) outlines the mechanisms for cholesterol reduction. The other 25% reduction of cholesterol is due to the favourable ratio of the amino acids, arginine and lysine, plant sterols, tocotrienols and monounsaturated fatty acids. In addition, β-glucan was shown to be the active ingredient that lowers serum glucose and insulin levels.

The function of β-glucan in the upper gastrointestinal tract is associated with molecular weight, structure and interaction with other dietary constituents (Rimsten et al., 2003). The molecular weight distribution of β-glucan changes with food processing (Frank et al., 2004). Åman et al. (2004) analysed the molecular weight distribution of β-glucan in raw oat materials, experimental foods and commercial oat products. The Calcofluor average molecular weight for rolled oats and oat bran ranged from 230 to 215 ($\times 10^{-4}$) g/mol whereas the Calcofluor average molecular weight for bread loaf and pancake batter was 63 and 19 ($\times 10^{-4}$) g/mol (Rimsten et al., 2003). They concluded that raw oat materials, commercial porridge, and breakfast cereals contained intact β-glucan with high molecular weight. However, processing such as baking or fermentation steps may result in extensive degradation of β-glucan reducing the molecular weight distribution. The degradation is a result of enzymatic reactions including β-glucanase, which lowers the molecular weight distribution. β-glucan's effectiveness is reduced with lower molecular weight (Tosh et al., 2004).

Many studies show that increased dietary fibre intake promotes satiety, decreasing hunger (Slavin and Green, 2007). Oats contain more protein on a per-gram basis than other cereals and in conjunction with dietary fibre, oats contribute a more sustained feeling of fullness and a decreased intake of foods high in fat or sugar.

10.3.5 Antioxidants

Oats are rich in antioxidants including vitamin E, avenanthramides, which are unique to oats, phenolic acids, flavonoids, sterols and phytic acid. This section will focus on vitamin E and avenanthramides. Consult Peterson (2001) for information on phenolic acids, flavonoids, sterols and phytic acid. Antioxidants protect the body from membrane damage, cancers, heart disease and age-related deterioration (Tsao and Akhtar, 2005).

Vitamin E is a generic term for compounds that show biological activity linked to α-tocopherol. Among the cereals, oats and barley have the highest concentrations. The primary oat tocol is α-tocotrienol with smaller amounts of α-tocopherol. Tocopherols are located in the embryo and the source for tocotrienols is the starchy endosperm. The total concentration of tocols in 12 varieties grown in the USA at three locations was 19.0–30.3 mg/kg (Peterson, 2001).

A group of amides unique to oats with antioxidant activities were found by Collins and published in 1989. About 24 different avenanthramides were identified (Collins, 2011). However there are three avenanthramides that occur in higher concentrations than all other avenanthramides. There are two different nomenclatures used by Collins (1989) and Dimberg et al. (1993) in the literature for the three dominant avenanthramides A=2f, B=2p and C=2c. Peterson (2001) compared the levels of avenanthramides in whole groats and pearling fractions. Avenanthramides A, B and C were highest in the pearling fractions suggesting they are located near the surface of the groat. In a study conducted by Peterson (2002) the three avenanthramides were synthesised and the antioxidant activities examined in two in vitro systems. The researchers found that avenanthramide C contributed significantly more antioxidant activity in oat groat extracts than avenanthramides A and B.

Peterson et al. (2005) found that oat varieties responded differently to environmental factors affecting avenanthramide concentration. The average concentration for three varieties grown in seven USA locations varied from 12 to 22.5 ppm for avenanthramide A, 22–42.1 ppm for avenanthramide B and 37.1–65.7 ppm for avenanthramide C (Emmons and Peterson, 2001). Levels of avenanthramides were found to vary due to the effects of variety, environment and tissue type. Overall avenanthramides A, B and C were dominant in European and North American oat varieties, whereas this was not the case in Chinese and Asian varieties (Collins, 2011).

Avenanthramides are associated with resistant responses to defend the plant against pathogens by reducing hyphae growth in host-pathotype interactions (Myama et al., 1995). Uchihashi et al. (2011) found that when the oat variety Shokan 1 was infected with an incompatible pathotype (resistant reaction) of crown rust (*Puccinia coronata* f. sp. *avenae*) there was a significant increase of avenanthramide A from 36 to 72 h after inoculation in the leaves. When the variety was inoculated with a compatible pathotype (susceptible reaction) only low levels of avenanthramide A were detected at 36 h after inoculation.

Antioxidant activity as well as other potential health benefits is discussed by Collins (2011). Bioavailability is the key to deliver these health benefits. Chen et al. (2007) conducted a human study and showed that avenanthramides were bioavailable in healthy adults.

10.3.6 Flavour

Oat flavour is a complex trait reliant on heat, which induces volatile flavour components. Steam treatment alone does not result in the characteristic oat flavour. The main flavour volatiles are aldehydes, ketones and alcohols and the non-volatiles are phenolic compounds, sugars, amino acids and small peptides, free fatty acids and lipids (Salmenkallio-Marttila et al., 2011). Phenolic acids, lignans and avenanthramides are also important, influencing oxidation rate by acting as antioxidants and adding directly to flavour (Dimberg et al., 1996).

Raw oat groats do not naturally have an 'oaty' taste. Precursors in oat groats are activated by heat during the commercial processing to produce oat flavour. When organic volatiles were extracted from toasted oats and separated into basic and neutral fractions, the basic fraction had strong nutty, earthy, toasted, slightly harsh and burnt flavours, whereas the neutral fraction had flavours of burnt grain, browned, heavy and not 'oaty'.

Sensory assessment was conducted by panels trained for basic tastes and taste threshold levels for fruit flavours, aromas, savoury flavours, herbs and spices, essences and texture (Zhou et al., 2000). The panellists evaluated sensory traits in oat porridge, such as oat and starch aroma, several types of flavour: overall, metallic, oat, bitter, creamy, raw, starch, soapy and metallic, oat and starch aftertaste flavours. Sensory assessment is important as there are varietal differences in flavour.

10.4 NUTRITION

The healthful benefits of oats have been demonstrated over the last 30 years. Recent data show the cardiovascular benefits are more than just cholesterol reduction with oat consumption favourably affecting blood pressure, blood glucose and body weight. Fig. 10.3 shows several health benefits from oat consumption.

10.4.1 Functional Food and Health Claims

Cardiovascular diseases (coronary heart disease and stroke) are the number one cause of death globally. An estimated 17.5 million people died from cardiovascular diseases in 2012 (WHO, 2015a). The global presence of diabetes was 9% in 2014 and 90% of those individuals have type 2 diabetes (WHO, 2015b). There were an estimated 1.9 billion adults over the age of 18 years and 42 million children under the age of 5 years that were overweight in 2014 (WHO, 2015c). Cardiovascular diseases and type 2 diabetes are prevented by a healthy diet, regular physical activity, avoiding tobacco and maintaining normal body weight.

Diet and lifestyle are the preferred avenues for the prevention and management of hypertension and cardiovascular diseases. Studies have shown diets rich in whole grains, especially oats, protect against cardiovascular diseases (lower cholesterol levels), high blood pressure and type 2 diabetes, whereas refined grain has no protective association (Truswell, 2002; Pins et al., 2002; Fung et al., 2002). Consumption of 3 g of β-glucan lowers serum cholesterol by 5–15%. A 1% decrease in serum cholesterol is estimated to reduce the observed rate of heart disease by 2–3% (Behall and Hallfrisch, 2011). Hence a diet rich in soluble fibre is effective as a functional food.

Health claims are an effective means for consumers to identify products with health value. Truthful labelling and advertising enable consumers to manage their own health. Government regulations ensure the nutrient specifications in products are met in order to advertise health claims. Clinical research into

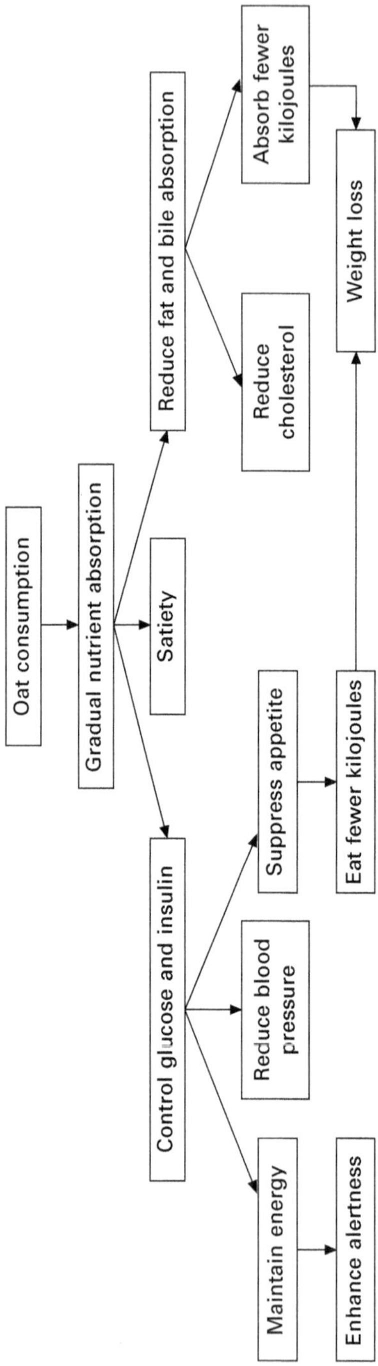

FIGURE 10.3 Health benefits from oat consumption.

functional foods and ingredients is required to substantiate claims and the consumer benefits from this additional research. Manufacturers producing healthy products gain a competitive edge, but also results in greater availability of healthy food.

10.4.2 Coeliac Disease

Individuals that have coeliac disease cannot tolerate foods containing gluten. This is an autoimmune disease that leads to chronic inflammation of the small intestine (see Chapter 14). Prolamins are responsible for the toxicity and are known as gliadin in wheat, hordein in barley, secalin in rye and avenin in oats. The prolamin content in wheat, barley and rye constitute about 40% of the protein, whereas it comprises about 15% in oats (Haboubi et al., 2006). People with coeliac disease must follow a life time gluten free diet. A number of long term studies show that oats are safe for the majority of people affected by coeliac disease because the prolamin content in oats is low compared to wheat, barley and rye (Ellis and Ciclitira, 2008). Hardy et al. (2015) in a study with 73 participants with coeliac disease found that 6 people had an immune response when oats were eaten. The specific oat peptides were identified, so in the future there may be ways to determine who is sensitive to oat consumption. Their research shows oat consumption may be safe for most people with coeliac disease. Due to a lack of fibre in coeliac diets, those that can tolerate oats would welcome this addition to their diet.

Oat products available on the grocery store shelf may be contaminated with wheat, barley and rye (Jones et al., 2016). Hernando et al. (2008) analysed 109 commercial oat products produced in Europe, Canada and the United States for wheat, barley and rye contamination using the Sandwich R5 ELISA. They found that 80% of the samples had some level of contamination and the primary cereal was barley. Oat products free from other cereal contamination is essential to ensure a place in coeliac diets.

Europe, UK, Canada and USA adopted the Codex Standard for gluten free foods. This defines food gluten free with gluten levels not exceeding 20 ppm in total. Currently, Australia does not accept the Codex Standard for gluten free foods.

10.5 PRODUCTION AND MAJOR USES OF OAT

Oats are currently ranked seventh in world cereal grain production after maize, rice, wheat, barley, sorghum and millet (FAO Statistics, 2009). The top five oat producing countries averaged from 2010 to 2013 were the Russian Federation, Canada, Poland, Australia and Finland, ranked in that order. Other countries with substantial oat production are Sweden, Ukraine, China, Germany, and the United Kingdom (FAO Statistics, 2009).

Most oat varieties have hulls that tightly enclose the groat and remain with the grain after harvest. In some countries, such as China and the United

Kingdom, naked oat varieties predominate. Naked oats have the free-threshing character like wheat, so when harvested, the hull is removed from the grain.

10.5.1 Food

Products produced from oats are generally made with the whole grain after the hull is removed, whereas other cereal products are produced from grain with the germ and bran removed. The unique composition of whole oats results in healthy products. Whole oats have the highest protein content and a favourable ratio of unsaturated to saturated fatty acids compared to other cereals. Whole oats also provide vitamins, minerals and antioxidants. The heat treatment used to stop rancidity also enhances the unique sensory characters of oats.

The main product produced from oats is porridge or oatmeal. Other food products processed from oats are cold cereals, infant foods, muesli or granola bars, breads, biscuits or cookies, thickeners and specialty flour. New processes were recently developed to manufacture non-dairy products such as milk, ice cream and yoghurt using oats. Antioxidants in oat flour can be used to stabilise some milk and meat products sensitive to fat oxidation during storage. Oat gum primarily composed of β-glucan is used to stabilise ice cream. Oat proteins have been used in many food products including heat-resistant chocolate, because of viscosity and emulsification properties.

10.5.2 Feed

Oats have been a traditional feed grain for centuries. Recent advances in corn, wheat, barley, soybean and canola as feed grains have resulted in a worldwide decline of oat production for feed. Despite the decline, the primary use of oats remains as a feed grain. Oats were the primary feed for horses that powered farm equipment prior to mechanisation. Oats are used primarily as a feed for dairy and beef cattle, sheep and horses, but can also be used for cats, dogs, birds, rabbits, bison, deer and fish. Naked oats are suited as a feed for weaner and grower pigs, poultry, racehorses and birds.

The nutrient value for animal feed is based on the proportion of groat to hull. The ratio varies with variety, but the growing environment can also influence groat percent, which varies from about 60–80%. The high oil content, comprised primarily of unsaturated fatty acids, can alter the fatty-acid composition of the animal fat. High protein content, with a greater proportion of lysine compared to other cereals, is also beneficial to animals.

Lignin is the major fibre fraction of the hull; it reduces grain digestibility in animals. Lignin content also varies in different varieties with most low-lignin varieties possessing improved grain digestibility compared to high-lignin varieties. Increased grain digestibility results in faster animal-weight gain. The hull plays an important role in feed as it reduces digestive problems such as acidosis.

The hull is a major constraint as a feed grain for poultry and pigs, because non-ruminants do not possess the enzymes necessary to digest the cellulose, hemicellulose and lignin. Because naked oats do not have the hull, the grain provides a good energy source for grower and weaner pigs, broilers, laying hens and turkeys.

Oats are the preferred feed for horses due to the palatability, digestibility and nutritive value of the grain. Naked oats are also used for racehorses, because they have limited grain intake and need a good source of energy.

10.5.3 Non-Food

There are many potential uses of oat grain fractions and hulls in products such as adhesives, cosmetic products, pharmaceuticals and oil-spill cleanup.

The unique chemical composition of the oat groat is responsible for oat-starch gels having more elastic, adhesive and translucent properties than wheat or maize starches. Oat starch could replace wheat or maize, with the waxy starches the exception. Products currently produced from starch include brown paper and cardboard products and coating agents for tablets in the pharmaceutical industry.

Oats have been used in cosmetic products for some time. Starch and lipid characteristics make oats a useful ingredient in bath products to relieve itching and nappy rash. Other products manufactured from oats include cleansers to replace soap in facial masks, soap for cleansing dry and oily skin, shampoo, lotions, bath additives and skin care.

The hull can also be used to produce industrial products. Oat hulls are used to produce furfural and furan compounds. These compounds are used in the manufacturing process of: crude petroleum; nylons; formaldehyde furfural resins; solvents for dyes, resins, lacquers, paints and varnish; phenolic resin glues and plywood adhesives; and cellulose pulp. A very different use of hulls is the cariostatic properties of tooth protection.

10.6 QUALITY REQUIREMENTS FOR OATS

Grain quality for the two major uses of oats, food and feed, require different levels of the main constituents in the groat.

10.6.1 Food

Grain quality characters important for oat varieties used to produce food products are both biochemical and physical.

Physical quality characters important for milling include groat percent, grain plumpness and hectolitre weight. Groat percent or the proportion of groat to total grain weight including the hull is one of the most important quality characters for millers. High groat percent is desired and values greater than 72% are acceptable, but 75% or higher is optimum. Ease of dehulling is also important and is often associated with varieties possessing high groat percent. High groat percent and ease of dehulling increase milling yield and milling efficiency. Varieties with high groat percent also produce large oat flakes often used in premium porridge products. Grain plumpness and uniformity reducing the proportion of small grains (screenings percent) also improves milling productivity, because small grain must be removed prior to processing, resulting in lost tonnage of the purchased grain. Hectolitre weight is a volume measurement of the grain. Historically, it was used to ensure that grain quality would be acceptable for milling. It is still used today, but groat percent and screenings percent reflect milling quality more precisely. Grain weight is an important quality trait for growers and is measured by weighing 500 or 1000 kernels. Heavier seeds often have improved germination and emergence.

The biochemical characters important to food products are the constituents discussed in Section 10.3. Low oil content is desired for food products, because it reduces processing problems and the potential for rancidity. In addition low oil may meet labelling regulations for health claims. High β-glucan is desirable to increase the health benefits for lowering blood cholesterol levels, so less porridge is required to achieve desirable results. The oat product must have a certain β-glucan content to meet labelling regulations for health claims. High protein content is also desirable. The newest research area is avenanthramides; breeding for high levels of avenanthramide and Vitamin E should increase antioxidant levels in oat products thereby improving nutritive value.

One other important quality requirements for oat products is contamination of grain by mycotoxins caused by *Fusarium* species. *Fusarium* species and mycotoxins change by region and year, and are favoured by wet humid conditions at flowering. The pathogen enters the developing grain after pollination and is protected by the hull. Common mycotoxins produced by *Fusarium* species are deoxynivalenol (DON), zearalenone, nivalenol (NIV), HT2 and T2 (Edwards, 2006). The European Commission has set legislative limits for DON and zearalenone levels. Adler et al. (2003) compared mycotoxin levels between conventional and naked oats. They found conventional oats were more vulnerable to fungal and bacterial infection due to the tightly enclosed groat by the hull. At harvest, the fungal and mycotoxin contamination remains with the grain. However, naked oats have a thinner loosely adhering hull resulting in improved ventilation and less microorganism growth. The authors concluded that removing the hull from conventional oats significantly reduced mycotoxin content and compared to naked oats was superior (Adler et al., 2003).

10.6.2 Feed

As with food products, grain quality characters important for oat varieties used as animal feed products are both biochemical and physical.

Physical quality characters important for feed include groat percent, grain plumpness and hull brightness. High groat percent is desirable as the groat will have more starchy endosperm and the grain will have less hull, thus improving energy values for feed. Grain plumpness is also associated with the starchy endosperm. Hull brightness is primarily a requirement for the horse-racing industry where bright, plump oats are sought.

The biochemical characters important to animal feed are the components discussed in Section 10.3. High oil or lipid content is desired to deliver increased energy to animals. High protein is also important for feed varieties. Unlike food quality characters, low β-glucan content is required in feed varieties to reduce the possibility of sticky stools. Hull lignin content is associated with grain digestibility. Generally animals digest oat grain with low hull lignin more effectively than oat grain with high lignin content. Although the association of low hull lignin and improved grain digestibility has been shown, there are exceptions. Research has shown live weight gain tripled in animals fed low lignin and highly digestible oat grain.

A combination of high oil and protein content in the groat, along with low hull lignin, increases the nutritive value of oats for feed by increasing grain digestibility and metabolisable energy. The flow-on effect of improving feed quality will be reflected in feed-test results for grain digestibility, metabolisable energy, acid- and neutral-detergent fibre and protein; these attributes are routinely measured by commercial companies using near infrared reflectance spectroscopy (NIR).

10.6.3 Non-Food

There is little information about quality characters important for non-food products. Certainly high lipid and starch content would be important for cosmetics, soap, shampoo and conditioners and lotions. Oats are used in facial masks, because of the film produced on the face; this is reported to be the result of β-glucan. The use of hulls for industrial uses is mostly a by-product of processing.

10.7 MAINTAINING AND IMPROVING OAT QUALITY

Oat breeding programs and the milling industry for food have been proactive in developing linkages to identify essential quality traits. Consequently oat varieties have been developed with specific quality characters required by the milling industry. Once varieties have been developed with enhanced grain quality, proper storage is essential to maintain the integrity of the grain for processing.

10.7.1 Breeding for Quality

Developing improved milling-oat varieties requires identifying breeding priorities not only for quality traits, but also agronomic traits such as yield potential and disease resistance. Production constraints also need to be addressed in variety development, so growers will select the variety for production so that milling-quality grain will be readily available to industry.

Improving grain quality starts with identifying quality traits important to the industry using the grain. Although the primary use of oats is for feed, the greatest effort for improving grain quality is for the milling industry. As feed lots grow in importance, more emphasis may be placed on improving grain quality for feed. However, quality traits important for milling varieties are also important for feed varieties, but differing in the levels desired for the two end uses. The biochemical traits of interest are protein, oil, β-glucan and antioxidants. Physical traits important to enhanced quality are groat percent, plumpness and hectolitre weight.

In order to improve quality, there must be genetic variation for the traits of interest and the heritability of the traits must be reasonably high. Fortunately heritabilities are good for most of the quality traits discussed. Identification of parental lines possessing the desirable quality traits to generate populations for selection is the beginning of the breeding cycle. Development of reproducible evaluation methods for each of the traits is essential to allow selection of the individuals in the population possessing the optimum combination of quality traits. The methods used in the earlier generation material require non-destructive techniques for large numbers of breeding lines.

The NIR instrument has permitted the measurement of protein and oil contents, groat percent, hull brightness, hull lignin content, grain digestibility and metabolisable energy. NIR is a high throughput, non-destructive method where thousands of samples can be accurately assessed. Physical quality traits are measured using several different instruments. Hectolitre weight is measured using the equipment shown in Fig. 10.4. Plumpness is measured by the Sortimat screenings instrument shown in Fig. 10.5. A grain counter is used to count 500 grains and the sample is weighed to determine thousand-grain weight. A ground sample is required to determine β-glucan content using either a flow-injection instrument or a Megazyme kit.

10.7.2 Molecular Tools for Quality Improvement

Quantitative trait loci (QTL) identify chromosomal regions associated with significant genetic effects for traits of interest. Molecular markers are differences in the DNA between two individuals and are linked to genes or QTLs used to select for the desired trait. Molecular markers are useful for assessing large numbers of plants efficiently for traits that are difficult to phenotype. This is the basis for marker assisted selection.

Quantitative trait loci analysis was done for oil, β-glucan and protein content. QTL analysis for protein content was done in four mapping populations and 3 or 4 QTLs were found in each population (Molnar et al., 2011). β-glucan content was investigated in two mapping populations, but the QTLs only explained 20% of the phenotypic variation, indicating other QTLs exist that had not been

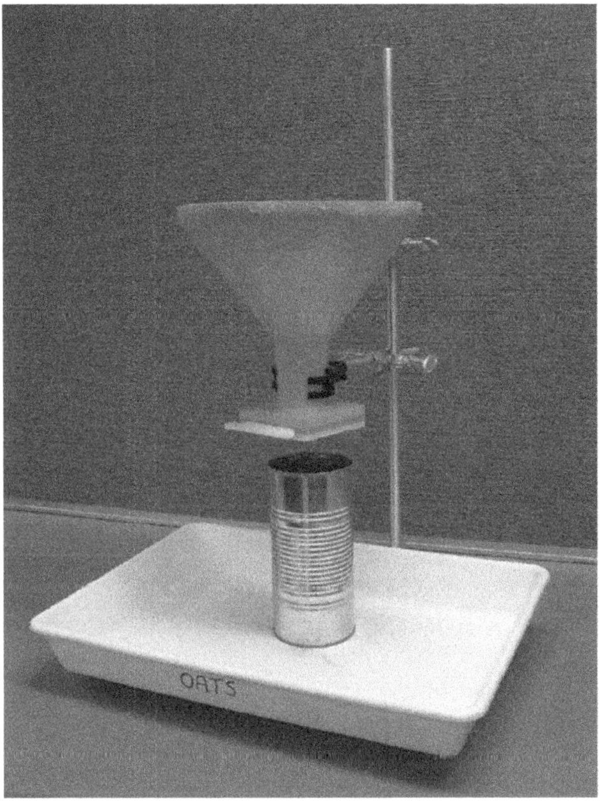

FIGURE 10.4 Equipment used to measure hectolitre weight.

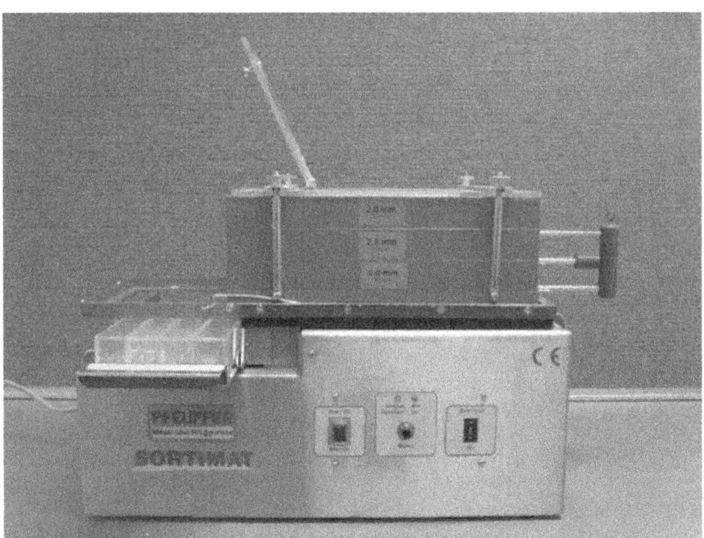

FIGURE 10.5 Equipment used to measure screenings percent and grain plumpness.

found (Molnar et al., 2011). Five to eight significant QTLs were found in the Ogle/MAM17-5 mapping population describing 30–42% of the phenotypic variation for oil (Zhu et al., 2004).

New molecular tools are being developed that offer rapid methods to discover and sequence thousands of genes, resulting in greater advances in oat quality. The expense associated with these tools is decreasing due to robotics and bioinformatics.

10.7.3 Storage Conditions

Oats can be stored for about 1 year in good facilities maintained under storage conditions of 20°C and 12–14% moisture with adequate protection against insects and rodents. Storage at moisture levels higher than that will cause the development of free fatty acids resulting in rancidity. In locations where moisture levels exceed 14%, there is a need for artificial drying before storage.

10.7.4 Processing

In order for oats to be processed, contaminants must be removed, the hull must be removed from the groat, and the manufacturing process must ensure that a product with quality appearance and taste is created. There are specifications for milling oats that include hull-to-groat ratio, colour and hectolitre weight.

Milling and processing operations consist of cleaning, grading, hulling, hull-and-fine separation, groat separation and kilning (Fig. 10.6). Grading occurs when groats are separated into two or three streams, based on groat length or thickness. The hulling process is the next step when groats are separated from the hull using either impact or stone-hulling systems. Impact hulling is more common than stone hulling. Hulls and fines are separated from whole groats, broken groats and unhulled oats. Further refinement occurs when groats are separated based on physical characters, such as groat size and weight. Because oat groats are rich in oil, a heat treatment is required to inactivate enzymes that cause rancidity and bitterness in the final product. Kilning is a process that heats the groat at a certain temperature and moisture content to inactivate lipase. Kilning can occur before grading or after groat separation.

Commercial processors generally can produce 100 kg of product from 175 kg of oats. Milling efficiency varies according to the variety and mill operating efficiency. Products produced include steel-cut groats, rolled oats, quick oats, baby oats, instant oat flakes, oat flour and oat bran.

10.8 FUTURE

The future holds many opportunities for oats used as food, feed and non-food products. Oat consumption increased 130% from 1980 to 2001 even before the health benefits were claimed. Eight years later, production capacities were

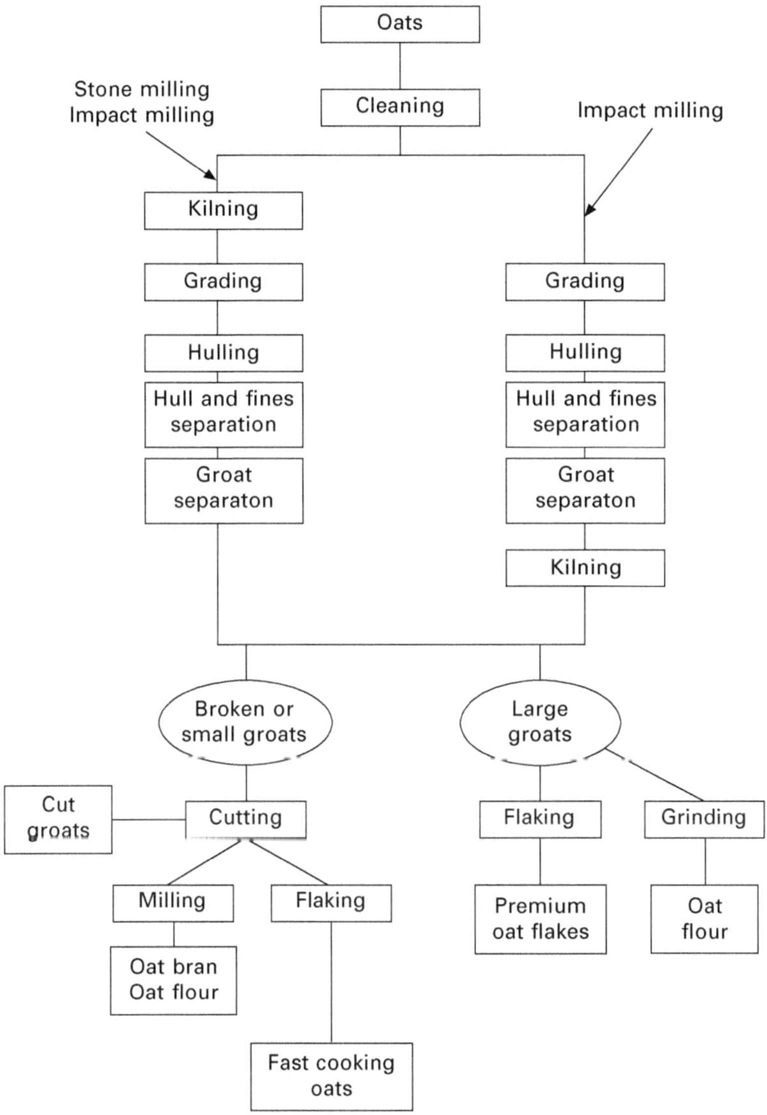

FIGURE 10.6 The milling process of the oat kernel.

increasing as much as 10% a year in some parts of the world. Oat is nature's functional food providing numerous health benefits including combating obesity. As research progresses on the healthy and unique constituents in the oat groat, new food products that are fun to eat will be developed. The area of nutraceuticals, where components are isolated and purified from grain and sold in medicinal forms not usually associated with food products, offers great opportunities for oats. Improved oat varieties, with increased levels of β-glucan and avenanthramides/antioxidants, will underpin many of the new potential uses. The increase of feed lots and the development of improved oat varieties for feed value will also increase the demand for oats. Increased consumption in countries such as India and China are also anticipated to increase demand for oats. The future looks bright for oats.

SOURCE OF FURTHER INFORMATION

Coffman, F.A. (Ed.), 1961. Oats and Oat Improvement, vol. 8, American Society of Agronomy, Madison, Wisconsin.

Chinese Academy of Agricultural Sciences, 2012. Ninth International Oat Conference: Proceedings at Beijing, China, 20–23 June 2012.

Cross, R.J., 2000. Sixth International Oat Conference: Proceedings at Lincoln University, Lincoln, NZ, 13–16 November 2000, Christchurch, New Zealand, New Zealand Institute for Crop and Food Research Limited.

Marshall, H.G., Sorrells, M.E. (Eds.), 1992. Oat Science and Technology. American Society of Agronomy, Madison, Wisconsin, Agronomy no. 33, 820 pp.

Peltonen-Sainio, P., Topi-Hulmi, M. (Eds.), 2004. Seventh International Oat Conference Proceedings, MTT Agrifood Research Finland, Helsinki, 17–22 July 2004, MTT Agrifood Research Finland, Data and Information Services, Jokioinen, Finland.

Scoles, G., Rossnagel, R. (Eds.), 1996. Fifth International Oat Conference and Seventh International Barley Genetics Symposium Proceedings, University of Saskatchewan, Saskatoon, 30 Jul–6 Aug 1996, University Extension Press, Saskatoon, Canada, University of Saskatchewan.

Stuthman, D., (Ed.), 2008. Eighth International Oat Conference Proceedings, University of Minnesota, Minneapolis, 28 June–2 July 2008.

Webster, F.H., Wood, P.J. (Eds.), 2011. Oats: Chemistry and Technology. second ed. AACC International, Inc, St. Paul, MN, p. 376.

Welsch, R.W. (Ed.), 1995. The Oat Crop: Production and Utilization. Chapman and Hall, London, UK, p. 560.

REFERENCES

Adler, A., Lew, H., Moudry, J., Sterba, Z., Vratilova, K., Edinger, W., Brodacz, W., Kiendler, E., 2003. Microbiological and mycotoxicological quality parameters of naked and covered oats with regard to the production of bran and flakes. Die Bodenkultur 54, 41–48.

Åman, P., Rimsten, L., Andersson, R., 2004. Molecular weight distribution of β-glucan in oat-based foods. Cereal Chem. 81, 356–360.

Banks, W., Greenwood, C.T., 1967. The fractionation of laboratory isolated cereal starches using dimethyl Sulfoxide. Starch/Staerke 19, 394.

Behall, K.M., Hallfrisch, J., 2011. Oats as a functional food for health. In: Webster, F.H., Wood, P.J. (Eds.), Oats: Chemistry and Technology. AACC International, St. Paul, MN.

Chen, O.C.-Y., Milbury, P.E., Collins, F.W., Blumberg, J.B., 2007. Avenanthramides are bioavailable and have antioxidant activity in humans after acute consumption of an enriched mixture from oats. J. Nutr. 137, 1375–1382.

Collins, F.W., 1989. Oat phenolics: avenanthramides novel substituted *N*-cinnamoylanthranilate alkaloids from oat groats and hulls. J. Agric. Food Chem. 37, 60–66.

Collins, F.W., 2011. Oat phenolics: biochemistry and biological functionality. In: Webster, F.H., Wood, P.J. (Eds.), Oats: Chemistry Technology. AACC International, St. Paul, MN, pp. 157–217.

Davidson, M.H., Dugan, L., Burns, J.H., Bova, J., Story, K., Drennan, K.B., 1991. The hypocholesterolemic effects of β-glucan in oatmeal and oat bran: a dose controlled study. J. Am. Med. Assoc. 265, 1833–1839.

Dimberg, L.H., Theander, O., Lingnert, H., 1993. Avenanthramides—A group of phenolic antioxidants in oats. Cereal Chem. 70, 637–641.

Dimberg, L.H., Molteberg, E.L., Solheim, R., Frølich, W., 1996. Variation in oat groats due to variety, storage, and heat treatment. 1: phenolic compounds. J. Cereal Sci. 24, 263–272.

Edwards, S., 2006. Investigation of Fusarium mycotoxins in UK barley and oat production', Harper Adams University College Final Report, CO4033 and CO4034.

Ellis, H.J., Ciclitira, P.J., 2008. Should coeliac sufferers be allowed their oats? Eur. J. Gastroenterol. Hepatol. 20, 492–493.

Ellis, R.P., Cochran, M.P., Dale, M.F., Dufus, C.M., Morrison, I.M., Prentice, R.D., Swanson, I.S., Tiller, S.A., 1998. Starch production and industrial use. J. Sci. Food Agric. 77, 289–311.

Emmons, C.L., Peterson, D.M., 2001. Antioxidant activity and phenolic content of oat as affected by cultivar and location. Crop Sci. 41, 1676–1681.

FAO Statistics, 2009. www.faostats.org.

Frank, J., Sundberg, B., Kamal-Eldin, A., Vessby, B., Åman, P., 2004. Yeast-leavened oat breads with high or low molecular weight β-glucan do not differ in their effects on blood concentrations of lipids, insulin, or glucose in humans. J. Nutr. 134, 1384–1388.

Fulcher, R.G., 1986. Morphological and chemical organization of the oat kernel. In: Webster, F.H. (Ed.), Oats: Chemistry and Technology. American Association of Cereal Chemists, St. Paul, MN.

Fung, T.T., Hu, F.B., Pereria, M.A., Liu, S., Stampfer, M.J., Colditz, G.A., Willett, W.C., 2002. Whole-grain intake and the risk of type 2 diabetes: a prospective study in men. Am. J. Clinical Nutr. 76, 535–540.

Haboubi, N.Y., Taylor, S., Jones, S., 2006. Coeliac disease and oats: a systematic review. Postgrad. Med. J. 82, 672–678.

Hardy, M.Y., Tye-Din, J.A., Stewart, J.A., Schmitz, F., Dudek, N.L., Hanachapola, I., Purcell, A.W., Anderson, R.P., 2015. Ingestion of oats and barley in patients with celiac disease mobilizes cross-reactive T cells activated by avenin peptides and immune-dominant hordein peptides. J. Autoimmun. 56, 56–65.

Hernando, A., Mujico, J.R., Mena, M.C., Lombardia, M., Méndez, E., 2008. Measurement of wheat gluten and barley hordeins in contaminated oats from Europe, the United States and Canada by Sandwich R5 ELISA. Eur. J. Gastroenterol. Hepatol. 20, 545–554.

Jones, J.M., Zevallos, V., Wrigley C.W., 2016. Appendix 3. Grains, foods, and ingredients suiting gluten-free diets for ceoliac disease. In: Wrigley, C., Corke, H., Seetharaman, K., Faubion, J. (Eds.), Encyclopedia of Food Grains. second ed. Academic Press, Oxford, pp. 460–466.

Kapica, C., 2001. Oats—nature's functional food. Nutr. Today 36 (2), 56–60.

Lockhart, H.B., Hurt, H.D., 1986. Nutrition of oats. In: Webster, F.H. (Ed.), Oats: Chemistry and technology. American Association of Cereal Chemists, St. Paul, MN.

Miller, S.S., Fulcher, R.G., Sen, A., Arnason, J.T., 1995. Oat endosperm cell walls: I. Isolation, composition, and comparison with other tissues. Cereal Chem. 72, 421–427.

Molnar, S.J., Tinker, N.A., Kaeppler, H.F., Rines, H.W., 2011. Molecular genetics of quality in oats. In: Webster, F.H., Wood, P.J. (Eds.), Oats: Chemistry and Technology. AACC International, St. Paul, MN.

Myama, S., Bordin, A.P.A., Morikawa, T., Tanpo, H., Kato, H., 1995. Association between avenalumin accumulation, infection hypha length and infection type in oat crosses segregating for resistance to *Puccinia coronata* f. sp. *avenae* race 226. Physiol. Mol. Plant Pathol. 46, 255–261.

Paton, D., 1977. Oat starch. I. Extraction, purification and pasting properties. Starch/Staercke 29, 149–153.

Paton, D., 1986. Oat starch; physical, chemical, and structural properties. In: Webster, F.H. (Ed.), Oats: Chemistry and Technology. American Association of Cereal Chemists, St. Paul, MN.

Peterson, D.M., 2001. Oat antioxidants. J. Cereal Sci. 33, 115–129.

Peterson, D.M., 2002. Oat avenanthramides exhibit antioxidant activities in vitro. Food Chem. 79, 473–478.

Peterson, D.M., Brinegar, A.C., 1986. Oat storage proteins. In: Webster, F.H. (Ed.), Oats: Chemistry and Technology. American Association of Cereal Chemists, St. Paul, MN.

Peterson, D.M., Wesenberg, D.M., Burrup, D.E., Erickson, C.A., 2005. Relationships among agronomic traits and grain composition in oat genotypes grown in different environments. Crop Sci. 45, 1249–1255.

Pins, J.J., Geleva, D., Keenan, J.M., Frazel, C., O'Connor, P.J., Cherney, L.M., 2002. Do whole-grain cereals reduce the need for antihypertensive medications and improve blood pressure control? J. Fam. Pract. 51, 353–359.

Rimsten, L., Stenberg, T., Andersson, R., Andersson, A., Åman, P., 2003. Determination of Åman P molecular weight using SEC with Calcofluor detection in cereal extracts. Cereal Chem. 80, 485–490.

Salmenkallio-Marttila, Heiniö, R.L., Kaukovirta-Norja, A., Poutanen, K., 2011. Flavor and texture in processing of new oat foods. In: Webster, F.H., Wood, P.J. (Eds.), Oats: Chemistry and Technology. AACC International, St. Paul, MN.

Shewry, P.R., Halford, N.G., 2002. Cereal seed storage proteins; structures, properties and role in grain utilization. J. Exp. Bot. 53, 947–958.

Slavin, J., Green, H., 2007. Dietary fibre and satiety. Br. Nutr. Found. Nutr. Bull. 32, 32–42.

Tester, R.F., Karkalas, J., 1996. Swelling and glatinization of oat starches. Cereal Chem. 73 (2), 271–277.

Tosh, S.M., Wood, P.J., Wang, Q., Weisz, J., 2004. Structural characteristics and rheological properties of partially hydrolysed oat β-glucan: the effects of molecular weight and hydrolysis method. Carbohydr. Polym. 55, 425–436.

Truswell, A.S., 2002. Cereal grains and coronary heart disease. Eur. J. Clin. Nutr. 56, 1–14.

Tsao, R., Akhtar, M.H., 2005. Nutraceuticals and functional foods: I. Current trends in phytochemical antioxidant research. J. Food Agric. Environ. 3, 10–17.

Uchihashi, K., Nakayashiki, H., Okamura, K., Ishihara, A., Tosa, Y., Park, P., Myama, S., 2011. In situ localization of avenanthramide A and its biosynthetic enzyme in oat leaves infected with the crown rust fungus. Physiol. Mol. Plant Pathol. 76, 173–181.

World Health Organisation, 2015a. Cardiovascular diseases, Fact sheet No 317, www.who.int/mediacentre/factsheets/fs317/en.

World Health Organisation, 2015b. 'Diabetes', Fact sheet No 312, www.who.int/mediacentre/factsheets/fs312/en.

World Health Organisation, 2015c. 'Obesity and overweight', Fact sheet No 311, www.who.int/mediacentre/factsheets/fs311/en.

Zhou, M.X., Glennie-Holmes, M., Robards, K., Helliwell, S., 1998. Fatty acid composition of lipids of Australian oats. J. Cereal Sci. 28, 311–319.

Zhou, M., Robards, K., Glennie-Holmes, M., Helliwell, S., 1999. Oat lipids. J. Am. Oil Chem. Soc. 76, 159–169.

Zhou, M., Robards, K., Glennie-Holmes, M., Helliwell, S., 2000. Contribution of volatiles to the flavour of oatmeal. J. Sci. Food Agric. 80, 247–254.

Zhu, S., Rossnagel, B.G., Kaeppler, H.F., 2004. Genetic analysis of quantitative trait loci for groat protein and oil content in oat. Crop Sci. 44, 254–260.

Chapter 11

Corn: Grain-Quality Characteristics and Management of Quality Requirements

Zoltán Győri
Institute of Human Nutrition, University of Debrecen, Debrecen, Hungary

11.1 INTRODUCTION: THE IMPORTANCE OF CORN (MAIZE) PRODUCTION IN THE INTERNATIONAL ARENA

Corn (*Zea mays* L.) is one of the most important grains in the world. The total world production of corn was over a billion tons in 2013, with United States, China, Brazil contributing to 31, 24 and 8% of the total production, respectively. The oldest corn from archaeological findings was found in the valley of Tehuacán in Mexico. There exist a number of theories as regards its origin, the most acceptable one being that the teosinte (*Zea mexicana*) is the early progenitor of corn. The latest knowledge can be found in Lee's paper (2016). Before its discovery by Europeans, corn spread northward into Canada and southward into Argentina. This period was followed by its spread into Europe, Africa and Asia. Nowadays it is grown in large areas in all the five continents.

Z. mays is known in some English-speaking countries as corn. In British English, 'corn' can mean any type of 'grain' especially 'wheat'. The word 'maize' was originally Spanish, and comes from the word 'mahiz' in the Arawak language of Haiti. The settlers called it 'Indian corn' (http://english.stackexchange.com/questions/96522/why-does-corn-mean-maize-in-american-english.epure.org/resources/statistics/, 2014). *The name in* French-speaking area *is maiz or* blé d'Espagne or blé de Turquie or blé d'Inde. The name in Italy is granturco.

The sowing areas, average yield and total yield amounts for the most important corn-producing countries are found in the tables in Chapter 2. Its human utilisation has a history of thousands of years in a number of regions of the world, while in other locations its spread is the result of the globalisation currently underway. On the basis of the colour, shape and hardness of the kernel, the crop is classified into several general categories. Its chemical composition provides versatile uses for

Cereal Grains. http://dx.doi.org/10.1016/B978-0-08-100719-8.00011-5

both the grain crop and its components as well as its total use as a green fodder. Its cultivation can safely be mechanised and it can readily be stored. Its main and secondary products may also be utilised outside the agricultural industry. In addition to being employed as animal feed, the grain crop is an important raw material for the food, starch, fermentation and chemical industries while the by-products are used by the cellulose, energy and chemical industries in addition to being used as feedstuff (Godon and Williams, 1994; Ranum et al., 2014). The harvested and stored green crop is an important feedstuff for ruminant animals used in milk production.

The production, storage and processing of corn is of great importance from breeding activities to industrial use. One of the new challenges of our days is the increase in the value of renewable energy sources and the increase in the CO_2 content of the atmosphere, which has put this crop in a new light. In this chapter we do not deal with green or sweet corn. Readers can find further information about these aspects of corn in a review by Hallauer (2001).

11.2 UTILISATION OF CORN

The utilisation of corn is significant both in locations where it is produced as well as in places that are distant from the sites of production. Corn is a major bulk commodity in the international grain trade. Its industrial utilisation in the USA is shown in Fig. 11.1. Overall, 34.3% of corn production in the USA is used as feedstuff for animals, 34.9% finds its way into ethyl alcohol production and a further 9.3% is used as food and as seed-corn. In addition, exports account for 13.1%, and 8.4% is the closing stock (www.usda.gov). In Europe, the ratio is different because the use of animal feed represents a higher proportion than ethanol production. Share of feedstock used to produce renewable ethanol in the EU are as follows: maize (42%), wheat (34%), beet-based (sugar equivalent) (17%), other cereals (7%) and others (0.3%) (www.epure.org).

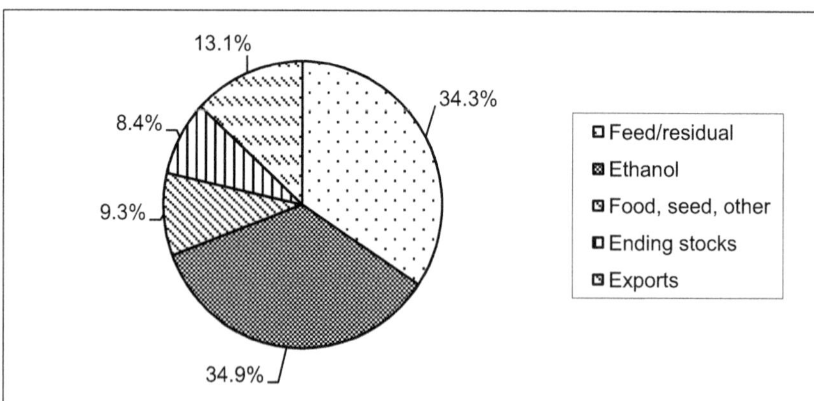

FIGURE 11.1 Field corn, 2013/2014, percentage of supply. *(Data taken from USDA, 2015. www.usda.gov.)*

TABLE 11.1 Human Consumption of Maize in Certain Countries (2011–12) (Annual kg per Capita)

Africa		America		Europe		Asia	
Cameroon	42	Brazil	28	Austria	13	China	8
Egypt	63	Canada	20	Bosnia	83	Indonesia	34
Ethiopia	43	Columbia	30	Bulgaria	41	Japan	11
Kenya	77	Guatemala	85	Croatia	8	Turkey	18
Mozambique	58	Honduras	80	Germany	10	Nicaragua	67
Namibia	40	Mexico	116	France	12	Nigeria	31
South-Africa	100	USA	13	Moldova	94	Pakistan	15
Zambia	115	Venezuela	57	Romania	39	Peru	20
Zimbabwe	108			Serbia	26		
				Slovenia	33		

Source: Data taken from www.faostat.fao.org.

Its use in human nutrition is considerable as a staple food in the diet of countries in Latin America, certain parts of Africa, and also in Romania, Moldova and Bosnia-Herzegovina on the European continent. The major data for the human consumption of corn in countries are found in Table 11.1.

Direct use of corn-on-the-cob for human nutrition includes boiling the whole cob (generally immature sweet corn) in water and flavouring with salt, butter and jam. Different kinds of tortilla are important foods in Central and South America. They are prepared in various ways using corn that has been cooked in slightly alkaline water containing lime, a process called 'nixtamalisation'. Then the mass is ground and 'masa' is obtained. The disk prepared from this mass is then baked. Several varieties of tortilla constitute important foods for human nutrition since they can be eaten in combination with legumes, different meat products, cheeses and vegetables. In North America, the consumption of hominy has long-standing traditions; this food is produced from corn cooked in an alkaline solution. The immigration and cultural diversification have created a high demand for ethnic foods in the United States and Europe. In Africa and Asia, corn is used for producing flat breads, beverages and different fermented and non-fermented porridges. The modern food industry produces various corn-based foods: tortillas, snacks and corn cakes are examples of the traditional foods described earlier. A comprehensive description of these products is found in the book *Speciality Corns* by Hallauer (2001). In Europe, the annual per capita consumption both for processed products and sweet corn averages over 35 kg. In Romania and on the Balkan Peninsula, where the consumption of hominy has long-standing traditions, it is favoured when boiled either

TABLE 11.2 Ethanol Yield of Various Crops

Plant	Yield (t/ha)	Average bioethanol yield (L/ha)
Sugar beet	40	4000
Sweet sorghum	35	3500
Potato	20	2000
Winter wheat	5	1500
Corn	6	2300
Jerusalem artichoke	50	4200

Source: Data taken from Lakner, Z., Kóbor, K., Pozsonyi, F., Pándi, F., 1993. The possibilities and chances of a Hungarian Bioethanol Program. Acta Agron. Hung. 42. 424–428.

in water or milk. In Hungary, rice, used to make the filling in stuffed cabbages, is replaced in certain locations by corn mush. In Italy, a variety of milled corn called 'polenta' is a popular dish. In different parts of the world the consumption of various morning dishes has become widespread and several of these contain corn grits, corn flour or corn bran produced from white or yellow corn.

With the increase in the value of renewable energy sources over the past few years, corn has also been receiving more and more attention. Ethyl alcohol (ethanol) used as an engine fuel is the product of a technology that has the potential to be implemented in large-scale industrial size. The amount of ethanol obtainable from plant material grown on 1 ha is compared in Table 11.2 for a few crops. The importance of this technology is well illustrated by the fact that in the USA, for example the volume of alcohol produced in this way has increased by 10 times of its original value over the past 10 years. The efficiency of this process has improved considerably in recent years and, considering the prices for crude oil, the trend is envisaged to remain the same in the future as well. This use of corn is on the increase in other countries where corn is grown in large areas, for example China, Thailand, Brazil, France and Hungary.

11.3 THE STRUCTURE AND CHEMICAL COMPOSITION OF CORN

The precise structure and chemical composition of the kernel depend on the type the kernel itself belongs to. Kernel types can be grouped as follows: dent, flint, flour, sweet, pop and pod (Watson and Ramstad, 1991). Modern dent hybrids are grown mostly in the US Corn Belt and some other places around the world (Argentina, China, France, Italy, Hungary, etc.). In these places the dominant varieties have a yellow endosperm (97%) with the remainder having a white endosperm. White maize is used in the food industry. Flint kernels are smooth and rounded without denting. Flints are predominantly grown in South America and northern Europe. Floury corn is one of the oldest types of corn, which the ancient Aztecs and Incas used to grow and consume. It is widely grown in the Andean region of South America. Sweet corn is an important crop for human food, being consumed directly in fresh, canned or frozen products. The main growing areas in the world are in the United States, France, Hungary, Ukraine and China. A very important corn variety, waxy corn (wx), contains starch consisting almost entirely of amylopectin. The diversity of corn species is so great that special nutritional and feed uses are possible (O2, opaque; ae, amylose extender; su2, sugary 2).

Corn (maize) is similar to other cereal grains in that its grain consists of the three main parts shown in Fig. 11.2, namely, the pericarp, the endosperm and the germ. Some authors include the tip cap as a fourth constituent (Lásztity, 1999). As regards its dimensions, the average corn grain is 8–17 mm long, its width is 4–6 mm and the weight of 1000 grains is 250–400 g.

As regards its composition, the main groups of constituents are: water (moisture content), carbohydrates including fibre as non-starch polysaccharides, proteins, amino acids, lipids, as well as vitamins, enzymes, mineral substances, coloured components and compounds that produce the flavour and aroma characteristics. The proportions of these constituents differ from one grain type to another. The types of endosperm variation include dent, flint, floury, pop, sweet and waxy (Zuber and Darrah, 1991). Grains are characterised by having a high starch content, considerable protein and a low fat content overall, but a lot of the fat is in the germ. There are high-oil hybrids, and these genotypes are useful in places where maize oil is a common culinary ingredient.

It is the endosperm that constitutes the major part of a corn kernel, a large part of which is starch. At the same time, over 70% of the protein is found in the endosperm. The distribution of the main chemical components of the major parts of the corn grain is provided in Table 11.3. The germ is rich in lipids but 70% of the ash and sugar content of the grain can also be found here. As regards the oils contained in the germ, the most important fatty acid is linoleic acid (unsaturated), which enhances its favourable dietary effects (Roderuck and Fox, 1987).

11.3.1 Moisture Content

Moisture is present in corn grains, as in all living organisms. The water is present in both free and bound forms. Bound water, which is bound physicochemically in colloidal materials (proteins, carbohydrates) has characteristics different from those of free water. It can only be removed from the grain by applying chemical methods and it does not play a role in the course of processing. 'Free water' refers to physically bound moisture, which has all the properties of water, that is it freezes and evaporates relatively easily.

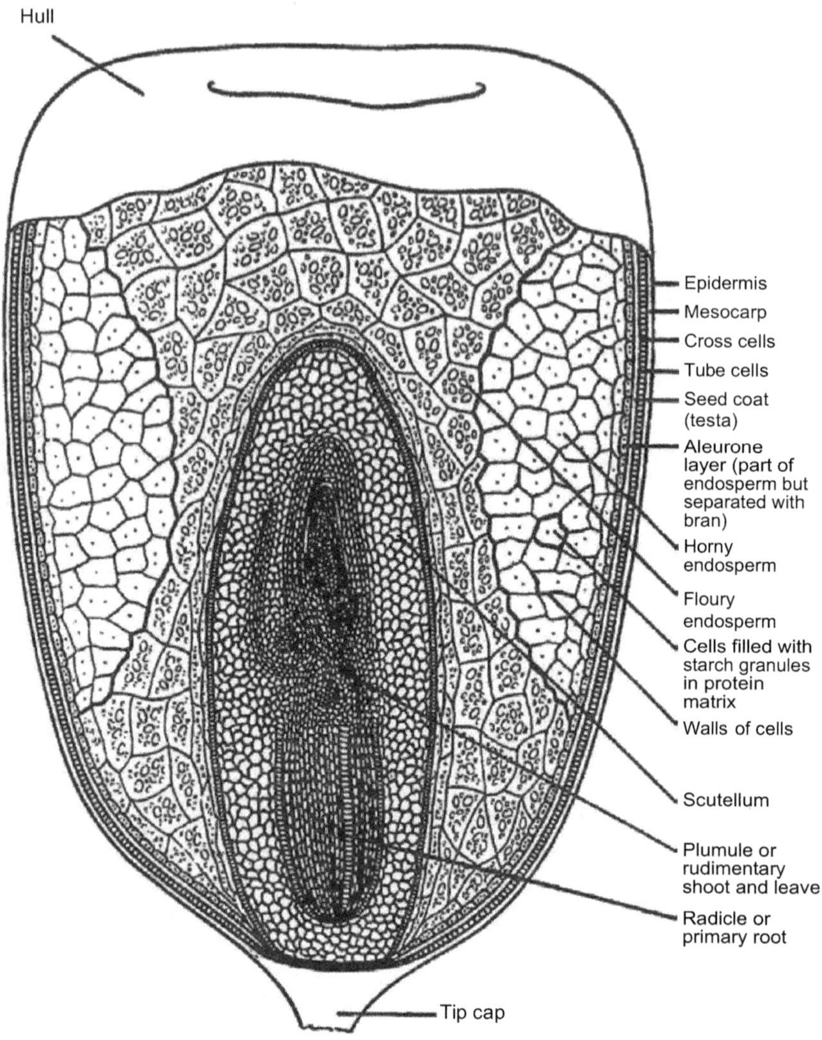

FIGURE 11.2 Longitudinal section of maize kernel. *(Data taken from Lásztity, R., 1999. Cereal Chemistry. Akadémiai Kiadó, Budapest.)*

In determining moisture content, it is only this free water that is determined. Since vital processes such as germination are initiated by the presence of free water; one of the aims of safe storage is regulating the moisture content of the kernels in a way that the amount of this moisture should not increase above the critical level that is characteristic for certain plants (14–15%). According to its origin, free water may be 'endogenous' and 'exogenous' (physiological and physical). The former gets into the kernel through physiological processes,

TABLE 11.3 The Distribution of the Main Components in the Different Parts of Corn Grain (%)

Botanical parts	Starch	Fat	Protein	Ash	Sugar
Endosperm	98.1	15.4	73.8	17.9	28.9
	97.8–98.7	13.3–17.4	69.5–78.9	12.6–23.3	23–37.3
Germ	1.5	82.6	26.2	78.4	69.3
	0.7–1.7	80.9–85.0	18.4–27.8	72.4–83.3	60.8–75.1
Bran	0.6	1.3	2.6	2.9	1.2
	0.4–0.7	0.8–1.7	1.4–2.6	0.9–3.6	0.7–1.7

Source: Data taken from Győri, Z., Győriné Mile, I., 2002. A kukorica minősége és feldolgozása. Budapest, Hungary, Szaktudás Kiadó (in Hungarian).

while the latter is a result of the hygroscopic properties of the kernel. As a result, the grain takes up or loses water without any intervention, depending on the humidity and temperature of its environment (Fig. 11.3). These aspects of the storage environment play an important role during storing since kernels take up or lose as much water as is necessary for creating the steady-state characteristic

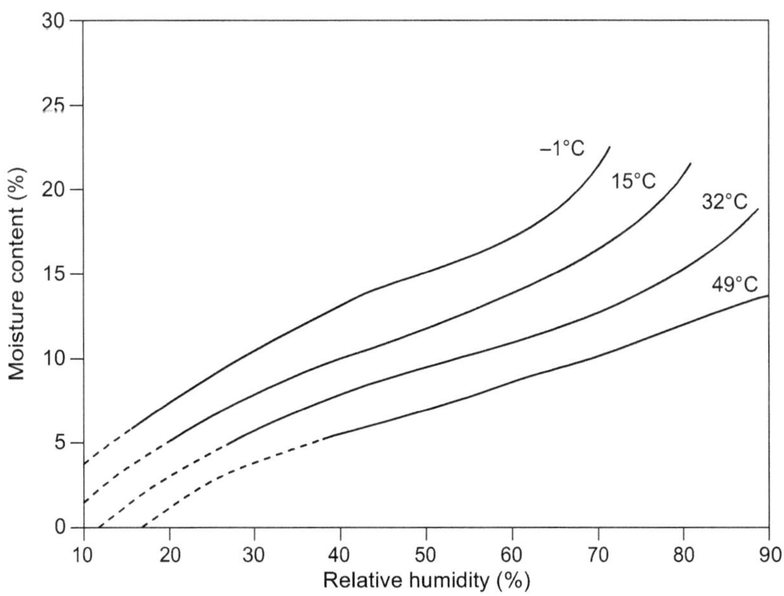

FIGURE 11.3 The relative moisture content of the kernel of corn (%). *(Data taken from Carl, W., Hall, P.E., 1980. Drying and Storage of Agricultural Crops. AVI Publishing Company Inc, Westport, CT, USA.)*

for a given temperature. This state is called hygroscopic equilibrium or balanced moisture content. Its value depends on a number of conditions, such as variety, chemical composition and also on whether the grain was dried naturally or in a heated air dryer, since the moisture content of the heat-dried grain is usually lower than that of naturally dried grain.

11.3.2 Carbohydrates

In cereal grains, it is the carbohydrates that occur in the largest amounts in the forms of starch, sugar, cellulose and non-cellulosic polysaccharides. Sugar is present in corn kernels at a level of 1–3%. In the embryo, sugars occur to a greater extent as disaccharides and to a lesser extent as monosaccharides. Cellulose, hemicellulose and pectic substances are primarily found in the pericarp (bran), while glucose and starch take part in forming cellulose. Starch is the most important polysaccharide found in cereal grains, constituting 60–80% of the dry matter in kernels. Most starch is found in the cells of the endosperm in the form of granules whose shape and size are characteristic of the grain species. The differing starch granule shapes (ball, lenticular, kidney, rectangular or multi-angular) and sizes (in μm) allow the microscopic differentiation of flour types with respect to origin (e.g. wheat or rye). The bigger corn starch particles are ball or kidney shaped, with diameters of 15–40 μm, whereas the smaller ones are ball-shaped and have diameters of 1–10 μm.

Starch consists of two components—amylose and amylopectin, which are present in the ratio of approximately 1:3 in starches of various origins. For ordinary corn, the proportions of amylose and amylopectin are 26–27% and 74–73%, respectively. The starch in waxy corn consists of amylopectin alone and its amylose content is lower than 1%. The amylose content of amylocorn is between 51% and 65%. Owing to the different chemical structures of these various starches, they have different characteristics. Amylose can be dissolved in warm water, yielding a viscous solution, while amylopectin can only be dissolved in water under pressure, when it yields a highly viscous solution. Amylose has a long non-branching chain consisting of 100–200 glucose molecules, whereas the glucose molecules in amylopectin form a structure with a branching chain (200–1000 molecules). Starch will decompose when boiled in acids, the end product being glucose. Starch syrup (thick dextrose solution) and glucose (starch sugar) are produced in this way.

Hemicelluloses have a mechanical role on the one hand, while acting as reserve nutrients on the other. This latter role is important when they are mobilised during germination. In these ways, hemicelluloses form a transition between celluloses and starches. When they decompose, pentoses are produced in large amounts, so hemicelluloses are also known as 'pentosans' (the terms hemicelluloses and pentosans are often used interchangeably). The pentosan content of the kernel of corn is 5.8–6.7%. It can also be one possible feedstock of the lignocellulosic biomass conversion for ethanol (Dionisi et al., 2015).

There are several different corn genotypes that are endosperm mutants (wy, ae, su, sh, bt, du); these grains contain sugars at widely varying levels (Salunkhe et al., 1985).

11.3.3 Nitrogenous Compounds

The nitrogenous substances in the corn kernel can initially be divided into two groups, namely, organic and inorganic compounds. The amounts of inorganic N compounds in the mature corn kernel are very small, mainly consisting of ammonia derivatives and nitrate compounds. Proteins play important roles in shaping the processes of germination, maturation and storage for the kernel. Proteins are important constituents with respect to the utilisation of corn. The products made from corn and their digestibility, nutritional values and aesthetic properties primarily depend on proteins.

As regards the organic nitrogenous compounds contained in the corn kernel, the amounts of free amino-acids are negligible, and proteins can be classified into two major groups, that is functional and storage proteins. The former proteins include the important enzymes; the latter are contained mainly in the endosperm, but they are also in the aleurone layer and the germ.

The most widely accepted classification of grain proteins in the technical literature is based on their solubility, namely, the Osborne fractionation procedure, which groups grain proteins into the categories of albumins, globulins, prolamins and glutelins.

The simple proteins—albumins (water soluble) and globulins (soluble in salt solution)—occur in grains in relatively minor amounts. For albumins, leucosin is the most important and the best known one for the corn grain. The globulin of corn is called mayzin.

Prolamins are found in the endosperm of all cereal grains. On hydrolysis, the prolamins yield a large amount of proline. Prolamins are soluble in 70% alcohol but are insoluble in water and salt solution. The prolamin contained in corn is called 'zein', which is concentrated in the endosperm. This fraction is poor in tryptophan and lysine. Glutelins are insoluble in water, alcohol and neutral salt solution, but are soluble in weak acids and alkali solutions. The glutelin contained in corn is called 'zeanin'.

The lysine content of normal corn protein is 2.1–2.8%, whereas that of opaque hybrids, which are rich in protein, is 4.0–6.5% (Sharobeem et al., 1985). Modified opaque hybrids are called 'quality protein maize' (Mertz, 1992). Unfortunately, their production potential is still below that of normal hybrids and so they are being introduced into general production more slowly (Table 11.4).

The non-protein nitrogenous compounds contained in corn are amino acids and amides, most of which are concentrated in the aleurone layer and the germ. They are found in small amounts, due to the fact that when the kernel matures, these small molecules are transmitted to the kernels and turned into proteins

TABLE 11.4 Amino Acid Composition of Whole Grain of Some Corn Varieties From USA

Amino acid (g per 100 g protein)	White flint corn	Yellow flint corn	K811 dent corn	K812 Opaque-2	K813 Opaque-2	K814 Opaque-2
Aspartic acid	6.9	5.89	7.59	11.73	13.88	18.98
Threonine	2.56	2.9	4.58	4.58	3.71	3.68
Serine	5.19	3.95	4.62	4.92	5.67	5.24
Glutamic acid	32.07	21.9	20.31	20.23	25.37	25.55
Proline	6.53	12.2	7.45	6.97	4.37	3.41
Glycine	2.72	4.22	5.74	4.03	3.14	3.51
Alanine	6.28	7.01	5.68	7.05	8.55	7.06
Cysteine	1.8	1.65	1.61	1.64	2.13	1.28
Valine	2.75	4.14	3.76	2.68	2.4	2.92
Methionine	1.66	2.25	1.23	1.62	1.4	1.49
Isoleucine	1.77	2.77	2.3	2.07	1.96	2.01
Leucine	11.44	14.77	8.68	7.1	8.65	6.69
Tyrosine	3.6	3.37	3.38	3.11	2.87	2.85
Phenylalanine	4.24	4.71	4.16	3.08	2.45	3.3
Lysine	2.77	2.17	5.96	6.21	4.69	4.81
Histidine	2.83	2.38	3.4	4.12	3.17	2.34
Tryptophan	0.23	0.31	0.36	0.43	0.43	0.41
Arginine	4.4	3.63	8.68	7.73	5.4	4.7

Source: Data taken from Sharobeem, S.F., Lásztity, R., Hidvégi, M., Salgó, A., Sarkadi, L., 1985. Amino acid content and in vitro protein quality of different corn varieties. In: Lásztity, R., Hidvégi, M. (Eds.), Amino Acid Composition and Biological Value of Cereal Proteins. Reidel Publishing, pp. 421–437.

unless there is some disorder occurring in nutrient uptake and deposition. At germination, the process is just the opposite when, under the influence of proteolytic enzymes, the proteins are broken down in the endosperm into polypeptides and amino acids.

Although enzymes (as metabolically active proteins) are found only in small amounts, they play an important role in the formation of starch, proteins and fats in corn. It is also the enzymes that break down these components on germination and when flour and coarse meal are processed. Enzyme activity depends on the moisture content of the developing grain, decreasing with the advancement of maturation. Then a static condition follows, but the activity increases again as germination starts.

11.3.4 Lipids

The endosperm and the bran are low in fat, while the aleurone layer and the germ are rich in oils. Corn oil belongs to the group of unsaturated oils, since its main fatty acid component is linoleic acid (C 18:2), which is followed in abundance by oleic acid (C 18:1), palmitic acid (C 16:0) and stearic acid (C 18:0). These compounds may be extracted with organic solvents (ether or petroleum ether); the so-called 'crude fat' so obtained contains phosphatides, sterols, mixed vitamins and pigments, in addition to fats and oils (fatty substances, greases). The individual germ tissues of the major cereal species contain the following amounts of crude fat:

Wheat	12–15%
Rye	11–13%
Barley	20–23%
Corn	20–22%

Under the influence of the enzyme lipase, the different types of fat break down into glycerine and free fatty acid during storage. When exposed to light and air, these fatty acids become rancid very quickly, producing unpleasant flavours and odours. The phosphatides (lecithin, cephalin) are primarily found in the germ and the aleurone layer. Lipases break down these compounds too, and the resulting phosphoric acid causes the acidification of the stored grain. Most phosphatides are present in grain kernels in complex form, bound to proteins and carbohydrates. In addition to these lipids, corn oil contains sterols (sitosterol, campesterol and stigmasterol).

11.3.5 Other Substances

The kernels of different corn varieties and hybrids vary in colour, from the common, yellow and brown through to red, blue and even metallic silver. All these variations can be brought about by genetic differences in the pericarp, the aleurone layer, the germ and the endosperm. Of the various colouring components, it is carotenoids that occur in the largest amounts in mature kernels. The most important representative of the carotenoids is carotene. Human and animal organisms are capable of converting most of β-carotene into vitamin A. The carotenoid content of corn is in the range of 5–80 mg/kg. In addition to the substances mentioned earlier, grains contain several pigments (xanthophyll, flavonoids) (FAO, 1992). The aleurone layer and the germ, especially, contain several vitamins. The most important ones are vitamins A and E and certain B vitamins.

Corn kernels produce 1.5–3% ash on combustion. The specific amount of ash, just like other components, is influenced by climatic, soil and fertilisation conditions. Certain grains make use of ash components (P, K, Si, Na, Ca, Mg, S) to varying degrees. Most of the mineral substances are contained in the husk and the germ, while the endosperm is poor in ash-producing substances. As regards calcium, manganese and copper, the mineral element content of corn is lower than that of wheat (Bódi et al., 2008). In contrast to earlier production findings, the amount of calcium is lower (Sager and Hoesch, 2005; Gwirtz and Garcia-Casal, 2014) in the hybrids grown today (70–150 mg/kg) than it used to be (300 mg/kg).

With the implementation of new pieces of equipment (plasma emission spectrometer ICP), there is now the opportunity to analyse elements that are tested less frequently. One of these elements, sulfur (Salunkhe et al., 1985), which belongs to the class of macro-elements, is an important nutrient and its average amount in the corn kernel is 1000–1300 mg/kg (dry-matter basis). Under adequate nutrient supply, the N–S ratio is constant and has a value of 12–16 (Tabatabai, 1984; Győri, 1998; Camberato et al., 2012; Győri and Boros 2013).

The other element to be highlighted is strontium, which is a concomitant element in calcareous minerals and rocks. Since the Chernobyl nuclear reactor disaster, the question of strontium in plant products is an issue of concern. Its level in corn kernels is below 1 mg/kg.

We have to examine the heavy metals (Hg, Cd, Pb, Zn) in the soil-plant system only where corn is grown in contaminated soil. This can occur if the floodplain is covered by mining sludge (Győri et al., 2003; Stanovič et al., 2015).

The microflora of corn has gained an ever-increasing importance over the past decades since both field and warehouse moulds produce toxic mycotoxins. Fungi (F-toxins) of the *Aspergillus* species (aflatoxins), which belong to the genus *Fusarium* (DON, zearalenone, T-toxins, fumonisin) produce mycotoxins that are poisonous even in small amounts (Miller, 2008; Russell et al., 2010; EFSA, 2012). When corn contaminated by aflatoxins was fed to cows, there was detectable Aflatoxin M1 present in milk (Trevisani et al., 2014). Nowadays, a new risk has been discovered regarding masked mycotoxins (Berthiller et al., 2013; De Saeger and Van Egmond, 2012).

11.4 DETERMINATION OF CORN QUALITY

11.4.1 Physical Properties of Kernel and Bulk Corn

In the real world, cereal grains are not dealt with individually, that is one by one, but as a bulk consignment of a multitude of grains. This is why we have to speak of mass properties, and why we must differentiate between individual properties that are characteristic of particular kernels. The most important mass properties include purity (broken corn and foreign material—BCFM), the evenness of kernels, hectolitre weight, hundred-kernel weight and roundness (Watson and Ramstad, 1991). These characteristics are the basis of classification. Kernel soundness, moisture content, hardness and density or vitreousness may also be included here. Other properties include heat conductivity and combustion heat. In commercial trade, the effects of pests and pathogens need to be considered.

Also important are identification of species and variety (hybrid), based on the properties of kernel shape, the type of endosperm (popcorn, sweet corn, floury, dent corn) and colour (white, yellow, red, purple). These aspects of quality are subject to visual assessment, but grain quality is so complex that laboratory tests may also be needed. Apart from physical properties, knowledge about nutritional value (e.g. for animal feed) is also important, especially when there are several processing objectives involved. This is the reason why a detailed quality assessment is necessary to ensure a uniform and high-quality supply for the processing industry. This information is also needed in anticipation of storage and shipment. Various methods are used to determine corn quality as no single indicator is sufficient. Moreover, there is no such thing as 'absolute quality'. 'Fitness for purpose' is probably the best definition for grain quality.

Quality as such always has to be defined in accordance with the methods applied in processing and utilisation, since the quality requirements differ from one use to another—flour versus invert sugar versus feedstuff. Accordingly, a system of indicators best suited to processing needs must be worked out, preferably on the basis of objective measurements. Such procedures may involve the sense organs or be based on physical, chemical, microbiological, radiological and biological methods. Thus the determination of quality is nothing but the judgement regarding suitability based on parameters defined by different methods. The different grades and standards in this respect are contained in the regulations of national authorities and international organisations. A prominent set of regulations are the US Grade Requirements (Table 11.5).

TABLE 11.5 Grades and Grade Requirements for Corn

		Maximum limits of:		
		Damaged kernels		
Grade	Minimum test weight per bushel (lbs)	Heat damaged kernels (%)	Total (%)	Broken corn and foreign material (%)
US No. 1	56.0	0.1	3.0	2.0
US No. 2	54.0	0.2	5.0	3.0
US No. 3	52.0	0.5	7.0	4.0
US No. 4	49.0	1.0	10.0	5.0
US No. 5	46.0	3.0	15.0	7.0

US sample grade.
US sample grade is corn that:
a. Does not meet the requirements for the US grade Nos. 1, 2, 3, 4 or 5; or
b. contains stones with an aggregate weight in excess of 0.1% of the sample weight, two or more pieces of glass, three or more crotalaria seeds (*Crotalaria* spp.), two or more castor beans (*Ricinus communis* L.), four or more particles of an unknown foreign substance(s) or a commonly recognised harmful or toxic substance(s), eight or more cockleburs (*Xanthium* spp.), or similar seeds singly or in combination, or animal filth in excess of 0.20% in 1000 g; or
c. has a musty, sour or commercially objectionable foreign odour; or
d. is heating or otherwise of distinctly low quality.
Source: Data taken from USDA, 2008. www.usda.gov.

The four major quality attributes are BCFM, damaged kernels and test weight. The US grain grades and standards are administered by the Federal Grain Inspection Service (FGIS), which specifies three classes of corn: Yellow Corn, White Corn and Mixed Corn. In other major corn producing and exporting countries (e.g. Argentina and the EU), there are equivalent standards and food-standard specifications for the quality of corn and the foods made from it (EC 824/2000). They fall into two different categories:

1. Establishing nutritional and feed values.
2. Testing industrial-processing standards.

The classification systems are now suitable for demonstrating not only changes resulting from natural influences but also quality modifications resulting from factors related to intensive agricultural management. As production becomes more intensive, testing is needed for more and more quality attributes, such as starch composition, micro-elements, amino-acid composition, mycotoxins and genetically modified (GM) status. Suitable methods for these further requirements are the subject of ongoing research.

11.4.2 Standard Analyses of Corn Quality

The main analyses needed to assess corn quality are listed in Table 11.6. Despite the range of tests needed for these analyses, they are far fewer than are needed for establishing the objective classification of winter wheat. The first and critical step in conducting any test is representative sampling. If this initial step is not conducted with the necessary diligence, there may be serious mistakes. When determining cleanness or mixed state (BCFM) all the components that are foreign to the corn kernels are isolated from the corn sample, especially material that will have unfavourable effects on utilisation. The mass of this material is weighed. In the USA, sieves having 4.8 mm (14/64 in.) and 2.4 mm (6/64 in.)

TABLE 11.6 Standard Analyses of Corn Quality

Hectolitre weight	Ash content
Hundred kernel weight	Protein content
Purity	Amino acid composition
Moisture content	Vitamins
Starch content	Mycotoxins
Oil content	Pesticide remains
Fibre content	Radioactivity

Source: Data taken from Győri, Z., Boros, N., 2013. Analysis of the sulfur content in corn in field experiments. In: Marton, L.C.s., Spitkó, T., (eds.), Proceeding of 60 Years of Hungarian Hybrid Maize (1953–2013). Pannonian Plant Biotechnology Association, Martonvásár, pp. 47–54.

round-hole are used for this purpose. In EU intervention purchases and stocks, the standard is a 4.5 mm round-hole sieve and a 1.0 mm slit sieve. In Hungary, 4.5 and 2.5-mm round-hole sieves are in use for this purpose.

As has already been indicated, moisture content is an important consideration as regards the storability of the grains to be received either when wet storage (lactic acid fermentation) (28–32% moisture content) is used or when corn is stored in bulk. The desired quality must be maintained during storage, transportation and delivery, involving long periods of time. Moisture content affects both the test weight and bulk density of maize and, consequently, these measures should only be quoted on a standardised moisture content. As the specific gravity of corn is 1.3 and that of water is 1.0, when the moisture content increases it is the value of the former that decreases. In theory, there can be three groups of methods applied in the determination of the moisture content of corn (Watson, 1986):

1. Basic reference methods (e.g. Karl Fischer method or drying a ground sample of corn under vacuum in the presence of a strong desiccating agent).
2. Practical reference methods, which according to the relevant regulations are by the air-oven method (AACC, 2000a, No. 44-15/A and ICC, 2006a, No. 110-1, 90 min 130–135 °C from a ground sample) (Williams et al., 2008). Despite being precise these methods are often slow in relation to demands. They do not help with avoiding storing extremely wet corn or determining the need for heated air drying.
3. Empirical methods used in practice ensure quick measurement based on different measurement principles. These are: dielectric constituents, quick air oven, nuclear magnetic resonance (NMR) and near-infrared spectroscopy (NIR). In order to obtain more precise readings for the varieties produced, it is worthwhile calibrating these devices with the oven-drying method every year. The great advantage of these devices is that they are portable, thus providing the opportunity to conduct the testing of samples taken directly in the field. The more widely used devices operating on the principle of NIR and NMR are also suitable for determining the levels of other important components—protein, fat and starch.

The determination of moisture content is especially important in the case of corn as there must be a special emphasis on drying to avoid kernels being broken. In the course of handling (placing into storage), kernels that have microscopic cracks (checking) in their pericarp will break up into several parts and the original biological value will decrease. These broken kernels may later become the starting points for decomposition processes. This general requirement is well illustrated by the fact that EU corn lots, which are compulsorily withheld for intervention purchase and stocks (to buy up the over-production on fixed price by the EU), have quality requirements where the regulation for kernels of irregular quality comes directly after the one for moisture content. In addition to broken kernels, levels of 3% heat damaged and 6% germinated kernels are allowed.

As a basic test, the crude protein content is most frequently conducted using the Kjeldahl digestion method, which dates back to 1883 (AACC, 2000b, Approved Method No. 46-12 and ICC, 2006b, Standard No. 105-2), followed by distilling the ammonia obtained and determining the content with an acid-alkaline based titration. With the exception of the nitrate–nitrite method, this method can determine all the substances containing nitrogen (proteins, peptides, free amino acids, amines and ammonia) as so-called 'total nitrogen', and by multiplying the value obtained by 6.25, the crude protein content is obtained.

The so-called 'nitrogen determination by combustion', based on the Dumas principle which dates back to 1848 (AACC, 2000c, Approved Method No. 46-30 and ICC, 2006c, Standard No. 167) is becoming popular worldwide. This method measures the nitrate–nitrite content in combination with the total nitrogen content. As a result, this method yields higher readings than the Kjeldahl method, especially for green corn samples, but not with mature kernels. Despite the high investment required, the utilisation of this method is spreading because it requires fewer chemicals, less workforce and puts a smaller load on the environment. Instrumental methods such as NIR and near-infrared transmission (NIT) are also widely used.

The main corn protein is zein, which contains less of the limiting amino acids lysine and tryptophan than, for example winter wheat. This is the reason why it is also important to be aware of the protein content and composition of new hybrids too.

'Crude fat content' is a collective name for substances that can be removed with petroleum ether or ether. Fats, oils, waxes, chlorophylls, xanthophylls, carotenoids, essential oils and sterols all belong here. The determination of the different fats and oils is possible with gas chromatography, which yields precise readings as regards the nutritional-physiological and industrial qualities of fats.

Of the organic components, crude fibre is important. It is composed of substances constituting the pectic substances, hemicellulose, cellulose and lignin, which constitute the cell wall and other non-structural materials (plant gums and mucilage).

The mineral elements are determined with titrimetry, photometry or spectrophotometry after the organic matter has been removed (incineration). Furthermore, non-destructive testing methods may also be applied (X-ray fluorescence). In choosing an adequate testing method the important consideration is the expected concentration of a particular substance in the material. For the purpose of determining the ash content, the ground sample must be burnt to ash in the furnace overnight at 550–600°C where the mineral substances get completely burnt in 6 h or overnight. The results are given in percentage points.

Starch may be determined with a polarimeter on the basis of the specific rotation power of the solution obtained following acidic hydrolysis and the precipitation of proteins (ICC, 2006d, Method 122/1, which dates back to 1976). However, enzymic methods of glucose determination following enzymic digestion of the starch have found a wider acceptance recently (ICC, 2006e,

Method 128/1; AACC International (2000d), Method 76-13.01). To determine processing suitability, the determination of the amylose–amylopectin ratio is more important. A ratio higher than 95% of waxy-to-non-waxy kernels is a basic requirement. The iodine colour test, the promatest and the floating number serve to determine this ratio.

Several quality indicators of corn and other grains can be determined without the use of chemical processes in the near-infrared range with devices operating in the so-called NIR or NIT (near infrared reflection or transmittance) modes. These devices open the door to perform more measurements in one step in numerous grain species even without grinding. The amount of crude protein, crude fat, starch, ash, moisture and in a number of cases that of fibre, may be determined with calibrations obtained by using traditional methods. Portable devices of this kind are available for deployment on-the-spot (in the field or warehouse) so that quick decisions can be made with respect to harvesting, grain receival and when determining prices that are especially advantageous.

Under unfavourable weather (wet, steamy weather during ripening) and storage conditions, certain fungi (e.g. *Fusarium*) may infest corn with mycotoxins. The determination of these extremely toxic compounds also constitutes part of corn-quality assessment. The most frequently occurring mycotoxins in corn are Zearalenone (F2) and Deoxynivalenol (DON). There are several methods available for detecting them and determining their quantities (thin-layer chromatography, mycotoxin test kits, gas chromatography, high-pressure liquid chromatography). This is covered in more detail in Chapter 15.

Several plant and grain protectants are employed in corn production. The relevant regulations limit values for these substances, but the determination of the chemicals and their residues fall out of the scope of the frequently conducted quality tests. Additional testing involves the determination of residual values of compounds used for fumigation.

Radio-active contamination of agricultural products has become an issue following major incidents such as at Chernobyl or at Fukushima. Most countries in the world have set critical limits for many isotopes. For sweet-corn quality, the deep-freezing industry puts a special emphasis on observing microbiological limits.

11.5 EFFECTS OF AGRONOMY AND STORAGE ON QUALITY

11.5.1 Major Factors Determining Corn Quality

There are several factors that interact to influence the quality of plant products. The major factors are listed in Table 11.7. Genetic properties determine quality fundamentally. The protein contents of old Hungarian varieties exceeded 12% whereas those of the modern ones are below 10% (dry-matter basis). It is a fact, however, that their productivity is several times higher than that of old varieties.

TABLE 11.7 Main Factors Determining Plant Quality

Genetic properties of the plant	Environmental influences	Crop management practice
Biological resources	Enviromental resources	Technical resources
Varieties	*Climate*	*Plant sequence*
Hybrids	*Soil*	*Soil cultivation*
	Water	*Sowing time*
		Amelioration
		Plant population
		Plant protection
		Plant growth regulators
		Fertilisation
		Irrigation
		Drying at harvest
		Transportation
		Storage

Source: Data taken from Győri, Z., 1999. Tápanyagellátás hatása a növényi termékek minőségére. In: Füleky Gy (Ed.), Tápanyag-gazdálkodás. Budapest, Hungary, Mezőgazda Kiadó (in Hungarian).

Breeding activity has produced a weak negative correlation between average yields and protein content. Corn proteins are poor in lysine and tryptophan and as a consequence animals, especially monogastric ones, have to receive supplementary proteins, soy, in their rations. Data for the so-called 'high-lysine hybrids' (Opaque-2), are shown in Table 11.4. Production of such hybrids on large areas is currently limited by their low average yields. As a result of the breeding activities currently underway, further changes may be possible to the composition of corn. The amylose–amylopectin ratio can be given as an example of the basis for the production of a raw material that has expanded the range of favourable properties for use in the food industry.

The cropping year may have a measurable effect on the crude protein content of corn. Temperature and precipitation data serve to characterise production years. In addition, their distribution over the vegetation period determines the effects on protein content. In drier years, the crude protein content of corn is 1.5% higher in Hungary than in wet years. In addition, mineral fertilisation is more effective in increasing crude protein content in dry years. This claim is confirmed by the fact that under the same plant-density conditions, irrigated corn has lower protein content than when it is grown under dry conditions. Individual hybrids react to increasing plant density in different ways and there may be changes not only in their dry-matter contents

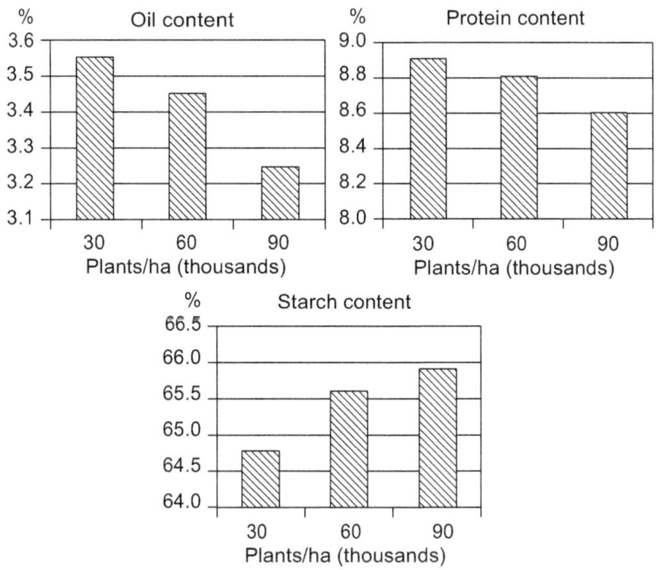

FIGURE 11.4 Effect of plant density on the quality of corn. *(Data taken from Sárvári, M., El Hallof, N., Molnár, Zs., 2007. Effect of determining factors on maize yield with special regards to plant density. Cereal Res. Commun. 35 (2), 1037–1040.)*

(Nagy, 2008) but also in their quality (Matz, 1991). Changes in protein content under different plant-density conditions are demonstrated in Fig. 11.4 (Sárvári et al., 2007). A quality-improving effect of fertilisation (NPK) can be observed in corn grown under intensive management. We can get more thoroughly acquainted about different corn cultivation methods in the book by Nafziger (2016).

As in all cereals, starch assimilation in maize is more sensitive than protein synthesis to variations in mineral supply, temperature, moisture and radiation. As a consequence, factors that reduce yield will tend to increase protein content (Fig. 11.5). The effect that a protein content increase of 0.2% per 10 kg/ha nitrogen (up to max. 120 kg/ha) is obtained in continuous growing (monoculture). This increase in a corn–wheat rotation is only 0.15%.

When high doses of NPK are applied, there are changes observed in the proteins too. However, it is not significant for the user, especially in the feedstuff industry, whether or not higher protein contents, resulting from the application of mineral fertilisers, have unfavourable effects on the crop's amino-acid composition. Studies over several years have shown that the fertiliser treatments (max. N_{240} + PK) did not decrease the lysine content of the dry matter in a statistically significant way since its ratios within protein did not change; nor did the proportion of the two sulfur-containing essential amino acids, methionine and cysteine, change significantly. It is also important to consider whether the

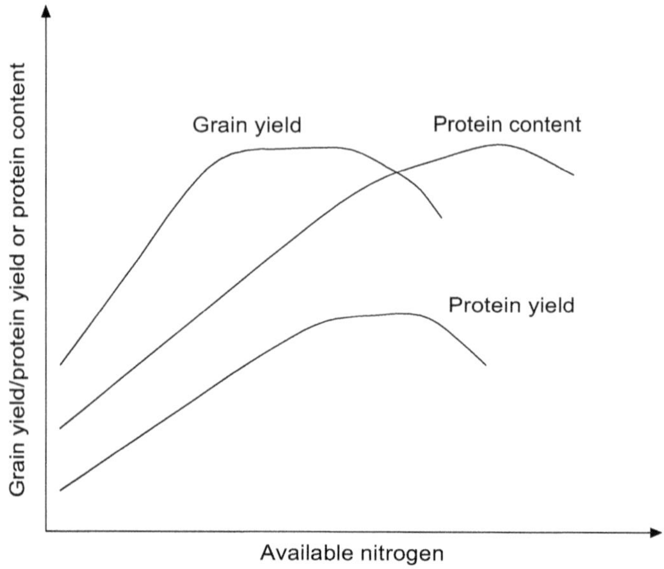

FIGURE 11.5 Effect of nitrogen supply on yield and protein content. *(Data taken from Győri, Z., 1999. Tápanyagellátás hatása a növényi termékek minőségére. In: Füleky Gy (Ed.), Tápanyag-gazdálkodás. Budapest, Hungary, Mezőgazda Kiadó (in Hungarian).)*

uptake of zinc may be impeded (high amount of available phosphorus in the soil) with a decrease in the tryptophan content. Experimental results show that increased NPK doses did decrease the zinc content but at the same time tryptophan contents increased as the protein content of the kernels also increased. It will be essential that new hybrids are tested for such considerations too (Fig. 11.6).

Different cropping locations and agro-ecological regions also make their effects felt on the quality of identical corn hybrids. These effects are strong, primarily, in locations where the use of manure is modest and the application of natural or mineral fertilisers moderates this change. Tests carried out over past years have proved that a decreasing sulfur supply in the soil (the decrease of atmospheric deposition and the decrease of the utilisation of superphosphate) has not yet influenced the sulfur content of corn. However, a decreasing sulfur supply may bring about a considerable deterioration in the quality of wheat (reducing the opportunity for disulfide-bond interactions). As an outcome of our analyses, an N–S correlation has been determined in the kernel. According to this finding the sulfur content grows in direct ratio to the nitrogen content of the kernel. Despite this fact and considering, for example frequent corn–wheat crop rotations, the sulfur fertilisation of corn (as the preceding crop) has to be considered seriously.

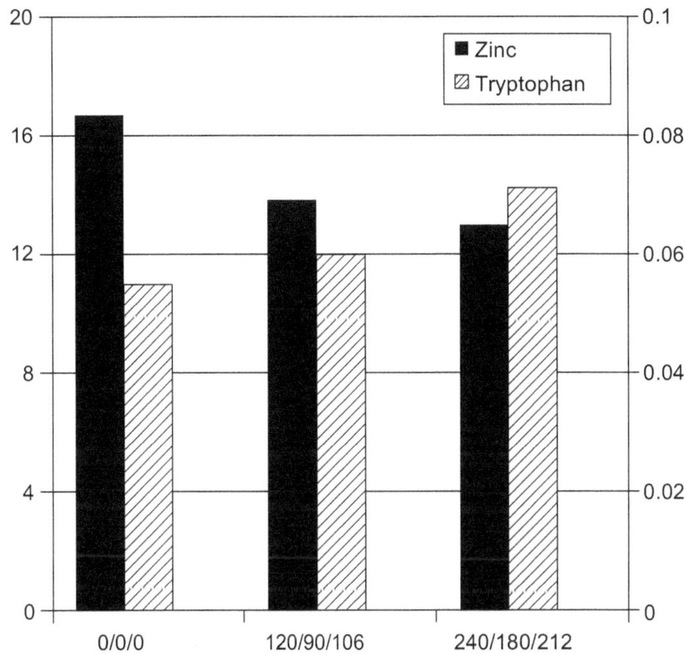

FIGURE 11.6 Effect of mineral (NPK) fertilisation on the zinc and tryptophan contents of corn. *(Data taken from Elek, Á., Borbély, M., Győri, Z., 2007. Tryptophan content of some food raw materials. Cereal Res. Commun. 35, 373–376.)*

11.6 MILLING TECHNOLOGIES

Corn-processing plants may adopt either dry or wet milling. The food industry employs dry milling as it can produce grits, flour and corn germ of an excellent quality. Thus, in this case the main objective is the separation of the main components of the kernel and the processing of the endosperm.

The technology of dry milling consists of the following operations (Kent, 1975):

1. hulling,
2. conditioning,
3. removing the germ,
4. grinding,
5. sifting,
6. purifying and aspirating of grits, and
7. packaging.

Foreign materials contained in the corn are removed in the cleaning unit with sieves and dry-stone removing devices. Contaminants of this kind may be metals, pieces of stem, stones and broken kernels, insects, rodents and birds (or anything else one can imagine). For germ removal, the moisture content of the kernel is very important. A level of 16% is preferred to keep the germ in a properly resilient state that enables a clean separation from the endosperm. For this purpose, after cleaning there is an automatic moistening device regulating the amount of water dispensed into the moistening screw, which delivers the corn into the tempering bin. It is here that the moisture penetrates into the internal part of the corn. The period for temporary storage is 6–8 h.

Germ removal comes after this temporary storage. The operation is mostly carried out with a Beall-type de-germinating device. In Europe, it is mostly the so-called ‘turbo-crushers’ that are employed for this purpose. Next, the milled material is cooled down, if necessary, and it is classified into large- or medium-sized granule fractions, germ and meal. The medium and large fractions are the result of the primary grinding, while fine hominy and germ are the products of secondary grinding. Separation of the corn germ is extremely important for two reasons:

1. Oil can be pressed from the germ part with a high efficiency.
2. Due to the high fat content of the corn germ, the grits and flour produced cannot be stored safely unless they are germ-free.

The grinding of germ-free ‘broken’ corn kernels, whose fat content may be up to 2.5%, is done in roller mills. There are flat sieves employed for classification, which will yield the grits, meal and flour products of the desired particle sizes. The grits leaving the sieves are further classified into purifying and aspirating machines. It is only in this way that corn grits, very low in fat (below 1%), can be obtained.

The products of an American mill are (in mm): flaking grits (between 3.4 and 5.8, 12%), coarse grits (between 1.4 and 2.0, 15%), medium grits (between 1.0 and 1.4), fine grits (between 0.65 and 1.0) (medium- and fine grits are 23% together), coarse meal (between 0.3 and 0.65, 10%), fine meal (between 0.17 and 0.3, 10%), corn flour (below 0.17, 5%), germ (between 6.7 and 0.5, 14%) and hominy feed (11%). The characteristics of the products from a Hungarian mill are: edible corn grits (between 1.450 and 0.400 mm, 45–48%), edible flour (below 0.400 mm, 5–8%), germ (8–10%) and flour for animal feed (34–42%).

Wet milling is traditionally the first step in the process of starch production. Initially, the two main by-products are the hull and the germ. The hulled corn is steeped in acidic water containing 0.2–0.4% SO_2 for 24–50 h, where the water-uptake is very rapid until 40% moisture is reached. The process then slows down until 45–48% moisture has been reached. This process makes the separation of the three parts of the kernel (hull, germ and endosperm) easier. In the course of steeping, the S—S bonds of the protein are broken, thus facilitating the separation of the starch and the protein. By removing the water used for steeping, material containing 5–7% dry matter, with 35–45% protein content on dry basis, is obtained. This material is ground in a cutting machine, which produces a starch–water suspension from which the germs can be removed with a hydrocyclone. The protein is separated by filtration from the suspension, and water is removed in a centrifuge at high speed. Then the protein is dried down to a moisture content of 10–12%. The starch is obtained by centrifuging and drying. Processes that combine the dry-wet grinding methods are also employed in the first phase of starch production.

11.7 COMPLEX UTILISATION OF CORN

In countries where corn is widely used, it is an important means of providing mass nutrition (as flour and coarse meal). In these cases, corn is a feedstuff high in energy but its industrial utilisation is also increasing because of its varied uses of the starch that can be obtained. In countries with modern industries and agriculture, an advanced corn industry has developed over the past years. As a result, the components (carbohydrates, proteins, fats, etc.) are separated and extracted in a concentrated form. In this way, over 1000 products—from foundry-press powder to invert sugar—have been produced from corn. In the course of industrial processing, a great variety of materials are manufactured: products for human consumption (e.g. isoglucose syrup, glucose syrup, germ oil, dextrose), industrial raw materials (starch, dextrin, rectified alcohol, corn jam, enzymes, vitamins, etc.) as well as raw materials for mixed feeds (feed yeast, oil-meal, gluten, maize bran, corn-returns).

There have been several processing plants constructed for the complex utilisation of corn all over the world especially in the United States, China and Europe, and there has been one such plant in operation in Hungary since 1982. Prior to construction, liquid sugar and rectified alcohol were considered to be the main products but many other products have also been produced satisfactorily. Today in the United States, the same amount of field corn is processed into ethanol as is used up as animal feed. There were 212 ethanol plants in the United States in 2013 (Johnson and Tolman, 2013).

Further processing is the same for both dry and wet milling procedures, although there are some differences in the actual details. For sugar production, a starch suspension is liquefied at pH 6.5 and at 85–89°C in 2–3 h by using α-amylase enzyme. This is when the breaking down of starch containing big molecules takes place. Protein and fibre materials are removed by straining at pH 4.8. To the resulting clarified fluid is added amyloglucosidase (glucoamylase) enzyme at a temperature of 60°C. This is the process of saccharification, which takes places over 24 h so that the starch breaks down almost completely to glucose (over 97% dextrose equivalent). The sugar solution thus obtained has 30% dry-matter content, after being purified in separators. It is then condensed to 38% of its previous volume in a vacuum evaporator and, in a further step, may be isomerised enzymatically to fructose (high-fructose corn syrup, HFCS) at 70°C over 60 h. The resulting product up to 50% fructose and the sugar solution obtained in this way can be regarded as invert sugar because its composition is the same as is produced by the hydrolysis and inversion of beet sugar. It also contains a small amount of maltose and isomaltose, and is also called isoglucose.

With the dry preparatory processing method, rectified alcohol is produced from the mealy fraction; while with the wet one, it is produced from part of the starch milk. As a matter of fact after the potato, the second most important raw material for rectified alcohol production in the traditional way is corn. In this procedure, saccharification occurs under the influence of amylase and then it is yeast that brings about the alcoholic fermentation. Rausch and Eckhof (2016) present the different milling technologies in detail in their book.

The plant in Hungary, which basically operates by using the wet technology, uses 1 million metric tons of corn every year, producing various finished products: starch, liquid sugar, rectified alcohol, corn germ, gluten and animal feedstuff. Out of the products prepared, liquid sugar is used for industrial purposes, although it is also in circulation for human consumption in a number of countries. This is the largest isoglucose producer in Europe comprising 27% of the total European isoglucose output. In addition to corn starch and isoglucose, 450 m^3 of bioethanol per day are also produced. In the course of industrial processing (for sweets, soft drinks, canned fruit, beer, wine, liquor), this product may be needed in large amounts as the dissolved state is an advantage since granulated sugar can be used only after being dissolved in water. In this way, the problem of granulating and dissolving again can be eliminated. Transportation of the product for this latter purpose is relatively inexpensive as it can be delivered in tanker trucks, although the construction of suitable receiving and storing systems is costly.

The utilisation of rectified alcohol is well-known since current modern technologies (pharmaceutical industry, vinegar production, the cosmetics industry, the production of domestic chemicals, eluting solvents, ethanol as a motor fuel) require larger volumes of rectified alcohol than the beverage industry. Recently its use as a fuel for engines, either on its own or in a mixed form, is spreading in an increasing number of countries (e.g. USA, Canada, Brazil and EU). Ethanol production from corn has risen to 14.3 billion gallons (54.5 billion litres) for 2014 in the USA (www.bioproducts bioenergy.gov). A schematic chart of producing ethyl alcohol from corn is seen in Fig. 11.7. Plants operating with the latest technologies do not combine the two types of milling.

Fine cooking oil can be obtained from the germ of corn, whose properties equal the quality (favourable fatty acid composition) of olive oil. It has another important characteristic being free from cholesterol.

Corn gluten is a concentrated protein feed, which is utilised by feed-formulation plants. It is available in two types: corn gluten meal (at 60% protein content) and corn gluten feed (20% protein). In either the wet or dry process, further by-products are produced: 1 bushel (35 L) of corn in the dry-grind process generates 17 lbs (7.7 kg) DDGS (distillers' dried grains with solubles), 1.6 lbs (0.73 kg) corn oil and 2.6 lbs (1.18 kg) gluten meal. The wet process generates 13.5 lbs (6.12 kg) of gluten meal at 60% protein content per bushel

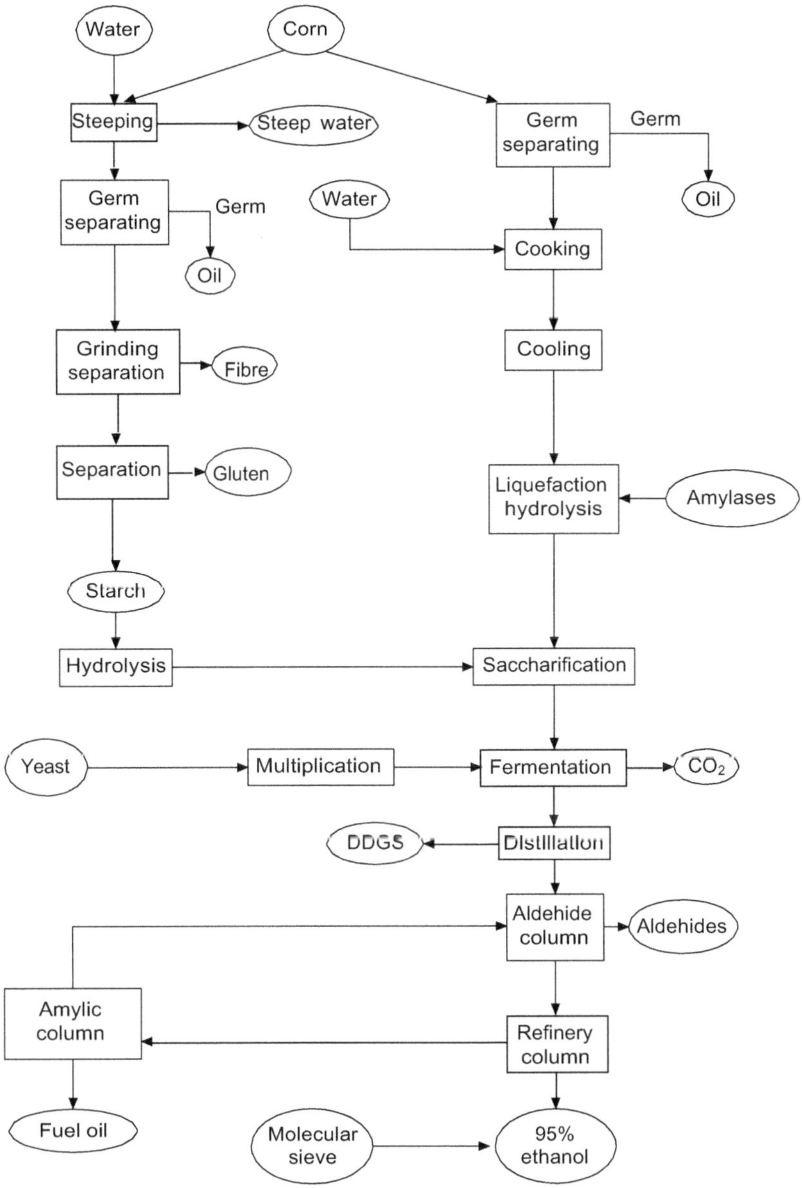

FIGURE 11.7 Ethyl alcohol production from corn. *(Data taken from Győri, Z., Győriné Mile, I., 2002. A kukorica minősége és feldolgozása. Budapest, Hungary, Szaktudás Kiadó (in Hungarian).)*

of corn (Bothast and Schlicher, 2005). A favourable aspect of processing into ethanol is that the resulting yeast proteins augment the protein content of the original corn gluten (Lásztity, 1996).

The values of these products are increased by the fact that this material contains relatively high amounts of mineral material, fermentation factors and water-soluble vitamins. This is why there is intensive research on corn gluten utilisation in different foods (Rosentrater and Krishnan, 2006). The economic efficiency of ethanol production can be further increased by the utilisation of carbon-dioxide in soft-drink production as well as by the fact that phytosterol and tocopherol components may be extracted from corn distillers' dried grain (Winkler et al., 2009).

By modifying the technologies to a lesser or greater extent and by taking away the necessary mass flow in certain locations, several more products may be obtained, which are also important industrial raw materials. Thus glucose is an important base material for sorbitol and vitamin C production while the remains of steep-water products are important raw materials for producing antibiotics and enzymes.

Highly complex industries based on corn processing are developing quickly all over the world (e.g. China and Thailand) with consideration for strict environmental regulations that produce only minimal amounts of waste material. These so-called 'biological industries' are suitable for processing various materials (from by-products to single-cell protein). Enzymes, such as α-amylase and glucoamylase, as well as biologically active substances, are utilised in an ever-increasing number of locations. The products of so-called biorefining include: sweeteners, polysaccharides, pharmaceuticals, nutraceuticals, fibres, biodegradable films, organic acids, pigments, polyols, vitamins and many others (Bothast and Schlicher, 2005).

11.8 TORTILLA PRODUCTION

From the US corn production 3–4% is used for human nutrition. The base of masa production is the dry corn milling industry. Corn masa is used in Mexican foods.

In the traditional tortilla preparation, corn is lime-cooked in pots over a fire and steeped, to form the nixtamal. The nixtamal is washed by hand or with mechanical washers to remove the skin, or pericarp, and then ground on a stone grinder (called 'metate') to form the masa (FAO, 2007). A schematic chart of producing masa from corn is shown in Fig. 11.8.

11.9 THE STATE OF AFFAIRS WITH GM CORN IN THE WORLD

When projecting future processes, one has to consider the changes already taking place since the past 20 years in crop production around the world (Fig. 11.9). In addition to improvements in the management of the production of several crop varieties (soybeans, corn, cotton, canola), it is the use of genetically engineered seeds that has come to the forefront in a number of developed and developing countries (USA, The Argentine Republic, Brazil, China).

This process can be traced from production to the utilisation of seeds but due to regulations the two may follow separate product cycles. James (2014) gives an excellent overview of the state of affairs as regards production and his compilation says that since 1996 the worldwide sowing areas of biotech soy, corn, cotton and canola have increased to 91 , 51 , 25 and 9 million ha, respectively. By this time the sowing areas of biotech crops have increased to 181 million ha,

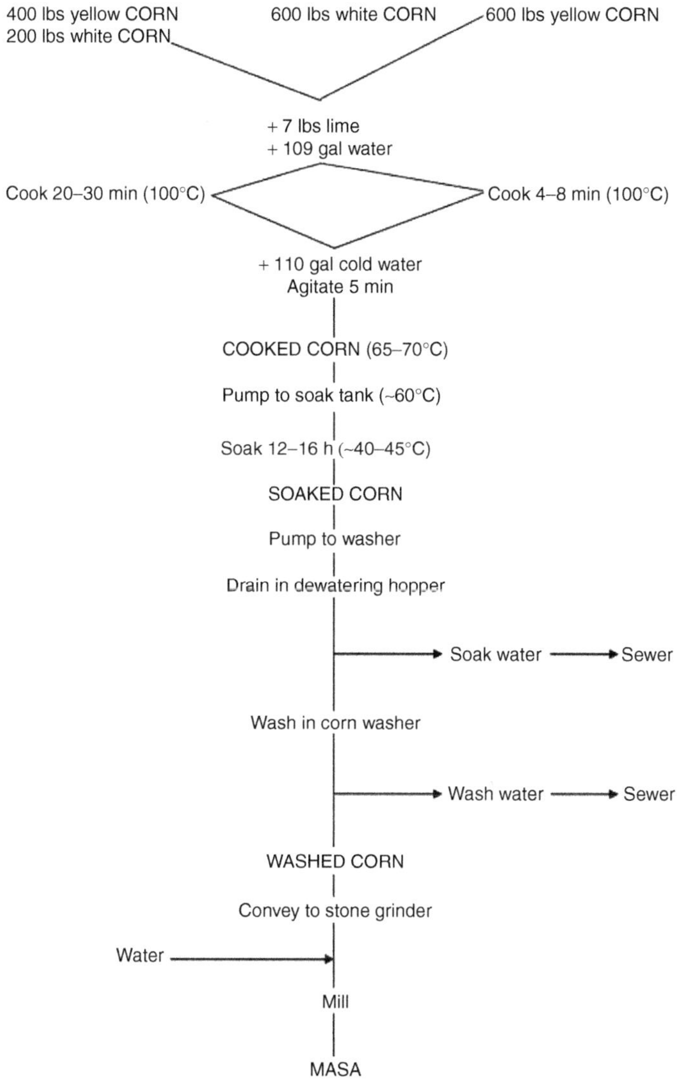

FIGURE 11.8 Typical commercial corn and tortilla chip processing. *(Data taken from Watson, S.A., Ramstad, P.E., 1991. Corn: Chemistry and Technology. American Association of Cereal Chemists Inc, St. Paul, MN, USA.)*

of which the area in developing countries is c. 85 million ha. The distribution of the different crop species is shown in Fig. 11.10, which clearly indicates that the ratios of biotech to conventional corn varieties is 30:70%.

As regards the research findings of quality tests, which compare the data for conventional and GM corn hybrids, it is the volume by Blair and Regenstein (2015) that provides detailed figures.

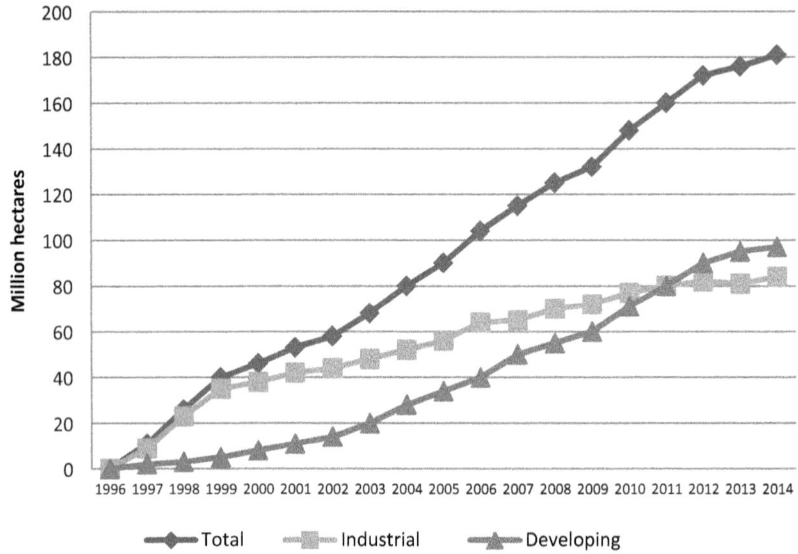

FIGURE 11.9 Global area of biotech crops 1996–2014: industrial and developing countries. *(Data taken from James, C., 2014. ISAAA Report on Global Status of Biotech/GM Crops. http://www.isaaa.org.)*

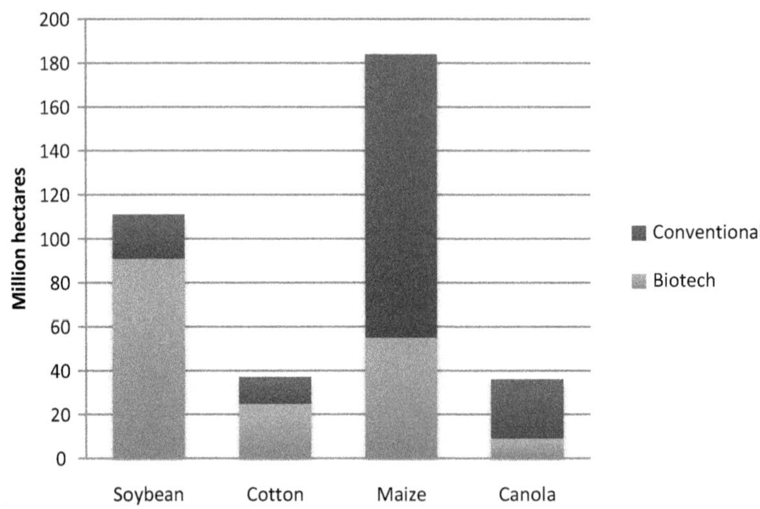

FIGURE 11.10 Global adoption rates (%) for principal biotech crops (2014). *(Data taken from James, C., 2014. ISAAA Report on Global Status of Biotech/GM Crops. http://www.isaaa.org. Hectarage based on FAO preliminary data for 2012.)*

Despite its spectacular growth, however, even today, the production of GM crops is the subject of heated debates about the crop production of the world, especially in situations in individual countries where discussions are characterised by split opinions between scientific and political interests. Consequently, the governing regulations are different, for which the best example is the difference between the USA and the EU (Lynch and Vogel, 2001; Plan and Van den Eede, 2010; EFSA, 2015). Furthermore, there are countries where GM production is banned and is accompanied by a strict observance of the rules. In Europe, politically tinted disputes have recently multiplied as regards the outcomes of the free-trade agreement between the EU and the USA (Grossman and Endres, 2000).

11.10 FUTURE TRENDS

The importance of corn production and processing could increase in the coming decades since both its food and feed uses are likely to increase. It should be noted that in the United States the Hispanic population is growing rapidly and this fact increases the demand for white-eyed hybrids. In the meantime, the endeavours made by certain countries to become energy-independent also set up new claims. In countries where the objective is to increase meat consumption (e.g. China), it is the increase in the demand for feed for monogastric animals (pork and poultry) that puts new claims on corn. This in turn will require the production of new hybrids with higher production potentials, whose chemical composition (protein and starch contents) is also favourable. The role of so-called production-processing integration will undoubtedly grow in the production of corn products. Other than conventional (feed) uses of the corn stalk, that is its utilisation in boilers and power stations may become a more significant use. The changes in the demands for utilisation are already sending important messages to breeders about quality targets.

The use of corn for food purposes will also increase as a result of globalisation. Corn products are expected to appear in foods and in locations where they have not been widely used previously. In Europe, products like tortillas, or gluten-free flakes and corn beer are being consumed more, and the consumption of sweet corn is also expected to increase with higher vegetable consumption in some countries. The development of different corn hybrids containing modified starch (amylose, amylopectin) required for the production of the different starches used in the food industry will also increase.

A major change is expected in the industrial utilisation of corn, showing changes similar to the trends in the United States for bio-alcohol produced from corn. This increase is expected to occur in the major corn-producing countries of Europe (France, Italy, Hungary and Romania). Projects of this nature are considerably influenced by the price of crude oil in the world market and international policies, such as EU attitudes. On this basis, the demand for and price of ethanol, as a fuel, will also become a factor. Addressing the raw-material requirements of this process also raises several questions for breeders, for example changing the protein:starch ratio in the corn kernel for increased starch. This would be desirable in locations where animal feed-lots are not available to utilise the protein by-product as a feedstuff. The growing and processing facilities to be established for the purposes of corn-ethanol production will require changes in the management technologies in countries and regions with intensive corn production (considerable application of mineral fertilisers and irrigation), changes in the protein:starch ratio and employing new hybrids that respond well to intensive management technologies.

Improvement in technologies for fermentation of alcohol are also expected, increasing the economic efficiency of the process yielding about 10% more alcohol from 1 tonne of corn. Wider application should decrease transportation and handling costs. The use of high-temperature yeast is also relevant since it may increase the alcohol content and decrease the cost per unit of production. Another possibility is the quick steeping process, allowing the separation of germ in dry grinding so that the oil will not finish up in the by-product at the end of the process. In addition to the alcohol obtained from the kernel, technologies currently in the experimental stages are expected to enable the production of alcohol from green crops. Thus, there are new perspectives opening up for whole-plant utilisation in addition to making silage, the current method (Dorr, 2008). The production of ethanol from cellulosic biomass will develop too (Lewis et al., 2010; Wettstein et al., 2012; Kang et al., 2014).

The previously mentioned facts put new demands on quality tests as well, and these require even faster and more precise publication of the data obtained. In this area, improving the potential for starch determination with NIR/NIT-based devices will allow the application of calibrations for starch and alcohol yield (www.foss.us). Detection of mycotoxins by rapid tests will also become more widespread, as their presence causes a deterioration of feed conversion in animals as well as in the efficiency of fermentation. Thus, quick results will become ever more important in handling goods and the importance of measurements with large equipment is expected to diminish, although they will always be required to provide the standards used to calibrate rapid tests.

In the course of these changes, it will be necessary to avoid the diversion of corn for animal feed and for industrial processing at the expense of populations who depend on corn for sustenance. Perhaps this is the major challenge facing the corn industry in the future.

Many of the changes described earlier are made more complicated by the fact that the regulations of a number of countries in the world prevent the growing of GM corn (Davison and Berthau, 2008; Popp and Lakner, 2013).

SOURCES OF FURTHER INFORMATION

http://www.agpm.com/prestation_service/en/grains.php
http://www.cemp.org
http://www.gafta.com

Belitz, H.D., Grosch, W., 1987. Food Chemistry. Springer-Verlag, Berlin, Heidelberg.
Caballero, B., Finglas, P.M., Toldrá, F., 2016. Encyclopedia of Food and Health. Elsevier, Amsterdam, Tokyo.
Fast, R.B., Caldwell, E.F., 1990. Breakfast Cereals and How They Are Made. American Association of Cereal Chemists Inc, St. Paul, MN, USA.
Fennema, O.R., 1985. Food Chemistry. CRC Press, Boca Raton, FL, USA.
Hamer, R.J., Hoseney, R.C., 1998. Interactions: The Keys to Cereal Quality. American Association of Cereal Chemists Inc, St. Paul, MN, USA.
O'Brien, L., Henry, R.J., 2000. Transgenic Cereals. American Association of Cereal Chemists Inc, St. Paul, MN, USA.
Parekh, S.R., 2004. The GMO Handbook. Humana Press Inc, Totowa, NJ.
Pomeranz, Y., 1978. Advances in Cereal Science and Technologyvol. IIAmerican Association of Cereal Chemists Inc, St. Paul, MN, USA.
Pomeranz, Y., 1980. Advances in Cereal Science and Technologyvol. IIIAmerican Association of Cereal Chemists Inc, St. Paul, MN, USA.
Spelman, C.A., 1994. Non-Food Uses of Agricultural Raw Materials. Cab International, Wallingford, CT, USA.
Wrigley, C.W., 1995. Identification of Food Grain Varieties. American Association of Cereal Chemists Inc, St. Paul, MN, USA.

REFERENCES

AACC, 2000a. Method 44-15A Moisture—Air-Oven Methods, AACC Methods, tenth ed. AACC International, St. Paul, MN, USA.
AACC, 2000b. Method 46-12.01 Crude Protein—Kjeldahl Method, Boric Acid Modification, AACC Methods, tenth ed. AACC International, St. Paul, MN, USA.
AACC, 2000c. Method 46-30.01 Crude Protein—Combustion Method, AACC Methods, tenth ed. AACC International, St. Paul, MN, USA.
AACC, 2000d. Method 76-13.01 Total Starch Assay Procedure (Megazyme Amyloglucosidase/alpha-Amylase Method), AACC Methods, tenth ed. AACC International, St. Paul, MN, USA.
Berthiller, F., Crews, C., Dall'Asta, D., De Saeger, S., Haesaert, G., Karlovsky, P., Oswald, I.P., Seefelder, W., Speijers, G., Stroka, J., 2013. Masked mycotoxins: a review. Mol. Nutr. Food Res. 57, 165–186.
Blair, R., Regenstein, J.M., 2015. Genetic Modification and Food Quality: A Down to Earth Analysis. Wiley Online Library, pp. 276.
Bódi, Z., Pepó, P., Kovács, A., Széles, É., Győri, Z., 2008. Macro- and microelement contents of blue and red kernel corns. Cereal Res. Commun. 36 (1), 147–155.
Bothast, R.J., Schlicher, M.A., 2005. Biotechnological process for conversion of corn into ethanol. Appl. Microbiol. Biotechnol. 67, 19–25.
Camberato J., Maloney S., Casteel S., 2012. Sulfur Deficiency in Corn. www.kingcorn.org/news/timeless/SulfurDeficiency.pdf
Davison, J., Berthau, Y., 2008. The theory and practice of European traceability regulations for GM food and feed. Cereal Foods World 53, 186–196.
De Saeger, S., Van Egmond, H., 2012. Masked mycotoxins. World Mycotoxin J. 5 (3), 203–206, (Special issue).
Dionisi, D., Anderson, J.A., Aulenta, F., McCue, A., Paton, G., 2015. The potential of microbial processes for lignocellulosic biomass conversion to ethanol: a review. J. Chem. Technol. Biotechnol. 90 (3), 366–383.

Dorr, T.C., 2008. Biofuels and food. Cereal Foods World 53, 76–77.

EC, 2000. European Commission, Commission Regulation 824/2000/EC of 19 April 2000 establishing procedures for the taking-over of cereals by intervention agencies and laying down methods of analysis for determination the quality of cereals. Off. J. Eur. Commun. L100, 31–50.

EFSA, 2012. Modeling, Predicting and Mapping the Emergence of Aflatoxins in Cereals in the EU due to Climate Change. http://www.efsa.europa.eu/en/supporting/pub/223e.htm

EFSA, 2015. Regulation of Genetically Modified Organisms in the European Union. https://www.efsa.europa.eu

FAO, 1992. Maize in human nutrition. Food and Nutrition Series No. 25, Rome.

FAO, 2007. Tortilla-Tortilla Chips. www.fao.org

Godon, B., Williams, C., 1994. Primary Cereal Processing: A Comprehensive Sourcebook. John Wiley & Sons, Inc, Cambridge, UK.

Grossman, M.R., Endres, A.B., 2000. Regulation of genetically modified organisms in the European Union. Am. Behav. Sci. 44 (3), 378–434.

Gwirtz, J.A., Garcia-Casal, M.N., 2014. Processing maize flour and corn meal food products. Ann. NY Acad. Sci. 1312 (1), 66–75.

Győri., Z., 1998. A termesztési tényezők hatása egyes gabonafélék és maghüvelyesek minőségére. Hungarian Academy of Sciences, Doctoral Thesis, Budapest, Hungary.

Győri, Z., Alapi, K., Sipos, P., Zubor, Á., 2003. Effects of heavy metals on floodplain soils and pastures of the River Tisza. Hungary II. Examination of soil and herbaceous plants in the Upper Tisza. In: Adriano, D., Németh, T., Győri, Z. (Eds.), Natural Attenuation of Metals Along the Tisza River–Floodplain Wetlands Continuum. University of Debrecen, Debrecen, pp. 161–173.

Győri, Z., Boros, N., 2013. Analysis of the sulphur content in corn in field experiments. In: Marton, L.C.s., Spitkó, T. (Eds.), In: Proceeding of 60 Years of Hungarian Hybrid Maize (1953–2013). Pannonian Plant Biotechnology Association Budapest, Martonvásár, Hungary, pp. 47–54.

Hallauer, A.R., 2001. Speciality Corns. CRC Press, Boca Raton, FL, USA.

ICC, 2006a. Method No. 110-1, Determination of Moisture Content of Cereals and Cereal Products (Practical Method). ICC Standard Methods (Complete ed. 2006). ICC International Association for Cereal Science and Technology, Vienna, Austria.

ICC, 2006b. Standard No. 105-2, Determination of Crude Protein in Cereals and Cereal Products for Food and Feed. ICC Standard Methods (Complete ed. 2006). ICC International Association for Cereal Science and Technology, Vienna, Austria.

ICC, 2006c. Method 167, Determination of Crude Protein in Grain and Grain Products for Food and Feed by the Dumas Combustion Principle. ICC Standard Methods (Complete ed. 2006). ICC International Association for Cereal Science and Technology, Vienna, Austria.

ICC, 2006d. Method 122/1, Determination of Starch Content by Calcium Chloride Dissolution. ICC Standard Methods (Complete ed. 2006). ICC International Association for Cereal Science and Technology, Vienna, Austria.

ICC, 2006e. Method 128/1, Procedure for the Determination of Starch After Enzymic Decomposition. ICC Standard Methods (Complete ed. 2006). ICC International Association for Cereal Science and Technology, Vienna, Austria.

James, C., 2014. ISAAA Report on Global Status of Biotech/GM Crops. http://www.isaaa.org

Johnson, P., Tolman, R., 2013. World of Corn Unlimited Possibilities. NCGA, Washington, DC. www.ncga.com

Kang, Q., Appels, L., Tan, T., Dewil, R., 2014. Bioethanol from lignocellulosic biomass: current findings determine research priorities. Sci. World J. 2014, 13.

Kent, N.L., 1975. Technology of Cereals. Pergamon Press, New York, USA.

Lásztity, R., 1996. The Chemistry of Cereal Proteins, second ed. CRC Press Inc, Boca Raton, FL, USA.

Lásztity, R., 1999. Cereal Chemistry. Akadémiai Kiadó, Budapest.
Lee, E.A., 2016. Maize genetics. Wrigley, C., Corke, H., Seetharaman, K., Faubion, J. (Eds.), Encyclopedia of Food Grains, 4, Elsevier, Oxford, UK, pp. 407–419.
Lewis, M.F., Lorenzana, R.E., Jung, H.-J.G., Bernado, R., 2010. Potential for simultaneous improvement of corn grain yield and stover quality for cellulosic ethanol. Crop Sci. 50, 516–523.
Lynch, D., Vogel, D., 2001. The Regulation of GMOs in Europe and the United States: A Case-Study of Contemporary European Regulatory Politics.' CFR.org. Council on Foreign Relations. Web. 3 Nov. 2015.
Matz, S.A., 1991. The Chemistry and Technology of Cereals as Food and Feed. AVI Publishing Company Inc, New York, USA.
Mertz, E.T., 1992. Quality Protein Maize. American Association of Cereal Chemists Inc, St. Paul, MN, USA.
Miller, J.D., 2008. Mycotoxins in small grains and maize: old problems, new challenges. Food Addit. Contamin. Part A 25 (2), 219–230.
Nafziger, E.D., 2016. Maize agronomy. Wrigley, C., Corke, H., Seetharaman, K., Faubion, J. (Eds.), Encyclopedia of Food Grains, 4, Elsevier, Oxford, UK, pp. 194–200.
Nagy, J., 2008. Maize Production: Food, Bioenergy, Forage. Hungary Akadémiai Kiadó, Budapest, p. 391.
Plan, D., Van den Eede, G., 2010. The EU Legislation on GMOs Overview European Commission Joint Research Centre Institute for Health and Consumer Protection EUR 24279 EN—2010 An.
Popp, J., Lakner, Z., 2013. Global socio-economic and environmental dimensions of GM maize cultivation. Food Nutr. Sci. 4, 8–20.
Ranum, P., Pena-Rosas, J.P., Garcia-Casal, M.N., 2014. Maize production, utilization, and consumption. Ann. N.Y. Acad. Sci. 1312 (1), 105–112.
Rausch, K.D., Eckhof, S.R., 2016. Wet milling. Wrigley, C., Corke, H., Seetharaman, K., Faubion, J. (Eds.), Encyclopedia of Food GrainsElsevier, Oxford, UK.
Roderuck, C.E., Fox, H., 1987. Nutritional value of cereal grains. In: Olson, R.A., Frey, K.J. (Eds.), Nutritional Quality of Cereal Grains. American Society of Agronomy, Madison, WI, USA.
Rosentrater, K.A., Krishnan, P.G., 2006. Incorporating distillers' grains in food products. Cereal Foods World 51, 52–60.
Russell, R., Paterson, M., Lima, N., 2010. How will climate change affect mycotoxins in food? Food Res. Int. 43, 1902–1914.
Sager, M., Hoesch, J., 2005. Macro- and micro element levels in cereals grown in lower Austria. J. Cent. Eur. Agric. 6 (4), 461–472.
Salunkhe, D.K., Chavan, J.K., Kadam, S.S., 1985. Postharvest Biotechnology of Cereals. CRC Press Inc, Boca Raton, FL, USA.
Sárvári, M., El Hallof, N., Molnár, Zs., 2007. Effect of determining factors on maize yield with special regards to plant density. Cereal Res. Commun. 35 (2), 1037–1040.
Sharobeem, S.F., Lásztity, R., Hidvégi, M., Salgó, A., Sarkadi, L., 1985. Amino acid content and in vitro protein quality of different corn varieties. In: Lásztity, R., Hidvégi, M. (Eds.), Amino Acid Composition and Biological Value of Cereal Proteins. Reidel Publishing, Dordrecht, pp. 421–437.
Stanovič, R., Kujovsky, M., Vollmannová, A., Árvay, J., Musilová, J., 2015. The content of Cd, Pb and Hg in the grain of maize (*Zea mays*, L.) harvested in the alluvia soils of the upper reaches of the River Nitra. J. Microbiol. Biotechnol. Food Sci. 4 (3), 142–144.
Tabatabai, M.A., 1984. Importance of sulphur in crop production. Biogeochemistry 1, 45–62.
Trevisani, M., Farkas, Z., Serraino, A., Zambrini, A.V., Pizzamiglio, V., Giacometti, F., Ámbrus, A., 2014. Analysis of industry-generated data. Part 1: a baseline for the development of a tool to

assist the milk industry in designing sampling plans for controlling aflatoxin M1 in milk. Food Addit. Contam. 31 (7), 1246–1256.

Watson, S.A., 1986. Corn moisture methods used by wet millers. NC-151. Grain Quality Newsletter. Ohio Agriculture Reservation and Development Centre, Wooster 8 (1), p. 20.

Watson, S.A., Ramstad, P.E., 1991. Corn: Chemistry and Technology. American Association of Cereal Chemists Inc, St. Paul, MN, USA.

Wettstein, S.G., Alonso, D.M., Gürbüz, E.I., Dumesic, J.A., 2012. A roadmap for conversion of lignocellulosic biomass to chemicals and fuels. Energy Environ. Eng. 1 (3), 218–224.

Williams, P., Landhauer, M.G., Poms, R.E., Weheing, R.L., Bergthaller, W., Gaines, C.S., 2008. The Joint AACC International-ICC Methods Harmonization Project. Cereal Foods World 53, 99–102.

Winkler, J.K., Rennick, K.A., Eller, F.J., Vaughan, S.F., 2009. Phytosterol and tocopherol components in extracts of corn distiller's dried grain. J. Agric. Food. Chem. 55, 6482–6486.

Zuber, M.S., Darrah, L.L., 1991. Breeding, genetics, and seed corn production. In: Watson, S.A., Ramstad, P.E. (Eds.), Corn: Chemistry and Technology. American Association of Cereal Chemists Inc, St. Paul, MN, USA, pp. 31–53.

Chapter 12

Rice: Grain-Quality Characteristics and Management of Quality Requirements

Melissa Fitzgerald
School of Agriculture and Food Science, University of Queensland, St. Lucia, QLD, Australia

12.1 INTRODUCTION

Rice is the major staple food crop of the world, providing about a quarter of the population's calories and up to 75% of calories consumed by poor people (FAO, 2002). Different varieties of rice are adapted to grow in many types of environments, in both temperate and tropical areas, in flood- and drought-prone areas, and in cool and warm regions (Yoshida, 1981). Food security poses real challenges to rice improvement programs. On one axis, the global population is steadily increasing, on another, arable rice-growing land is being claimed by increasing urbanisation accompanied by growth in megacities, and on top of these global warming is likely to place downward pressure on the total yield of rice (Peng et al., 2004), and on the yield of acceptable rice due to quality issues of broken grains and chalkiness (Fitzgerald and Resurreccion, 2009). Over the last decades, yield growth has been of the order of about 1–2% (Zhao and Fitzgerald, 2013). In order to continue to feed a growing population with less arable land, in a warming world, yields must increase by an annual rate of about 50 kg/ha (www.irri.org), and to ensure consumer acceptance, the higher yielding varieties must be of the quality consumers require. A clear illustration of the importance of this is the non-adoption of the first generation of hybrid rice, which offered a 15% yield advantage. Farmers were quick to adopt hybrid rice, but the lack of acceptance in the market quickly communicated to farmers that there was not a significant market for the grain from their super high-yielding rice. Examples such as this illustrate the importance of consumer acceptance as the defining factor affecting the adoption and persistence of a new variety.

The value of the rice market chain is largely determined by quality traits of the grains, both for rice traded domestically and internationally. The cooking

Cereal Grains. http://dx.doi.org/10.1016/B978-0-08-100719-8.00012-7

and sensory properties of different varieties determine the reputation of the rice, and its physical properties determine its economic value. Rice sold in domestic markets is first graded by varietal class, and then assessed on physical quality within that class. The two properties used to value polished rice are the proportion of translucent grains and of whole grains. Samples with many broken grains and/or chalky grains (opaque regions of the grain) are given a low grading in domestic markets, and fetch lower prices.

Rice sold in the international market is subject to criteria that define six grades (USDA); these criteria are defined by proportions of broken and/or chalky grain. Samples with too many broken or chalky grains can be completely excluded from international trade. Unfortunately, these two major traits are linked, and for each percentage increase in chalkiness, there is an equivalent decrease in whole grains (Zhao and Fitzgerald, 2013).

Tropical rice accounts for 90% of world rice trade; developing countries are the main players, accounting for 83% of exports and 85% of imports (FAO, 2002). When volume, not quality, drives a nation's market chain, the rice rarely reaches the higher grades, so farmers can face a situation of making little or no profit from their crop (http://www.ifc.org/ifcext/mekongpsdf.nsf/AttachmentsByTitle/MDB1/$FILE/MDB1-EN.pdf.).

Most rice is grown, traded and consumed in developing countries and those in or approaching economic transition. As an economy develops and incomes rise, demand increases for higher quality rice (Gerard et al., 2001), thus rice improvement programs must be equipped with knowledge and capacity to ensure that rice varieties released meet the quality required by target markets. Previously these were traits of physical, sensory and cooking quality, but the scope has been broadened in response to current challenges to human health, and now includes nutritional quality.

Our current knowledge of how the components of the grain interact to give different physical or sensory traits has reached a plateau. This is demonstrated by the persistence of benchmark varieties over many decades despite the yield gains achieved over the same decades (Fitzgerald et al., 2009a).

Current phenotyping methods provide data to enable breeders to select for a basic package of quality traits, but two varieties with the same scores can be differentiated by consumers (Fitzgerald, 2004; Calingacion et al., 2014). An example of this is the Thai fragrant rice Khao Dawk Mali 105 (KDML105). This variety was selected in 1959, and is still the most popular rice of Thailand, however it is susceptible to many biotic stresses (Toojinda et al., 2004). An IRRI-Thailand project sought to introduce tolerance to the Rice Blast fungus, which attacks the aerial parts of the plant. A variety resulted from this work, Pathumthani 1 (PTT1), which captured all the quality traits of KDML105, as measured with the current tools, as well as tolerance to Rice Blast. However, consumers are able to detect differences between PTT1 and KDML (Laohakunjit and Kerdchoechuen, 2007), and this is demonstrated by the lower price that PTT1 fetches in the market place (Supanee Jongdee, personal communication).

Significant gains have been made recently in understanding the genetic basis of several of the current traits of quality (Calingacion et al., 2014; Concepcion et al., 2015), but there is a continuing need to delve deeper into the grain and make use of new technology to identify adjectives that describe flavour, taste and texture and identify the structural and genetic basis of these so that high quality can be combined with high yield to meet the future needs of a growing population of rice consumers coupled with the constraints of changing growing environments.

12.2 UNIQUE CHARACTERISTICS OF RICE

Rice diverged from its Gramineae relatives of wheat, maize and barley before leaving the damp shade of the forest floor, and did not evolve the capacity to grow and thrive in dry open habitats (Kellogg, 2001). Therefore rice is unique amongst the staple food grasses in that it is adapted to flooded conditions. Research programs are in place to increase the water use efficiency of rice and to search for varieties that are able to grow aerobically. There has been significant progress in the identification of genes that confer drought tolerance, and breeding programs now have pipelines to introgress genes that enable tolerance to abiotic stresses.

It is commonly perceived that rice requires more water than irrigated wheat, but the water productivity of rice is higher than for irrigated wheat in countries with high rice yields per hectare (Ali and Talukder, 2008). This relationship between rice and water makes the cultivation of rice more environmentally profitable than wheat. In addition, the development of rice varieties that are able to grow in water-restricted conditions will substantially increase the water productivity of rice in relation to other cereals.

The way rice is consumed differs from its major food-crop relatives. Wheat and barley are mostly consumed after the grain is ground to flour, but most rice is consumed as polished grains. This elevates the importance of being able to predict the quality traits of the rice grain. If a sample of wheat flour does not meet desired quality standards, it is possible to blend the wheat with another sample of flour to achieve the correct quality. However, rice cannot be blended to achieve a particular standard of eating quality, because one type might be much firmer or take longer to cook. The particle size of a rice grain means that a consumer will be more able to detect two different qualities in a sample of two varieties of steamed rice much more easily than in a loaf of bread or a noodle made from blended wheat.

12.3 TRAITS OF QUALITY

The traits of quality include physical, sensory, cooking and more recently, nutritional value. Science has delivered functional, structural and genetic information about several of the traits, but not all are fully understood. Furthermore, some of the more esoteric traits are yet to be defined. The umbrella description of 'good' or 'bad' quality needs to be dissected to reveal a new collection of adjectives that properly describe the traits of sensory quality.

12.3.1 Physical Traits

The dimensions of the grain often define the quality of rice for various markets. Several famous examples illustrate the importance of grain dimensions. The Japanese variety Koshihikari is shorter than a medium grain, and all breeding efforts to capture Koshihikari quality must pay attention to grain shape. Basmati varieties are known to be extra long and slender. A very popular variety of rice in India, Samba Mahsuri, is a short, slender grain, and millers consider that the shape of this grain is the most important trait of quality for this variety because they can achieve high head rice yield (HRY) and the grain shape is considered elegant (Sundaram et al., 2008). When developing stress-tolerant versions of Samba Mahsuri, the shape of the grain was always a primary selection target (Sundaram et al., 2008). Grains can come in many different lengths and widths (Fitzgerald et al., 2009a), but breeding programs tend to focus primarily on obtaining the following combinations of length and width: short-bold, medium-bold, medium-slender or long-slender types.

One of the difficulties of dialogue on grain shape is that different major trading countries have different standards for grain length (Calingacion et al., 2014). This leads to difficulties in assigning grain type during trade. Several genes have been identified that associate with parameters of grain shape, and as these become validated, it should be possible to use genotyping to sort out the definitions in each standard.

A source of confusion regarding dialogue on grain shape derives from the terms used. Many people express grain dimensions as a shape, which is defined by the ratio of length and width. Three classes are defined: bold (<2), medium ($2.1-3$) and slender (>3). Confusingly, these terms could also describe width. The expression of length and shape, for example short and slender, when describing grains, illustrates to people exactly what type of grain is being described.

Most rice is sold as polished grain, so physical quality is immediately obvious to consumers. The two major traits of physical quality are chalky grain and broken grain (Fig. 12.1). These both affect the price of the rice along the

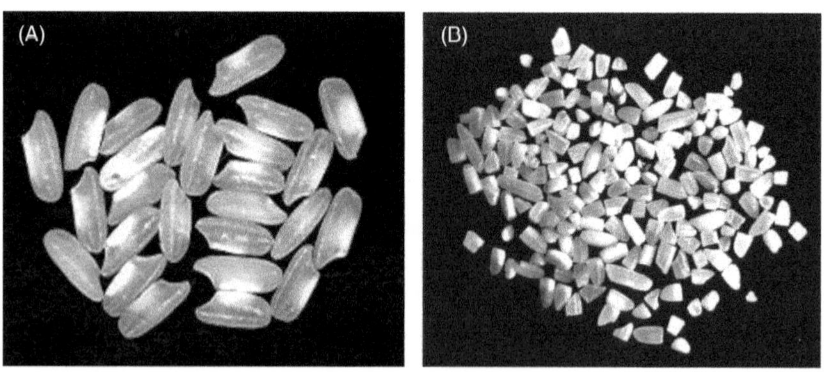

FIGURE 12.1 **Chalk (A) and broken grain (B)—the two traits of physical quality that devalue rice in all markets.**

whole spectrum of rice marketing, from the smallest rice farmer to the largest international rice exporter. Chalky regions weaken the grain and cause it to break or pulverise during polishing (Swamy and Bhattacharya, 1979). Small landholders who grow rice crops to feed their family will generally take the harvest to the local miller to de-husk and polish the grains. If the crop is very chalky, most of the grains will break or turn to powder. This means that the rice cannot be eaten and cannot be sold. Chalk affects farmers in every socio-economic scale and no breeding or research program has been able to eliminate it.

Environmental conditions play a large role in the incidence of chalk; high temperatures during grain-filling lead to chalk in many varieties of rice (Lisle et al., 2000). A recent study using a number of varieties in a field trial has shown that high temperature in combination with high humidity leads to chalk (Zhao and Fitzgerald, 2013). The study showed that high temperature without high humidity did not necessarily increase chalk.

Several factors demonstrate that there is a considerable genetic component to the occurrence of chalk. Several varieties consistently produce chalky grain, and the most notable of these is IR8—the rice of the green revolution (Peng et al., 1999). IR8 survived selection because its yield was quadruple that of the popular varieties at the time. Heritability is also illustrated by the identification of a number of QTLs for chalk (Wan et al., 2004, 2005; Zhou et al., 2009), several of which are stable in different environments (Wan et al., 2005).

Chalky areas of rice grains are characterised by airspaces and small starch granules (Lisle et al., 2000). Consequently, many studies have turned towards the processes of starch synthesis (Lisle et al., 2000; Tashiro and Wardlaw, 1991; Bhattacharya et al., 1979; Patindol and Wang, 2003) to search for genetic explanations of chalk. However, no enzyme of starch synthesis, nor any allelic variant of any gene of starch synthesis has been associated with the incidence of chalk. In one study using temperate japonica varieties, chalky parts of grains were separated from translucent parts and no difference was found in the molecular weight distribution of the starches from chalky and translucent sections of the grains (Lisle et al., 2000). This suggests that enzymes directly responsible for elongating the chains of starch are not the cause of chalk. The key difference between starch from chalky and translucent parts of the grain is that starch granules from chalky parts of the grain are much smaller than those found in translucent parts (Lisle et al., 2000). Moreover, chalky grains are generally found on secondary branches of the panicle (Lisle et al., 2000), which are known to begin development later than grains in primary positions on the panicle (Mohapatra et al., 1993; Yang et al., 2006). It has been shown that grain-filling is terminated before synthesis of starch reaches completion, resulting in small and immature starch granules in grains of lower sink dominance, with the phenotype being chalk (Fitzgerald and Resurreccion, 2009). Chalk is a trait of high scientific priority because it is a key factor that determines the yield of acceptable rice.

Two studies have recently revealed genetic loci that associate with chalk. In the first study, a vacuolar H^+ pyrophosphatase (VPPT) gene on chromosome

5 was discovered in a doubled haploid population of Chinese rice that significantly reduces chalkiness (Li et al., 2014). However, in the second recent study (Zhao et al., 2015), it is clearly demonstrated that the VPPT gene is not sufficient to protect against chalkiness. In that work, four other QTLs were discovered, which when combined in one variety, lead to negligible chalk in any environment (Zhao et al., 2015). These four QTLs and one gene must be tested more widely across various germplasm groups. Once validated, these QTLs must be converted to markers for use in breeding programs.

Broken grain destroys the visual beauty of a sample of rice, and because breaks can occur through cells (Swamy and Bhattacharya, 1982), starch leaches from the broken grains during cooking, forming a 'glue' around the cooked grains and a sticky layer on the bottom of the pan that is difficult to remove. The proportion of whole milled grain, head rice yield (HRY), is a key trait of quality measured in every quality evaluation program and also contributes to the yield of marketable and acceptable rice. There is a genetic component to HRY and QTLs have been mapped to regions on chromosome 3 in one study (Tan et al., 2001) and to chromosomes 1 and 5 in another (Septiningsih et al., 2003). The QTL on chromosome 3 is located in the same position as a major gene for grain length (Tan et al., 2001; Fan et al., 2006; Takano-Kai et al., 2009) and grain length is a factor that affects HRY. The population used by Tan et al. (2001) was derived from parents that differ in grain length and the allele from the shorter parent improved HRY. The population used by Septiningsih et al. (2003) was derived from parents of similar grain length, which reasonably explains why this QTL was found by Tan et al. (2001) and not by Septiningsih et al. (2003). It further suggests that the QTLs on chromosome 1 and 5 might directly affect HRY whereas the one on chromosome 3 is an indirect contribution.

Possibly the largest factor that affects the proportion of broken grain is post-harvest drying and storage of the rice. If the grain goes through several cycles of wetting and drying, rapid drying, or is not equilibrated prior to polishing, HRY decreases (Banaszek and Siebenmorgen, 1990; Cnossen et al., 2003; Jindal and Siebenmorgen, 1994; Kocher et al., 1990). Wetting and drying can occur if mature grain remains on the panicle in an environment where dewy nights are followed by hot, dry days. Modelling studies indicate that if grain dries to below a moisture content of 20% on a dry basis just once, HRY will be high, but if moisture content cycles several times above and below 20%, then HRY will fall significantly (Clampett et al., 2004). This has been shown previously using several varieties of rice that were cycled between 14% and 25% moisture or between 14% and 18% moisture. After 4 cycles between 14% and 25% moisture, almost all the grains were cracked or broken in both long and medium grain varieties whereas very few were broken or cracked when the moisture cycles did not exceed 20% (Fig. 12.2) (Fitzgerald et al., 2000).

Grains in a single panicle will flower over an interval of about seven days and grains on primary positions of a panicle undergo grain-filling faster than grains on secondary branches of the panicle (Mohapatra et al., 1993, 2009). The more

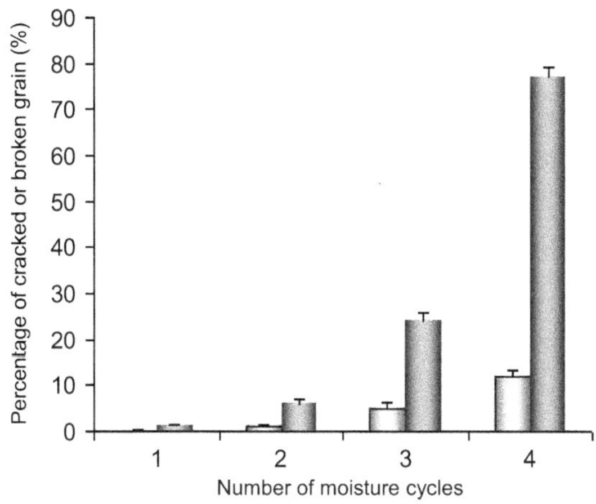

FIGURE 12.2 Percentage of cracked grain after each cycle from 14% moisture to either 18% (light grey) or 25% (dark grey) moisture and back to 14%. *(Reprinted with permission from Rural Industries Research and Development Corporation.)*

tillers on a plant, then the longer the interval of flowering for each plant (Kunze and Calderwood, 1985). This means that when grains are harvested, the length of time between reaching physiological maturity and harvest differs for each grain, resulting in variable moisture content of freshly harvested grain (Kocher et al., 1990). If rice is harvested at an average of about 22% moisture, the range in moisture content can be from 15% to 45% (Kunze and Calderwood, 1985). Moisture content of grains should be about 12–14% for optimal HRY, any higher, the grains are too soft, and lower they are too brittle (Kunze and Calderwood, 1985). If grain is dried slowly to optimal moisture content, and given time for moisture contents to equilibrate the range in moisture content is also narrowed, and HRY is optimised (Cnossen et al., 2003; Siebenmorgen et al., 1990).

Grains break because fissuring occurs within the grain (Swamy and Bhattacharya, 1982). These fissures are lines of weakness and during the milling process, the grain is exposed to mechanical forces that produce fractures along existing fissures (Cnossen et al., 2003). The glass transition of the starch is the process identified that contributes the most to fissuring (Cnossen and Siebenmorgen, 2000; Perdon et al., 2000). Glass transition (T_g) is the stage when a crystal begins to go rubbery, in the first stages of melting. A rice grain is primarily starch, and starch is semi-crystalline. If moisture content decreases at a given temperature, then the rubbery starch reverts to crystalline starch once more (Slade and Levine, 1988). T_g of native starch at 20% moisture is about room temperature. Moving in and out of T_g, as would occur if the moisture content transitioned through 20% a number of times, alters the perfection of the crystal (Slade and Levine, 1988). It is suggested that this altered crystalline structure establishes regions of stress within the starch granules, causing fissuring (Cnossen and Siebenmorgen, 2000; Perdon et al., 2000).

The colour of the polished grain is another physical trait of quality that breeders select for and that influences decisions made by consumers (Monsoor et al., 2004; Bergman et al., 2004; Fitzgerald and Reinke, 2006). Colour varies from white to yellow, and the yellowness index is used to score the colour. Consumers prefer rice that is white rather than yellower types (Bergman et al., 2004). There are a number of reasons why the rice grain could appear yellow. Protein content associates positively with yellowness and generally higher protein is not desired from a sensory point of view (Matsue et al., 1994; Champagne et al., 2009). Rice also develops a yellow tinge when it is stored poorly (Swamy et al., 1971; Tran et al., 2005). This is due to oxidation of fatty acids and results in the accumulation of compounds like 2-nonenal that give a rancid flavour (Lam and Proctor, 2003). The flavour of freshly harvested rice can be easily differentiated from rice stored for some time at room temperature, but not when the rice was stored at low temperature (Tran et al., 2005). This is consistent with previous work showing that low temperature decreases the activity of the odour-producing compounds (Tsugita et al., 1983). Storage also increases the ratio of high molecular weight protein sub-units (Chrastil and Zarins, 1992) which could play some role in changing the light absorbing characteristics of the rice, leading to the yellowish tinge.

A rice grain cannot just taste nice, it also has to look nice. Thus the traits of physical quality are important. The physical beauty of the grain is immediately seen by consumers and governs the value of the rice in the marketplace. Thus it is very important that the traits of physical quality are measured in rice improvement programs and the methods are standardised between programs.

12.3.2 Cooking and Sensory Traits

Sensory properties of the cooked rice include visual factors of gloss and shininess, as well as flavour, taste, aroma and textural features (Champagne et al., 2004). Cooking and sensory properties of grains determine the reputation of different varieties (Wada et al., 2006). With aroma as the exception, sensory traits are not well enough defined to be useful as selection tools in breeding programs. Illustrating this, the sensory quality of the best variety from various regions was compared with that of a higher yielding replacement variety. The sensory panel found significant differences between each of the pairs in terms of the sensory properties (Champagne et al., 2010) suggesting that sensory properties need more investigation, and techniques need to be developed that can screen for them in a breeding program.

Since rice is cooked in water, the hydration properties of different varieties are intuitively a topic of interest. Water uptake during cooking has been studied extensively over the decades and a recent review concludes that water uptake does not provide information about the quality of rice (Bhattacharya, 2009).

Elongation and width-wise expansion of rice grains during cooking is considered to be an important trait of cooking quality. Several Burmese varieties are prized for the elongation of the grains during cooking, and length-wise elongation during cooking is a prime quality trait of basmati varieties (Hirannaiah et al., 2001). Using a recombinant inbred population, QTLs for these traits were found on several chromosomes, including the short arm of chromosome 6, where the *Waxy* locus is found (Ge et al., 2005). In addition, QTLs on chromosomes 3, 8 and 11 have been associated with length-wise elongation of basmati varieties (Amarawathi et al., 2008).

Most rice is consumed in developing countries so cooking time and the cost of cooking, is a most important trait of cooking quality. The cooking time of rice is influenced by the gelatinisation temperature (GT) of the amylopectin fraction of the starch (Singh et al., 2003; Cuevas et al., 2010). Rice with a high GT takes approximately 4 min more to cook compared with rice of a low GT (Halick and Kelly, 1959). A recent calculation taking account of the number of times rice is cooked in one day—about 1.4 billion times—reports that global savings of about 10,000 years of cooking time each day can be made if all rice was low GT, which saves 4–5 min of cooking time (Fitzgerald et al., 2009a), representing huge savings to fuel and carbon stocks.

GT also affects texture and is a trait measured in every rice-breeding program to select or discard material early in varietal development. GT provides information about the temperature that rice begins to cook, the time it takes to cook and the texture of the cooked rice (Juliano, 1985). The GT of rice starch can fall anywhere between about 55°C and 80°C (Juliano, 1992). Most rice is cooked or processed at temperatures at or below 100°C. With an upper limit of 100°C, the lower the GT, the larger the window for the processes of cooking to occur that follow gelatinisation. The temperature of gelatinisation is likely to affect the rate of processes following gelatinisation, with effects on the sensory properties of the rice. Therefore GT serves as a link between cooking and sensory properties.

Gelatinisation describes the melting of the crystalline lamellae of amylopectin molecules (Slade and Levine, 1989). In these crystalline lamellae, chains of degrees of polymerisation (DP) between 6 and 24 are clustered in planes of double helices (Durrani and Donald, 2000; Mouille et al., 1996; Lelievre et al., 1986; Jane et al., 1999). Double helices form between outer chains on adjacent molecules (Hizukuri, 1986), leading to long-distance arrays of crystalline order. A higher proportion of shorter chains (DP 6–12), crystallised within the array, decreases the perfection of the crystals, which decreases the amount of heat required to melt them (Slade and Levine, 1988). Crystalline lamellae alternate with amorphous lamellae (Donald et al., 2001) and chains of DP 24 – 35 span both lamellae (Castro et al., 2005), forming one cluster.

In the crystalline lamellae, short amylopectin chains with DP of 6–12 are elongated by starch synthase (SS) I (Fujita et al., 2006), and those chains are subsequently elongated by SSIIa to lengths ranging from DP 12–24 (Umemoto et al., 2004). A higher proportion of chains remaining as DP 6–12 in rice is due to polymorphisms in the *SSIIa* gene which decrease the binding and catalytic capacity of the SSIIa protein (Nakamura et al., 2002, 2005, 2006; Umemoto and Aoki, 2005; Waters et al., 2006). The higher proportion of short chains destabilises the crystal, which drives GT down. It is unknown whether SSIIa activity is partially or completely abolished. The genetic findings suggest that there are two classes of GT, but rice improvement programs recognise three. Further, many varieties that are currently popular are intermediate GT (Juliano, 2007). It is therefore likely that other genes contribute to GT. Mutants lacking SSIIIa activity have significantly lower GT (Ryoo et al., 2007), so if allelic variation exists at that locus, different combinations of *SSIIa* and *SSIIIa* alleles could reveal the genetic definition of intermediate classification.

The amylose fraction of the starch is a very important trait of rice quality and associates with many traits of sensory quality (Juliano, 1972). However, amylose content is not a direct predictor of quality, since varieties with the same amylose content can differ in quality (Champagne et al., 1999). The enzyme that is responsible for amylose content is encoded by the *Waxy* locus (Sano and Katsumata, 1984). A number of different alleles of the *Wx* gene have been described (Chen et al., 2008a; Mikami et al., 2008), and the activity of the encoded protein, Granule Bound Starch Synthase (GBSS)I differs in different genetic backgrounds (Fitzgerald et al., 2009a).

Amylose content is described by the following classifications: 0–2% (waxy); 2–12% (very low); 13–19% (low); 20–24% (intermediate); and >25% (high). The International Network for Rice Quality carried out a number of ring tests in order to improve the method used to quantify amylose and to ensure that all rice improvement programs were using the same type of standards (Fitzgerald et al., 2009b). Genetic studies have revealed allelic variation at the *Wx* locus that defines each of the classes of amylose content. The waxy class is due to a null mutation caused by a duplication of 23 bp in exon 2 (Wanchana et al., 2003), which causes a frame-shift and leads to a premature translation termination codon in exon 2 (Isshiki et al., 2001) leading to no production of amylose. The next mutation found was a substitution from G to T at the splice site of exon 1, defining the Wx^b allele, which leads to inefficient splicing of the first intron (Wang et al., 1995). Retention of intron 1 in the messenger transcript means that GBSSI protein cannot be translated (Isshiki et al., 1998; Larkin and Park, 1999; Ayres et al., 1997). This mutation separates low amylose varieties from intermediate and high (Ayres et al., 1997). Amylose does accumulate in varieties carrying the Wx^b allele because about 20% of splicing occurs at one of four non-consensus splice sites nearby (Larkin and Park, 1999; Prathepha, 2007; Cai et al., 1998). The Wx^a allele was previously defined as not having the substitution at the splice site of exon 1, and carrying a G in that position (Isshiki et al., 1998; Cai et al., 1998). The presence of G at the splice site of exon 1 is found in varieties of both high amylose and intermediate amylose. A SNP was found on exon 6 (from A to C), that separates varieties of intermediate amylose from those of high amylose (Chen et al., 2008a; Mikami et al., 2008). The SNP causes an amino acid change from tyrosine to serine, and the allele has been designated the Wx^{in}

allele (Mikami et al., 2008). An SNP on exon 4 of the *Wx* gene is found in varieties with very low amylose and very low GBSSI activity (Mikami et al., 2008). This SNP, A to G, changes the amino acid from aspartate to glycine, and has only been found in varieties carrying the Wx^a allele (Mikami et al., 2008), suggesting that it occurs in an important region for either binding or catalysis of the protein. The SNPs in exons 4 and 6 are found in conserved regions of GBSSI in maize, wheat, barley, pea and potato, as well as GBSSII and SSI in rice (Mikami et al., 2008). Such wide conservation indicates the functional importance of that region for the synthesis and accumulation of amylose.

The general association between amylose content and sensory properties is a positive correlation with the hardness and a negative correlation with stickiness of the cooked rice (Li et al., 2016). However, there are many examples of varieties with the same amount of amylose but different firmness when the rice is cooked (Champagne et al., 1999). A classic example is found in the high amylose classification. Rice with high amylose can be either soft or firm after cooking. This difference is termed the gel consistency of the rice (Cagampang et al., 1973). It is measured by a simple test that essentially determines how quickly the gel sets. As the gel sets during and after the cooking process, chains of amylose aggregate, embedding amylopectin molecules and remnants of starch granules (Gidley and Bulpin, 1989). The arrangement of molecules inside the granules and in the grains determines whether the gel continues to flow for some time, or whether it sets rapidly (Gidley, 1989). The classic illustration of gel consistency is the comparison between two high amylose varieties IR5, which is soft-textured with a flowing gel, and IR8, which is firm-textured with a gel that sets rapidly (Fig. 12.3). One of the features of the starch that explains differences in gel consistency is the proportion of amylose that is soluble in hot water relative to the proportion that is insoluble (Bhattacharya et al., 1978). Studies show that varieties with a higher proportion of insoluble amylose will have a less mobile, faster setting gel (Bhattacharya, 2009; Bhattacharya et al., 1978; Bhattacharya et al., 1982). Consistent with a role for amylose, a number of studies, using different mapping populations, all find an association between the *Wx* locus and gel consistency (Tang et al., 1991; He et al., 1999, 2006; Tan et al., 1999; Bao et al., 2000; Lanceras et al., 2000; Bao, 2002; Bao et al., 2002, 2003; Li et al., 2003; Fan et al., 2005; Tian et al., 2005; Wang et al., 2007; Kiani et al., 2008). The validation of all these associations with the *Wx* locus, in a recent study using a mapping population, demonstrated that a SNP in exon 10 of the *Wx* gene associates with both gel consistency and with the proportion of insoluble amylose (Tran et al., 2011). Markers have been developed to screen for this SNP to enable direct incorporation into grain quality evaluation for high amylose and soft texture, such as is required in Burma, India and Uruguay (Calingacion et al., 2014).

There are many traits of texture; examples are cohesiveness, adhesiveness, stickiness and gumminess (Champagne et al., 2004). These traits of texture are poorly understood in terms of the composition of the grain or the genotype of the variety. Starch is not the only component of the grain and protein content

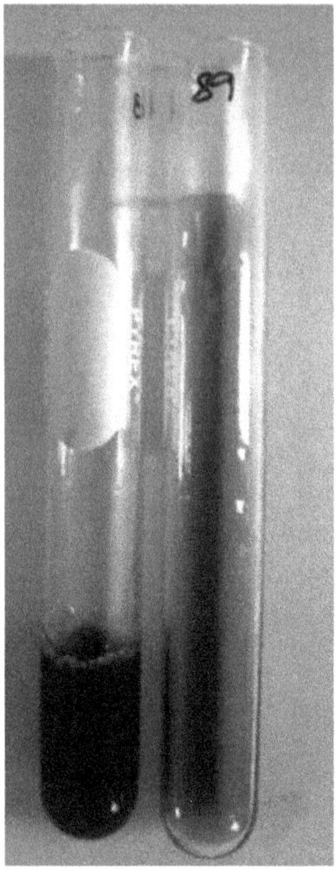

FIGURE 12.3 Gel consistency test showing rapidly set gel of IR8 (left), and the slower setting gel of IR5 (right) which flows further before setting.

can affect some of these textural traits. In a recent study, protein content was found to affect stickiness between grains, roughness and hardness of the cooked rice (Champagne et al., 2009). Protein content correlated negatively with stickiness in that study (Champagne et al., 2009). Protein also affects the pasting properties of the rice and the amount of water required to cook the rice (Sun et al., 2008). Korean and Japanese rice improvement programs actively breed for low protein, presumably because consumers in those countries prefer soft and sticky rice (Wada et al., 2006). These other features of texture are not screened routinely in rice improvement programs, presumably because our understanding of these descriptors from structural, biochemical and genetic perspectives is, at best, in its infancy.

Flavour and taste are regularly referred to by consumers as they describe rice they are eating. The most commonly considered compound for taste is the aromatic compound 2-acetyl 1-pyrroline (2AP). Recently much has been revealed about the genetics and the biochemistry of the synthesis of this compound. In almost every fragrant variety, 2AP accumulates because of mutations in the betaine aldehyde dehydrogenase 2 (BADH2) gene (Chen et al., 2008b; Kovach et al., 2009). A mutation in exon 7 of the gene is the most common mutation leading to the accumulation of 2AP (Fitzgerald et al., 2008), but at least 10 other mutations can be found in the coding region of the gene that also lead to the accumulation of aroma (Kovach et al., 2009). The location of the mutation and the nature of the mutation (SNP, insertion or deletion) associates with the amount of 2AP that accumulates (Kovach et al., 2009).

The effect of the mutations in the BADH2 gene causing 2AP to accumulate has been shown (Bradbury et al., 2008). BADH2 is thought to oxidise 4-aminobutyraldehyde, produced from putrescine, to γ-amino butyric acid (GABA). However, varieties carrying *badh2* cannot carry out the oxidation step, and accumulate both 4-aminobutyraldehyde and delta-1-pyrroline, which is proposed to activate the pathway of 2AP synthesis (Bradbury et al., 2008).

The same mutation of the BADH2 gene can be found in both the jasmine and basmati style of rice, and interestingly, the fragrance gene seems to have been introgressed into indica germplasm from japonica during domestication (Kovach et al., 2009). However the aroma of jasmine rice is considerably different from that of basmati rice (Champagne et al., 2010) and consumers can easily discriminate between the two. A recent study involving a number of metabolomics platforms and rice scientists involved in the development of new varieties of jasmine and basmati rices has shown that there are many compounds that discriminate between high quality basmati and high quality jasmine rices. Most of these discriminating compounds are volatile (Mumm et al., 2016), but several of them are unable to be identified due to the technologies of detection and separation outstripping that of correct annotation. Many of the compounds that can be identified as discriminatory are those that can be placed into pathways of lipid oxidation (Mumm et al., 2016).

Metabolomics technologies offer a suite of platforms for unravelling important components of taste, flavour and nutritional value (Fitzgerald et al., 2009a; Calingacion et al., 2012, 2015; Hall, 2011), introducing a host of new information that could be delivered as genetic tools for selection. A recent study of the volatile compounds from cooking popular varieties of rice identified indole as an unpleasant compound that can be detected as soon as the lid is lifted from a rice cooker containing the variety IR64 (Calingacion et al., 2015). This unpleasant aroma is well known to occur in cooking of IR64 and perhaps by identifying the gene for its presence, breeding could remove it from the aura of future varieties.

12.3.3 Nutritional Traits

The nutritional properties of rice have recently joined the suite of quality traits. Rice grains contain mineral elements and high levels of B vitamins and tocopherols (Bergman and Xu, 2003; Goffman and Bergman, 2004; Berardo et al., 2009; Villareal et al., 1991). Rice also contains a number of phytochemicals with potential health benefits (Bergman and Xu, 2003). However, almost all of these vitamins and minerals are found in the bran layers and embryo of the grains. If the grain is consumed with the bran layers intact, the amount of vitamins and minerals in the grains can contribute to meeting the daily requirement of each. For example the amount of thiamine in the bran layers of rice can meet the daily requirement for thiamine; this has been demonstrated by the high levels of mortality due to thiamine deficiency in populations that consumed well-polished rice compared with populations that consumed partially polished rice (Butterworth, 2001). After removal of the bran layers, the grain contains almost no micronutrients (Butterworth, 2001). Before the arrival of friction mills, rice was pounded and the bran layers removed manually (van Ruiten, 1985). Manual polishing is strenuous work and rarely produces completely polished white grains (van Ruiten, 1985). As electricity penetrated deeper into developing countries, electrical friction mills replaced manual pounding methods; these mills were able to remove the bran layers completely, to produce polished white grains. The polished grains contain only trace amounts of micronutrients, vitamins or phytochemicals. The reason so much effort is taken to remove bran completely is because the lipids in the bran layer are easily oxidised and lead to off-flavours and rancidity of rice (Shirasawa et al., 2008).

Millions of children die every year because of hidden hunger resulting from insufficient amounts of micronutrients like iron, zinc and vitamin A. Rice bran can contain reasonably high levels of both iron and zinc (Batten et al., 1992; Zhang et al., 2004), but very little pro-Vitamin A (www.harvestplus.com); with present milling practices, these nutrients are fed to animals and not to humans. Diverse diets could provide minerals and vitamins but people who derive most of their calories from a staple, often do not always have access to dietary diversity. The major reasons that consumers prefer polished grain is that unpolished grain has a short shelf life and cooking it consumes more resources. One way to address this is through biofortication and an example of this is Golden Rice (Paine et al., 2005), which is predicted to alleviate vitamin A deficiency (Stein et al., 2006) once it is released.

The major component of the polished grain is starch and variability in the structure of the starch has potential for preventing or managing chronic diseases like Type II diabetes (Marsh and Brand-Miller, 2008). Type II diabetes can be managed with diet and drugs, but untreated, it leads to a plethora of health problems such as coronary artery and peripheral vascular disease, stroke, diabetic neuropathy, non-traumatic amputations, renal failure and blindness (http://www.eatlas.idf.org/webdata/docs/Atlas2003-Summary.pdf.).

Eighty percent of people with Type II diabetes live in developing countries, and for most, rice is the staple. The number of people with Type II diabetes is predicted to increase by 150% over the next 25 years, and the greatest impact will be felt in developing countries (http://www.eatlas.idf.org/webdata/docs/Atlas2003-Summary.pdf.; Walgate, 2008). In 2007, more people died from diabetes than from HIV/AIDS, and 4 times more people died from diabetes than from malaria (http://www.eatlas.idf.org/webdata/docs/Atlas2003-Summary.pdf.).

Much diabetes research in developed countries has centred around the potential of the glycaemic index (GI) of food—a measure of the relative capacity of different dietary carbohydrates to raise blood glucose levels—to protect against the development of Type II diabetes, or to provide a tool to manage the condition (Marsh and Brand-Miller, 2008; Jenkins et al., 1981; Wolever and Mehling, 2002; Brand-Miller et al., 2003; Opperman et al., 2004; Livesey et al., 2008; Barclay et al., 2008; Halton et al., 2008). Low GI diets reduce the amount of glycated proteins (a measure of diabetic damage), increase insulin sensitivity after fasting, and lead to lower glucose levels in the blood (Marsh and Brand-Miller, 2008; Wolever and Mehling, 2002; Brand-Miller et al., 2003; Opperman et al., 2004; Livesey et al., 2008). A recent meta-analysis of 37 prospective observational studies concluded that carbohydrates of high GI are strongly associated with increased risk of developing Type II diabetes as well as a number of other chronic diseases (Barclay et al., 2008). Another recent study spanning 20 years, not included in the previous one, confirms that the type of carbohydrate, rather than the amount, associates strongly with development of Type II diabetes (Halton et al., 2008). In one study, mice were fed diets that differed in GI only due the *type* of carbohydrate in the feed, and at the end of the trial the mice on the high GI diet contained many of biochemical indicators that signal the onset of Type II diabetes in humans (Pawlak et al., 2004).

Over the past 30 years, a significant body of data has accumulated from epidemiological studies, animal models, and clinical trials indicating that a low GI diet is beneficial in lessening susceptibility to developing Type II diabetes, and that low GI foods can be achieved with different types of carbohydrates (Marsh and Brand-Miller, 2008). For many developing countries currently in, or poised to enter economic transition, rice is the staple, and rice is primarily carbohydrate. The structure of the starch components of rice is highly variable (Choi et al., 1999; Kim et al., 2005; Mizukami et al., 1999; Ramesh et al., 1999; Umemoto et al., 1999; Mizukami and Takeda, 2000; Reddy et al., 1993; Ong and Blanshard, 1995; Shu et al., 2006, 2007; Horibata et al., 2004; Kang et al., 2006; Nakamura, 2002; Wang et al., 2002), and exploring the relationship between different structures and the digestibility of rice could provide a pathway for preventing the onset or progression of Type II diabetes in developing countries. A survey of the GI of 300 popular varieties, both traditional and improved, demonstrated that the GI of rice ranges from about 50 to 95 (Fitzgerald et al., 2011). The study also showed that the strongest association was between GI and amylose content. This study demonstrates diversity in GI and also provides information for personalised nutrition by rice consumers.

12.4 ENVIRONMENTAL EFFECTS ON RICE QUALITY

Many of the traits of quality are altered by environmental conditions. Higher temperatures during grain-filling pose the largest challenge to maintaining the quality traits of rice. Chalk is a trait that is particularly susceptible to elevated temperatures during grain-filling (Lisle et al., 2000; Tashiro and Wardlaw, 1991; Ward, 2007). Another study shows that high temperature shortens the interval of substrate supply to the panicle (Fitzgerald and Resurreccion, 2009). Further, there are genotypic differences in the ways that varieties respond to the decreased supply of substrates. High temperature causes yield of paddy rice to fall (Peng et al., 2004), but the way the plant responds to the high temperature can have a dramatic effect on the yield of edible rice. In some varieties, the plant responds by sacrificing a proportion of the grains and the remaining grains are translucent and high quality, whereas varieties that try to fill all spikelets produce small, chalky grains that turn to powder in the mill (Fitzgerald and Resurreccion, 2009). Finding the genetic explanation of this difference in physiological response to high temperature will immediately provide a selection tool for breeding programs to develop varieties that are able to produce acceptable grain as the world becomes warmer.

Amylose content decreases in many varieties in high temperature and GT increases. Amylose content decreases in varieties carrying the Wx^b (Larkin and Park, 1999; Cheng et al., 2005; Zhong et al., 2005) or the Wx^{int} (Chen et al., 2008c) allele. For varieties carrying the Wx^b allele, high temperature alters the efficiency of splicing at the non-consensus splice sites (Larkin and Park, 1999) but the mechanism is unknown for the Wx^{int} varieties, though in one study using a variety carrying the Wx^{int} allele, expression of the GBSSl protein was lowered in high temperature (Ward, 2007).

In high temperature, the chain length distribution of amylopectin chains is altered, leading to higher GTs (Asaoka et al., 1984, 1985). Studies show that expression of branching enzyme (BE) llb is lowered when grain-filling occurs in high temperature (Jiang et al., 2003; Yamakawa et al., 2007). The activity of BEllb certainly contributes to GT, and this is shown by the elevated GT of mutants that lack BEllb activity (Juliano et al., 1990; Mizuno et al., 2001; Nishi et al., 2001). If mutations can be found in *BEllb*, that would provide opportunities to maintain GT at the preferred level.

It is generally accepted now that the world has become warmer and that we are facing some of the hottest years on record (Elmendorf et al., 2015). Such findings mean that rice scientists need to understand how to work with the changing environment to cultivate rice in the future to ensure yield and quality.

12.5 FUTURE CHALLENGES

The major challenges facing rice improvement programs are future food security, climate change and obesity and its consequences. There are a number of factors that define the challenge, and these are population growth, the rise in obesity and chronic disease, decreasing proportion of arable land and the effect of global warming on environmental sustainability; these all impact the amount of acceptable/marketable rice that is produced and the types of rice that are likely to be consumed.

As rice improvement programs develop varieties that can overcome these social and nutritional challenges, it is important that the tools enabling selection for grain quality are available to ensure that the new varieties produce quality grain. One example of rice breeding that has improved yield by introducing tolerance to an abiotic stress is the finding of the *sub1* gene (Xu et al., 2006). This gene provides submergence tolerance to the rice which allows yield to be maintained even after a period of submergence (Xu et al., 2006). Current breeding efforts at IRRI involve introgressing submergence tolerance into many varieties of rice (Septiningsih et al., 2009), ensuring that the quality of the grain is maintained (www.irri.org). Another example of lowering the digestibility of rice is the use of genetic silencing tools to alter the structure of the starch to ensure it is digested more slowly, and that more of it is fermented in the large intestine (Butardo et al., 2011).

Ensuring that grain quality is maintained, as higher-yielding, climate-ready varieties replace lower-yielding varieties, requires a deeper understanding of how the processes of grain-filling affect the synthesis of the components of the grain and how these components interact to define the quality required by consumers. Our understanding of the traits of quality is far from complete, and new traits must be identified. In particular, the traits of flavour, taste and texture are not well understood, and technology is now at a point where research could reveal these traits. For example it is becoming possible to obtain a full molecular weight distribution of starch at its multiple levels of organisation (Hoang et al., 2008; Baker et al., 2001; Ayoub et al., 2006; Peng et al., 2007). Such information could be used to determine causative effects on texture. Further, technologies for detecting and quantifying compounds to examine the flavour and taste of foods are rapidly undergoing development (Hall, 2007; Hall et al., 2008). With the development of such technology, and the strengthening capacity of bioinformatics, it is highly likely that previously unknown factors of quality will be revealed, and in time, developed for delivery to rice improvement programs.

REFERENCES

Ali, M.H., Talukder, M.S.U., 2008. Increasing water productivity in crop production-A synthesis. Agric. Water Manag. 95 (11), 1201–1213.

Amarawathi, Y., et al., 2008. Mapping of quantitative trait loci for basmati quality traits in rice (*Oryza sativa* L.). Mol. Breed. 21 (1), 49–65.

Asaoka, M., et al., 1984. Effect of environmental temperature during development of rice plants on some properties of endosperm starch. Starch-Starke 36 (6), 189–193.

Asaoka, M., Okuno, K., Fuwa, H., 1985. Effect of environmental temperature at the milky stage on amylose content and fine structure of amylopectin of waxy and nonwaxy endosperm starches of rice (*Oryza sativa* L.). Agric. Biol. Chem. 49 (No. 2), 373–379.

Ayoub, A., Ohtani, T., Sugiyama, S., 2006. Atomic force microscopy investigation of disorder process on rice starch granule surface. Starch-Starke 58 (9), 475–479.

Ayres, N.M., et al., 1997. Microsatellites and a single-nucleotide polymorphism differentiate apparent amylose classes in an extended pedigree of US rice germ plasm. Theor. Appl. Genet. 94 (6–7), 773–781.

Baker, A.A., Miles, M.J., Helbert, W., 2001. Internal structure of the starch granule revealed by AFM. Carbohydr. Res. 330 (2), 249–256.

Banaszek, M.M., Siebenmorgen, T.J., 1990. Head rice yield reduction rates caused by moisture sorption. Arkansas Farm Res. 39, 4–8.

Bao, J., 2002. Analysis of the relationship between Wx alleles and some starch quality parameters of rice (*Oryza sativa* L.). Cereal Res. Commun. 30 (3-4), 397–402.

Bao, J.-S., et al., 2000. Analysis of genetic covariances between rice eating and cooking quality and plant height at different stages. J. Agric. Zhejiangensis 12 (1), 6–10.

Bao, J.-S., et al., 2002. QTL for rice grain quality based on a DH population derived from parents with similar apparent amylose content. Euphytica 128 (3), 317–324.

Bao, J., et al., 2003. Analysis of quantitative trait loci for starch properties of rice based on an RIL population. Acta Botanica Sinica 45 (8), 986–994.

Barclay, A.W., et al., 2008. Glycemic index, glycemic load, and chronic disease risk--a meta-analysis of observational studies. Am. J. Clin. Nutr. 87 (3), 627–637.

Batten, G.D., et al., 1992. Mineral content of shoots of Australian rice crops. Commun. Soil Sci. Plant Anal. 23 (11–12), 1195–1208.

Berardo, N., et al., 2009. Characterization of maize germplasm for the chemical composition of the grain. J. Agric. Food Chem. 57 (6), 2378–2384.

Bergman, C.J., Xu, Z., 2003. Genotype and environment effects on tocopherol, tocotrienol, and gamma-oryzanol contents of Southern US rice. Cereal Chem. 80 (4), 446–449.

Bergman, C.J., Bhattacharya, K.R., Ohtsubo, K.I., 2004. Rice end-use quality analysis. In: Champagne, E.T. (Ed.), Rice Chemistry and Technology. The American Association of Cereal Chemists, St. Paul, Minnesota.

Bhattacharya, K.R., 2009. Cereal Food World. Physicochemical basis of eating quality of rice 54 (1), 18–28.

Bhattacharya, K.R., Sowbhagya, C.M., Indudhara Swamy, Y.M., 1978. Importance of insoluble amylose as a determinant of rice quality. J. Sci. Food Agric. 29, 359–364.

Bhattacharya, K.R., Indudhara Swamy, Y.M., Sowbhagya, C.M., 1979. Varietal difference in equilibrium moisture content of rice and effect on kernel chalkiness. J. Food Sci. Technol. 16, 214–215.

Bhattacharya, K.R., Sowbhagya, C.M., Indudhara Swamy, Y.M., 1982. Quality profiles of rice: a tentative scheme for classification. J. Food Sci. 47 (2), 564–569.

Bradbury, L., et al., 2008. Inactivation of an aminoaldehyde dehydrogenase is responsible for fragrance in rice. Plant Mol. Biol. 68 (4), 439–449.

Brand-Miller, J., et al., 2003. Low-glycemic index diets in the management of diabetes: A meta-analysis of randomized controlled trials. Diabetes Care 26 (8), 2261–2267.

Butardo, V., et al., 2011. Impact of down regulation of starch branching enzyme IIb in rice by artificial microRNA and Hairpin-RNA-Mediated RNA silencing. J. Exp. Bot. 62 (14), 4927.

Butterworth, R.F., 2001. Maternal thiamine deficiency: still a problem in some world communities. Am. J. Clin. Nutr. 74 (6), 712–713.

Cagampang, G.B., Perez, C.M., Juliano, B.O., 1973. A gel consistency test for eating quality in rice. J. Sci. Food Agric. 24, 1589–1594.

Cai, X.L., et al., 1998. Aberrant splicing of intron 1 leads to the heterogeneous 5 ' UTR and decreased expression of waxy gene in rice cultivars of intermediate amylose content. Plant J. 14 (4), 459–465.

Calingacion, M.N., et al., 2012. A genomics and multi-platform metabolomics approach to identify new traits of rice quality in traditional and improved varieties. Metabolomics 8 (5), 771–783.

Calingacion, M., et al., 2014. Diversity of global rice markets and the science required for consumer-targeted rice breeding. PLoS ONE 9 (1), e85106.

Calingacion, M., et al., 2015. Delving deeper into technological innovations to understand differences in rice quality. Rice 8 (1), 6.

Castro, J.V., et al., 2005. Mechanistic information from analysis of molecular weight distributions of starch. Biomacromolecules 6 (4), 2248–2259.

Champagne, E.T., et al., 1999. Correlation between cooked rice texture and rapid visco analyser measurements. Cereal Chem. 76 (5), 764–771.

Champagne, E.T., et al., 2004. Sensory characteristics of diverse rice cultivars as influenced by genetic and environmental factors. Cereal Chem. 81 (2), 237–243.

Champagne, E.T., et al., 2009. Unraveling the impact of nitrogen nutrition on cooked rice flavor and texture. Cereal Chem. 86 (3), 274–280.

Champagne, E.T., et al., 2010. Important sensory properties differentiating premium rice varieties. Rice 3, 270–281.

Chen, M.-H., et al., 2008a. Waxy gene haplotypes: associations with pasting properties in an international rice germplasm collection. J. Cereal Sci. 48 (3), 781–788.

Chen, S., et al., 2008b. Badh2, encoding betaine aldehyde dehydrogenase, inhibits the biosynthesis of 2-acetyl-1-pyrroline, a major component in rice fragrance. Plant Cell 20 (7), 1850–1861.

Chen, M.-H., et al., 2008c. Waxy gene haplotypes: associations with apparent amylose content and the effect by the environment in an international rice germplasm collection. J. Cereal Sci. 47 (3), 536–545.

Cheng, F., et al., 2005. Temperature induced changes in the starch components and biosynthetic enzymes of two rice varieties. Plant Growth Regul. 46 (1), 87–95.

Choi, H.C., et al., 1999. Varietal variations in physicochemical characteristics and amylopectin structure of grain in glutinous rice. Korean J. Crop Sci. 44 (3), 207–213.

Chrastil, J., Zarins, Z.M., 1992. Influence of storage on peptide subunit composition of rice oryzenin. J. Agric. Food Chem. 40 (6), 927–930.

Clampett, W.S., Williams, R.L., Lacy, J.M., 2004. Improvement of rice grain quality. Rural Industries Research and Development Corporation: City, p. 44.

Cnossen, A.G., Siebenmorgen, T.J., 2000. The glass transition temperature concept in rice drying and tempering: Effect on milling quality. Trans. Am. Soc. Agric. Eng. 43 (6), 1661–1667.

Cnossen, A.G., Jimenez, M.J., Siebenmorgen, T.J., 2003. Rice fissuring response to high drying and tempering temperatures. J. Food Eng. 59 (1), 61–69.

Concepcion, J.C.T., et al., 2015. The need for new tools and investment to improve the accuracy of selecting for grain quality in rice. Field Crops Res. 182, 60–67.

Cuevas, R., et al., 2010. Melting the secrets of gelatinisation temperature. Funct. Plant Biol. 37, 439–447.

Donald, A.M., et al., 2001. Scattering studies of the internal structure of starch granules [Review]. Starch-Starke 53 (10), 504–512.

Durrani, C.M., Donald, A.M., 2000. Shape, molecular weight distribution and viscosity of amylopectin in dilute solution. Carbohydr. Polym. 41, 207–217.

Elmendorf, S.C., et al., 2015. Experiment, monitoring, and gradient methods used to infer climate change effects on plant communities yield consistent patterns. Proc. Natl. Acad. Sci. 112 (2), 448–452.

Fan, C.C., et al., 2005. The main effects, epistatic effects and environmental interactions of QTLs on the cooking and eating quality of rice in a doubled-haploid line population. Theor. Appl. Genet. 110 (8), 1445–1452.

Fan, C., et al., 2006. GS3, a major QTL for grain length and weight and minor QTL for grain width and thickness in rice, encodes a putative transmembrane protein. Theor. Appl. Genet. 112 (6), 1164–1171.

FAO Rice Information, Vol. 3., 2002. Rome: Food and Agriculture Organization of the United Nations. 222. Available from: http://www.fao.org/DOCREP/006/Y4751E/y4751e03.htm

Fitzgerald, M.A., 2004. Starch. In: Champagne, E.T. (Ed.), Rice Chemistry and Technology. AACC, St. Paul, Minnesota.

Fitzgerald, M.A., Reinke, R.F., 2006. Rice Grain Quality III, Rural Industries Research and Development Corporation, pp. 1–78.

Fitzgerald, M.A., Resurreccion, A.P., 2009. Maintaining the yield of edible rice in a warming world. Funct. Plant Biol. 36 (12), 1037–1045.

Fitzgerald, M.A., Lewin, L.G., Reinke, R.F., 2000. Evaluation of rice grain quality II. Rural Industries Research and Development Corporation, Canberra, 47.

Fitzgerald, M.A., et al., 2008. Is there a second gene for fragrance in rice? Plant Biotechnol. J. 6 (4), 416–423.

Fitzgerald, M.A., McCouch, S.R., Hall, R.D., 2009a. More than just a grain of rice, the global quest for quality. Trends Plant Sci. 14 (3), 133–139.

Fitzgerald, M.A., et al., 2009b. Addressing the dilemmas of measuring amylose in rice. Cereal Chem. 86 (5), 492–498.

Fitzgerald, M., et al., 2011. Identification of a major genetic determinant of glycaemic index in rice. Rice 4, 66–74.

Fujita, N., et al., 2006. Function and characterization of starch synthase I using mutants in rice. Plant Physiol. 140 (3), 1070–1084.

Ge, X.J., et al., 2005. QTL analysis of cooked rice grain elongation, volume expansion, and water absorption using a recombinant inbred population. Plant Breed. 124 (2), 121–126.

Gerard, C., et al., 2001. A multi-stages biosynthetic pathway in starch granules revealed by the ultrastructure of maize mutant starches. J. Cereal Sci. 34 (1), 61–71.

Gidley, M.J., 1989. Molecular mechanisms underlying amylose aggregation and gelation. Macromolecules 22, 351–358.

Gidley, M.J., Bulpin, P.V., 1989. Aggregation of amylose in aqueous systems: the effect of chain length on phase behavior and aggregation kinetics. Macromolecules 22, 341–346.

Goffman, F.D., Bergman, C.J., 2004. Rice kernel phenolic content and its relationship with antiradical efficiency. J. Sci. Food Agric. 84 (10), 1235–1240.

Halick, J.V., Kelly, V.J., 1959. Gelatinization and pasting characteristics of rice varieties as related to cooking behavior. Cereal Chem. 39, 91–98.

Hall, R.D., 2007. Food metabolomics: META-PHOR a new European research initiative. Agro Food Ind. Hi-Tech 18 (1), 14–16.

Hall, R.D., 2011. Plant metabolomics in a nutshell: potential and future challenges. In: Hall, R.D. (Ed.), Biology of Plant Metabolomics. Wiley Blackwell, Oxford, pp. 1–24.

Hall, R.D., Brouwer, I.D., Fitzgerald, M.A., 2008. Plant metabolomics and its potential application for human nutrition. Physiologia Plantarum 132, 162–175.

Halton, T.L., et al., 2008. Low-carbohydrate-diet score and risk of type 2 diabetes in women. Am. J. Clin. Nutr. 87 (2), 339–346.

He, P., et al., 1999. Genetic analysis of rice grain quality. Theor. Appl. Genet. 98 (3–4), 502–508.

He, Y., et al., 2006. Functional analysis of starch-synthesis genes in determining rice eating and cooking qualities. Mol. Breed. 18 (4), 277–290.

Hirannaiah, B.V., Bhashyam, M.K., Ali, S.Z., 2001. An improved cooking quality test for basmati rice. J. Food Sci. Technol. 38 (2), 116–119.

Hizukuri, S., 1986. Polymodal distribution of the chain lengths of amylopectins, and its significance. Carbohydr. Res. 147, 342–347.

Hoang, N.-L., et al., 2008. Toward a full characterization of native starch: separation and detection by size-exclusion chromatography. J. Chromatogr. A 1205 (1–2), 60–70.

Horibata, T., et al., 2004. Structure and physicochemical characteristics of endosperm starches of rice cultivars recently bred in Japan. J. Appl. Glycosci. 51, 303–313.

Isshiki, M., et al., 1998. A naturally occurring functional allele of the rice waxy locus has a GT to TT mutation at the 5' splice site of the first intron. Plant J. 15 (1), 133–138.

Isshiki, M., et al., 2001. Nonsense-mediated decay of mutant waxy mRNA in rice. Plant Physiol. 125 (3), 1388–1395.

Jane, J.-L., et al., 1999. Effects of amylopectin branch chain length and amylose content on the gelatinization and pasting properties of starch. Cereal Chem. 76 (5), 629–637.

Jenkins, D.J., et al., 1981. Glycemic index of foods: a physiological basis for carbohydrate exchange. Am. J. Clin. Nutr. 34 (3), 362–366.

Jiang, H., Dian, W., Wu, P., 2003. Effect of high temperature on fine structure of amylopectin in rice endosperm by reducing the activity of the starch branching enzyme. Phytochemistry 63 (1), 53–59.

Jindal, V.K., Siebenmorgen, T.J., 1994. Effects of rice kernel thickness on head rice yield reduction due to moisture adsorption. Trans. Am. Soc. Agric. Eng. 37 (2), 487–490.

Juliano, B.O., 1972. Physicochemical properties of starch and protein in relation to grain quality and nutritional value of rice. In: IRRI Rice Breeding, IRRI, Los Banos, Philippines, pp. 389–405.

Juliano, B.O., 1985. Polysaccharides, proteins and lipids of rice. In: Juliano, B.O. (Ed.), Rice, Chemistry and Technology. American Association of Cereal Chemists, St Paul, MN, pp. 59–60.

Juliano, B.O., 1992. Structure, chemistry, and function of the rice grain and its fractions. Cereal Food World 37 (10), 772.

Juliano, B.O., 2007. Rice Chemistry and Quality. Philippine Rice Research Institute, Munoz, Nueva Ecija, 402.

Juliano, B.O., et al., 1990. Some grain properties of IR36-based starch mutants. Starch-Stärke 42, 256–260.

Kang, H.-J., et al., 2006. Comparison of the physicochemical properties and ultrastructure of Japonica and Indica rice grains. J. Agric. Food Chem. 54 (13), 4833–4838.

Kellogg, E.A., 2001. Evolutionary history of the grasses. Plant Physiol. 125, 1198–1205.

Kiani, S., et al., 2008. Inheritance of gelatinization temperature and gel consistency in rice (*Oryza sativa* L.). J. Appl. Sci. 8 (8), 1503–1510.

Kim, K.S., et al., 2005. Ultrastructure of individual and compound starch granules in isolation preparation from a high-quality, low-amylose rice, ilpumbyeo, and its mutant, G2, a high-dietary fiber, high-amylose rice. J. Agric. Food Chem. 53 (22), 8745–8751.

Kocher, M.F., et al., 1990. Rice kernel moisture content variation at harvest. Trans. Am. Soc. Agric. Eng. 33 (2), 541–548.

Kovach, M.J., et al., 2009. The origin and evolution of fragrance in rice (*Oryza sativa* L.). Proc. Natl. Acad. Sci. 106 (34), 14444–14449.

1985. Rough rice drying. In: Kunze, O., Calderwood, D., Juliano, B.O. (Eds.), Rice Chemistry and Technology. third ed. AACC, St Paul, Minnesota, pp. 233–255.

Lam, H.S., Proctor, A., 2003. Milled rice oxidation volatiles and odor development. J. Food Sci. 68 (9), 2676–2681.

Lanceras, J.C., et al., 2000. Mapping of genes for cooking and eating qualities in Thai jasmine rice (KDML105). DNA Res. 7 (2), 93–101.

Laohakunjit, N., Kerdchoechuen, O., 2007. Aroma enrichment and the change during storage of non-aromatic milled rice coated with extracted natural flavor. Food Chem. 101 (1), 339–344.

Larkin, P.D., Park, W.D., 1999. Transcript accumulation and utilization of alternate and non-consensus splice sites in rice granule-bound starch synthase are temperature-sensitive and controlled by a single-nucleotide polymorphism. Plant Mol. Biol. 40 (4), 719–727.

Lelievre, J., Lewis, J.A., Marsden, K., 1986. The size and shape of amylopectin: a study using analytical ultracentrifugation. Carbohydr. Res. 153 (2), 195–203.

Li, Z.F., et al., 2003. Mapping of quantitative trait loci controlling physico-chemical properties of rice grains (*Oryza sativa* L.). Breed. Sci. 53 (3), 209–215.

Li, Y., et al., 2014. Chalk5 encodes a vacuolar H(+)-translocating pyrophosphatase influencing grain chalkiness in rice. Nat. Genet. 46 (4), 398–404.

Li, H., et al., 2016. The importance of amylose and amylopectin fine structure for textural properties of cooked rice grains. Food Chem. 196, 702–711.

Lisle, A.J., Martin, M., Fitzgerald, M.A., 2000. Chalky and translucent rice grains differ in starch composition and structure and cooking properties. Cereal Chem. 77 (5), 627–632.

Livesey, G., et al., 2008. Glycemic response and health a systematic review and meta-analysis: relations between dietary glycemic properties and health outcomes. Am. J. Clin. Nutr. 87 (1), 258S–268S.

Marsh, K., Brand-Miller, J.C., 2008. State of the art reviews: glycemic index, obesity, and chronic disease. Am. J. Lifestyle Med. 2 (2), 142–150.

Matsue, Y., Odahara, K., Hiramatsu, M., 1994. Differences in protein content, amylose content and palatability in relation to location of grains within rice panicle. Jpn. J. Crop Sci. 63 (2), 271–277.

Mikami, I., et al., 2008. Allelic diversification at the wx locus in landraces of Asian rice. Theor. Appl. Genet. 116 (7), 979–989.

Mizukami, H., Takeda, Y., 2000. Chewing properties of cooked rice from new characteristic rice cultivars and their relation to starch molecular structures. J. Appl. Glycosci. 47 (1), 61–65.

Mizukami, H., Takeda, Y., Hizukuri, S., 1999. The structure of the hot-water soluble components in the starch granules of new Japanese rice cultivars. Carbohydr. Polym. 38 (4), 329–335.

Mizuno, K., et al., 2001. Characterization of an isoform of rice starch branching enzyme, RBE4, in developing seeds. Plant Cell Physiol. 42 (4), 349–357.

Mohapatra, P.K., Patel, R., Sahu, S.K., 1993. Time of flowering affects grain quality and spikelet partitioning within the rice panicle. Aust. J. Plant Physiol. 20, 231–241.

Mohapatra, P.K., Sarkar, R.K., Kuanar, S.R., 2009. Starch synthesizing enzymes and sink strength of grains of contrasting rice cultivars. Plant Sci. 176, 256–263.

Monsoor, M.A., Proctor, A., Siebenmorgen, T.J., 2004. Surface lipid and free fatty acids (FFA) content of head and broken rice produced by milling after different drying treatments. Cereal Chemistry 81 (6), 705–709.

Mouille, G., et al., 1996. Preamylopectin processing—a mandatory step for starch biosynthesis in plants. Plant Cell 8 (8), 1353–1366.

Mumm, R., et al., 2016. Multi-platform metabolomics analyses of a broad collection of fragrant and non-fragrant rice varieties reveals the high complexity of grain quality characteristics. Metabolomics 12, 38.

Nakamura, Y., 2002. Towards a better understanding of the metabolic system for amylopectin biosynthesis in plants: Rice endosperm as a model tissue [Review]. Plant Cell Physiol. 43 (7), 718–725.

Nakamura, Y., et al., 2002. The fine structure of amylopectin in endosperm from Asian cultivated rice can be largely classified into two classes. Starch-Stärke 54 (3-4), 117–131.

Nakamura, Y., et al., 2005. Essential amino acids of starch synthase IIa differentiate amylopectin structure and starch quality between japonica and indica rice varieties. Plant Mol. Biol. 58 (2), 213–227.

Nakamura, Y., Sato, A., Juliano, B.O., 2006. Short-chain-length distribution in debranched rice starches differing in gelatinization temperature or cooked rice hardness. Starch-Starke 58 (3-4), 155–160.

Nishi, A., et al., 2001. Biochemical and genetic analysis of the effects of amylose-extender mutation in rice endosperm. Plant Physiol. 127 (2), 459–472.

Ong, M.H., Blanshard, J.M.V., 1995. Texture determinants in cooked, parboiled rice. I. Rice starch amylose and fine structure of amylopectin. J. Cereal Sci. 21 (3), 251–260.

Opperman, A.M., et al., 2004. Meta-analysis of the health effects of using the glycaemic index in meal-planning. Br. J. Nutr. 92, 367–381.

Paine, J.A., et al., 2005. Improving the nutritional value of Golden Rice through increased pro-vitamin A content. Nat. Biotech. 23 (4), 482–487.

Patindol, J., Wang, Y.J., 2003. Fine structures and physicochemical properties of starches from chalky and translucent rice kernels. J. Agric. Food Chem. 51 (9), 2777–2784.

Pawlak, D., Kushner, J., Ludwig, D.S., 2004. Effects of dietary glycaemic index on adiposity, glucose homoeostasis, and plasma lipids in animals. Lancet 364, 778–785.

Peng, S., et al., 1999. Yield potential trends of tropical rice since the release of IR8 and the challenge of increasing rice yield potential. Crop Sci. 39 (6), 1552–1559.

Peng, S., et al., 2004. Rice yields decline with higher night temperature from global warming. Proc. Natl. Acad. Sci. USA 101 (27), 9971–9975.

Peng, L., Liu, Z., Kennedy, J.F., 2007. The study of starch nano-unit chains in the gelatinization process. Carbohydr. Polym. 68 (2), 360–366.

Perdon, A., Siebenmorgen, T.J., Mauromoustakos, A., 2000. Glassy state transition and rice drying: development of a brown rice state diagram. Cereal Chem. 77 (6), 708–713.

Prathepha, P., 2007. Identification of variant transcripts of waxy gene in non-glutinous rice (*O. sativa* L.) with different amylose content. Pak. J. Biol. Sci. 10 (15), 2500–2504.

Ramesh, M., Ali, S.Z., Bhattacharya, K.R., 1999. Structure of rice starch and its relation to cooked-rice texture. Carbohydr. Polym. 38 (4), 337–347.

Reddy, R.K., Ali, Z.S., Bharracharya, K.R., 1993. The fine structure of rice-starch amylopectin and its relation to the texture of cooked rice. Carbohydr. Polym. 22, 267–275.

Ryoo, N., et al., 2007. Knockout of a starch synthase gene OsSSIIIa/Flo5 causes white-core floury endosperm in rice (*Oryza sativa* L.). Plant Cell Rep. 26, 1083–1095.

Sano, Y., Katsumata, M., 1984. Differential regulation of waxy gene-expression in rice. Jpn. J. Genet. 59 (6), 652.

Septiningsih, E.M., et al., 2003. Identification of quantitative trait loci for grain quality in an advanced backcross population derived from the *Oryza sativa* variety IR64 and the wild relative O-rufipogon. Theor. Appl. Genet. 107 (8), 1433–1441.

Septiningsih, E.M., et al., 2009. Development of submergence-tolerant rice cultivars: the Sub1 locus and beyond. Ann. Bot. 103 (2), 151–160.

Shirasawa, K., et al., 2008. Identification of gene for rice (*Oryza sativa*) seed lipoxygenase-3 involved in the generation of stale flavor and development of SNP markers for lipoxygenase-3 deficiency. Breed. Sci. 58 (2), 169–176.

Shu, X., et al., 2006. Starch structure and digestibility of rice high in resistant starch. Starch-Stärke 58 (8), 411–417.

Shu, X., et al., 2007. The influences of chain length of amylopectin on resistant starch in rice (*Oryza sativa* L.). Starch-Stärke 59 (10), 504–509.

Siebenmorgen, T.J., Banaszek, M.M., Kocher, M.F., 1990. Kernel moisture content variation in equilibrated rice samples. Trans. Am. Soc. Agric. Eng. 33 (6), 1979–1983.

Singh, N., et al., 2003. Physico-chemical, morphological, thermal, cooking and textural properties of chalky and translucent rice kernels. Food Chem. 82 (3), 433–439.

Slade, L., Levine, H., 1988. Non-equilibrium melting of native granular starch: part I. Temperature location of the glass transition associated with gelatinisation of A-type cereal starches. Carbohydr. Polym. 8, 183–208.

Slade, L., Levine, H., 1989. A food polymer science approach to selected aspects of starch gelatinization and retrogradation. In: Millane, R., BeMiller, J.N., Chandrasekaran, R. (Eds.), Frontiers in Carbohydrate Research 1. Food Applications. Elsevier, London, pp. 215–270.

Stein, A.J., Sachdev, H.P.S., Qaim, M., 2006. Potential impact and cost-effectiveness of golden rice. Nat. Biotech. 24 (10), 1200–1201.

Sun, J., Hou, C., Zhang, S., 2008. Effect of protein on the rheological properties of rice flour. J. Food Process. Preserv. 32 (6), 987–1001.

Sundaram, R., et al., 2008. Marker assisted introgression of bacterial blight resistance in Samba Mahsuri, an elite indica rice variety. Euphytica 160 (3), 411–422.

Swamy, Y.M.I., Bhattacharya, K.R., 1979. Breakage of rice during milling—effect of kernel defects and grain dimension. J. Food Proc. Eng. 3, 29–42.

Swamy, Y.M.I., Bhattacharya, K.R., 1982. Breakage of rice during milling 1. Types of cracked and immature grains. J. Food Sci. Technol. 19, 106–111.

Swamy, Y.M.I., Zakiuddin Ali, S., Bhattacharya, K.R., 1971. Relationship of moisture content and temperature to discolouration of rice during storage. J. Food Sci. Technol. (8), 150–152.

Takano-Kai, N., et al., 2009. Evolutionary history of GS3, a gene conferring grain length in rice. Genetics 182 (4), 1323–1334, 182.

Tan, Y.F., et al., 1999. The three important traits for cooking and eating quality of rice grains are controlled by a single locus in an elite rice hybrid, Shanyou 63. Theor. Appl. Genet. 99 (3–4), 642–648.

Tan, Y.F., et al., 2001. Mapping quantitative trait loci for milling quality, protein content and color characteristics of rice using a recombinant inbred line population derived from an elite rice hybrid. Theor. Appl. Genet. 103 (6–7), 1037–1045.

Tang, X., Khush, G.S., Juliano, B.O., 1991. Genetics of gel consistency in rice (*Oryza sativa* L.). J. Genet. 70 (2), 69–78.

Tashiro, T., Wardlaw, I.F., 1991. The effect of high-temperature on kernel dimensions and the type and occurrence of kernel damage in rice. Aust. J. Agric. Res. 42 (3), 485–496.

Tian, R., et al., 2005. Mapping quantitative trait loci underlying the cooking and eating quality of rice using a DH population. Mol. Breed. 15 (2), 117–124.

Toojinda, T., et al., 2004. Molecular breeding for rainfed lowland rice in the Mekong Region. In: Proceedings for the Fourth International Crop Science Congress, Brisbane Australia.

Tran, N.A., et al., 2011. A single nucleotide polymorphism on the waxy gene explains gel consistency. Theor. Appl. Genet. 123 (4), 519–525.

Tran, T.U., et al., 2005. Detection of changes in taste of japonica and indica brown and milled rice (*Oryza sativa* L.) during storage using physicochemical analyses and a taste sensing system. J. Agric. Food Chem. 53 (4), 1108–1118.

Tsugita, T., Ohta, T., Kato, H., 1983. Cooking flavor and texture of rice stored under different conditions. Agric. Biol. Chem. 47 (3), 543–549.

Umemoto, T., Aoki, N., 2005. Single-nucleotide polymorphisms in rice starch synthase IIa that alter starch gelatinisation and starch association of the enzyme. Funct. Plant Biol. 32 (9), 763–768.

Umemoto, T., et al., 1999. Differences in amylopectin structure between two rice varieties in relation to the effects of temperature during grain-filling. Starch-Stärke 51 (2–3), 58–62.

Umemoto, T., et al., 2004. Natural variation in rice starch synthase IIa affects enzyme and starch properties. Funct. Plant Biol. 31, 671–684.

USDA: http://archive.gipsa.usda.gov/reference-library/standards/ricestandards-milled_rice.pdf

van Ruiten, H.T., 1985. Rice milling: an overview. In: Juliano, B.O. (Ed.), Rice, Chemistry and Technology. Americal Association of Cereal Chemistry Inc, St. Paul, Minnesota, pp. 349–379.

Villareal, C.P., Maranville, J.W., Juliano, B.O., 1991. Nutrient content and retention during milling of brown rices from the International Rice Research Institute. Cereal Chem. 68 (4), 437–439.

Wada, T., et al., 2006. Evaluation and use of physicochemical properties as index traits for selecting rice cultivars with extremely high palatability. Jpn. J. Crop Sci. 75 (1), 38–43.

Walgate, R., 2008. Diabetes research for developing countries. N. Biotechnol. 25 (2–3), 111–116.
Wan, X.Y., et al., 2004. QTL detection for eating quality of cooked rice in a population of chromosome segment substitution lines. Theor. Appl. Genet. 110 (1), 71–79.
Wan, X.Y., et al., 2005. Stability of QTLs for rice grain dimension and endosperm chalkiness characteristics across eight environments. Theor. Appl. Genet. 110 (7), 1334–1346.
Wanchana, S., et al., 2003. Duplicated coding sequence in the waxy allele of tropical glutinous rice (*Oryza sativa* L.). Plant Sci. 165 (6), 1193–1199.
Wang, Z.Y., et al., 1995. The amylose content in rice endosperm is related to the post-transcriptional regulation of the waxy gene. Plant J. 7 (4), 613–622.
Wang, L.-F., Wang, Y.-J., Porter, R., 2002. Structures and physicochemical properties of six wild rice starches. J. Agric. Food Chem. 50 (9), 2695–2699.
Wang, L.Q., et al., 2007. Genetic basis of 17 traits and viscosity parameters characterizing the eating and cooking quality of rice grain. Theor. Appl. Genet. 115 (4), 463–476.
Ward, R.M., 2007. Potential Impact of Temperature and Carbon Dioxide Levels on Rice Quality. University of Sydney, School of Chemistry, Sydney.
Waters, D.L.E., et al., 2006. Gelatinization temperature of rice explained by polymorphisms in starch synthase. Plant Biotechnol. J. 4 (1), 115–122.
Wolever, T.M.S., Mehling, C., 2002. High-carbohydrate-low-glycaemic index dietary advice improves glucose disposition index in subjects with impaired glucose tolerance. Br. J. Nutr. 87, 477–487.
Xu, K., et al., 2006. Sub1A is an ethylene-response-factor-like gene that confers submergence tolerance to rice. Nature 442 (7103), 705–708.
Yamakawa, H., et al., 2007. Comprehensive expression profiling of rice grain filling-related genes under high temperature using DNA microarray. Plant Physiol. 144 (1), 258–277.
Yang, J.C., et al., 2006. Post-anthesis development of inferior and superior spikelets in rice in relation to abscisic acid and ethylene. J. Exp. Bot. 57 (1), 149–160.
Yoshida, S., 1981. Fundamentals of Rice Crop Science. International Rice Research Institute, Los Banos, Laguna.
Zhang, M.W., Guo, B.J., Peng, Z.M., 2004. Genetic effects on Fe, Zn, Mn and P contents in Indica black pericarp rice and their genetic correlations with grain characteristics. Euphytica 135 (3), 315–323.
Zhao, X., Fitzgerald, M., 2013. Climate change: implications for the yield of edible rice. PLoS ONE 8 (6), e66218.
Zhao, X., et al., 2015. Identification of stable QTLs causing chalk in rice grains in nine environments. Theor. Appl. Genet. 129, 1–13.
Zhong, L., et al., 2005. The deterioration of eating and cooking quality caused by high temperature during grain filling in early-season indica rice cultivars. J. Agron. Crop Sci. 191 (3), 218–225.
Zhou, L., et al., 2009. Fine mapping of the grain chalkiness QTL qPGWC-7 in rice (*Oryza sativa* L.). Theor. Appl. Genet. 118 (3), 581–590.

Chapter 13

Sorghum and Millets: Grain-Quality Characteristics and Management of Quality Requirements

John R.N. Taylor and Kwaku G. Duodu
Institute for Food, Nutrition and Well-being and Department of Food Science, University of Pretoria, Hatfield, Pretoria, South Africa

13.1 INTRODUCTION

Sorghum and the millets are normally considered together as they are all tropical (C4) type cereal species that originated in Africa and Eurasia and are cultivated in the same agro-ecological zones. The term millet does not refer to a single species but is used collectively for tiny-grained cereals. Historically, sorghum and millets have not been used for industrial processing to near the same extent as the other major cereals. As a consequence of this, methods of assessment and management of quality requirements for processing and end-use have not been developed to the same extent. Since the late 1980s, however, industrialisation of sorghum and to a lesser extent millet processing has taken place rapidly (Kleih et al., 2000; Taylor and Dewar, 2001; Haggblade and Holzapfel, 2004; Taylor et al., 2006, 2010; Taylor and Emmambux, 2008a,b; Saleh et al., 2013). Several drivers of sorghum and millet industrialisation can be identified.

In developing regions, notably Africa, India and China where these grains are traditionally cultivated for food and beverages, there is rapid urbanisation and expansion of the urban middle class. These consumers have created a demand for high-quality convenience-type foods and beverages. These are often based on traditional home- or village-made products made from sorghum and millets. Related to this, the cost of imported commodity cereals, specifically wheat and barley used to make bread and beer, respectively, can be prohibitively expensive for many developing countries. Hence, processors are increasingly substituting locally produced grains such as sorghum in their manufacture

Cereal Grains. http://dx.doi.org/10.1016/B978-0-08-100719-8.00013-9

(Taylor et al., 2006). Cost is not the only factor driving this change. It is now being increasingly recognised that using grains produced by local farmers, instead of importing grains, is highly beneficial for economic development in developing countries. This is driving government–private sector partnerships to promote industrial utilisation of locally grown grains, for example sorghum in lager brewing (Mackintosh and Higgins, 2004).

In developed countries, there is growing recognition that sorghum and millets are rich in health-promoting phytochemicals, especially phenolics (Taylor and Duodu, 2015) and that their consumption may have benefits in preventing or alleviating the so-called lifestyle diseases of type-II diabetes and cardiovascular disease, and possibly some types of cancer. Additionally, they are 'gluten-free' grains (Ciacci et al., 2007). This is creating demand for sorghum and millets in specialty multigrain foods and for gluten-free products.

This chapter firstly describes sorghum and millets with respect to their taxonomy, grain production, grain structure and chemistry. It then discusses traditional, emerging and novel end-uses and the quality requirements and methods of assessment for these uses. Next, international and national quality management systems for sorghum and millets are examined. Lastly, future trends in sorghum and millet quality evaluation are considered, especially with respect to quality management needs.

13.2 TAXONOMY AND GRAIN PRODUCTION

Table 13.1 shows the taxonomy of sorghum and major cultivated millet species. Sorghum and most millets belong to the Paniceae tribe, the exceptions being finger millet and teff (alternatively spelled tef) which belong to the Eragrostideae tribe. A very important characteristic of sorghum and millets is that they are hardy and can be cultivated in low agricultural potential environments. Sorghum is notably water efficient and drought-tolerant and requires a minimum seasonal rainfall of 400 mm, compared to for example maize, which requires 500–600 mm (Muchow, 1989; Dendy, 1995). The millets are generally even hardier than sorghum. Pearl millet, for example can produce a crop where the seasonal rainfall is as little as 300 mm.

In terms of world production, sorghum is the fifth most important cereal after wheat, maize, rice and barley with an annual production in 2013 of approximately 58 million tonnes (FAOSTAT, 2013). Collectively, the millets are the sixth most important with an annual production of >54 million tonnes. There are some eight major cultivated millets, with pearl millet being the major millet, accounting for some 50% of total millet production (Table 13.1). The major sorghum producing countries and regions are the USA, Northern Africa (Nigeria, Sudan Ethiopia, Niger, Burkina Faso), Argentina, China, Australia and Brazil. The major regions of pearl millet production are West and Central Africa, India, Eastern and Southern Africa, USA and Brazil. The other millets

TABLE 13.1 Sorghum and the Major Millet Species

Common English name	Other common and vernacular names	Tribe, genus and species	World production (tonnes × 1000)
Sorghum	Milo, Jowar, Mabele, Mtama	Paniceae, *Sorghum bicolor* (L.) Moench	58
Finger millet	Ragi, Wimbi	Eragrostideae, *Eleusine coracana* L. Gaertn.	7
Teff	Tef	Eragrostideae, *Eragrostis tef* (Zuccagni) Trotter	>4
White fonio	Fonio, Acha	Paniceae, *Digitaria exilis* (Kippist) Stapf	0.5 (both fonio species combined)
Black fonio	Black acha	Paniceae, *Digitaria iburua* Stapf	
Japanese millet	Japanese barnyard millet	Paniceae, *Echinochloa esculenta* (A. Braun) H. Scholz	No data
Proso millet	Common millet	Paniceae, *Panicum miliaceum* L. subsp. *miliaceum*	5
Foxtail millet	Italian millet	Paniceae, *Setaria italica* (L.) P. Beauv. subsp. *italica*	5
Pearl millet	Babala, Bajra/Bajira, Mahangu	Paniceae, *Pennisetum glaucum* (L.) R. Br.	>24

Information mainly from the USDA Germplasm Resources Information Network (undated) and FAOSTAT (2013).

are cultivated widely in Eurasia (especially India, China, Korea, Eastern and Southern Europe) and Eastern Africa (Ethiopia, Kenya and Uganda).

Cultivation practices for sorghum and the millets are highly variable. In countries like the USA, Argentina, Australia, Brazil, Mexico and South Africa, sorghum is mainly produced by large-scale, mechanised farming. High-yielding hybrid cultivars are used. However, most sorghum production is at a small-scale using traditional manual farming practices, especially in Africa. Here open pollinated varieties (OPVs) are used. However, improved higher-yielding varieties are now rapidly replacing traditional landraces. The difference in production practices has a large impact on yield. For example concerning the two leading sorghum producing countries, the USA and Nigeria, in the former mechanised farming and hybrids result in a yield of 3.7 t/ha, whereas in the latter with OPVs and traditional farming practices the yield is only 1.2 t/ha (FAOSTAT, 2013). With regard to the millets, they are produced almost exclusively by traditional, low-input small-scale farming. Mostly, traditional landraces are cultivated, although improved varieties and hybrids are increasingly available. As a consequence, worldwide millet yields are generally very low, only 0.9 t/ha (FAOSTAT, 2013). However, where mechanised farming of high-yield hybrids is taking take place in Argentina, as an example, yields are substantially higher at 1.8 t/ha, as is the case in China where cultivation is generally non-mechanised but is intensive with yields of 2.4 t/ha. Up until now, there are no genetically modified (GM) sorghum or millets under commercial cultivation.

13.3 GRAIN STRUCTURE AND CHEMISTRY

Sorghum and some millets, pearl millet, fonio and teff, are regarded as true caryopses. A caryopsis is a single-seeded fruit (e.g. cereals) in which the fruit coat or pericarp surrounds the seed and adheres tightly to the seed coat. Finger millet, foxtail millet and proso millet on the other hand are regarded as utricles, where the pericarp is loosely attached to the kernel and not completely fused to the seed coat or testa.

Table 13.2 shows some structural features of sorghum and some millets. Compared to sorghum, the millets are relatively smaller grains, as can be seen from 1000 kernel weight measures. Sorghum for instance, has a 1000 kernel weight of 25–35 g compared to 3–15 g for pearl millet (Abdelrahman et al., 1984; Serna-Saldivar and Rooney, 1995). For both sorghum and the millets, the basic kernel structure is similar, the principal anatomical parts being the pericarp, germ and the endosperm. The endosperm forms the largest anatomical component of the kernel. Pearl millet stands out in that it has a proportionally larger germ, about 16.5% of the kernel compared to an average of 9.4% for sorghum (Abdelrahman et al., 1984).

The pericarp of sorghum and millets consists of three layers with varying thicknesses, namely, the epicarp (outermost layer), followed by the mesocarp and finally an inner endocarp layer (Earp and Rooney, 1982). Sorghums with a thick pericarp tend to have small starch granules in the mesocarp. Sorghum and millets contain a testa layer just beneath the endocarp with varying thickness. In sorghums, the testa is pigmented in types that contain condensed tannins. Sorghums with pigmented testa are divided into type II and type III, with the type III sorghums containing the higher amount of tannins. Finger millet has an especially thick testa consisting of five layers with varying degrees of pigmentation from red to purple (McDonough et al., 1986). The occurrence of tannins in finger millet is a varietal property as in sorghum (Siwela et al., 2007). These workers showed that tannins in finger millet are located in the testa layer, which tends to be thicker than that of finger millet grain types that do not contain tannins.

Underneath the testa layer is the endosperm, which consists of a one-layered aleurone in sorghum and millets and peripheral, corneous and floury endosperm

TABLE 13.2 Some Structural Features of Sorghum and Some Millets

	Sorghum	Pearl millet	Foxtail millet	Finger millet	Proso millet	Fonio	Teff
Seed type	Caryopsis	Caryopsis	Utricle	Utricle	Utricle	Caryopsis	Caryopsis
1000 kernel weight (g)	30	8	5	2.5	6.1	0.5	0.3–0.4
Testa (layers)	1	1	Remnant	5	1	1	1
Pigmented	Sometimes	Probably no	No	Sometimes	No	No	No
Aleurone (layers)	1	1	1	1	1	1	1
Starch granule size (μm)	2–30	6–12	5–25	8–21	1–17	7–8	6–22
Starch granule type	Simple	Simple	Simple Yes	Compound and simple	Simple	Simple	Compound and simple
Waxy type	Yes	No	Yes	No	Yes	No	No
Protein body size (μm)	0.4–2.0	0.6–1.2	1.0–2.0	2.0	0.5–2.5	1.2–1.4	1.2–1.4

Source: Data from: Taylor, J.R.N., Novellie, L., Liebenberg, N.v.d.W., 1984. Sorghum protein body composition and ultrastructure. Cereal Chem. 61, 69–73; McDonough, C.M., Rooney, L.W., Earp, C.F., 1986. Structural characteristics of *Eleusine coracana* (finger millet) using scanning electron and fluorescence microscopy. Food Microstruct. 5, 247–256; McDonough, C.M., Rooney, L.W., 1989. Structural characteristics of *Pennisetum americanum* using scanning electron and fluorescence microscopies. Food Microstruct. 8, 137–149; FAO, 1995. Sorghum and Millets in Human Nutrition. Available from: http://www.fao.org; Serna-Saldivar, S., Rooney, L.W., 1995. Structure and chemistry of sorghum and millets. In: Dendy, D.A.V. (Ed.), Sorghum and Millets: Chemistry and Technology. American Association of Cereal Chemists, St Paul, MN. pp. 69–124; and Taylor, J.R.N., Kruger, J., 2016. Millets. In: Caballero, B., Finglas, P., Toldrá, F. (Eds.), The Encyclopedia of Food and Health, vol. 3. Academic Press, Oxford. pp. 748–757.

TABLE 13.3 Typical Proximate Composition (g/100 g Dry Basis) of Sorghum and the Major Millets

	Protein	Lysine (g/100 g protein)	Fat	Dietary fibre	Ash
Sorghum[a]	12.1	2.1	3.8	7.6	1.6
Millets[b]					
Pearl	12.4	2.8	5.1	13.8	2.8
Finger	8.5	2.9	1.4	15.2	2.7
Proso	12.6	1.5	2.6	13.1	2.7
Fonio	8.4	2.2	2.6	18.2	3.4
Teff	10.0	2.7	2.1	8.0	2.6

[a]*USDA (2011).*
[b]*Taylor and Kruger (2016).*

portions. The cells of the aleurone contain protein bodies, enzymes, mineral deposits and oil in the form of spherosomes (Serna-Saldivar and Rooney, 1995). The peripheral endosperm contains a dense protein matrix and small starch granules. The corneous (also known as vitreous or horny) and floury endosperm portions contain starch granules and protein bodies embedded in a protein matrix. The packing of the starch granules and protein bodies within the protein matrix in the floury endosperm is of a more loose arrangement compared to the continuous and tightly packed corneous endosperm. The floury endosperm has air voids, which diffract incoming light, giving an opaque appearance (Hoseney et al., 1974). The cell walls of the endosperm fluoresce strongly, indicating the presence of phenolic compounds such as ferulic acid. The germ consists of an embryonic axis and a scutellum and it is a reserve tissue containing oil, protein, enzymes and minerals.

Table 13.3 shows the proximate compositions of sorghum and the millets. Notably, pearl millet is generally richer in protein and fat compared to sorghum and the other millets. Starch is the most abundant chemical component in sorghum and millets. Starches from regular sorghum types contain 23–30% amylose, while waxy (high amylopectin type) sorghums contain up to 5% amylose (Serna-Saldivar and Rooney, 1995). Sorghum starch granules range from 2 to 30 μm (average 15–20 μm) in diameter. Starch granules in the corneous endosperm tend to be smaller and angular while those of the floury endosperm tend to be larger and round. Starch granules of the millets may be simple or compound and tend to be angular or polygonal in the corneous endosperm and more spherical in the floury endosperm.

Prolamin proteins, mainly located within the protein bodies of the endosperm are the major protein fraction in sorghum and millets, followed by glutelins that are mainly located within the protein matrix. Sorghum and millet proteins are lysine-deficient because of the predominance of the prolamin fraction. Although the albumin, globulin and glutelin fractions are lysine-rich, they are minor protein fractions (Taylor and Schüssler, 1986). Most lipids of sorghum and millets are located in the scutellar tissue of the germ and thus may be greatly reduced in quantity as a result of debranning during milling. Non-polar lipids (triglycerides) are the most abundant. Polar lipids in sorghum and millets include phospholipids and glycolipids and the unsaponifiable matter of the lipid fraction contains components such as phytosterols, carotenoids and tocopherols. The pericarp, aleurone and germ are rich in minerals and therefore refined products are low in minerals. Finger millet and teff appear to be richer in calcium than sorghum and other millets (Serna-Saldivar and Rooney, 1995). Sorghum and millets are also important sources of B-vitamins, which are concentrated in the aleurone and the germ.

Phenolic compounds and phytates are important non-nutritional components of sorghum and millets. They are normally regarded as anti-nutritional factors. Tannins and phytates are able to bind minerals and dietary protein making them unavailable, or they may bind and inhibit digestive enzymes (Serna-Saldivar and Rooney, 1995). However, there is now increasing focus on anti-oxidant properties of phenolic compounds (Awika and Rooney, 2004) and phytates (Graf and Eaton, 1990) and their potential health benefits. Condensed tannins are the most important phenolic compounds in sorghum and millets. As stated, only types II and III sorghums contain condensed tannins, which are located in the testa and of the millets, only finger millet has been reliably reported to contain condensed tannins. The aleurone and germ are the areas of location for phytates. Decortication can therefore remove up to about 50% of phytate and total phosphorus (Doherty et al., 1982).

13.4 TRADITIONAL, NOVEL AND EMERGING USES

Table 13.4 summarises the end uses for sorghum and millets. As can be seen, these are rapidly increasing in diversity. Concerning animal feed, in the USA sorghum has been used as an animal feed grain for more than 100 years and it is also widely grown for this purpose in Argentina and Brazil. In Australia, it is now also majorly used as poultry feed as the country is too arid to produce sufficient maize. In the USA, sorghum is now also used in formulated animal feeds, often in the form of distillers dry grains and solubles (DDGS), the co-product of grain bioethanol production (Urriola et al., 2009). The millets have long been popular as companion bird food in the form of heads of seeds. Pearl millet is now gaining popularity as poultry feed on account of its high energy and good-quality protein (Taylor, 2016). It is also increasingly being cultivated as forage crop in many countries (Newman et al., 2010).

With regard to food applications, the major use for sorghum and millets is still in the production of traditional products such as porridges, flatbreads

TABLE 13.4 Traditional, Novel and Emerging Uses for Sorghum and Millets

Traditional	Novel	Emerging
Foods		
Flours and meals (Africa, Asia)	Instant porridges and infant foods (Africa)	Biofortified grains (India, Africa)
Dumplings, porridges and gruels (Africa, Asia)	Expanded snack foods (Africa)	Phytochemical rich foods (USA)
Rice (Africa and Asia) Popcorn (Africa and India)	Ready-to-eat breakfast cereals (USA, Australia, Africa)	Gluten-free foods (USA, Europe, Australia)
Flatbreads (Africa, Central America)	Noodles (Japan)	
Couscous (Africa)		
Malt (Africa, India)		
Beverages		
Non-alcoholic fermented beverages (Africa, Europe, Asia)	Lager beers and stouts (Africa)	Gluten-free beers (USA, Europe, Australia)
Cloudy and opaque beers (Africa, Asia)		
Wines (Taiwan) Spirits (China)		
Animal feed		
Animal feed (USA, South America)	Formulated cattle feed (USA, South America)	Formulated dog food (South Africa)
Bird food (Asia, USA, Africa)		
Poultry feed (Australia)		
Other		
Syrup from sweet sorghum (USA) Vinegar (China)	Forage pearl millet (many countries)	
Industrial uses		
	Starch (USA, Africa) Bioethanol from starch (USA, Australia, Uruguay)	Bioethanol from sweet sorghum (USA, Africa, India) Cellulosic bioethanol from sweet and forage sorghums (USA)

and fermented beverages (beers and non-alcoholic beverages) in developing countries. Food processing, except for milling, is still mainly carried out at household level. Brewing traditional opaque and cloudy beers, however, is a major small- and industrial-scale commercial enterprise. The largest novel food and beverage application for sorghum is for Western-type lager and stout beer brewing and unfermented non-alcoholic 'malt' beverages. In Nigeria, Africa's second largest beer market, virtually all beers and malt beverages are produced with at least some sorghum (Ogbonna, 2011). Sorghum lager beer brewing is now also growing rapidly in East and Central Africa. As indicated, these developments are driven by economics. In developed countries, there is also a small but growing market for so-called 'gluten-free' beer made from sorghum, aimed at the coeliac market.

Regarding novel food applications, in Africa a rapidly expanding range of ready-to-eat products made from sorghum and also pearl millet is being manufactured, particularly porridges and infant foods. In developed countries, there are now many multigrain and 'gluten-free' baked goods such as breads and muffins and some breakfast cereal products contain sorghum and millets, especially teff. In the USA, one line in particular that has attracted much media interest has added sorghum bran as source of phenolic phytochemical antioxidants (Grain Berry, undated).

In the USA, sorghum grain is widely used as a feedstock for bioethanol production, although on a much smaller scale than maize. Regarding potential large-scale emerging applications, there is also great interest in using sweet sorghums, varieties with high levels of sucrose in the stem, for bioethanol production and forage sorghums for cellulosic bioethanol (Han et al., 2012).

The diversity of applications for sorghum grain in particular is very large, even within a country. For example in South Africa which has a total utilisation of less than 250,000 tonnes, its usage is divided approximately as follows: 50% of commercially grown sorghum is used for malting for home and industrial opaque beer brewing and breakfast cereal manufacture, 30% is milled into meal which is sold for home porridge making, 10% is processed into instant porridge and baby food powders, and 10% is used in various animal feed applications including poultry, pigeon, ostrich and, when price competitive, formulated dog food. (Sorghum, undated). Grain bioethanol production from sorghum is also now being planned in detail. For each of these applications, grain with different quality parameters is required. This poses complex challenges with regard to managing grain quality throughout the agricultural value chain (Table 13.5), especially when small quantities of grain are involved.

13.5 SORGHUM AND MILLET END-USE QUALITY REQUIREMENTS AND THEIR ASSESSMENT

13.5.1 Animal Feed

The single most important sorghum quality parameter for animal feed, human food and industrial applications is whether the variety contains tannins. The tannins bind with dietary proteins and with digestive enzymes reducing the feed and feed and food nutritional quality of the grains. For poultry and pig feeding, the feed efficiency and nitrogen digestibility is typically reduced by 10–20%

TABLE 13.5 Products and Issues Associated With the Various Stages of the Sorghum and Millets Grain Value Chain

Stages		Products	Issues for analysis and management
Breeding	→	Hybrids and improved varieties	Yield factors, biotic and abiotic stress tolerance, herbicide tolerance, grain mould resistance Molecular technologies, marker-assisted selection End-use quality: tannins, grain colour, biofortification, grain hardness, amylase activity potential
Seed multiplication		Pure seed	Germinability, phenotypic and genotypic identification
↓			
Sowing	→		Soil analysis, time to sow, seed treatment fertiliser
↓			
Growing	→	Forage and fodder for ruminants	On-farm management, precision agriculture
↓			
Harvesting	→	Stalks for construction, fuel and bioethanol straw for animal feed	Fibre and grain moisture Fibre sugar content
↓			
Delivery to silo	→	Grain for testing	Moisture, protein, contaminants, defects Quality-based segregation and binning
↓			
Grain storage	→		Fumigation, test and maintain grain quality: germinability
↓			
Transport to processor	→		Identity preservation
↓			

Processing			Pesticide residues, heavy metals, mycotoxins, GMO-free, gluten-free
Feed formulation	→	Animal feed	Energy content, protein, lysine
Wet milling	→	Starch	Starch yield and quality
Dry grind process	→	Bioethanol	Starch content
Malting	→	Malt	Grain size, germination, extract, amylase, free amino nitrogen
↓			
Brewing	→	Beers and non-alcoholic malt beverages	Grain size, sensory quality
Dry milling	→	Flour	Moisture, grain size, grain hardness, milling yield, flour colour, phytochemicals, starch damage, porridge texture
Extrusion cooking	→	Snacks, ready-to-eat porridges, gluten-free products	Degree of cook, texture, sensory quality
Baked goods processing	→	Gluten-free baked products	Texture, shelf-life, sensory quality
↓			
Retailing	→	Food product labelling	Nutrient composition analysis: energy, protein, carbohydrates, glycaemic carbohydrates, dietary fibre, micronutrients –vitamins and minerals Presence of allergens such as gluten
↓			
Consumer choice	→	Food product ethics and consumer satisfaction	Non-GMO, organic, fair trade, carbon footprint Sensory acceptability
Consumer nutrition and health	→	Functional foods and nutraceuticals Allowed health claims, approval by non-governmental health-promoting associations	Whole grain, glycaemic response, antioxidant activity, heart healthy, anticancer, phytochemicals

when animals are fed diets based on tannin sorghum compared to non-tannin sorghum (Hancock, 2000). Hence, it is critical to identify whether a consignment of sorghum is of the tannin type or has been accidentally or intentionally adulterated with tannin sorghum.

Contrary to popular belief, tannin sorghums cannot easily be identified on the basis of grain colour. Most red sorghum varieties do not contain tannins and some white sorghum varieties are tannin containing. The simplest and most unequivocal method to detect tannin sorghum grains is by use of the Bleach test. The Bleach test determines the presence or absence of a pigmented testa (presence or absence of condensed tannins) in intact sorghum kernels (Waniska et al., 1992; ICC, 2015). The kernels are soaked in the bleaching reagent (an alkaline solution of sodium hypochlorite) and left to stand at room temperature for about 10–30 min. With tannin sorghums (types II and III), the reagent dissolves away the pericarp and oxidises the constituents of the testa, yielding black pigments on the surface of the grain. Non-tannin sorghums (type I) do not turn black. The method is inexpensive and easy to perform. However, it does not give a quantitative measure of tannin content and also does not always distinguish between types II and III sorghums (Waniska et al., 1992). The Bleach test can also give false positives with type I sorghum grains that have been moulded or undergone weathering or insect damage (Waniska et al., 1992; Dykes and Rooney, 2006). The Bleach test has also been applied successfully to finger millet (Siwela et al., 2007). The difference was that the finger millet grains had to be incubated for a longer time (overnight) in the bleaching reagent.

Methods used for quantification of condensed tannins in sorghum include the butanol-HCl, vanillin-HCl and protein precipitation and enzyme inhibition assays (Schofield et al., 2001). The butanol-HCl method, also called the proanthocyanidin method, involves depolymerisation of condensed tannin in butanol catalysed by hydrochloric acid to yield a red anthocyanin product that can be detected by colorimetry (Watterson and Butler, 1983; Schofield et al., 2001). The vanillin-HCl method is based on the ability of vanillin to form coloured complexes with condensed tannin polymers (Waterman and Mole, 1994; Schofield et al., 2001). Non-tannin phenolics are a source of interference in the vanillin-HCl assay and as a result type I sorghums and millets that do not contain condensed tannins give apparent low tannin values with this assay (Dykes et al., 2005; Siwela et al., 2007). Assays based on precipitation of proteins with tannin have been used to measure tannin content (Hagerman and Butler, 1978). The ability of tannins to bind and inhibit enzymes may be used as a means of quantifying tannins. Various enzymes have been used in such assays including trypsin, amylase, cellulase and alkaline phosphatase (Schofield et al., 2001).

Often the level of total phenolic compounds in sorghum and millets is determined. With sorghum the level of total phenolics generally correlates with the tannin content. However, total phenolics actually comprise three main groups of phenolic compounds: phenolic acids, flavonoids (including anthocyanin and anthocyanidin pigments) and condensed tannins. The first two groups are present in all sorghum and millets but in varying quantities, whereas, as explained, condensed tannins are only present in some sorghum and finger millet varieties (Dykes and Rooney, 2006). Some common colorimetric methods used to determine total phenolics include the Prussian blue, Folin-Denis, Folin-Ciocalteu and ferric ammonium citrate (ISO, 1988) assays. A limitation of all these methods is that they quantify the total concentration of phenolic hydroxyl groups in an extract being assayed, irrespective of the specific molecules in which the phenolic hydroxyl groups occur (Waterman and Mole, 1994).

In contrast to the presence of tannins, pericarp colour in sorghum, which is related to the content of non-tannin polyphenols especially the anthocyanins/anthocyanidins, does not seem to influence animal feed value (Hancock, 2000). However, sorghum endosperm texture may slightly influence its feed value. In pig feeding, grain with intermediate endosperm texture was found to have somewhat higher dry matter and energy digestibility than corneous (hard) or floury (soft) grains (Hancock, 2000). Methods for estimating sorghum and millet grain hardness are discussed in Section 13.5.2.

13.5.2 Dry Milling

To produce meal (coarse particles) and flour (fine particles) for porridge and baked goods production, sorghum and millet grains must be dry milled. Sorghum and millet grain meal and flour colour are typically measured using a tristimulus reflectance-type colour meter (Kebakile et al., 2008).

Milling, in addition to reducing the grain into small particles, also invariably involves removal of the outer components of the grain, the germ and pericarp, commonly referred to as the bran. With sorghum and millets, bran removal is performed by mechanical abrasion, a process generally referred to as dehulling or decortication, whereby the outer layers are progressively abraded off. To obtain a clean separation of the bran from the endosperm, grain with a hard endosperm is required. With soft endosperm grain, the kernel breaks up during dehulling, resulting in the endosperm flour becoming contaminated with pieces of bran. It is noteworthy that tannin sorghums are invariably soft. Hence, the tannins, which as stated are located in the testa, will contaminate the flour.

The biochemical basis of endosperm hardness in sorghum and maize (and presumably the millets) is explained in a review by Chandrashekar and Mazhar (1999). In principle, grain hardness, or more strictly speaking grain strength, involves adhesion of the starch and protein components of the endosperm. Sorghum and millet grain hardness is generally related to kernel bulk density (Chiremba et al., 2011). Kernel bulk density can simply be roughly assessed by determining test weight (hectolitre weight). In the USA, test weight is used to grade sorghum (GIPSA, 2013). The International Crops Research Institute for the Semi-Arid Tropics (ICRISAT) (Gomez et al., 1997) gives detailed instructions for a test to determine the true density of batches of sorghum and pearl millet grain. It is a flotation ‘Percent Floaters’ test employing a sodium nitrate solution of specific gravity 1.3 g/cm^3.

A common test method for sorghum and millet grain hardness itself is using the Tangential Abrasive Dehulling Device (TADD) developed by Reichert et al. (1986). The TADD has a motor driving a vertical axle, which is connected to a horizontal carborundum or abrasive paper-covered disc of approximately 30 cm in diameter. Open ended, small vertical metal cylinders (normally eight) are mounted on a fixed plate on top of the disc. Weighed amounts of grain samples (approximately 10 g) are put into the cylinders and the motor is turned on. The rotating abrasive disc progressively abrades off the outer layers of the grains as they roll around in the cylinders; the harder the grain the slower the material is abraded off. Grain hardness is calculated either in terms of time to remove a certain weight percentage of the grain, or percentage removed over a fixed time. ICRISAT (Gomez et al., 1997) gives full details for determining what is referred to as Percent Dehulling Loss in sorghum and pearl millet. For sorghum, the TADD is recommended but for pearl millet, a laboratory-scale rice polisher (dehusker) is recommended. An alternative instrument that is increasingly being applied to estimate sorghum grain hardness is the Single Kernel Characterization System (SKCS) (Bean et al., 2006). The instrument measures the hardness of individual kernels by crushing them between a toothed rotor and a progressively narrowing crescent, and measuring the force (Perten Instruments, undated.).

As described, sorghum and pearl millet grains all have a hard outer endosperm, generally referred to as corneous and an inner softer floury endosperm. The relative proportion of corneous and floury endosperm is highly variable and is genetically controlled (Rooney and Miller, 1982). Grains with a high proportion of the corneous endosperm are harder. Hence, sorghum and pearl millet grain hardness can be estimated in terms of the proportion of the corneous to floury endosperm. Probably the most straightforward of such methods is the International Association for Cereal Science and Technology (ICC) Standard—Estimation of Sorghum Grain Endosperm Texture (ICC, 2015). In this test, kernels are sectioned longitudinally and then visually categorised as corneous (hard), intermediate or floury (soft) against diagrammatic standards.

13.5.3 Traditional Foods

As mentioned, probably the major traditional food use for sorghum and millets is for making porridges and gruels. Grain colour and whether the sorghum or finger millet variety contains tannins are important consumer quality criteria but usage is very much related to community and individual preference. Porridge texture is a very important quality parameter, which can vary from pourable, through spoonable to mashed potato-like. It is influenced by many factors, including meal particle size, method of milling and meal chemical composition (Kebakile et al., 2008) as well as flour–water ratio. The role of grain chemical composition is very incompletely understood, but factors such as starch amylose–amylopectin ratio certainly play a role (Fliedel, 1994). The porridge-making quality of sorghum and millets can be determined by preparing a porridge under standardised conditions and then measuring the texture of the porridge. The Southern African Grain Laboratory (SAGL, 2015) and ICRISAT (Gomez et al., 1997) both give standard methods to prepare cereal porridges. Instrumental methods such as the Bostwick Consistometer (CSC Scientific Company, undated), the Brookfield Viscometer (Brookfield Engineering, undated) and texture analysers such as the Stable Micro Systems TA-XT series (Stable Micro Systems, undated) can then be used to measure porridge texture. Alternatively, porridge quality can be more comprehensively evaluated by descriptive type (trained panel) sensory evaluation. Sorghum porridge sensory parameters are given by Kebakile et al. (2008).

Another alternative is to cook and measure the texture of the porridge using a single instrument. This can be done using an amylograph-type stirring viscometer such as the Brabender Visco Amylograph or Rapid Visco Analyser as described in AACC International (2000) Methods.

With regard to grain-quality criteria for traditional flatbreads such as injera (Ethiopia), roti (India) and tortilla (Central America), it has, for example been found that sorghum endosperm softness was associated with the positive injera quality attributes of softness, rollability and fluffiness (Yetneberk et al., 2004). The authors describe a standard method for preparing injera and an instrumental texture analysis procedure. Grain colour and hence flatbread colour is an important quality criterion. White tan-plant-type (non-tannin, white endosperm, pale pericarp) sorghums have been found to be strongly preferred for roti (Vannalli et al., 2010) and tortilla making (Bedolla et al., 1983). Similarly, white teff is much preferred over red or brown types for making injera (Dr. Kebede Abegaz, Food Scientist and Technologist, Hawassa University, Ethiopia, personal communication).

13.5.4 Malting, Brewing and Grain Bioethanol

13.5.4.1 Malting and Brewing

For any grain to produce malt of good quality, it is essential that a very high percentage of the grains must germinate uniformly under malting conditions. The percentage of grains that will germinate under such conditions is called Germinative Energy (GE). Germination in sorghum and millets is adversely affected by poor storage conditions, particularly high humidity and temperature. For sorghum and millets, GE can be determined by ICC Standard 174 (ICC, 2015). The method involves germinating sound grains in Petri dishes on moistened filter paper under specific conditions. A GE at 72 h of $\geq$90% is recommended. To determine the malting quality of a batch of grain, it is necessary to malt the grain under controlled conditions in the laboratory. For sorghum there is an approved South African Council for Scientific and Industrial Research (CSIR) laboratory malting procedure (Dewar et al., 1995). The malting procedure involves steeping (soaking) cleaned grain in nylon bags at controlled temperature (25°C) and for specified times over 24 h, and then germinating the grain at controlled temperature (25°C) under essentially 100% relative humidity for up to 5 days. The malt is then dried at low temperature (50°C) to conserve its enzymic activity. The laboratory malting procedure can also be applied to the millets but the steeping time has to be greatly reduced as these smaller grains germinate much more rapidly (Pelembe et al., 2002).

The major quality measure of sorghum and millet malt quality for opaque beer brewing is diastatic power (DP), the joint α- and β-amylase activity of the malt. Both South Africa and Botswana have standards for malting sorghum defined in terms of the DP potential of particular cultivars (see Section 13.6.4). In Southern Africa, sorghum and millet malt DP is determined according to the South African Bureau of Standards (SABS) Method 235 (SABS, 1970), and expressed as Sorghum Diastatic Units (SDU)/g malt. Sometimes, for millets the units are stated as Millet Diastatic Units or similar but they are quantitatively the same. In principle, the method is the same as that of other malt DP assays, such as the European Brewery Convention (EBC) Method 4.12 (EBC, 1998). An enzyme extract is made from the malt and this is used to hydrolyse soluble starch under controlled conditions. DP is quantified in terms of the quantity of sugars produced, measured by redox titration. Outside southern Africa, sorghum malt DP is generally determined using the EBC method. An approximate conversion factor is 1 SDU = 0.6 Windisch–Kolbach units (the EBC DP units) + 8.

The SABS method incorporates one important difference in that the malt enzymes are extracted with a solution of peptone (partially hydrolysed protein). The function of the peptone is to react with the tannins in sorghum malts made from tannin sorghums, thus preventing the tannins inactivating the malt enzymes. The SABS method therefore measures the potential DP of the sorghum sample, which is important in breeding programmes.

A fundamental difference between sorghum and millet malts and barley malt is that the level of β-amylase is several times higher in barley malt (Dufour et al., 1992; Taylor and Robbins, 1993). β-amylase is the enzyme that produces the maltose, which is fermented by the yeast into ethanol and carbon dioxide. Hence, maltsters and brewers of sorghum and millet malts are interested in optimising β-amylase activity. The most specific method of measuring malt β-amylase activity is by the Megazyme Beta-amylase Assay Procedure (Betamyl-3 Method) (Megazyme International, 2010). In this assay, the substrate is a dye-linked maltotrioside, which is essentially only attacked by the β-amylase enzyme. Thus, any hydrolysis of the substrate by α-amylase activity in the enzyme extract is avoided, preventing overestimation of β-amylase activity. Hydrolysis of the maltotrioside results in release of the dye, which is measured colorimetrically.

Another very important measure of sorghum, millet and barley malt quality is Free Amino Nitrogen (FAN). FAN comprises individual amino acids and short peptides, and is the product of hydrolysis of the malt protein by proteolytic enzymes. Its major function in brewing is to act as a source of nitrogen for yeast growth during fermentation. FAN is particularly important in commercial sorghum opaque beer and beer brewing as the malt (sorghum or barley malt) generally comprises only a small proportion of the cereal grist, and a high percentage being unmalted sorghum grain adjunct, which is very low in FAN. The most common method used to measure FAN is using the EBC ninhydrin method 4.10 (EBC, 1998). Reduced ninhydrin reacts with ammonia from the amino group of the free amino acids to produce a blue–purple complex, which is measured colorimetrically. In the EBC method, the malt is mashed at laboratory scale under standardised conditions and then the FAN is measured. Thus, the FAN in the malt and that produced by proteolysis during brewing is measured. In Southern Africa, there is a CSIR approved method for measuring sorghum malt FAN, Method 5 Determination of Free Amino Nitrogen in Sorghum Malt (Dewar et al., 1995). This is an adaptation of the EBC method. The major difference is that in the CSIR method, the FAN is extracted from the malt under strongly acidic conditions to prevent any proteolysis during extraction so that just the FAN present in the malt is measured. Thus, as with DP, this procedure is foremost a cultivar selection tool.

In barley malt brewing, the most important malt quality parameter is extract. Extract is the proportion of malt or cereal grain that can be expected to go into solution when the malt or cereal grain is mashed during brewing. It is essentially a measure of how much of the starch will be solubilised and hydrolysed during brewing, although other chemical components, in particular preformed dextrins and sugars and also protein peptides and amino acids, contribute to extract. Malt extract is determined by subjecting the malt to a laboratory-scale mashing procedure under standardised conditions. During mashing, the malt starch is gelatinised and then hydrolysed into soluble dextrins and sugars by the malt amylases. The percentage of the malt solids that has gone into solution is measured by specific gravity. Probably the most commonly used malt extract assay is the EBC Congress mashing procedure Method 4.5.1 (EBC, 1998). This method can be applied to measure sorghum malt extract where sorghum malt is used to brew lager and stout beer.

In East and Central Africa, sorghum lager beer brewing is done using unmalted sorghum grain and commercial enzymes. To determine the quality of sorghum for this type of brewing, the milled grain is first cooked to gelatinise the starch, and then mashed under standardised conditions with commercial amylase or barley malt, for example using EBC Method 6.6 (EBC, 1998).

13.5.4.2 Sorghum Grain Bioethanol

The process to produce bioethanol from sorghum grain is in principle the same as that for brewing with unmalted grain, whereby the grain is cooked to gelatinise the starch, hydrolysed into fermentable sugars using commercial amylases and fermented into ethanol and carbon dioxide by yeast. It would appear logical that starch is the most important sorghum grain quality parameter for bioethanol production. In fact, it has been shown that ethanol yield is highly significantly correlated with sorghum grain starch content (Wang et al., 2008). In view of the large scale and automation of bioethanol production, NIR spectroscopy is the preferred technique for measuring grain quality parameters. Recently Fourier Transform (FT) NIR spectroscopy has been found to give excellent rapid and non-destructive measurement of sorghum starch content and prediction of ethanol yield (Li et al., 2015).

However, sorghums with similar starch content can vary by as much as 7% in terms of ethanol yield (Wang et al., 2008). Factors that influenced ethanol yield included protein digestibility, level of extractable protein, protein–starch interactions, levels of phenolic compounds (including tannins) and starch amylose–amylopectin ratio. The protein effects were thought, most likely, to be due to small starch granules being embedded in the endosperm protein matrix, and hence inaccessible to amylases.

Concerning tannins, the tannins can inhibit the amylases and reduce ethanol yield and hence consignments need to be assayed for tannins. Thus, the use of tannin sorghum would appear to be undesirable. However, since bioethanol production is very large scale and driven by economics there could be trade-offs which can be evaluated; lower grain cost versus higher enzyme usage and lower ethanol yields, and possibly distillers dried grain and solubles (DDGS) animal feed co-product of lower nutritional quality.

13.5.5 Baked Goods—Focus on Gluten-Free Foods

Flour from white tan-plant type sorghum is preferred for gluten-free baked goods. For this and related applications, the United States Department of Agriculture (USDA) has a White sorghum class (see Section 13.6.2). This is not simply because of the light colour of the flour, but also because this type of sorghum is less astringent and bitter than tannin-containing and red flavonoid-rich non-tannin type sorghums (Kobue-Lekalake et al., 2007). The same applies to millets, where white teff flour is a flour of choice for gluten-free products.

Baked goods produced from sorghum tend to have a rather dry and gritty consistency. The dryness is probably due to a combination of the more hydrophobic nature of kafirin prolamin proteins compared to wheat gluten, the high gelatinisation temperature of sorghum starch and insoluble nature of its glucurono-arabinoxylan endosperm cell wall polysaccharides (Taylor et al., 2006).

The grittiness is presumably due to flour particles from the corneous (hard) endosperm. Soft, floury endosperm is better for baked goods making than hard, corneous endosperm. Munck (1995) refers to work that showed that soft endosperm sorghum flour had substantially lower water absorption and higher loaf volumes in wheat–sorghum composite breads than hard endosperm. Corneous endosperm sorghum seems to be disadvantageous for baked goods making due to the endosperm protein matrix envelope which limits starch granule expansion (Ezeogu et al., 2008), thus producing a product with gritty texture. This is the opposite situation to milling where hard endosperm sorghum is advantageous, as described. To date, endosperm texture does not seem to have been used as selection criterion for sorghum for baked goods. However, there seems to be considerable potential, as recent work has shown that high protein digestibility and waxy (high amylopectin) traits in sorghum, which influence endosperm texture, dramatically affect flour water solubility and pasting properties (Elhassan et al., 2015).

Teff flour seems to be advantageous for gluten-free baked goods as it does not impart a gritty texture. This is probably due to the fact that its starch granules have a different structure to those of maize and sorghum (Bultosa et al., 2002) and its prolamin storage proteins are less polymerised and hydrophobic than those of sorghum (Adebowale et al., 2011).

13.6 SORGHUM AND MILLET QUALITY MANAGEMENT SYSTEMS

Grain quality means different things to the various partners in the grain supply chain. At least three different concepts of quality can be identified: grain food safety, conformation to trade specifications and conformation to end-use requirements. The various existing different sorghum and millet quality management systems emphasise different aspects of these three concepts. Thus, representative international and national standards for sorghum and millet quality will be explained.

13.6.1 Codex Alimentarius Commission Standards

There are Codex (2008) standards (last modified in 1995) for sorghum grains, sorghum flour, pearl millet grains and pearl millet flour. These standards are mainly about food safety and cover issues such as toxic seeds, heavy metals, pesticide residues and mycotoxins. Table 13.6 summarises the Codex standards for sorghum grains and sorghum flour. The others are similar in content.

In the case of sorghum grains and flour, there is a quality specification with respect to tannin content. Maximum levels for tannins of 0.5% for whole sorghum grain, 0.3% for decorticated sorghum grain and 0.3% for sorghum flour are specified, although strangely the method for determining tannin content is not stated. For sorghum and pearl millet there are also proximate chemical compositional standards. They are, however, very loose standards. Perhaps the most useful of them is for crude fibre content in pearl millet grains, where maxima of 3.0–4.5% and 2.0% are specified for whole and decorticated grain, respectively. The lower level of decorticated grain should help with detection of cases where whole grain is passed off as refined grain. There is also a crude fibre specification for pearl millet flour, maximum 1.5%, which is somewhat illogical since flour is normally simply milled decorticated grain. Also, strangely, there are no similar Codex crude fibre specifications for whole and decorticated sorghum grain.

TABLE 13.6 Summary of the Codex Food Safety Standards for Sorghum Grain and Sorghum Flour (Codex, 2008)

Factor	Sorghum grain (CODEX STAN 172-1989)	Sorghum flour (CODEX STAN 173-1989)
General	Sorghum grains shall be safe and suitable for human consumption. Sorghum grains shall be free from abnormal flavours, odours and living insects. Sorghum grains shall be free from filth (impurities of animal origin, including dead insects) in amounts that may represent a hazard to human health.	Sorghum flour shall be safe and suitable for human consumption. Sorghum flour shall be free from abnormal flavours, odours and living insects. Sorghum flour shall be free from filth (impurities of animal origin, including dead insects) in amounts that may represent a hazard to human health.
Moisture content	14.5% m/m maximum	15.0% m/m maximum
Defects	Not more than 8.0% total defects. Not more than 2.0% extraneous matter of which not more 0.5% shall be inorganic matter. Filth—0.1% m/m maximum.	None specified
Toxic or noxious seeds	Sorghum grains shall be free from the following in amounts that may represent a hazard to human health: Crotolaria (*Crotolaria* spp.), Corn cockle (*Agrostemma githago* L.), Castor bean (*Ricinus communis* L.), Jimson weed (*Datura* spp.) and other seeds that are commonly recognised as harmful to health.	None specified
Tannin content	For whole sorghum grains, tannin content shall not exceed 0.5% on a dry matter basis. For decorticated sorghum grains, tannin content shall not exceed 0.3% on a dry matter basis.	Tannin content of sorghum flour shall not exceed 0.3% on a dry matter basis.

Heavy metals	Sorghum grains shall be free from heavy metals in amounts that may represent a hazard to human health	Sorghum flour shall be free from heavy metals in amounts that may represent a hazard to human health.
Pesticide residues	Sorghum grains shall comply with the maximum residue limits established by the Codex Alimentar us Commission.	Sorghum flour shall comply with those maximum residue limits established by the Codex Alimentarius Commission.
Mycotoxins	Sorghum grains shall comply with those maximum mycotoxin limits established by the Codex Alimentar us Commission.	Sorghum flour shall comply with those maximum mycotoxin limits established by the Codex Alimentarius Commission.
Hygiene	To the extent possible in good manufacturing practice, the product shall be free from objectionable matter. Shall be free from microorganisms in amounts that may represent a hazard to health. Shall be free from parasites in amounts that may represent a hazard to health. Shall not contain any substance originating from microorganisms in amounts that may represent a hazard to health.	To the extent possible in good manufacturing practice, the product shall be free from objectionable matter. Shall be free from microorganisms in amounts that may represent a hazard to health. Shall be free from parasites in amounts that may represent a hazard to health. Shall not contain any substance originating from microorganisms in amounts that may represent a hazard to health.

TABLE 13.7 USDA GIPSA Sorghum Classes (GIPSA, 2013)

Class	Definition
Sorghum	Sorghum which lacks a pigmented testa (subcoat) and contains less than 98.0% White sorghum and not more than 3.0% Tannin sorghum. The pericarp colour of this class may appear white, yellow, red, pink, orange or bronze.
Tannin sorghum	Sorghum which has a pigmented testa (subcoat) and contains not more than 10.0% of kernels without a pigmented testa.
White sorghum	Sorghum which lacks a pigmented test (subcoat) and contains not less than 98.0% of kernels with a white pericarp, and contains not more than 2.0% of other classes. This class includes sorghum containing spots that, singly or in combination, cover 25.0% or less of the kernel.
Mixed sorghum	Sorghum which does not meet the requirements for any of the classes Sorghum, Tannin sorghum or White sorghum.

13.6.2 United States Sorghum Standards

There are United States Department of Agriculture Grain Inspection, Packers and Stockyards Administration (GIPSA) standards for sorghum (GIPSA, 2013). The US standards for sorghum include food safety standards, but are primarily trading standards, which relate somewhat to end-use quality. Sorghum is divided into four classes: Sorghum, Tannin sorghum, White sorghum and Mixed sorghum. Hence, sorghum is essentially classified according to whether it is of tannin type or white in colour. The complete definitions of the US sorghum classes are given in Table 13.7. Sorghum colour is determined visually and the presence of a pigmented testa is determined using a version of the Bleach test described earlier. Each of the four classes of sorghum is divided into four numerical grades and a US Sample grade, with the highest grade being 1 and the lowest 4. The highest grade has the highest test weight and lowest percentage of heat- and other-damaged kernels, broken kernels and foreign material. The Sample grade is for grain that does not meet the requirements for the four grades.

13.6.3 Australian Sorghum Standards

In Australia, standards for sorghum grain have been published by Grain Trade Australia. This is a body set up to standardise grain trading standards, trade rules and grain contracts across the Australian grain industry. The standards state that there are no varietal restrictions regarding grain sorghum which can be regarded as either red, white or yellow (Grain Trade Australia, 2015). Two grades of sorghum are identified, Sorghum No. 1 and Sorghum No. 2 with set limits for various parameters such as moisture, test weight, foreign material, defective

TABLE 13.8 Australian Sorghum Standards (Grain Trade Australia, 2015)

Parameter	Sorghum No. 1	Sorghum No. 2
Moisture maximum (%)	13.5	13.5
Test weight minimum (kg/hL)	71	62
Foreign material maximum (% by weight) (all matter other than sorghum)	2.0	4.0
Screenings maximum (% by weight) (all matter passing through a 2-mm slotted screen after 40 shakes	11.0	25.0
Total defective grain maximum (% by count)	5% (15 grains)	25% (75 grains)
Of which Field fungi maximum (% by count)	3% (9 grains)	10% (30 grains)
Of which Bin burnt, heat damaged maximum (% by weight per half litre)	0.5%	1.0%
Of which Storage mould maximum (% by weight per half litre)	0.05%	0.1%
Sprouted maximum (% by count)	3% (9 grains)	10% (30 grains)

grain, seed contaminants and various other types of contaminants. Methods and procedures are also specified for determination of moisture, protein, test weight, trash and screenings, defective grains and contaminants. Table 13.8 gives some set parameters for Australian sorghum standards.

13.6.4 South African and Botswana Sorghum Standards

The South African Department of Agriculture (2008) and the Botswana Bureau of Standards (BOBS, 2002) have regulations and standards, respectively for sorghum, which emphasise end-use quality requirements as well as being safety trading standards. The South African regulations divide sorghum into four classes: GM (Grain sorghum—Malting), GL (Grain sorghum Low tannin—not malting) and GH (Grain sorghum High tannin—malting) and Other. It is a cultivar-based system. Sorghum cultivars are categorised according to whether, when their grain is malted, it produces high DP, as measured by the SABS method described earlier. The classification system is based on the fact that the major commercial use for sorghum in South Africa is for malting for opaque beer brewing. The complete definitions of the South African sorghum classes are given in Table 13.9. Within each of the classes there are two grades, which are similar in scope to those of the US sorghum grades.

The BOBS (2002) sorghum standards system is similar to the South African system in that it is a cultivar-based system with a class of malting sorghum. However, the BOBS sorghum standards further emphasise end-use quality

TABLE 13.9 South African Department of Agriculture Sorghum Standards (2008)

Class	Definition
GM (Grain sorghum—malting)	Consists of malt sorghum that does not have a dark testa Is of a GM cultivar as specified in the cultivar list Complies with the standards for one of the grades of Class GM sorghum
GL (Grain sorghum Low tannin—not malting)	Consists of sorghum that does not have a dark testa Is of a GL cultivar as specified in the cultivar list Complies with the standards for one of the grades of Class GL sorghum
GH (Grain sorghum High tannin—malting)	Consists of sorghum that has a dark testa Is of a GH cultivar as specified in the cultivar list Complies with the standards for one of the grades of Class GH sorghum
Other	Does not comply with the standards for Class GM, GL or GH sorghums

requirements with additional classes of sorghum intended for food and feed use. This is related to the fact that sorghum is Botswana's major cereal crop and staple food. In the standards there are four classes of sorghum: Food class sorghum (FOC), Feed class sorghum (FEC), Malting class sorghum (MC) and Other class sorghum (OC). The complete definitions of the Botswana sorghum classes are given in Table 13.10. As seen, the Food class sorghum has sub-classes in terms of grain, which is assessed visually. It is also defined in terms of endosperm hardness, which is assessed by the ICC (2015) method described earlier. For malting sorghum, there is a GE specification, also measured using the ICC (2015) method. Also of interest is the fact that a maximum tannin level of 0.5% is specified, as determined by the ISO (1988) method described earlier. As with the South African sorghum regulations, there are two grades within each of the classes.

13.6.5 Namibian Pearl Millet Standards

Namibia has standards for pearl millet grain (Namibian Ministry of Agriculture, Water and Forestry, 2004/5) and regulations for flour/meal (Namibian Ministry of Agriculture, Water and Forestry, 2008). Pearl millet is the major cereal crop and staple food in Namibia, where it is commonly called mahangu. The grain standards are primarily concerned with food safety and currently simply divide pearl millet into two classes: (1) cleaned mahangu grain, and (2) graded mahangu grain. There are two grades with differing permissible levels of screenings, impurities and defective grains. The flour/meal standards are also primarily about food safety but also concern product identity. There are three

TABLE 13.10 Botswana Bureau of Standards Sorghum Standards (BOBS, 2002)

Class	Definition
Food class sorghum (FOC)	Consists of sorghum that does not have a dark testa The grains are of intermediate to hard endosperm texture Consists of either white grains (sub-class FOCW); red grains (sub-class FOCR); or is of grain colour other than white or red, or mixture of grain colours (sub-class FOCO) Is of a FOC cultivar, as specified by the Department of Crop Production of the Ministry of Agriculture Complies with the standards for one of the grades of Class FOC sorghum
Feed class sorghum (FEC)	Consists of sorghum that does not have a dark testa Is of a FOC cultivar, that cannot be graded as of Class FOC, or is of a FEC cultivar as specified by the Department of Crop Production of the Ministry of Agriculture Complies with the standards for one of the grades of Class FEC sorghum
Malting class sorghum (MC)	Shall comprise two sub-classes: Malting Class sorghum with a dark testa (MCDT) Malting Class sorghum without a dark testa (MCCT). Consists of sorghum that either has a dark testa (sub-class MCDT); or that does not have a dark testa (sub-class MCCT) The grains have high germinability (GE > 90%) Is of a MCDT cultivar, or of a MCCT cultivar as specified by the Department of Crop Production of the Ministry of Agriculture Complies with the standards for one of the grades of Class MC sorghum
Other class sorghum (OC)	Does not comply with the standards for Class FOC, Class FEC or Class MC sorghum

classes: fermented product (max. pH 5.0), unfermented product (min. pH 6.0) and unspecified product (pH between 5.0 and 6.0). These pH-based flour/meal classes are based on the fact that in Namibia pearl millet grain is traditionally steeped and then semi-wet milled during which time it undergoes a non-alcoholic (lactic acid) fermentation. Additionally, according to the standards, the pearl millet product is graded as either flour or meal depending on particle size (granulation) sieving specifications.

13.6.6 Organic Food Standards

Due to the growing production and international trade in organic foods, various certification bodies and governments have put guidelines in place to regulate various aspects of the production and trade in organic foods. Guidelines for labelling seem to be common to all. The guidelines also cover both plant and animal food sources. Therefore, one would not find organic standards specifically for sorghum and millets.

The Codex Alimentarius Commission has published such guidelines for the production, processing, labelling and marketing of organically produced foods (Codex, 2007). Elements of these guidelines cover aspects such as labelling and claims, rules for production and preparation of organic foods, inspection and certification systems and imports. IFOAM—Organics International (International Federation of Organic Agriculture Movements) is a worldwide umbrella organisation for the organic agriculture movement with about 800 affiliates in more than 100 countries. They have published comprehensive standards for organic production and processing which cover a broad range of issues including accreditation programmes (IFOAM, 2014).

Various countries also have established guidelines for production and trade in organic foods. For example in Australia the most important body for the organic industry is the Organic Federation of Australia. This body has published a National Standard for Organic and Bio-dynamic Produce (Organic Federation of Australia, 2015) to regulate the organic food industry. The standard identifies general principles that apply to organic activities, stipulate conditions to be met by operators of organic units and outlines exceptions that are allowed to the standard. The USDA has comprehensive rules, regulations and standards that govern production and trade in organic foods, which are captured under a National Organic Program (NOP) (USDA, 2015).

13.7 FUTURE DIRECTIONS

13.7.1 Impact of Molecular Biology on Sorghum and Millet End-Use Quality

As mentioned, GM sorghum and millets are not as yet cultivated commercially. Notwithstanding this, improvement in the end-use quality of sorghum through genetic modification continues to be studied. For example genes coding for wheat high-molecular weight glutenin subunits have been expressed in sorghum using a sorghum protein promoter, with the aim of improving its bread-making quality (Mishra et al., 2008; Kumar et al., 2012). To improve sorghum protein digestibility and lysine content, RNA interference (RNAi) technology has been used to suppress the synthesis of specific combinations of sub-classes of the kafirin storage protein (Jung et al., 2008; Da Silva et al., 2011) (See Section 13.7.2). Nevertheless, it is the authors' opinion that GM sorghum and millets will not be approved for cultivation in the foreseeable future. This is because there are concerns about gene transfer between such GM crops and their weedy and wild relatives (Chapman and Burke, 2006), and also the fact that sorghum and millets are not GM is a valuable consumer attribute.

However, molecular biology has the potential to revolutionise 'conventional' breeding of these cereals and improve grain quality assessment. Notably, the sequencing by Mace et al. (2013) of the genomes of 44 sorghum representative varieties of cultivated sorghum, exhibiting valuable traits such as stay-green drought resistance, insect resistance, grain size and grain quality, should reveal many useful single nucleotide polymorphisms (SNPs) associated with such traits (Mace et al., 2013). As an example of this potential, in 2014 the first sorghum varieties were released for cultivation in sub-Saharan Africa where DNA marker-assisted selection had been used in breeding (Mohamed et al., 2014). The selected trait was resistance to *Striga,* a parasitic weed that can devastate sorghum and maize crops.

Molecular techniques are starting to find application in breeding programmes to detect germplasm with grain quality desired traits. For example with regard to biofortification (see Section 13.7.2) it has been found that Fe and Zn contents in pearl millet are largely under additive genetic control (Govindaraj et al., 2013). Related to this, it has been shown that in maize there are quantitative trait loci (QTLs) associated with high iron bioaccessibility (in vitro bioavailability) (Bouis and Welch, 2010). QTLs associated various grain-quality traits in sorghum, including amylose content, tannin content and endosperm texture, were first identified in the late 1990s (Rami et al., 1998). More recently, QTLs for the yellow endosperm trait and carotenoids (Salas Fernandez et al., 2008) and for protein digestibility (Winn et al., 2009) have been identified in sorghum. Concerning millets, QTLs for the opaque endosperm texture modifier (*Opm*) and the related trait of tryptophan content have been identified in finger millet (Babu et al., 2014). From this work, expressed sequence tag-simple sequence repeat (EST-SSR) primers have been developed for detection of high protein and high calcium finger millet genotypes (Nirgude et al., 2014).

13.7.2 Biofortification of Sorghum and Millets

Biofortification aims to combat malnutrition in communities in developing countries that do not majorly consume manufactured food products (Bouis and Welch, 2010). Biofortification involves increasing the density and bioavailability of key essential nutrients, especially micronutrients, in staple foods such as grains. It is most commonly achieved through breeding, either conventional or genetic engineering. Worldwide, there are several on-going sorghum and millet biofortification projects. ICRISAT is leading projects to increase iron and zinc levels in pearl millet and sorghum through conventional breeding (Rai et al., 2012; Ashok Kumar et al., 2013). The Africa Biofortified Sorghum project, led by the Africa Harvest Foundation in Kenya, used genetic engineering approaches to substantially increase the content of provitamin A, improve protein lysine content and protein digestibility and improve essential mineral bioavailability by reducing phytate levels in sorghum (Taylor et al., 2014).

A significant problem with biofortification is the cost and time required to assay the nutrients of interest in the large number of experimental lines generated in breeding programmes. Hence, rapid assay methods are having to be developed. With respect to micronutrients, it has been found that energy-dispersive X-ray fluorescence (EDXRF) gave high correlations for both iron and zinc contents ($r \geq 0.97$) in pearl millet with the standard inductively coupled plasma optical emission spectroscopy (ICP-OES) method (Rai et al., 2012). Not only was EDXRF much more rapid than ICP-OES, it was also only about 11% the cost.

13.7.3 Health-Promoting Foods—Functional Foods and Nutraceuticals

There is increasing interest in sorghum and millets as foods because they have potential long-term health-promoting properties. Their health-promoting properties have been linked to their high content of various bioactive compounds, especially phenolics. The prevailing hypothesis is that these bioactive phenolics may reduce and/or prevent oxidative stress and hence have anti-diabetic, anti-inflammatory, anti-hypertensive and cardiovascular disease preventative and anti-cancer properties (Taylor et al., 2014; Taylor and Duodu, 2015). In this regard, a significant future trend is expected to be the development of sorghum- or millet-based functional foods and nutraceutical-type products.

The health-promoting properties of sorghum and millet grains and also some sorghum and millet foods have been indicated using a large number of in vitro and in vivo studies. The in vitro radical scavenging properties of sorghum, millets and their foods is well established (Awika et al., 2004; Dicko et al., 2005; Ragaee et al., 2006; Awika et al., 2009; Chandrasekara and Shahidi, 2010, 2011). It has become normal practice in this area of research for radical scavenging properties of these grains and their foods to be shown against synthetic radicals such as ABTS [2,2'-azino-bis(3-ethylbenzothiazoline-6-sulfonic acid)], DPPH (2,2-diphenyl-1-picrylhydrazyl) and oxygen radical. Animal models have also been used to determine health-promoting properties in vivo (Hegde et al., 2005; Burdette et al., 2010; Moraes et al., 2012). Anti-diabetic effects of sorghum and millets have been shown by determining the ability of phenolic extracts from these grains to inhibit starch hydrolysing enzymes (Kim et al., 2011; Kunyanga et al., 2012; Links et al., 2015).

In the future development of biofortified sorghum and millets and of bioactive-rich functional foods and nutraceuticals from them, there will be increasing focus on the content of particular micronutrients and phenolics and on their physiological actions. This has important implications for quality evaluation and management of sorghum and millets (see Table 13.5). An increasingly more multi-disciplinary approach is required, bringing together different types of expertise such as plant breeders, agriculturists, food scientists, biochemists, medical scientists, nutritionalists, food processors and most importantly consumers.

SOURCES OF FURTHER INFORMATION

www.aaccnet.org: AACC International (formerly the American Association of Cereal Chemists) publishes standard methods of analysis and holds an annual meeting which frequently features sorghum and millet science and technology.

www.codexalimentarius.net: The Codex Alimentarius Commission is a joint body of the Food and Agriculture Organization and World Health Organization. It prepares internationally accepted food safety and quality standards and publishes them in the Codex Alimentarius. There are Codex Alimentarius standards for sorghum and pearl millet and their flours and for many other grains and their products.

www.icrisat.org: The International Crops Research Institute for the Semi-Arid Tropics (ICRISAT) is the Consultative Group on International Agricultural Research (CGIAR) organisation responsible for sorghum and millets and tropical legumes. ICRISAT's primary concerns are cultivar breeding and agricultural development in the developing world. ICRISAT also has a site called EXPLOREit@ICRISAT (http://exploreit.icrisat.org) which is particularly useful as it provides easily comprehensible information about its mandate crops.

www.icc.or.at: ICC (the International Association for Cereal Science and Technology) publishes standard methods of analysis, including methods specifically on sorghum and has a study group devoted specifically to sorghum, millets, legumes and composite flours.

http://crsps.net/2013/04: INTSORMIL (the International Sorghum, Millets and Other Grains Collaborative Research Support Program). This USAID supported network of sorghum and millet experts in the USA and developing countries ran from 1979 to 2013. INTSORMIL was involved in all aspects of sorghum and millet development from breeding, processing and marketing. This website archives a large resource of useful technical information, including sorghum quality determination, and sorghum and millet processing.

www.k-state.edu/smil: The Sorghum and Millet Innovation Laboratory (SMIL), a USAID Feed the Future Project, is the successor to INTSORMIL. SMIL, which is hosted by Kansas State University, is a collaborative programme between US and African scientists and is focussed on countries in East and West Africa where sorghum and millets are critically important crops. SMIL has two related research and development focuses: creating entrepreneurial advances to reduce poverty and hunger, and creating new sorghum and millet feed and food products to stimulate demand to economic benefits.

Encyclopedia of Food Grains: Wrigley, C.R., Corke, H., Seetharaman, K., Faubion, J. (Eds.), 2016. Encyclopedia of Food Grains, second ed. Elsevier-Academic Press, Oxford, UK. This second edition of the standard reference on cereals and other grains comprises some 210 chapters covering all aspects of grain science and technology, including 4 chapters on sorghum and 2 on millets.

REFERENCES

AACC International, 2000. Approved Methods of the American Association of Cereal Chemists. Methods: 61-01 Amylograph Method for Milled Rice; 61-02 Determination of the Pasting Properties of Rice With the Rapid Visco Analyser. tenth ed AACC International, St. Paul, MN.

Abdelrahman, A., Hoseney, R.C., Varriano-Marston, E., 1984. The proportions and chemical compositions of hand-dissected anatomical parts of pearl millet. J. Cereal Sci. 2, 127–133.

Adebowale, A.A., Emmambux, M.N., Beukes, M., Taylor, J.R.N., 2011. Fractionation and characterization of teff proteins. J. Cereal Sci. 54, 360–366.

Ashok Kumar, A., Reddy, B.V.S., Ramaiah, B., Sahrawat, K.L., Pfeiffer, W.H., 2013. Gene effects and heterosis of grain iron and zinc concentration in sorghum [*Sorghum bicolor* (L.) Moench]. Field Crops Res. 146, 86–95.

Awika, J.M., Rooney, L.W., 2004. Sorghum phytochemicals and their impact on human health. Phytochemistry 65, 1199–1221.

Awika, J.M., Rooney, L.W., Waniska, R.D., 2004. Anthocyanins from black sorghum and their antioxidant properties. Food Chem. 90, 293–301.

Awika, J.M., Yang, L., Browning, J.D., Faraj, A., 2009. Comparative antioxidant, antiproliferative and phase II enzyme inducing potential of sorghum (*Sorghum bicolor*) varieties. LWT—Food Sci. Technol. 42, 1041–1046.

Babu, B.K., Agrawal, P.K., Pandey, D., Kumar, A., 2014. Comparative genomics and association mapping approaches for opaque2 modifier genes in finger millet accessions using genic, genomic and candidate gene-based simple sequence repeat markers. Mol. Breed. 34, 1261–1279.

Bean, S.R., Chung, O.K., Tuinstra, M.R., Pederson, J.F., Erpelding, J., 2006. Evaluation of the Single Kernel Characterization System (SKCS) for measurement of sorghum grain attributes. Cereal Chem. 83, 108–113.

Bedolla, S., De Palacios, M.G., Rooney, L.W., Diehl, K.C., Khan, M.N., 1983. Cooking characteristics of sorghum and corn for tortilla preparation by several cooking methods. Cereal Chem. 60, 263–268.

Botswana Bureau of Standards (BOBS), 2002. Botswana Standard, BOS63:2002, Cereals—Classification and Grading of Sorghum Grains Intended for Sale in Botswana—Specification. BOBS, Gaborone.

Bouis, H.E., Welch, R.M., 2010. Biofortification—a sustainable agricultural strategy for reducing micronutrient malnutrition in the global south. Crop Sci. 50 (Suppl. 1), S-20.

Brookfield Engineering, undated. Viscometers Available from: www.brookfieldengineering.com

Bultosa, G., Hall, A.N., Taylor, J.R.N., 2002. Physico-chemical characterization of grain tef (*Eragrostis tef* (Zucc.) Trotter) starch. Starch/Stärke 54, 461–468.

Burdette, A., Garner, P.L., Mayer, E.P., Hargrove, J.L., Hartle, D.K., Greenspan, P., 2010. Anti-inflammatory activity of select sorghum (*Sorghum bicolor*) brans. J. Med. Food 13, 879–887.

Chandrasekara, A., Shahidi, F., 2010. Content of insoluble bound phenolics in millets and their contribution to antioxidant capacity. J. Agric. Food Chem. 58, 6706–6714.

Chandrasekara, A., Shahidi, F., 2011. Inhibitory activities of soluble and bound millet seed phenolics on free radicals and reactive oxygen species. J. Agric. Food Chem. 59, 428–436.

Chandrashekar, A., Mazhar, H., 1999. The biochemical basis and implications of grain strength in sorghum and maize. J. Cereal Sci. 30, 193–207.

Chapman, M.A., Burke, J.M., 2006. Letting the gene out of the bottle: the population genetics of genetically modified crops. New Phytol. 170, 429–443.

Chiremba, C., Rooney, L.W., Taylor, J.R.N., 2011. Relationships between simple grain quality parameters for the estimation of sorghum and maize hardness in commercial hybrid cultivars. Cereal Chem. 88, 570–575.

Ciacci, C., Maiuri, L., Caporaso, N., Bucci, C., Del Giudice, L., Massardo, D.R., Pontieri, P., Di Fonzo, N., Bean, S.R., Ioeger, B., Londei, M., 2007. Celiac disease: in vitro and in vivo safety and palatability of wheat-free sorghum food products. Clin. Nutr. 26, 799–805.

Codex, 2007. Organically Produced Foods: Guidelines for the Production, Processing, Labelling and Marketing of Organically Produced Foods (GL 32-1999). Available from: ftp://ftp.fao.org/docrep/fao/010/a1385e/a1385e00.pdf

Codex, 2008. Codex Alimentarius International Food Standards. Whole and Decorticated Pearl Millet Grains (169-1989), Pearl Millet Flour (170-1989), Sorghum Grains (172-1989), Sorghum Flour (173-1989). Available from: http://www.codexalimentarius.org/standards/list-of-standards

CSC Scientific Company, undated. Bostwick Consistometer. Available from: www.cscscientific.com

Da Silva, L.S., Jung, R., Zhao, Z., Glassman, K., Grootboom, A.W., Mehlo, L., O'Kennedy, M.M., Taylor, J., Taylor, J.R.N., 2011. Effect of suppressing the synthesis of different kafirin sub-classes on grain endosperm texture, protein body structure and protein nutritional quality in improved sorghum lines. J. Cereal Sci. 54, 160–167.

Dendy, D.A.V., 1995. Sorghum and millets: production and importance. In: Dendy, D.A.V. (Ed.), Sorghum and Millets: Chemistry and Technology. American Association of Cereal Chemists, St. Paul, MN, pp. 11–26.

Dewar, J., Taylor, J.R.N., Joustra, S.M., 1995. Accepted Methods of Sorghum Malting and Brewing Analyses, Methods: 3 Laboratory Malting of Sorghum; 4 Determination of Diastatic Power of Sorghum Malt; 5 Determination of Free Amino Nitrogen in Sorghum Malt. CSIR Food Science and Technology, Pretoria.

Dicko, M.H., Gruppen, H., Traore, A.S., Van Berkel, W.J.H., Voragen, A.G.J., 2005. Evaluation of the effect of germination on phenolic compounds and antioxidant activities in sorghum varieties. J. Agric. Food Chem. 53, 2581–2588.

Doherty, C., Faubion, J.M., Rooney, L.W., 1982. Semiautomated determination of phytate in sorghum and sorghum products. Cereal Chem. 59, 373–377.

Dufour, J.P., Melotte, L., Srebrnik, S., 1992. Sorghum malts for the production of a lager beer. J. Am. Soc. Brew. Chem. 111, 110–119.

Dykes, L., Rooney, L.W., 2006. Sorghum and millet phenols and antioxidants. J. Cereal Sci. 44, 236–251.

Dykes, L., Rooney, L.W., Waniska, R.D., Rooney, W.L., 2005. Phenolic compounds and antioxidant activity of sorghum grains of varying genotypes. J. Agric. Food Chem. 53, 6813–6818.

Earp, C.F., Rooney, L.W., 1982. Scanning electron microscopy of the pericarp and testa of several sorghum varieties. Food Microstruct. 1, 125–134.

European Brewery Convention (EBC), 1998. Analytica EBC, 5th ed. Methods: 4.5.1 Extract of Malt: Congress Mash; 4.10 Free Amino Nitrogen of Malt by Spectrophotometry; 4.12 Diastatic Power of Malt; 6.6 Extract Content of Maize: Enzymatic Method. Fachverlag Hans Carl, Nuremberg, Germany.

Elhassan, M.S., Emmambux, M.N., Hays, D.B., Peterson, G.C., Taylor, J.R.N., 2015. Novel biofortified sorghum lines with combined waxy (high amylopectin) starch and high protein digestibility traits: effects on endosperm and flour properties. J. Cereal Sci. 65, 132–139.

Ezeogu, L.I., Duodu, K.G., Emmambux, M.N., Taylor, J.R.N., 2008. Influence of cooking conditions on the protein matrix of sorghum and maize endosperm flours. Cereal Chem. 85, 397–402.

Faostat, 2013. Production. Available from: http://faostat3.fao.org

Fliedel, G., 1994. Evaluation de la qualité du sorgho pour la fabrication du tô. Agric. Dév. 4, 12–21.

Germplasm Resources Information Network, undated. GRIN Taxonomy for Plants. Available from: www.ars-grin.gov

Grain Inspection, Packers and Stockyards Administration (GIPSA), 2013. Sorghum, Grain Inspection Handbook II, United States Department of Agriculture, Washington, DC (Chapter 9). Available from: www.gipsa.usda.gov

Gomez, M.I., Obilana, A.B., Martin, D.F., Madzvamuse, M., Monyo, E.S., 1997. Manual of Laboratory Procedures for Quality Evaluation of Sorghum and Pearl Millet. ICRISAT, Patancheru, India.

Govindaraj, M., Rai, K.N., Shanmugasundaram, P., Dwivedi, S.L., Sahrawat, K.L., Muthaiah, A.R., Rao, A.S., 2013. Combining ability and heterosis for grain iron and zinc densities in pearl millet. Crop Sci. 53, 507–517.

Graf, E., Eaton, J.W., 1990. Antioxidant functions of phytic acid. Free Radical Biol. Med. 8, 61–69.

Grain Berry, undated. Grain Berry Antioxidant Foods. Available from: http://www.grainberry.com

Grain Trade Australia, 2015. Grain Trade Australia Section 2—Sorghum Trading Standards, 2015/16 Season. Available from: http://www/graintrade.org.au/commodity_standards

Hagerman, A.E., Butler, L.G., 1978. Protein precipitation method for the quantitative determination of tannins. J. Agric. Food Chem. 26, 809–812.

Haggblade, S., Holzapfel, W.H., 2004. Industrialization of Africa's indigenous beer brewing. In: Steinkraus, K.H. (Ed.), Industrialization of Indigenous Fermented Foods, Revised and Expanded. Marcel Dekker, New York, pp. 271–361.

Han, K.-J., Pitman, W.D., Kim, M., Day, D.F., Alison, M.W., McCormick, M.E., Aita, G., 2012. Ethanol production potential of sweet sorghum assessed using forage fiber analysis procedures. Global Change Bioenergy, doi 10.1111/j.1757-1707.2012.01203.x.

Hancock, J.D., 2000. Value of sorghum and sorghum coproducts in diets for livestock. In: Wayne Smith, C., Fredericksen, R.A. (Eds.), Sorghum: Origin, History, Technology and Production. John Wiley and Sons, New York, pp. 731–750.

Hegde, P.S., Rajasekaran, N.S., Chandra, T.S., 2005. Effects of the antioxidant properties of millet species on oxidative stress and glycemic status in alloxan-induced rats. Nutr. Res. 25, 1109–1120.

Hoseney, R.C., Davis, A.B., Herbers, L.H., 1974. Pericarp and endosperm structure of sorghum grain shown by scanning electron microscopy. Cereal Chem. 51, 552–558.

International Association for Cereal Science and Technology (ICC), 2015. Standard 174 Determination of Germinative Energy of Sorghum Grain; Standard 176 Estimation of Sorghum Grain Endosperm Texture; Standard 177 Detection of Tannin Sorghum Grain by the Bleach Test. ICC, Vienna.

International Federation of Organic Agriculture Movements (IFOAM), 2014. The IFOAM Norms for Organic Production and Processing, Version 2014. Available from: http://www.ifoam.bio/sites/default/files/ifoam_norms_version_july_2014.pdf

International Organization for Standardization (ISO), 1988. ISO/DIS 9648 Sorghum—Determination of Tannin Content. ISO, Paris.

Jung R., Hu, W.-N., Meeley, R.B., 2008. Grain quality improvement through altered expression of seed proteins. European Patent EP 2,003,205.

Kebakile, M.M., Rooney, L.W., De Kock, H.L., Taylor, J.R.N., 2008. Effects of sorghum type and milling process on the sensory characteristics of sorghum porridge. Cereal Chem. 85, 307–313.

Kim, J.S., Hyun, T.K., Kim, M.J., 2011. The inhibitory effects of ethanol extracts from sorghum, foxtail millet and proso millet on α-glucosidase and α-amylase activities. Food Chem. 124, 1647–1651.

Kleih, U., Bala Ravi, S., Dayakar Rao, B., Yoganand, B., 2000. Industrial Utilization of Sorghum in India. ICRISAT, Patancheru, India, pp. 1–44.

Kobue-Lekalake, R.I., Taylor, J.R.N., De Kock, H.L., 2007. Effects of phenolics in sorghum grain on its bitterness, astringency and other sensory properties. J. Sci. Food Agric. 87, 1940–1948.

Kumar, T., Dweikat, I., Sato, S., Ge, Z., Nersesian, N., Chen, H., Elthon, T., Bean, S., Ioerger, B.P., Tilley, M., Clemente, T., 2012. Modulation of kernel storage proteins in grain sorghum (*Sorghum bicolor* (L.) Moench). Plant Biotechnol. J. 10, 533 544.

Kunyanga, C.N., Imungi, J.K., Okoh, M.W., Biesalski, H.K., 2012. Total phenolic content, antioxidant and antidiabetic properties of methanolic extract of raw and traditionally processed Kenyan indigenous food ingredients. LWT—Food Sci. Technol. 45, 269–276.

Li, J., Danao, M.G.C., Chen, S.F., Li, S., Singh, V., Brown, P.J., 2015. Prediction of starch content and ethanol yields of sorghum grain using near infrared spectroscopy. J. Near Infrared Spectrosc. 23, 85–92.

Links, M.R., Taylor, J., Kruger, M.C., Taylor, J.R.N., 2015. Sorghum condensed tannins encapsulated in kafirin microparticles as a nutraceutical for inhibition of amylases during digestion to attenuate hyperglycaemia. J. Funct. Foods 12, 55–63.

Mace, E.S., Tai, S., Gilding, E.K., Li, Y., Prentis, P.J., Bian, L., et al., 2013. Whole-genome sequencing reveals untapped genetic potential in Africa's indigenous cereal crop sorghum. Nat. Commun. 4, 2320.

Mackintosh, I., Higgins, B., 2004. The development of a sorghum-based lager beer in Uganda: a model of co-operation between industry and government in the development of local ingredients for the production of quality lager beer and consequential benefits for the parties involved. Aspects Appl. Biol. 72, 235–245.

McDonough, C.M., Rooney, L.W., Earp, C.F., 1986. Structural characteristics of *Eleusine coracana* (finger millet) using scanning electron and fluorescence microscopy. Food Microstruct. 5, 247–256.

Megazyme International, 2010. Beta-amylase Assay Procedure (Betamyl-3 Method). Available from: www.megazyme.com

Mishra, A., Pandey, R., Bansal, S., Tomar, A., Khanna, V.K., Garg, G.K., 2008. Construction of gene cassette harboring HMW glutenin gene of wheat driven by γ-kafirin promoter of sorghum. Acta Biol. Szeged. 52, 277–282.

Mohamed, A., Ali, R., Elhassan, O., Suliman, E., Mugoya, C., Masiga, C.W., Elhusien, A., Hash, C.T., 2014. First products of DNA marker-assisted selection in sorghum released for cultivation by farmers in sub-saharan Africa. J. Plant Sci. Mol. Breed. 3, 3.

Moraes, E.A., Natal, D.I.G., Quieroz, V.A.V., Schaffert, R.E., Cecon, P.R., de Paula, S.O., dos Anjos Benjamin, L., Ribeiro, S.M.R., Martino, H.S.D., 2012. Sorghum genotype may reduce low-grade inflammatory response and oxidative stress and maintains jejunum morphology of rats fed a hyperlipidic diet. Food Res. Int. 49, 553–559.

Muchow, R.C., 1989. Comparative productivity of maize, sorghum and pearl millet in a semi-arid tropical environment. II. Effect of water deficits. Field Crops Res. 20, 207–219.

Munck, L., 1995. New milling technologies and products: whole plant utilization by milling and separation of the botanical and chemical components. In: Dendy, D.A.V. (Ed.), Sorghum and Millets: Chemistry and Technology. American Association of Cereal Chemists, St. Paul, MN, pp. 223–281.

Namibian Ministry of Agriculture, Water and Forestry, 2004/5. Quality, Classification and Grading Standards for Whole Pearl Millet (Mahangu) Grains. Namibian Ministry of Agriculture, Water and Forestry, Windhoek.

Namibian Ministry of Agriculture, Water and Forestry, 2008. Regulations Relating to Composition and Quality of Pearl Millet (Mahangu) Products: Classification, Grading and Quality Standards of Pure Mahangu Flour/Meal and Mixed Products. Namibian Ministry of Agriculture, Water and Forestry, Windhoek.

Newman, Y., Jennings, E., Vendramini, J., Blount, A., 2010. Pearl Millet (*Pennisetum glaucum*): Overview and Management. University of Florida, Florida, Available from: http://edis.ifas.ufl.edu.

Nirgude, M., Babu, B.K., Shambhavi, Y., Singh, U.M., Upadhyaya, H.D., Kumar, A., 2014. Development and molecular characterization of genic molecular markers for grain protein and calcium content in finger millet (*Eleusine coracana* (L.) Gaertn.). Mol. Biol. Rep. 41, 1189–1200.

Ogbonna, A.C, 2011. Current developments in malting and brewing trials with sorghum in Nigeria: a review'. J. Inst. Brew. 117, 394–400.

Organic Federation of Australia, 2015. National Standard for Organic and Bio-dynamic Produce. Available from: http://www.ofa.org.au/national_standard

Pelembe, L.A.M., Dewar, J., Taylor, J.R.N., 2002. Effect of malting conditions on pearl millet malt quality. J. Inst. Brew. 108, 7–12.

Perten Instruments, undated. SKCS 4100. Available from: www.perten.com

Ragaee, S., Abdel-Aal, E.M., Noaman, M., 2006. Antioxidant activity and nutrient composition of selected cereals for food use. Food Chem. 98, 32–38.

Rai, K.N., Govindaraj, M., Rao, A.S., 2012. Genetic enhancement of grain iron and zinc content in pearl millet. Qual. Assur. Saf. Crops Foods 4, 119–125.

Rami, J.F., Dufour, P., Trouche, G., Fliedel, G., Mestres, C., Davrieux, F., Blanchard, P., Hamon, P., 1998. Quantitative trait loci for grain quality, productivity, morphological and agronomical traits in sorghum (*Sorghum bicolor* L. Moench). Theor. Appl. Genet. 97, 605–616.

Reichert, R.D., Tyler, R.T., York, A.E., Schwab, D.J., Tatarynovich, J.E., Mwasaru, M.A., 1986. Description of a production model of the tangential abrasive dehulling device and its application to breeders' samples. Cereal Chem. 63, 201–207.

Rooney, L.W., Miller, F.R., 1982. Variation in the structure and kernel characteristics of sorghum. In: Mertin, J.V. (Ed.), International Symposium on Sorghum Grain Quality. ICRISAT, Patancheru, India, pp. 143–162.

South African Bureau of Standards (SABS), 1970. Method 235 Standard Test Method for the Determination of the Diastatic Power of Malts Prepared From Kaffircorn (Sorghum) Including Bird-Proof Varieties, and Millet. South African Bureau of Standards, Pretoria.

Southern African Grain Laboratory (SAGL), 2015. Industry Accepted Method: 005 Pap Cooking Test. Available from: www.sagl.co.za

Salas Fernandez, M.G., Hamblin, M.T., Li, L., Roone, W.L., Tuinstra, M.R., Kresovich, S., 2008. Quantitative trait loci analysis for endosperm color and carotenoid content in sorghum grain. Crop Sci. 48, 1732–1743.

Saleh, A.S.M., Zhang, Q., Chen, J., Shen, Q., 2013. Millet grains: nutritional quality, processing, and potential health benefits. Compr. Rev. Food Sci. Food Saf. 12, 281–295.

Schofield, P., Mbugua, D.M., Pell, A.N., 2001. Analysis of condensed tannins—a review. Anim. Feed Sci. Technol. 91, 21–40.

Serna-Saldivar, S., Rooney, L.W., 1995. Structure and chemistry of sorghum and millets. In: Dendy, D.A.V. (Ed.), Sorghum and Millets: Chemistry and Technology. American Association of Cereal Chemists, St Paul, MN, pp. 69–124.

Siwela, M., Taylor, J.R.N., de Milliano, W.A.J., Duodu, K.G., 2007. Occurrence and location of tannins in finger millet grain and antioxidant activity of different grain types. Cereal Chem. 84, 169–174.

Sorghum SA, undated. Products. Available from: www.sorghumsa.co.za

South African Department of Agriculture, 2008. Regulations Relating the Grading, Packing and Marking of Sorghum Intended for Sale in the Republic of South Africa. Available from: http://www.daff.gov.za/daffweb3/Branches/Agricultural-Production-Health-Food-Safety/Food-Safety-Quality-Assurance/Local-and-Import-Regulations

Stable Micro Systems, undated. Texture Analysis. Available from: www.stablemicrosystems.com

Taylor, J.R.N., 2016. Millet pearl. In: Wrigley, C., Corke, H., Seetharaman, K., Faubion, J. (Eds.), Encyclopedia of Food Grains, second ed. Academic Press, Oxford, pp. 190–198.

Taylor, J.R.N., Barrion, S.C., Rooney, L.W., 2010. Pearl millet—New developments in an ancient food grain. Cereal Foods World 55, 16–19.

Taylor, J.R.N., Belton, P.S., Beta, T., Duodu, K.G., 2014. Review: increasing the utilisation of sorghum, millets and pseudocereals: developments in the science of their phenolic phytochemicals, biofortification and protein functionality. J. Cereal Sci. 59, 257–275.

Taylor, J.R.N., Dewar, J., 2001. 'Developments in sorghum food technologies'. Taylor, S. (Ed.), Advances in Food and Nutrition Research, vol. 43, Academic Press, San Diego, CA, pp. 217–264.

Taylor, J.R.N., Duodu, K.G., 2015. Effects of processing sorghum and millets on their phenolic phytochemicals and the implications of this to the health-enhancing properties of sorghum and millet food and beverage products. J. Sci. Food Agric. 95, 225–237.

Taylor, J.R.N., Emmambux, M.N., 2008a. Millets. In: Arendt, E., Dal Bello, F. (Eds.), Gluten free Cereal Products and Beverages. Elsevier, London, pp. 119–148.

Taylor, J.R.N., Emmambux, M.N., 2008b. Products containing other speciality grains: sorghum, the millets and pseudocereals. In: Hamaker, B.R. (Ed.), Technology of Functional Cereal Products. Woodhead Publishing, Abington, UK, pp. 281–335.

Taylor, J.R.N., Kruger, J., 2016. Millets. Caballero, B., Finglas, P., Toldrá, F. (Eds.), The Encyclopedia of Food and Health, vol. 3, Academic Press, Oxford, pp. 748–757.

Taylor, J.R.N., Robbins, D.J., 1993. Factors influencing beta-amylase activity in sorghum malt. J. Inst. Brew. 99, 413–416.

Taylor, J.R.N., Schober, T.J., Bean, S.R., 2006. Novel food and non-food uses for sorghum and millets. J. Cereal Sci. 44, 252–271.

Taylor, J.R.N., Schüssler, L., 1986. The protein compositions of the different anatomical parts of sorghum grain. J. Cereal Sci. 4, 361–369.

Urriola, P.E., Hoehler, D., Pedersen, C., Stein, H.H., Shurson, G.C., 2009. Amino acid digestibility of distillers dried grains with solubles, produced from sorghum, a sorghum-corn blend, and corn fed to growing pigs. J. Anim. Sci. 87, 2574–2580.

United States Department of Agriculture (USDA), 2011. National Nutrient Database for Standard Reference, Release, 28. Available from: http://ndb.nal.usda.gov/ndb/food

United States Department of Agriculture (USDA), 2015. Organic Standards. Available from: http://www/ams.usda.gov/grades-standards/organic-standards

Vannalli, S., Kasturiba, B., Naik, R.K., Yenagi, N., 2010. Nutritive value and quality characteristics of sorghum genotypes. Karnataka J. Agric. Sci. 20, 586–588.

Wang, D., Bean, S., McLaren, J., Seib, P., Madl, R., Tuinstra, M., Shi, Y., Lenz, M., Wu, X., Zhao, R., 2008. Grain sorghum is a viable feedstock for ethanol production. J. Ind. Microbiol. Biotechnol. 35, 313–320.

Waniska, R.D., Hugo, L.F., Rooney, L.W., 1992. Practical methods to determine the presence of tannins in sorghum. J. Appl. Poultry Res. 1, 122–128.

Waterman, P.G., Mole, S., 1994. Analysis of Phenolic Plant Metabolites. Alden Press, Oxford, UK, pp. 66–103.

Watterson, J.J., Butler, L.G., 1983. Occurrence of an unusual leucoanthocyanidin and absence of proanthocyanidins in sorghum leaves. J. Agric. Food Chem. 31, 41–45.

Winn, J.A., Mason, R.E., Robbins, A.L., Rooney, W.L., Hays, D.B., 2009. QTL mapping of a high protein digestibility trait in *Sorghum bicolor*. Int. J. Plant Genomics, Article ID 471853.

Yetneberk, S., De Kock, H.L., Rooney, L.W., Taylor, J.R.N., 2004. Effects of sorghum cultivar on injera quality. Cereal Chem. 81, 314–327.

Chapter 14

Ancient Wheats and Pseudocereals for Possible use in Cereal-Grain Dietary Intolerances

Ferenc Békés*, Regine Schoenlechner and Sándor Tömösközi†**
**FBFD PTY LTD, North Parramatta, NSW, Australia; **Institute of Food Technology, Department of Food Science and Technology, University of Natural Resources & Technology, Vienna, Austria; †Department of Applied Biotechnology and Food Sciences, BUTE, Budapest, Hungary*

14.1 INTRODUCTION

Several grain species have been recommended as being suitable for people with various forms of dietary intolerance to the gluten proteins of wheat and related cereal species. These 'alternative' or specialty grains include the ancient progenitors of modern wheats and the grains classed as pseudocereals. These grains have attracted the attention of the food industry out of proportion to their traditional agricultural significance, based on the possibility that they would be suitable for gluten-free foods and for people who have adverse reactions to fermentable oligosaccharides, disaccharides, monosaccharides and polyols (FODMAPs). Before describing these grain species, it is necessary to explain the significance of disorders related to gluten and FODMAPs, and thus the criteria necessary for a grain-based food to be genuinely suitable in these cases.

Wheat gluten is a protein-lipid-carbohydrate complex formed as a result of specific covalent and non-covalent interactions among flour components during dough making as the components are hydrated and energy from mechanical input from the mixing process is provided (Islam et al., 2011). However, nowadays the terms 'gluten' and 'gluten-free' are used more widely (and wrongly) in everyday practice, the latter loosely referring to food products not containing cereal prolamin proteins, that is wheat gliadin and glutenin analogues.

Consumption of 'gluten-containing' food causes disorders for a significant minority of people who consume foods derived from wheat, rye, barley and possibly

Cereal Grains. http://dx.doi.org/10.1016/B978-0-08-100719-8.00014-0

oats. The consideration of oat as a 'harmless cereal' is one of the oldest and open problems among experts. While several publications explain its potential danger as a result of contamination by other cereal, others claim that a certain proportion of coeliac patients are sensitive to epitopes found in avenins, the storage proteins of oat (Pulido et al., 2009). The fact is however, that in several types of intolerances related to the consumption of 'gluten-containing' cereals, the trigger compounds are not components of the gluten. There is a lack of related knowledge about the causes of different symptoms or disorders and about the terms 'gluten', 'prolamins', 'gliadin', 'glutenin' not only in the minds of the consumer but also in medical practice (Branchi et al., 2015).

The general public in most Western countries is now aware of the potential adverse effects of cereals containing gluten with reports appearing in the lay press (Braly and Hoggan, 2002; Ford, 2008; Wangen, 2009; Davis, 2011) promoting gluten-free diets. Many of these reports fail to draw attention to the importance of appropriate diagnosis or defining the nature of the gluten 'intolerance' an individual may have, hence posing a significant public relations threat to the grain industry.

In a recent survey (Gluten-Free Foods in the US, 5th ed.) of those who eat gluten-free foods, 36% do so for reasons other than sensitivity, 65% because they think it is healthier, 27% because they think it aids weight loss, 7% do so to help reduce inflammation and 4% do so to combat depression, only 5.7% claimed a formal medical diagnosis.

Gluten-free business has been booming in the last 5 years in western countries. In the USA retails sales reached $10.5 billion in 2013 and more than $15 billion is predicted for 2016. In Australia, where more than 23% of new bakery products have been marketed as gluten-free in 2014, it is expected that the gluten-free retail sales exceed $100 million in the next 5 years (Jargon, 2014).

As the basic knowledge about the mechanisms of certain disorders is improving, we learn that the causing components of cereal products are not always gluten proteins but often soluble proteins (albumins and globulins), even soluble oligosaccharides (often called FODMAPs). This fact underlines the need to better understand what type of diet is needed for different sensitive individuals instead of simply recommending 'gluten-free' diet for them.

Recently the most successful innovation area in the baking industry is governed by the large demand on 'gluten-free' products. Because of their chemical, nutritive and functional properties, the application of two groups of often neglected grain species, the so-called pseudocereals (buckwheat, amaranth, etc.) and some primitive wheat relatives such as Einkorn, Emmer end spelt play an important role in these developments.

This chapter provides a short overview on four separate but interrelated topics. After a brief summary of cereal-related health issues, the quality aspects of pseudocereals and primitive wheats are discussed, finally the specific quality related issues of the production and commercialisation of 'gluten-free' and/or low FODMAP products will be described. Further details on these topics can be found on these issues in excellent publications such as review articles and/or books of Anderson and Wieser (2006), Arendt and Dal Bello (2008), Belton and Taylor (2002), Faris (2014), Haraszi et al. (2011), Thompson and Mendez (2008) and Wieser (2008).

14.2 CEREAL-RELATED HEALTH DISORDERS

14.2.1 Protein-Related Disorders

The spectrum of 'gluten'-related disorders can be divided into three main groups: autoimmune, allergic and not autoimmune-allergic disorders (Sapone et al., 2012) (Fig. 14.1).

Two of these disorder types, the autoimmune and the allergic disorders are relatively well documented; however further studies are needed to reveal mechanisms related to the diverse symptoms of allergic reactions. The autoimmune disorders include coeliac disease (CD) (Anderson and Wieser, 2006) gluten ataxia and dermatitis herpetiformis (Lauriere et al., 2006). The allergic reactions involve respiratory allergy (Amano et al., 1998), food allergy (FA) (Mills et al., 2004), wheat-dependent exercise-induced anaphylaxis (Armentia et al., 2012) and contact urticaria. Symptoms of CD are triggered by an abnormal response of

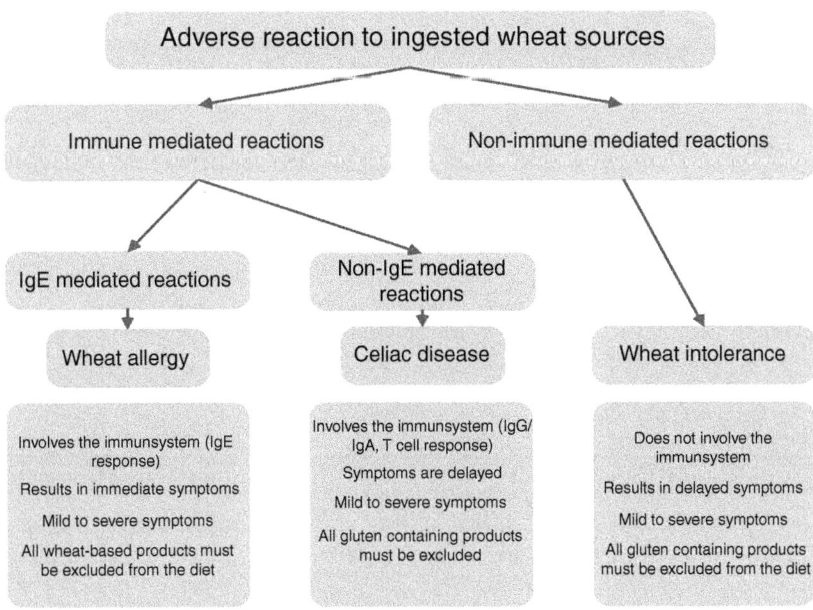

FIGURE 14.1 Spectrum of gluten-related disorders. *(Based on Sapone, A., Bai, J.C., Ciacci, C., Dolinsek, J., Green, P.H.R., Hadjivassiliou, M., Kaukinen, K., Rostami, K., Sanders, D.S., Schumann, M., Ullrich, R., Villaltam D., Voltam U., Catassi, C., Fasano, A., 2012. Spectrum of gluten-related disorders: consensus on new nomenclature and classification. BMC Med. 10, 13.)*

auto-antibodies, such as antibodies to tissue-transglutaminase, or antibodies against proline- and glutamine-rich wheat gluten proteins or their barley and rye homologs (Green and Cellier 2007). CD shows strong association with HLA–DQ alleles and carriage of certain HLA-DQ molecules is the most important genetic factor to develop detrimental symptoms (Sollid et al., 2012). HLA-DQ heterodimers function as surface-type receptor proteins bound to the surface of the antigen-presenting cells. The presence of HLA-DQ molecules, such as DQ2.5, DQ8, DQ2.2, is a clear marker for developing autoimmune symptoms, with modifying effects coming from genetic and environmental factors (Anderson et al., 2000). The different serotypes recognise different peptides. A set of criteria can be defined for the structure of an active CD epitope in a cereal protein: (1) a size (nine amino acids), (2) the presence of a tissue transglutaminase 2 (tTG) enzyme-binding site, (3) surrounded by amino acids with defined charge and hydrophobicity (Sollid et al., 2012). Patients suffering from CD produce a range of autoimmune responses to practically all gliadins and LMW glutenins and to several HMW glutenin sub-units as well as their analogous polypeptides in the consumed cereal products (Juhász et al., 2012).

The development of wheat allergies is mediated more directly by the recognition of allergens by specific IgE epitopes bound to mast cells (Catassi and Fasano, 2008). Depending on the route of the wheat allergen exposure, symptoms can be divided into symptoms of classic food-allergy; wheat-dependent exercise-induced anaphylaxis (WDEAI); occupational asthma and urticaria (Sapone et al., 2012). Classic FA symptoms may affect the skin, gastrointestinal or respiratory tract. While atopic dermatitis is mainly characteristic in children, WDEIA or urticaria were mostly identified in adults. Bakers' asthma, one of the most prevalent respiratory allergies, is a significant symptom among adults working with wheat flour. In wheat allergy, allergen presenting B-cells and T-cells have the same level of impact and, similar to other food allergies allergic reactions, are the result of special cross-links between specific immunoglobulin E and short peptides rich in glutamine and proline that result from the breakdown of wheat-seed proteins by endogenous proteases (Sapone et al., 2012). This interaction stimulates basophils and mast cells to release chemical mediators, such as histamines resulting in different categories of inflammatory reactions. The allergenic regions of protein recognised by the binding sites of IgE molecules are called IgE-binding epitopes. These epitopes can be classified into two categories, conformational and linear epitopes and both types are involved in the development of allergic reactions (Akagawa et al., 2007).

Wheat-related allergy is caused by two group of proteins: primary triggers, that is linear and structural immune reactive peptides of wheat grain proteins and cross allergens—proteins with structural similarity to known allergens. Allergen proteins of wheat are not only prolamins (α, β, γ and ω gliadins and LMW glutenin sub-units) but proteins which are not components of the gluten complex—amylases, the lipid transfer proteins (LTP), peroxidase, thioredoxin and serine proteinase inhibitors (Tatham and Shewry, 2008).

There remains uncertainty regarding the prevalence of FA, particularly to fruits, vegetables, nuts and other edible plants. Although diagnostic studies have objectively verified allergic reactions caused by ingestion of plant food, the extent of the problem on the population level remains unclear for patients and their families, schools, the catering industry, food producers and retailers, health professionals and policy makers (Zuidmeer et al., 2008). Therefore research activities determining the prevalence of allergy caused by certain foods, differentiating the causes of symptoms are extremely important.

Recently a large set of blood samples from randomly selected individuals ($n = 1145$) (Pasco et al., 2012) was analysed by the IgE RAST method against wheat and milk antigens, and FA questionnaires were used with statistical analyses to establish relationships between individuals with clinical symptoms and immune-responses (Vu et al., 2014). Results indicated that the prevalence of wheat allergy was 2.5% where both a positive IgE immune response and symptoms against wheat were observed. These results are in full agreement with similar investigations on wheat where the prevalence of wheat IgE sensitisation in European countries is 2.9% (Zuidmeer et al., 2008; Siles and Hsieh, 2013). It was postulated that the remaining 12.8% ($n = 125$) of individuals who showed raised IgE antibody levels without symptoms may have a latent wheat sensitivity with the potential of developing symptoms sooner or later. The large proportion (12.9%) of the investigated population who have symptoms associated with consumption of wheat products but did not demonstrate raised IgE may suffer from other wheat-related disorders (i.e. not IgE mediated) such as CD, non-coeliac gluten sensitivity (NCGS) or reaction to fructans (FODMAPs) for those with irritable bowel syndrome (IBS).

NCGS has been described about 40 years ago, being defined as mainly abdominal symptoms related to ingestion of gluten-containing cereals, after CD and wheat (Schuppan et al., 2015). There is strong evidence that NCWS is caused by an innate immune response to wheat proteins (different from the adaptive, T cell-mediated response to gluten in CD, or to gluten and non-gluten proteins in wheat allergy) (Sapone et al., 2012; Catassi et al., 2013). NCGS can usually be separated clinically from food intolerances (mainly lactase deficiency and FODMAP intolerance)—conditions that lead to increased fermentation of carbohydrates in the intestine and that do not cause intestinal or general inflammation.

In a study carried out by Junker et al. (2012), gliadin peptides, including alpha-gliadin peptide p31-49, were not found as an activator of innate immunity in intestinal epithelia of myeloid cells, whereas a water-soluble fraction and a peptic-tryptic digest of commercial gliadin strongly activated mouse and human myeloid cells in vitro, most potently dendritic cells. This activity was not due to, for example lipopolysaccharide contamination and not associated with alpha- or gamma-gliadins, or synthetic gliadin peptides, leading to the identification of the family of amylase-trypsin inhibitors as triggers of innate immunity in wheat.

NCGS has become a topic of increasing interest for the scientific community. The results of the survey of Branchi et al. (2015) demonstrate a certain degree of awareness of the clinical relevance of gluten-related disorders among gastroenterology specialists. Despite this, 40% of the survey responders are not able to give the correct definition of NCGS. Efforts should be made to improve the knowledge about these disorders, as well as to promote the circulation of unambiguous and scientifically relevant information. The establishment of new guidelines approved by the international scientific community would help towards the proper management of patients with gluten-related disorders.

14.2.2 Carbohydrate-Related Disorders

IBS is a common gastrointestinal condition contributing to considerable financial burden and impact on quality of life. Until recently therapeutic strategies, however, had disappointing efficiency but it improved significantly by the findings of the research group led by Peter Gibson, Melbourne, Australia (Gibson et al., 2007, 2015). Because many abdominal symptoms might originate from bowel digestion, altering factors that contribute to luminal digestion, particularly the osmotic load within the lumen, and the fermentative gas content might offer symptomatic benefit. Candidate substrates that are highly fermentable exerting an osmotic effect are dietary, poorly absorbed, short-chain carbohydrates collectively termed FODMAPs (Shepherd and Gibson, 2006). FODMAPs can include fructose and lactose in patients in whom these are malabsorbed, polyols such as sorbitol because they are poorly absorbed by humans, as well as fructo-oligosaccharides (fructans) and galacto-oligosaccharides (such as raffinose) for which humans do not express suitable hydrolases and are always poorly absorbed (Shepherd et al., 2008).

Major dietary FODMAPs include fructose and fructans. Common dietary sources of fructose are fruits and honey; sources of fructans are cereals and onion (Muir et al., 2009; Biesiekierski et al., 2011). FODMAPs are substrates for fermentation by bacteria not considered to be probiotic. Bacterial fermentation of carbohydrates yields short-chain fatty acids, including butyrate, the major energy substrate for the colonic epithelium. Butyrate is also a key regulator of colonocyte proliferation and apoptosis and has immunomodulatory effects. In these ways, fermentable carbohydrates delivered to the colon have potential anti-carcinogenic and anti-inflammatory actions. Restriction of FODMAPs delivered to the colon to reduce gas production, subsequent luminal distension and GI symptoms consequently have adverse effects on colonic health.

The evidence-base for efficacy of the low FODMAP diet is building (Shepherd and Gibson, 2006; Shepherd et al., 2008) with a recent blinded placebo-controlled cross-over study confirming previous studies that about 75% of patients gain clinically significant benefit (Halmos et al., 2014). The low-FODMAP diet is increasingly being applied by health professionals in patients with IBS as first-line therapy (Halmos et al., 2014, 2015). Given the high prevalence and chronic nature of IBS, it is possible that many people will restrict intake of FODMAPs over months to years (Shepherd and Gibson, 2006).

14.3 ANCIENT WHEAT RELATIVES

Some ancient cereal crops of the Mediterranean region (einkorn, emmer and spelt) played critical roles in the history of human civilisation through the centuries. Here and in the neighbouring Near East, wild ancestors of very unattractive appearance (brittle rachis, small seeds, etc.) were domesticated by farmers who in their simple breeding approaches produced plants possessing more useful agrobotanic traits. These cereals were popular within the region for hundreds of years and long represented a staple food. At a certain point in history, however, the introduction of higher-yielding, free-threshing wheats caused hulled wheats to fall into a state of neglect to such an extent that they have become a relic crop, as in the case of einkorn (Galterio et al., 2001).

For social, cultural or simply economic reasons, hulled wheats are becoming popular once again. Today, they are no longer seen as the 'food of the poor' as they were in the past. On the contrary, they have become an exclusive and fashionable food for which discerning consumers are prepared to pay a higher price than for any other wheat product.

Einkorn (*T. monococcum*) is genetically the most basic wheat type with a diploid chromosome number ($2n = 14$) and contains two sets of a single genome (designated A). Emmer (*T. dicoccum*), another ancient wheat, is tetraploid ($2n = 28$) and combines two distinct genomes (AB), whereas spelt (*T. spelta*) and common wheat (*T. aestivum*) are hexaploid ($2n = 42$) and contain two sets of three different genomes (ABD) (Fig. 14.2.)

14.3.1 Einkorn

Einkorn (*Triticum monococcum* ssp. *monococcum* L.) is a diploid ancestral wheat, related to durum (*T. turgidum* ssp. *durum*) and to bread (*T. aestivum* ssp. *aestivum*) wheats. It was instrumental in the rise and spread of agriculture and a significant food source for several thousand years, before it was replaced by the more productive polyploid wheats during the Endolithic period (Nesbitt and Samuel, 1996). Einkorn originated in the mountainous areas of Turkey and its wild progenitor (*T. boeoticum* Boiss.) can still be found growing spontaneously. It was one of the first wheats to be domesticated some 10, 000 years ago in the Near East. After the Bronze age, einkorn was replaced by emmer, spelt and barley and its cultivation was dramatically reduced in Europe. However, it is now sporadically grown in western Turkey, the Balkan countries, Germany, Switzerland, Spain and Italy and it was still grown on the Swedish island of Gotland in the mid-1960s (Luje et al., 2003).

Einkorn is a potential crop for environmentally friendly organic farming (Luje et al., 2003). Compared to common wheat, einkorn is generally more resistant to diseases and has the ability to withstand drought.

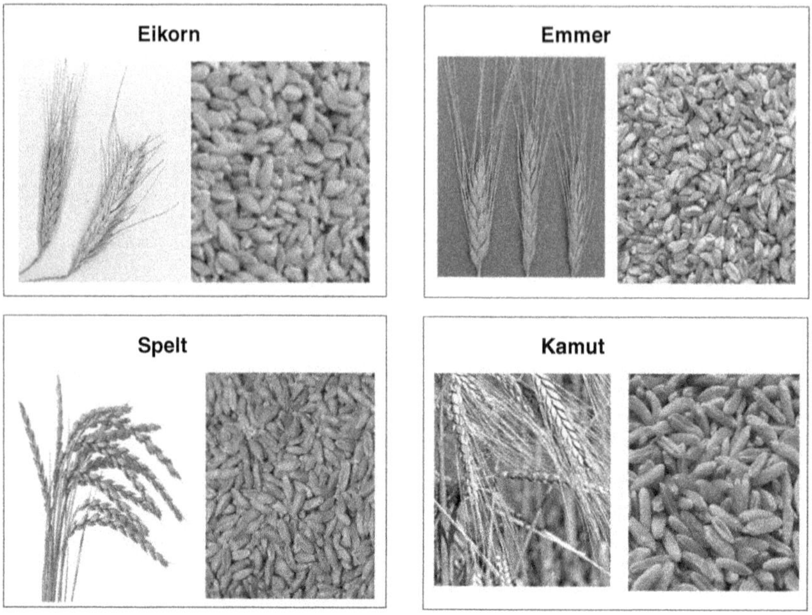

FIGURE 14.2 **Spikes and grains of some ancient wheat relatives.**

The renewed nutritional interest in this cereal is related to its high protein and yellow pigment contents (Abdel-Aal et al., 1995; Borghi et al., 1997; Corbellini et al., 1999; D'Egidio and Vallega, 1994), as well as its putative low allergenicity (De Vincenzi et al., 1996; Molberg et al., 2005). Einkorn may also be an excellent source of traits such as disease resistance, yellow pigment content, and so on, for durum and bread wheats. The yellow pigments, as in other wheats, consist mostly of lutein (Abdel-Aal et al., 2002); in whole einkorn flours, lutein content is around 8.5 mg/kg (dry matter), with an average value 4 times higher than bread wheat (Abdel-Aal et al., 2002).

The grains of einkorn are enclosed by tough glumes, which protect them against insects and pollutants, but before milling they must be dehulled by passing through rubber-coated rollers. The dehulled grain is typically smaller than grains of common wheat varieties (~ 18 and 34 mg/kernel, respectively). The grain texture of einkorn has been characterised as very soft compared to other wheat varieties (Corbellini et al., 1999). Flour yield from milling of einkorn has been reported to be similar or higher than wheat, possibly due to lack of the deep crease typical of polyploid wheats (D'Egidio and Vallega, 1994) (see Chapter 22). A high content of both ash (approx. 2.2%) and protein (12.5–25.2%) has been reported in dehulled einkorn, while total dietary fibre in einkorn (8.7%) was found to be considerably lower than in common wheat (12.5%) (Corbellini et al., 1999). In general the baking quality of einkorn is poor (LeClerc et al., 1978); however certain einkorn lines with acceptable gluten strength (alveograph values) and farinograph stability have been also reported (Borghi et al., 1997).

Einkorn was traditionally used for porridge or cooked whole grains (D'Egidio and Vallega, 1994). The main focus of the end-use quality of einkorn has been on its baking quality and only a few studies have investigated other possible applications. More recently, cookies and flakes made from einkorn for breakfast cereals have been investigated by D'Egidio and Vallega (1994). The very soft endosperm characteristics of einkorn and its high gelatinisation viscosity suggest specific applications for cake and pastry products. Another application could be weaning foods due to its high gelatinisation viscosity, high protein content and low content of dietary fibre. In addition, whole boiled kernels of einkorn can be used as a substitute for rice due to its good consistency and pleasant flavour (D'Egidio and Vallega, 1994).

14.3.2 Emmer

Emmer (*Triticum turgidum* ssp. *dicoccum*) also known as farro (Italian for hulled wheat) has survived in Italy only in a few mountainous marginal areas of central and south Italy where it is cultivated according to organic farming procedures or using very low N-fertilisation in competition for yield with other cereals (Vallega and Waines, 1987). Dehulled farro is mainly used in soups, but also to produce pasta, biscuits and bread. It is used in the treatment of several disorders such as colitis, allergies and high blood cholesterol (Borghi et al., 1997) and even of CD; however treatment of CD is completely wrong because farro contains gluten polypeptides with toxic epitopes. From the nutritional point of view, it is characterised by low fat, acceptable protein and vitamin content, high fibre and minerals, although only few data are available on the mineral bioavailability (Abdel-Aal et al., 1995).

14.3.3 Kamut

Kamut is a registered trademark cereal (an organic crop) whose origin is much older; it has Egyptian origin. The history of its reinvention and commercialisation is a very colourful story (Abdel-Aal et al., 1998). Kamut is cultivated in mainly North America but recently there is increasing production in Thailand and Australia under exclusive licence conditions. Because of the market interest, European, especially Italian, farmers have been attracted by Kamut cultivation because the market is increasing. The references on the agronomic, nutritional and technological traits of Kamut cultivated in European environments are still rare and elusive (Grausgruber et al., 2005; Stagnari et al., 2008).

Kamut brand wheat can be found in breakfast cereals, breads, cookies, snacks, waffles, pancakes, bread mixes, baked goods and prepared and frozen meals. Because of the inherent sweetness of this grain (referred to by some as 'the sweet wheat'), no sugar is required to hide the subtle bitterness associated with most wheats and whole-wheat products. Many are utilising the natural firmness of the kernels to produce tasty pilafs, cold salads, soups, or a substitute for beans in chili. Kamut brand bulgur and couscous are also popular in Europe. Kamut brand wheat also makes an outstanding pasta which is superior to all other whole grain pastas in texture and flavour. Because of the strong gluten in the protein, Kamut pasta can be frozen and reheated without losing its firmness. Recently, Green Kamut was introduced. It is becoming the rage of the green foods market because of its concentrated health benefits and mild, fresh taste when compared to other wheat grass juices. Thus the leaves as well as the grain of this remarkable plant are proving to be valuable.

The complete nutritional analysis of Kamut brand grain substantiates that it is higher in energy than other wheats. Compared to common wheat, it is higher in eight out of nine minerals; contains up to 65% more essential amino acids; and boasts more lipids and fatty acids. The most striking superiority of Kamut brand wheat is found in its protein level—up to 40% higher than the national average for wheat. Because of its higher percentage of lipids, which produce more energy than carbohydrates, Kamut brand can be described as a 'high energy grain'. Athletes, people with busy lives and anyone looking for quality nutrition will find Kamut brand products a valuable addition to their diet (Abdel-Aal et al., 1998).

14.3.4 Spelt

Spelt (*T. aestivum* ssp. *spelta* (L.) Thell.) ($2n = 6x = 42$, genome = AABBDD) is one of the oldest cereals of the modern world. Molecular evidence indicates that the spelts are derived from hybridisation events between non-free threshing tetraploid wheat ($2n = 4x = 28$, genome = AABB) and free-threshing hexaploid wheat (Blatter et al., 2004). Most of the hulled wild wheats and hexaploid subspecies such as *spelta, macha* and *vavilovii* have a recessive allele (q) at the Q locus that conditions major domestication traits in wheat (Faris, 2014). Extensive systematic breeding and selection led to the development of free-threshing and high-yielding varieties of spelt wheat having wide adaptability to different agro-climatic conditions and led to the replacement of most of the low-yielding traditional landraces and subspecies. Now, in most parts of the world, spelt has been superseded by common wheat (*T. aestivum* ssp. *aestivum*) and is being grown as a marginal or niche crop with low inputs. After many years of marginal cultivation, recent interest in the use of spelt for ecologically grown food has led to a resurgence in its cultivation. In fact, the food use of spelt has been a tradition in Europe, particularly in Germany where it is widely used in many grain-based foods (Cubadda and Marconi, 2002).

Spelt demonstrates a higher resistance to environmental factors than common wheat and can be cultivated in harsh ecological conditions, without the use of pesticides (Raman et al., 2008). Moreover, spelt can grow on poorly-drained and low-fertility soils and therefore is a suitable crop for organic farming

(Bonafaccia et al., 2002). In the past, spelt has not been targeted for systematic breeding, the objectives for improvement have only recently been identified (Neeson et al., 2008).

More recently, spelt is becoming a valuable crop due to its reputation as a healthy food especially for patients with wheat allergies (Onishi et al., 2006) and its suitability for speciality products such as bread, pastry, noodles (Galova and Knoblochova, 2001; Elía et al., 2004) and in confectionary. As a functional food, spelt flour is also reputed to possess a unique flavour and to be more nutritious than bread wheat (Campbell, 1997). Generally, spelt products such as bread attract a higher price in the marketplace, especially organic products.

The chemical, nutritive and functional properties of spelt wheats compared to *aestivum* wheats have been reported in numerous publications in the last decade, reviewed recently by Escarnot et al. (2012). Results showed that, on average, spelt whole meals and milling fractions were higher in lipids and unsaturated fatty acids as compared to wheat, whereas tocopherol content was lower in spelt, suggesting that the higher lipid content of spelt may not be related to a higher germ proportion. Although milling fractionation produced similar proportions of flour and brans in spelt and wheat, it was found that ash, copper, iron, zinc, magnesium and phosphorus contents were higher in spelt samples, especially in aleurone-rich fine bran and in coarse bran. Even though phosphorus content was higher in spelt than in wheat brans, phytic acid content showed the opposite trend and was 40% lower in spelt versus wheat fine bran, which may suggest that spelt has either a higher endogenous phytase activity or a lower phytic acid content than wheat. The results of the study of Ruibal-Menieta et al. (2005) give important indications on the real nutritional value of spelt compared to wheat. Moreover, it was shown by Gomez-Becerra et al. (2010) that the Ca/Fe ratio, combined with that of oleate/palmitate, provides a highly discriminating tool to authenticate spelt from wheat flours and to face the growing issue of spelt-flour adulteration. The results indicated that spelt is a highly promising source of genetic diversity for grain protein and mineral nutrients, particularly for Zn and Fe.

The spelt-based products are easily digestible and have a slightly higher protein content than wheat. The main difference in the nutritive value of spelt flours is the variation in the amount and type of grain proteins, especially the prolamins (Galli et al., 2000; Shewry, 2002). There has been anecdotic clinical evidence for a long time that a large proportion of non-CD patients suffering from wheat-related health disorders can tolerate products made from certain spelt varieties. Recently the first research paper with robust experimental and statistical results also demonstrated this important observation (Armentia et al., 2012).

14.3.5 Pseudocereals

Besides the 'true' cereals (monocotyledonous grasses) there is a group of plants that produce starch-rich grains, which can be used like cereals (Fletcher, 2016).

FIGURE 14.3 Spikes and grains of some pseudocereals.

Botanically they are assigned to the dicotyledonous-plant family and according to most researchers they are summarised under the term pseudocereals. Although there are several plants under this group only three have gained some importance, namely amaranth (*Amaranthus spp.* L), quinoa (*Chenopodium quinoa* Willd) and buckwheat (*Fagopyrum esculentum* Moench) (Fig. 14.3.). According to several phylogenetic classifications the *Amaranthus* and *Chenopodium* genera belong together to the *Caryophyllales*, whereas buckwheat (*Fagopyrum*) belongs to the *Polygonales*. *Polygonales* and *Caryphyllales* are closely related and are combined together in *Caryophyllidae*.

Worldwide more than 60 species of amaranth exist (Corke et al., 2016). The main grain amaranth species used today are *Amaranthus caudatus* L. (syn. *edulis* Spegazzini), *Amaranthus cruentus* L. (syn. *paniculatus* L.) and *Amaranthus hypochondriacus* L. Among quinoa sweet and bitter varieties exist, which is dependent on the content of saponins; that is if the saponin content is below 0.11% the variety is considered as being a sweet variety (Koziol, 1991). A wide range of quinoa varieties can be found, including coloured varieties ranging from white, yellow, orange and red to black coloured. Mostly used worldwide are the white species. For buckwheat two varieties are commonly cultivated: common buckwheat (*Fagopyrum esculentum*) and tartary buckwheat (*Fagopyrum tataricum*). Amaranth seeds are lentil-shaped and measure about 1 mm in diameter. The weight of 1000 kernels is only 0.5–1.4 g. Quinoa seeds are slightly bigger than amaranth, the 1000 kernel weight is approximately 1.9–4.3 g. In contrast to cereals, in amaranth and quinoa the embryo surrounds the starch-rich tissue (perisperm) in the form of a ring and makes about 25% of the total seed weight. The buckwheat seed is a three-angled achene, 6–9 mm long. The fruit of *F. tataricum* is smaller (4–5 mm) and more rounded at the edges. The 1000 kernel weight (10–20 g) depends mainly on the hull thickness. Structurally and chemically, the endosperm resembles that of a cereal grain consisting of a non-starchy aleurone layer and large cells packed with starch granules constituting the majority of the endosperm.

While the origin of amaranth and quinoa is in Latin America, buckwheat originally was spread from Asia (Cai et al., 2016). In pre-Columbian times amaranth and quinoa were important staple foods for the indigenous population as maize was the only other cereal available at those times. After the Spanish Conquest the cultivation of both plants was suppressed but since the 1970s both crops are experiencing a renaissance due to their favourable nutritional composition. Yet, it has to be mentioned that until today amaranth and quinoa have not been a commercial success worldwide, which is mainly due to the dominance of cereals like wheat, rice and maize. The origin of buckwheat is middle Asia from where it was transferred by nomadic people and reached middle Europe (Germany, Austria, Italy) in the 13th century. However, the importance of buckwheat was lost due to the cultivation of cereals.

Today the world production for amaranth and quinoa is increasing, while for buckwheat it is decreasing. In the year 2013, the amount for quinoa produced was 103,418 tonnes (FAOSTAT, 2014), which is almost 3 times as much as 20 years earlier. For buckwheat, the amount was 234,7558 tonnes in the year 2013, only about half the amount produced in 1993 (FAOSTAT, 2014). For amaranth production, no worldwide data are available. The main amaranth-producing countries are the tropical regions of South America, but also of Africa, Central and Southeast Asia (especially India), as well as the warm regions of North America. Quinoa is mainly grown in the Andean region, Peru, Bolivia and to a minor extent Ecuador, Chile, Argentina and the main producers of buckwheat are China, Russia and Ukraine, to some extend also North America and only a little in Europe.

The main reason for the increase in production and utilisation of the pseudocereals can be found in two factors: their good nutritional composition and their gluten-free status.

14.3.6 The Chemical and Nutritional Composition of Pseudocereals

All three pseudocereals show excellent nutritional properties, which is one of the main reasons why they have attracted so much interest in recent years. In Table 14.1, the main chemical composition of amaranth, quinoa, buckwheat and wheat (as representative for the true cereals) is given. In comparison to wheat, the composition of pseudocereals is distinct, but more important is the fact that their single components are sometimes of extraordinary quality.

Protein can be found in similar amounts to that of cereals, although differences between species exist as was determined by Tömösközi et al. (2009) for

TABLE 14.1 Chemical Composition of Amaranth, Quinoa and Buckwheat in Comparison to Wheat (% dm)

(% dm)	Starch	Fat	Protein	Ash	TDF	TPC[a]
Wheat	52.20	1.68	13.04	1.76	11.99	505.1
Buckwheat	77.37	2.19	11.45	2.34	6.65	860.9
Amaranth	58.38	6.33	14.74	2.95	12.67	95.8
Quinoa	57.18	7.73	12.18	2.72	12.75	284.5

dm, Dry matter; TDF, total dietary fibre; TPC, total phenolic content.
[a]*(mg ferulic acid equivalents/100 g dm).*
Source: Schoenlechner and Tömösközi, 2014, own measurements, unpublished data.

amaranth. Protein content ranged from 14.2% to 17.4%. Compared to cereals, the nutritional quality of the protein in pseudocereals is generally higher and is more comparable to legume protein (Drzewiecki et al., 2003). In particular amaranth and quinoa have a very high amount of essential amino acids. Lysine, arginine, tryptophan and sulfur-containing amino acids are found in higher amounts than in the true cereals (Matuz et al., 2000; Gorinstein et al., 2002; Vega-Galvez et al., 2010). The balanced amino-acid composition of amaranth and quinoa is close to the optimum protein reference pattern in the human diet according to FAO/WHO requirements (Montoya-Rodríguez et al., 2015). Unlike in cereals, the main storage proteins in pseudocereals are globulins and albumins. Prolamins can only be found in very small amounts (Gorinstein et al., 2002; Drzewiecki et al., 2003; Muchova et al., 2000; Prakash and Pal, 1998; Thanapornpoonpong et al., 2008).

The amount of fat in buckwheat is comparable to the true cereals, but in amaranth and quinoa it is about 3–4 times higher (Table 14.1). In amaranth the main fatty acids are linoleic, palmitic and stearic acid (Becker, 1994; Leon-Camacho et al., 2001; Cai et al., 2003). Additionally amaranth contains a sginifcant amount of squalene (Becker, 1994). The main fatty acids in buckwheat and quinoa are linoleic, oleic and palmitic acids (Bonafaccia et al., 2003; Repo-Carrasco et al., 2003), Overall, the content of essential fatty acids is very high in three pseudocereals.

Starch in pseudocereals is found in comparable amounts to cereals, which allows them to be used for starch-based products. Starch granules in amaranth and quinoa are extraordinary small and their content of amylose is very low (below 10%) (Atwell et al., 1983; Qian and Kuhn, 1999; Lindeboom et al., 2005). In contrast, buckwheat contains a high amount of amylose (Wolter et al., 2013). In pseudocereals starch is stored in the perisperm, unlike in cereal starch, where it is stored in the endosperm (Berghofer and Schoenlechner, 2002). The different starch properties of pseudocereals influence their processing properties to a great extent.

The content of dietary fibre in pseudocereals is similar to that of common cereals, but they show great variation between species, which is also common for true cereals. Some authors report relatively low dietary fibre content for pseudocereals (Alvarez-Jubete et al., 2009), others found higher amounts. Repo-Carrasco (1992) found that amaranth (*A. caudatus*) was a rich source of dietary fibre, especially the insoluble fraction. La Mothe et al. (2015) studied the dietary fibre content and composition in quinoa and amaranth. Total dietary fibre content was 10% for quinoa and 11% for amaranth. For both pseudocereals, 78% of the dietary fibre was insoluble. Insoluble fibre from quinoa and amaranth was mainly composed of galacturonic acid, arabinose, galactose, xylose and glucose. In buckwheat, Steadman et al. (2001) determined a total dietary fibre content of 7%. Dziedzic et al. (2012) studied the components of dietary fibre in buckwheat, which were 6.8% of lignin, 2.2% of hemicellulose and 10.6% of cellulose.

Pseudocereals are excellent sources of micronutrients such as minerals and vitamins. The content of minerals is about twice as much as in cereals and they are particularly rich in calcium, magnesium, iron, potassium and zinc. Regarding vitamins, riboflavin (vitamin B_2) and tocopherols (vitamin E) are the ones that are found abundantly. But most impressive is their content in folic acid, a vitamin that is deficient in a significant portion of the world's population. Schoenlechner et al. (2010a) determined the total folate content in the pseudocereals amaranth, quinoa and buckwheat in comparison to four cereal species and found that in amaranth the folate content was about 4 times higher than in cereals and in quinoa the amount reached even values that were 10 times higher. Some quinoa species can even contain about 250 μg/100g (unpublished data). Only in buckwheat, folic acid was in the range of other cereals.

Pseudocereals can also contain significant amounts of phenolic compounds, which are phytochemicals present in plants that can promote health benefits due to their antioxidant capacity. Great variation exists between species, also dependent on growing conditions. Among the pseudocereals, amaranth seems to contain rather low amounts of phenolic compounds (Repo-Carrasco-Valencia et al., 2009), in contrast to quinoa and buckwheat (Table 14.1). Miranda et al. (2012), Pasko et al. (2009) and Repo-Carrasco-Valencia and Astuahuaman (2011) determined the total phenolic compound in quinoa and found values of 1.59–374 mg gallic acid equivalents/100 g in different varieties. Buckwheat was shown to have 2–5 times more phenolic compounds than oats or barley (Zdunczyk et al., 2006). The primary antioxidants in buckwheat are rutin, quercetin, hyperin and catechins (Morishita et al., 2007).

14.3.7 Functional Properties of Pseudocereals

Pseudocereals have different physical properties compared to cereals, which is mainly a result of the different seed morphology of the pseudocereals (ring embryo) and their distinct chemical composition.

In amaranth and quinoa, the very small starch granules and their high content of amylopectin (usually at least 90%) are responsible for the higher viscosity, good freeze-thaw stability, higher water-binding capacity and higher swelling power; in addition, their starch shows less retrogradation than cereals. Due to these properties, they have excellent ability as thickening agents (Atwell et al., 1983; Lorenz, 1990; Ahamed et al., 1996; Hunjai et al., 2004). In buckwheat the amylose content is high and lies between 18.3% and 47% (Qian et al., 1998; Yoshimoto et al., 2004; Wolter et al., 2013). Qian et al. (1998) found that buckwheat amylose forms complexes with lipids. These complexes can decrease the swelling power and solubility of buckwheat starch thus affecting the functional properties of the starch.

Regarding protein, research has revealed that all pseudocereal proteins are highly soluble and are thus a good source to be applied in functional foods (Segura-Nieto et al., 1999; Bejosano and Corke, 1999; Kovacs et al., 2001; Salcedo-Chavez et al., 2002). Amaranth protein isolates have shown good foaming and film forming properties (Fidantsi and Doxastakis, 2001). Films from amaranth native protein isolates (Condés et al., 2013) showed low water-vapour permeability but poor mechanical properties, which were improved by thermal treatment (protein denaturation) of the proteins. Depending on protein and thermal conditions, amaranth proteins are able to form self-supporting gels that could be applied in different gel-like foods (Avanza et al., 2005). Bolontrade et al. (2016) found that foams from amaranth proteins at acidic pH were flexible and elastic and thus suggested the application of amaranth proteins into acid foam-type foods like dessert stuffing or ice cream.

Quinoa protein solubility is increased under alkaline conditions (Abugoch, 2009). Its emulsification capacity and stability is higher compared to pearl millet or wheat, whereas foaming capacity seems to be lower (Oshodi et al., 1999). Protein isolates of quinoa have a water-holding capacity of 2.8–4.5 mL water/g sample depending on extraction conditions, which is similar to soy isolates (Abugoch, 2009).

14.3.8 Utilisation of Pseudocereals for Food Products

14.3.8.1 General Processing Aspects

Principally all three pseudocereals can be used for a wide range of products, similar to cereals. Breakfast cereals, cereal beverages (similar to soy beverage), extruded snack foods are examples for products from pseudocereals, which can be found on the market.

The biggest limitation though for functional properties is that pseudocereals do not contain gluten and they thus do not possess any dough-forming or baking properties. For the production of bread and bakery products they cannot be used without addition of further ingredients or without adaptation of processing conditions. Up to a certain amount they can be added to wheat-based products, thereby improving the nutritional properties of the resulting product.

Sanz-Penella and Tamayo-Ramos (2012), for example supplemented wheat bread with amaranth (20 and 40%) and found an increased iron concentration in the resulting bread. The production of bakery products like cookies from 100% pseudocereals does not cause great difficulties, especially when buckwheat or popped amaranth is used. This was shown by trials, where cookies were produced from amaranth, quinoa or buckwheat (Schoenlechner et al., 2006). According to Omary et al. (2012), germinating pseudocereals may improve the contents of nutrients, vitamins, minerals, total polyphenols and antioxidant activity, while decreasing anti-nutrients. Thus germinated gluten-free cereals and pseudocereals have potential to be used as natural means of fortification and enrichment in (gluten-free) foods. Their substitution may affect texture and taste (Makinen et al., 2013), which needs further investigation.

The local and/or international suppliers usually offer their products with declared purity and quality parameters, like those of the websites of Scalzo and Canadian Grains Commission, respectively (http://www.scalzofoods.com.au/products/ancient_grains and https://www.grainscanadagc.ca/oggg-gocg/13/oggg-gocg-13d-eng.htm). However, till now there are no generally accepted and used purity/quality specifications when buying the grains and flours of pseudocereals.

14.3.8.2 Bread Baking With Pseudocereals

Due to the lack of gluten in pseudocereals, their addition, especially in high amounts, to wheat dough results in a change of the processing conditions and quality of the end product. These changes include increased dough yield, moist dough, decreased fermentation tolerance, lower volume, tense and non-elastic crumb and various flavour changes (Sanz-Penella et al., 2013; Poutanen et al., 2014; Stikic et al., 2012; Iglesias-Puig et al., 2015). Traditional bread making therefore has to be adapted, when higher amounts (more than 20%) are to be included.

Bread making from pseudocereals has been studied by several research groups. Generally they have all found that the addition of pseudocereals has increased the nutritional value of the resulting end product. Depending on the amount of flour substitution (12–50%), the protein, mineral (in particular Cu, Mn, Zn, Fe, Ca, Mg and K), dietary fibre, lipid, squalene and folate content were enhanced (Sanz-Penella et al., 2013; Tosi et al., 2002; Schoenlechner et al., 2010a). On the other hand, it was also found that after pseudocereal flour addition the phytate content was significantly increased (Bilgicli and İbanoglu, 2015; Iglesias-Puig et al., 2015), but this can be easily avoided by the used of exogenous phytases from bifidobacteria (Iglesias-Puig et al., 2015). The use of flours from germinated seeds could also minimise the phytate content in bakery products.

Pseudocereals are also rich sources of some bioactive compounds like flavonoids (e.g. rutin in buckwheat) and it was found that some of them were not altered after baking, for example total phenolic content of bread containing quinoa was retained (Brend et al., 2012), but in bread containing buckwheat, rutin concentration was decreased (Vogrincic et al., 2010).

If the purpose of pseudocereal inclusion in bread making is to increase protein amount and quality, protein isolates could be preferred in order to avoid quality deterioration of the end product (Tömösközi et al., 2011). Another possibility to improve the quality of bread from/with pseudocereals, is the application of sourdough fermentation. Amaranth and quinoa flour seemed to be suitable for various species of Lactobacillus (Jekle et al., 2010; Houben et al., 2010; Coda et al., 2010; Dallagnol et al., 2015). In wheat bread with added buckwheat flour sourdough fermentation improved bread volume and crumb structure (Moroni et al., 2011).

14.3.8.3 Gluten-Free Foods From Pseudocereals

Important in a gluten-free diet is not only to exclude all gluten-containing components (i.e. wheat, rye, barley and subspecies and possibly also oats) and avoidance of gluten-contamination of foods, but also to replenish deficient nutrients, especially when gut absorption is still impaired. It is important to ensure nutritional adequacy as well as excluding gluten, as people suffering from any gluten intolerance have to adhere to such a diet life-long. Until recent times, gluten-free processing was based on a narrow range of raw materials, mainly rice and maize. The consequence was a low nutritional quality (high energy density, but low nutrient density including a low content of vitamins, minerals, trace elements, dietary fibre, secondary plant metabolites) and low sensory quality, in particular for bread. The challenges for production of gluten-free foods is thus to improve the nutritional quality and to improve the sensory properties. Considering this, pseudocereals have a very high potential for the development of high-quality gluten-free foods.

The task for gluten-free processing is to replace the functionality of gluten (formation of a three dimensional network in bread, agglutination and elasticity in pasta, etc.) by other means like using specific ingredients or adapting processing conditions. The simulation of the unique properties of gluten is a great technological challenge that makes the production of sensory appealing, gluten-free bakery products very difficult. In most cases it is not the use of one single gluten-free raw material that meets this aim, but the use of a mixture of components in combination with additional chemical or physical methods and treatments. For gluten-free bread baking, the addition of a system hydrocolloid-emulsifier-protein (e.g. egg, soybean or milk protein) has emerged to be best suitable for acceptable gluten-free bread quality within the research in the last decade.

Apart from recipe parameters, emphasis is also taken to modify or adapt the baking process. Sourdough fermentation has come into interest in this respect as it offers textural and sensory advantages. According to Houben et al. (2010), sourdough fermentation was able to produce gluten-free dough with viscosity and elasticity similar to that found in pure wheat flours. Alvarez-Jubete et al. (2010) compared the baking properties of all three pseudocereals (amaranth, quinoa and buckwheat), each blended with rice flour (50:50), with a control bread (rice flour and potato starch 50:50). They could show that pseudocereal flour may be introduced into a gluten-free bread formulation to enhance crumb softness and cohesiveness and without adversely affecting sensory properties of the bread loaves.

In gluten-free pasta, the network formation properties of gluten are not responsible for volume; rather, gluten is mainly responsible for the typical pasta texture. The lack of gluten represents a challenge because the available gluten-free raw materials have low functional qualities to provide pasta with good structure.

The use of amaranth, quinoa and buckwheat for the production of gluten-free pasta was studied by Schoenlechner et al. (2010b). The results showed that pasta produced from amaranth had decreased texture firmness and cooking time, while pasta from quinoa mainly showed increased cooking loss. In buckwheat pasta, the least negative effects were observed. By combination of all three raw materials into one flour blend in the ratio of 60% buckwheat, 20% amaranth and 20% quinoa, the dough matrix was improved. Dough moisture had to be lowered (30% vs. 34.5% in wheat pasta). The addition of isolated protein (the most suitable was egg albumen) had the highest effect on improving pasta firmness.

For the production of gluten-free pasta, the role of starch and its properties are important (Marti et al., 2011). Phenomena related to starch retrogradation were found to play a central role for the final texture of the products (Mariotti et al., 2011). Thus more attention is being paid on the technology of pasta making, utilising pre-gelatinised flours or dough, or applying different temperature regimes (Mastromatteo et al. 2011; Marti and Pagani, 2013). Both approaches seem to be promising for improving gluten-free pasta quality.

14.4 GLUTEN-FREE MANAGEMENT AND GLUTEN ANALYSIS

Food-related disorders due to food components (allergy, sensitivity, CD, NCGS, etc.) affect a significant proportion of the population. The main feature of these types of food safety problems is that they are caused by small amounts of natural food components (certain epitopes in food proteins) and the sensitivity level of patients can be very different depending on the individuals and also on the health status of persons (Anderson, 1986; Skypala, 1999). The safest or even the only effective approach to handling these disorders is to avoid the consumption of food products containing the causative proteins.

Food allergen management has become a special area of food-safety systems. For establishing the threshold of gluten-free products in the food chain, three types of approaches can be identified in industrial practice—Analytical Methods, Safety Assessment and Risk Assessment (FoodDrinkEurope, 2013; The Threshold Working Group, 2006).

Patients suffering from these symptoms have to identify somehow the components or food products in their diet causing their problems. To avoid these products, proper labelling of foods is the simplest way to inform the consumers about the possible risk factors of consuming that product. So, the first and most general legislation level in the related food safety area is regulation on the provision of food information to consumers (Regulation (EU) No 1169, 2011; Food Allergen Labeling and Consumer Protection Act, 2004).

The development of labelling instructions is a complex issue because of (1) the great number of foodstuffs potentially causing adverse reactions, (2) the wide range of sensitivities among patients and (3) the variety of foods that contain such food ingredients. Identification of toxic components is therefore needed by responsible national and/or international authorities and regulatory agencies.

Food-allergen labelling regulation exists in most countries but the labelling regulatory frameworks differ significantly around the world (Gendel, 2012; Lampidonis and Siragakis, 2014). Because of the globalised food production and commercialisation system, the lack of uniformity among frameworks from country to country is a very big problem. One of the main difficulties in the regulations is how to define and how to determine the threshold levels. In most cases, there is no clinical evidence for the minimum concentrations of the different allergenic proteins. Therefore, very often the zero tolerance concept is applied, but this approach is not realistic and not applicable for the different stages of the whole food chain and not interpretable in the area of food analysis (Kerbach et al., 2009; Madsen et al., 2011).

Wheat and cereals are included in 'allergen' lists of all countries, although the definitions for this food group are different. The common description of cereals is 'cereals containing gluten,' including wheat, rye, barley, oats, spelt, kamut or their hybridised strains and products. However, in the US regulation only wheat and triticale are included while in Japan and Korea wheat and buckwheat are also mentioned. Regarding oats, it seems on the basis of the latest scientific results that most gluten-intolerant people can tolerate non-contaminated oat-based products (Scherf and Poms, 2015; Londono et al., 2013).

After a long-term process, the Codex Alimentarius Committee has completed the first regulation for foodstuffs suitable for gluten-intolerant people (Codex Alimentarius, 1979). In this regulation gluten is defined 'as a protein fraction from wheat, rye, barley, oats or their crossbred varieties and derivatives thereof, to which some persons are intolerant and that is insoluble in water and 0.5 M NaCl'. Naturally this definition is highly simplified, the nutritional, clinical and proteomic background of cereal protein-related disorders is much more complex, but partly because of the lack of generally accepted, scientific evidence and more importantly the lack of routinely applicable analytical solutions, this approach is still applied today.

The fundamental basis of the existing system is that it defines and measures the 'gluten' content in the product, instead of monitoring the amount(s) of the causative compounds (toxic/allergenic epitopes) in the food product. Both the concept of defining gluten and the way it is quantified are a compromise and huge activities in both, basic and applied research are in progress for further fine-tuning and replacement with more precise methodologies.

Today food products with less than 20 ppm gluten (10 ppm gliadin) can be labelled as gluten-free and foods with less than 100 ppm gluten (50 ppm gliadin) content are defined as very low gluten level products (Codex Alimentarius, 1979; Catassi et al., 2007). Naturally this 1:1 glutenin/gliadin ratio is a rough approximation of the reality. The real gliadin (prolamin) to glutenin ratio in different common wheat varieties can vary between 0.7 and 3, for example in emmer this ratio can increase up to 7 (Wieser and Koehler, 2009). This situation foreshadows the difficulties to serve this regulation with exact analytical methods. There are some minor differences in regional or national regulations or standards for gluten-free products (Regulation (EU) No 609/2013 of the European Parliament; Health Canada's Position on Gluten-free Claims, 2012; Food and Drug Administration, 2013; Australia New Zealand Food Standards Code, 2014) but all of the internationally used and accepted legislation is in harmony with the basic Codex standard.

The cited Codex document also recommends the use of the Enzyme-linked Immunoassay (ELISA) R5 Mendez Method, which became Type I method. This method is based on the very specific immune reaction between the potentially coeliac-toxic epitope QQPFP, which occurs repeatedly in α-, γ- and ω-gliadins, hordeins and secalins and the monoclonal antibody R5 raised against a secalin extract. It uses the sandwich ELISA principle to quantify low levels of gluten (Codex Alimentarius, 1999; Valdes et al., 2003).

The ELISA-based laboratory kits and the lateral-flow rapid tests (Valdes et al., 2003) still remain the most frequently used analytical methods in routine laboratories and food safety practice. The main advantages of immune based methods are their high specificity and sensitivity. However, there are some well-known but only partly or absolutely unanswered questions around the application or even the validation of immunoanalytical procedures (Gabrovska et al., 2006; Slot et al., 2015). For example, different protein molecules/protein fractions/epitopes are targeted and their antibodies (monoclonal R5, G12, 401/21, α-20, or polyclonal) are used, respectively, in the commercially available analytical methods. The presence (the concentration) of target proteins depends on genetic and/or environmental factors, like varieties of plants, environmental stresses, year of harvesting. The different target proteins require different calibration procedures—including calibrators and calculation methods. The results of comparison studies show clearly the degree of analytical errors originating from the abovementioned reasons (Bugyi et al., 2013; Bruins et al., 2015). Suitable reference materials—mainly in real food matrices—are not available, therefore the analytical conditions for method validation and for quality assurance of results are insufficient (Tömösközi et al., 2014).

The physical (i.e. solubility) and chemical (i.e. immuno-reactivity) behaviour of target proteins depends on the food matrices and can change considerably during food processing (Geng et al., 2008; Török et al., 2015a). Both primary structure and conformation of storage proteins of wheat are significantly altered during food processing, mostly by the heat treatment during baking. The denaturation, structural changes and complex formation modify the solubility properties and the immunochemical reactivity of proteins (Gomaa and Boye, 2013). Protein denaturation and the structure modification processes during heat treatment were studied mainly in baking trials (Singh and MacRitchie, 2004) but insufficient information is available concerning the immuno-analytical reactions in simplified model systems (Rumbo et al., 2001). Processed and heat treated model food matrices with defined gliadin content were studied by Bugyi et al. (2013) and by Török et al, 2015b) using commercially available ELISA assays. Results of ELISA measurements showed a decrease of measurable gliadin protein content throughout the process, which can be caused by the changes in antibody binding activity and protein solubility. This underlines the need of developing standardised methodologies for end products using heat treated model food matrices.

The detection and especially the quantification of gluten proteins are extremely important not only due to their direct effect on end-use quality but also for food safety reasons. Grain composition varies between cereal genotypes and therefore leads to methodological problems in food allergen research and genotype selection in breeding for quality. The high sequence similarity and multi-species origin of prolamins coupled with limitations in the available methodologies (Haraszi et al., 2011) make the exact identification of proteins that trigger health problems and their genotypic frequency, variability and stability difficult to determine. High-resolution methods such as mass spectrometry (MS) require accurate molecular quantitative relationships between prolamin peptide biomarkers and the final gluten/prolamin content to relate the detection of peptide mass to their protein sources. These quantitative relationships however are difficult to establish due to genotypic and environmental variability. In addition, the significantly higher portion of proline and glutamine residues has led to poor digestibility by trypsin, one of the most commonly used enzyme in MS-based proteomics. Other enzymes (e.g. chymotrypsin, thermolysin) and enzyme combinations (e.g. LysCþtrypsin) were found to work better to obtain prolamin peptides (Salplachta et al., 2005; Sealey-Voyksner et al., 2010; Rombouts et al., 2013) and thus introduced further challenges such as the optimisation of enzymatic digestion prior to liquid chromatography mass spectrometry (LC-MS) analyses or processing of mass spectra with bioinformatics software that are usually optimised for the use of trypsin.

To assist peptide biomarker searches, epitope mapping, protein selection and medical studies, a database (ProPepper, https://propepper.net) was developed to contain members of the prolamin superfamily proteins identified from Poaceae species, peptides obtained with single and multi-enzyme in silico digestion as well as linear epitopes responsible for wheat-related food disorders (Juhász et al., 2015).

A complete proof of the unfound and false character of one of the main criticisms of the anti-gluten lobby can be given by using ProPepper: gluten proteins of all of the diploid ancestors of the nowadays bread-wheat contain the same amount of toxic epitopes as their hexaploid derivate (Békés, 2015). So, the increased prevalence of wheat-related health disorders is not the result of wheat breeding. As it was shown by Kasarda (2013), it is also not true that modern wheat varieties contain a larger gluten content, producing a larger exposure to gluten, with the suggested consequence of responsibility for the increasing number of health problems. On the other hand, as Kasarda (2013) showed, gluten exposure has indeed increased in the last 50 years in the developed countries, which is caused by the increased application of vital gluten in many processed foods.

Knowledge of the composition and amounts of epitopes present in a single wheat sample represents a significant gap and the detailed wheat proteome datasets now available can provide the necessary information to carry out an estimation of allergen prediction for a single cultivar. The combined use of genome sequence and allergen databases, prediction methodology and cereal chemistry results in a better understanding of the level of toxicity present in the end-products from wheat flour. The workflow presented in the review of Juhász et al. (2012) provides information about the number and distribution of epitopes for a single protein or for a protein fraction. In addition, epitopes present in the highest frequency and for the most harmful proteins can be identified. The 'epitope toxicity' value obtained in this way is a significant research output from the analysis of large datasets that can be applied to the food industry.

14.5 ANALYSIS AND QUALITY MANAGEMENT OF LOW FODMAP FOOD PRODUCTS

Realisation of the crucial importance of FODMAP components in relation to food-related health disorders initiated active research and development in analysing FODMAP-content, as well as utilising food materials with low FODMAP content. The current view of medical experts is that, excluding people suffering with CD, the majority of people who are feeling better on the 'wheat-free' or 'gluten-free' diet are automatically selecting a low FODMAP diet by choosing 'gluten-free' products.

Compared to the 'gluten-free' phenomena, where the problematic food sources are a relative small group of certain cereals, the establishment a low FODMAP diet involves considering a much wider range of plant-origin foodstuffs. In the last 5 years, excellent printed publications (Muir et al., 2007, 2009; Biesiekierski et al., 2011) and even computer and mobile phone apps (http://www.med.monash.edu.au/cecs/gastro/fodmap/iphone-app.html) have been published to help developing custom-designed low FODMAP diets.

One of the often underestimated nutritive values of the pseudocereals is that, beyond being gluten-free, their FODMAP content is very low, compared to traditional cereals. It is the reason why the pseudocereals play a very important role in low FODMAP diets, not only utilising them as ingredients in different processed foods or using them in their traditional forms, but also as source materials for novel dishes, specifically developed by expert chefs (Wilson, 2015).

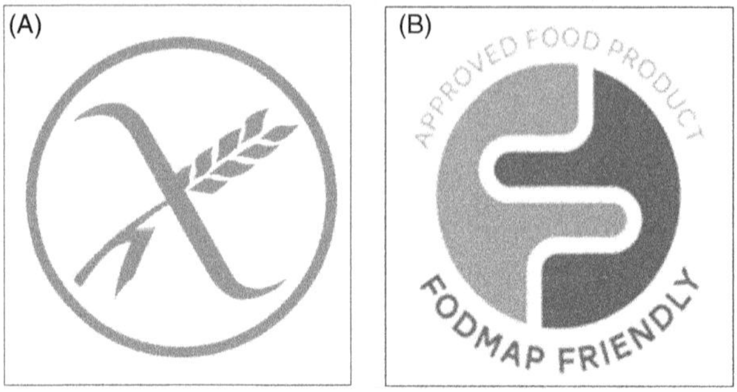

FIGURE 14.4 (A) The internationally accepted logo for gluten free food products (B) Logo indicating low FODMAP level in Australian food products.

The FODMAP analytical data in these data sources have been mostly determined by the two-column based HPLC method of Muir et al. (2009) and by applying the commercially available enzymatic Megazyme Fructan HK Assay kit (McCleary and Blakeney, 1999). In Australia several service laboratories have the legislative right to provide the LOW FODMAP logo on the label of a product containing less than 1.2 g/100 g (Fig. 14.4.).

Traditional cereals (wheat, barley and in particular rye) are relatively rich in FODMAPs and therefore any food product made from them are not recommended for low FODMAP diets.

Because of the fact that FODMAP components are located mostly in the endosperm of the grains, reducing the milling yield, producing less bran in the flour, does not improve the FODMAP level. The baking technology on the other hand has a significant effect, because yeast metabolises a certain amount of soluble carbohydrates during fermentation. So, FODMAP content in the final product can be altered both by the amount of yeast applied in the bread formulation and/or the length of fermentation. Compared to yeasted dough products, sour dough technology has an even more significant effect in reducing the FODMAP level.

Compared to *aestivum* wheat, spelt contains significantly lower amounts of fructose and fructans. But even more important is that there is around a fivefold larger inter-varietal difference among spelt varieties compared to those found among bread wheats (Békés and Suter, unpublished results). It provides the basis for screening the spelt cultivars for low FODMAP content.

An Australian grown spelt line (GWF spelt) was successfully selected with significantly lower FODMAP content. The FODMAP content of the baking products of this line–using an optimised formulation and technology—was significantly lower than the threshold defined for low FODMAP products. Interestingly, the same line was found to have significantly less immune reactivity than any wheat and other spelt cultivars (Vu et al., 2014). It was also found that the albumin-globulin protein composition of GWF spelt is different from other wheat and wheat-related species. One of these differences has been identified by comparing the structure of genes coding the allergenic expansin proteins in different wheat and spelt germplasms. The expansins are a family of closely related non-enzymatic proteins found in plant cell walls. The expansin from GWF spelt has an insertion of three amino acids in the primary sequence, subsequently altering the conformation of the entire protein (Breen et al., 2010). This alteration in the soluble protein composition could be one of the reasons for the significantly lower immune reactivity of GWF spelt.

The potential candidate source for making cereal products with dual health benefits has been recognised by several bakeries in Australia. A closed production line has been developed, including contract growing of GWF spelt with strict guidelines to avoid any contamination of the grain. A small milling capacity has been established with special care of cleaning the whole technology prior to GWF spelt milling.

A PCR-based procedure has been developed to detect the above mutation in the expansin gene (Suter and Békés, 2012) to screen for any contamination in the grain or flour samples caused by any other cereals with the wild type expansin protein. This procedure is utilised for quality assurance purposes. FODMAP content of flour and their products (sour dough bread with special formulation) is monitored by an expert laboratory. Permission has been granted to use the 'LOW FODMAP' logo on the final products.

Preliminary experiences provide some evidence to support the dual health benefits of this spelt line. The question remains as to whether this finding is associated with the apparent lower allergenicity of spelt for those with wheat allergy. An answer requires a double-blind randomised placebo-controlled clinical trial, comparing the allergic symptoms and IgE levels caused by the consumption of wheat and GWF spelt products, involving both, those that eat predominantly wheat and those who have replaced spelt for wheat in their diet.

The above procedure of producing a cereal line with health benefits can be used as an example, indicating how to manage (protect, police) the quality of the final product from contamination by any other cereal material from the field to the bakery. This type of issues will hopefully appear in the near future when germplasm with altered gluten and/or non-gluten protein composition and proven reduced allergenicity will be selected or produced. Genetic engineering is capable of producing such germplasm; the best example for this approach is the work of Altenbach et al. (2014a,b), where the most allergenic gluten protein, the omega-5 gliadin has been successfully eliminated from the samples tested. It is a different issue however, whether or not this and similar, genetically modified wheats will be allowed to be commercialised for human consumption.

14.6 FURTHER SAFETY-RELATED QUALITY ASPECTS

No doubt, major food-safety problems have to be solved in the whole food chain of gluten-free products to ensure the elimination of harmful cereal proteins and to keep the GF status of the processed food under control. However, some other nutritionally important aspects have to be taken into consideration for cereal-based GF foods as well. Some anti-nutritive factors, like phytates, saponins, tannins or heat-labile trypsin inhibitors and lectins can be identified in pseudocereals, mainly in buckwheat. Generally, the concentrations of these compounds are lower than in other minor cereals (i.e. sorghum) or legume seeds (i.e. soy, beans and peas) but sometimes higher than in the main cereals, like wheat, maize or rice. In the case of buckwheat, some allergenic protein sub-units between 9 and 100 kDa were identified (Berghofer and Schoenlechner, 2002; Ruales and Nair, 1993; Linh et al., 2014).

14.7 CONCLUSIONS

Commercial activity related to 'gluten toxicity' and 'wheat allergy' is largely focused upon 'gluten-free' food. However, as more and more information is collected in the medical research, for medium or long term, the solution to solve this problem and satisfy what the different consumer groups really need is not simply having traditional and 'gluten-free' products on the shelf of the supermarkets. Gluten-free diet is not healthier but incomparably more expensive than the consumption of gluten-containing products (Missbach et al., 2015). Beyond those individuals who suffer coeliac disease ($\sim$ 1% of the Caucasian population), consumers with different cereal-related health disorders can tolerate products containing reduced and/or altered gluten content and reduced amounts of certain soluble carbohydrates (FODMAPs) (Halmos et al., 2015).

The development, manufacturing and commercial availability of such products require further detailed basic and applied research in relation to developing reliable, cost-effective and high-throughput analytical tools to monitor the presence/absence and quantity of certain key components of the source material, the intermediates and the final product through the whole production chain from the farm to the supermarket. However, the development of the infrastructure for the production and commercialisation of a more versatile, healthier cereal product basis is one of the several requirements needed to deal with the complex issue of providing suitable food for people suffering from cereal-based health disorders. The legislative environment must be altered to allow the production/commercialisation of products with low/altered gluten content, defining thresholds of certain key components, giving instructions of analytical procedures, labelling, regulating and policing the very often misleading media-based marketing, etc. But the ultimate need is the better communication, collaboration among plant science, medical research and practice, as well as grain industry and the permanent, honest and up-to-date promotion of the customer.

REFERENCES

Abdel-Aal, E.-S.M., Hucl, P., Sosulski, F.W., 1995. Compositional and nutritional characteristics of spring einkorn and spelt wheats. Cereal Chem. 72, 621–624.

Abdel-Aal, E.S.M., Sosulski, F.W., Hucl, P., 1998. Origins, characteristics and potentials of ancient wheats. Cereal Food World 43, 708–715.

Abdel-Aal, E.-S.M., Young, J.C., Wood, P.J., Rabalski, I., Hucl, P., Falk, D., Fregeau-Reid, J., 2002. Einkorn: a potential candidate for developing high lutein wheat. Cereal Chem. 79, 455–457.

Abugoch, L.E., 2009. Quinoa (*Chenopodium quinoa* Willd.): composition, chemistry, nutritional and functional properties. Adv. Food Nutr. Res. 58, 1–31.

Ahamed, N., Singhal, R., Kulkarni, P., Pal, M., 1996. Physicochemical and functional properties of *Chenopodium quinoa* starch. Carbohydr. Polym. 31, 99–103.

Alvarez-Jubete, L., Arendt, E.K., Gallagher, E., 2010. Nutritive value of pseudocereals and their increasing use as functional gluten-free ingredients. Trends in Food Sci. Technol. 21, 106–113.

Alvarez-Jubete, L., Holse, M., Hansen, A., Arendt, E.K., Gallagher, E., 2009. Impact of baking on vitamin E content of pseudocereals amaranth, quinoa, and buckwheat. Cereal Chem. 86, 511–515.

Akagawa, M., Handoyo, T., Ishii, T., Kumazawa, S., Morita, N., Suyamam, K., 2007. Proteomic analysis of wheat flour allergens. J. Agric. Food Chem. 55, 6863–6870.

Altenbach, S.B., Tanaka, C.T., Allen, P.V., 2014a. Quantitative proteomic analysis of wheat grain proteins reveals differential effects of silencing of omega-5 gliadin genes in transgenic lines. J. Cereal Sci. 59, 118–125.

Altenbach, S.B., Tanaka, C.T., Seabourn, B.W., 2014b. Silencing of omega-5 gliadins in transgenic wheat eliminates a major source of environmental variability and improves dough mixing properties of flour. BMC Plant Biol. 14 (1), 1–13.

Amano, M., Ogawa, H., Kojima, K., Kamidaira, T., Suetsugu, S., Yoshihama, M., Satoh, T., Samejima, T., Matsumoto, I., 1998. Identification of the major allergens in wheat floor responsible for baker's asthma. Biochem. J. 330, 1229–1234.

Anderson, J.A., 1986. The establishment of common language concerning adverse reactions to foods and food additives. J. Allergy Clin. Immunol. 78, 140–144.

Anderson, R.P., Degano, P., Godkin, A.J., Jewell, D.P., Hill, A.V.S., 2000. In vivo antigen challenge in celiac disease identifies a single transglutaminase-modified peptid as the dominant A gliadin T-cell epitope. Nature (Med) 6, 337–342.

Anderson, R.P., Wieser, H., 2006. Medical applications of gluten-composition knowledge. In: Wrigley, C.W., Békés, F., Bushuk, W. (Eds.), Gliadin and Glutenin. The Unique Balance of Wheat Quality. AACCI Press, St Paul, Min., USA, pp. 387–409.

Arendt, E.K., Dal Bello, F., 2008. Gluten Free Cereal Products and Beverages. Academic press, San Diego, USA.

Armentia, A., Martín, S., Araceli Diaz-Perales, A., Arantxa Palacín, A., Tordesillas, L., Herrero, M., Martín-Armentia, M., 2012. A possible hypoallergenic cereal in wheat food allergy and baker's asthma. Am. J. Plant Sci. 3, 1779–1781.

Atwell, W.A., Patrick, B.M., Johnson, L.A., Glass, R.W., 1983. Characterization of quinoa starch. Cereal Chem. 60, 9–11.

Australia New Zealand Food Standards Code, 2014. Standard 2.9.5 Food for Special Medical Purposes, vol. 30, Federal register legislative instruments F2014C01202.

Avanza, M.V., Puppo, M.C., Anon, M.C., 2005. Rheological characterisation of amaranth protein gels. Food Hydrocolloids 19, 889–898.

Becker, R., 1994. Amaranth oil: composition, processing, and nutritional qualities. In: Paredes-Lopez, O. (Ed.), Amaranth—Biology, Chemistry, and Technology. CRC Press, London, pp. 133–142.

Bejosano, F.P., Corke, H., 1999. Properties of protein concentrates and hydrolysates from amaranthus and buckwheat. In: Proceedings of Sixth Symposium on Renewable Resources for the Chemical Industry, 23–25 March, 1999 Bonn, pp. 53.

Békés, F., 2015. Health related quality attributes of wheat. In: Proceedings of Third Conference of Cereal Biotechnology and Breeding/CBB3, Pauk, J. (Ed.). Academiai Kiado, Budapest, Hungary, pp. 1.

Belton, P., Taylor, J., 2002. Pseudocereals and less common cereals. Grain properties and utilization potential. Springer, Berlin, Heidelberg.

Berghofer, E., Schoenlechner, R., 2002. Grain amaranth. In: Belton, P.S., Taylor, J.R.N. (Eds.), Pseudocereals and less common cereals. grain properties and utilization potential. Springer-Verlag, Berlin Heidelberg, pp. 219–253.

Bilgicli, N., İbanoglu, S., 2015. Effect of pseudo cereal flours on some physical, chemical and sensory properties of bread. J. Food Sci. Technol. 52, 7525–7527.

Biesiekierski, J.R., Rosella, O., Rose, R., Liels, K., Barrett, J.S., Shepherd, S.J., Gibson, P.R., Muir, J.G., 2011. Quantification of fructans, galacto-oligosacharides and other short-chain carbohydrates in processed grains and cereals. J. Hum. Nutr. Diet 24, 154–176.

Blatter, R.H.E., Jacomet, S., Schlumbaum, S., 2004. About the origin of European spelt (*Triticum spelta* L.): allelic differentiation of the HMW Glutenin B1-1 and A1-2 subunit genes. Theor. Appl. Genet. 108, 360–367.

Bolontrade, A.J., Scilingo, A.A., Anon, M.C., 2016. Amaranth proteins foaming properties: film rheology and foam stability—Part 2. Colloids Surf. B Biointerfaces 141, 643–650.

Bonafaccia, G., Galli, V., Francisci, R., Mair, V., Skrabanja, V., Kreft, I., 2002. Characteristics of spelt wheat products and nutritional value of spelt wheat-based bread. Food Chem. 68, 437–441.

Bonafaccia, G., Gambelli, L., Fabjan, N., Kreft, I., 2003. Trace elements in flour and bran from common and tartary buckwheat. Food Chem. 80, 1–5.

Borghi, B., Castagna, R., Corbellini, M., Heun, M., Salamini, F., 1997. Breadmaking quality of einkorn wheat (*Triticum monococcum* ssp. monococcum). Cereal Chem. 73, 208–214.

Braly, J., Hoggan, R., 2002. Dangerous grains. Why gluten cereal grains may be hazardous to your health. Penguin Group, New York, N.Y., USA.

Branchi, F., Ferretti, F., Norsa, L., Roncoroni, L., Conte, D., Bardella, M.T., Elli, L., 2015. Management of nonceliac gluten sensitivity by gastroenterology specialists: data from an italian survey. BioMed Res. Int. 2015, 6.

Breen, M., Li, D., Dunn, D.S., Bekes, F., Kong, X., Zhang, J., Jia, J., Wicker, T., Mago, R., Ma, W., Bellgard, M., Appels, R., 2010. Wheat beta-expansin (EXPB11) genes: identification of the expressed gene on chromosome 3BS carrying a pollen allergen domain. BMC Plant Biol. 10, 99, (http://www.biomedcentral.com/1471-2229/10/99).

Brend, Y., Galili, L., Badani, H., Hovav, R., Galili, S., 2012. Total phenolic content and antioxidant activity of red and yellow quinoa (*Chenopodium quinoa* Willd.) seeds as affected by baking and cooking conditions. Food Nutr. Sci. 3, 1150–1155.

Bruins, I.D., Slot, I.D., Bremer, M.G.E.G., Fels-Klerx, I., Hamer, R.J., 2015. Evaluating the performance of gluten ELISA test kits: the numbers do not tell the tale. Cereal Chem. 92, 513–521.

Bugyi, Z., Török, K., Hajas, L., Adonyi, Z., Popping, B., Tömösközi, S., 2013. Comparative study of commercially available gluten ELISA kits using an incurred reference material. Qual. Assur. Saf. Crop. Foods 5, 79–87.

Cai, Y.Z., Corke, H., Wang, D., Li, W.D., 2016. Buckwheat: overview. Wrigley, C., Corke, H., Seetharaman, K., Faubion, J. (Eds.), Encyclopedia of Food Grains, vol. 1, second ed. Elsevier Ltd, Oxford, UK, p. 307.

Cai, Y., Sun, M., Corke, H., 2003. Antioxidant activity of betalains from plants of the amaranthaceae. J. Agric. Food Chem. 51, 2288–2294.

Campbell, K.G., 1997. Spelt: agronomy, genetics and breeding. Plant Breed. Rev. 15, 187–213.

Catassi, C., Fabiani, E., Iacono, G., D'Agate, C., Francavilla, R., Biagi, F., Volta, U., Accomando, S., Picarelli, A., Vitis, I., de Pianelli, G., Gesuita, R., Carle, F., Mandolesi, A., Bearzi, I., Fasano A., 2007. A prospective, double-blind, placebo-controlled trial to establish a safe gluten threshold for patients with celiac disease. Am. J. Clin. Nutr. 85, 160–166.

Catassi, C., Fasano, A., 2008. Celiac disease. In: Arendt, E.K., Dal Bello, F. (Eds.), Gluten-free cereal products and beverages. Academic Press, San Diego, USA, pp. 1–28.

Catassi, C., Bai, J.C., Bonaz, B., Bouma, G., Calabro, A., Carroccio, A., 2013. Non-celiac gluten sensitivity: the new frontier of gluten related disorders. Nutrients 5, 3839–3853.

Coda, R., Rizzello, C.G., Gobbetti, M., 2010. Use of sourdough fermentation and pseudo-cereals and leguminous flours for the making of a functional bread enriched of γ-aminobutyric acid (GABA). Int. J. Food Microbiol. 137, 236–245.

Codex Alimentarius Commission, 1999. Recommended Methods of Analysis and Sampling. Codex Stan 234.

Codex Alimentarius Commission, 1979. Standard for Foods for Special Dietary Use for Persons Intolerant to Gluten. Codex Stan 118.

Condés, M.C., Anon, M.C., Mauri, A.N., 2013. Amaranth protein films from thermally treated proteins. J. Food Eng. 119, 573–579.

Corbellini, M., Empillli, S., Vacdno, P., Brandolini, A., Borghi, I., Heun, J., Salamini, F., 1999. Einkorn characterization for bread and cookie production in relation to protein subunit composition. Cereal Chem. 76, 727–733.

Corke, H., Cai, Y.Z., Wu, H.X., 2016. Amaranth: overview. Wrigley, C., Corke, H., Seetharaman, K., Faubion, J. (Eds.), Encyclopedia of Food Grains, vol. 1, second ed. Elsevier Ltd, Oxford, UK, p. 287.

Cubadda, R., Marconi, E., 2002. Spelt wheat. In: Belton, P., Taylor, J. (Eds.), Pseudocereals and less common cereals. Grain properties and utilization potential. Springer, Berlin, Heidelberg, pp. 153–177.

Dallagnol, A.M., Pescuma, M., Rollán, G., Torino, M.I., de Valdez, G.F., 2015. Optimization of lactic ferment with quinoa flour as bio-preservative alternative for packed bread. Appl. Microbiol. Biotechnol. 99, 3839–3849.

Davis, W., 2011. Wheat Belly: Loose the Wheat and Find Your Path Back to Health. Rodale Books Inc., New York.

D'Egidio, M.G., Vallega, V., 1994. Bread baking and dough mixing quality of diploid wheat *Triticum monococcum* L. Italian. Food Beverage Technol. 4, 6–9.

De Vincenzi, M., Luchetti, R., Dal Belin Peruffo, A., Curioni, A., Pogna, N.E., Gasbarrini, G., 1996. In vitro assessment of acetic-acid-soluble proteins (glutenin) toxicity in celiac disease. J. Biochem. Toxicol. 11, 205–210.

Drzewiecki, J., Delgado-Licon, E., Haruenkit, R., Pawlezik, E., Martin-Belloso, O., Park, Y.S., Jung, S.T., Trakhtenberg, S., Gorinstein, S., 2003. Identification and differences of total proteins and their soluble fractions in some pseudocereals based on electrophoretic pattern. J. Agric. Food Chem. 51, 7798–7804.

Dziedzic, K., Górecka, D., Kucharska, M., Przybylska, B., 2012. Influence of technological process during buckwheat groats production on dietary fibre content and sorption of bile acids. Food Res. Int. 47, 279–283.

Elía, M., Moralejo, M., Rodríguez-Quijano, M., Molina-Cano, J.-L., 2004. Spanish spelt: a separate gene pool within the spelt germplasm. Plant Breed. 123, 297–299.

Escarnot, E., Jacquemin, J.-M., Agneessens, R., Paquot, M., 2012. Comparative study of the content and profiles of macronutrients in spelt and wheat, a review. Biotechnol. Agron. Soc. Environ. 16, 243–256.

FAOSTAT, 2014: FAO Statistics Division 2013, Available from: http://faostat.fao.org/

Faris, J.D., 2014. Wheat domestication: key to agricultural revolutions past and future. In: Tuberosa, R. (Ed.), Genomics of Plant Genetic Resources. Springer Science + Business Media, Dordrecht, Germany, pp. 439–464, Chapter 18.

Fidantsi, A., Doxastakis, G., 2001. Emulsifying and foaming properties of amaranth seed protein isolates. Colloids Surf. B Biointerfaces 21, 119–124.

Fletcher, R.J., 2016. Pseudocereals: overview. Wrigley, C., Corke, H., Seetharaman, K., Faubion, J. (Eds.), Encyclopedia of Food Grains, 1, second ed. Elsevier Ltd, Oxford, UK, p. 274.

Ford, R., 2008. The Gluten Syndrome. Is wheat causing you harm? RRS Global LT, Christchurch, New Zealand.

Food Allergen Labeling and Consumer Protection Act of 2004 (Public Law 108-282, Title II), U.S. Food and Drug Administration.

Food and Drug Administration, 2013. Food labeling; gluten-free labeling of foods. Fed. Reg. 78 (150), 47154–47179.

FoodDrinkEurope, 2013. Guidance on food allergen management for food manufacturers. http://www.fooddrinkeurope.eu/uploads/press-releases_documents/temp_file_FINAL_Allergen_A4_web1.pdf

Gabrovska, D., Rysova, J., Filova, V., Plicka, J., Cuitra, P., Kubik, M., Barsova, S., 2006. Gluten determination by gliadin enzyme-linked immunosorbent assay kit: interlaboratory study. J. AOAC Int. 89, 154–160.

Galli, G., Francisci, V., Mair, R., Skrabanja, V., Kreft, I., 2000. Characteristics of spelt wheat products and nutritional value of spelt wheat-based bread. Food Chem. 68, 437–441.

Galova, Z., Knoblochova, H., 2001. Biochemical characterstics of five spelt wheat cultivars (*Triticum spelta* L.). Acta Fytotechn. Zootechn. 4, 85–86.

Galterio, G., Cardarilli, D., Codianni, P., Acquistucci, R., 2001. Evaluation of chemical and technological characteristics of new lines of *Triticum turgidum* ssp *dicoccum*. Nahrung/Food 45, 263–266.

Gendel, S.M., 2012. Comparison of international food allergen labeling regulations. Regul. Toxicol. Pharmacol. 63, 279–285.

Geng, C., Westphal, D., Yeung, J.M., 2008. Detection of gluten by commercial test kits: effects of food matrices and extraction procedures. In: Food Contaminants ACS Symposium Series, vol. 1001, pp. 462–475, Chapter 30.

Gibson, P.R., Newnham, R., Barrett, J.S., Shepherd, S.J., Muir, J.G., 2007. Review article: fructose malabsorption and the bigger picture. Aliment. Pharmacol. Ther. 25, 349–363.

Gibson, P.R., Varney, J., Malakar, S., Muir, J.G., 2015. Food components and irritable bowel syndrome. Gastroenterology 148, 1158–1174.

Gomaa, A., Boye, J.I., 2013. Impact of thermal processing time and cookie size on the detection of casein, egg, gluten and soy allergens in food. Food Res. Int. 52, 483–489.

Gomez-Becerra, H.F., Erdem, H., Yazici, A., Tutus, Y., Torun, B., Ozturk, L., Cakmak, I., 2010. Grain concentrations of protein and mineral nutrients in a large collection of spelt wheat grown under different environments. J. Cereal Sci. 52, 342–349.

Gorinstein, S., Pawelzik, E., Delgado-Licon, E., Haruenkit, R., Weisz, M., Trakhtenberg, S., 2002. Characterisation of pseudocereal and cereal proteins by protein and amino acid analyses. J. Sci. Food Agric. 82, 886–891.

Grausgruber, H., Oberforster, M., Ghambashidze, G., Ruckenbauer, P., 2005. Yield and agronomic traits of khorasan wheat (*Triticum turanicum* Jakubz.). Field Crops Res. 91, 319–327.

Green, P.H.R., Cellier, M.D.B., 2007. Celiac disease. New England J. Med. 357, 1731–1743.

Halmos, E.P., Power, V.A., Shepherd, S.J., Gibson, P.R., Muir, J.G., 2014. A diet low in FODMAPs reduces symptoms of irritable bowel syndrome. Gastroenterology 146, 67–75.

Halmos, E.P., Christophersen, C.T., Bird, A.R., Shepherd, S.J., Gibson, P.R., Muir, J.G., 2015. Diets that differ in their FODMAP content alter the colonic luminal microenvironment. Gut 64, 93–100.

Haraszi, R., Chassaigne, H., Maquet, A., et al., 2011. Analytical methods for detection of gluten in food—method developments in support to the legislations on labelling of foodstuffs. J. AOAC Int. 94, 1006–1025.

Health Canada's Position on Gluten-free Claims, 2012. Available from: http://www.hc-sc.gc.ca/fn-an/alt_formats/pdf/securit/allerg/cel-coe/gluten-position-eng.pdf

Houben, A., Götz, H., Mitzscherling, M., Becker, T., 2010. Modfication of the rheological behavior of amaranth (*Amaranthus hypochondriacus*) dough. J. Cereal Sci. 51, 350–356.

Hunjai, C., Wansoo, K., Malshik, S., 2004. Properties of Korean amaranth starch compared to waxy millet and waxy sorghum starches. Starch/Staerke 56, 469–477.

Iglesias-Puig, E., Monedero, V., Haros, M., 2015. Bread with whole quinoa flour and bifidobacterial phytases improve contribution to dietary mineral intake and their bioavailability without substantial loss of bread quality. LWT-Food Sci. Technol. 60, 71–77.

Islam, S., Ma, W., Yan, G., Békés, F., Appels, R., 2011. Modifying processing and health attributes of wheat bread through changes in composition, genetics and breeding. In: Cauvain, S.P., Tran, B. (Eds.), Bread Making. Improving Quality. second ed. Woodhead Publishing Limited, Cambridge, pp. 259–296.

Jargon, J., 2014. The gluten-free craze: is it healthy? Wall St. J 14 (06), 9–11.

Jekle, M., Houben, A., Mitzscherling, M., Becker, T., 2010. Effects of selected lactic acid bacteria on the characteristics of amaranth sourdough. J. Sci. Food Agric. 90, 2326–2332.

Juhász, A., Gell, Gy., Békés, F., Balázs, E., 2012. The epitopes in wheat proteins for defining toxic units relevant to human health. Funct. Integr. Genomics 12, 585–598.

Juhász, A., Haraszi, R., Maulis, Cs., 2015. ProPepper: a curated database for identification and analysis of peptide and immune-responsive epitope composition of cereal grain protein families. Database 2015, 1–16.

Junker, Y., Zeissig, S., Kim, S.J., Barisani, D., Wieser, H., Leffler, D.A., 2012. Wheat amylase trypsin inhibitors drive intestinal inflammation via activation of toll-like receptor 4. J. Exp. Med. 209, 2395–2408.

Kasarda, D.D., 2013. Can an increase in celiac disease be attributed to an increase in the gluten content of wheat as a consequence of wheat breeding? J. Agric. Food Chem. 61, 1155 1159.

Kerbach, S., Alldrick, A.J., Crevel, R.W.R., Dömötör, L., DunnGalvin, A., Mills, E.N.C., Pfaff, S., Poms, R.E., Popping, B., Tömösközi, S., 2009. Managing food allergens in the food supply chain—viewed from different stakeholder perspectives. Qual. Assur. Saf. Crop Foods 1, 50–60.

Kovacs, E.T., Maraz-Szabo, L., Varga, J., 2001. Examination of the protein-emulsifier-carbohydrate interactions in amaranth based pasta products. Acta Alimentaria 30, 173–187.

Koziol, M.J., 1991. Afrosimetric estimation of threshold saponin concentration for bitterness in quinoa (*Chenopodium quinoa* Willd). J. Sci. Food Agric. 54, 211–219.

Lampidonis, A., Siragakis, G., 2014. Allergens and food industry: customer and legislation demands. In: Siragakis, G., Kizis, D. (Eds.), Food Allergen Testing. John Wiley &Sons, Chichester, UK, pp. 197–212.

La Mothe, L., Srichuwong, S., Reuhs, B., Hamaker, B., 2015. Quinoa (*Chenopodium quinoa* W.) and amaranth (*Amaranthus caudatus* L.) provide dietary fibres high in pectic substances and xyloglucans. Food Chem. 167, 490–496.

Lauriere, M., Pecquet, C., Bouchez-Mahiout, I., Snegaroff, J., Bayrou, O., Raison-Peyron, N., Vigan, M., 2006. Hydrolyzed wheat proteins present in cosmetics can induce immediate hypersensitivities. Contact Dermatitis 54, 283–289.

LeClerc, J.A., Bailey, L.H., Vessling, H.L., 1978. Milling and baking tests of einkorn, emmer, spelt and polish wheat. J. Amer. Soc. Agron. 10, 215–217.

Leon-Camacho, M., Garcia-Gonzalez, D.L., Aparicio, R., 2001. A detailed and comprehensive study of amaranth (*Amaranthus cruentus* L.) oil fatty profile. Eur. Food Res. Technol. 213, 349–355.

Lindeboom, N., Chang, P.R., Falk, K.C., Tyler, R.T., 2005. Characteristics of starch from eight quinoa lines. Cereal Chem. 82, 216–222.

Linh, N.T.N., Khoa, D.V.A., Halas, V., 2014. Buckwheat as valuable feed and food resource. Nova J. Med. Biol. Sci. 3, 1–6.

Londono, D.M., van't Westende, W.P.C., Goryunova, S., Salentijn, E.M.J., van den Broeck, H.C., van der Meer, I.M., Visser, R.G.F., Gilissen, L.J.W.J., Smulders, M.J.M., 2013. Avenin diversity analysis of the genus *Avena* (oat). Relevance for people with celiac disease. J. Cereal Sci. 58, 170–177.

Lorenz, K., 1990. Quinoa *(Chenopodium quinoa)* starch—physico-chemical properties and functional characteristics. Starch/Staerke 42, 81–86.

Luje, H., Muller, B., Laustsen, A.M., Hansen, A.E., 2003. Chemical composition, functional properties and sensory profiling of einkorn (*Triticum monococcum* L.). J. Cereal Sci. 37, 231–240.

Madsen, C.B., Hattersley, S., Allen, K.J., Beyer, K., Chan, C.-H., Godefroy, S.B., Hodgson, R., Mills, E.N.C., Munoz-Furlong, A., Schnadt, S., Ward, R., Wickman, M., Crevel, R., 2011. Can we define a tolerable level of risk in food allergy? Clin. Exp. Allergy 42, 30–37.

Makinen, O.E., Zannini, E., Arendt, E.K., 2013. Germination of oat and quinoa and evaluation of the malts as gluten free baking ingredients. Plant Foods Hum. Nutr. 68, 90–95.

Mariotti, M., Iametti, S., Cappa, C., Rasmussen, P., Lucisano, M., 2011. Characterisation of gluten-free pasta through conventional and innovative methods: evaluation of the uncooked products. J. Cereal Sci. 53, 319–327.

Marti, A., Pagani, M.A., Seetharaman, K., 2011. Understanding starch organisation in gluten-free pasta from rice flour. Carbohydr. Polym. 84, 1069–1074.

Marti, A., Pagani, M.A., 2013. What can play the role of gluten in gluten free pasta? Trends Food Sci. Technol. 31, 63–71.

Mastromatteo, M., Chillo, S., Iannetti, M., Civica, V., Del Nobile, M.A., 2011. Formulation optimisation of gluten-free functional spaghetti based on quinoa, maize and soy flours. Int. J. Food Sci. Technol. 46, 1201–1208.

Matuz, J., Bartok, T., Morocz-Salomon, K., Bona, L., 2000. Structure and potential allergenic character of cereal proteins I. Protein content and amino acid composition. Cereal Res. Comm. 28, 263–270.

McCleary, B.V., Blakeney, A.B., 1999. Measurement of inulin and oligofructan. Cereal Food World 44, 398–406.

Mills, E.N.C., Jenkins, J.A., Alcocer, M.J.C., Shewry, P.R., 2004. Structural, biological, and evolutionary relationships of plant food allergens sensitizing via the gastrointestinal tract. Crit. Rev. Food Sci. Nutr. 44, 379–407.

Miranda, M., Vega-Gálvez, A., Martinez, E., López, J., Rodríguez, M.J., Henríquez, K., Fuentes, F., 2012. Genetic diversity and comparison of physicochemical and nutritional characteristics of six quinoa (*Chenopodium quinoa* Willd.) genotypes cultivated in Chile. Food Sci. Technol. 32, 835–843.

Missbach, B., Schwingshackl, L., Billmann, A., Mystek, A., Hickelsberger, M., Bauer, G., Konig, J., 2015. Gluten-free food database: the nutritional quality and cost of packaged gluten-free foods. Peer J. 3, e1337.

Molberg, O., Uhlen, A.K., Jensen, T., Flaete, N.S., Fleckenstein, B., Arentz-Hansen, H., Raki, M., Lundin, K.E., Sollid, L.M., 2005. Mapping of gluten T-cell epitopes in the bread wheat ancestors: implications for celiac disease. Gastroenterology 128, 393–401.

Montoya-Rodríguez, A., Gómez-Favela, M.A., Reyes-Moreno, C., Milán-Carrillo, J., González de Mejía, E., 2015. Identification of bioactive peptide sequences from amaranth (*Amaranthus hypochondriacus*) seed proteins and their potential role in the prevention of chronic diseases. Comp. Rev. Food Sci. Food Safety 14, 139–158.

Morishita, T., Yamaguchi, H., Degi, K., 2007. The contribution of polyphenols to antioxidative activity in common buckwheat and tartary buckwheat grain. Plant Prod. Sci. 10, 99–104.

Moroni, A.V., Dal Bello, F., Zannini, E., Arendt, E.K., 2011. Impact of sourdough on buckwheat flour, batter and bread: Biochemical, rheological and textural insights. J. Cereal Sci. 54, 195–202.

Muchova, Z., Cukova, L., Mucha, R., 2000. Seed protein fractions of amaranth (*Amaranthus sp*). Rostl. Vyroba 46, 331–336.

Muir, J.G., Rose, R., Rosella, O., Liels, K., Barrett, J.S., Shepherd, S.J., Gibson, P.R., 2009. Measurement of short-chain carbohydrates in common australian vegetables and fruits by high-performance liquid chromatography (HPLC). J. Agric. Food Chem. 57, 554–565.

Muir, J.G., Shepherd, S.J., Rosella, O., Rose, R., Gibson, P.R., 2007. Fructan and free fructose content of common Australian vegetables and fruit. J. Agric. Food. Chem. 55, 6619–6627.

Neeson, R., Evans, J., Burnett, V., Luckett, D., Wellings, C., Taylor, H., Raman, H., Van Meeuwen, E., Bowden, P., 2008. Optimising the quality and yield of spelt under organic production in SE Australia. In: Proceedings of Fourteenth Australian Society of Agronomy Conference, ASAC, Adelaide, SA, Australia, pp. 24–31.

Nesbitt, M., Samuel, D., 1996. From staple crop to extinction? The archaeology and history of the hulled wheats. In: Paludosi, S., Hammer, K., Heller, J. (Eds.), Hulled Wheats, Proceedings of the First International Workshop on Hulled Wheats, Castelvechio Pascoli, Roma, Italy, pp. 1–21.

Omary, M.B., Fong, C., Rothschild, J., Finney, P., 2012. Effects of germination on the nutritional profile of gluten-free cereals and pseudocereals: a review. Cereal Chem. 89, 1–14.

Onishi, I., Hongo, A., Sasakuma, T., Kawahara, T., Kato, K., Miura, H., 2006. Variation and segregation for rachis fragility in spelt wheat, *Triticum spelta* L. Genet. Resour. Crop Evol. 53, 985–992.

Oshodi, A.A., Ogungbenle, H.N., Oladimeji, M.O., 1999. Chemical composition, nutritionally valuable minerals and functional properties of benniseed (*Sesamum radiatum*), pearl millet (*Pennisum typhoides*) and quinoa (*Chenopodium quinoa*) flours. Int. J. Food Sci. Nutr. 50, 325–331.

Pasco, J.A., Nicholson, G.C., Kotowicz, M.A., 2012. Cohort profile: geelong osteoporosis study. Int. J. Epidemiol. 41, 1565–1575.

Pasko, P., Barton, H., Zagrodzki, P., Gorinstein, S., Folta, M., Zachwieja, S., 2009. Anthocyanins, total polyphenols and antioxidant activity in amaranth and quinoa seeds and sprouts during their growth. Food Chem. 115, 994–998.

Poutanen, K., Sozer, N., Della-Valle, G., 2014. How can technology help to deliver more of grain in cereal foods for a healthy diet? J. Cereal Sci. 59, 327–336.

Prakash, D., Pal, M., 1998. *Chenopodium*: seed protein, fractionation and amino acid composition. Int. J. Food Sci. Nutr. 49, 271–275.

Pulido, O.M., Gillespie, Z., Zarkadas, M., Dubois, S., Vavasour, E., Rashid, M., Switzer, C., Godefroy, S.B., 2009. Introduction of oats in the diet of individuals with celiac disease: a systematic review. Adv. Food Nutr. Res. 57, 235–825.

Qian, J.Y., Rayas-Duarte, P., Grant, L., 1998. Partial characterization of buckwheat (*Fagopyrum esculentum*) starch. Cereal Chem. 75, 365–373.

Qian, J.Y., Kuhn, M., 1999. Characterization of *Amaranthus cruentus* and *Chenopodium quinoa* starch. Starch/Staerke 51, 116–120.

Raman, H., Rahman, R., Luckett, D., Raman, R., Békés, F., Láng, L., Bedő, Z., 2008. Characterisation of genetic variation for aluminium resistance and polyphenol oxidase activity in genebank accessions of spelt wheat. Breed. Sci. 59, 373–381.

Regulation (EU) No 1169/2011 of the European Parlament and the council on on the provision of food information to consumers, Official Journal of the European Union.

Regulation (EU) No 609 2013. European Parlament and the on food intended for infants and young children, food for special medical purposes, and total diet replacement for weight control.

Repo-Carrasco, R., Espinoza, C., Jacobsen, S.-E., 2003. Nutritional value and use of the Andean crops quinoa (*Chenopodium quinoa*) and kañiwa (*Chenopodium pallidicaule*). Food Rev. Int. 19, 179–189.

Repo-Carrasco-Valencia, R., Astuahuaman, L., 2011. Quinoa (*Chenopodium quinoa* Willd.) as a source of dietary fiber and other functional components. Food Sci. Technol. 31, 225–230.

Repo-Carrasco-Valencia, R., Peña, R.J., Kallio, H., Salminen, S., 2009. Dietary fiber and other functional components in two varieties of crude and extruded kiwicha (*Amaranthus caudatus*). J. Cereal Sci. 49, 219–224.

Repo-Carrasco, R., 1992. Andean Crops and Infant Nourishment. Report B 25. Institute of Development Studies, University of Helsinki, Finland.

Rombouts, I., Lagrain,B., Brunnbauer, M. et al., 2013. Improved identification of wheat gluten proteins through alkylation of cysteine residues and peptide-based mass spectrometry. Sci. Rep., 3, Article number: 2279.

Ruales, J., Nair, B.M., 1993. Saponins, phytic acid, tannins and protease inhibitors in quinoa, (*Chenopodium qumoa*, Willd) seeds. Food Chem. 48, 137–143.

Ruibal-Menieta, N.L., Delacroix, D.L., Mignelot, E., Picke, J.M., Marques, C., Meurens, M., Delzenne, N.M., Larondelle, Y., 2005. Spelt (*Triticum aestivum* ssp.spelta) as a source of breadmaking flours and bran naturally enriched in oleic acid and minerals but not phytic acid. J. Agric. Food Chem. 53, 2751–2759.

Rumbo, M., Chirdo, F.G., Fossati, C.A., Anon, M.C., 2001. Analysis of the effects of heat treatment on gliadin immunochemical quantification using a panel of anti-prolamin antibodies. J. Agric. Food Chem. 49, 5719–5726.

Salcedo-Chavez, B., Osuna-Castro, J.A., Guevara-Lara, F., Dominguez-Dominguez, J., RParedes-Lopez, O., 2002. Optimisation of the isoelectric presicpitation method to obtain protein isolates from amaranth (*Amaranthus cruentus*) seeds. J. Agric. Food Chem. 50, 6515–6520.

Salplachta, J., Marchetti, M., Chmelik, J., et al., 2005. A new approach in proteomics of wheat gluten: combining chymotrypsin cleavage and matrix-assisted laser desorption/ionization quadrupole ion trap reflectron tandem mass spectrometry. Rapid. Commun. Mass Spectrom. 19, 2725–2728.

Sanz-Penella, J.M., Tamayo-Ramos, J.A., 2012. Application of bifidobacteria as starter culture in whole wheat sourdough breadmaking. Food Bioprocess Technol. 5, 2370–2380.

Sanz-Penella, J.M., Wronkowska, M., Soral-Śmietana, M., Haros, M., 2013. Effect of whole amaranth flour on bread properties and nutritive value. LWT-Food Sci. Technol. 50, 679–685.

Sapone, A., Bai, J.C., Ciacci, C., Dolinsek, J., Green, P.H.R., Hadjivassiliou, M., Kaukinen, K., Rostami, K., Sanders, D.S., Schumann, M., Ullrich, R., Villaltam, D., Voltam, U., Catassi, C., Fasano, A., 2012. Spectrum of gluten-related disorders: consensus on new nomenclature and classification. BMC Med. 10, 13.

Scherf, K.A., Poms, R.E., 2015. Recent developments in analytical methods for tracing gluten. J. Cereal Sci. 61, 1–11.

Schoenlechner, R., Linsberger, G., Kaczyk, L., Berghofer, E., 2006. Production of gluten-free short dough biscuits from the pseudocereals amaranth, quinoa and buckwheat with common bean. Ernaehrung/Nutrition 30, 101–107.

Schoenlechner, R., Wendner, M., Siebenhandl-Ehn, S., Berghofer, E., 2010a. Pseudocereals as alternative sources for high folate content in staple foods. J. Cereal Sci. 52, 475–479.

Schoenlechner, R., Drausinger, J., Ottenschlaeger, V., Jurackova, K., Berghofer, E., 2010b. Functional properties of gluten-free pasta produced from amaranth, quinoa and buckwheat. Plant Foods Hum. Nutr. 65, 339–349.

Schuppan, D., Pickert, G., Ashfaq-Khan, M., Zevallos, V., 2015. Non-celiac wheat sensitivity: differential diagnosis, triggers and implications. Best Pract. Res. Clin. Gastroenterol. 29, 469–476.

Sealey-Voyksner, J.A., Khosla, C., Voyksner, R.D., et al., 2010. Novel aspects of quantitation of immunogenic wheat gluten peptides by liquid chromatography-mass spectrometry/mass spectrometry. J. Chromatogr. A 1217, 4167–4183.

Segura-Nieto, M., Shewry, P.R., Paredes-Lopez, O., 1999. Globulins of the pseudoereals: amaranth, quinoa, and buckwheat. In: Shewry, P.R., Casey, R. (Eds.), Seed Proteins. Kluwer Academic Publishers, Dordrecht, The Netherlands, pp. 453–475.

Shepherd, S.J., Gibson, P.R., 2006. Fructose malabsorption and symptoms of IBS: guidelines for effective dietary management. J. Am. Diet. Assoc. 106, 1631–1639.

Shepherd, S.J., Parker, F.C., Muir, J.G., Gibson, P.R., 2008. Dietary triggers of abdominal symptoms in patients with IBS: randomized placebo-controlled evidence. Clin. Gastroenterol. Hepatol. 6, 765–771.

Shewry, P.R., 2002. The major seed storage proteins of spelt wheat, sorgum, millets and pseudocereals. In: Belton, P., Taylor, J. (Eds.), Pseudocereals and Less Common Cereals. Grain Properties and Utilization Potential. Springer, Berlin, Heidelberg, pp. 1–24.

Siles, R.I., Hsieh, F.H., 2013. Allergy blood testing: a practical guide for clinicians. Cleve. Clin. J. Med. 78, 585–592.

Singh, H., MacRitchie, F., 2004. Changes in proteins induced by heating gluten dispersions at high temperature. J. Cereal Sci. 39, 297–301.

Skypala, I., 1999. Adverse food reactions—an emerging issue for adults. J. Am. Diet. Assoc. 111, 1877–1891.

Slot, I.D.B., Bremer, M.G.E.G., ven der Fels-Klerx, I., Hamer, R.J., 2015. Evaluating the performance of gluten ELISA test kits: the numbers do not tell the tale. Cereal Chem. 92, 513 521.

Sollid, L.M., Qiao, S.W., Andersonm, R.P., Gianfrani, C., Koning, F., 2012. Nomenclature and listing of celiac disease relevant gluten T-cell epitopes restricted by HLA-DQ molecules. Immunogenetics 64, 455–460.

Stagnari, F., Codianni, P., Pisante, M., 2008. Agronomic and kernel quality of ancient wheats grown in central and southern Italy. Cereal Res. Commun. 36, 313–326.

Steadman, K.J., Burgoon, M.S., Lewis, B.A., Edwardson, S.E., Obendorf, R.L., 2001. Buckwheat seed milling fractions: description, macronutrient composition and dietary fibre. J. Cereal Sci. 33, 271–278.

Stikic, R., Glamoclija, D., Demin, M., Vucelic-Radovic, B., Jovanovic, Z., Milojkovic-Opsenica, D., Jacobsen, S.E., Milovanovic, M., 2012. Agronomical and nutritional evaluation of quinoa seeds (*Chenopodium quinoa* Willd.) as an ingredient in bread formulations. J. Cereal Sci. 55, 132–138.

Suter, D.A.I., Békés, F., 2012. Wheat immunoreactivity. Australian Patent http://v3.espacenet.com/publicationDetails/ biblio?CC=HU&NR= AU2011000468

Tatham, A.S., Shewry, P.R., 2008. Allergy to wheat and related cereals. Clin. Exp. Allergy 38, 1712–1726.

The Threshold Working Group, Food and Drug Administration, 2006. Approaches to establish thresholds for major food allergens and for gluten in food, Available from: http://www.fda.gov/downloads/Food/Ingredients PackagingLabeling/UCM192048.pdf

Thanapornpoonpong, S.N., Vearasil, S., Pawelzik, E., Gorinstein, S., 2008. Influence of various nitrogen applications on protein and amino acid profiles of amaranth and quinoa. J. Agric. Food Chem. 56, 11464–11470.

Thompson, T., Mendez, E., 2008. Commercial assays to assess gluten content of gluten-free foods: why they are not created equal. J. Am. Diet. Assoc. 108, 1682–1687.

Tömösközi, S., Baracskai, I., Schoenlechner, R., Berghofer, E., Lasztity, R., 2009. Comparative study of composition and technological quality of amaranth I. Gross chemical composition, amino acid and mineral content. Acta Alimentaria 38, 341–347.

Tömösközi, S., Gyenge, L., Pelcéder, Á., Abonyi, T., Schönlechner, R., Lásztity, R., 2011. Effects of flour and protein preparations from amaranth and quinoa seeds on the rheological properties of wheat-flour dough and bread crumb. Czech J. Food Sci. 29, 109–116.

Tömösközi, S., Török, K., Bugyi, Zs., Hajas, L., 2014. Reference materials for food allergen testing. In: Siragakis, G., Kizis, D. (Eds.), Food Allergen Testing. John Wiley &Sons, Chichester, UK, pp. 215–233.

Török, K., Hajas, L., Bugyi, Zs., Balázs, G., Tömösközi, S., 2015a. Characterization of an incurred reference material: effects of food processing and matrix components on the analytical results. Acta Alimentaria Hungarica 44, 390–399.

Török, K., Hajas, L., Horváth, V., Schall, E., Bugyi, Zs., Kemény, S., Tömösközi, S., 2015b. Identification of the factors affecting the analytical results of food allergen ELISA methods. Eur. Food Res. Technol. 241, 127–136.

Tosi, E.A., Re, E.D., Masciarellim, R., Sanchez, H., Osella, C., de la Torre, M.A., 2002. Whole and defatted hyperproteic amaranth flours tested as wheat flour supplementation in mold breads. Lebensm. Wissensch. Technol. 35, 472–475.

Valdes, I., Garcia, E., Llorente, M., Mendez, E., 2003. Innovative approach to low-level gluten determination in foods using a novel sandwich enzyme-linked immunosorbent assay protocol. Eur. J. Gastroenterol. Hepatol. 15, 465–474.

Vallega, V., Waines, J.G., 1987. High molecular weight glutenin subunit variation in *Triticum turgidum* var. *dicoccum*. Theor. Appl. Genet. 74, 706–710.

Vega-Galvez, A.M., Miranda, J., Vergara, E., Uribe, L., Puente, X., Martinez, E.A., 2010. Nutrition facts and functional potential of quinoa (*Chenopodium quinoa* Willd.), an ancient Andean grain: a review. J. Sci. Food Agric. 90, 2541–2547.

Vogrincic, M., Timoracka, M., Melichacova, S., Vollmannova, A., Kreft, I., 2010. Degradation of rutin and polyphenols during the preparation of tartary buckwheat bread. J. Agric. Food Chem. 58, 4883–4887.

Vu, N.T., Chin, J., Pasco, J.A., Kovács, A., Wing, L.W., Békés, F., Suter, D.A.I., 2014. The prevalence of wheat and spelt sensitivity in a randomly selected Australian population. Cereal Res. Comm. 43, 97–107.

Wangen, S., 2009. Healthier Without Wheat A New Understanding of Wheat Allergies Celiac Disease and Non-Celiac Gluten Intolerance. Innate Health Publ Sieattle, WS, USA.

Wieser, H., 2008. Detection of gluten. In: Arendt, E.K., Dal Bello, F. (Eds.), Gluten-Free Cereal Products and Beverages. Academic Press, San Diego, USA, pp. 47–80.

Wieser, H., Koehler, P., 2009. Is the calculation of the gluten content by multiplying the prolamin content by a factor of 2 valid? Eur. Food Res. Technol. 229, 9–13.

Wilson, L.A., 2015. Grains as Mains. Modern Recipes Using Ancient Grains. Penguin Random House Company, London.

Wolter, A., Hager, A.-S., Zanini, E., Arendt, E.K., 2013. In vitro starch digestibility and predicted glycaemic indexes of buckwheat, oat, quinoa, sorghum, teff and commercial gluten-free bread. J. Cereal Sci. 58, 431–436.

Yoshimoto, Y., Egashira, T., Hanashiro, I., Ohinata, H., Takase, Y., Takeda, Y., 2004. Molecular structure and some physicochemical properties of buckwheat starches. Cereal Chem. 81, 515–520.

Zdunczyk, Z., Flis, M., Zielinski, H., Wroblewska, M., Antoszkiewicz, Z., Juskiewicz, J., 2006. In vitro antioxidant activities of barley, husked oat, naked oat, triticale, and buckwheat wastes and their influence on the growth and biomarkers of antioxidant status in rats. J. Agric. Food Chem. 54, 4168–4175.

Zuidmeer, L., Goldhahn, K., Rona, R.J., Gislason, D., Madsen, C., Summers, C., Sodergren, E., Dahlstrom, J., Lindner, T., Sigurdardottir, S.T., McBride, D., Keil, T., 2008. The prevalence of plant food allergies: a systematic review. J. Allergy Clin. Immunol. 121, 1210–1218.

Part III

Grain-Quality Analysis and Management

Chapter 15

Food Safety Aspects of Grain and Cereal Product Quality

Anton J. Alldrick
Campden BRI, Chipping Campden, United Kingdom

15.1 INTRODUCTION

At a basic level, food-safety management requires: identification of the hazard; factors (risk factors) which influence the frequency and degree to which the hazard occurs and; the processes necessary to reduce the risk of the hazard occurring. As with any other food stuff, hazards associated with cereals can be chemical, physical or microbiological in nature and can be introduced either through the effects of nature and/or the acts of man. Given that events at any one stage of the supply chain can have an impact on consumer health; international best practice (Schlundt et al., 2003) in respect of food-safety management considers the supply chain as a continuum. Recognising this through adoption of a field to table approach enables an understanding of both, where hazards can be introduced and the attendant risk factors.

For the purposes of this chapter the grain supply chain is described in a simplified form (Fig. 15.1). It begins with cultivation of the plant, followed by harvesting and storage of the grain. As required, the grain undergoes some form of primary processing, which may be as simple as removing foreign material or more sophisticated, for example mechanical reduction through milling. The end product (e.g. whole grains or flour) may then either be sold directly to the consumer or passed on to a second processor to be used as an ingredient in another food (e.g. flour in bread), before being sold.

In terms of successful risk management, consideration of standard BS ISO 31000:2009 (British Standards Institute, 2009) indicates that within a (food-) business context, risk management takes into account not only the aspects of the hazard and the probability of it occurring; but also all business operations and the mechanisms necessary to ensure that risk can be kept to an acceptable level. As will be discussed later in this chapter, this requires not only an appreciation of operational issues but also the potential for the introduction of hazards as a result of other (e.g. criminal) activities and the adoption

Cereal Grains. http://dx.doi.org/10.1016/B978-0-08-100719-8.00015-2

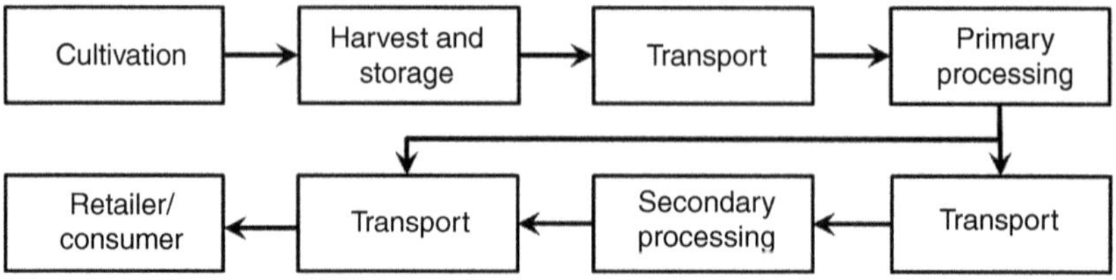

FIGURE 15.1 **'Field to table' overview of the production of grain-based foods.**

of appropriate food defense programs with particular regard to the application of 'CARVER + Shock' (US Food & Drug Administration, 2009) and TACCP principles (British Standards Institute, 2014).

The purpose of this chapter is therefore to consider the possible hazards associated with cereal-based foods, factors which contribute to their occurrence and the management processes to remove or mitigate the consequential risk. It uses a field to table approach to consider hazards, where they are introduced and the risk they present. An overview of food-safety issues associated with each stage of the supply chain is provided in Table 15.1. As will be discussed

TABLE 15.1 Types of Food-Safety Hazards Potentially Associated With Cereal-Based Food

Origin	Types	Examples	References
Inherent (produced in raw material)	Phytotoxins	Tannins in sorghum	Masha and Svanberg (1983)
Bioaccumulation	Heavy metals	Arsenic in rice and rice products	Meharg et al. (2008)
Agronomic and/or post-harvest practices	Fungal (mycotoxins)	Aflatoxicosis associated with contaminated maize	Azziz-Baumgartner et al. (2005)
Primary and secondary processing	Microbiological	Hepatitis contamination due to poor worker health screening	Schoenbaum et al. (1976)
	Pesticides	Presence of pesticides in excess of maximum residue limits due to cross contamination	Weeks (1967)
	Process-induced food toxicants	3-Monochloropropane diol; acrylamide	Velišek (2009); Svensson et al. (2003)
Criminal acts	Fraudulent addition	Melamine in wheat bran	
	Sabotage	Methamidophos contamination of frozen dumplings	British Broadcasting Corporation (2014)

subsequently, dependent on which stage of the supply chain is being considered, particular hazards have to be managed. Some of these reflect natural phenomena, while others are a consequence of agronomic/manufacturing practices or a result of criminal acts.

Given the different species of cereals cultivated and the diverse range of foods made from them, this chapter will focus principally on the chemical and microbiological hazards associated with the production of wheat and its subsequent conversion into flour and ultimately bread. Where appropriate, reference will also be made to other cereals and/or products to extend or emphasise a particular point.

15.2 THE ROLE OF AGRONOMY

The processes of cultivating and harvesting grain are described in more detail in Fig. 15.2. In terms of impact on food safety, both environmental phenomena and agronomic practices influence the nature of the hazard and the probability of it occurring. As discussed later in the chapter, this begins with varietal selection, which determines the crop's propensity to produce inherent toxicants (phytotoxins) as well as to bioaccumulate particular contaminants such as heavy metals. During cultivation, soil composition is a determinant in terms of bioaccumulation of heavy metals and climate affects the ability of mycotoxigenic moulds to successfully infect the crop. Agronomic practices can mitigate the risks presented by some of these hazards or can introduce further hazards, for example through misapplication of pesticides, while post-harvest activities in particular grain cleaning and drying strategies can influence the occurrence of phytotoxins associated with certain co-harvested seeds as well as mycotoxins due to the growth of particular mycotoxigenic fungi.

15.2.1 Phytotoxins

Many plants produce secondary metabolites which may be toxic to the consumer (phytotoxins), these include many staple crops. Examples include lectins in legumes, glucosinolates in brassicas and alkaloids in potatoes and tomatoes [reviewed in D'Mello et al. (1991)]. Although cereals are unusual in that for most of the human population, the levels of endogenous toxic materials produced by them are low, two compounds which merit further interest are phytate and tannins. In addition the issue of phytotoxins will be extended to consider

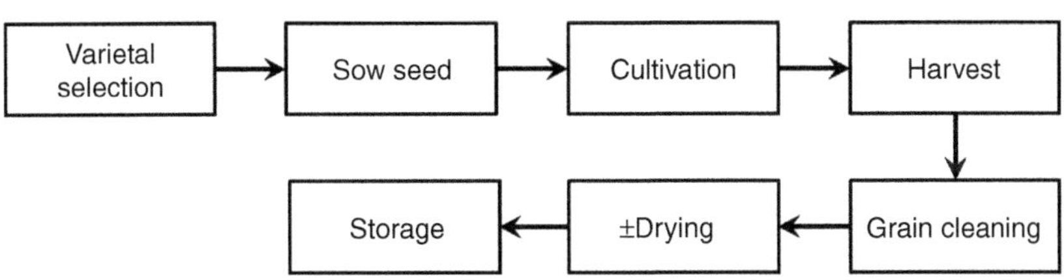

FIGURE 15.2 Schematic description of cereal cultivation and post-harvest handling.

the question of compounds found in the seeds of other plants which may be co-harvested with cereals and have public health consequences (pyrrolizidine and tropane alkaloids).

15.2.1.1 Phytates

Phytic acid (CAS Registry No. 83-86-3) is unstable in its acid form and is usually found as its calcium or calcium magnesium salt (phytate) within the kernel. As phytate, this compound is thought to account for up to 80% of the phosphorus present within the cereal tissue. Its distribution differs depending on the species of plant considered (Reddy et al., 1989). Thus while it is mainly concentrated in the germ of maize, in wheat it is found mainly in the aleurone layers and in pearl millet it appears to be uniformly distributed through the whole kernel. Phytic acid is resistant to the endogenous enzymes of the mammalian digestive tract and has been considered by some to be an anti-nutrient. This is either by virtue of it being an unavailable source of dietary phosphorus—of particular concern in the case of livestock, or as a sequestering agent for nutritionally significant metal ions (in particular calcium and iron). The nutritional significance of phytate depends very much on both the population and the product they eat. Phytate-sequestration of iron is certainly regarded to be an issue in the case of complementary foods intended for the relief of malnourished individuals [reviewed by Hurrel (2003)]. In terms of bread, removal of the bran layers and germ from the wheat kernel to produce white bread-making flour effectively removes most of the phytate present in the kernel. It should, however, be noted that in more affluent populations, phytate, even when present in wholemeal (100% extraction) flour products is not generally considered to be an issue in terms of mineral nutrition (UK Department of Health, 1991).

15.2.1.2 Tannins

The term tannin describes a group of high molecular weight compounds of plant origin, with a sufficiently high number of phenolic hydroxyl groups (1–2 per 100 Da) to form cross-links between proteins and/or other complex macromolecules (Griffiths, 1991). The tannin content of cereals varies depending on the species and variety. Values (14% moisture basis) range from 0.1% in brown rice through 0.4% for wheat and rye (Juliano, 1985) to 5% for certain varieties of sorghum (Collins, 1986). Condensed tannins are dimers or oligomers of diverse flavon-3-ols and have been considered to have anti-nutritious attributes, mainly as a consequence of their ability to reduce protein digestibility [reviewed by Griffiths (1991)]. Although of significance to livestock nutrition, their significance as anti-nutrients in terms of human nutrition in the Developed World has been questioned (Bender and Bender, 1996). In other parts of the world, this may not be the case; for example where tannin-rich cereals such as sorghum play a role in weaning foods [e.g. parts of Africa; Masha and Svanberg (1983)]. In the case of sorghum, there are a number of strategies available for reducing its tannin content. These include cultivation of low tannin varieties, characterised by a reduced presence/absence of pigment in the seed coat testa and pericarp (Taylor and Duodu, 2010; Chapter 13). Over and above varietal selection, other methods of reducing tannin content in foods containing sorghum include one or a combination of: physical removal (decortication) of the tannin-containing seed coat material prior to milling (Youssef et al., 1988) and malting (Lefyedi and Taylor, 2006).

15.2.1.3 Pyrrolizidine and Tropane Alkaloids

These two groups of compounds are not produced by cereals themselves but are considered to be an issue in that they are found in the seeds of plants which grow in the vicinity of the crop and form part of the admixture of the harvested grain. Thus in the case of 1,2 unsaturated pyrrolizidine alkaloids, these have been identified as the causative principle of hepatic veno-occlusive disease, which in its acute form has a high mortality incidence and in the chronic form exhibits itself as cirrhosis of the liver (World Health Organization, 1988). Poisonings attributed to these compounds have been associated with contamination of cereal crops with seeds of *Heliotropium* species in various parts of the world, for example Afghanistan (Mohabbat et al., 1976). However, the issue of phytotoxins associated with admixed plant material is not restricted to the developing world. This is exemplified by the issue of tropane alkaloids [e.g. atropine/(−)-hyoscyamine and (−)-scopolamine] which are associated with at least seven Angiosperm (flowering plant) families: Brassicaceae, Convolvulaceae, Erythroxylaceae Euphorbiaceae, Proteaceae, Rhizophoraceae and Solanaceae (Griffin and Lin, 2000). A review of the literature undertaken by Adamse and van Egmond (2010) suggests that a major source of tropane alkaloid poisoning in humans is the seed of *Datura stramonium* (Jimson weed or thorn apple) present within the admixture of harvested grain. Other sources of poisoning in humans reported by Adamse and van Egmond include seeds from other *Datura* species, the berries of *Atropa belladonna* (deadly nightshade) and seeds of *Hyoscyamus niger* (henbane). In terms of the quantity of contaminating seeds necessary to elicit an adverse response, Perharič et al. (2013) discussed a mass outbreak of tropane alkaloid poisoning in Slovenia, where individuals had eaten foods made from whole buckwheat containing 190 *D. stramonium* seeds per kilogram grain. The ability of relatively low levels of contamination is further exemplified by a report concerning an incident in Austria (Fretz et al., 2007) associated with millet-based foods which contained 50 *D. stramonium* seeds per kilogram of grain.

Incidents such as those have led to a reconsideration of the issue in various jurisdictions including the European Union. A recent risk assessment undertaken by the European Food Safety Authority (2013a) considered the results of a surveillance exercise where food samples collected primarily from the Netherlands but also Germany (total number of samples, 124) were tested for the presence of the tropane alkaloids (−)-scopolamine and (−)-hyoscyamine. In most cases neither alkaloid was detected, however, one or both alkaloids were found in

21 samples, of which 19 were foods intended for infants and small children and categorised as '*Simple cereals which are or have to be reconstituted with milk or other appropriate nutritious liquids*'. In its hazard and risk characterisation exercise, the Authority considered these alkaloids jointly in terms of their acute toxicity and determined an acute reference dose (ARfD) for the sum of (−)-scopolamine and (−)-hyoscyamine of 0.016 μg/kg bodyweight per day. Using information abstracted from the EFSA European Comprehensive Food Consumption Data Base (European Food Safety Authority, 2015) the Authority estimated that, when considering high consuming (95 percentile) toddlers (children aged 1–3 years old) of '*Simple cereals which are or have to be reconstituted with milk or other appropriate nutritious liquids*', it was probable that their combined exposure to these two alkaloids exceeded the group ARfD between 11% and 18% of consumption days. In terms of risk management the regulatory response has been to extend monitoring programmes within Member States for these alkaloids (European Commission, 2015a) and to introduce legislation setting maximum limits (atropine, 1.0 μg/kg; scopolamine 1.0 μg/kg) for their presence in '*Processed cereal-based foods and baby foods for infants and young children, containing millet, sorghum, buckwheat or their derived products*' (European Commission, 2016a).

15.2.2 Heavy Metals

The term 'heavy metals' embraces a number of metallic elements with known toxic effects. Of particular relevance to cereals are the elements arsenic, cadmium and lead. The dietary toxicological significance of these metals has been the subject of recent reviews (e.g. European Food Safety Authority, 2012, 2013b, 2014). Consideration of these reviews suggests that for the European population at least, cereals and cereal-based foods make a significant contribution to the dietary exposure of these metals. As discussed later in the chapter, the occurrence of these compounds is determined by the geological conditions where the crops are cultivated as well as the activities of man himself.

15.2.2.1 Arsenic

Arsenic occurs in cereals in both organic and inorganic molecular forms, with inorganic arsenic (in either the As (III) or As (V) forms) considered to be the more toxic form. Inorganic arsenic is classified as a category I (known human) carcinogen (International Agency for Research on Cancer, 1987) and consequently more attention has been paid to these molecular forms both in terms of dietary exposure and in risk management and a similar approach has been adopted here. Dietary exposure to arsenic (either inorganic or organic) is effected through drinking water and/or food itself. Given the contribution made by cereal-based foods to the human diet, they have the potential to act as a significant source of dietary arsenic. This applies particularly to rice (Agusa et al., 2009; Chatterjee et al., 2010; Liang et al., 2010), which has a greater propensity than other cereals to accumulate the element (US Environmental Protection Agency, 1982). Arsenic contamination of cereals is effected by uptake through the roots and, in some plants (e.g. rice), contamination is highly dependent on the amount of the metal present in the irrigation water (Baig et al., 2011). Levels of arsenic found in rice are therefore very much dependent on levels present in the environment where the crop is grown (Zhao et al., 2013). In parts of Asia in particular where rice is cultivated and environmental levels of arsenic are high, dietary exposure to arsenic can be sufficiently high to be a cause of toxicological concern in terms of cancer incidence (Li et al., 2011). In contrast in the United States, a retrospective analysis of data from 45,231 men and 160,408 women failed to show any linkage between total, white or brown rice consumption and cancer incidence (Zhang et al., 2016a). Nevertheless, cereal-linked arsenic consumption is not only an issue for indigenous populations eating home-produced cereals. It can also be a factor for those populations who purchase and import such commodities. A case in point is rice imported into the European Union, where the European Food Safety Authority (European Food Safety Authority, 2009a) estimated that in the case of high rice consumers (in particular certain ethnic groups), inorganic arsenic exposure could be in the region of 1 μg/kg bodyweight per day. This compares with an index of safe intake (Bench Mark Dose of 1% extra risk, BMD01) of 0.3–8 μg/kg bodyweight per day. Another population found to be at risk included children in the process of weaning and who often have a heavy reliance on rice-based foods. This has been highlighted in a number of countries including the United Kingdom. Between 2006 and 2009 the United Kingdom Food Standards Agency published the results of surveys to determine the significance (if any) of the presence of heavy metals (including arsenic) in weaning-foods and infant formulae (Food Standards Agency, 2006a, 2007a) and rice milk (Food Standards Agency, 2009). Although none of the products contained arsenic at levels which exceeded the then statutory maximum limit for arsenic in any food (1 mg/kg), a risk assessment based on consumption data indicated that children consuming such products might be at risk of adverse exposure (Meharg et al., 2008). This was particularly true of rice milk preparations (often used as a substitute in cases of milk intolerance) and led to the Agency advising that rice milk should not be used as a substitute when feeding babies or young children (Food Standards Agency, 2009). Internationally, regulatory risk management concerning levels of arsenic has led to the Codex Alimentarius Commission proposing a maximum residue limit of 0.2 mg/kg inorganic arsenic in polished rice (Codex Alimentarius Commission, 2014). These proposals have been acted on in a number of jurisdictions within the European Union where limits of 0.20, 0.25 and 0.1 mg inorganic arsenic per kg have been set for polished rice, parboiled or husked rice and rice intended as an ingredient for foods to be consumed by young children respectively (European Commission, 2015b).

In the medium to long term, the most effective route for dealing with the issue is probably to breed varieties which bioaccumulate arsenic at lower levels. However, an understanding of the factors leading to the bioaccumulation of arsenic within the plant is, at the time of writing, still incomplete (Tuli et al., 2010). Nevertheless progress is being made, for example Ma et al. (2008) and Panda et al. (2010) have described at biochemical and genetic levels some of the mechanisms responsible for the uptake and distribution of arsenic within the rice plant. Other groups such as Smith et al. (2009) and Carey et al. (2010) have described how inorganic arsenic (arsenate) and organic arsenic are taken up and distributed throughout the whole plant. Panda et al. (2010) have suggested that the ability of the plant to produce phytochelatins (metal-binding thiol peptides) may be one factor to be investigated in relation to reducing the occurrence of the metal in rice.

15.2.2.2 Cadmium

Salts of this metal occur naturally in the environment as a consequence of erosion, other natural phenomena and also as a consequence of the activities of man. It is a nephrotoxin due to its accumulation in the proximal tubules of the kidney and it can also promote bone demineralisation either through its effect on kidney function or by direct action on the tissue itself. Like arsenic, cadmium is considered to be a class I carcinogen and has been associated with increased incidences of cancers of the lung, endometrium, breast and bladder (European Food Safety Authority, 2009b). In terms of cereals, the metal has been identified as the causative factor of the condition Itai Itai related to cadmium contamination of rice in Japan (Ogawa et al., 2004; Kobayashi et al., 2009), while studies in China have identified cereal-based products (rice and wheat) as significant sources of dietary cadmium in children living in Jiangsu province and at risk of significant risk through chronic exposure to the metal (Zhang et al., 2016b). In terms of agricultural practice, as in the case of arsenic (discussed earlier), irrigation is a major contributory factor for the occurrence of cadmium in rice. For example studies both in Italy (Cattani et al., 2008) and Pakistan (Hussain et al., 2010) have demonstrated that use of waste water to irrigate rice can lead to increases in levels of cadmium contamination. Fertiliser practices are another factor which can lead to increased bioaccumulation, for example through intensive use of cadmium contaminated phosphorus fertilisers, as seen in rice grown in Sri Lanka (Bandara et al., 2010). It has also been observed that other practices might also contribute to cadmium bioaccumulation in cereals. Dahlin et al. (2016) have raised the question of the mobilisation of indigenous cadmium as a consequence of soil amendments through addition of chloride ions associated with urine and biogas digestates and subsequent bioaccumulation in wheat. Others have shown interactions between environmental factors and nitrogen application practices with cadmium bioaccumulation in both winter (Wångstrand et al., 2007) and durum wheats (Perilli et al., 2010).

As already discussed in the case of arsenic—in many parts of the world, the only approaches that will lead to significant reductions in cereals-related cadmium intake will be through the development of appropriate agronomic practices as well as cereal varieties which do not accumulate the metal into the kernel in the first place. Achieving the latter objective requires knowledge of the genetics and biochemistry of factors leading to amelioration. At the level of molecular genetics, progress has been in identifying those genes responsible for suppressing cadmium bioaccumulation in rice (Frank et al., 2009; Ueno et al., 2010). Nevertheless, in this case there is a problem of trying to maintain yields while reducing plant cadmium concentrations (Liu et al., 2009). With regard to other cereals, commercial varieties with a lower propensity to bioaccumulate cadmium have already been described [e.g. durum wheat, Clarke et al. (2010)].

15.2.2.3 Lead

In terms of human toxicology, the primary targets of this heavy metal are the nervous and cardiovascular systems as well as the kidneys (European Food Safety Authority, 2013b). In contrast to arsenic and cadmium contamination, contamination of cereals with lead is first and foremost a consequence of man's activities, in particular through industrial emissions and the use (now discontinued in many jurisdictions) of leaded petrol (Kabata-Pendias and Mukherjee, 2007). Lead accumulation in plants appears to be primarily a function of atmospheric deposition and although some plants have the ability to take up lead through the roots (United States Agency for Toxic Substances and Disease Registry, 2007). In the case of wheat both atmospheric pollution and contamination of soil have been shown to be risk factors with regard to lead bioaccumulation (Yang et al., 2015; Fang et al., 2015). Adoption of more stringent environmental protection regulations has led to a reduction of levels of lead in the atmosphere (e.g. Central, Europe, UN Environment Programme, 2010), which have been associated with a concurrent reduction in the amounts of lead found in plant-derived foods such as cereals (Kabata-Pendias and Mukherjee, 2007).

15.2.3 Mycotoxins

Mycotoxins are poisonous metabolites produced by certain filamentous fungi. Having a range of diverse chemical structures their toxicological properties are similarly varied. As in the case of heavy metals, depending on levels of exposure they can exert both acute and chronic toxicity. It has been estimated that approximately 25% of the world's crop production are, to one degree or another, contaminated with them (Charmley et al., 1995). Given their toxicity and ubiquity, mycotoxins present the risk of a significant hazard to both man and his livestock. With regard to cereals, mycotoxin contamination has historically been of public health significance. The earliest references to human mycotoxicoses probably relate to ergot alkaloid (e.g. ergometrine) poisoning associated with grain infected with *Claviceps purpurea* and can be traced as far back to references in Assyrian cuneiform tablets dating from approximately 600 BC (van Dongen and de Groot, 1995). Mycotoxin contamination of grain can still have catastrophic consequences as exemplified by the aflatoxicosis outbreak in Kenya during 2004, during which over 100 persons died from aflatoxin-induced liver failure following consumption of contaminated maize (Azziz-Baumgartner et al., 2005). It is therefore unsurprising that in many jurisdictions the presence of mycotoxins is legislated for. Arguably the most extensive regulations for materials intended for human consumption are those in force within the European Union (European Commission, 2006b). In the case of mycotoxins associated with cereals, these regulations currently set limits for aflatoxins, ochratoxin A, fumonisins, deoxynivalenol (a trichothecene), zearalenone and citrinin. The significance of the generic risk presented by mycotoxins can be seen when considering regulatory enforcement activity in relation to cereals and cereal-based foods. Consideration of regulatory enforcement notifications made within the European Union RASFF [Rapid Alert System for Food and Feed (European Commission, 2016b)] reveals that contravention of mycotoxin control legislation is the single largest reason for notifications (Fig. 15.3).

Broadly speaking, mycotoxins found in cereals can be classified on the basis of when they are produced—either before or after harvest. Miller (1995) referred to these two groupings of mycotoxins as 'field' and 'storage' respectively. Field mycotoxins (e.g. deoxynivalenol) are generally produced during crop

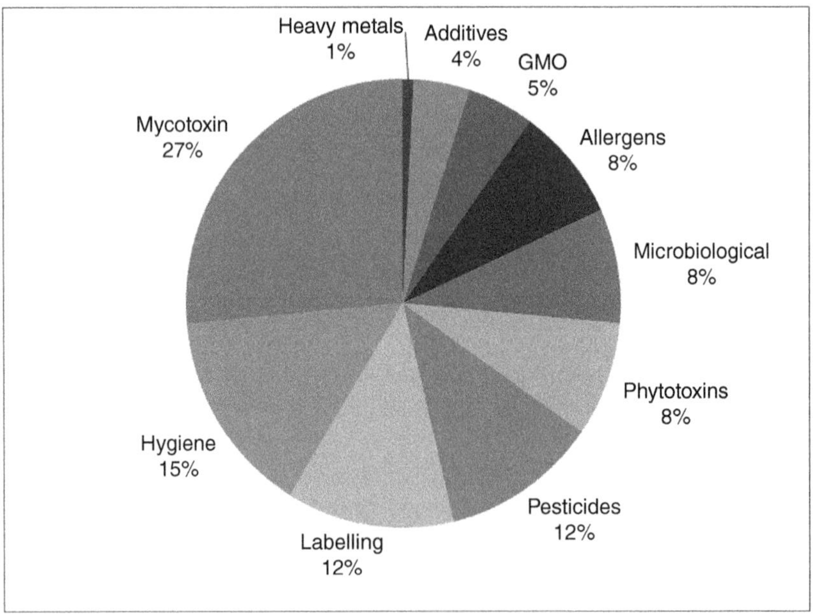

FIGURE 15.3 Analysis of notifications (n = 122) concerning 'Cereals and Bakery Products' communicated through the European Rapid Alert System for Food and Feed during the calendar year 2015.

development, while storage mycotoxins (e.g. ochratoxin A) are formed post-harvest. However the classification is not absolute. Fumonisins and zearalenone (which like deoxynivalenol are produced by *Fusarium* spp. and were originally considered to be 'field mycotoxins') have been demonstrated to be formed both post- as well as pre-harvest (Ono et al., 2003; Alldrick and Hajšelová, 2004). Similarly, aflatoxins, although considered to be storage mycotoxins, have also been shown under some circumstances to be formed pre-harvest (Hill et al., 1983). As will be discussed further in the chapter, the now commonly accepted strategy for the management of both field and storage mycotoxin contamination of cereals is one of *prevention is better than cure* (Battalglia et al., 1996). The tactics necessary to realise this strategy are based essentially on HACCP principles (Codex Alimentarius Commission, 2012; European Commission, 2006b). Consequently management of 'Field' mycotoxins primarily revolves around events at pre-harvest, and 'Storage' mycotoxins post-harvest.

Generally speaking, those field mycotoxins of importance to cereals are the ergot alkaloids produced by *Claviceps purpurea* together with mycotoxins produced by *Fusarium* spp. (e.g. trichothecenes, fumonisins and zearalenone). In many cases agronomic control of these compounds revolves around reducing the inoculum available to infect the cereal during growth, for example weed control, adoption of appropriate tillage practices and crop rotation practices through to, where permissible, suppressing fungal growth through the use of appropriate fungicide application. The importance of the correct use of fungicidal regimes in managing fusarial mycotoxin formation in cereals is increasingly being recognised. One of the original motivations for fungicide application was prevention or suppression of fusarium ear blight (FEB). Early on, some fungicides [e.g. triadimefon and propiconazole; Boyacioglu et al. (1992)] were observed to be effective in managing both FEB and mycotoxin contamination. Subsequent studies showed that other compounds effective in controlling FEB (e.g. azoxystrobulin) were actually associated with increased levels of mycotoxin contamination (Simpson et al., 2001). In this case, the increased level of contamination may have been attributed to the selective action of the fungicide towards the, apparently, non-toxigenic, plant pathogen *Microdochium nivale* also associated with FEB (Jennings et al., 2000). Specific tools have also been developed to optimise these management approaches. For example Canadian farmers have access to a weather prediction tool called DONcast (Weather Innovations Incorporated, 2016). This assists farmers in deciding whether or not to apply appropriate fungicide treatments at anthesis to reduce both the risk of FEB and eventual deoxynivalenol contamination. Other grain markets have developed risk assessment tools to assist purchasers of grain to assess the risk of fusarium mycotoxin contamination in grain. For example in the United Kingdom, shipments of home-produced grain are accompanied by a 'grain passport' which includes a risk assessment concerning *Fusarium* spp. mycotoxin contamination and is produced in a prescribed format (Agriculture and Horticulture Development Board, 2015).

In terms of managing the risk of mycotoxin contamination of stored cereals, of key concern are:

- The hygienic state of harvesting and grain storage facilities.
- Time taken to transfer grain with a high moisture content from field to the dryer.
- Thermodynamics of the drying process itself to reduce the moisture content.
- Maintenance of the eventual reduced moisture content.

Reducing the risk of storage mycotoxins again begins with reducing the probability of the grain being inoculated with mycotoxigenic fungi in the first place. Cleaning of harvesting and storage equipment to remove debris from the year's previous harvest and thus potential inocula and the possibility of mycotoxin contamination as in the case of ochratoxin A, are recommended in codes of practice (Food Standards Agency, 2007b). A further factor to consider is grain moisture at harvest and the questions of when and how to reduce moisture to minimise the risk of mycotoxin production. The importance of the interval between harvest and drying is highlighted by work such as that of Ono et al. (2003), working with Brazilian maize. They observed that even short post-harvest holding intervals (days) before drying could lead to over 10-fold increase in contamination with fumonisin mycotoxins. However, speedy transfer of wet grain to the dryer is not in itself, sufficient to assure that mycotoxin contamination does not become excessive; a second factor to consider is the actual process of drying the grain. Studies addressing the thermal kinetics of grain drying, such as those by Jonsson (1996) in small-grain crops, have demonstrated that both the geographical location and the methods of grain drying are of significance to mycotoxin (ochratoxin A) formation. Jonsson's work in Sweden demonstrated that in some years, even when drying with heated air, average grain moistures of greater than 14% after 5 months' storage were observed. In the case of grain dried with near-ambient air or near-ambient air plus added heat, average grain moistures rising to 20% were found. Drying with ambient air therefore can be disadvantageous in terms of eventual mycotoxin contamination. This does not appear to be a problem unique to small-grain crops; similar results have been reported for aflatoxin in maize by Tuite et al. (1986).

Once grain has been dried to an appropriate moisture content, it is essential that this status be maintained during storage. Control of grain moisture while in storage has long been identified as a critical factor in preventing spoilage and by implication mycotoxin contamination (Christensen et al., 1992). A key issue is the formation of localised zones of relatively high moisture content that can facilitate fungal growth and consequent mycotoxin production. Factors leading to this type of problem include poor storage practices leading to temperature differentials and moisture migration due to convection as well as blending of grain with different moisture contents to obtain satisfactory average moisture content (Christensen et al., 1992). The situation can be exacerbated by insufficient drying, since grain metabolism of carbohydrates itself leads to the generation of both moisture and heat. As discussed by Pomeranz (1992), wheat held below 15% moisture and an ambient temperature of 15°C, exhibits little metabolic activity. Raising the moisture content above this level leads to increases in metabolic activity and consequent localised increases in grain moisture activity, with the consequential risk of increased mould activity.

15.2.4 Plant Protection Products (Pesticides)

Pesticides are arguably one of the most extensively regulated group of compounds in terms of food-based chemical hazards. Such regulations, for example European Union Regulation 396/2005 (European Parliament and Council, 2005) specify which pesticides may be applied to which crops and the amounts of residues permitted in food intended for consumption. Given the large number of possible pesticide crop/pesticide combinations this is supported by a data base to assist in determining compliance (European Commission, 2016c). The risk presented by the use of pesticides is in part reduced by the advice provided by responsible pesticide manufacturers to end-users. This will include details of which crops the pesticide is permitted to be used for and recommended rates of application. The consequent treatment regimes are therefore designed not only to achieve the appropriate level of crop-protection but also to ensure that any residues are below the maxima set in local legislation. Compliance with such advice is further enhanced through primary producers complying with commercial standards of agricultural best practice which require those certified to them to apply pesticides in an appropriate manner. For example within the United Kingdom, flour-millers require their grain suppliers to be certified under an approved scheme, for example the 'Red Tractor Scheme' (Red Tractor Assurance, 2014) and that deliveries of individual grain lots should be accompanied by a 'grain passport', which details its pesticide application history (National Association of British and Irish Millers, 2014).

A second area relates to use of pesticides to manage infestation during grain storage or transport. For many years the fumigant of choice was methyl bromide. However this compound has been identified as an ozone-depleting agent and its use has been discontinued in developed countries since 2005 and in third world countries by 2015 in accordance with the Montreal Protocol (Environment Agency, 2008). Alternative methods of control have been proposed; however, the fumigant often used in both temperate and tropical zones is now phosphine (Brice and Golob, 2000; Agriculture and Horticulture Development Board, 2011).

From a food-safety point of view it should be noted that most reports relating to illnesses arising as a result of exposure to pesticides in food do not reflect the presence of residues following their use in agriculture. Rather, they reflect issues relating to, improper transport and storage of food intended for direct consumption (Weeks, 1967), misapplication in food-production facilities (Chaudhry et al., 1998), consumption of treated material (e.g. seed stock) not intended for human consumption (Bakir et al., 1973) and as a vector for sabotage (Kavalci et al., 2009).

15.2.5 Man-Made Contaminants

As discussed earlier, in temperate climatic zones and depending on conditions at harvest, recourse to warm-air drying may well be necessary. Warm-air drying of grain can, however, be associated with additional food-safety linked challenges. One of these is potential contamination with fuel combustion by-products and in particular dioxins. Generically the term dioxin describes a diverse range of heterocyclic organochlorine pollutants released into the environment from a variety of human activities. The archetypal compound is 2,3,7,8-tetrachlorodibenzo-*p*-Dioxin (TCDD). Dioxins are considered to be human carcinogens (International Agency for Research on Cancer, 1997) and have also been associated with a diverse range of toxicoses, including effects on the endocrine and immune systems.

Dioxins are characterised by their ability to be bioaccumulated within the organism (plant or animal) and in particular within the lipid fraction. One Japanese study (Otani et al., 2006) reviewed survey data gathered over a 4-year period for locally grown cereals and legumes. Levels of dioxins present in the crops were not considered to give cause for concern. However, the capacity of dioxins to bioaccumulate through the supply chain means that the meat or other products from animals fed dioxin-contaminated feed can be a public health issue. Grain drying using inappropriate fuels of incorrectly designed flues can be a route by which such contamination can occur, a point recognised in relevant Codex Alimentarius codes of practice concerning risk minimisation of dioxin contamination of foods (Codex Alimentarius, 2006). Use of inappropriate fuel to dry feed was considered to be the reason for the recall of Irish pork and beef suspected of being contaminated with dioxins during December, 2008 (Tlustos, 2009).

15.3 GRAIN PROCESSING (FLOUR MILLING)

Grain processing as exemplified by flour milling (Fig. 15.4) is essentially a physical process whereby the kernel is cleaned, adjusted to an appropriate moisture content and then mechanically reduced to the desired particle size to produce

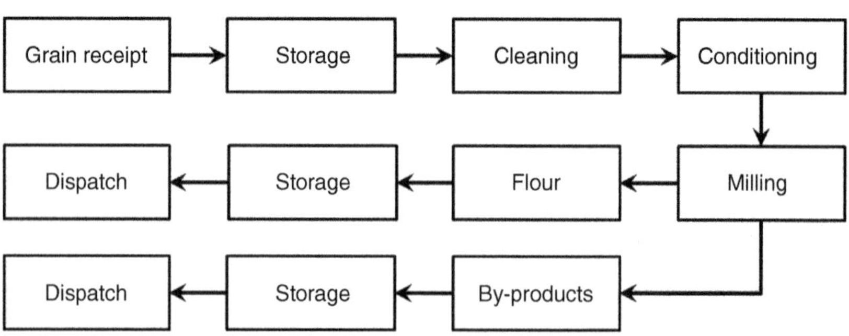

FIGURE 15.4 Primary processing (e.g. flour milling).

a flour and by-products (e.g. bran). Where appropriate, flour production also involves fractionation—not only to separate bran, germ and endosperm from each other but also to assure the correct particle size of the milled endosperm.

The process only involves physical processes in terms of food-safety steps, such as grain cleaning. For example the use of gravity tables can be effective in removing admixtures such as seeds of toxic plants or shrivelled grains which might be local sources of high levels of *Fusarium* mycotoxins. The risk to the consumer is therefore to a large degree predicated by the level of contamination of the grain supplied to the miller. It is therefore incumbent on the miller not only to set specifications detailing limits for the presence of particular compounds but also to have the appropriate supplier quality assurance (SQA) systems in place to ensure compliance. In developing these specifications attention has to be paid not only to any statutory levels set for a particular contaminant in grain but also to levels which are compatible with the needs and obligations of the miller's customers. In this case, the term 'customer' applies not only to those purchasing the flour but also the milling by-products (e.g. animal-feed compounders). The complexity of the situation is shown by consideration of current EU regulations concerning permitted levels of deoxynivalenol in wheat (European Commission, 2006a). Depending on the food, limits are set at 250 mg/kg for cereal-based foods intended for babies, infants and young children; 500 mg/kg for bakery wares, breakfast cereals etc.; 750 mg/kg for grains and flour intended for human consumption and 1250 mg/kg for unprocessed grains. Given the mechanically reductive nature of the milling process, this will have implications in terms of the final specification set by the miller to the vendor for the grain. It is therefore likely that the miller will have to pass on the consequences of his customers' requirements to his grain supplier by specifying maximum levels of mycotoxin contamination for the grain lower than the limit set for unprocessed grain in the regulations. The margin of difference required will depend on the flour extraction rate used (discussed later in the chapter), and the requirements of the miller's customers. With regard to SQA, for it to be effective, it is necessary for the purchaser to have both the appropriate knowledge and information to set relevant specifications as well as the tools to verify compliance. Acquisition of knowledge and information requires an integrated appreciation of both the grain's provenance and associated food-safety hazards, together with the risk of them occurring. From this base, appropriate verification programmes can be developed. At a generic level these could involve some form of supply chain inspection and audit process, which could be undertaken either by the purchaser or through a third party supplier certification scheme. An example of the latter approach can be found in the standard terms and conditions of trade adopted by the National Association of British and Irish Millers (National Association of British and Irish Millers, 2014). These specify that grain grown in the United Kingdom will be purchased from third-party certified (Red Tractor Assurance, 2014) suppliers.

Any management system requires that it be validated, verified and reviewed as necessary [see HACCP principle 6, Codex Alimentarius (2003)]. At one level this could be requiring the vendor to supply an appropriate certificate of analysis. Ideally information provided in the certificate, should be based on analyses performed in an appropriately accredited [e.g. ISO 17025:2005, International Organisation for Standardisation (2005)] laboratory on a sample collected in an appropriate manner. Ultimately, however, verification will also include periodic analysis commissioned by the grain processor. Strategies for the development of appropriate analytical programmes for food-safety management purposes within the cereals industry have been described by Poms et al. (2009). Essentially the intensity of testing depends on the degree of confidence that can be placed on the functionality of the SQA system in respect of a particular contaminant. Where confidence is high, the frequency of analyses may be reduced to as little as once per harvest. Under certain circumstances, for example in years where the risk of mycotoxin contamination is unacceptably high it may be necessary to positively accept at-mill intake on the basis of results from the use of rapid test kits [discussed by Alldrick (2014)].

Notwithstanding this observation the milling process can bring about changes in the distribution of contaminants when comparing amounts within the grain and the resultant mill fractions. Contaminants are not only heterogeneously distributed within any particular lot of grain but also within the kernel. Modern milling processes can effectively achieve contaminant redistribution in the flour and by-products produced. This is achieved at two levels. Prior to milling, grain is 'cleaned' (Fig. 15.4) by the physical removal of extraneous matter and sub-standard grains. Cleaned grain may also be subject to further 'cleaning' by the removal of surface bran layers by pearling or scouring. A second factor is that milling involves processes that bring about both mechanical reduction and also fractionation. Certain mill streams (e.g. endosperm) will therefore often contain lower concentrations of a particular contaminant while others (e.g bran layers) may contain higher levels. The significance of the milling process can be demonstrated by consideration of its role in the quality of flour in terms of microbiology and mycotoxin content.

In terms of the microbiological quality of flour, the major contributory factor is the level of microbiological contamination of the grain from which it is produced (Richter et al., 1993; Berghofer et al., 2003). However, mill hygiene may also be a contributory factor. Australian studies have associated microbiological loading, in part, with the conditioning stage of the flour milling process due to the presence of highly contaminated wheat residues in the damper (Eyles et al., 1989). In common with a number of other types of contamination, microbial contamination appears to occur on the surface of the kernel. Consequently during the milling process microbial loading is differentiated between the bran and the endosperm fractions, with higher levels of loading associated with those fractions associated with the grain surface (e.g. bran), as opposed to the endosperm (Berghofer et al., 2003). Weather conditions at harvest may also be a factor. Surveys and analyses of UK flour produced from new harvest wheats by Legan and co-workers (Legan, 1994; Legan and Curtis, 1990) correlated wetter harvests with both increased total bacterial viable counts as well as total number of moulds and yeasts.

As shown in Fig. 15.3, mycotoxin contamination of cereals and cereal products is a significant issue with regards to notifications made by enforcement agencies within the European Union. Milling can contribute to a reduction in mycotoxin loading of the finished flour. This reduction is particularly dependent on facilities within the screen room and the flour extraction rate. In the case of *Fusarium* spp. mycotoxins, physical sorting methods such as the use of specific-gravity tables can lead to the removal of certain types of contaminated (shrivelled) grains and hence reduce the mycotoxin content of the resultant flour (Tkachuk et al., 1991). Reliance on such systems pre-supposes a strong linkage between grain damage and mycotoxin contamination [which may not necessarily be the case, Edwards et al. (2001)] and that not only the capital costs of the equipment but also of handling the rejected grain have to be met. The milling process itself can also contribute to a reduction in the final levels of mycotoxin contamination seen in white flour. Separation of endosperm from the bran layers has been reported as contributing to reductions in levels of both ochratoxin A (Alldrick, 1996) and deoxynivalenol (Abbas et al., 1985; Lepschy-von Gleissenthal and Suess, 1996) contamination of the final flour. Fractionation of grain in this manner can be a two-edged source since the process does not destroy the mycotoxin, it simply removes the fraction containing it, with a consequent increase in mycotoxin concentration within that fraction. This can become an issue as within the European Union following the wet 2008 wheat harvest which led to higher than normal levels of zearalenone in the crop and eventually higher levels of zearalenone in wheat bran intended for breakfast cereals. This ultimately led to a derogation in the regulations and a temporary increase in the maximum permitted levels for this mycotoxin (Edwards, 2011).

15.4 FOOD PROCESSING

An overview of the process whereby a primary product (wheat flour) is converted to a food product (bread) which is eventually purchased by the consumer is described in Fig. 15.5. In terms of food-safety issues, these can relate to how

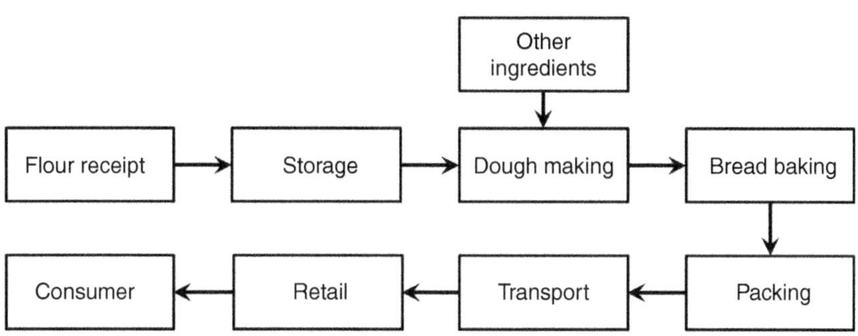

FIGURE 15.5 Food processing (bread baking).

the food business operates, the choice of other ingredients (in particular food additives), employee activities and the health needs of the consumer, particularly with regards to food hypersensitivity.

15.4.1 Microbiological Safety

From a food-safety hazard analysis viewpoint, the microbiological loading of flour presents a potential food-safety issue. One reason why microbiological issues do not feature so prominently is that in order to be palatable, cereal-based foods usually have to be subjected to some form of thermal treatment (cooking) before being eaten [discussed by Kent and Evers (1994)]. Cooking usually results in significant microbiological kill—effectively resulting in a pasteurised product. In terms of microbiological safety, incidents where consumer health has been compromised often represent issues with outgrowth of bacterial spores not destroyed under normal processing conditions. In the case of cereal products the most significant indigenous bacteria of concern are *Bacillus* spp., where spore germination following thermal shock leads to outgrowth in the cooled cooked product. Reports of food poisoning (sometimes fatal) associated with these bacteria and involving cooled rice and pasta (usually salads based on them) are well documented (Raevuori et al., 1976; Dierick et al., 2005). Incidents relating to food-poisoning associated with *Bacillus* spp. in bread are rare, this probably lies in its significance as a spoilage organism and in particular with the phenomenon of 'Rope'. Consumption of microbiologically compromised product is less likely if the product's taste and/or appearance are not what would be expected. Furthermore strategies for controlling *Bacillus* spp. outgrowth have been known for some considerable time (Henderson, 1918) including the use of vinegar and/or propionate in dough recipes to suppress bacterial growth (Kaur, 1983; Thompson et al., 1998).

Where reports of food poisoning linked to bread itself (rather than other ingredients; e.g. sandwich contents) have been made, the cause has either been unknown or reflective of staff hygiene. Steinberg et al. (2006) described a series of 16 food poisoning outbreaks taking place in 7 states in the United States and associated with burritos made by two separate companies. Although the food responsible for the incidents was identified, as being the burritos themselves, the actual causative agent was not described. Other reports of microbiological food poisoning associated with bakery products appear to be related to failures in worker health surveillance programmes leading to transfer of pathogenic bacteria (*Salmonella* spp.) or viruses (hepatitis) from production workers (Schoenbaum et al., 1976; Kimura et al., 2005) or food handlers (Milazzo and Rose, 2001) to the product itself. Management of microbiological hazards therefore reflects on good manufacturing practice (GMP), which includes sanitation, appropriate process controls and worker health/hygiene etc. GMP not only relates to microbiological hazards but also to those of chemical or physical origin as discussed later in the chapter.

15.4.2 Food Additives

The Codex Alimentarius Commission (2015) defined a food additive as:

> *Any substance not normally consumed as a food by itself and not normally used as a typical ingredient of the food, whether or not it has nutritive value, the intentional addition of which to food for a technological (including organoleptic) purpose …*

Their use is usually governed by local legislation, which varies from jurisdiction to jurisdiction. Nevertheless such legislation shares some common features, in particular that for any permitted additive it will specify the purpose(s) for which the additive can be used and in which foods. The legislation will go on to specify limits for the amount of residues that are allowed to remain in the finished product. It is important to note that the use of a food additive in one jurisdiction might be more limited or even prohibited in another. This is particularly the case with regard to food additives in cereal-based foods and concerns the use of 'flour-treatment agents' (e.g. chlorine and chlorine dioxide), dough conditioners (e.g. ascorbic acid, azodicarbonamide and potassium bromate), certain pH regulators and leavening agents (e.g. aluminium phosphates). The use of these chemicals has been the subject of significant scrutiny and in many jurisdictions, in particular the European Union, their use is often restricted or banned. The reasons for this change in the European Union relate partially to the harmonisation of relevant legislation within Member States and also to developments in the science of toxicology. Thus, following a review of toxicological data concerning the safety of cake baked with chlorinated flour, concerns were raised over the possible bioaccumulation of the chlorinated fatty acid 9,10-dichloro-octadeconoic acid, in mice fed lipid-fractions of the cake baked with chlorinated flour. The industry chose not to respond to a call for more evidence concerning safety in use and permission for chlorine and chlorine dioxide to be used as flour-treatment agents was consequently not included in the ensuing harmonised legislation (CoT, 1996). A similar situation arose in connection with the use of potassium bromate and azodicarbonamide. In the case of potassium bromate, the compound had been identified as a genotoxic carcinogen by Kurokawa et al. (1983); however, it was considered that it was completely degraded during the baking process. However, the advent of more sensitive methods of analysis with lower levels of detection led to bromate being discovered in baked bread (Dennis et al., 1994) in the United Kingdom. Partly as a result of such findings the use of this compound is now prohibited in many jurisdictions. A similar fate befell azodicarbonamide when it was shown to be associated with the occurrence of the carcinogen ethyl carbamate in bread (Dennis et al., 1997).

Use of non-approved additives can be a significant issue, particularly in terms of food imported from one jurisdiction into another. One example concerns the use of aluminium-based chemical leavening agents. Dietary aluminium exposure within the European Union in a number of population groups is considered to exceed the tolerable weekly intake (European Food Safety Authority, 2013c) and consequently the use of aluminium food additives is curtailed. Thus in the case of sodium aluminium phosphate (E541), its use is permitted in '*only sponge cakes produced from contrasting coloured segments held together by jam or spreading jelly and encased by a flavoured sugar paste*' (European Commission, 2012). Aluminium-based leavening agents are used elsewhere in the world in the production of noodles, this was highlighted as a potential issue in 2010/2011 (Laboratory of the Government Chemist, 2011) and although the issue has been resolved to a large degree, consideration of RASFF data indicates that during the calendar year 2015, at least two alerts related to the presence of unacceptable levels of aluminium in noodles imported from the Far East.

15.4.3 Process Contaminants

Paradoxically the very processes responsible for converting cereal-based ingredients into palatable and microbiologically safe foods can also lead to the formation of compounds of toxicological significance. Probably the most significant issue at the moment relates to acrylamide. Acrylamide is a known carcinogen and was first detected in 2002 by Swedish researchers within a number of high carbohydrate foods prepared at high temperatures (Svensson et al., 2003). The significance of these findings is still the subject of considerable debate (JECFA, 2005); however, there is a general consensus that the occurrence of this compound should be minimised. The mechanism by which acrylamide is formed during the baking process has been elucidated and the levels formed are to some degree related to the amounts of free asparagine present in the raw materials (JECFA, 2005). Food manufacturers have responded by collaborating with governments to further understand the problem and identify appropriate acrylamide reduction strategies for carbohydrate-rich products. In the European Union these have been consolidated into the, 'Acrylamide Tool Box' (Food Drink Europe, 2014). In the case of bread, most of the acrylamide appears to be formed in the crust. The 'Tool Box' suggests strategies contributing to a reduction in acrylamide formation at either the dough mixing or baking stages. In the case of dough mixing, these include possible use of asparaginase, limiting the use of reducing sugars, using flours with lower asparagine contents, the addition of calcium salts and extending the period of fermentation. In terms of baking, the key recommendation is to reduce baking temperatures to produce a lighter crust (Food Drink Europe, 2014). It is important to note that it is probable that for any one product, it would not be possible to adopt all of the strategies because other factors, in particular relating to consumer acceptability or technical feasibility might be unacceptably altered.

15.4.4 Food Allergy and Coeliac Disease

Some individuals are known to exhibit adverse reactions to food which most others can eat with no ill-effect, these have been referred to as 'non-toxic' due to the limited number of victims (Bruijnzeel-Koomen et al., 1995). The underpinning physiological mechanisms leading to such reactions are various but of relevance to this chapter are the conditions of food allergy and coeliac disease and the obligations of food businesses towards these groups of consumers.

15.4.4.1 Food Allergy

In excess of 170 foods have been identified as eliciting food allergy in at least one individual (Burks et al., 2012) and the latest estimates are that in the developed world food allergy can affect 5% of adults and 8% children (Sicherer and Sampson, 2014). The types of foods considered to be significant causes of food allergy vary geographically and legislation now exists in many jurisdictions concerning the provision of information to food allergic consumers. Codex Alimentarius (2010) identified six foods or groups of food as being significant causes of food allergy whose presence as ingredients in packaged food should be declared. These were: cereals containing gluten; that is wheat, rye, barley, oats, spelt or their hybridised strains and products of these; crustacea and products of these; eggs and egg products; fish and fish products; peanuts, soybeans and products of these; milk and milk products (lactose included); tree nuts and nut products; and sulfite in concentrations of 10 mg/kg or more. The regulatory requirements for this form of risk communication have been further developed in different jurisdictions, for example within the European Union the list extends to 14 foods or groups of foods and also places requirements on businesses selling unwrapped foods (European Parliament and Council, 2011). Food businesses therefore have a clear obligation to ensure that operations in connection with these foods are accurately reflected in the ingredients information provided to the consumer and that reasonable steps are taken to minimise cross-contamination [discussed by Alldrick (2009)]. Where, despite application of GMP, cross-contamination might occur, consideration may have to be given to precautionary labelling (e.g. 'May Contain'). Where used indiscriminately, the use of precautionary labelling can be controversial (discussed by Dunn-Galvin et al., 2015); however, guidance and tools are available to food businesses in terms of when and how to determine the need for such labelling as well as appropriate wording (Allergen Bureau, 2014; Food Standards Agency, 2006b).

15.4.4.2 Coeliac Disease

This, 'is a lifelong autoimmune disease caused by a reaction to gluten', (Coeliac UK, 2016) and generally represents an intolerance to particular proteins found in small grain cereals. As a consequence, food businesses have developed analogue foods which are compatible with sufferers from this condition (so called 'gluten free'). There is a Codex Alimentarius standard (Codex Alimentarius Commission, 2008) for gluten-free products which sets a maximum limit or 20 mg gluten per kg in 'gluten free' products that has been adopted in the legislation of some (but not all) jurisdictions. Given that these products are targeted at a particularly sensitive sub-set of the population, food businesses producing them must operate exemplary 'Good Manufacturing Practice' (GMP), which may have to be verified by additional measures including positive release on receipt of appropriate analytical data from an external laboratory for each production run. It is important to note that coeliac disease and allergy to a particular cereal are not the same and the threshold value of 20 mg/kg would not apply to foods specifically targeted at cereal-allergic individuals.

15.4.5 Criminal Activities

Food-safety hazards can also be introduced into food as a result of criminal activities. As discussed later in the chapter, motivation can be either economic (fraud) or grievance (sabotage) and the actual act may be perpetrated at any point within the supply chain.

15.4.5.1 Fraud

The susceptibility of food generally and cereal products in particular to fraud and the development of laboratory methods to detect such events was described as early as the 1820s (Accum, 1820). Fraud usually involves the addition of something to enhance a food or food ingredient's desirable characteristics; this can be in terms of visual appearance, taste or its perceived nutritional content. Recent examples include the addition of melamine to wheat bran to increase its nitrogen and by implication protein content. In this case the wheat bran was used as an ingredient in food for companion animals; in 2007, following a number of deaths of both cats and dogs, over 150 brands of pet-food were withdrawn from the US market (US Food & Drug Administration, 2014). In addition to health consequences, such activities can also have substantial financial costs, thus following one incident relating to use of the dye Sudan I to adulterate chilli powder, the consequent uninsured loss to one food business was 2.4 million pounds sterling (Premier Foods plc., 2006).

15.4.5.2 Sabotage

A second criminal route by which illegal activities can contribute to the presence of chemical hazards is as a consequence of pre-meditated sabotage. Motivations include extortion, terrorism and employee-grievance. The potential for chemical hazards to be introduced into the food chain by illegal activities has been considered at an international level (World Health Organization, 2002); while the potential for cyanide and other chemical compounds including ricin (a lectin derived from castor beans) and the paralytic shellfish poison, saxitoxin, to be weoponised and delivered through the food chain has also been discussed (Lee et al., 2003). The potential of the employee to be an agent of sabotage cannot be underestimated. It has been estimated that in some businesses, between 75% and 90% of employees behave in a manner which can be described as deliberately dysfunctional or deviant to business operations (Harris and Ogbonna, 2006). Cases of sabotage within the food and more specifically the cereals industry have been reported and which at their most extreme have had public health implications. Such was the case in 2008 when 10 Japanese consumers were poisoned after eating frozen dumplings contaminated with the organophosphorus pesticide methamidophos and originally produced in China. Subsequent investigation revealed that the pesticide had been injected into the dumplings by a production worker as a result of frustration with his wages and colleagues (British Broadcasting Corporation, 2014). A second case relates to an outbreak of poisoning following consumption of bagels in Ankara, Turkey. These were contaminated with 2,2-dichlorovinyl dimethyl phosphate which was attributed to an employee from a competitor company adding the pesticide to the dough mix (Kavalci et al., 2009).

In common with food fraud, risk management of these issues should be integrated into more holistic approaches to manage threats in terms of prevention, response and recovery. Advances in management of risk from diverse sources has led to the development of risk-mitigation tools such as Threat Analysis Critical Control Point (TACCP). This has been defined as the 'systematic management of risk through the evaluation of threats, identification of vulnerabilities, and implementation of controls to materials and products, purchasing, processes, premises, distribution networks and business systems by a knowledgeable and trusted team with the authority to implement changes to procedures' (British Standards Institute, 2014). Another complementary tool is the CARVER + SHOCK (Criticality, Accessibility, Recuperability, Vulnerability, Effect, Recognisability + Shock or impact) approach developed by the US Food and Drug Administration, and described as, 'an offensive targeting prioritization tool that has been adapted for use in the food sector. This tool can be used to assess the vulnerabilities within a system or infrastructure to an attack' (US Food & Drug Administration, 2009).

15.5 CONCLUSIONS

The relative safety of bread compared with other cereal-based foods is at least partly attributable to the technologically related quality attributes of grain and flour demanded by food businesses (e.g. low moisture content, absence of mould growth and foreign matter). Combining those attributes required of the raw material with the temperatures used to produce the finished product means that bread, in particular, has been a food with a relatively lower risk of contributing a significant hazard than many other foods. However, this is not a reason for complacency, since, as discussed in this chapter, there are significant food-safety hazards. For the food-safety manager, like commentators in other fields of management, inspiration might come from considering some of the ideas developed in ancient texts on military strategy. Two appropriate to this context are those written by the Chinese general Sun Tzu (544–496): '*If you know the enemy and know yourself, you need not fear the result of a hundred battles*' and the fourth century Roman strategist Vegetius: '*Let him who desires peace prepare for war*'. Good food-safety management rests on a continuing understanding of the hazards presented throughout the grain chain from cereal to finished product and the development of appropriate strategies to deal with them.

15.6 DISCLAIMER

Mention of any trademark, product or service is no indication of endorsement or any other form of comment by either the author or his organisation.

REFERENCES

Abbas, H.K., Mirocha, C.J., Pawlosky, R.J., Pusch, D.J., 1985. Effect of cleaning, milling and baking on deoxynivalenol in wheat. Appl. Environ. Microbiol. 50, 482–486.

Accum, F., 1820. A Treatise on Adulterations of Food, and Culinary Poisons, Exhibiting the Fraudulent Sophistications of Bread, Beer, Wine, Spirituous Liquors, Tea, Coffee, Cream, Confectionery, Vinegar, Mustard, Pepper, Cheese, Olive Oil, Pickles, and Other Articles Employed in Domestic Economy; and Methods of Detecting Them. Longman, Hurst, Rees, Orme, and Brown, London.

Adamse, P., van Egmond, H.P., 2010. Report Number 2010.011, Tropane Alkaloids in Food. Available from: http://edepot.wur.nl/160741

Agusa, T., Kunito, T., Minh, T.B., Trang, P.T.K., Iwata, H., Viet, P.H., Tanabe, S., 2009. Relationship of urinary arsenic metabolites to intake estimates in residents of the Red River Delta, Vietnam. Environ. Pollut. 157, 396–403.

Agriculture and Horticulture Development Board, 2011. HGCA Grain Storage Guide for Cereals and Oilseeds, third ed. Available from: http://cereals.ahdb.org.uk/media/178349/g52_grain_storage_guide_3rd_edition.pdf

Agriculture and Horticulture Development Board, 2015. Risk Assessment for Fusarium Mycotoxins in Wheat. Available from: http://cereals.ahdb.org.uk/media/418930/is40-risk-assessment-for-fusarium-mycotoxins-in-wheat.pdf

Alldrick, A.J., 1996. The effects of processing on the occurrence of ochratoxin A in cereals. Food Addit. Contam. 13 (Suppl.), 27–28.

Alldrick, A.J., 2009. Risk management—operational implications. In: Coutts, J., Fielder, R. (Eds.), Management of Food Allergens. Wiley-Blackwell, Oxford.

Alldrick, A.J., 2014. Looking for the best compromise in rapid food mycotoxin tests: speed, sensitivity, precision and accuracy. World Mycotoxin J. 7, 407–415.

Alldrick, A.J., Hajšelová, M., 2004. Zearalenone. In: Magan, N., Olsen, M. (Eds.), Mycotoxins in Foods: Detection & Control. Woodhead Publishing Ltd, Cambridge, pp. 351–366.

Allergen Bureau, 2014. VITAL®. Available from: http://allergenbureau.net/vital/

Azziz-Baumgartner, E., Lindblade, K., Gieseker, K., Schurz Rogers, H., Kieszak, S., Njapau, H., Schleicher, R., McCoy, L.F., Misore, A., DeCock, K., Rubin, C., Slutsker, L., The Aflatoxin Investigative Group, 2005. Case–control study of an acute aflatoxicosis outbreak, Kenya, 2004. Environ. Health Perspect. 113, 1779–1783.

Baig, J.A., Kazi, T.G., Shah, A.Q., Afridi, H.I., Kandhro, G.A., Khan, S., Kolachi, N.F., Wadhwa, S.K., Shah, F., Arain, M.B., Jamali, M.K., 2011. Evaluation of arsenic levels in grain crops samples, irrigated by tube well and canal water. Food Chem. Toxicol. 49, 265–270.

Bakir, F., Damluji, S.F., Amin-Zaki, L., Murtadha, M., Khalidi, A., al-Rawi, N.Y., Tikriti, S., Dahahir, H.I., Clarkson, T.W., Smith, J.C., Doherty, R.A., 1973. Methylmercury poisoning in Iraq. Science 181, 230–241.

Bandara, J.M.R.S., Wijewardena, H.V.P., Liyanege, J., Upul, M.A., Bandara, J., 2010. Chronic renal failure in Sri Lanka caused by elevated dietary cadmium: Trojan horse of the green revolution. Toxicol. Lett. 198, 33–39.

Battalglia, R., Hatzold, T., Kroes, R., 1996. Guest editorial: conclusions from the workshop on ochratoxin in food, organised by ILSI Europe in Aix-en-Provence (10–12 January 1996). Food Addit. Contam. 13 (Suppl.), 1–3.

Bender, D.A., Bender, A.E., 1996. Nutrition: A Reference Handbook. Oxford University Press, Oxford.

Berghofer, L.K., Hocking, A.D., Miskelly, D., Jansson, E., 2003. Microbiology of wheat and flour milling in Australia. Int. J. Food Microbiol. 85, 137–149.

Boyacioglu, D., Hettiarachchy, N.S., Stack, R.W., 1992. Effect of three systemic fungicides on deoxynivalenol (vomitoxin) production by *Fusarium graminearum* in wheat. Can. J. Plant Sci. 72, 93–101.

Brice, J.R., Golob, P., 2000. Fumigation with phosphine and carbon dioxide in metal grain tanks for on-farm use in Swaziland. J. Agric. Eng. Res. 76, 17–26.

British Broadcasting Corporation, 2014. China-Japan Poisoned Dumplings: Chinese Man Jailed. Available from: www.bbc.co.uk/news/world-asia-china-25805306

British Standards Institute, 2009. Risk Management—Principles and Guidelines (BS ISO 31000:2009). British Standards Institute, London.

British Standards Institute, 2014. PAS 96:2014—Guide to Protecting and Defending Food and Drink From Deliberate Attack. BSI Group, London.

Bruijnzeel-Koomen, C., Ortolani, C., Aas, K., Bindslev-Jensen, C., Björkstén, B., Moneret-Vautrin, D., Wüthrich, 1995. Adverse reactions to food. Allergy 50, 623–635.

Burks, A., Tang, M., Sicherer, S., Muraro, A., Eigenmann, P., Ebisawa, M., Fiocchi, A., Chiang, W., Beyer, K., Wood, R., Hourihane, J., Jones, S., Lack, G., Sampson, H., 2012. ICON: food allergy. J. Allergy Clin. Immunol. 129, 906–920.

Carey, A.M., Scheckel, K.G., Lombi, E., Newville, M., Yongseong, C., Norton, G.J., Charnock, J.M., Feldmann, J., Price, A.H., Meharg, A.A., 2010. Grain unloading of arsenic species in rice. Plant Physiol. 152, 309–319.

Cattani, I., Romani, M., Boccelli, R., 2008. Effect of cultivation practices on cadmium concentration in rice grain. Agron. Sustain. Dev. 28, 265–271.

Charmley, L.L., Trenholm, H.L., Prelusky, D.B., Rosenberg, A., 1995. Economic losses and decontamination. Nat. Toxins 3, 199–203.

Chatterjee, D., Halder, D., Majumder, S., Biswas, A., Nath, B., Bhattacharya, P., Bhowmick, S., Mukherjee Goswami, A., Saha, D., Hazra, R., Maity, P.B., Chatterjee, D., Mukherjee, A., Bundschuh, J., 2010. Assessment of arsenic exposure from groundwater and rice in Bengal Delta Region, West Bengal, India. Water Res. 44, 5803–5812.

Chaudhry, R., Lall, S.B., Mishra, B., Dhawan, B., 1998. A foodborne outbreak of organophosphate poisoning. Br. Med. J. 317, 268–269.

Christensen, C.M., Miller, B.S., Johnston, J.A., 1992. Moisture and its measurement. In: Sauer, D.B. (Ed.), Storage of Cereal Grains and Their Product. American Association of Cereal Chemists, St. Paul, NM, pp. 39–54.

Clarke, J.M., Knox, R.E., Depauw, R.M., Clarke, F.R., Fernandez, M.R., McCaig, T.N., Singh, A.K., 2010. Brigade durum wheat. Can. J. Plant Sci. 89, 505–509.

Codex Alimentarius Commission, 2003. Recommended International Code of Practice: General Principles of Food Hygiene CAC/RCP 1-1969, Rev. 4-2003. Available from: http://www.fao.org/fao-who-codexalimentarius/sh-proxy/en/?lnk=1&url=https%253A%252F%252Fworkspace.fao.org%252Fsites%252Fcodex%252FStandards%252FCAC%2BRCP%2B1-1969%252FCXP_001e.pdf

Codex Alimentarius Commission, 2006. Code of Practice for the Prevention and Reduction of Dioxin and Dioxin-Like pcb Contamination in Foods and Feeds CAC/RCP 62-2006+D112. Available from: http://www.fao.org/fao-who-codexalimentarius/sh-proxy/en/?lnk=1&url=https%253A%252F%252Fworkspace.fao.org%252Fsites%252Fcodex%252FStandards%252FCAC%2BRCP%2B62-2006%252FCXP_062e.pdf

Codex Alimentarius Commission, 2008. Standard for Foods for Special Dietary use for Persons Intolerant to Gluten: CODEX STAN 118-1979. Available from: http://www.fao.org/fao-who-codexalimentarius/sh-proxy/en/?lnk=1&url=https%253A%252F%252Fworkspace.fao.org%252Fsites%252Fcodex%252FStandards%252FCODEX%2BSTAN%2B118-1979%252FCXS_118e_2015.pdf

Codex Alimentarius Commission, 2010. General Standard for the Labelling of Prepackaged Foods, CODEX STAN 1-1985. Available from: http://www.fao.org/fao-who-codexalimentarius/sh-proxy/en/?lnk=1&url=https%253A%252F%252Fworkspace.fao.org%252Fsites%252Fcodex%252FStandards%252FCODEX%2BSTAN%2B1-1985%252FCXS_001e.pdf

Codex Alimentarius Commission, 2012. Prevention and Reduction of Food and Feed Contamination. Available from: ftp://ftp.fao.org/codex/Publications/Booklets/Contaminants/CCCF_2012_EN.pdf

Codex Alimentarius Commission, 2014. Report of the Eighth Session of the Codex Committee on Contaminants in Foods, The Hague, the Netherlands, 31 March–4 April 2014. Available from: ftp://ftp.fao.org/codex/Reports/Reports_2014/REP14_CFe.pdf

Codex Alimentarius Commission, 2015. Codex General Standard for Food Additives, CODEX STAN 192-1995. Available from: http://www.fao.org/fao-who-codexalimentarius/sh-proxy/en/?lnk=1&url=https%253A%252F%252Fworkspace.fao.org%252Fsites%252Fcodex%252FStandards%252FCODEX%2BSTAN%2B192-1995%252FCXS_192_2015e.pdf

Coeliac UK, 2016. About coeliac disease and dermatitis herpetiformis. Available from: https://www.coeliac.org.uk/coeliac-disease/about-coeliac-disease-and-dermatitis-herpetiformis/D72

Collins, F.W., 1986. Oat phenolics: structure occurrence and function. In: Webster, F.H. (Ed.), Oats: Chemistry & Technology. American Association of Cereal Chemists, St. Paul, MN, pp. 227–295.

CoT—Committee on Toxicity of Chemicals in Food, Consumer Products and The Environment, 1996. Statement on the safety-in-use of chlorine and chlorine dioxide as flour treatment agents. Available from: http://www.archive.official-documents.co.uk/document/doh/toxicity/chap-1a.htm

Dahlin, A.S., Eriksson, J., Campbell, C.D., Öborn, I., 2016. Soil amendment affects Cd uptake by wheat—are we underestimating the risks from chloride inputs? Sci. Total Environ. 554–555, 349–357.

Dennis, M.J., Burrell, A., Mathieson, K., Willetts, P., Massey, R.C., 1994. The determination of the flour improver potassium bromate in bread by gas chromatographic and ICP-MS methods. Food Addit. Contam. 11, 633–639.

Dennis, M.J., Massey, R.C., Ginn, R., Willetts, P., Crews, C., Parker, I., 1997. The contribution of azodicarbonamide to ethyl carbamate formation in bread and beer. Food Addit. Contam. 14, 101–108.

Dierick, K., Van Coillie, E., Swiecicka, I., Meyfroidt, G., Devlieger, H., Meulemans, A., Hoedemaekers, G., Fourie, L., Heyndrickx, M., Mahillon, J., 2005. Fatal family outbreak of *Bacillus cereus*-associated food poisoning. J. Clin. Microbiol. 43, 4277–4279.

D'Mello, J.P.F., Duffus, C.M., Duffus, J.H., 1991. Toxic Substances in Crop Plants. Royal Society of Chemistry, Cambridge.

Dunn-Galvin, A., Chan, C.-H., Crevel, R., Grimshaw, K., Poms, R., Schnadt, S., Taylor, S.L., Turner, P., Allen, K.J., Austin, M., Baka, A., Baumert, J.L., Baumgartner, S., Beyer, K.L., Bucchini, K.L., Fernandez-Rivas, M., Grinter, K., Houben, G.F., Hourihane, J., Kenna, F., Kruizinga, A.G., Lack, G., Madsen, C.B., Mills, E.N.C., Papadopoulos, N.G., Alldrick, A., Regent, L., Sherlock, R., Wal, J.-M., Robert, G., 2015. Precautionary allergen labelling: perspectives from key stakeholder groups. Allergy 70, 1039–1051.

Edwards, S., 2011. Zearalenone risk in European wheat. World Mycotoxin J. 4, 443–448.

Edwards, S.G., Pirgozliev, S.R., Hare, M.C., Jenkinson, P., 2001. Quantification of trichothecene-producing Fusarium species in harvested grain by competitive PCR to determine efficacies of fungicides against Fusarium head blight of winter wheat. Appl. Environ. Microbiol. 67, 1575–1580.

Environment Agency, 2008. Methyl Bromide. Available from: http://www.environment-agency.gov.uk/business/topics/pollution/470.aspx

European Commission, 2006a. COMMISSION REGULATION (EU) No 1881/2006 of 19 December 2006 setting maximum levels for certain contaminants in foodstuffs. Off. J. Eur. Union L364, 5–24.

European Commission, 2006b. Commission Recommendation of 17 August 2006 on the prevention and reduction of Fusarium toxins in cereals and cereal products (2006/583/EC). Off. J. Eur. Union L234, 35–40.

European Commission, 2012. COMMISSION REGULATION (EU) No 380/2012 of 3 May 2012 amending Annex II to Regulation (EC) No 1333/2008 of the European Parliament and of the Council as regards the conditions of use and the use levels for aluminium-containing food additives. Off. J. Eur. Union L119, 14–38.

European Commission, 2015a. COMMISSION RECOMMENDATION (EU) 2015/976 of 19 June 2015 on the monitoring of the presence of tropane alkaloids in food. Off. J. Eur. Union L157, 97–98.

European Commission, 2015b. COMMISSION REGULATION (EU) 2015/1006 of 25 June 2015 amending Regulation (EC) No 1881/2006 as regards maximum levels of inorganic arsenic in foodstuffs. Off. J. Eur. Union L161, 14–16.

European Commission, 2016a. COMMISSION REGULATION (EU) 2016/239 of 19 February 2016 amending Regulation (EC) No 1881/2006 as regards maximum levels of tropane alkaloids in certain cereal-based foods for infants and young children. Off. J. Eur. Union L45, 3–5.

European Commission, 2016b. RASFF—Food and Feed Safety Alerts. Available from: http://ec.europa.eu/food/safety/rasff/index_en.htm

European Commission, 2016c. EU—Pesticides Database. Available from: http://ec.europa.eu/food/plant/pesticides/eu-pesticides-database/public/?event=homepage&language=EN

European Food Safety Authority, 2009a. Scientific opinion on arsenic in food. EFSA J. 7 (10), 1–191, 1351.

European Food Safety Authority, 2009b. Scientific opinion on cadmium in food. EFSA J. 980, 1–139.

European Food Safety Authority, 2012. Cadmium dietary exposure in the European population. EFSA J. 10 (1), 2551.

European Food Safety Authority, 2013a. Scientific opinion on Tropane alkaloids in food and feed. EFSA J. 11, 3386, 113 pp.

European Food Safety Authority, 2013b. Scientific opinion on lead in food. EFSA J. 8 (4), 1570.

European Food Safety Authority, 2013c. Dietary exposure to aluminium-containing food additives. Supporting Publications 2013, EN-411, 17 pp.

European Food Safety Authority, 2014. Dietary exposure to inorganic arsenic in the European population. EFSA J. 12 (3), 3597.

European Food Safety Authority, 2015. EFSA European Comprehensive Food Consumption Data Base. Available from: http://www.efsa.europa.eu/en/food-consumption/comprehensive-database

European Parliament and Council, 2005. REGULATION (EC) No 396/2005 of The European Parliament And of The Council of 23 February 2005 on maximum residue levels of pesticides in or on food and feed of plant and animal origin and amending Council Directive 91/414/EEC foodstuffs. Off. J. Eur. Union L70, 1–16.

European Parliament and Council, 2011. REGULATION (EU) No 1169/2011 of The European Parliament and of the Council of 25 October 2011 on the provision of food information to consumers, amending Regulations (EC) No 1924/2006 and (EC) No 1925/2006 of the European

Parliament and of the Council, and repealing Commission Directive 87/250/EEC, Council Directive 90/496/EEC, Commission Directive 1999/10/EC, Directive 2000/13/EC of the European Parliament and of the Council, Commission Directives 2002/67/EC and 2008/5/EC and Commission Regulation (EC) No 608/2004. Off. J. Eur. Union L304, 18–63.

Eyles, M.J., Moss, R., Hocking, A.D., 1989. The microbiological status of Australian flour and the effects of milling procedures on the microflora of wheat and flour. Food Aust. 41, 704–709.

Fang, T., Liu, G., Zhou, C., Lu, L., 2015. Lead in soil and agricultural products in the Huainan Coal Mining Area, Anhui, China: levels, distribution, and health implications. Environ. Monit. Assess. 187, 152–162.

Food Drink Europe, 2014. Acrylamide Toolbox 2013. Available from: http://www.fooddrinkeurope.eu/uploads/publications_documents/AcrylamideToolbox_2013.pdf

Frank, T., Habernegg, R., Yuan, F.-J., Shu, Q.-Y., Engel, K.H., 2009. Assessment of the contents of phytic acid and divalent cations in low phytic acid (lpa) mutants of rice and soybean. J. Food Compos. Anal. 22, 278–284.

Fretz, R., Schmid, D., Brueller, W., Girsch, L., Pichler, A.M., Riediger, K., Safer, M., Allerberger, F., 2007. Food poisoning due to Jimson weed mimicking Bacillus cereus food intoxication in Austria, 2006. Int. J. Infect. Dis. 11, 557–558.

Food Standards Agency, 2006a. Survey of metals in weaning foods and formulae for infants. Food Safety Information Sheet 17/06. Available from: www.food.gov.uk/science/surveillance/fsisbranch2006/fsis1706

Food Standards Agency, 2006b. Guidance on allergen management and consumer. Available from: http://www.food.gov.uk/sites/default/files/multimedia/pdfs/maycontainguide.pdf

Food Standards Agency, 2007a. Survey of metals in weaning foods and formulae for infants—additional information on inorganic arsenic and methyl mercury levels. Food Safety Information Sheet 07/03. Available from: www.food.gov.uk/science/surveillance/fsisbranch2007/fsis0307

Food Standards Agency, 2007b. The UK code of good storage practice to reduce ochratoxin a in cereals. Available from: http://www.food.gov.uk/sites/default/files/multimedia/pdfs/ochratoxinacop.pdf

Food Standards Agency, 2009. Arsenic in rice drinks. Food Safety Information Sheet 02/09. Available from: www.food.gov.uk/multimedia/pdfs/fsisarsenicinrice.pdf

Griffiths, D.W., 1991. Condensed tannins. In: D'Mello, J.P.F., Duffus, C.M., Duffus, J.H. (Eds.), Toxic Substances in Crop Plants. Royal Society of Chemistry, Cambridge, pp. 180–201.

Griffin, W.J., Lin, G.D., 2000. Chemotaxonomy and geographical distribution of tropane alkaloids. Phytochemistry 53, 623–637.

Harris, L.C., Ogbonna, E., 2006. Service sabotage: a study of antecedents and consequences. J. Acad. Mark. Sci. 34, 543 558.

Henderson, L.J., 1918. The prevention of rope in bread. Science 48, 247–248.

Hill, R.A., Blankenship, P.D., Cole, R.J., Sanders, T.H., 1983. Effects of soil moisture and temperature on preharvest invasion of peanuts by the Aspergillus flavus group and subsequent aflatoxin development. Appl. Environ. Microbiol. 45, 628–633.

Hurrel, R.F., 2003. Influence of vegetable protein sources on trace element and mineral bioavailability. J. Nutr. 133, 2973S–2977S.

Hussain, A., Murtaza, G., Ghafoor, A., Basra, A.S.M., Qadir, M., Sabir, M., 2010. Cadmium contamination of soils and crops by long term use of raw effluent, ground and canal waters in agricultural lands. Int. J. Agric. Biol. 12, 851–856.

International Agency for Research on Cancer, 1987. IARC Monographs on the Evaluation of the Carcinogenic Risks to Humans, Suppl. 7, Overall Evaluations of Carcinogenicity: An Updating of IARC Monographs Volumes 1 to 42. International Agency for Research on Cancer, Lyon.

International Agency for Research on Cancer, 1997. Polychlorinated dibenzo para-dioxins and polychlorinated dibenzofurans. IARC Monographs on the Evaluation of Carcinogenic Risks to Humans Vol. 69. International Agency for Research on Cancer, Lyon.

International Organisation for Standardisation, 2005. ISO, 2005, ISO/IEC 17025:2005 General Requirements for the Competence of Testing and Calibration Laboratories. ISO, Geneva.

JECFA—Joint FAO/WHO Expert Committee on Food Additives, 2005. Report Of The 64th Meeting, Rome 8–17 February 2005. Available from: www.Who.Int/Ipcs/Food/Jecfa/Summaries/Summary_Report_64_Final.Pdf

Jennings, P., Turner, J.A., Nicholson, P., 2000. Overview of Fusarium ear blight in the UK—effect of fungicide treatment on disease control and mycotoxin production. Proceedings of the BCPC Conference Pests and Diseases, 2000 2, 707–712.

Jonsson, N., 1996. Konserva och lagra spannmål på rätt sätt! Jordbrukstekniska Institutet, Uppsala.

Juliano, B.O., 1985. Production and utilisation of rice. In: Juliano, B.O. (Ed.), Rice: Chemistry and Technology. American Association of Cereal Chemists, St. Paul, MN, pp. 1–16.

Kabata-Pendias, A., Mukherjee, A., 2007. Trace Elements From Soil to Human. Springer, Berlin, 550 pp.

Kaur, P., 1983. Survival and growth of Bacillus cereus in bread. J. Appl. Microbiol. 60, 513–516.

Kavalci, C., Durukan, P., Ozer, M., Cevik, Y., Kavalci, G., 2009. Organophosphate poisoning due to a wheat bagel. Intern. Med. 48, 85–88.

Kent, N.L., Evers, A.D., 1994. Kent's Technology of Cereals, fourth ed. Pergamon, Oxford.

Kimura, A.C., Palumbo, M.S., Meyers, H., Abbott, S., Rodriguez, R., Werner, S.B., 2005. A multi-state outbreak of Salmonella serotype Thompson infection from commercially distributed bread contaminated by an ill food handler. Epidemiol. Infect. 133, 823–828.

Kobayashi, E., Suwazono, Y., Dochi, M., Honda, R., Kido, T., Nakagawa, H., 2009. Influence of drinking and/or cooking with Jinzu river water on the development of itai-itai disease. Biol. Trace Elem. Res. 129, 46–57.

Kurokawa, Y., Hayashi, Y., Maekawa, A., Takahashi, M., Kokubo, T., Odashima, S., 1983. Carcinogenicity of potassium bromate administrated orally to F344 rats. J. Natl. Cancer Inst. 71, 965–972.

Laboratory of the Government Chemist, 2011. Aluminium in Imported Noodles. Available from: https://www.gov.uk/government/uploads/system/uploads/attachment_data/file/317870/Aluminium_in_imported_noodles.pdf

Lee, R.V., Harbison, R.D., Draughon, F.A., 2003. Food as a weapon. Food Prot. Trends 23, 664–674.

Lefyedi, M.L., Taylor, J.R.N., 2006. Effect of dilute alkaline steeping on the microbial contamination, toxicity and diastatic power of sorghum malt. J. Inst. Brew. 112, 108–116.

Legan, J.D., 1994. The microbiological condition of 1994 new harvest flours, Chorleywood Digest (December 1994), Chorleywood, UK, Flour Milling & Baking Research Association, pp. 112–113.

Legan, J.D, Curtis, P.S., 1990. Predicting wheat and flour quality from sunspot activity: fact or fantasy? FMBRA Bulletin (April, 1990), Chorleywood, UK, Flour Milling & Baking Research Association, pp. 89–93.

Lepschy-von Gleissenthal, J., Suess, A., 1996. Verteilung des Trichothecenmykotoxins Deoxynivalenol bei der Vermahlung von Weizen. Getreide Mehl Brot. 50, 340–342.

Li, G., Sun, G.X., Williams, P.N., Nunes, L., Zhu, Y.G., 2011. Inorganic arsenic in Chinese food and its cancer risk. Environ. Int. 37, 1219–1225.

Liang, F., Li, Y., Zhang, G., Tan, M., Lin, J., Liu, W., Li, Y., Lu, W., 2010. Total and speciated arsenic levels in rice from China. Food Addit. Contam. 27, 810–816.

Liu, J., Zhu, Q., Zhang, Z., Xu, J., Yang, J., Wong, M.H., 2009. Variations in cadmium accumulation among rice cultivars and types and the selection of cultivars for reducing cadmium in the diet. J. Sci. Food Agric. 85, 147–153.

Ma, J.F., Yamaji, N., Mitan, I.N., Xu, X.-Y., Su, Y.-H., McGrath, S.P., Zhao, F.-J., 2008. Transporters of arsenite in rice and their role in arsenic accumulation in rice grain. Proc. Natl. Acad. Sci. USA 105, 9931–9935.

Masha, A.C., Svanberg, U., 1983. Preparation of weaning foods with high nutrient density using flour of germinated cereals. Food Nutr. Bull. 5, 10–14.

Meharg, A.A., Sun, G., Williams, P.N., Adomako, E., Deacon, C., Zhu, Y.-G., Feldmann, J., Raab, A., 2008. Inorganic arsenic levels in baby rice are of concern. Environ. Pollut. 152, 746–749.

Milazzo, A., Rose, N., 2001. An outbreak of Salmonella Typhimurium phage type 126 linked to a cake shop in South Australia. Commun. Dis. Intell. 25 (2), 73.

Miller, J.D., 1995. Fungi and mycotoxins in grain. J. Stored Prod. Res. 31, 1–16.

Mohabbat, O., Younos, M.S., Merzad, A.A., Srivastava, R., Sediq, G.G., Aram, G.N., 1976. An outbreak of hepatic veno-occlusive disease in north-western Afghanistan. Lancet 308, 269–271.

National Association of British and Irish Millers, 2014. **nabim** Recommended code of practice for mill intake. Available from: www.nabim.org.co.uk

Ogawa, T., Kobayashi, E., Okubo, Y., Suwazono, Y., Kido, T., Nogawa, K., 2004. Relationship among prevalence of patients with Itai-itai disease, prevalence of abnormal urinary findings, and cadmium concentrations in rice of individual hamlets in the Jinzu River basin, Toyama prefecture of Japan. Int. J. Environ. Health Res. 14, 243–252.

Ono, E.Y.S., Sasaki, E.Y., Hashimoto, E.H., Hara, L.N., Corrêa, B., Itano, E.N., Sugiura, T., Ueno, Y., Hirooka, E.Y., 2003. Post-harvest storage of corn: effect of beginning moisture on mycoflora and fumonisin contamination. Food Addit. Contam. 19, 1081–1090.

Otani, T., Seike, N., Miwa, T., 2006. Levels of dioxins in rice, wheat, soybean, and adzuki bean cultivated in 1999 to 2002 in Japan and estimation of their intake. J. Food Hygiene Soc. (Japan) 47, 182–188.

Panda, S.K., Upadhyay, R.P., Nath, S., 2010. Arsenic stress in plants. J. Agron. Crop Sci. 196, 161–174.

Perharič, L., Koželj, G., Družina, B., Stanovnik, L., 2013. Risk assessment of buckwheat flour contaminated by thorn-apple (*Datura stramonium* L.) alkaloids: a case study from Slovenia. Food Addit. Contam. Part A 30, 321–330.

Perilli, P., Mitchell, L.G., Grant, C.A., Pisante, M., 2010. Cadmium concentration in durum wheat grain (*Triticum turgidum*) as influenced by nitrogen rate, seeding date and soil type. J. Sci. Food Agric. 90, 813–822.

Pomeranz, Y., 1992. Biochemical, functional and nutritive changes during storage. In: Sauer, D.B. (Ed.), Storage of Cereal Grains and Their Products. American Association of Cereal Chemists, St. Paul, MN, pp. 55–141.

Poms, R.E., Alldrick, A., Grazina, J., Ermolli, M., Binder, E.-M., Reynolds, S., Goekmen, V., Haas-Lauterbach, S., 2009. Testing for food safety. In: Cauvain, S.P., Young, L.S. (Eds.), Applications of Cereal, Flour and Dough Testing. D E Stech Publications Inc., Lancaster, PA, pp. 237–240.

Premier Foods plc., 2006. Premier Foods plc Annual Report 2005. Available from: www.premierfoods.co.uk/shadomx/apps/fms/fmsdownload.cfm?file_uuid=EA640BAC-F1FF-886B-4B8B-0C0342C46EC8&siteName=premierfoods

Raevuori, M., Kiutamo, T., Niskanen, A., Salminen, K., 1976. An outbreak of food-poisoning in Finland associated with boiled rice. J. Hygiene (Camb.) 76, 316.

Reddy, N.R., Pierson, M.D., Sathe, S.K., Salunkhe, D.K., 1989. Phytates in Cereals and Legumes. CRC Press, Boca Raton, FA.

Red Tractor Assurance, 2014. Combinable crops and sugar beet standards Vs.03. Available from: http://assurance.redtractor.org.uk/contentfiles/Farmers-5611.pdf

Richter, K.S., Dorneanu, E., Eskridge, K.M., Rao, C.S., 1993. Microbiological quality of flours. Cereal Foods World. 38, 367–369.

Schlundt, J., van Eck, W., Vallanjon, M., 2003. WHO and FAO have a recipe for safer food. Bull. World Health Org. 81, 315.

Schoenbaum, S.C., Baker, O., Jezek, Z., 1976. Common-source outbreak of hepatitis due to glazed and iced pastries. Am. J. Epidemiol. 10, 74–80.

Sicherer, S.H., Sampson, H.A., 2014. Food allergy: epidemiology, pathogenesis, diagnosis, and treatment. J. Allergy Clin. Immunol. 133, 291–307.

Simpson, D.R., Weston, G.E., Turner, J.A., Jennings, P., Nicholson, P., 2001. Differential control of head blight pathogens of wheat by fungicides and consequences for mycotoxin contamination of grain. Eur. J. Plant Pathol. 107, 421–431.

Smith, E., Kempson, I., Juhasz, A.L., Weber, J., Skinner, W.M., Grafe, M., 2009. Localization and speciation of arsenic and trace elements in rice tissues. Chemosphere 76, 529–535.

Steinberg, E.B., Henderson, A., Karpati, A., Hoekstra, M., Marano, N., Souza, J.M., Simons, M., Kruger, K., Giroux, J., Rogers, H.S., Hoffman, M.K., Kadry, A.R., Griffin, P.M., 2006. Mysterious outbreaks of gastrointestinal illness associated with burritos supplied through school lunch programs. J. Food Protect. 69, 1690–1698.

Svensson, K., Abramsson, L., Becker, W., Glynn, A., Hellenäs, K.-E., Lind, Y., Rosén, J., 2003. Dietary intake of acrylamide in Sweden. Food Chem. Toxicol. 41, 1581–1586.

Taylor, J.R.N., Duodu, K.G., 2010. Sorghum and millets; characteristics and quality requirements. In: Wrigley, C.W., Batey, I.L. (Eds.), Cereal Grains: Assessing and Managing Quality. Woodhead Publishing Limited, Oxford, pp. 237–263.

Thompson, J.M., Waites, W.M., Dodd, C.E.R., 1998. Detection of rope spoilage in bread caused by Bacillus species. J. Appl. Microbiol. 85, 481–486.

Tkachuk, R., Dexter, J.E., Tipples, K.H., Nowicki, T.W., 1991. Removal by specific gravity table of tombstone kernels and associated trichothecenes from wheat infected with Fusarium head blight. Cereal Chem. 68, 428–431.

Tlustos, C., 2009. The dioxin contamination event in Ireland 2008. Organohalogen Compounds 71, 1155 1159.

Tuite, J., Foster, G.H., Eckhoff, S.R., Shotwell, L., 1986. Sulfur dioxide treatment to extend corn drying time. Cereal Chem. 63, 462–464.

Tuli, R., Chakrabarty, D., Trivedi, K.P., Tripathi, D.R., 2010. Recent advances in arsenic accumulation and metabolism in *rice*. Mol. Breed. 26, 307–323.

Ueno, D., Yamaji, N., Kono, I., Huang, C.-F., Ando, T., Yano, M., Ma, J.F., 2010. Gene limiting cadmium accumulation in rice. Proc. Natl. Acad. Sci. USA 107, 16500–16505.

UK Department of Health, 1991. Dietary reference values for food energy and nutrients for the United Kingdom. Report on Health & Social Subjects Number 41, HMSO, London.

UN Environment Programme, 2010. Final review of scientific information on lead. Version of December 2010. Available from: http://www.unep.org/chemicalsandwaste/Portals/9/Lead_Cadmium/docs/Interim_reviews/UNEP_GC26_INF_11_Add_1_Final_UNEP_Lead_review_and_apppendix_Dec_2010.pdf.

US Agency For Toxic Substances and Disease Registry, 2007. Toxicological Profile for lead. U.S. Department of Health and Human Services. 1–582. Available from: http://www.atsdr.cdc.gov/toxprofiles/tp13.pdf

US Environmental Protection Agency, 1982. Exposure and risk assessment for arsenic. U.S. Environmental Protection Agency. Office of Water Regulations and Standards. PB85221711. EPA-440/4-85-005.

US Food & Drug Administration, 2009. CARVER+Shock Primer. Available from: www.fda.gov/Food/FoodDefense/FoodDefensePrograms/ucm376791.htm

US Food & Drug Administration, 2014. Melamine Pet Food Recall of 2007. Available from: http://www.fda.gov/AnimalVeterinary/SafetyHealth/RecallsWithdrawals/ucm129575.htm+D180

van Dongen, P.W.J., de Groot, A.N.J.A., 1995. History of ergot alkaloids from ergotism to ergometrine. Eur. J. Obstetr. Gynecol. Reprod. Biol. 60, 109–116.

Velíšek, J., 2009. Chloropropanols. In: Stadler, R.H., Lineback, D.R. (Eds.), Process Induced Food Toxicants. Wiley, Hoboken, NJ, pp. 539–562.

Wångstrand, H., Eriksson, J., Öborn, I., 2007. Cadmium concentration in winter wheat as affected by nitrogen fertilization. Eur. J. Agron. 26, 209–214.

Weather Innovations Incorporated, 2016. Available from: http://www.weatherinnovations.com/models.cfm

Weeks, D.E., 1967. Endrin food-poisoning. A report on four outbreaks caused by two separate shipments of endrin-contaminated flour. Bull. World Health Organ. 37, 499–512.

World Health Organization, 1988. Pyrrolizidine Alkaloids, International Programme on Chemical Safety, Environmental Health Criteria 80. Available from: www.inchem.org/documents/ehc/ehc/ehc080.htm

World Health Organization, 2002. Terrorist Threats to Food: Guidance for Establishing and Strengthening Prevention and Response Systems. Available from: http://apps.who.int/iris/bitstream/10665/42619/1/9241545844.pdf?ua=1

Yang, J., Chen, T., Lei, M., Zhou, X., Huang, Q., Ma, C., Gu, R., Guo, G., 2015. New isotopic evidence of lead contamination in wheat grain from atmospheric fallout. Environ. Sci. Pollut. Res. 22, 14710–14718.

Youssef, A.M.M., Bolling, H., Moustafa, E.K., Moharram, Y.G., 1988. Extraction, determination and fractionation of sorghum polyphenols. Food Chem. 30, 103–111.

Zhang, R., Zhang, X., Wu, K., Wu, H., Sun, Q., Hu, F.B., Han, J., Willett, W.C., Giovannucci, E.L., 2016a. Rice consumption and cancer incidence in US men and women. Int. J. Cancer 138, 555–564.

Zhang, Y., Liu, P., Wang, C., Wu, Y., 2016b. Human health risk assessment of cadmium via dietary intake by children in Jiangsu Province, China. Environ. Geochem. Health, Available from: http://link.springer.com/article/10.1007%2Fs10653-016-9805-5.

Zhao, F.J., Zhu, Y.-G., Meharg, A.A., 2013. Methylated arsenic species in rice: geographical variation, origin, and uptake mechanisms. Environ. Sci. Technol. 47, 3957–3966.

Chapter 16

Breeding for Grain-Quality Traits

Zoltán Bedő, László Láng and Mariann Rakszegi
Agricultural Institute, Centre for Agricultural Research, Hungarian Academy of Sciences, Martonvásár, Hungary

16.1 INTRODUCTION

Diversity in wheat use requires breeding efforts to focus on several quality attributes and their interactions, such as grain hardness, gluten characteristics, starch and lipid properties, and so on. Cereal products are based on distinct types of quality (Lookhart and Wrigley, 1997). When setting goals for the future, breeders will need to comply with the new requirements of a changing society, a new lifestyle and its consequences for the human diet and health. Classical quality-oriented breeding has a long tradition, with a focus mainly on the improvement of yield, protein content and composition. This is the most important wheat quality criterion, especially for bread making. A new challenge for small grain cereal breeders in our changing world is to develop novel cereals with modified nutritional composition. To achieve new developments in quality traits, breeders need new tools to meet the future challenges. For this purpose they study the natural variation in different quality traits, develop pre-breeding programs, widen the genetic variability by introducing genes from wild and cultivated relatives, combine classical selection methods with molecular marker technologies, utilise the advances of gene technology to develop genetically modified (GM) cereals or apply reverse genetic approach in mutation breeding, etc., in order to improve grain-quality traits.

16.2 CLASSICAL BREEDING FOR QUALITY IMPROVEMENT

The method most widely used in classical breeding is cross-breeding, which is divided into several stages. The first step is to develop populations with broad genetic variability. After crossing, the diverse properties of the parents recombine in the progeny, producing genetically variable populations from which stable, homozygous lines can be selected. In practice, plant selection usually means ear selection, giving enough seeds to allow the sowing of ear progeny

Cereal Grains. http://dx.doi.org/10.1016/B978-0-08-100719-8.00016-4

rows in the following year. The third stage of breeding is testing. After several years of experiments in different environments, it is possible to identify lines that have significantly better agronomical properties than existing cultivars, combined with satisfactory end-use quality.

The parental partners most frequently used by cereal breeders are adapted lines or varieties, exotic lines or varieties, land races or old varieties, and wild and cultivated relatives. The crossing method depends to a great extent on the type of parental partner. Smaller genetic gains in quality traits can be hoped for when crossing adapted lines or varieties, since in this case the adaptability does not change substantially, providing a greater chance of rapidly selecting a new commercial variety. When using non-adapted genotypes (e.g. land races or exotic lines) the backcross method is generally applied, where breeders endeavor to retain the agronomic traits of the recurrent, adapted parent to achieve the desired quality.

In order for every possible recombination in a segregating population to occur at least once, a large population size would be needed, enabling desirable genotypes occurring at low frequency to be selected. Large population size is important because of the polygenic nature of many quality traits. This is why selection is generally begun at the time when homozygous lines develop (in the F_5–F_6 generation). If preliminary selection in early generations is to be successful, several conditions must be satisfied. In many regions, the window between harvest and sowing is very narrow, whereas time is required for the analyses on which selection decisions can be based. There is thus a need for rapid methods with which large numbers of lines can be tested easily on small samples. When testing samples from early generations it is also necessary to ensure that a sufficient quantity of seed is retained after the quality analysis for sowing and raising the next generation. For this reason, a number of micro-analytical methods has been elaborated, for instance for wheat gluten analysis or wheat dough analysis. The first significant quality analyser of this type was the 2 g Mixograph (Rath et al., 1990), developed in Australia, which gave well-replicated results and was more objective due to the automatic evaluation (Békés et al., 2001). The micro-extension tester reported by Rath et al. (1995) is valuable in early-generation selection for wheat quality, where the size of the available sample does not allow testing by the standard method (Grasgruber et al., 2002). As little as 10 g flour is required for this analysis. The micro Z-arm mixer (Haraszi et al., 2004), the mixing action of which is analogous to that of the traditional valorigraph and farinograph, has the potential to become widely used for the estimation of dough rheological traits, particularly among breeders in the countries of Central and Northern Europe, where the farinograph has traditionally been used. These analyses could be complemented by a micro-baking test (Békés et al., 2003). Near infrared spectroscopy (NIR) is widely used to provide rapid analysis of large numbers of small samples for various quality traits, and can be used to group genotypes based on specific characters (Osborne, 2006), including endosperm structure or protein content. It gives a good prediction of the Zeleny sedimentation index, which is a good indicator of loaf volume. NIR analysis of the whole grain will accelerate the breeding process by reducing the preparation time required and by decreasing the cost of selection, while the intact samples that are analysed can be planted for further selection.

Among the genetic resources, there is a long tradition of using populations of land races and old varieties, as in many countries the genetic background for increasing protein content and improving rheological quality comes from improved land races produced by mass selection from local land races, or by varieties with primitive, heterogeneous populations. With the cultivation of old varieties selected from land races, the Canadian spring wheat Red Fife, for instance, laid the foundations for the breeding of wheat with good bread-making quality in the United Kingdom, while another Canadian variety, Marquis, served the same purpose in Finland and Hungary. In Central Europe, the Bánkúti wheat varieties represented a transitional stage between land races and modern varieties.

Despite the fact that the genetic variability within *Triticum aestivum* (L.) is nowhere near as greatly exploited as would be desirable, varieties containing genes or gene complexes originating from wild and cultivated relatives of *T. aestivum* are increasingly of interest to plant breeders. One of the obstacles faced when transferring genes from related species to wheat via direct hybridisation is the incompatibility between species, as the male and female gametes from different species are unable to unite to form a zygote. Distant hybridisation has been given particular attention in regions where natural populations of wild species have great genetic variability, and where agronomically valuable, locally adapted forms are available to breeders. One such program, involving *T. dicoccoides*, is underway in Israel. In addition to its favourable disease resistance, this species is also a promising source for broadening the genetic basis of *T. aestivum* in terms of increased kernel mass and protein content. The results reported by Feldman et al. (1994) indicated that common wheat lines with both competitive yield potential and increased protein content had been produced.

Selection specifically for increased protein content and technological quality was carried out by Sharma and Gill (1983). Plainsman V, Plainsman IV, Encore and Frontiersman contain genes from *Aegilops ovata,* while addition lines containing the 1E chromosome from *Thinopyrum elongatum* have improved seed storage protein quality (Garg et al., 2009). Similar results were obtained with *Thinopyrum intermedium* when Cao et al. (2007) examined substitution for chromosome 1D of wheat. The V genome of *Dasypyrum villosum* (L.) Candargy (Dv), also known as *Haynaldia villosa* (L.) Schur, an annual wild diploid relative of common wheat, contains genes for elevated seed protein content and also has a positive effect on gluten strength (Shewry et al., 1987, 1991). Zhang et al. (2014b) suggested that the *Glu-V1* and *Gli-V1/Glu-V3* loci, located on chromosome 1V of *D. villosum* (L.), could have a positive effect on grain quality traits, like protein content, Zeleny sedimentation value and the rheological characteristics of wheat flour dough.

Barley is one of the most important β-glucan sources. A significantly higher β-glucan level was detected in the leaves and grains of the wheat/barley 1HS

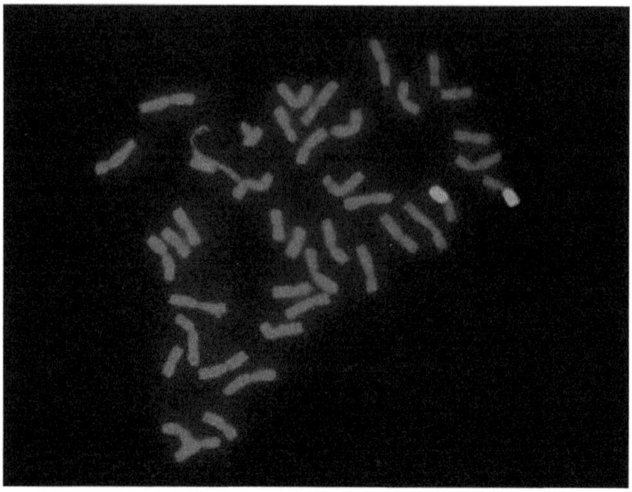

FIGURE 16.1 Genomic in situ hybridisation (GISH) on mitotic chromosomes of the 'Asakaze komugi' (wheat)—'Manas' (barley) 4BS.7HL translocation line. Labelled genomic DNA of barley was used as probe and highlighted barley chromosome arm 7HL in light grey (red in web version). The wheat chromosomes were counterstained with DAPI in dark grey (blue in web version). *(Photo by Klaudia Kruppa.)*

and 7H addition lines compared to the control wheat line. The expression of the HvCslF9 and HvCslF6 genes in the genetic background of wheat was also determined by quantitative RT-PCR, and the HvCslF9 gene was mapped to the short arm of the 1H chromosome (Cseh et al., 2013). An increased β-glucan level was also detected in the 4BS.7HL translocation line (Fig. 16.1), demonstrating that the HvCslF6 gene is of potential relevance for the manipulation of wheat (1,3;1,4)-β-D-glucan levels (Cseh et al., 2011).

16.3 MOLECULAR BREEDING TECHNOLOGIES

The US Plant Breeding Research Priorities Survey was carried out in 2009 to identify the research areas most important for continued progress in plant breeding. The top three categories based on plant breeders' needs were connected with the characterisation, collection and exchange of germplasm, molecular marker development and mapping, including sequencing, and the development of plant phenotyping databases. The choice of these research areas proves that breeding technologies have become more complex and that molecular breeding tools now play an important role.

Cereal breeding programs for processing quality require the selection of new germplasm, widening the genetic variability and increasing the efficiency of selection. In the past, the selection of breeding lines with enhanced performance relied mainly on the direct measurement of the traits of interest, though the use of markers was always important, and morphological, cytogenetic and biochemical markers have long proved to be useful tools for breeders. Biochemical markers have found wide application in wheat breeding for technological quality, since wheat endosperm storage proteins exhibit a high level of polymorphism between genotypes. DNA markers have now become routinely used in selection technology. Plant breeders apply these markers to identify, introduce and select genetic variation in breeding programs. The most frequent application is molecular marker-assisted selection (MAS), which offers a tool for improving the methodology of the selection process.

The use of molecular markers for the development of new germplasm and to improve the efficiency of pre-breeding for processing quality is a great step forward in cereal improvement. It enables breeders to accelerate the introgression and backcrossing of genes into diverse genetic backgrounds, while allowing genes with similar phenotypic effects to be pyramided. Marker-assisted backcross breeding is a useful tool to accelerate the introgression of genes into an adapted background. By simulating recombination during meiosis, breeders are now able to recover the recurrent parent using molecular markers more efficiently than by traditional backcrossing (Frisch et al., 2000). The agronomic traits of BC_5 and BC_6 lines are very similar to those of the recurrent parents. Marker-assisted backcross breeding provides efficient positive foreground selection for the donor trait, positive background selection for the recurrent parental genome and negative background selection against undesirable donor parent alleles.

The identification of DNA sequences for various alleles of the genes encoding the proteins of high and low molecular weight glutenins, puroindolines, granule-bound starch synthase (GBSS), etc., and the wide application of the polymerase chain reaction have enabled diagnostic DNA markers to be designed for quality traits (Gale, 2005). Such markers are now being used by wheat breeders to select lines with the required quality attributes. Marker-assisted breeding offers the advantage of being applicable in any developmental stage for any type of plant tissue, while being independent of environmental conditions. MAS is a useful method for the selection of traits in cases where phenotypic evaluation is difficult or environmental effects are strong.

The identification of DNA markers linked to quantitative trait loci (QTL) is an important tool for use in MAS aimed at improving quality traits in breeding programs that are designed to select new genetic resources for the development of varieties with the desired end-use quality. QTL mapping in plants is usually conducted using a population derived from a cross between two inbred lines. The power of such QTL detection and the estimation of the effects greatly depend on the choice of the two parental lines.

QTL analysis, carried out by Rousset et al. (2001), revealed that bread-making quality is subject to complex control and that the *Glu-1* locus is only one component in the genetic control of this trait. Other authors came to a similar conclusion concerning the complexity of bread-making quality. Tsilo et al. (2013) detected six QTLs for farinograph water absorption on chromosomes

1A, 1B, 4B, 4D and 5A. These QTLs confirmed earlier studies suggesting that flour water absorption is a function of protein content, starch damage and gluten strength.

QTL mapping was also undertaken in a CD87/Katepwa wheat population that was phenotyped for grain oil and protein concentration in 2 years. Both the total grain lipid and protein concentrations varied significantly across the progeny, ranging from 3.87 to 5.77 and 11.3–15.6%, respectively. Nine and twelve QTLs were identified for grain lipid and protein concentrations, respectively (Moore et al., 2015).

A recombinant inbred line population developed from a cross between an elite wheat line (WCB414) and an exotic genotype was used to identify QTLs and new alleles for eight quality traits. Composite interval mapping for thousand-kernel weight, kernel-volume weight, grain-protein content, percentage flour extraction and four mixograph-related traits identified a total of 69 QTLs, 19 of which were stable. These QTLs were located on 18 different chromosomes (Echeverry-Solarte et al., 2015).

Exploring the genetic diversity in existing germplasm by means of extensive genomic characterisation and phenotypic evaluation is one of the priorities of plant breeders. Studies of allelic diversity can produce datasets to assess the sequence polymorphism of collections by evaluating them for associations between observed polymorphism and phenotypic variability. An alternative approach for QTL detection, commonly referred to as 'association mapping' or genome-wide association study, is based on the detection of correlations between genotypes and phenotypes in a group of individuals represented by a germplasm collection or natural populations (Zondervan and Lon, 2004). This method exploits linkage disequilibrium and the higher amount of recombination typically available in a population used for association mapping compared to a biparental mapping population to identify regions of the genome associated with the trait of interest.

16.4 TRANSGENIC TECHNOLOGY

Nowadays transgenesis is an important breeding technology to introduce genes for various agronomical traits into agricultural crops. The worldwide use of GM crops in agricultural production is close to 200 million hectares in the mid-2010s. Most of the global soybean production is GM (81%), whereas 81% of cotton, 35% of maize and 30% of oilseed rape production consists of transgenic crops (James, 2012).

Breeding for transgenic varieties is a two-step process, involving the integration of foreign DNA and the regeneration of fertile plants, followed by the selection of a GM variety. The first results in transgenic plant development were obtained using model plants and a few important agricultural crops like maize and rice, while the transformation protocol for small-grain cereals was developed later. The first fertile transgenic wheat was developed by Vasil et al. (1992)

using a biolistic method. *Agrobacterium*-mediated transformation protocols were developed a decade later for all the major cereal species (Lazzeri and Jones, 2009). Well-established transformation protocols are available for wheat and barley with increasingly efficient *Agrobacterium*-mediated transformation techniques, better integration patterns and improved co-transformation. The increasing number of traits tested in field trials is a proof of the rapid development of the transgenic breeding technology for small grain cereals. Several transgenic wheat and barley transgenic genotypes with agronomically important traits have been submitted for registration (Dunwell, 2008).

It would take many generations to introgress new quality traits or widen the genetic diversity in existing germplasm with classical breeding methods. The transgenic technology is a precise and efficient tool to introduce genes from any source directly into the local varieties used by subsistence farmers, to increase the levels of organic and/or mineral nutrients (Perez-Massot et al., 2013).

The first world-famous success achieved in the improvement of crop nutritional composition using the transgenic technology was the development of Golden Rice. Vitamin A deficiency is one of the most prevalent diseases in developing countries, affecting more than 4 million children each year, up to 500,000 of whom become partially or totally blind. One possible solution to this problem would be to distribute GM varieties that contain vitamin A or its precursor, beta-carotene, which can be converted into vitamin A. The GM rice producing beta-carotene was crossed with another line engineered to contain multiple genes improving iron availability. The high beta-carotene/high available iron hybrid was called Golden Rice (Potrykus, 2003). Co-transformation, when several genes are transferred simultaneously using *Agrobacterium*-mediated transformation in order to develop enhanced quality traits, has resulted in a growing number of commercial transgenic crops containing 'stacked' genes to modify multiple traits.

Two strategies are realistic ways of handling the amino acid deficiency of wheat. One is the transformation of plants to produce proteins containing essential amino acids, and the other is to regulate the amino acid metabolism to increase the availability of essential amino acids. One of the earliest attempts to overcome the poor nutritional value of cereal proteins was the expression of lysine-rich pea legumin in cereal endosperm, which resulted in transgenic wheat grain protein containing elevated lysine content (Stöger et al., 2001). When Tamás et al. (2009) transformed an amaranth (*Amaranthus hypochondriacus*) albumin gene into the Cadenza wheat variety, the lysine content increased by 6.4% compared to the non-transgenic control.

Attempts to increase dough strength have focussed on high-molecular-weight (HMW) glutenin sub-units (GS). HMW glutenin sub-unit genes have been used to transform a range of genotypes, resulting in increased amounts of HMW sub-unit protein and changes in the grain functional properties of the wheat grain (Barro et al., 1997). The expression of transgenes encoding single sub-units may have significantly different effects on grain processing properties

TABLE 16.1 Comparison of Protein Composition of the Transformed Line, B73-6-1, and the Control, L88-6

	L88-6	B73-6-1
Glutenin/gliadin	0.82	0.95[b]
HMW/LMW	0.40	0.66[b]
Dx/Dy	2.58	14.67[b]
Bx/By	4.14	5.24[b]
HMW-GS[a]	100%	177.53%[b]
LMW-GS[a]	100%	128.56%[b]

GS, Glutenin sub-units; HMW, high-molecular-weight; LMW, low-molecular-weight.
[a]*Amount of individual protein classes expressed as the percentage of amounts found in L88-6.*
[b]*Values are significantly different from the control line, L88-6 at 0.5 probability level.*
Source: Rakszegi, M., Békés, F., Láng, L., Tamás, L., Shewry, P.R., Bedő, Z., 2005. Technological quality of transgenic wheat expressing an increased amount of a HMW sub-unit of glutenin. J. Cereal Sci. 42, 15–23.

(Darlington et al., 2003; Rakszegi et al., 2005; Shewry et al., 2006). For example, the expression of the sub-unit 1Ax1 transgene led to increased dough strength, while the expression of the homoeologous sub-unit 1Dx5 transgene led to 'overstrong' dough with low extensibility and poor water absorption. According to the results of field experiments on GM spring wheats over-expressing HMW sub-units 1Ax1 and 1Dx5, significant changes in the structure of the glutenin polymers were detected in flour functional properties (Table 16.1), caused by the altered ratio of x-type to y-type HMW sub-units (Rakszegi et al., 2008).

The use of transgenic flours with increased 1Dy10 glutenin content in commercial blends could provide advantages in sponge-and-dough baking applications. Dough properties were improved by the transgene-encoded increases in 1Dy10 in both unblended and blended flours. The mean farinograph stability of transgenic samples was twice that of the control, and blends with transgenic samples demonstrated increases in stability proportional to the amount of transgenic flour included (Graybosch et al., 2013).

GM plants can be created using additional copies of naturally occurring genes from the species of interest, or from a closely related species, which are called cisgenes. Cisgenesis differs from transgenesis in that DNA fragments from wheat or cross-compatible species are incorporated into the genome, so cisgenic plants do not contain foreign or modified genetic material. Cisgenic plants are produced by the same transformation techniques as transgenic plants. As wheat has many wild and cultivated relatives, the prospects for developing precision breeding via the cisgenesis technology are promising. This technology would be a modern alternative to classical breeding. The cisgenic technology makes it possible to transfer genes more accurately than with classical breeding methods such as introgression, translocation breeding or mutation breeding, as it is not necessary to eliminate the inferior germplasm of the wild relatives or exotic varieties used to transfer a single agronomically important trait. Cisgenesis has great potential to overcome many problems of traditional breeding, such as linkage drag, crossing barriers, the introduction of many deleterious genes linked with one useful gene, and so on. The first results, published by Lusser et al. (2011), were achieved for potato and apple.

The transgenic technology is the most widely known new plant breeding method, and has been much criticised by the general public, especially in Europe. To avoid any potential dangers GM crops must undergo a risk assessment to assess their potential impact on human health and the environment before they receive market authorisation. This involves detailed molecular characterisation, comparison to conventional crops of the same species, assessment of potential toxicity/allergenicity and nutritional analysis, etc. (EFSA, 2011).

16.5 MUTATION BREEDING

Mutation breeding had a significant impact on the improvement of crop varieties in the second-half of the last century. However, selection for mutations induced using radiation or chemical agents was usually based on visible phenotyping, an approach not easily applied to polyploid species such as wheat, as most mutations are recessive and complemented by functional copies in the additional genomes. A recent development has contributed significantly to the efficiency of mutant discovery and is applicable to polyploid as well as diploid species. This is called targeted induced local lesions in genomes (TILLING) and allows the identification of DNA polymorphisms in large numbers of genotypes, based on DNA sequence differences. Selection can therefore be made at the molecular level for mutant candidate genes which can be combined to develop new varieties. In TILLING, the genomic DNA of the mutant plant in question is mixed with a reference DNA, and PCR is used to amplify the target region of DNA with asymmetrically-labelled fluorescent primers. Mismatches in heteroduplex DNA formed by the annealing of wild-type and mutant PCR products are then recognised and cleaved by cell nuclease, after which the cleavage products are separated by gel electrophoresis. Putative polymorphisms detected in one fluorescence channel can therefore be verified by the corresponding cut strand in the other channel.

TILLING is an ideal technique for bread wheat as it allows the detection of recessive mutations in polyploid species (Comai et al., 2004). This technology provides the direct identification of the mutation induced and allows the rapid, parallel screening of multiple genes. It also allows good predictions to be made about the number of alleles that will be identified, on the basis of the mutation frequency and population size. The EMS-mutagenised population of bread wheat variety Cadenza was studied in the M_3 generation, and several morphological mutant lines were observed (Rakszegi et al., 2010). Selection for white kernel mutants was carried out after treatment with NaOH solution (Fig. 16.2).

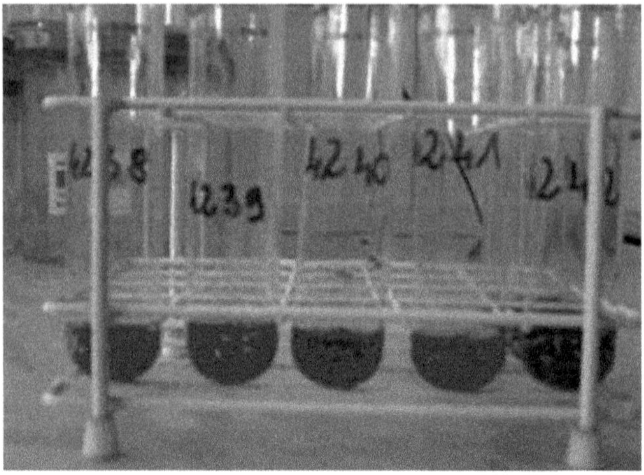

FIGURE 16.2 White kernel mutants of the EMS-mutagenised wheat variety Cadenza in NaOH solution.

Slade et al. (2005) used the TILLING technology to identify 246 alleles of the waxy starch genes in a mutagenised population of bread wheat, encoding enzymes ranging in activity from near wild type to null.

16.6 SELECTION FOR PROTEIN CONTENT AND BREAD-MAKING QUALITY

Wheat breeders have long faced the challenge of resolving the conflict between improving the yield or the grain protein concentration, as there is a negative correlation between these parameters (Fig. 16.3), the physiological basis of which is that every gram of glucose produced during photosynthesis can be utilised to produce either 0.83 g of carbohydrate or 0.40 g of protein, when NO_3 is the N source (Penning de Vries et al., 1974).

Simultaneous selection for both grain yield and grain protein concentration was carried out based on Canadian Western Red Spring (CWRS) germplasm. In the F_9 generation both the mean grain protein of the highest-yielding family and the mean yield of the highest protein family were significantly higher than in the checks, which included the parents. This demonstrates that it is possible to select correlation-breaking, high protein/high yielding gene recombinants, though the negative relationship between the quantity of protein and grain yields was not broken in the whole breeding material (DePauw et al., 1998). Sears (1998) concluded that there are two basic criteria for selecting high protein/high-yielding wheat genotypes: selection for germplasm with high protein content that will complement selection efforts for grain yield and access to a rapid protein screening system which allows measurement before planting. As a result of these efforts he selected the varieties Karl, Karl 92 and Jagger, which represent increases in both

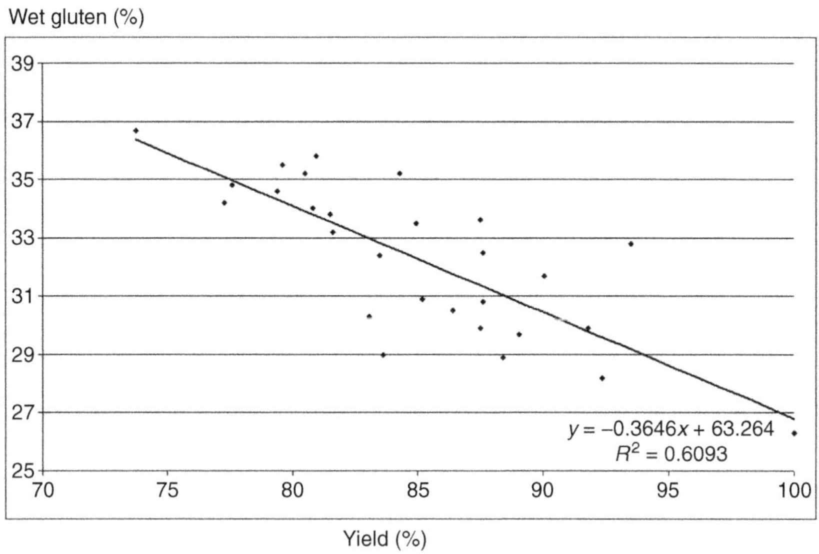

FIGURE 16.3 Negative correlation between the grain yield and wet gluten content of wheat varieties grown under Hungarian environmental conditions (2014–15).

grain yield and grain protein content. Irrespective of the fact that both yielding ability and protein content have polygenic inheritance, the heritability index is 0.42 for protein content and 0.19 for grain yield in a North American environment (DePauw and Townley-Smith, 1988). This suggests that selection for protein content should be commenced in an earlier generation than for grain yield. An increase in protein content was the primary aim when the varieties Nap Hal and Atlas 66 were included in international selection programs (Johnson et al., 1978).

Grain protein content is controlled by a complex genetic system and is strongly influenced by several environmental factors, making genetic analysis much more difficult. Two QTLs for grain protein content were detected on the 3B and 5B chromosomes. The QTL identified by barc77 on chromosome 3B had an additive effect of 0.17, while the other QTL, identified by gwm499 on chromosome 5B, had an additive effect of 0.19 (Heo and Sherman, 2013).

The efficiency of breeding for bread-making quality was greatly promoted by the spread of biochemical markers in breeding practice. In particular, the comprehensive investigation of HMW glutenins brought significant advances in breeding for technological quality based on the points system elaborated by Payne et al. (1979). The HMW sub-units can be found at three loci on the 1A, 1B and 1D chromosomes. The 1 and 2* sub-units, at the *Glu-A1* locus on chromosome 1A, both have a positive effect on flour quality (Payne et al. 1981; Branlard and Dardevet 1985). According to Payne et al. (1984) the 7 + 8, 13 + 16 and 17 + 18 sub-unit combinations on chromosome 1B all improve bread-making quality. Various authors have demonstrated, however, that sub-units at the *Glu-D1* locus on chromosome 1D have the greatest effect on bread-making quality (Payne et al., 1987; Lagudah et al., 1988), and it has been proved that the 5 + 10 sub-units at the *Glu-D1* locus improve flour quality compared with sub-units 2 + 12 and 3 + 12.

D'Ovidio et al. (1997) underlined the fact that HMW glutenins have the greatest influence on dough strength, which is affected not only by the composition, but also by the quantities of individual HMW GS. Among the naturally occurring over-expressed forms, outstanding importance is attached to the *Glu-B1*-encoded sub-unit Bx7, which leads to greater protein expression. Over-expressed Bx7 sub-units have been detected in several wheat genotypes, in both old landraces and modern varieties, including Red River 68, Bánkúti 1201, etc. (D'Ovidio et al., 1997; Juhász et al., 2003; Butow et al., 2004). However, the old Hungarian wheat, Bankuti 1201, which has good bread-making properties, has sub-units 2 + 12 on chromosome 1D (Bedő et al., 1995), which normally indicates poor bread-making quality. A similar exception is the UK wheat variety Hereward, which has 0, 7 + 9, 3 + 12 HMW composition.

These results confirm that although the HMW sub-units are clearly correlated with technological traits, they are not the only factors responsible for bread-making quality. Besides HMW GS the remaining genetic variance could be attributed to low-molecular-weight (LMW) glutenins, gliadins and lipids. An analysis of 216 genotypes from the population of Bánkúti 1201, for example revealed 6 HMW glutenin and 19 different gliadin types (Juhász et al., 2003). Various analyses have indicated a correlation between technological quality and LMW GS (Gupta and Shepherd, 1990; Juhász et al., 2003). Disulfide bonds, which influence dough elasticity, play an important role in the final bread-making quality (Wrigley, 1996), and continue to be significant during dough raising and baking (Wrigley et al., 1998).

Efforts have been made to model grain dry mass and protein concentration, nitrogen uptake and protein composition, especially for gluten-forming proteins (Porter, 1993; Jamieson and Semenov, 2000; Martre et al., 2003). The Sirius Quality 1 model simulates the accumulation of biomass and total grain protein accumulation. The results support the hypothesis that the synthesis of the main protein fractions is determined by the quantity of N per grain and by the rate of N transfer to the grain (Martre et al., 2006). Based on the individual and interactive contribution of HMW and LMW glutenins to dough quality parameters, a prediction method was developed using 3000 samples of Australian bread wheat varieties. Direct measurements covering a wide range of glutenin allelic combinations revealed correlations with this bread-making quality trait (Békés et al., 2006).

A major objective in wheat breeding is to improve the stability of bread-making quality traits, but in this respect breeders are faced with a conflict between improving yield or bread-making quality. One example is the widespread use of wheat-rye chromosome substitutions and translocations. Although these were not developed to broaden the genetic base for technological quality, their widespread cultivation has had a considerable influence on bread-making quality and quality stability. The gene complex originating from rye is well known to carry numerous disease resistance genes (Heun and Fischbeck, 1987), but this is combined with an unfavourable effect on the technological quality of wheat, partly due to the presence of gamma secalin instead of omega and gamma gliadins, and partly to the lack of the LMW GS found on wheat chromosome 1BS (Dhaliwal et al., 1988). Wheat-rye translocation forms were developed in Weihenstephan and Salzmünde in Germany (Zeller and Fuchs, 1983), and also in the USA (1RS.1DL Amigo) and Japan (1RS.1BL Salmon). The undesirable technological traits include small loaf volume and sticky crumb (Zeller, 1973), together with poor rheological traits.

Breeders have made considerable efforts in recent years to eliminate unfavourable technological quality by incorporating the translocated fragment into chromosomes of homoeologous group 1 or by altering the present genetic background. In the first case, Zeller and Fuchs (1983) incorporated the rye translocation into chromosome 1A and Koebner and Shepherd (1986) into 1D. Knackstedt et al. (1994) reported a new wheat–rye translocation on the long arm of chromosome 2B, carrying a resistance gene against Hessian fly, which did not influence the wheat storage proteins or the technological quality.

Another example is the *Lr19* leaf rust resistance gene, which was transferred to common wheat from *Thinopyrum ponticum* by Sharma and Knott (1966). *Lr19* is an important gene not only due to the rust resistance conferred by this gene, but also because of the yield increases produced in different backgrounds when alien chromatin carrying *Lr19* is introgressed in wheat. Unfortunately, the *Lr19* translocation also carries a gene(s) for yellow endosperm pigmentation that renders the resistance useless in countries where this trait is regarded as undesirable because of quality considerations. Several modifications were made to construct a shorter form of the translocation, lacking the yellow pigment genes (*Lr19-149*). Marais et al. (2001) carried out a yield trial with near isogenic lines of both the original and shortened translocations, which suggested that *Lr19* may cause a small reduction in kernel size and an increase in loaf volume, effects which are not associated with *Lr19-149*.

16.7 BREEDING FOR STARCH IMPROVEMENT IN CEREALS

Cereal carbohydrates play a significant role in the human diet, as they are the most important sources of energy. Starch is a major component of wheat flour, accounting for 65–70% of the dry weight of the wheat grain. Wheat endosperm also contains starch as polysaccharide, which could also have a positive effect on human health. Starch is composed of two types of polymers: the mainly linear amylose (20–30%) and the highly branched amylopectin (70–80%). The ratio of amylose to amylopectin determines many of the physico-chemical properties of the starch and thereby affects the end-use of the wheat flour. The 60 kDa GBSS is the key enzyme involved in the production of amylose and is the product of the waxy gene. In bread wheat, GBSS is encoded by three *waxy* loci (*Wx-A1*, *Wx-D1*, *Wx-B1*), located on the short arms of chromosomes 7A and 7D and on the long arm of chromosome 4A. The quantity of amylose is determined by the presence of these *waxy* alleles. When the starch is amylose-free it is referred to as *waxy*, while lines with one or two null-alleles are called 'partially waxy' lines and have reduced amylose content.

The relative amounts of amylose and amylopectin are known to influence both nutritional and technological properties such as susceptibility to enzymatic hydrolysis, gelling and pasting behaviour. The major part of the dietary fibre consists of carbohydrates which are not digested and absorbed in the human small intestine and are thus not responsible for an increase in the post-prandrial blood glucose content (Mann et al., 2007). Dietary fibre is one of the major food constituents that influence the rate and extent to which blood glucose increases after the ingestion of a carbohydrate food. The type of starch is important in this respect, as amylose is slowly digested by the alpha-amylases present in the human duodenum, while amylopectin is very rapidly digested after the ingestion of a carbohydrate food because the branched structure provides multiple sites for enzymatic hydrolysis (Lafiandra et al., 2014).

Because of its different chemical composition, some of the processing characteristics of waxy wheat flour (water-holding capacity, dough development time, extensibility, swelling power and setback) are better than those of normal wheat flour. However, it also has some defects in rheological characteristics, including a higher degree of softening, a lower farinograph quality number and a smaller resistance-to-extensibility ratio. Blending waxy wheat flour with normal flour is a promising way to improve product quality in baked foods and to prolong the shelf-life of these products (Zhang et al., 2014a).

To modify the amylose/amylopectin ratio, and therefore starch properties, Nakamura et al. (1995) developed the first waxy wheat variety with 0% amylose content. Several breeding programs have since selected full waxy and partial waxy wheat varieties. For example Yamamori (2009) and Yamamori and Yamamoto (2011) used diverse wheat germplasm to develop several partial waxy wheats, each carrying two null and one modified-activity allele to generate amylose contents between 7% and 20%. Partial waxy whcats look morc promising than the full waxy ones for various noodle quality types. The partial waxy allele (*Wx-B1b*) was found in all Australian Standard White (ASW) varieties in Japan in 2009 and two-thirds of ASW seeds in 2011. Most US Western White (WW) seeds had a null allele (*Glu-A1c*) or alleles that lacked one of the two *Glu-B1* sub-units. Most hard red winter (HRW) seeds had *Glu-B3b* or *Glu-B3g* (Ikeda et al., 2014).

Whole barley grain consists of about 65–68% starch, 10–17% protein, 4–9% β-glucan, and 2–3% free lipids. Based on grain starch composition, barley has wide diversity, and can be classified as normal, waxy or high-amylose starch types.

Durum wheat contains two waxy genes, which encode the *Wx-A1* and *Wx-B1* proteins. In order to investigate the effect of the over-expression of the waxy genes on amylose content and starch properties, durum wheat transgenic lines over-expressing the *Wx-B1* gene were produced. The amylose content of the transgenic lines showed no increase compared to the untransformed cultivar. One of the transgenic lines was crossed with a mutant waxy genotype of durum wheat lacking all the waxy protein isoforms. The screening of F_2 segregant plants permitted the identification of lines possessing various combinations of waxy genes, including a line containing only the construct used in the transformation. In this line, the presence of amylose restored the normal vitreous seed phenotype, confirming the correct functioning of the transgene (Sestili et al., 2012).

Selection for high amylose content wheat using classical breeding methods is a difficult task as natural variation for this trait is limited. One solution for the selection of new germplasm with increased amylose is the application of the transgenic technology. GM wheat with a significantly higher content of amylose has been developed to produce combined SBE IIa/SBE IIb knockouts (Regina et al., 2006). The transgenic line exhibited a 25–70% increase in amylose content, but further breeding efforts (backcrossing) will be necessary to combine high amylose content with other agronomic traits to produce cultivars adapted to specific climates and regions (Patton, 2006).

Another way of inducing genetic variation for amylose content is mutation breeding. Work in the HEALTHGRAIN EU FP6 project has combined such mutations in the genes encoding an enzyme involved in amylopectin synthesis to produce triple mutant lines in which the content of amylose is increased to about 40% (Sestili et al., 2010). Furthermore, these elevated amylose lines have been produced in European wheat backgrounds to facilitate their exploitation by breeders (Rakszegi et al., 2015). Although the results show that it is possible to increase the amylose content, new lines with higher amylose have lower thousand-kernel weight, which is a negative agronomic character, reducing yield potential and adaptability features (Table 16.2).

Mutations in one or more starch synthases lead to decreased amylopectin and increased amylose content in durum wheat. Hogg et al. (2013) identified two lines lacking SGP-A1, which were crossed with the durum variety 'Mountrail'. SGP-A1 null progeny lines derived from the F_5 generation were then treated with ethyl-methane-sulfonate (EMS). SGP-B1 null mutations were identified, each being a mis-sense mutation. The selected high amylose durum mutant lines will be useful in the development of pasta products with a low glycaemic index.

16.8 SELECTION FOR CEREALS WITH INCREASED DIETARY FIBRE AND PHYTOCHEMICAL CONTENT

Breeders are showing increasing interest in epidemiological studies that show the protective role of dietary fibre and phytochemicals against many chronic diseases. Dietary fibre comprises non-starch polysaccharides and oligosaccharides, resistant starch and lignin. Because of the health benefits, plant breeders have elaborated selection programs which include studies on natural variation, breeding for widening genetic variability in pre-breeding, and variety

TABLE 16.2 Amylose Content and Thousand-Kernel Weight (TKW) of Lines Selected in the High Amylose Breeding program (2012–14)

Line	Amylose (%)	TKW (g)
SOLSTICE (parental line)	23.44	35.48
KORELI (parental line)	25.33	41.67
UKRAINKA (parental line)	26.14	39.64
993-11-SGP1 (parental line)	26.19	33.79
SGP/UKRAINKA5	26.81	39.19
SGP/UKRAINKA6	26.86	41.67
SGP/SOLSTICE/LONA	27.69	38.53
SGP/UKRAINKA2	27.76	39.92
SGP/SOLSTICE/UKRAINKA	27.77	40.50
SGP/UKRAINKA4	27.78	40.25
SGP/UKRAINKA1	28.02	37.29
SGP/UKRAINKA3	28.34	32.88
SGP/LONA5	29.22	32.53
SGP/LONA4	31.37	32.07
SGP/KORELI2	33.79	26.58
SGP/UKRAINKA7	35.95	31.40
SGP/KORELI1	36.37	30.70
SGP/LONA3	36.59	31.12
SGP/UKRAINKA9	37.56	29.55
SGP/SOLSTICE/UKRAINKA	37.70	26.13
SGP/LONA1	37.73	29.58
SGP/LONA6	37.90	29.23
SGP/LONA2	39.95	29.61
SGP/LONA7	39.98	31.02
SGP/UKRAINKA8	41.05	27.08
$LSD_{5\%}$	4.16	3.08

development. For efficient selection it is important to know the extent of natural variation, the heritability of traits, that is the proportion of variance that is due to the genotype, and the distribution of these components within the mature grain.

According to Ward et al. (2008) the estimated dietary fibre contents ranged from about 85–270 mg/g DM [dry matter] and were greatest in barley, rye and oats. The major dietary fibre components in flour and bran fractions are β-glucans and arabinoxylans (AX) (Fig. 16.4). Barley and oats are rich in

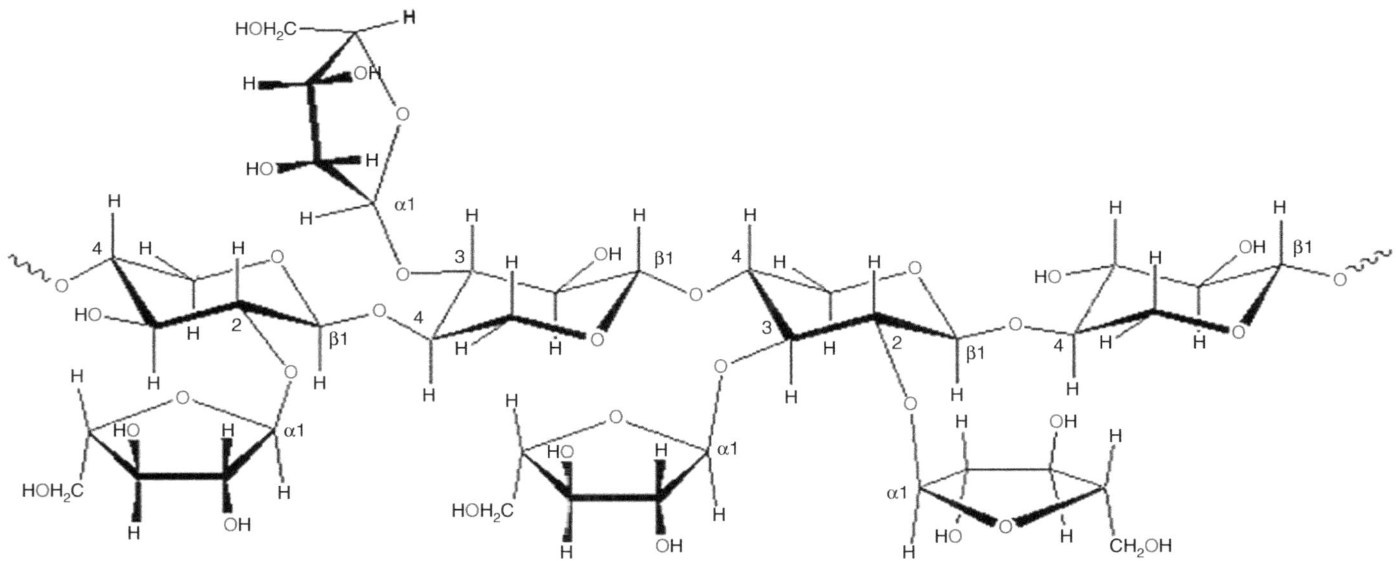

FIGURE 16.4 **The major constituents of dietary fibre in wheat are the arabinoxylans (AX), consisting of a xylose backbone with arabinose side chains hemicellulose.**

β-glucans, whereas wheat and rye are richer in AX. The soluble dietary fibre fraction, known as water-extractable arabinoxylan (WE-AX), has a special health protective role, and the highest concentration was measured in rye, followed by wheat. The largest variation in WE-AX was observed in wheat and ranged from 0.30% to 1.40% DM in the HEALTHGRAIN EU project. This is important, as AX is quantitatively the most abundant polysaccharide in the dietary fibre in wheat, making up 60–70% of the dry matter weight of the starchy endosperm cell wall (Gebruers et al., 2008).

Ward et al. (2008) compared the ranges of natural variation and the amounts of arabinoxylans in different cereal species, including *T. monococcum, T. dicoccum, T. durum* cultivated spring and winter hexaploid common wheat, rye, oats and barley, and concluded that the highest amounts of total arabinoxylans (TOTAX) and water-extractable arabinoxylans could be measured in rye in both the bran and flour fractions. However, TOTAX was higher in the bran fraction of common wheat genotypes, while *T. monococcum and T. dicoccum* had similar or lower contents of the different arabinoxylan fractions than common or durum wheats.

The genetics of WE-AX in hexaploid common wheat were studied by Charmet et al. (2009). In two recombinant populations derived from crosses between WEAX-high and WEAX-low parents, two QTL regions were identified using the viscosity of the flour extract (WEAX-viscosity) as a predictor of WEAX content. One of these was located on chromosome 6B and had a major effect. This single QTL explained as much as 59% of the phenotypic variation. Chromosome 1R has also been reported as influencing arabinoxylans in rye (Boros et al., 2002), and wheat varieties bearing the well-known 1B/1R wheat-rye translocation are often richer in WEAX and give more viscous flour water extracts (Selanere and Anderson, 2002). Shewry and Hey (2015) summarised heritability studies on dietary fibre components and concluded that both AX and β-glucan show high heritability in the white flour and bran fractions, making their amount and composition amenable to manipulation using classical plant breeding.

The genetic variability in AX content was investigated in a set of 104 tetraploid wheat genotypes, and regions attributable to AX content were identified using genome-wide association studies (Marcotuli et al., 2015). This revealed a total of 37 significant marker-trait associations (MTA), identifying 19 QTLs associated with AX content. Seven MTAs were identified on chromosome 5A, where three QTL regions were associated with AX content, while the lowest number of MTAs was detected on chromosomes 2B and 4B, where only one MTA identified a single locus.

The natural variation and heritability of phytochemicals was studied in the frame of the HEALTHGRAIN EU FP6 project in six different environments in Europe using 26 wheat genotypes. Wide variation was measured for almost all the phytochemical compounds (Table 16.3). The largest variation was found for phenolic compounds, such as tocols, alkylresorcinols and folates, where the environmental variance was found to be dominant in most cases. Heritability

TABLE 16.3 Range of Variation in the Phytochemical Content of 26 Wheat Genotypes Grown in six Environments, and the Calculated Proportion of the Variance due to Genotype (Heritability)

Phytochemicals (wholemeal)	Range	Heritability
Total sterols (including stanols) (μg/g DW)	645–1039	0.57
Stanols (% sterols + stanols)	15–29	0.22
Tocols (μg/g DW)	32–89	0.77
Folates (ng/g DW)	323–889	0.24
Alkylresorcinols (μg/g DW)	281–981	0.63
Total phenolic acids (μg/g DW)	456–1171	0.28

DW, Dry weight.
Source: Shewry, P.R., Hey, S.J., 2015. Improving wheat grain quality for processing. Strategic Objectives 1 and 4 (also 9), Rothamsted Research, 248, UK.

was the highest for tocols and alkylresorcinols, which could be interesting for cereal breeders aiming to improve their local germplasm. By contrast, some other components, such as folates and phenolic acids, exhibited low levels of heritability (Shewry and Hey, 2015).

Interest in barley for food uses focuses on the effects of β-glucans in lowering blood cholesterol levels and the glycaemic index. Total dietary fibre ranges from 11% to 34% and soluble dietary fibre from 3% to 20% (Baik and Ullrich, 2008; Fastnaught, 2001). Barley grains contain much greater amounts of phenolic compounds (0.2–0.4%) than other cereal grains (Bendelow and LaBerge, 1979). Barley is one of the best sources of tocols, containing both a high concentration of total tocols and a favourable distribution of the major biologically active forms, such as tocopherols and tocotrienols (Panfili et al., 2008).

Rye is another important cereal species for improving the phytochemical content. In general rye genotypes have higher folate content, sterols and alkylresorcinols than hexaploid common wheat. The 10 genotypes studied, including old land races and modern varieties, showed high natural variation in phytochemical and dietary fibre composition (Nyström et al., 2008).

The most prevalent forms of mineral malnutrition in the world are iron and zinc deficiencies. Cereal grains are poor sources of zinc, but because this mineral is more soluble than iron and is easier to take up from the soil, strategies to increase the zinc content of plants have concentrated on transport and accumulation (Palmgren et al., 2008). Plant breeders paid little attention to the improvement of micronutrient concentration, so the grain concentrations of iron and zinc were lower in modern semi-dwarf cultivars than in older tall cultivars (Shewry and Hey, 2015). Amiri et al. (2015) came to a similar conclusion when they studied varieties selected over the last 70 years and found that iron and zinc concentrations had decreased. They identified a wide range of genetic diversity in the uptake of micronutrients, particularly in that of iron and zinc, within the wheat genotypes studied, suggesting that selection for improved micronutrient efficiency is possible.

16.9 ROLE OF ENVIRONMENT AND CLIMATE CHANGE ON BREEDING FOR QUALITY TRAITS

Cereal quality characteristics are influenced by genotype, environment and genotype × environment interactions. The achievement of the desired end-use quality is a complex process and has become an even more critical issue due to the increasing frequency of drought, heat, heavy rainfall, cold, etc., due to climate change. This means that the frequency of extreme weather events is increasing worldwide, causing severe problems for the production of crops with stable quality traits. The role of environmental effects depends on the developmental stage of the plant, and may influence both yield and grain quality properties. Breeding for improved heat and drought stress tolerance during meiosis and anthesis remains a significant challenge. Dough properties and baking quality vary among wheat varieties and may be influenced by abiotic stresses such as high temperature, especially during the grain-filling period (Semenov et al., 2014).

Understanding the magnitude of environmental effects is a key component in breeding quality traits, such as bread-making properties. Peterson et al. (1997) reported that the environment contributed 51 and 49% of the total variance measured in mixograph peak time and loaf volume, respectively. According to Robert et al. (2001) the genetic variability was significant for both the rate and duration of nitrogen grain filling, though variability was considerably higher for the rate. The environment had the major effect on the duration, while genotype and environment had similar effects on the rate.

In studies performed in different environments, including one that was sown late in order to expose the lines to high temperatures during grain filling and one in a year in which conditions were unusually cool and wet, genomic regions containing photoperiod sensitivity loci were found to affect grain protein content, while the *Ha* (puroindoline) locus on chromosome 5D was associated with loaf quality traits. Other QTLs (on chromosomes 2B, 3B and 5A) were novel and not associated with any known quality or phenology genes (Maphosa et al., 2015).

Extreme weather conditions, like heavy rainfall during the harvest period, can cause the premature germination of seeds, known as pre-harvest sprouting (PHS). This phenomenon has become a serious problem, and wheats with white-coloured grain are particularly susceptible. PHS reduces bread-making quality through the breakdown of starch reserves in the endosperm by the increased enzyme activity of alpha-amylase. Lines from a cross between the PHS-resistant variety Cayuga and the PHS-susceptible variety Caledonia were used for QTL mapping. A total of 16 environments were used to detect 15 different PHS QTLs including a major QTL, QPhs.cnl-2B.1, which was significant in all the environments tested and explained from 5% to 31% of the trait variation in a given environment (Munkvold et al., 2009).

Analysis of the contents of bioactive components (tocols, sterols, alkylresorcinols, folates, phenolic acids and fibre components) in 26 wheat cultivars grown in six site × year combinations showed that the extent of variation due to variety and environment differed significantly between the components. Significant correlations were also observed between the contents of bioactive components and environmental factors (precipitation and temperature) during grain development (Shewry et al., 2010).

In the search for new germplasm, synthetic hexaploid lines and their derivatives were intensively used in breeding programs to provide new genetic variability for better adaptability under various environments. Using SSR markers, Warburton et al. (2006) found a significant increase in the genetic diversity of recently developed synthetic hexaploid wheat derivatives. The inherent diversity of the new synthetic derivatives is comparable to that of the land races; however, they also exhibit improved stress tolerance and better end-use quality.

One of the most widely known effects of climate change is the expected increase in the atmospheric CO_2 concentration. Crops that follow the C3 metabolic pathway for carbon fixation during photosynthesis, including wheat and other small grain cereals, have lower concentrations of zinc and iron when grown under field conditions at elevated atmospheric CO_2 concentration. C3 crops also have lower concentrations of protein, whereas C4 crops seem to be less affected. Differences between varieties suggest that breeding for decreased sensitivity to atmospheric CO_2 concentration could partly address these new challenges to global health (Myers et al., 2014).

Agronomic management is one of the most important factors influencing the growing environment of wheat intended for quality production. Recently organic farming has received ever greater attention and international experiments have been established in order to examine conventional bread wheat varieties under many different environmental growing conditions. Because of the specific nature of growing environments the need has been felt to start studies on the necessity of breeding cereals for organic agriculture. Organic cereal breeding has mainly resulted in varieties with taller plant height and higher grain protein content compared to modern varieties, so they are better suited for sites with low nitrogen availability (Wolfe et al., 2008). The limited, less controlled availability of nitrogen is one of the major problems in organic farming. Cereals bred for organic farming should have vigorous early plant growth, resulting in better weed suppression, the better utilisation of nutrients in early growth stages and better resistance to pests. One of the results of these breeding efforts is hexaploid tritordeum, a new cereal species that arose from a cross between *T. durum* and *Hordeum chilense* (Martın and Sanchez-Monge Laguna, 1980). This new species is suitable for organic production and was found to have a protein content of 19–24% (Martın and Cubero, 1981), combined with a high level of bioactive components, such as dietary fibre.

16.10 CONCLUSIONS

Plant breeders have contributed significantly to improvements in the quality of wheat and other small grain cereals during the 20th century. This success story was possible as a result of the green revolution, when the grain yield tripled while the growing area did not change significantly worldwide. Today we face different challenges in the cereal supply chain: changing diets and food safety requirements in developed countries, and an increasing demand for cereal foods, especially in developing countries, including regions that are only marginally suitable for wheat production but which have a great need for a healthy diet with increasing protein and calorie consumption. These complex phenomena will influence future trends in cereal breeding for quality traits. Wheat quality research is not only about identifying ways to meet customer demands for protein quality and quantity. The protein-oriented research strategy has changed during the last 2 decades, as wheat is an important source not only of protein but also of various types of starches, dietary fibre, vitamins, micronutrients, etc., which exert positive physiological effects on human health. New molecular breeding tools, cereal chemistry methods, and milling and baking industry technologies have all come to the aid of plant breeders, who have also established close working relationships with experts in diverse fields, aimed at improving the efficiency of the breeding process and shortening the time required to develop new varieties. The participation of plant breeders in this worldwide collaboration, involving the joint efforts of public and private sectors, will be necessary to establish a quality-oriented whole cereal supply chain for the future.

REFERENCES

Amiri, R., Bahraminejad, S., Sasani, S., Jalali-Honarmand, S., Fakhri, R., 2015. Bread wheat genetic variation for grain's protein, iron and zinc concentrations as uptake by their genetic ability. Europ. J. Agron. 67, 20–26.

Baik, B.K., Ullrich, S.E., 2008. Barley for food: characteristics, improvement, and renewed interest. J. Cereal Sci. 48, 233–242.

Barro, F., Rooke, L., Békés, F., Gras, P., Tatham, A.S., Fido, R., Lazzeri, P.A., Shewry, P.R., Barcelo, P., 1997. Transformation of wheat HMW subunit genes results in improved functional properties. Nat. Biotechnol. 15, 1295–1299.

Bedő, Z., Kárpáti, M., Vida, Gy., Kramarikné Kissimon, J., Láng, L., 1995. Good breadmaking quality wheat (*Triticum aestivum*) genotypes with 2 + 12 subunit composition at the Glu-D1 locus. Cer. Res. Comm. 23, 283–289.

Békés, F., Gras, P., Appels, R., 2001. Small-Scale dough testing as a breeding and research tool. In: Bedő, Z., Láng, L. (Eds.), Wheat in a Global Environment. Proceedings of the Sixth International Wheat Conference. Kluwer Academic Publishers, Budapest, Hungary, pp. 285–290.

Békés, F., Lukow, O.M., Uthayakumaran, S., Mann, G., 2003. Small-scale quality measurements. In: Wheat Gluten Protein Analysis, American Ass. Cer. Chem. St Paul, USA, 173–198.

Békés, F., Kemény, S., Morell, M., 2006. An integrated approach to predicting end-product quality of wheat. Europ. J. Agron. 25, 155–162.

Bendelow, V.M., LaBerge, D.E., 1979. Relationship among barley, malt and beer phenolics. J. Am. Soc. Brewing Chem. 37, 89–90.

Boros, D., Lukaszewski, A.J., Aniol, A., Ochodzki, P., 2002. Chromosome location of genes controlling the content of dietary fibre and arabinoxylans in rye. Euphytica 128, 1–8.

Branlard, G., Dardevet, M., 1985. Diversity of grain protein and bread wheat quality. II. Correlation between high molecular weight subunits of glutenin and flour quality characteristics. J. Cereal Sci. 3, 345–354.

Butow, B.J., Gale, K.R., Ikea, J., Juhász, A., Bedő, Z., Tamás, L., Gianibelli, M.C., 2004. Dissemination of the highly expressed Bx7 glutenin subunit (Glu–B1 allele) in wheat as revealed by novel PCR markers and RP-HPLC. Theor. Appl. Genet. 109, 1525–1535.

Cao, S., Xu, H., Li, Z., Wang, X., Wang, D., Zhang, A., Jia, X., Zhang, X., 2007. Identification and characterization of a novel *Ag. intermedium* HMW-GS gene from *T. aestivum* to *Ag. intermedium* addition lines TAI-I series. J. Cereal Sci. 45, 293–301.

Charmet, G., Masood-Quraishi, C.U., Ravel, C.C., Romeuf, C.I., Balfourier, C.F., Perretant, C.M.R., Joseph, C.J.L., Rakszegi, M., Guillon, C.F., Sado, C.P.E., Bedő, Z., Saulmier, C.L., 2009. Genetics of dietary fibre in wheat. Euphytica 170, 155–168.

Comai, L., Young, K., Till, B.J., Reynolds, S.H., Green, E.A., Codomo, C.A., Enns, L.C., Johnson, J.E., Burtner, C., Odden, A.R., Henikoff, S., 2004. Efficient discovery of DANN polymorphism in natural population by ecotilling. Plant J. 34, 778–786.

Cseh, A., Kruppa, K., Molnar, I., Rakszegi, M., Dolezel, J., Molnar-Lang, M., 2011. Characterization of a new 4BS.7HL wheat–barley translocation line using GISH, FISH, and SSR markers and its effect on the b-glucan content of wheat. Genome 54, 795–804.

Cseh, A., Soos, V., Rakszegi, M., Turkosi, E., Balazs, E., Molnar-Lang, M., 2013. Expression of HvCslF9 and HvCslF6 barley genes in the genetic background of wheat and their influence on the wheat b-glucan content. Ann. Appl. Biol. 163, 142–150.

Darlington, H., Fido, R., Tatham, A.S., Jones, H., Salmon, S.E., Shewry, P.R., 2003. Milling and baking properties of field grown wheat expressing HMW subunit transgenes. J. Cereal Sci. 38, 301–306.

DePauw, R.M., Townley Smith, T.F., 1988. Patterns of response for genotype grain yield and protein content in seven environments. In: Miller, T.E., Koebner, R.M.D., (Eds.), Proceedings of the Seventh International Wheat Genetics Symposium, Cambridge, UK, 993–997.

DePauw, R.M., Clarke, J.M., McCaig, T.N., Townley-Smith, T.F., 1998. Opportunities for the improvement of Western Canadian Wheat concentration, grain yield and quality through plant breeding. In: Fowler, D.B., Geddes, W.E., Johnston, A.M., Preston, K.R., (Eds.), Proceedings of Wheat Protein Symposium, Saskatoon, Canada, 75–93.

Dhaliwal, A.S., Mares, D.J., Marshall, D.R., Skeritt, J.H., 1988. Protein composition and pentosan content in relation to dough stickiness of 1B/1R translocation wheats. Cereal Chem. 62, 143–149.

D'Ovidio, R., Masci, S., Porceddu, E., Kasarda, D.D., 1997. Duplication of the Bx7 high –molecular-weight glutenin subunit gene in bread wheat (*Triticum aestivum*, L.) cultivar 'Red River 68'. Plant Breed. 116, 525–531.

Dunwell, J.M., 2008. Transgenic wheat, barley, and oats: future prospects. In: Jones, H.D., Shewry, P.R., (Eds.), Methods in Molecular Biology; Transgenic wheat, barley and oats: production and characterization protocols, 478, 333–346.

Echeverry-Solarte, M., Kumar, A., Kianian, S., Simsek, S., Alamri, M.S., Mantovani, E.E., McClean, P.E., Deckard, E.L., Elias, E., Schatz, B., Xu, S.S., Mergoum, M., 2015. A population developed from an exotic line with supernumerary spikelets was genetically dissected for eight quality traits, discovering new genes/alleles with potential use in wheat breeding programs. Theor. Appl. Genet. 128, 893–912.

EFSA Annual Report 2011, Eurpean Food Safety Authority, Parma, Italy, 1–56.

Fastnaught, C.E., 2001. Barley fibre. In: Cho, S., Dreher, M. (Eds.), Handbook of Dietary Fibre. Marcel Dekker, New York, pp. 519–542.

Feldman, M., Millet, E., Abbo, S., 1994. Exploitation of wild emmer wheats to increase yield and protein content in durum and common wheat. In: Proceedings of Genetic Resources Eucarpia Meeting, Clermont Ferrand, France, 151–162.

Frisch, M., Bohn, M., Melchinger, A.E., 2000. PLABSIM: software for simulation of marker-assisted backcrossing. J. Hered. 91, 86–87.

Gale, K.R., 2005. Diagnostic DNA markers for quality traits in wheat. J. Cereal Sci. 41, 181–192.

Garg, M., Tanaka, H., Ishikawa, N., Takata, K., Yanaka, M., Tsujimoto, H., 2009. *Agropyron elongatum* HMW-glutenins have a potential to improve wheat end-product quality through targeted chromosome introgression. J. Cereal Sci. 50, 358–363.

Gebruers, K., Dornez, E., Boros, D., Fras, A., Dynkowska, W., Bedő, Z., Rakszegi, M., Delcour, J.A., Courtin, C.M., 2008. Variation in the content of dietary fibre and components thereof in wheats in the HEALTHGRAIN diversity screen. J. Agric. Food Chem. 56, 9740–9750.

Grasgruber, H., Schoggl, G., Ruckenbauer, P., 2002. Investigations on the validity of the micro-extensigraph method to measure rheological properties of wheat doughs. Eur. Food Res. Techn. 214, 79–82.

Graybosch, R.A., Seabourn, B., Chen, Y.H.R., Blechl, A.E., 2013. Transgenic enhancement of high-molecular-weight glutenin subunit 1Dy10 concentration: effects in wheat flour blends and sponge and dough baking. Cereal Chem. 90, 164–168.

Gupta, R.B., Shepherd, K.W., 1990. Two-step one-dimensional SDS-PAGE analysis of LMW subunits of glutelin I. Variation and genetic control of the subunits in hexaploid wheats. Theor. Appl. Genet. 80, 65–74.

Haraszi, R., Gras, P.W., Tomoskozi, S., Salgó, A., Békés, F., 2004. Application of a micro Z-arm mixer to characterise mixing properties and water absorption of wheat flour. Cereal Chem. 81, 555–560.

Heo, H., Sherman, J., 2013. Identification of QTL for grain protein content and grain hardness from winter wheat for genetic improvement of spring wheat. Plant Breed. Biotech. 1, 347–353.

Heun, M., Fischbeck, G., 1987. Identification of wheat powdery mildew resistance genes by analysing host-pathogen interactions. Plant Breed. 98, 124–129.

Hogg, A.C., Gause, K., Hofer, P., Martin, J.M., Graybosch, R.A., Hansen, L.E., Giroux, M.J., 2013. Creation of a high-amylose durum wheat through mutagenesis of starch synthase II (SSIIa). J. Cereal Sci. 57, 377–383.

Ikeda, T.M., Yanaka, M., Takata, K., 2014. Comparison of quality-related alleles among Australian and North American wheat classes exported to Japan. Cereal Chem. 91, 616–622.

James, C., 2012. Global Status of commercialized biotech/GM crops: 2012. ISAAA Brief No. 44. Ithaca: ISAAA.

Jamieson, P.D., Semenov, M.A., 2000. Modelling nitrogen uptake and redistribution in wheat. Field Crop Res. 68, 21–29.

Johnson, V.A., Wilhelmi, K.D., Kuhr, S.L., Mattern, P.J., Schmidt, J.W., 1978. Breeding progress for protein and lysin in wheat. In: Ramanujam, S. (Ed.), Proceedings of the Fifth International Wheat Genetics Symposium, New Delhi, India, 825–835.

Juhász, A., Larroque, O.R., Tamás, L., Hsam, S.L.K., Zeller, F.J., Békés, F., Bedő, Z., 2003. Bánkúti 1201—an old Hungarian wheat variety with special storage protein composition. Theor. Appl. Genet. 107, 697–704.

Knackstedt, M.A., Sears, R.G., Rogers, D.E., Lookhart, G.L., 1994. Effects of T2BS-2RL wheat-rye translocation on breadmaking quality in wheat. Crop Sci. 34, 1066–1070.

Koebner, R.M.D., Shepherd, K.W., 1986. Controlled introgression to wheat of genes from rye chromosome arm 1RS by induction of allosyndesis. 1. Isolation of recombinants. Theor. Appl. Genet. 7, 197–208.

Lafiandra, D., Riccardi, G., Shewry, P.R., 2014. Improving cereal grain carbohydrates for diet and health. J. Cereal Sci. 59, 312–326.

Lagudah, E.S., O'Brien, L., Hannoran, G.M., 1988. Influence of gliadin composition and high-molecular-weight subunits of glutenin on dough properties in an F_3 population of a bread wheat cross. J. Cereal Sci. 7, 33–42.

Lazzeri, P.A., Jones, H.D., 2009. Transgenic wheat, barley and oats: production and characterization. Jones, H.D., Shewry, P.R. (Eds.), Methods in Molecular Biology, vol. 478, Humana Press, New York City, pp. 3–22.

Lookhart, G.L., Wrigley, C.W., 1997. International Wheat Quality Conference, Session I. Biochemical Determinants for Quality Prediction, Introduction. In: Steele, J.L., Chung, O.K. (Eds.), Proceedings of the International Wheat Quality Conference. Manhattan, Kansas, USA, pp. 41–43.

Lusser, M., Parisi, C., Plan, D., Rodríguez-Cerezo, E., 2011. New plant breeding techniques: state-of-the-art and prospects for commercial development. JRC Scientific and Technical Reports, JRC European Commission, EUR 24760, pp. 24–25.

Mann, J., Cummings, J.H., Englyst, H.N., Key, T., Liu, S., Riccardi, G., Summerbell, C., Uauy, R., van Dam, R.M., Venn, B., Vorster, H.H., Wiseman, M., 2007. FAO/WHO scientific update on carbohydrates in human nutrition: conclusions. Eur. J. Clin. Nutr. 61, 132–137.

Marais, G.F., Marais, A.S., Groenewald, J.Z., 2001. Evaluation and reduction of Lr19-149, a recombined form of the *Lr19* translocation of wheat. Euphytica 121, 289–295.

Maphosa, L., Langridge, P., Taylor, H., Emebiri, L.C., Mather, D.E., 2015. Genetic control of grain protein, dough rheology traits and loaf traits in a bread wheat population grown in three environments. J. Cereal Sci. 64, 147–152.

Marcotuli, I., Houston, K., Waugh, R., Fincher, G.B., Burton, R.A., Blanco, A., Gadaleta, A., 2015. Genome wide association mapping for arabinoxylan content in a collection of tetraploid wheats. PLoS ONE 10, 1–14.

Martın, A., Sanchez Monge Laguna, E., 1980. A hybrid between Hordeum chilense and *Triticum turgidum*. Cer. Res. Comm. 8, 349–353.

Martın, A., Cubero, J.I., 1981. The use of Hordeum chilense in cereal breeding. Cer. Res. Comm. 9, 317–323.

Martre, P., Porter, J.R., Jamieson, P.D., Triboi, E., 2003. Modelling grain nitrogen accumulation and protein composition to understand the sink/source regulation of nitrogen remobilization for wheat. Plant Physiol. 133, 1959–1967.

Martre, P., Jamieson, P.D., Semenov, M.A., Zyskowski, R.F., Porter, J.R., Triboi, E., 2006. Modelling protein content and composition in relation to crop nitrogen dynamics for wheat. Europ. J. Agron. 25, 138–154.

Moore, C.M., Richards, R.A., Rebetzke, G.J., 2015. Phenotypic variation and QTL analysis for oil content and protein concentration in bread wheat (*Triticum aestivum* L.). Euphytica 204, 371–382.

Munkvold, J.D., Tanaka, J., Benscher, D., Sorrells, M.E., 2009. Mapping quantitative trait loci for preharvest sprouting resistance in white wheat. Theor. Appl. Genet. 119, 1223–1235.

Myers, S.S., Zanobetti, A., Kloog, I., Huybers, P., Leakey, A., Bloom, A.J., Carlisle, E., Dietterich, L.H., Fitzgerald, G., Hasegawa, T., Holbrook, N.M., Nelson, R.L., Ottman, M.J., Raboy, V., Sakai, H., Sartor, K.A., Schwartz, J., Seneweera, S., Tausz, M., Usui, Y., 2014. Increasing CO_2 threatens human nutrition. Nature 510, 139–142.

Nakamura, T., Yamamori, M., Hirano, H., Hidaka, S., Nagamine, T., 1995. Production of waxy (amylose-free) wheats. Mol. General Genet. 248, 253–259.

Nyström, L., Lampi, A.M., Anderson, A.A.M., Kamal-Eldin, A., Gebruers, K., Courtin, C.M., Delcour, J.A., Li, L., Ward, J.L., Fras, A., Boros, D., Rakszegi, M., Bedő, Z., Shewry, P.R., Piironen, V., 2008. Phytochemicals and dietary fibre components in rye varieties in the HEALTHGRAIN diversity screen. J. Agric. Food Chem. 56, 9758–9766.

Osborne, B.G., 2006. Application of near infrared spectroscopy in quality screening of early generation material in cereal breeding programmes. J. NIR Spectr. 14, 93–101.

Palmgren, M.G., Clemens, S., Williams, L.E., Krämer, U., Borg, S.U.K., Schjørring, J.K., Sanders, D., 2008. Zinc biofortification of cereals: problems and solutions. Trends Plant Sci. 13, 464–473.

Panfili, G., Fratianni, A., DiCriscio, T., Marconi, E., 2008. Tocol and beta-glucan levels in barley varieties and in pearling by-products. Food Chem. 107, 84–91.

Patton, D., 2006. Limagrain, GRDC to bring GM healthy wheat to market. Available from: http://foodnavigator.com/news/ng.asp?id=71757

Payne, P.I., Corfield, K.G., Blackman, J.A., 1979. Identification of a high-molecular-weight subunit of glutenin whose presence correlates with bread-making quality in wheats of related pedigree. Theor. Appl. Genet. 55, 153–159.

Payne, P.I., Corfield, K.G., Holt, L., Blackman, J.A., 1981. Correlation between the inheritance of certain high-molecular-weight subunits of glutenin and breadmaking quality in progenies of six crosses of bread wheat. J. Sci. Food. Agric. 32, 51–60.

Payne, P.I., Holt, L.M., Jackson, E.A., Law, C.N., 1984. Wheat storage proteins, their genetics and their potential for manipulation by plant breeding. Phil. Trans. R. Soc. Lond. 304, 359–371.

Payne, P.I., Lawrence, G.J., Nightingale, M.A., Krattiger, A.F., Holt, L.M., 1987. The relationship between HMW subunit composition and bread-making quality of British-grown varieties. J. Sci. Food Agric. 40, 51–60.

Penning de Vries, F.W.T., Brunsting, A.H.M., van Laar, H.H., 1974. Products, requirements and efficiency of biosynthesis: a quantitative approach. J. Theor. Biol. 45, 339–377.

Perez-Massot, E., Banakar, R., Gómez-Galera, S., Zorrilla-López, U., Sanahuja, G., Arjo, G., Miralpeix, B., Vamvaka, E., Farre, G., Rivera, S.M., Dashevskaya, S., Berman, J., Sabalza, M., Yuan, D., Bai, C., Bassie, L., Twyman, R.M., Capell, T., Christou, P., Zhu, C., 2013. The contribution of transgenic plants to better health through improved nutrition: opportunities and constraints. Genes Nutr. 8, 29–41.

Peterson, C.J., Sears, R.G., Graybosch, R.A., Shelton, D.R., 1997. Breeding for quality. In: Steele, J.L., Chung, O.K. (Eds.), Proceedings of the International Wheat Quality Conference, Manhattan, Kansas, USA, pp. 175–184.

Porter, J.R., 1993. AFRCWHEAT2: a model of the growth and development of wheat incorporating responses to water and nitrogen. Eur. J. Agron. 2, 69–82.

Potrykus, I., 2003. 'Nutritionally enhanced rice to combat malnutrition disorders of the poor. Nutr. Rev. 61, 101–104.

Rakszegi, M., Békés, F., Láng, L., Tamás, L., Shewry, P.R., Bedő, Z., 2005. Technological quality of transgenic wheat expressing an increased amount of a HMW subunit of glutenin. J. Cereal Sci. 42, 15–23.

Rakszegi, M., Kisgyörgy, B.N., Tearall, K., Shewry, P.R., Láng, L., Phillips, A., Bedő, Z., 2010. Diversity of the agronomical and morphological properties in a mutant population of bread wheat studied in the Healthgrain programme. Euphytica 174, 409–421.

Rakszegi, M., Pastori, G., Jones, H.D., Békés, F., Butow, B., Láng, L., Bedő, Z., Shewry, P.R., 2008. Technological quality of field grown transgenic lines of commercial wheat cultivars expressing the 1Ax1 HMW glutenin subunit gene. J. Cereal Sci. 47, 310–321.

Rakszegi, M., Kisgyorgy, B.N., Kiss, T., Sestili, F., Lang, L., Lafiandra, D., Bedő, Z., 2015. Development and characterization of high-amylose wheat lines. Starch-Starke 67, 247–254.

Rath, C.R., Gras, P.W., Wrigley, C.W., Walker, C.E., 1990. Evaluation of dough properties from two grams of flour using the Mixograph principle. Cereal Food World 35, 572–574.

Rath, C.R., Gras, P.W., Zhonglin, Z., Appels, R., Békés, F., Wrigley, C.W., 1995. A prototype extension tester for two-gram dough samples. In: Pannozo, J.F., Downie, P.G. (Eds.), Proceedings of the Fourty Fourth Annual RACI Conference, Melbourne, Australia, pp. 122–126.

Regina, A., Bird, A., Topping, D., Bowden, S., Freeman, J., Barsby, T., Kosar-Hashemi, B., Li, Z., Rahman, S., Morell, M., 2006. High amylose wheat generated by RNA interference improves indices of large-bowel health in rats. Proc. of the Natl. Acad. Sci. USA 103, 3546–3551.

Robert, N., Hennequet, C., Bérard, P., 2001. Dry matter and nitrogen accumulation in wheat kernel: genetic variation in rate and duration of grain filling. J. Genet. Breed. 55, 297–305.

Rousset, M., Brabant, P., Kota, R.S., Dubcovsky, J., Dvorak, J., 2001. Use of recombinant substitution lines for gene mapping and QTL analysis of bread making quality in wheat. Euphytica 119, 81–87.

Sears, R.G., 1998. Improving grain protein concentration and grain yield in USA hard winter wheat In: Fowler, D.B., Geddes, W.E., Johnston, A.M., Preston, K.R. (Eds.), Proceedings of Wheat Protein Symposium, Saskatoon, Canada, pp. 63–67.

Selanere, M.L., Anderson, R., 2002. Cell wall composition of 1B/1R translocation wheat grains. J. Sci. Food Agric. 82, 538–545.

Semenov, M.A., Stratonovitch, P., Alghabari, F., Gooding, M.J., 2014. Adapting wheat in Europe for climate change. J. Cereal Sci. 59, 245–256.

Sestili, F., Botticella, E., Bedo, Z., Phillips, A., Lafiandra, D., 2010. Production of novel allelic variation for genes involved in starch biosynthesis through mutagenesis. Mol. Breed. 25, 145–154.

Sestili, F., Botticella, E., Proietti, G., Janni, M., D'Ovidio, R., Lafiandra, D., 2012. Amylose content is not affected by overexpression of the Wx-B1 gene in durum wheat. Plant Breed. 131, 700–706.

Sharma, D., Knott, D.R., 1966. The transfer of leaf rust resistance from *Agropyron* to *Triticum* by irradiation. Can. J. Genet. Cytol. 8, 137–143.

Sharma, H.C., Gill, B.S., 1983. Current status of wide hybridization in wheat. Euphytica 32, 17–31.

Shewry, P.R., Hey, S.J., 2015. Improving wheat grain quality for processing. Strategic Objectives 1 and 4 (also 9), Rothamsted Research, 248, UK.

Shewry, P.R., Parmar, S., Pappin, D.J.C., 1987. Characterization and genetic control of the prolamins of *Haynaldia villosa*: relationship to cultivated species of the *Triticeae* (rye, wheat and barley). Biochem. Genet. 25, 309–325.

Shewry, P.R., Powers, S., Field, J.M., Fido, R.J., Jones, H.D., Arnold, G.M., West, J., Lazzeri, P.A., Barcelo, P., Barro, F., Tatham, A.S., Békés, F., Butow, B., Darlington, H., 2006. Comparative field performance over three years and two sites of transgenic wheat lines expressing HMW subunit transgenes. Theor. Appl. Genet. 113, 128–136.

Shewry, P.R., Piironen, V., Lampi, A.M., Edelmann, M., Kariluoto, S., Nurmi, T., Nyström, L., Ravel, C., Charmet, G., Andersson, A.A.M., Åman, P., Boros, D., Gebruers, K., Dornez, E., Courtin, C.M., Delcour, J.A., Rakszegi, M., Bedo, Z., Ward, J.L., 2010. The HEALTHGRAIN wheat diversity screen: effects of genotype and environment on phytochemicals and dietary fiber components. J. Agric. Food Chem. 58, 9291–9298.

Shewry, P.R., Sabelli, P.A., Parmar, S., Lafiandra, D., 1991. λ-type prolamins are encoded by genes on chromosomes 4Ha and 6Ha of *Haynaldia villosa* Schur (syn. *Dasypyrum villosum* L.). Biochem. Genet. 29, 207–211.

Slade, A.J., Fuerstenberg, S.I., Loeffler, D., Steine, M.N., Facciotti, D., 2005. A reverse genetic, nontransgenic approach to wheat crop improvement by TILLING. Nat. Biotech. 23, 75–81.

Stöger, E., Parker, M., Christou, P., Casey, R., 2001. Pea legumin overexpressed in wheat endosperm assembles into an ordered paracrystalline matrix. Plant Physiol. 125, 1732–1742.

Tamás, C., Kisgyörgy, B.N., Rakszegi, M., Wilkinson, M.D., Yang, M.S., Láng, L., Tamás, L., Bedő, Z., 2009. Transgenic approach to improve wheat (*Triticum aestivum* L.) nutritional quality. Plant Cell Rep. 28, 1085–1094.

Tsilo, T.J., Nygard, G., Khan, K., Simsek, S., Hareland, G.A., Chao, S., Anderson, J.A., 2013. Molecular genetic mapping of QTL associated with flour water absorption and farinograph related traits in bread wheat. Euphytica 194, 293–302.

Vasil, V., Castillo, A.M., Fromm, M.E., Vasil, I.K., 1992. Herbicide resistant fertile transgenic wheat plants obtained by microprojectile bombardment of regenerable embryogenic callus. Bio-technology 10, 667–674.

Warburton, M.L., Crossa, J., Franco, J., Kazi, M., Trethowan, R., Rajaram, S., Pfeiffer, W., Zhang, P., Dreisigacker, S., van Ginkel, M., 2006. Bringing wild relatives back into the family: recovering genetic diversity in CIMMYT improved wheat germplasm. Euphytica 149, 289–301.

Ward, J.L., Poutanen, K., Gebruers, K., Piironen, V., Lampi, A.M., Nyström, L., Andersson, A.A.M., Åman, P., Boros, D., Rakszegi, M., Bedő, Z., Shewry, P.R., 2008. The HEALTHGRAIN cereal diversity screen: concept, results and prospects. J. Agric. Food Chem. 56, 9699–9709.

Wolfe, M.S., Baresel, J.P., Desclaux, D., Goldringer, I., Hoad, S., Kovacs, G., Loschenberger, F., Miedaner, T., Østerga, H., Lammerts van Bueren, E.T., 2008. Developments in breeding cereals for organic agriculture. Euphytica 163, 323–346.

Wrigley, C.W., 1996. Giant proteins with flour power. Nature 381, 738–739.

Wrigley, C.W., Andrews, J.L., Békés, F., Gras, W.P., Gupta, R.B., MacRitchie, F., Skerritt, J.H., 1998. Protein-protein interactions - essential to dough rheology. In: Hamer, R.J., Hoseney, R.C. (Eds.), The Keys to Cereal Quality. AACC, St. Paul, Minnesota, USA, pp. 17–46.

Zeller, F.J., 1973. 1B/1R wheat-rye substitutions and translocations. In: Sears, E.R. (Ed.), Proceedings of the Fourth International Wheat Genetics Symposium, Missouri, USA, pp. 209–221.

Zeller, F.J., Fuchs, E., 1983. Citologie und Krankenheitsrezistenz einer 1A/1R und mehreren 1B/1R Weizen-Roggen Translokationssorten. Zeitschrift für Pflanzenzüchtung 90, 285–296.

Zhang, H., Zhang, W., Xua, Ch., Zhou, X., 2014a. Studies on the rheological and gelatinization characteristics of waxy wheat flour. Int. J. Biol. Macromol. 64, 123–129.

Zhang, R., Zhang, M., Wang, X., Chen, P., 2014b. Introduction of chromosome segment carrying the seed storage protein genes from chromosome 1V of *Dasypyrum villosum* showed positive effect on bread-making quality of common wheat. Theor. Appl. Genet. 127, 523–533.

Zondervan, K.T.C., Lon, R., 2004. The complex interplay among factors that influence allelic association. Nat. Rev. Genet. 5, 89–100.

Yamamori, M., 2009. Amylose content and starch properties generated by five variant Wx alleles for granule-bound starch synthase in common wheat (*Triticum aestivum* L.). Euphytica 165, 607–614.

Yamamori, M., Yamamoto, K., 2011. Effects of two novel Wx-A1 alleles of common wheat (*Triticum aestivum* L.) on amylose and starch properties. J. Cereal Sci. 54, 229–235.

Chapter 17

Identification of Grain Variety and Quality Type

Crispin A. Howitt* and Diane Miskelly**
*CSIRO Agriculture and Food, Canberra, ACT, Australia; **Westcott Consultants, Goulburn, NSW, Australia

17.1 INTRODUCTION

Cereals constitute a major component of human diets worldwide (Chapter 1) and are traded as a commodity globally. Maize is predominantly traded as a feed grain, barley for feed and malting, while wheat is predominantly traded for human consumption. The latter two categories (food and feed) have price differentials depending on the suitability of the grain, the desired end use or quality. This price differential for different classes of end-product suitability means that it is important for the buyer to be able to distinguish varieties and/or classes of grains that are suitable for use for a specific end product. This in turn provides different targets for researchers to understand the genetic basis of quality and thus allow breeders to develop varieties suitable for specific end products. This chapter will outline the methods available for variety identification, and provide a brief overview of some of the research into the understanding of the genetic basis of quality and development of markers suitable for use in marker-assisted selection (MAS) in breeding programs.

17.2 VARIETY IDENTIFICATION

In crops where quality is a differentiator in value and there is significant variation in end-product quality between different cultivars, there is a need to accurately identify genotype in a rapid and efficient manner. This need is important at all stages in the value chain and starts in the breeding process where the breeder must be able to distinguish new varieties from existing ones already available. In many countries before release, a breeder must be able to demonstrate distinctness, uniformity and stability (DUS) (Cooke, 1995; Cooke and Wrigley, 2004). This can be demonstrated using morphological traits alone but where there is little variation in these characteristics, analysis of biochemical

Cereal Grains. http://dx.doi.org/10.1016/B978-0-08-100719-8.00017-6

or molecular markers may also be required. Once a variety is registered it must be maintained as a pure line so that the farmer is assured of getting what was paid for, or in the case of seed stored on the farm from the previous year, to provide assurance that varieties have not been mixed, or mis-labelled. Finally when the farmer delivers grain to the receival site, there needs to be some way to rapidly determine if the grain is actually the variety declared, to prevent mixing of different varieties or quality classes, and for end-point royalty collection. This need has long been recognised and scientists have developed methods to accurately identify different varieties. While there are many techniques available, the issue has become one of providing cost-effective methods that are both rapid and accurate.

The methods in use today take advantage of three distinct characteristics and can be divided into visual, protein-based and DNA-based methods. The visual methods rely on morphological traits (see Chapter 4 and species-specific chapters) being bred into varieties to allow distinct classes to be segregated. The biochemical methods take advantage of the diversity in storage protein composition with each variety effectively having its own unique fingerprint of storage proteins. The genotypic methods utilise knowledge of the variation in DNA sequences between the cultivars to develop a unique 'barcode' that can be used for variety identification. The first two of these methods have been examined in detail in numerous treatises [e.g. see Uthayakumaran et al. (2006b) and Wrigley (1995) and references therein]; thus they will only be covered briefly with the focus on recent developments in variety identification.

17.2.1 Visual Classification Into Classes

For variety identification at the receival point to be effective, it must be accomplished rapidly, as the time frame for delivery of grain is measured in minutes. If a variety is mis-classified it will effectively contaminate the other grain in the bin where it has been stored. For example the maximum admixture of bread-wheat tolerated in a delivery of durum wheat may be as low as 3%, and many single variety segregations are required to be at least 95% pure. At present, grain morphology using the characteristics described in Section 4.7 is the only rapid test that satisfies the time frame available at a receival point during harvest delivery. Although this may seem like a daunting task, in practice only 5–6 varieties comprise the majority of each grain type delivered to a particular receival site in any one season. The other option is to sub-sample the delivery and test it as soon as possible, often using tests that require at least an overnight time frame. While this does not alleviate the contamination that has already occurred, it allows future deliveries from that farm to be quarantined until the variety can be accurately determined. One way to circumvent this issue is to classify the grain into classes of similar quality type rather than single varieties, thus minimising the need for accurate on-the-spot variety identification. This system is used in Canada, Australia and the United States. The US system of wheat classification takes into account the time of the year the crop was planted and harvested, the location, seed colour (red or white) and the hardness of the grain.

17.2.2 Biochemical Characterisation for Variety Identification Based on Protein Composition

17.2.2.1 Polyacrylamide Gel Electrophoresis

These methods take advantage of the stability and complexity of the storage protein composition providing a unique fingerprint for each variety. Two different systems are utilised: SDS-polyacrylamide gel electrophoresis (PAGE) of fully reduced flour extracts (Fig. 17.1) and acid-PAGE of the gliadins (Fig. 17.2) (Lookhart and Wrigley, 1995; Uthayakumaran et al., 2006b). Neither of these methods involves major technical challenges, utilising relatively inexpensive equipment; however, both are labour intensive and relatively slow, requiring 3 h or more for sample analysis. Historically acid-PAGE of gliadins, ICC Standard No. 143, (ICC, 2011) was the predominant method used as the gliadins are readily extracted and the patterns produced are simpler to characterise than those of the total protein extracts used in SDS-PAGE. Well-established acid-PAGE and SDS-PAGE electrophoresis methods are available for wheat, rye, triticale, barley, oats and rice (AGSA, 2003).

17.2.2.2 Reversed-Phase High-Performance Liquid Chromatography

Reversed-phase high-performance liquid chromatography (RP-HPLC) analysis of gliadins is a routine method in variety analysis of wheat. It was previously

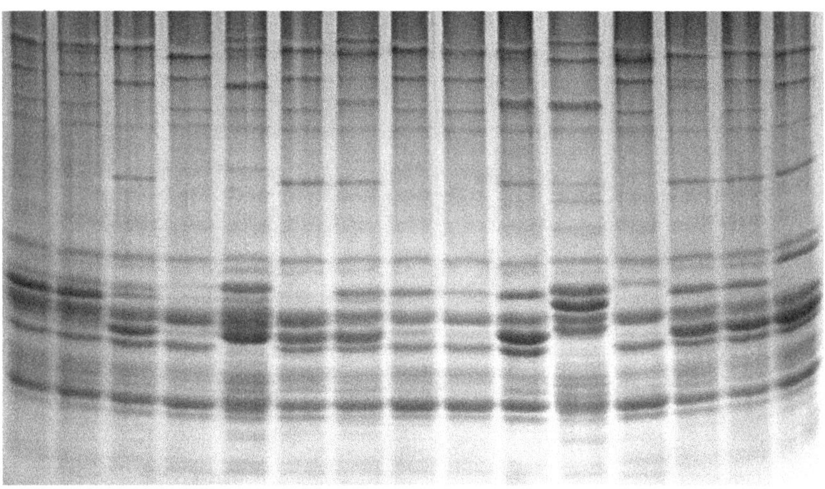

FIGURE 17.1 SDS-PAGE analysis of the storage protein composition of lines in a breeding population. Each lane represents a different line in the population.

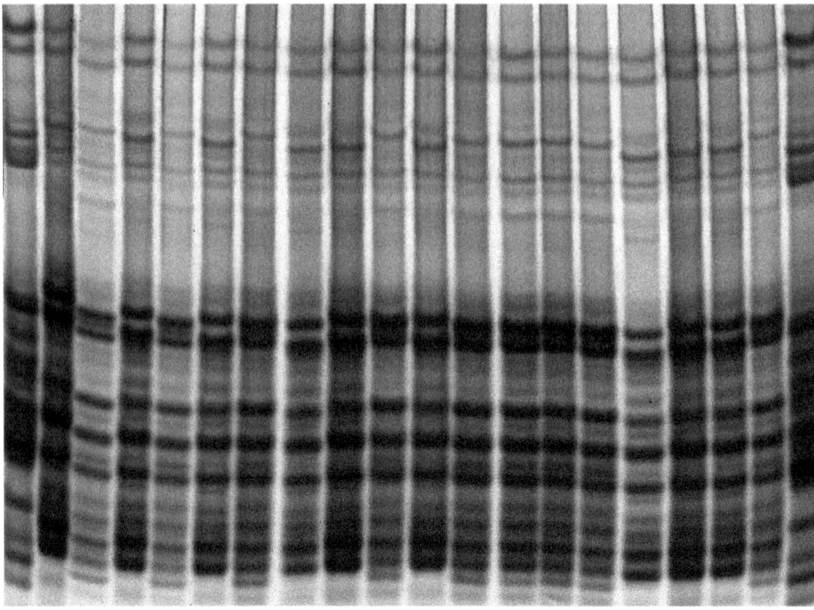

FIGURE 17.2 Acid-PAGE analysis of gliadins to determine the integrity of a line. Gliadins were extracted from 20 individual seeds from the line and analysed using Acid-PAGE. From this analysis, it is clear that the variety is not pure and consists of a mixture of different lines.

popular for variety identification, but now finds application in discrimination of glutenin sub-units. It provides a number of advantages over gel-electrophoretic methods in that it can be automated, the run time of an individual sample is relatively quick in comparison and there is the potential for the use of software to interpret the results. The method of separation of the gliadins is based upon hydrophobicity, as opposed to charge and size in acid-PAGE, thus it is likely to allow for distinction of varieties that are difficult to separate using acid-PAGE (Lookhart et al., 2003).

17.2.2.3 Capillary Electrophoresis

The use of capillary electrophoresis (CE) for variety identification utilises the principle of charged-based separation. The run times for CE are relatively rapid, from 4 to 20 min, and the process including data capture and interpretation can be automated. A number of studies have demonstrated the utility of CE for variety identification, based upon gliadin profiles (Bean and Lookhart, 2000, 2001; Siriamornpun et al., 2001, 2002), or identification of HMW-GS (Uthayakumaran et al., 2006a). It has also been shown that CE can be utilised for variety identification using the glutenin sub-units or proteins extracted from the bran or germ (Siriamornpun et al., 2002). Applications to barley, rye, maize and sorghum have been reviewed by Kaisoon et al. (2008).

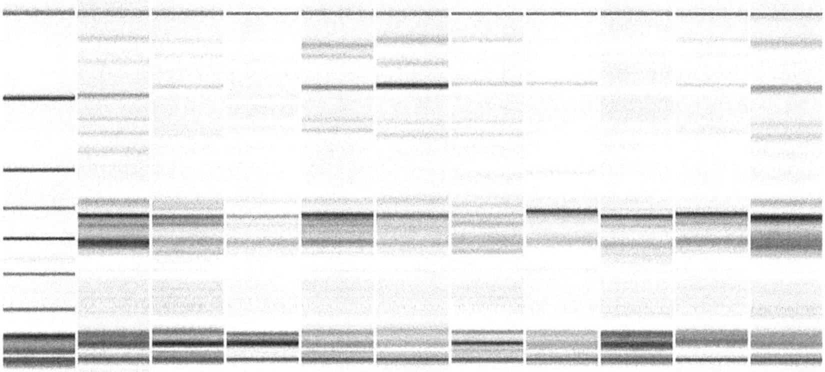

FIGURE 17.3 Lab-on-a-chip analysis CE of wheat flour proteins extracted under reducing conditions. Image displayed is a simulated gel-electrophoresis pattern. The left-hand lane shows the molecular weight standards. *(Image courtesy of S. Uthayakumaran and C. Wrigley.)*

17.2.2.4 Microfluidic Lab-on-a-Chip CE

While all the biochemical methods described earlier have become routine in laboratories, with skilled staff, it is not feasible for them to be performed at the point of grain receival; thus research has been carried out on methods that can give 'real time' feed back to the operators. The most promising of these is based upon the Agilent 2100 Bioanalyzer and has been dubbed 'Lab-on-a-chip' (http://www.agilent.com/cs/library/applications/5989-7735EN.pdf). This equipment has a small footprint, is relatively inexpensive and is relatively quick with the ability to analyse a single sample, including the protein extraction, in 20 min. Ten samples can be completed in less than 1 h (Uthayakumaran et al., 2005). Utilising this method Uthayakumaran et al. (2005) were able to effectively distinguish a set of 40 Australian wheat cultivars (Fig. 17.3). The time frames of this technique are close to those required at the point of grain receival to ensure grains are not mis-classified and assigned to the wrong storage bin. The potential for use in 'on-the-spot' verification of grower-declared wheat variety has been demonstrated at harvest-receival sites in Australia (Miskelly and Wrigley, 2016).

17.2.2.5 Matrix-Assisted Laser Desorption/Ionisation Time-of-Flight Mass Spectrometry

Adaptation of techniques such as matrix-assisted laser desorption/ionisation time-of-flight mass spectrometry (MALDI-TOF) for variety identification provides the potential for analysis of very large sample sets in a relatively short space of time, with the throughputs in the order of 1000 samples per day becoming feasible. The rate-limiting step becomes the sample preparation time. Throughputs on this scale could allow a breeder to analyse the storage-protein composition of every line in a breeding population, provided they can afford the initial outlay for the equipment. Bloch et al. (1999) successfully demonstrated that MALDI-TOF analysis of gliadins could be used to unambiguously distinguish 10 Danish wheat varieties from one another, while Ens et al. (2000) utilised it to distinguish 16 Canadian varieties. Subsequent studies using Danish cultivars were able to distinguish 30 varieties based upon gliadin composition, when neural networks were used; however, there were some issues with reproducibility and the classifications had an error rate of 13% (Sorensen et al., 2002). Similar results were obtained when a partial least-squares regression algorithm was used (Sorensen et al., 2004).

A study utilising MALDI-TOF to define 16 HMW-GS, resulted in an optimised extraction method, which provided a rapid, complete and precise profile of the sub-units without a separation step (Zhang et al., 2008). A method such as this is useful for the identification of lines in a breeding program, which have favourable combinations of alleles because it is simple, fast and precise.

17.2.3 Barley and Maize

Both barley and maize contain their own unique sets of grain storage proteins, hordeins and zeins respectively, and thus in theory any of the techniques described earlier can be adapted for these cereals. However, in practice they are not often utilised and do not seem to be able to provide the same level of discrimination as that seen in wheat when protein-based methods are used. A number of reasons for this lack of discrimination have been proposed and they include the lack of diversity in modern barley varieties and the fact that barley and maize are diploid, (whereas wheat is hexaploid) and thus there are fewer storage proteins for barley and maize (Lookhart, 1991). However, DNA-based methods are now proving useful for discriminating barley and maize varieties.

17.2.4 DNA-Based Variety Identification

Over the last 2 decades there have been significant efforts to understand the genetic basis of agronomic and quality traits of cereals, and as a result molecular markers for these traits have been developed. These in turn can be used effectively to provide a genetic 'barcode' for variety identification. As all of these methods require the extraction of genomic DNA (gDNA) from the sample, amplification of the marker (typically by PCR) and then analysis of the marker are needed. Common molecular markers used for variety identification are SNP (single nucleotide polymorphism), SSP (single sequence repeats) and DArT (Diversity Arrays Technology) using a variety of detection platforms. DNA methods do not provide 'real time' analysis and consequently they cannot be utilised at the point of grain receival. However, as gDNA can be extracted

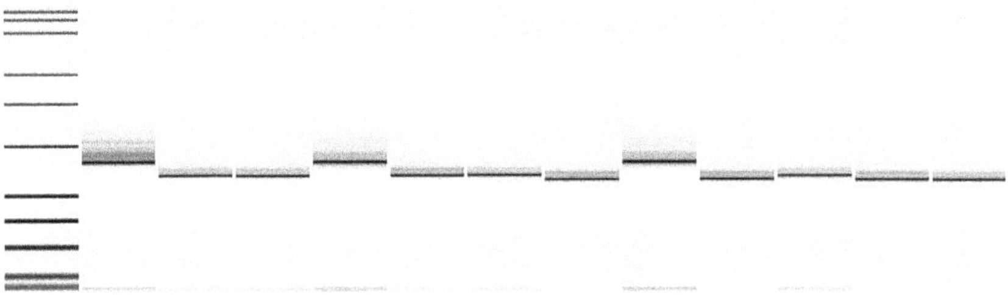

FIGURE 17.4 Lab-on-a-chip analysis of barley varieties utilising a microsatellite marker. DNA was extracted from single seeds of Australian barley varieties. Each lane represents a different grain. *(Image courtesy of S. Uthayakumaran and C. Wrigley.)*

from any tissue, there is a potential to identify which varieties the farmers are growing, at an early stage in the season, and thus minimise the potential for misclassification of a variety at receival. Molecular markers are also particularly useful in breeding programs as they allow the breeder to identify the lines most likely to be the best ones and focus their resources on those. Whichever markers are used, they must be able to distinguish between all the varieties of a particular grain that will be grown in a region and new varieties must be added to the barcode library as they are released for commercial cultivation. This requires considerable preliminary investment before routine screening can be undertaken at harvest delivery, and validation of markers if they are to be transferred for use on a different population.

Examples of the use of molecular markers for barley, wheat, maize and rice have been described in the literature. A set of 48 single-sequence repeats (SSR) that were useful for genotyping barley, and could be used to discriminate a reference set of 24 cultivars, have been reported (Macaulay et al., 2001). A sub-set of 5 of these markers has been utilised to discriminate Australian barley varieties and, of the 11 varieties tested, 9 could be unambiguously identified with these markers (Tran-Dinh et al., 2009). This study also demonstrated that the lab-on-a-chip system could be utilised to separate the PCR products, thus reducing the time frame required for the assay (Fig. 17.4). Similarly, MALDI-TOF was used to profile SNPs from barley; 33 different markers were identified which could distinguish all but 2 common Australian barley varieties (Pattemore et al., 2010). The minimum number of SSR markers needed to distinguish 80 Italian durum wheat varieties was found to be 2 (Mangini et al., 2010). More recently, a set of 10 SSR markers was used to distinguish 48 popular Canadian malting and feed barley varieties. As with previous studies, a number of biotypes were observed (Perry et al., 2014), a biotype being an individual distinguishable from others, all being valid members of the commercial variety. Similar outcomes were reported for the identification of 16 Chilean

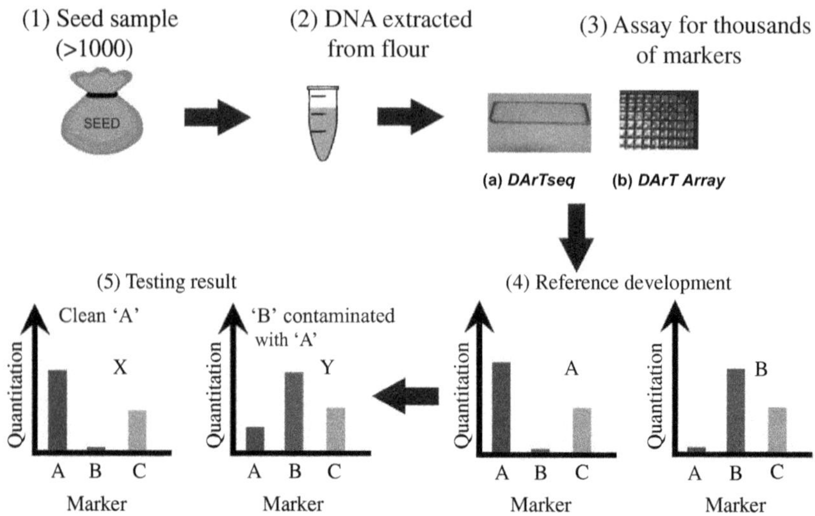

FIGURE 17.5 The mechanism by which the Diversity Arrays Technology (DArT) system can determine the proportions of mixtures of varieties. This simple case shows a mix of two varieties and only three markers. *(Image courtesy of Diversity Arrays Technology Pty Ltd. Reproduced with permission from Miskelly, D.M., Wrigley, C.W., 2016. Identification of varieties of food grains. In: Wrigley, C., Corke, H., Seetharaman, K., Faubion, J. (Eds.), Encyclopedia of Food Grains, vol. 4. Elsevier Ltd, Oxford, UK. pp. 29–41.)*

rice varieties (Becerra et al., 2015) and 36 Taiwanese and international rice varieties (Chuang et al., 2011).

SNP markers are replacing SSR markers as they occur in high frequency and have thus improved the efficiency of genetic diversity studies. Thirty-two selected SNP markers were used to profile 128 Canadian bread-wheat varieties; the application in varietal identification was verified using a further set of 119 varieties (Perry and Lee, 2015). Progress has been reported in the identification of maize hybrids using an SNP array, which could correctly identify 94% of a Chinese population of 276 varieties (Tian et al., 2015).

DArT is another DNA method which provides high throughput, whole-genome screening utilising a hybridisation microarray. A large number of markers have been developed, enabling a high degree of discrimination and the ability to determine the proportions of contamination varieties without the necessity for grain-by-grain analysis (Fig. 17.5). It is used routinely in Australia to verify declarations of wheat and barley varieties. In a recent audit of 400 individual farmers' wheat deliveries to grain-receival points in eastern Australia, 80% were found to be the nominated variety and correctly declared, with the remainder inconsistent with the farmer's declaration or a mixture of two different varieties (www.grdc.com.au).

17.2.5 Applications of Variety Identification Methods

17.2.5.1 *Plant Breeders' Rights*

Plant Breeders' Rights (PBR) is an internationally recognised system, which allows the breeder to hold intellectual property rights in the propagation of a new variety for commercial use. It enables the collection of end-point royalties (EPR), which can then be used to fund future breeding activities. New varieties must be DUS and this is based on a suite of visually-assessed morphological characteristics such as plant height, growth habit, presence/absence of awns etc., as described in Chapter 4. Although the potential of DNA profiling has been recognised by the International Union for the Protection of new Varieties of Plants (UPOV), it can only be used in certain situations such as where the link between the molecular marker and the morphological characteristic has a high level of reliability (www.upov.int). Perfect genetic markers are available for some morphological traits such as 'seasonal growth habit' in barley (Cockram et al., 2012) and the search continues for others, which have the potential to streamline the DUS process.

17.2.5.2 *Confirmation of Pure Seed Identity*

Although plant morphological characteristics are used to confirm the identity and purity of breeders' seed, commercial pure seed or farmer-retained seed, the methods are time consuming as seed must be sown and assessed over a growing season. Protein or DNA methods can provide a much faster result. An Australian survey of 23 commercial seed lots of the variety Sunvale (an older but still popular variety released in 1995) found that only 13 of the seed lot samples were 100% pure (www.grdc.com.au). In this instance, Sunvale had a unique DNA fingerprint, eliminating the possibility of biotypes confounding the results.

17.2.5.3 *Verification of Variety at Grain Receival*

It is important for grain buyers and millers to ensure that the grain that has been purchased not only satisfies physical criteria such as test weight and protein content, but is the correct variety. Some varieties may attract a premium (or a discount) relative to that paid for satisfying the minimum physical parameters and EPR may need to be deducted for payment to breeders. On-site identification will generally involve identification using morphological characteristics, but more likely, a protein- or DNA-based method will be applied post-harvest to a representative sample taken at the time of delivery. The method chosen for authentication will depend on the speed at which results are required as well as the equipment and expertise available. As more reliable DNA-based methods are becoming available at lower cost, it is likely that they will become the preferred methods for variety authentication.

If the delivery contains a mixture of varieties, or if adulteration is suspected, it may be necessary to carry out a grain-by-grain analysis rather than use a

TABLE 17.1 Statistical Interpretation of Results From the Identification of Individual Grains Sampled From a Total Consignment of Grain

Distinct identity found for given % of grains identified	Confidence limits[a] (%) depending on total number of grains identified			
	10 grains	20 grains	100 grains	500 grains
0	0–31	0–17	0–4	0–1
5	—	0–25	2–11	3–7
10	0–45	1–32	5–17	7–13
20	3–56	6–44	13–29	17–23
50	19–81	27–73	40–60	46–54

[a] *Limits at 95% confidence level.*
Source: Reproduced with permission from Miskelly, D.M., Wrigley, C.W., 2016. Identification of varieties of food grains. In: Wrigley, C., Corke, H., Seetharaman, K., Faubion, J. (Eds.), Encyclopedia of Food Grains, vol. 4. Elsevier Ltd, Oxford, UK. pp. 29–41.

ground sample. In this instance, a large number of grains must be tested to ensure the results are statistically significant and the confidence limits are within an acceptable range (Table 17.1).

17.2.5.4 Identification of GM Proteins

Considerable controversy about GM crops exists in some parts of the world and commercialisation is regulated. In some countries it is necessary to segregate and preserve the identity of GM and non-GM grain throughout the supply chain. It is also necessary to ensure that the maximum adventitious presence of GM contamination is not exceeded in non-GM grain. At present, maize is the only cereal grain with GM and non-GM (hybrid) varieties. Several methods are used to detect the marker proteins that are expressed by the modified gene. The simplest is a lateral flow test strip, which has specific protein antibodies in a membrane-based detection system. When the 'dip stick' is immersed in a homogenised solution of the test sample, a visible line is formed to indicate the presence of the particular GM. This method takes only a few minutes, so it is suitable for use at delivery or in a small regional laboratory. Another form of immunoassay is the enzyme-linked immunosorbent assay (ELISA) technique in which specific antibodies are adsorbed to microtiter plates, enabling high-volume screening due to automation. A testing time of 2–4 h makes this assay more suitable for use in a central laboratory. DNA testing can also be used to detect GM, and as multiple transformations may need to be recognised, qualitative screening tests based on real-time PCR are used routinely. Further development is required in terms of method validation. Guidelines for standardisation have been documented by Broeders et al. (2014).

17.2.6 Future Challenges for Variety Identification

A number of challenges still remain for variety identification. Choice of the most appropriate method will depend on a number of factors, including the speed at which results are required and the degree of discrimination required. While the lab-on-a-chip system has reduced processing times to near those required for 'real time' assessment, it has only been utilised in limited studies on wheat and rice. Systems based on storage protein alone do not have the power to differentiate some specialty traits. Two recent examples of these are Hi-Maize^{TM} and BARLEYmax^{TM}, which have health benefits as a result of mutations in starch biosynthetic enzymes, but in terms of storage protein composition, they are indistinguishable from equivalent varieties that do not contain these mutations. Thus molecular markers are required for unequivocal identification of these varieties. In the last few years, protein-based methods have been replaced by DNA-based methods as knowledge of the genetic basis of quality grows and with advances in DNA-marker characterisation.

On the other hand, some analytical methods of variety identification have become too discriminating. Both protein- and DNA-based methods have, in some cases, provided distinction between single grains of the same variety due to the presence of biotypes within that variety (Wrigley, 1995; Uthayakumaran et al., 2006b). These biotypes are a legitimate part of the variety, being descendants from the original cross, but they represent a degree of polymorphism that was not evident at the stage of establishing the original breeder's sample, which became the ultimate representation of the variety. The presence of multiple biotypes complicates the task of identification in that, for the relevant identification system, a specific variety is represented by more than one pattern and analysis of a milled sample will produce a pattern that is a mix of patterns. Thus it may be necessary to analyse grains individually (one-at-a-time) to obtain clearly identifiable patterns or to use a system such as DArT based on a large number of markers.

If the requirement to identify varieties after harvest relates mainly to the need to specify grain quality, as quality class is generally specified by the admission of a limited range of specified varieties into each quality class, specification of variety can be seen as a surrogate for quality type. If suitable test systems were available to determine key aspects of quality, these tests might be used to replace the need for variety specification, for variety identification and for annual updating of the range of varieties acceptable for each grade or class of grain quality. There is thus the challenge to elucidate markers for quality traits, thus to segregate for quality type in breeding, at grain receival and in monitoring identity preservation during grain transport and storage.

However, this would not satisfy the need to identify a variety in the event where EPR payments are assessed at the receival, or where individual varieties are segregated into a separate class, particularly where mis-declaration is suspected.

There is an emerging need to identify, authenticate and trace species, subspecies and varieties in food manufacture, from raw material intake through to the finished product. An example is the accidental or deliberate adulteration of durum pasta with bread-wheat flour or semolina, which is illegal in some countries. In food manufacturing, processing stages such as cooking can result in difficulties in extraction of intact DNA, so there is a need for the development of customised food diagnostics methods, which are readily available and simple to use.

17.3 IDENTIFICATION OF QUALITY TYPE USING MOLECULAR MARKERS

Cereal grains are the major component of the diet for a large proportion of the world's population (Chapter 1), and have been bred over thousands of years for improvements in agronomic performance and yield, to produce the crops we know today. A major challenge for today's breeders is to continue to increase the yield of these cereal crops while at the same time improving grain quality for end-product manufacture. This is particularly important as the world's population continues to grow, with it estimated to increase to 9.6 billion by 2050.

Traditionally breeding has relied on physical measurements to select for phenotypes (traits) such as plant height, grain number and grain size; this type of characterisation is relatively simple and quick to do. Many lines can be characterised at an early stage in the breeding program with only the best lines being carried forward. When breeding for quality, the process is not as straight forward because analysis of quality traits requires significant quantities of grain; thus it cannot be performed at an early stage. To compound this problem, grain-quality analysis is time and labour intensive, with quality assessments in wheat breeding programs representing up to 25% of the breeders' costs; thus only a limited number of lines can be analysed in a comprehensive manner. To overcome these limitations, efforts are underway to identify markers for quality traits that can be utilised in MAS programs, which will allow the breeders to identify lines that are likely to be of higher quality and focus their efforts on those lines. For comprehensive reviews of the principles of MAS, the reader is referred to Chapter 16, Collard and Mackill (2008) and Xu et al. (2013).

When utilising MAS, two types of markers are available to the breeder. The first is diagnostic (also known as 'perfect'), where the marker is directly associated with the gene that influences the trait, thus having absolute linkage with the trait being selected, and not requiring independent validation for each parental line used in a breeding program. The second type is linked markers, and they are generally associated with quantitative trait loci (QTL) identified, in a mapping population, for the trait of interest. They are not directly associated with the gene underpinning the trait, and thus they only have a probability of being co-inherited with the trait of interest. As linked markers are identified within a specific population, it is required that they be validated before being used in breeding programs. Markers have evolved from restriction fragment-length polymorphisms (RFPLs) to PCR-based SSRs and more recently SNPs and DArT. Techniques such as genome-wide association mapping allow simultaneous assessment at a higher resolution and many more markers are being discovered. Bioinformatics tools can be applied to study associations with previously detected QTLs.

The following sections provide a brief overview of some quality traits and markers or QTLs associated with these traits. Given the number of studies reported in recent years and the number of traits characterised, it is not possible to provide a comprehensive review here.

17.3.1 Markers for Quality Traits in Wheat

17.3.1.1 Gluten Polymers and Quality

The high molecular weight (HMW) and low molecular weight (LMW) glutenin sub-units (GS) are critical determinants of the structure of the glutenin macro-polymer. Thus they are a major determinant of end-product quality, with the HMW-GS contributing to dough strength and the LMW-GS to extensibility (Bushuk, 1998; Goesaert et al., 2005; Gras et al., 2001; Shewry et al., 2003). There are three loci for the HMW-GS (*Glu-A1, Glu-B1* and *Glu-D1*), each encoding an x- and a y-type HMW-GS, and three major loci for the LMW-GS (*Glu-A3, Glu-B3* and *Glu-D3*), with each encoding multiple proteins. Due to their influence on quality, and the relative simplicity of the loci encoding them, much effort has been put into molecular characterisation of these loci, which has enabled the development of relevant molecular markers. These markers were reviewed by Gale (2005), Howitt et al. (2006) and more recently by Goutam et al. (2013). A summary of the many markers developed is provided in Table 17.2. These include dominant and co-dominant markers as well as novel markers used to provide better discrimination and identification of mis-classified genotypes. According to Rasheed et al. (2014) more than 69 alleles at *Glu-B1* and 29 alleles at *Glu-D1* have been documented in bread wheat in the Gene Catalogue (http://www.shigen.nig.ac.jp/wheat/komugi/genes/download.jsp). Wide-scale application studies have been reported using allele-specific PCR markers on 718 wheat cultivars and advanced lines from 20 countries and the frequencies of occurrence quantified by country, identifying potential sources for use in future breeding (Jin et al., 2011).

LMW-GS account for approximately 60% of the total glutenin sub-units (Rasheed et al., 2014) and until recently have received far less attention than the HMW-GS. This deficiency can be directly attributed to the complexity of the *Glu-3* loci, making it difficult for both genetic characterisation and scoring of allelic variants on SDS-PAGE. The rapid advances in sequencing techniques and the reduction in cost, in recent years, have made characterisation of these loci

TABLE 17.2 PCR-Based Markers for HMW-GS Sub-Units

Gene	Alleles	References
Ax, Ay	Various	Lafiandra et al. (1997)
Ax null	*Glu-A1c*	Dovidio et al. (1995); Lafiandra et al. (1997)
Ax	*Various*	de Bustos et al. (2001)
Ax1, Ax2*	*Glu-A1a, b*	Ma et al. (2003); Radovanovic and Cloutier (2003)
Ax2*	*Various*	Liu et al. (2008)
Ax2*B		Dovidio and Anderson (1994); Juhasz et al. (2003), using marker of Dovidio and Anderson (1994)
Bx7/Bx17	Glu-B1b	Ahmad (2000); Ma et al. (2003)
Bx7/Bx7*	Glu-B1	Espí et al. (2012)
Bx7*, Bx7^{OE}	*B1al*	Butow et al. (2003); Radovanovic and Cloutier (2003); Ragupathy et al. (2008)
Bx	*B1 al, b, i*	Butow et al. (2004)
Bx6	*Glu-B1d*	Schwarz et al. (2004)
By8	*Various*	Lei et al. (2006)
By9	Various	Lei et al. (2006)
By16/By20		Lei et al. (2006)
Dx, Dy	*Various*	de Bustos et al. (2001); Dovidio et al. (1994)
Dx5	Glu-D1d	Ahmad (2000); Dovidio and Anderson (1994); Ma et al. (2003); Radovanovic and Cloutier (2003)
Dx2/Dx5	Glu-D1d/a	Liu et al. (2008)
Dy10/Dy12	Glu-D1d/a	Ahmad (2000); Liu et al. (2008); Smith et al. (1994)

and the development of molecular markers feasible. Markers are now available for the major alleles at *Glu-3A* and *Glu-3B* (Wang et al., 2009a, 2010; Zhang et al., 2004). A summary of the markers available is provided in Table 17.3. Combinations of PCR-based markers with SDS-PAGE, 2DE, CE and MALDI-TOF MS are being used to detect more allelic variations. Recently, 48 unique allele combinations were reported in a population of wheats and substitution lines from Australia, France, Canada and CIMMYT (Wang et al., 2015).

An international collaborative program is underway to share methods, unify the criteria and develop a uniform nomenclature (Ikeda et al., 2013). The Glutenin Composition Database contains information on HMW-GS and LMW-GS for thousands of genotypes and also includes data on dough strength and extensibility (Békés and Wrigley, 2013).

TABLE 17.3 PCR-Based Markers for LMW-GS and Gliadins

Loci	Alleles discriminated	References
Glu-A3		Devos et al. (1995)
Glu-A3	a–f	Zhang et al. (2004)
Glu-A3	a/c, b, d–f	Wang et al. (2015) using MALDI-TOF
Glu-A3,B3,D3		Ikeda et al. (2006); Long et al. (2005)
Glu-B3		Zhao et al. (2007a)
Glu-B3	a–i	Wang et al. (2009a)
Glu-B3	a–c, d/i, f–hi	Wang et al. (2010)
Glu-D3		Zhao et al. (2007a)
Glu-D3	a–e	Zhao et al. (2007b)
Glu-D3	c, h, and l	Appelbee et al. (2009) using markers of Zhao et al. (2007a) and Zhao et al. (2007b)
Glu-D3	a–f	Wang et al. (2010)
γ-Gliadins		Zhang et al. (2003)

17.3.1.2 Non-Glutenin Effects

While the glutenin alleles play a significant role in determining end-product quality, they are not entirely responsible for all of the variation and approximately 25% is attributable to other genes. Determining the cause of this variation has been the subject of many studies and multiple QTLs have been identified. As these studies have been carried out in multiple populations and often the genetic maps created have few markers in common, it is very difficult to directly compare QTLs from one study to the next. For the markers identified to be of use in breeding populations, they will need to be validated across all lines used in the breeding programs. The following section will briefly highlight some of the non-glutenin QTLs identified to date.

17.3.1.3 Grain Protein Content

Understanding the genetics of grain protein content (GPC) in wheat has proven to be difficult as the trait has a low heritability, is very sensitive to environmental conditions and increased GPC is correlated with decreased yield. Turner et al. (2004) could not identify any major genes underlying the trait, while several studies have shown the effects of chromosome 5 (Blanco et al., 2002; Campbell et al., 2001; Ma et al., 2007; Sun et al., 2008). A number of studies have identified QTLs for GPC on group 3 and 7 chromosomes, 7AL (Campbell et al., 2001), 7A and 7B (Blanco et al., 2002), 3A and 7D (Groos et al., 2004), 7B

(Blanco et al., 2006), 3A (Sun et al., 2008) and 3A and 7A (Mann et al., 2009). More recently, Bogard et al. (2013) found genomic regions on chromosomes 3A and 5D where protein content was independent of grain yield.

Studies on the lines in which chromosomal regions from *Triticum turgidum* var. *dicoccoides* were introgressed into tetraploid and hexaploid wheat have revealed that the presence of a region on chromosome 6B, *Gpc-6B1,* resulted in increased GPC without an associated yield penalty (Cantrell and Joppa, 1991). Linked markers for this region were developed and the loci was targeted for candidate gene identification by fine mapping (Olmos et al., 2003) and micro-co-linearity, using wheat–rice synteny (Distelfeld et al., 2004, 2006). These studies identified a 245-kb region that contained the gene of interest. The entire region was sequenced from both the *dicoccoides* and tetraploid parental line, Langdon. This identified an NAC transcription factor, *Nam-B1*, that contained a single base pair (bp) insertion in the Langdon allele (Uauy et al., 2006). Perfect molecular markers have been developed that can be used in MAS (Uauy et al., 2006). A sequence-tagged marker linked to *Gpc-6B1* is now used in Canadian wheat breeding programs. This has resulted in the release of high yielding, high protein commercial varieties such as Lillian and Burnside. These varieties retain high protein content without a yield penalty (Randhawa et al., 2013).

17.3.1.4 Grain Hardness

The major genetic locus controlling hardness is the *Ha* locus on chromosome 5D, which encodes *pinA*, *pinB* and *gsp-1*. The dominant 'soft' phenotype requires the presence of both the wild-type *pinA-D1a* and the *pinB-D1a* alleles. If either of these alleles is absent or contains variations in the coding sequence from the wild-type allele, then a hard phenotype is seen. These variations in the proteins produced directly affect the binding of puroindolines to starch granule surface lipids. This binding in turn influences endosperm texture, which feeds through to effects on properties such as milling energy, flour particle size, milling yield, damage to starch granules, water absorption of flour, dough development time and extensibility and dough foam stability. All these factors in turn influence end-product quality (Morris, 2002). As a result, these genes have been extensively studied [see Bhave and Morris (2008a,b) for reviews]. Simple and co-dominant PCR markers for grain hardness were developed for the *Pina-D1a, Pina-D1b, Pinb-D1a, Pinb-D1b, Pinb-D1c and Pinb-D1d* alleles and used to investigate 100 genotypes from 5 countries (Huang and Brûlé-Babel, 2011). Recently, Chen et al. (2013) characterised the pin *Pina-D1* and *Pinb-D1* in 493 cultivars from diverse locations and found the highest diversity of *Pin-D1* alleles in Chinese landraces. This complexity makes it difficult to develop PCR-based molecular markers to assist in MAS, unless all the alleles present in the breeding population are known. While the genetic basis distinguishing the two major hardness classes is now well established, little is known about the residual variation within each of the classes and their genetic components.

TABLE 17.4 Chromosomal Regions Identified for Wheat Endosperm Yellowness

Chromosome	References
3A, 7A	Parker et al. (1998)
3B, 7A	Mares and Campbell (2001)
7B	Kuchel et al. (2006)
7A, 7B	Elouafi et al. (2001) (durum); Zhang and Dubcovsky (2008) (durum)
5B, 5D	Mares and Campbell (2001)
4A, 5A	Hessler et al. (2002) (durum)
1A, 1B, 3B, 4A, 7D	Zhang et al. (2009)
1B, 7A	He et al. (2008)
3A, 3B, 5B, 7A	Howitt et al. (2009)
1B, 2A, 4B, 6A, 7A, 7B	Pozniak et al. (2007, 2012) (durum)
7A	Singh et al. (2009)
1A, 1B, 4A, 7A, 7B	Zhang et al. (2009)

17.3.1.5 Milling Yield

Milling yield is of significant value in the industry because an increase of as little as 1% represents a significant cost saving for the miller. The genetic basis of milling yield is still poorly understood and is influenced by hardness, grain size and morphology, and thickness and the colour of the seed coat. QTLs have been identified on multiple chromosomes from different populations, with QTLs identified on 3A, 5A, 7D (Parker et al., 1999), 1B, 3B 6B (Campbell et al., 2001), 5B, 7D (Smith et al., 2001), 1A, 2A, 6A, (Kuchel et al., 2006), 2BS, 4B, 5AL, 6BL (Lehmensiek et al., 2006), 1B, 3B and 7D (McCartney et al., 2006) and 1B, 5B and 5D (Tsilo et al., 2011). As yet, none of the markers associated with these QTLs have been validated in other populations; thus their use in MAS at present is limited.

17.3.1.6 Flour Colour

The yellow pigment of wheat endosperm and hence flour colour is due to carotenoids. In both durum and bread wheat, the major QTLs for flour colour are on group 7 chromosomes and many minor QTLs have been reported (Table 17.4). This topic has been reviewed for durum and bread wheat by Schulthess and Schwember (2013).

More recent studies have attempted to associate genes encoding carotenoid biosynthetic enzymes with these QTLs. Phytoene synthase is a rate-limiting enzyme in carotenoid biosynthesis. In durum, phytoene synthase2 (*PSY2*), phytoene desaturase (*PDS*) and zeta carotene desaturase (*ZDS*) were mapped to chromosomes 5, 4 and 2 respectively (Cenci et al., 2004). *PSY1* has been mapped to chromosomes 7A and 7B and shown to be tightly linked to the major QTLs identified (Pozniak et al., 2007; Singh et al., 2009). Similarly in bread wheat, different alleles of the 7A homeoform of *Psy1* have been shown to be tightly linked to the major QTLs for flour colour (He et al., 2008; Howitt et al., 2009; Zhang and Dubcovsky, 2008). All three of these studies developed dominant markers for the alleles identified, with (Howitt et al., 2009) developing a multiplexed co-dominant cleaved amplified polymorphic sequence (CAPS) marker that could distinguish all four alleles identified. Association studies have linked alleles of the 7B homeoform of PSY with variation of flour colour, and diagnostic molecular markers for the five alleles identified have been developed (He et al., 2009), while epsilon cyclase has been linked to the QTL on 3B (Howitt et al., 2009) and a co-dominant CAPS marker developed. *PYS1D* has also been shown to influence yellowness in bread wheat (Ravel et al., 2013).

17.3.1.7 Polyphenol Oxidase

The level of polyphenol oxidase (PPO) present in flour is the major factor contributing to the darkening of end products such as steamed breads and noodles, thus breeding for reduced PPO content will enhance flour quality for these products. QTLs associated with PPO activity have been identified on group 2 chromosomes (Beecher et al., 2012; Mares and Campbell, 2001; Raman et al., 2007; Simeone et al., 2002). Jukanti et al. (2004) cloned six PPO genes, and identified two groups, one of which was expressed during kernel development. Utilising these sequences, Sun et al. (2005) developed a sequence-tagged site (STS) marker PPO18 for wheat kernel PPO activity, associated with the gene located on chromosome 2A. Subsequently, He et al. (2007) developed a dominant marker, PPO29, for the gene on 2D. Wang et al. (2009b) have developed the markers STS1, which can discriminate two alleles on 2D, and PPO05 that could discriminate the low-activity PPO allele *PPO-2Ab* from the higher activity *PPO-2Aa* allele. These markers were validated in a large range of Chinese cultivars and can now be used in wheat breeding programs to select for lines with low PPO activity.

17.3.1.8 Water Absorption

Water absorption is a critical quality trait for bakers as higher water absorption means less flour needs to be used for each loaf of bread and thus input costs are lower, increasing profitability. The *pinb* gene at the *ha* locus has been associated with water absorption (Campbell et al., 2001). While this was the strongest QTL identified in this study, other QTLs have also been reported on chromosomes 1A, 1B, 1D, 2D, 3B (Mansur et al., 1990), 1DL, 3A 5AS, 5BS, 5DS (Campbell et al., 2001), 2A and 2D (Kuchel et al., 2006), 2B, 4D, 6B, 7D

(McCartney et al., 2006) 5A (Ma et al., 2007), 1A (Glu-A3), 3A, 5A and 5B (Mann et al., 2009) and 1A, 1B, 4B, 4D and 5A (Tsilo et al., 2013). This last study using SSRs and DArT marker loci provided validation of QTLs found by previous researchers and that water absorption is a function of protein content, starch damage and gluten strength.

17.3.1.9 Dough Rheological Traits

The contribution of the HMW- and LMW-GS to dough properties has been extensively studied [for reviews see Bushuk (1998); Goesaert et al. (2005); Gras et al. (2001); Shewry et al. (2003)]. These proteins alone do not explain all the variation observed in dough rheological parameters, such as dough strength or extensibility, and thus loci other than these are the targets of studies to understand the genetic basis of quality. A brief coverage of some of these studies is given later in the chapter.

QTLs for dough mixing time have been identified on 1B and 2A (Zemetra et al., 1987), 2B, 4AL, 7AS, 7DS (Campbell et al., 2001), 1B and 4D (McCartney et al., 2006). Law et al. (2005), using substitution lines, identified regions on 7A, and to lesser extent regions on chromosomes 5BS and 7BS, that influenced mixing time.

Zemetra et al. (1987) identified two significant QTLs on chromosomes 3B and 7B for mixing tolerance, while Campbell et al. (2001) detected QTLs on 1BL and 1A for mixing tolerance in a soft versus hard wheat cross. Kuchel et al. (2006) reported a QTL on 2A associated with both R_{max} and stability. In a five-site trial, using the Kukri X Janz doubled haploid population, QTLs for R_{max} (Farinogram maximum height) were identified on 1B and 1D corresponding to the *Glu B1* and *Glu D1* loci for all five sites, while a further QTL, on 1B (*Glu-B3*) and one on 2B were common to three of the five sites. In this study, significant QTLs were also identified for dough rheology traits that are measured by the Z-arm mixer, the majority of these were associated with the *Glu-B3*, *Glu-B1* and *Glu-D1* loci. Additional QTLs for various Z-arm parameters were also identified from one site, on 2B, 5A, 7A and 7D (Mann et al., 2009).

Studies using the Alveograph have identified QTLs on 1B and 3A for W (dough strength), on 1B, 2B, 3B, 6D for P (tenacity) and QTLs on chromosomes 2B, 3B, 5B for L (extensibility) (Groos et al., 2004). However, none of the QTLs for rheological characteristics had large effects. Perretant et al. (2000) identified three QTLs for W, two of these, on 1A and 5D, co-located with QTLs for hardness, and one on 3B. Tadesse et al. (2015) found three markers on 1D, 1B and 5B that accounted for 14% of the variation in W.

17.3.1.10 Loaf Volume

In recent years, studies analysing quality have revealed that dough rheological properties do not adequately reflect end-product quality, and that to fully understand a parameter such as loaf volume, baking studies must be performed.

TABLE 17.5 Chromosomal Regions Identified That Influence Loaf Volume, but do not Encode Glutenins or Gliadins

Method	Chromosome	References
Straight dough (AACC method 10-10B)	1A, 2A, 3A, 5D, 6B, 6D[a]	Rousset et al. (2001); Elangovan et al. (2008)
Chorleywood bread process	3A	Law et al. (2005)
No time spiral mix baking procedure	3A	Law et al. (2005)
Rapid dough bread-making	2A, 3A	Kuchel et al. (2006)
Long fermentation baking (AACC method 10-10)	4D	McCartney et al. (2006)
French bread-making (AFNOR method NF V03-016)	1A, 3A, 5B, 7A, 7B	Groos et al. (2007)
Sponge and dough	5D	Mann et al. (2009)
Long fermentation baking (AGSA method 07-02)	5D, 6A	Maphosa et al. (2013)

[a]*Only the major QTLs are reported here, 30 QTLs for loaf volume were identified in this study.*

These studies have shown that when loaf volume is measured, a number of QTLs that do not map to the glutenin loci can be identified. Despite different baking methods being used, a number of these studies have identified a QTL on 3A. These QTLs identified in these studies are summarised in Table 17.5. More recently, the *wbm* gene located on chromosome 7L, a region where no other quality genes are located, has been postulated to contribute to bread quality, and had no relationship with protein content (Furtado et al., 2015).

17.3.2 Barley Malting Quality

There are two predominant uses of barley: (1) as a feed stock and (2) in the brewing industry for the production of beer, whisky and shochu. Barley for the brewing market captures a higher price than that used for feed grain and thus is the focus of significant research efforts as a new malting barley must satisfy a diverse set of requirements including small-scale micro-malting evaluation. Additionally, some end users have different requirements depending on the type of beer and regional preferences. This section will briefly touch on some parameters that are important for malting quality for the production of beer, and on some QTLs and markers that have been associated with these traits. The reader is referred to Fox et al. (2003) for a comprehensive coverage of these topics. Recent work on markers is mainly concentrated on a number of breeding populations which have been established, principally in North America, Europe and Australia (http://wheat.pw.usda.gov/ggpages/bgn/30/SEU1txt_and_figs_color.htm).

17.3.2.1 Grain Size

Grain size is an important parameter in barley to be used for malting; it is influenced by both genetic and environmental factors. Smaller grains generally have reduced starch content and higher protein levels which limit the extract potential. To date, many of the studies on grain size have been conducted in populations that were not designed for analysis of this trait and many QTLs with small effects have been identified, even using molecular techniques. Interestingly, the most significant QTLs appear to be co-located with genes associated with plant development, such as earliness per se (*eps*) and semi-dwarfing (*sdw1*). They may be a result of plieotropic effects; thus utilising these loci for MAS may be of limited value [see Coventry et al. (2003) for a comprehensive review]. Recently, Pasam et al. (2012) used association mapping and found 36 markers yielding 21 QTLs significantly associated with thousand grain weight. These were present on all chromosomes.

17.3.2.2 Diastatic Power

Diastatic power is the term used to describe the collective activity of the starch-degrading enzymes α-amylase, β-amylase, limit dextrinase and α-glucosidase in malt. QTLs for diastatic power have been identified on 1H, 2H, 4H, 5H, 6H, 7H (Hayes et al., 1993), 4H, 5H, 7H (Oziel et al., 1996), 1H, 3H, 5H (Thomas et al., 1996), 1H, 5H, 6H, 7H (Mather et al., 1997), 2H, 7H (Marquez-Cedillo et al., 2000) and 1H, 4H, 5H (Collins et al., 2001). In order to validate some of these, Coventry et al. (2003) selected eight markers associated with these QTLs, validating them in unmapped breeders' populations. Two of the markers used, *Bmy1* on 4H and *EBmac501* on 1H, were validated as being useful in MAS breeding for diastatic power. In a more recent study, *EBmac501* was not found to be associated with diastatic power in the populations studied, suggesting this marker is only useful in certain backgrounds (Panozzo et al., 2007). QTLs for α-amylase and β-amylase activities have also been detected in different populations.

17.3.2.3 α-Amylase

α-Amylase (EC 3.2.1.1) is an endohydrolase that randomly cleaves α-(1-4) glucosidic bonds in starch, and is involved in the conversion of starch to glucose during malting. QTLs for α-amylase activity have been identified on all chromosomes with all studies identifying regions on chromosome 5 that were associated with this trait (Collins et al., 2001; Coventry et al., 2003; Han et al., 2004; Hayes et al., 1993; Marquez-Cedillo et al., 2000; Mather et al., 1997; Oziel et al., 1996; Panozzo et al., 2007). Ayoub et al. (2003) demonstrated the potential to use markers for a QTL on 5H in one population, to select for lines with increased α-amylase activity in a second population. Similarly Coventry et al. (2003) demonstrated the potential to use the marker *abg463*, on 5H, for increased α-amylase activity in six unmapped breeders' populations. Recently, sequencing of the *amy-1* gene, on 6H from 326 released cultivars, identified four haplotypes for the gene and molecular markers were developed. Further, association mapping using these markers revealed that within the spring barleys analysed, haplotype *amy1_H1* was associated with significantly higher friability values than *amy1_H4*. While haplotype *amy1_H2* explained 19% of the phenotypic variation of the malting quality index (MQI) for all varieties and 35% in a sub-set of 72 winter barleys (Matthies et al., 2009). Strong QTL × environment interactions were found using a genome-wide selection strategy which found a large number of minor QTLs contributing to malt α-amylase (Yang and Ham, 2012).

17.3.2.4 β-Amylase

β-Amylase (EC 3.2.1.2) is an exoenzyme that cleaves the disaccharide maltose from the non-reducing end of amylose and amylopectin, and is one of the key enzymes in the production of maltose, which is utilised by yeast during fermentation. This enzyme is found in mature grain in two forms, free and bound, and is encoded by a multi-genic locus (*Bmy1*, *2* and *3*) (Li et al., 2002).

Three major forms of β-amylase have been identified, Sd1, Sd2 and Sd3, with the first two predominant in commercial malting varieties, while the third, which has increased thermostability was identified in *Hordeum spontaneum* (Eglinton et al., 1998; Kihara et al., 1998). Upon the basis of thermostability, Sd2 can be split into two classes: Sd2L (low) and Sd2H (high). SNPs were identified in the coding sequences of *Bmy1*, encoded on 4H, allowing the development of molecular markers to distinguish these two alleles (Eglinton et al., 1998). As well as these SNPs, a 126-bp insertion in intron 3 that could be used to distinguish the alleles, was identified (Erkkila, 1999; Erkkila et al., 1998). Two independent studies used the presence of the 126-bp insertion to demonstrate it could be used to select for lines with higher thermostability of β-amylase, as it was not present in the Sd2L allele (Coventry et al., 2003; Gunkel et al., 2001). A more detailed analysis of the sequences encoding the four alleles of *Bmy1* identified two SNPs which, when used in combination, could be used to genotype barley varieties and unambiguously determine which allele was present (Paris et al., 2002). The utility of these markers has been demonstrated (Malysheva et al., 2004; Polakova et al., 2003) and they show great potential to be used in MAS for increased thermostability of β-amylases in malting barley.

17.3.2.5 Malt Extract

Malt extract refers to the amount of soluble material that can be extracted from the malt during mashing. Malt extract is a key parameter in assessing the quality of the grain, but assessment of this trait requires large amounts of grain and is destructive. Thus if suitable markers that have a positive correlation with malt extract can be identified, they will provide an invaluable tool for breeding using MAS. Assessing malt extract is difficult as it is influenced by environment, genetics and technology, both malting and mashing (Collins et al., 2003). Variations in any one of these between different studies can influence the outcome and make validation of QTLs between studies difficult. The use of different marker types in genome-wide association mapping has been shown to have an effect on the discovery rate for significant marker-trait associations for malt extract (Matthies et al., 2012).

To date, QTLs for malt extract have been identified on all the barley chromosomes (Bezant et al., 1997; Collins et al., 2001, 2003; Han et al., 2004; Hayes et al., 1993; Marquez-Cedillo et al., 2000; Mather et al., 1997; Oziel et al., 1996; Panozzo et al., 2007; Thomas et al., 1996), with the majority of studies reporting QTLs on 1H, 2H and 5H. Collins et al. (2003) investigated the suitability of markers for six of the regions identified in their studies in breeders' populations, which contained a parent that was a high malt-extract variety. Four of these markers, two on 2H and two on 5H, were found to be associated with the allele from the high malt-extract parent and thus have the potential to be utilised in a MAS for malt extract.

17.3.2.6 Wort Viscosity

Wort viscosity influences the processing of the beer, with higher viscosities impacting on filtration and thus reducing the efficiency of the brewing process. Like malt extract, this trait is influenced by genetics, environment and the technology used in the processing, thus it is a trait for which MAS could be used to accelerate the progress of breeding. QTLs for wort viscosity have been identified on chromosomes 2H, 3H, 4H, 5H (Thomas et al., 1996), 5H, 6H, 7H (Mather et al., 1997) and 1H, 2H and 3H (Panozzo et al., 2007). A wort viscosity QTL was mapped on chromosome 5H for a spring × winter doubled haploid (DH) cross (Laidò et al., 2009) and on chromosome 7H using European barley varieties (Matthies et al., 2014).

17.3.2.7 Haze Stability

Analysis of factors that influence haze stability identified a 12-kDa barley silica eluate (SE) protein, that when present, reduced haze stability in pilot brewing trials (Evans et al., 2003; Robinson et al., 2004). Subsequent analysis of the SE protein has identified it as a barley trypsin inhibitor of the chloroform/methanol type, and the gene encoding it has been cloned. Analysis of the gene sequence in SE+ and SE− lines has identified a consistent set of SNPs that can be used to distinguish the two alleles (Robinson et al., 2007a,b). While the mechanism of how this protein influences haze stability remains to be elucidated, the presence of these SNPs identified will allow the development of molecular markers to identify which allele is present. Recently, a novel protein qACH, which explained about 20% of phenotypic variations was identified on chromosome 4H (Ye et al., 2015).

17.3.2.8 Utilising MAS for Improving Malting Quality in Barley

There is great potential for utilising markers to improve quality, Karakousis et al. (2003) identified a set of 36 markers that were likely to be of use in MAS in Australian barley breeding programs for agronomic, disease resistance and quality traits. A study utilising two markers, one of which was identified by Karakousis et al. (2003), highlighted the potential of MAS in barley to improve malting quality, through the pyramiding of QTLs in a breeding program. In this program, two regions of the genome that contained multiple quality QTLs were selected for by MAS during the breeding program, and introgressed into a poor-quality malting line that had superior agronomic characteristics. At the end of the cycle, micro-malting results revealed that, compared to the recurrent parents, malt-extract levels were increased by 1.5–2%, diastase levels increased by 25–43%, α-amylase activity was up 36–53%, and wort viscosity was reduced by approximately 10% (Emebiri et al., 2009). Matthies et al. (2014) in a study of European barleys, found 140 marker trait associations on all seven chromosomes, but concentrated on 1H, 2H and 5H, corresponding with the previously identified QTL hotspots. New markers were found for malt-quality index and viscosity.

17.3.3 Markers for Enhancing the Nutritional Value of Maize

Significant resources, both public and private, have been devoted to the study of maize and the development of lines with improved agronomic characteristics and quality. The results of these efforts have meant that average yields in the United States have risen by 360% over the last 65 years (Fraley, 2009); at the same time, the quality of the grain produced has improved. It is beyond the scope of this chapter to go into these achievements in detail and the reader is referred to a review of factors that contribute to quality in maize (Balconi et al., 2007) and recent treatises on molecular genetic approaches to maize improvement (Kriz and Larkins, 2009; Wusirika et al., 2014) for comprehensive coverage of these topics. In this section, two aspects of the use of molecular markers to improving the nutrition quality of maize for the third world will be briefly discussed.

17.3.3.1 Quality Protein Maize

The proteins in corn endosperm are low in lysine and tryptophan, thus people with maize-based diets often have deficiencies in these amino acids, which must be obtained from other dietary sources. The discovery of the *opaque-2* (*o2*) mutation, which almost doubles the lysine content in maize endosperm by decreasing the synthesis of prolamin (zein) proteins and increasing the level of other types of endosperm proteins, raised expectations for the genetic improvement of cereal protein quality. However, the mutation also resulted in low yields, chalky-looking grain and susceptibility to pests and diseases, preventing the uptake of these lines. Extensive traditional breeding efforts at CIMMYT have introduced *opaque-2 modifiers* (*opm*) into these lines, which eliminate these defects, while the high lysine content is maintained. These modified *opaque-2* maize lines were designated 'Quality Protein Maize' or QPM (Prasanna et al., 2001). Introgression of *o2* and its modifiers into locally adapted cultivars is a slow process, as phenotypic evaluation for vitreousness is necessary. Research is on-going to develop markers for *opm* to allow MAS to be used in breeding programs.

Lopes et al. (1995) identified two chromosomal regions on the long arm of chromosome 7 that are associated with *o2* endosperm modification. Further studies have identified molecular markers for *o2* and a modifier (Bantte and Prasanna, 2003; Lin et al., 1997; Lopez et al., 1995), and these have been used to demonstrate that MAS can be used to accelerate the introgression of these traits into locally adapted germplasm (Danson et al., 2006). While this is a promising result, a more recent study has identified additional regions that are involved in modifying the *o2* phenotype. A major QTL, that explained 30% of the variance in endosperm texture, was identified near the centromere on chromosome 7, and six other QTLs that explained 5–12% of the variance were also identified (Holding et al., 2008). Further QTLs for tryptophan have been identified, including those on chromosomes 5, 7 and 9 (Babu et al., 2015). More detailed analysis of the regions identified is likely to result in the development of new markers that can be used in MAS to accelerate the introgression of the high lysine and tryptophan traits into locally adapted elite cultivars.

17.3.3.2 Enhancing the Pro-Vitamin A Content of Maize

In third world countries, more than 250,000 children go blind each year as a result of vitamin A deficiencies and more than half of those children die within a year of becoming blind. In addition, zeaxanthin and lutein play a role in the prevention of macular degeneration. Efforts have gone into metabolic engineering of the carotenoid biosynthetic pathway in cereals in order to increase the provitamin A content (Giuliano et al., 2008). In maize, initial efforts have focused on identification of QTLs for carotenoid content and the development of markers for MAS. The gene encoding *PSY1* has been shown to be associated with variation in endosperm colour (Palaisa et al., 2003). More comprehensive studies have identified QTLs for total carotenoid, β-carotene, β-cryptoxanthin, zeaxanthin and lutein content of the grain and shown *PSY1* is linked to these QTLs (Wong et al., 2004). This association was confirmed in a different population and a QTL for α-carotene content that was linked to *PSY1* was also identified (Chander et al., 2008). The gene encoding ζ-carotene desaturase was also linked to QTLs for the levels of these carotenoids in the grain (Wong et al., 2004). However, this association was not found in the population used by Chander et al. (2008). Analysis of four natural polymorphisms in epsilon cyclase revealed that these explained 58% of the variation in flux between the β- and ε-pathways and accounted for a three-fold difference in the level of pro-vitamin A compounds within the endosperm (Harjes et al., 2008). Lines containing 15 μg/g β-carotene have been developed and are being used as breeding parents (Yan et al., 2010). Development of more QTLs in conjunction with MAS is required to further enhance carotenoid content.

17.4 FUTURE TRENDS FOR MARKER-ASSISTED SELECTION OF QUALITY

The use of MAS in the breeding for quality in cereals is still developing. Numerous molecular markers and QTLs have been identified [see Liu et al. (2012) for a review of functional markers in wheat] and are beginning to be used routinely in breeding programs. In wheat, functional markers are being utilised in breeding programs for grain hardness, storage proteins, starch properties, dough strength, PPO, yellow pigment and lipoxygenase activity (Goutam et al., 2013). MAS in Canada has enabled wheat breeders to select genotypes with desirable qualities such as high protein content, improved gluten strength and improved pasta colour (Randhawa et al., 2013). Considerable progress has been made in China where a number of genes have been introduced into mega rice (Rao et al., 2014). This field is advancing rapidly as our knowledge of the genes that underpin cereal quality increases.

SNPs are the most common polymorphisms and the development of high throughput, automated genotyping and new detection platforms has enabled the construction of high-density molecular maps and significantly improved linkage mapping. SNPs have been identified in barley (Duran et al., 2009) and in the application of high-throughput screening in wheat (Akhunov et al., 2009; Berard et al., 2009; Paux et al., 2012). These applications are being combined with already established technologies such as DArT (Akbari et al., 2006; Wenzl et al., 2006) and multiplex-ready PCR systems for high-throughput screening of SSR markers in wheat and barley (Hayden et al., 2008a,b; Matthies et al., 2012).

Genome-wide selection studies utilise all available markers (not just a subset associated with a trait) and along with phenotypic data are used to predict the performance of the progeny [see Xu et al. (2012) for a review of whole-genome strategies]. This has led to the development of integrated genetic maps with SSR, SNP and DArT markers (Zhou et al., 2015). Genome-wide marker discovery has been assisted by reference sequencing such as the Wheat Genome Sequencing Consortium (http://www.wheatgenome.org) and access to the databases is expected to result in genetic tools suitable for use in breeding programs. Large marker and trait datasets (http://barleycap.org) are being established for barley quality improvement, which has been hindered by the use of small breeding populations and QTL × environment interactions (Mohammadi et al., 2015).

Rice is the first cereal to have its genome sequenced and knowledge of the sequence of elite cultivars is being used to design custom markers for amylose content (Duitama et al., 2015). These advances will flow through to sequencing studies in other cereals, enabling the screening of a greater number of lines at a lower cost. Quality evaluation will still be necessary, particularly at later stages in breeding programs, but it may become a rate-limiting step in early generation testing. Understanding the errors involved in quantitative testing as well as environmental impacts are essential components. Programs are being developed in the field of wheat quality to predict the genetic potential for quality (Békés et al., 2006, 2013; Eagles et al., 2002, 2004; Oury et al., 2010) and also to permit selection of the best candidate lines for quality analysis.

New areas of study have emerged, such as to identify gluten peptides in wheat associated with allergic responses and coeliac disease in wheat, with some researchers believing it may be possible to select for lines with lower levels of toxicity (Rasheed et al., 2014). The effect of climate change brings the potential for selection of drought and heat tolerant lines from genetically diverse populations (Lopes et al., 2015) and potential markers for elite lines grown under conditions of elevated temperatures and higher levels of carbon dioxide in the atmosphere (Ingvordsen et al., 2015).

17.5 CONCLUSIONS

There are now a number of options available to accurately identify variety at any stage throughout the value chain, with the majority now utilising DNA composition to assess variety. In the past for some grain-growing regions, visual identification has provided 'real time' identification, but this subjective approach has proven increasingly to be inadequate. Only one of the chemical analysis methods, lab-on-a-chip, offers close to 'real time' objective assessment of variety identification at the point of grain receival, where it is critical to determine quality class, payment basis and to minimise mixing of mis-matched grain types. Future developments will be in the area of automated DNA-based methods, which ideally could be utilised for on-the-spot receival testing, but are more likely to be used in a regional laboratory, so that costs can be kept relatively low. Development of these methods will benefit from improvements in marker identification and associated technologies being developed in breeding programs.

MAS is being adopted to a greater extent in breeding programs to produce lines that are differentiated on agronomic, disease resistance and/or quality traits. As sequencing costs decrease, rapid genotyping and density of markers is increasing towards the entire genome level. These genomics approaches are integrated with bioinformatics and knowledge databases and which will become valuable decision making tools for breeders. However, they are not expected to replace traditional phenotypic testing due to environmental effects on quality and the number of genes controlling many traits.

SOURCES OF FURTHER INFORMATION

GrainGenes comprehensive database for molecular and phenotypic information for wheat, rye and other cereals. Available from: http://wheat.pw.usda.gov/GG3/

Dhingani, R.M., Umrania, V.V., Tomar, R.S., Parakhia, M.V., Golakiya, B.A., 2015. Introduction to QTL mapping in plants. Ann. Plant Sci. 4, 1072–1079.

Henry, R.J., 2014. Wheat genomics for grain quality improvement. Cereal Food World 59, 22–24.

Miskelly, D.M., Wrigley, C.W., 2016. Identification of varieties of food grains. Wrigley, C., Corke, H., Seetharaman, K., Faubion, J. (Eds.), Encyclopedia of Food Grains, vol. 4, Elsevier Ltd, Oxford, UK, pp. 29–41.

REFERENCES

AGSA (formerly RACI), 2003. Official Testing Methods of the Australasian Grain Science Association, 4th ed. Australasian Grain Science Association, Narrabri, NSW, Australia.

Ahmad, M., 2000. Molecular marker-assisted selection of HMW glutenin alleles related to wheat bread quality by PCR-generated DNA markers. Theor. Appl. Genet. 101, 892–896.

Akbari, M., Wenzl, P., Caig, V., Carling, J., Xia, L., Yang, S.Y., Uszynski, G., Mohler, V., Lehmensiek, A., Kuchel, H., Hayden, M.J., Howes, N., Sharp, P., Vaughan, P., Rathmell, B., Huttner, E., Kilian, A., 2006. Diversity arrays technology DArT for high-throughput profiling of the hexaploid wheat genome. Theor. Appl. Genet. 113, 1409–1420.

Akhunov, E., Nicolet, C., Dvorak, J., 2009. Single nucleotide polymorphism genotyping in polyploid wheat with the Illumina GoldenGate assay. Theor. Appl. Genet. 119, 507–517.

Appelbee, M.J., Mekuria, G.T., Nagasandra, V., Bonneau, J.P., Eagles, H.A., Eastwood, R.F., Mather, D.E., 2009. Novel allelic variants encoded at the Glu-D3 locus in bread wheat. J. Cereal Sci. 49, 254–261.

Ayoub, M., Armstrong, E., Bridger, G., Fortin, M.G., Mather, D.E., 2003. Marker-based selection in barley for a QTL region affecting alpha-amylase activity of malt. Crop Sci. 43, 556–561.

Babu, B.K., Agrawal, P.K., Saha, S., Gupta, H.S., 2015. Mapping QTLs for opaque2 modifiers influencing the tryptophan content in quality protein maize using genomic and candidate gene-based SSRs of lysine and tryptophan metabolic pathway. Plant Cell Rep. 34, 37–45.

Balconi, C., Hartings, H., Lauria, M., Pirona, R., Rossi, V., Motto, M., 2007. Gene discovery to improve maize grain quality traits. Maydica 52, 357–373.

Bantte, K., Prasanna, B.M., 2003. Simple sequence repeat polymorphism in Quality Protein Maize (QPM) lines. Euphytica 129, 337–344.

Bean, S.R., Lookhart, G.L., 2000. Ultrafast capillary electrophoretic analysis of cereal storage proteins and its applications to protein characterization and cultivar differentiation. J. Agric. Food Chem. 48, 344–353.

Bean, S.R., Lookhart, G.L., 2001. Recent developments in high-performance capillary electrophoresis of cereal proteins. Electrophoresis 22, 1503–1509.

Becerra, V., Paredes, M., Gutiérrez, E., Rojo, C., 2015. Genetic diversity, identification and certification of Chilean rice varieties using molecular markers. Chilean J. Agric. Res. 75 (3), 267–274, Available from: http://www.scielo.cl/scielo.php?script=sci_arttext&pid=S0718-58392015000400001.

Beecher, B.S., Carter, A.H., See, D.R., 2012. Genetic mapping of new seed-expressed polyphenol oxidase genes in wheat (*Triticum aestivum* L.). Theor. Appl. Genet. 124, 1463–1473.

Békés, F., Kemeny, S., Morell, M., 2006. An integrated approach to predicting end-product quality of wheat. Eur. J. Agron. 25, 155–162.

Békés, F., Ma, W., Tömösközi, S., 2013. Estimating dough properties and end-product quality from flour composition. In: He, Z.H., Wang, D. (Eds.), In: Proceedings Eleventh International Gluten Workshop China, August 12-15, 2012. International Maize and Wheat Improvement Center CIMMYT, Mexico, DF, p. 114.

Békés, F., Wrigley, C.W., 2013. Gluten alleles and predicted dough quality for wheat varieties worldwide. Cereal Food World 58, 325–328.

Berard, A., Le Paslier, M.C., Dardevet, M., Exbrayat-Vinson, F., Bonnin, I., Cenci, A., Haudry, A., Brunel, D., Ravel, C., 2009. High-throughput single nucleotide polymorphism genotyping in wheat Triticum spp. Plant Biotechnol. J. 7, 364–374.

Bezant, J.H., Laurie, D.A., Pratchett, N., Chojecki, J., Kearsey, M.J., 1997. Mapping of QTL controlling NIR predicted hot water extract and grain nitrogen content in a spring barley cross using marker-regression. Plant Breed. 116, 141–145.

Bhave, M., Morris, C.F., 2008a. Molecular genetics of puroindolines and related genes: allelic diversity in wheat and other grasses. Plant Mol. Biol. 66, 205–219.

Bhave, M., Morris, C.F., 2008b. Molecular genetics of puroindolines and related genes: regulation of expression, membrane binding properties and applications. Plant Mol. Biol. 66, 221–231.

Blanco, A., Pasqualone, A., Troccoli, A., Di Fonzo, N., Simeone, R., 2002. Detection of grain protein content QTLs across environments in tetraploid wheats. Plant Mol. Biol. 48, 615–623.

Blanco, A., Simeone, R., Gadaleta, A., 2006. Detection of QTLs for grain protein content in durum wheat. Theor. Appl. Genet. 112, 1195–1204.

Bloch, H.A., Kesmir, C., Petersen, M., Jacobsen, S., Sondergaard, I., 1999. Identification of wheat varieties using matrix-assisted laser desorption/ionisation time-of-flight mass spectrometry and an artificial neural network. Rapid Commun. Mass Spectrom. 13, 1535–1539.

Bogard, M., Allard, V., Martre, P., Heumez, E., Snape, J.W., Orford, S., Griffiths, S., Gaju, O., Foulkes, Y., Le Gouis, J., 2013. Identifying wheat genomic regions for improving grain protein concentration independently of grain yield using multiple inter-related populations. Mol. Breed. 31, 587–599.

Broeders, S., Huber, I., Grohmann, L., Berben, G., Taverniers, I., Mazzara, M., Roosend, N., Morisset, D., 2014. Guidelines for validation of qualitative real-time PCR methods. Trends Food Sci. Technol. 372, 115–126.

Bushuk, W., 1998. Interactions in wheat doughs. In: Hamer, R.J., Hoseney, R.C. (Eds.), Interactions: The Keys to Cereal Quality. American Association of Cereal Chemists, St. Paul, MN, pp. 1–16.

Butow, B.J., Gale, K.R., Ikea, J., Juhasz, A., Bedo, Z., Tamas, L., Gianibelli, M.C., 2004. Dissemination of the highly expressed Bx7 glutenin subunit Glu-B1al allele in wheat as revealed by novel PCR markers and RP-HPLC. Theor. Appl. Genet. 109, 1525–1535.

Butow, B.J., Ma, W., Gale, K.R., Cornish, G.B., Rampling, L., Larroque, O., Morell, M.K., Békés, F., 2003. Molecular discrimination of Bx7 alleles demonstrates that a highly expressed high-molecular-weight glutenin allele has a major impact on wheat flour dough strength. Theor. Appl. Genet. 107, 1524–1532.

Campbell, K.G., Finney, P.L., Bergman, C.J., Gualberto, D.G., Anderson, J.A., Giroux, M.J., Siritunga, D., Zhu, J.Q., Gendre, F., Roue, C., Verel, A., Sorrells, M.E., 2001. Quantitative trait loci associated with milling and baking quality in a soft x hard wheat cross. Crop Sci. 41, 1275–1285.

Cantrell, R.G., Joppa, L.R., 1991. Genetic-analysis of quantitative traits in wild emmer *Triticum-turgidum* L. var dicoccoides. Crop Sci. 31, 645–649.

Cenci, A., Somma, S., Chantret, N., Dubcovsky, J., Blanco, A., 2004. PCR identification of durum wheat BAC clones containing genes coding for carotenoid biosynthesis enzymes and their chromosome localization. Genome 47, 911-917.

Chander, S., Guo, Y., Yang, X., Zhang, J., Lu, X., Yan, J., Song, T., Rocheford, T., Li, J., 2008. Using molecular markers to identify two major loci controlling carotenoid contents in maize grain. Theor. Appl. Genet. 116, 223–233.

Chen, F., Li, H., Cui, D., 2013. Discovery, distribution and diversity of Puroindoline-D1 genes in bread wheat from five countries (Triticum aestivum L.). BMC Plant Biol. 13 (1), 125, Available from: http://www.biomedcentral.com/1471-2229/13/125.

Chuang, H.Y., Lur, H., Hwu, K., Chang, M., 2011. Authentication of domestic Taiwan rice varieties based on fingerprinting analysis of microsatellite DNA markers. Bot. Stud. 52 (4), 393–405.

Cockram, J., Jones, H., Norris, C., O'Sullivan, D.M., 2012. Evaluation of diagnostic molecular markers for DUS phenotypic assessment in the cereal crop, barley *Hordeum vulgare* ssp. vulgare L. Theor. Appl. Genet. 125, 1735–1749.

Collard, B.C.Y., Mackill, D.J., 2008. Marker-assisted selection: an approach for precision plant breeding in the twenty-first century. Philos. Trans. R Soc. B Biol. Sci. 363, 557–572.

Collins, H.M., Logue, S.J., Jefferies, S.P., Barr, A.R. 2001. Validation of markers from malt extract, DP, alpha amylase and beta amylase. Tenth Australian Barley Technical Symposium, Canberra, Australia.

Collins, H.M., Panozzo, J.F., Logue, S.J., Jefferies, S.P., Barr, A.R., 2003. Mapping and validation of chromosome regions associated with high malt extract in barley *Hordeum vulgare* L. Aust. J. Agric. Res. 54, 1223–1240.

Cooke, R.J., 1995. Introduction: the reasons for variety identification. In: Wrigley, C.W. (Ed.), Identification of Food-Grain Varieties. American Association of Cereal Chemists, St. Paul, MN, pp. 1–18.

Cooke, R.J., Wrigley, C.W., 2004. Variety registration and breeders rights. In: Wrigley, C.W., Corke, H., Walker, C.E. (Eds.), Encyclopaedia of Grain Science. Elsevier Ltd, Oxford, pp. 314–321.

Coventry, S.J., Barr, A.R., Eglinton, J.K., McDonald, G.K., 2003. The determinants and genome locations influencing grain weight and size in barley *Hordeum vulgare* L. Aust. J. Agric. Res. 54, 1103–1115.

Danson, J.W., Mbogori, M., Kimani, M., Lagat, M., Kuria, A., Diallo, A., 2006. Marker assisted introgression of opaque2 gene into herbicide resistant elite maize inbred lines. Afr. J. Biotechnol. 5, 2417–2422.

de Bustos, A., Rubio, P., Soler, C., Garcia, P., Jouve, N., 2001. Marker assisted selection to improve HMW-glutenins in wheat. Euphytica 119, 69–73.

Devos, K.M., Bryan, G.J., Collins, A.J., Stephenson, P., Gale, M.D., 1995. Application of 2 microsatellite sequences in wheat storage proteins as molecular markers. Theor. Appl. Genet. 90, 247–252.

Distelfeld, A., Uauy, C., Olmos, S., Schlatter, A.R., Dubcovsky, J., Fahima, T., 2004. Microcolinearity between a 2-cM region encompassing the grain protein content locus Gpc-6B1 on wheat chromosome 6B and a 350-kb region on rice chromosome 2. Funct. Integr. Genomics 4, 59–66.

Distelfeld, A., Uauy, C., Fahima, T., Dubcovsky, J., 2006. Physical map of the wheat high-grain protein content gene Gpc-B1 and development of a high-throughput molecular marker. New Phytol. 169, 753–763.

Dovidio, R., Anderson, O.D., 1994. PCR analysis to distinguish between alleles of a member of a multigene family correlated with wheat bread-making quality. Theor. Appl. Genet. 88, 759–763.

Dovidio, R., Masci, S., Porceddu, E., 1995. Development of a set of oligonucleotide primers specific for genes at the Glu-1 complex loci of wheat. Theor. Appl. Genet. 91, 189–194.

Dovidio, R., Porceddu, E., Lafiandra, D., 1994. PCR analysis of genes encoding allelic variants of high-molecular-weight glutenin subunits at the Glu-D1 locus. Theor. Appl. Genet. 88, 175–180.

Duitama, J., Silva, A., Sanabria, Y., Cruz, D.F., Quintero, C., Ballen, C., Lorieux, M., Scheffler, B., Farmer, A., Torres, E., Oard, J., Tohme, J., 2015. Whole genome sequencing of elite rice cultivars as a comprehensive information resource for marker assisted selection. PLoS One 10 (4), e0124617, Available from: http://journals.plos.org/plosone/article?id=10.1371/journal.pone.0124617.

Duran, C., Appleby, N., Vardy, M., Imelfort, M., Edwards, D., Batley, J., 2009. Single nucleotide polymorphism discovery in barley using autoSNPdb. Plant Biotechnol. J. 7, 326–333.

Eagles, H.A., Eastwood, R.F., Hollamby, G.J., Martin, E.M., Cornish, G.B., 2004. Revision of the estimates of glutenin gene effects at the Glu-B1 locus from southern Australian wheat breeding programs. Aust. J. Agric. Res. 55, 1093–1096.

Eagles, H.A., Hollamby, G.J., Gororo, N.N., Eastwood, R.F., 2002. Estimation and utilisation of glutenin gene effects from the analysis of unbalanced data from wheat breeding programs. Aust. J. Agric. Res. 53, 367–377.

Eglinton, J.K., Langridge, P., Evans, D.E., 1998. Thermostability variation in alleles of barley beta-amylase. J. Cereal Sci. 28, 301–309.

Elangovan, M., Rai, R., Dholakia, B.B., Lagu, M.D., Tiwari, R., Gupta, R.K., Rao, V.S., Roder, M.S., Gupta, V.S., 2008. Molecular genetic mapping of quantitative trait loci associated with loaf volume in hexaploid wheat *Triticum aestivum*. J. Cereal Sci. 47, 587–598.

Elouafi, I., Nachit, M.M., Martin, L.M., 2001. Identification of a microsatellite on chromosome 7B showing a strong linkage with yellow pigment in durum wheat *Triticum turgidum* L. var. durum. Hereditas 135, 255–261.

Emebiri, L., Michael, P., Moody, D., Ogbonnaya, F., Black, C., 2009. Pyramiding QTLs to improve malting quality in barley: gains in phenotype and genetic diversity. Mol. Breed. 23, 219–228.

Ens, W., Preston, K.R., Znamirowski, M., Dworschal, R.G., Standing, K.G., 2000. Identification of wheat varieties using matrix-assisted laser desorption/ionization time-of-flight mass spectroscopy. In: Shewry, P.R., Tatham, A.S. (Eds.), Wheat Gluten. Royal Society of Chemistry, Cambridge, pp. 204–207.

Erkkila, M.J., 1999. Intron III-specific markers for screening of beta-amylase alleles in barley cultivars. Plant Mol. Biol. Rep. 17, 139–147.

Erkkila, M.J., Leah, R., Ahokas, H., Cameron-Mills, V., 1998. Allele-dependent barley grain beta-amylase activity. Plant Physiol. 117, 679–685.

Espí, A., Giraldo, P., Rodriguez- Quijano, M., Carrillo, J.M., 2012. A PCR-based method for discriminating between high molecular weight glutenin subunits Bx7 and Bx7* in *Triticum aestivum*. L. Plant Breed. 1315, 571–573.

Evans, D.E., Robinson, L.H., Sheehan, M.C., Tolhurst, R.L., Hill, A., Skerritt, J.S., Barr, A.R., 2003. Application of immunological methods to differentiate between foam-positive and haze-active proteins originating from malt. J. Am. Soc. Brew. Chem. 61, 55–62.

Fox, G.P., Panozzo, J.F., Li, C.D., Lance, R.C.M., Inkerman, P.A., Henry, R.J., 2003. Molecular basis of barley quality. Aust. J. Agric. Res. 54, 1081–1101.

Fraley, R.T., 2009. Molecular gentic approaches to maize improvement—an introduction. In: Kriz, a.L., Larkins, B.A. (Eds.), Molecular Genetic Approaches to Maize Improvement. Springer, Berlin, pp. 3–6.

Furtado, A., Bundock, P.C., Banks, P.M., Fox, G., Yin, X., Henry, R.J., 2015. A novel highly differentially expressed gene in wheat endosperm associated with bread quality. Sci. Rep. 5, 10446, Available from http://www.nature.com/articles/srep10446.

Gale, K.R., 2005. Diagnostic DNA markers for quality traits in wheat. J. Cereal Sci. 41, 181–192.

Giuliano, G., Tavazza, R., Diretto, G., Beyer, P., Taylor, M.A., 2008. Metabolic engineering of carotenoid biosynthesis in plants. Trends Biotechnol. 26, 139–145.

Goesaert, H., Brijs, K., Veraverbeke, W.S., Courtin, C.M., Gebruers, K., Delcour, J.A., 2005. Wheat flour constituents: how they impact bread quality, and how to impact their functionality. Trends Food Sci. Technol. 16, 12–30.

Goutam, U., Kukreja, S., Tiwari, R., Chaudhury, A., Gupta, R.K., Dholakia, B.B., Yadav, R., 2013. Biotechnological approaches for grain quality improvement in wheat: present status and future possibilities. Aust. J. Crop Sci. 7, 469–483.

Gras, P.W., Anderssen, R.S., Keentok, M., Bekes, F., Appels, R., 2001. Gluten protein functionality in wheat flour processing: a review. Aust. J. Agric. Res. 52, 1311–1323.

Groos, C., Bervas, E., Chanliaud, E., Charmet, G., 2007. Genetic analysis of bread-making quality scores in bread wheat using a recombinant inbred line population. Theor. Appl. Genet. 115, 313–323.

Groos, C., Bervas, E., Charmet, G., 2004. Genetic analysis of grain protein content, grain hardness and dough rheology in a hard X hard bread wheat progeny. J. Cereal Sci. 40, 93–100.

Gunkel, J., Voetz, M. and Rath F., 2001. Effect of the malting barley variety *Hordeum vulgare* L. on fermentability. EBC Twenty-Eighth International Congress, Budapest, Hungary, 355–361.

Han, F., Clancy, J.A., Jones, B.L., Wesenberg, D.M., Kleinhofs, A., Ullrich, S.E., 2004. Dissection of a malting quality QTL region on chromosome 1 7H of barley. Mol. Breed. 14, 339–347.

Harjes, C.E., Rocheford, T.R., Bai, L., Brutnell, T.P., Kandianis, C.B., Sowinski, S.G., Stapleton, A.E., Vallabhaneni, R., Williams, M., Wurtzel, E.T., Yan, J., Buckler, E.S., 2008. Natural genetic variation in lycopene epsilon cyclase tapped for maize biofortification. Science 319, 330–333.

Hayden, M.J., Nguyen, T.M., Waterman, A., Chalmers, K.J., 2008a. Multiplex-ready PCR: a new method for multiplexed SSR and SNP genotyping. BMC Genomics 9, 80.

Hayden, M.J., Nguyen, T.M., Waterman, A., McMichael, G.L., Chalmers, K.J., 2008b. Application of multiplex-ready PCR for fluorescence-based SSR genotyping in barley and wheat. Mol. Breed. 21, 271–281.

Hayes, P.M., Liu, B.H., Knapp, S.J., Chen, F., Jones, B., Blake, T., Franckowiak, J., Rasmusson, D., Sorrells, M., Ullrich, S.E., Wesenberg, D., Kleinhofs, A., 1993. Quantitative trait locus effects and environmental interaction in a sample of North-American barley germplasm. Theor. Appl. Genet. 87, 392–401.

He, X.Y., He, Z.H., Xia, W., Appels, R., Xia, X.C., 2009. Allelic variants of phytoene synthase 1 Psy1 genes in Chinese and CIMMYT wheat cultivars and development of functional markers for flour colour. Mol. Breed. 23, 553–563.

He, X.Y., He, Z.H., Zhang, L.P., Sun, D.J., Morris, C.F., Fuerst, E.P., Xia, X.C., 2007. Allelic variation of polyphenol oxidase PPO genes located on chromosomes 2A and 2D and development of functional markers for the PPO genes in common wheat. Theor. Appl. Genet. 115, 47–58.

He, X.Y., Zhang, Y.L., He, Z.H., Wu, Y.P., Xiao, Y.G., Xia, C.X., Xia, X.C., 2008. Characterization of phytoene synthase 1 gene Psy1 located on common wheat chromosome 7A and development of a functional marker. Theor. Appl. Genet. 116, 213–221.

Hessler, T.G., Thomson, M.J., Benscher, D., Nachit, M.M., Sorrells, M.E., 2002. Association of a lipoxygenase locus, Lpx-B1, with variation in lipoxygenase activity in durum wheat seeds. Crop Sci. 42, 1695–1700.

Holding, D.R., Hunter, B.G., Chung, T., Gibbon, B.C., Ford, C.F., Bharti, A.K., Messing, J., Hamaker, B.R., Larkins, B.A., 2008. Genetic analysis of opaque2 modifier loci in quality protein maize. Theor. Appl. Genet. 117, 157–170.

Howitt, C.A., Cavanagh, C.R., Bowerman, A.F., Cazzonelli, C., Rampling, L., Mimica, J.L., Pogson, B.J., 2009. Alternative splicing, activation of cryptic exons and amino acid substitutions in carotenoid biosynthetic genes are associated with lutein accumulation in wheat endosperm. Funct. Integr. Genomics 9, 363–376.

Howitt, C.A., Gale, K.R., Juhasz, A., 2006. Diagnostic markers for quality. In: Wrigley, C.W., Bekes, F., Bushuk, W. (Eds.), Gliadin and Glutenin the Unique Balance of Wheat Quality. AACC International, St. Paul, MN, pp. 333–361.

Huang, X.Q., Brûlé-Babel, A., 2011. Development of simple and co-dominant PCR markers to genotype puroindoline a and b alleles for grain hardness in bread wheat (*Triticum aestivum* L.). J. Cereal Sci. 53, 277–284.

ICC, 2011. Standard Methods of the International Association for Cereal Science and Technology. ICC, Vienna, Austria.

Ikeda, T.M., Araki, E., Fujita, Y., Yano, H., 2006. Characterization of low-molecular-weight glutenin subunit genes and their protein products in common wheats. Theor. Appl. Genet. 112, 327–334.

Ikeda, T.M., Rogers, W.J., Branlard, G., Peña, R.J., Lerner, S.E., Arrigoni, A., Xia, W.J., Appels, R., Lukow, O., Hurkman, W., Applebee, M., Sissons, M., Carrillo, J.M., He, Z.H., 2013. Update on low-molecular weight glutenin subunit identification. In: He, Z.H., Wang, D. (Eds.), In: Proceedings Eleventh International Gluten Workshop China, August 12-15, 2012. International Maize and Wheat Improvement Center CIMMYT, Mexico, DF, pp. 96–101.

Ingvordsen, C.H., Backes, G., Lyngkjær, M.F., Peltonen-Sainio, P., Jahoor, A., Mikkelsen, T.N., Jørgensen, R.B., 2015. Genome-wide association study of production and stability traits in barley cultivated under future climate scenarios. Mol. Breed. 35 (3), 1–14.

Jin, H., Yan, J., Peña, R.J., Xia, X.C., Morgounov, A., Han, L.M., Zhang, Y., He, Z., 2011. Molecular detection of high-and low-molecular-weight glutenin subunit genes in common wheat cultivars from 20 countries using allele-specific markers. Crop Pasture Sci. 62, 746–754.

Juhasz, A., Larroque, O.R., Tamaas, L., Hsam, S.L.K., Zeller, F.J., Bekes, F., Bedo, Z., 2003. Bankuti 1201—an old Hungarian wheat variety with special storage protein composition. Theor. Appl. Genet. 107, 697–704.

Jukanti, A.K., Bruckner, P.L., Fischer, A.M., 2004. Evaluation of wheat polyphenol oxidase genes. Cereal Chem. 81, 481–485.

Kaisoon, O., Siriamornpun, S., Meeso, N., 2008. Distinction between cereal genotypes based on the protein and DNA composition of the grain by capillary electrophoresis. World Appl. Sci. J. 4, 384–395.

Karakousis, A., Barr, A.R., Chalmers, K.J., Ablett, G.A., Holton, T.A., Henry, R.J., Lim, P., Langridge, P., 2003. Potential of SSR markers for plant breeding and variety identification in Australian barley germplasm. Aust. J. Agric. Res. 54, 1197–1210.

Kihara, M., Kaneko, T., Ito, K., 1998. Genetic variation of beta-amylase thermostability among varieties of barley, *Hordeum vulgare* L., and relation to malting quality. Plant Breed. 117, 425–428.

Kriz, A.L., Larkins, B.A. (Eds.), 2009. Molecular Genetic Approaches to Maize Improvement. Springer, Berlin.

Kuchel, H., Langridge, P., Mosionek, L., Williams, K., Jeffries, S.P., 2006. The genetic control of milling yield, dough rheology and baking quality of wheat. Theor. Appl. Genet. 112, 1487–1495.

Lafiandra, D., Tucci, G.F., Pavoni, A., Turchetta, T., Margiotta, B., 1997. PCR analysis of x- and y-type genes present at the complex Glu-A1 locus in durum and bread wheat. Theor. Appl. Genet. 94, 235–240.

Laidò, G., Barabaschi, D., Tondelli, A., Gianinetti, A., Stanca, A.M., Li Destri Nicosia, O., Di Fonzo, N., Francia, E., Pecchioni, N., 2009. QTL alleles from a winter feed type can improve malting quality in barley. Plant Breed. 128, 598–605.

Law, C.N., Bhandari, D.G., Salmon, S.E., Greenwell, P.W., Foot, I.M., Cauvain, S.P., Sayers, E.J., Worland, A.J., 2005. Novel genes on chromosome 3A influencing breadmaking quality in wheat, including a new gene for loaf volume, Lvl 1. J. Cereal Sci. 41, 317–326.

Lehmensiek, A., Eckermann, P.J., Verbyla, A.P., Appels, R., Sutherland, M.W., Martin, D., Daggard, G.E., 2006. Flour yield QTLs in three Australian doubled haploid wheat populations. Aust. J. Agric. Res. 57, 1115–1122.

Lei, Z.S., Gale, K.R., He, Z.H., Gianibelli, C., Larroque, O., Xia, X.C., Butow, B.J., Xia, W., 2006. Y-type gene specific markers for enhanced discrimination of high-molecular weight glutenin alleles at the Glu-B1 locus in hexaploid wheat. J. Cereal Sci. 43, 94–101.

Li, C.D., Langridge, P., Zhang, X.Q., Eckstein, P.E., Rossnagel, B.G., Lance, R.C.M., Lefol, E.B., Lu, M.Y., Harvey, B.L., Scoles, G.J., 2002. Mapping of barley *Hordeum vulgare* L. beta-amylase alleles in which an amino acid substitution determines beta-amylase isoenzyme type and the level of free beta-amylase. J. Cereal Sci. 35, 39–50.

Lin, K.R., Bockholt, A.J., Magill, C.W., Smith, J.D., 1997. Changes in soluble endosperm proteins associated with selection of quality protein maize lines. Maydica 42, 355–362.

Liu, S.X., Chao, S., Anderson, J., 2008. New DNA markers for high molecular weight glutenin subunits in wheat. Theor. Appl. Genet. 118, 177–183.

Liu, Y., He, Z., Appels, R., Xia, X., 2012. Functional markers in wheat: current status and future prospects. Theor. Appl. Genet. 125, 1–10.

Long, H., Wei, Y.M., Yan, Z.H., Baum, B., Nevo, E., Zheng, Y.L., 2005. Classification of wheat low-molecular-weight glutenin subunit genes and its chromosome assignment by developing LMW-GS group-specific primers. Theor. Appl. Genet. 111, 1251–1259.

Lookhart, G.L., 1991. Cereal proteins: composition of their major fractions and methods for identification. In: Lorenxe, K.J., Kulp, K. (Eds.), Hand Book of Cereal Science and Technology. Marcel Dekker, New York.

Lookhart, G.L., Bean, S.R., Bietz, J.A., 2003. HPLC of gluten monomeric proteins. In: Shewry, P.R., Lookhard, G.L. (Eds.), Wheat Gluten Protein Analysis. America Association of Cereal Chemists, St. Paul, MN, pp. 61–89.

Lookhart, G.L., Wrigley, C.W., 1995. Variety identification by electrophoretic analysis. In: Wrigley, C.W. (Ed.), Identification of Food-Grain Varieties. American Association of Cereal Chemists, St. Paul, MN, pp. 55–72.

Lopes, M.A., Takasaki, K., Bostwick, D.E., Helentjaris, T., Larkins, B.A., 1995. Identification of 2 opaque2 modifier loci in quality protein maize. Mol. Gen. Genet. 247, 603–613.

Lopes, M.S., El-Basyoni, I., Baenziger, P.S., Singh, S., Royo, C., Ozbek, K., Aktas, H., Ozer, E., Manickavelu, Ban, T., Vikram, P., 2015. Exploiting genetic diversity from landraces in wheat breeding for adaptation to climate change. J. Exp. Bot. 66, 3477–3486.

Lopez, M.A., Gloverson, L., Larkins, B.A., 1995. Genetic mapping of opaque2 modifier genes. Maize Newsl. 69, 125.

Ma, W., Zhang, W., Gale, K.R., 2003. Multiplex-PCR typing of high molecular weight glutenin alleles in wheat. Euphytica 134, 51–60.

Ma, W.J., Sutherland, M.W., Kammholz, S., Banks, P., Brennan, P., Bovill, W., Daggard, G., 2007. Wheat flour protein content and water absorption analysis in a doubled haploid population. J. Cereal Sci. 45, 302–308.

Macaulay, M., Ramsay, L., Powell, W., Waugh, R., 2001. A representative, highly informative genotyping set of barley SSRs. Theor. Appl. Genet. 102, 801–809.

Malysheva, L., Ganal, M.W., Roder, M.S., 2004. Evaluation of cultivated barley Hordeum vulgare germplasm for the presence of thermostable alleles of beta-amylase. Plant Breed. 123, 128–131.

Mangini, G., Taranto, F., Giove, S., Gadaleta, A., Blanco, A., 2010. Identification of durum wheat cultivars by a minimum number of microsatellite markers. Cereal Res. Commun. 38, 155–162.

Mann, G., Diffey, S., Cullis, B., Azanza, F., Martin, D., Kelly, A., Mcintyre, L., Schmidt, A., Xia, W.J., Nath, Z., Kutty, I., Leyne, P.E., Rampling, L., Quail, K.J., Morell, M.K., 2009. Genetic control of wheat quality: interactions between chromosomal regions determining protein content and composition, dough rheology, and sponge and dough baking properties. Theor. Appl. Genet. 118, 1519–1537.

Mansur, L.M., Qualset, C.O., Kasarda, D.D., Morris, R., 1990. Effects of Cheyenne chromosomes on milling and baking quality in Chinese Spring wheat in relation to glutenin and gliadin storage proteins. Crop Sci. 30, 593–602.

Maphosa, L., Langridge, P., Taylor, H., Chalmers, K.J., Bennett, D., Kuchel, H., Mather, D.E., 2013. Genetic control of processing quality in a bread wheat mapping population grown in water-limited environments. J. Cereal Sci. 57, 304–311.

Mares, D.J., Campbell, A., 2001. Mapping components of flour and noodle colour in Australian wheat. Aust. J. Agric. Res. 52, 1297–1309.

Marquez-Cedillo, L.A., Hayes, P.M., Jones, B.L., Kleinhofs, A., Legge, W.G., Rossnagel, B.G., Sato, K., Ullrich, E., Wesenberg, D.M., 2000. QTL. analysis of malting quality in barley based on the doubled-haploid progeny of two elite North American varieties representing different germplasm groups. Theor. Appl. Genet. 101, 173–184.

Mather, D.E., Tinker, N.A., Laberge, D.E., Edney, M., Jones, B.L., Rossnagel, B.G., Legge, W.G., Briggs, K.G., Irvine, R.B., Falk, D.E., Kasha, K.J., 1997. Regions of the genome that affect grain and malt quality in a North American two-row barley cross. Crop Sci. 37, 544–554.

Matthies, I.E., Malosetti, M., Röder, M.S., van Eeuwijk, F., 2014. Genome-wide association mapping for kernel and malting quality traits using historical European barley records. PLoS One 911, e110046.

Matthies, I.E., van Hintum, T., Weise, S., Röder, M.S., 2012. Population structure revealed by different marker types SSR or DArT has an impact on the results of genome-wide association mapping in European barley cultivars. Mol. Breed. 30, 951–966.

Matthies, I.E., Weise, S., Roder, M.S., 2009. Association of haplotype diversity in the alpha-amylase gene amy1 with malting quality parameters in barley. Mol. Breed. 23, 139–152.

McCartney, C.A., Somers, D.J., Lukow, O., Ames, N., Noll, J., Cloutier, S., Humphreys, D.G., McCallum, B.D., 2006. QTL analysis of quality traits in the spring wheat cross RL4452 × 'AC Domain. Plant Breed. 125, 565–575.

Miskelly, D.M., Wrigley, C.W., 2016. Identification of varieties of food grains. Wrigley, C., Corke, H., Seetharaman, K., Faubion, J. (Eds.), Encyclopedia of Food Grains, vol. 4, Elsevier Ltd, Oxford, UK, pp. 29–41.

Mohammadi, M., Blake, T.K., Budde, A.D., Chao, S., Hayes, P.M., Horsley, R.D., Olbert, D.E., Ullrich, S.E., Smith, K.P., 2015. A genome-wide association study of malting quality across eight US barley breeding programs. Theor. Appl. Genet. 128, 705–721.

Morris, C.F., 2002. Puroindolines: the molecular genetic basis of wheat grain hardness. Plant Mol. Biol. 48, 633–647.

Olmos, S., Distelfeld, A., Chicaiza, O., Schlatter, A.R., Fahima, T., Echenique, V., Dubcovsky, J., 2003. Precise mapping of a locus affecting grain protein content in durum wheat. Theor. Appl. Genet. 107, 1243–1251.

Oury, F.X., Chiron, H., Faye, A., Gardet, O., Giraud, A., Huemez, E., Rolland, B., Rousset, M., Trottet, M., Charmet, G., Branlard, G., 2010. The prediction of bread wheat quality: joint use of the phenotypic information brought by technological tests and the genetic information brought by HMW and LMW glutenin subunits. Euphytica 171, 87–109.

Oziel, A., Hayes, P.M., Chen, F.Q., Jones, B., 1996. Application of quantitative trait locus mapping to the development of winter-habit malting barley. Plant Breed. 115, 43–51.

Palaisa, K.A., Morgante, M., Williams, M., Rafalski, A., 2003. Contrasting effects of selection on sequence diversity and linkage disequilibrium at two phytoene synthase loci. Plant Cell 15, 1795–1806.

Panozzo, J.F., Eckermann, P.J., Mather, D.E., Moody, D.B., Black, C.K., Collins, H.M., Barr, A.R., Lim, P., Cullis, B.R., 2007. QTL analysis of malting quality traits in two barley populations. Aust. J. Agric. Res. 58, 858–866.

Paris, M., Jones, M.G.K., Eglinton, J.K., 2002. Genotyping single nucleotide polymorphisms for selection of barley beta-amylase alleles. Plant Mol. Biol. Rep. 20, 149–159.

Parker, G.D., Chalmers, K.J., Rathjen, A.J., Langridge, P., 1998. Mapping loci associated with flour colour in wheat *Triticum aestivum* L. Theor. Appl. Genet. 97, 238–245.

Parker, G.D., Chalmers, K.J., Rathjen, A.J., Langridge, P., 1999. Mapping loci associated with milling yield in wheat *Triticum aestivum* L. Mol. Breed. 5, 561–568.

Pasam, R.K., Sharma, R., Malosetti, M., van Eeuwijk, F.A., Haseneyer, G., Kilian, B., Graner, A., 2012. Genome-wide association studies for agronomical traits in a world wide spring barley collection. BMC Plant Biol. 12, 16, Available from: http://www.biomedcentral.com/1471-2229/12/16.

Pattemore, J.A., Rice, N., Marshall, D.F., Waugh, R., Henry, R.J., 2010. Cereal variety identification using MALDI-TOF mass spectrometry SNP genotyping. J. Cereal Sci. 52, 356–361.

Paux, E., Sourdille, P., Mackay, I., Feuillet, C., 2012. Sequence-based marker development in wheat: advances and applications to breeding. Biotechnol. Adv. 30, 1071–1088.

Perretant, M.R., Cadalen, T., Charmet, G., Sourdille, P., Nicolas, P., Boeuf, C., Tixier, M.H., Branlard, G., Bernard, S., Bernard, M., 2000. QTL analysis of bread-making quality in wheat using a doubled haploid population. Theor. Appl. Genet. 100, 1167–1175.

Perry, D.J., Fernando, U., Lee, S.J., 2014. Simple sequence repeat-based identification of Canadian malting barley varieties. Can. J. Plant Sci. 94, 485–496.

Perry, D.J., Lee, S.J., 2015. Identification of Canadian wheat varieties using OpenArray genotyping technology. J. Cereal Sci. 65, 267–276.

Polakova, K., Laurie, D., Vaculova, K., Ovesna, J., 2003. Characterization of beta-amylase alleles in 79 barley varieties with Pyrosequencing. Plant Mol Biol Rep 21, 439–447.

Pozniak, C.J., Clarke, J.M., Clarke, F.R., 2012. Potential for detection of marker-trait associations in durum wheat using unbalanced, historical phenotypic datasets. Mol. Breed. 30, 1537–1550.

Pozniak, C.J., Knox, R.E., Clarke, F.R., Clarke, J.M., 2007. Identification of QTL and association of a phytoene synthase gene with endosperm colour in durum wheat. Theor. Appl. Genet. 114, 525–537.

Prasanna, B.M., Vasal, S.K., Kassahun, B., Singh, N.N., 2001. Quality protein maize. Curr. Sci. 81, 1308–1319.

Radovanovic, N., Cloutier, S., 2003. Gene-assisted selection for high molecular weight glutenin subunits in wheat doubled haploid breeding programs. Mol. Breed. 12, 51–59.

Ragupathy, R., Naeem, H.A., Reimer, E., Lukow, O.M., Sapirstein, H.D., Cloutier, S., 2008. Evolutionary origin of the segmental duplication encompassing the wheat Glu-B1 locus encoding the overexpressed Bx7 Bx7OE high molecular weight glutenin subunit. Theor. Appl. Genet. 116, 283–296.

Raman, R., Raman, H., Martin, P., 2007. Functional gene markers for polyphenol oxidase locus in bread wheat *Triticum aestivum* L. Mol. Breed. 19, 315–328.

Randhawa, H.S., Asif, M., Pozniak, C., Clarke, J.M., Graf, R.J., Fox, S.L., Humphreys, D.G., Knox, R.E., DePauw, R.M., Singh, A.K., Cuthbert, R.D., Hucl, P., Spaner, D., 2013. Application of molecular markers to wheat breeding in Canada. Plant Breed. 132, 458–471.

Rao, Y., Li, Y., Qian, Q., 2014. Recent progress on molecular breeding of rice in China. Plant Cell Rep. 33, 551–564.

Rasheed, A., Xia, X., Yan, Y., Appels, R., Mahmood, T., He, Z., 2014. Wheat seed storage proteins: advances in molecular genetics, diversity and breeding applications. J. Cereal Sci. 60, 11–24.

Ravel, C., Dardevet, M., Leenhardt, F., Bordes, J., Joseph, J.L., Perretant, M.R., Exbrayat, F., Poncet, C., Balfourier, F., Chanliaud, E., Charmet, G., 2013. Improving the yellow pigment content of bread wheat flour by selecting the three homoeologous copies of Psy1. Mol. Breed. 31, 87–99.

Robinson, L.H., Evans, D.E., Kaukovirta-Norja, A., Vilpola, A., Aldred, P., Home, S., 2004. The interaction between malt protein quality and brewing conditions and their impact on beer colloidal stability. Tech. Quart. Master Brewers Assoc. Am. 41 (4), 353–362.

Robinson, L.H., Healy, P., Stewart, D.C., Eglinton, J.K., Ford, C.M., Evans, D.E., 2007a. The identification of a barley haze active protein that influences beer haze stability: the genetic basis of a barley malt haze active protein. J. Cereal Sci. 45, 335–342.

Robinson, L.H., Juttner, J., Milligan, A., Lahnstein, J., Eglinton, J.K., Evans, D.E., 2007b. The identification of a barley haze active protein that influences beer haze stability: cloning and characterisation of the barley SE protein as a barley trypsin inhibitor of the chloroform/methanol type. J. Cereal Sci. 45, 343–352.

Rousset, M., Brabant, P., Kota, R.S., Dubcovsky, J., Dvorak, J., 2001. Use of recombinant substitution lines for gene mapping and QTL analysis of bread making quality in wheat. Euphytica 119, 81–87.

Schulthess, A., Schwember, A.R., 2013. Improving durum wheat (*Triticum turgidum* L. var durum) grain yellow pigment content through plant breeding. Ciencia Invest. Agraria 40 (3), 475–490.

Schwarz, G., Felsenstein, F.G., Wenzel, G., 2004. Development and validation of a PCR-based marker assay for negative selection of the HMW glutenin allele Glu-B1-1d Bx-6 in wheat. Theor. Appl. Genet. 109, 1064–1069.

Shewry, P.R., Halford, N.G., Tatham, S., Popineau, Y., Lafiandra, D., Belton, P.S., 2003. The high molecular weight subunits of wheat glutenin and their role in determining wheat processing properties. In: Taylor, S.L. (Ed.), Advances in Food and Nutrition Research. Academic Press, San Diego, pp. 219–302.

Simeone, R., Pasqualone, A., Clodoveo, M.L., Blanco, A., 2002. Genetic mapping of polyphenol oxidase in tetraploid wheat. Cell. Mol. Biol. Lett. 7, 763–769.

Singh, A., Reimer, S., Pozniak, C.J., Clarke, F.R., Clarke, J.M., Knox, R.E., Singh, A.K., 2009. Allelic variation at Psy1-A1 and association with yellow pigment in durum wheat grain. Theor. Appl. Genet. 118, 1539–1548.

Siriamornpun, S., Wootton, M., Bekes, F., Wrigley, C.W., 2002. Capillary electrophoresis as an established method of variety identification for wheat and triticale, Cereals 2002 Proceedings of the Fifty-Second RACI Cereals Conference, Melbourne, 231–234.

Siriamornpun, S., Wootton, M., Cox, J.M., Bekes, F., Wrigley, C.W., 2001. Capillary electrophoresis of wheat gliadin proteins and its potential for wheat varietal identification using pattern matching software. Aust. J. Agric. Res. 52, 839–843.

Smith, A.B., Cullis, B.R., Appels, R., Campbell, A.W., Cornish, G.B., Martin, D., Allen, H.M., 2001. The statistical analysis of quality traits in plant improvement programs with application to the mapping of milling yield in wheat. Aust. J. Agric. Res. 52, 1207–1219.

Smith, R.L., Schweder, M.E., Barnett, R.D., 1994. Identification of glutenin alleles in wheat and triticale using PCR-generated DNA markers. Crop Sci. 34, 1373–1378.

Sorensen, H.A., Petersen, M.K., Jacobsen, S., Sondergaard, I., 2004. Mass spectrometry and partial least-squares regression: a tool for identification of wheat variety and end-use quality. J. Mass Spectrom. 39, 607–612.

Sorensen, H.A., Sperotto, M.M., Petersen, M., Kesmir, C., Radzikowski, L., Jacobsen, S., Sondergaard, I., 2002. Variety identification of wheat using mass spectrometry with neural networks and the influence of mass spectra processing prior to neural network analysis. Rapid Commun. Mass Spectrom. 16, 1232–1237.

Sun, D.J., He, Z.H., Xia, X.C., Zhang, L.P., Morris, C.F., Appels, R., Xia, W.J., Wang, H., 2005. A novel STS marker for polyphenol oxidase activity in bread wheat. Mol. Breed. 16, 209–218.

Sun, H.Y., Lu, J.H., Fan, Y.D., Zhao, Y., Kong, F.M., Li, R.J., Wang, H.G., Li, S.S., 2008. Quantitative trait loci QTLs for quality traits related to protein and starch in wheat. Prog. Nat. Sci. 18, 825–831.

Tadesse, W., Ogbonnaya, F.C., Jighly, A., Sanchez-Garcia, M., Sohail, Q., Rajaram, S., Baum, M., 2015. Genome-wide association mapping of yield and grain quality traits in winter wheat genotypes. PLoS One 10 (10), e0141339.

Thomas, W.T.B., Powell, W., Swanston, J.S., Ellis, R.P., Chalmers, K.J., Barua, U.M., Jack, P., Lea, V., Forster, B.P., Waugh, R., Smith, D.B., 1996. Quantitative trait loci for germination and malting quality characters in a spring barley cross. Crop Sci. 36, 265–273.

Tian, H.L., Wang, F.G., Zhao, J.R., Yi, H.M., Wang, L., Wang, R., Yang, Y., Song, W., 2015. Development of maize SNP3072, a high-throughput compatible SNP array, for DNA fingerprinting identification of Chinese maize varieties. Mol. Breed. 35 (6), 1–11.

Tran-Dinh, N., Carole, M., Uthayakumaran, S., Batey, I.L., Wrigley, C.W., 2009. Rapid DNA-based identification of wheat and barley varieties. J. Cereal Sci. 50, 388–391.

Tsilo, T.J., Hareland, G.A., Chao, S., Anderson, J.A., 2011. Genetic mapping and QTL analysis of flour color and milling yield related traits using recombinant inbred lines in hard red spring wheat. Crop Sci. 51, 237–246.

Tsilo, T.J., Nygard, G., Khan, K., Simsek, S., Hareland, G.A., Chao, S., Anderson, J.A., 2013. Molecular genetic mapping of QTL associated with flour water absorption and farinograph related traits in bread wheat. Euphytica 194, 293–302.

Turner, A.S., Bradburne, R.P., Fish, L., Snape, J.W., 2004. New quantitative trait loci influencing grain texture and protein content in bread wheat. J. Cereal Sci. 40, 51–60.

Uauy, C., Distelfeld, A., Fahima, T., Blechl, A., Dubcovsky, J., 2006. A NAC gene regulating senescence improves grain protein, zinc, and iron content in wheat. Science 314, 1298–1301.

Uthayakumaran, S., Batey, I.L., Wrigley, C.W., 2005. On-the-spot identification of grain variety and wheat-quality type by Lab-on-a-chip capillary electrophoresis. J. Cereal Sci. 41, 371–374.

Uthayakumaran, S., Listiohadi, Y., Baratta, M., Batey, I.L., Wrigley, C.W., 2006a. Rapid identification and quantitation of high-molecular-weight glutenin subunits. J. Cereal Sci. 44, 34–39.

Uthayakumaran, S., Wrigley, C.W., Batey, I.L., Bushuk, W., Lookhard, G.L., 2006b. Genotype identification. In: Wrigley, C.W., Bekes, F., Bushuk, W. (Eds.), Gliadin and Glutenin The Unique Balance of Wheat Quality. AACC International, St. Paul, MN, pp. 307–332.

Wang, L.H., Zhao, X.L., He, Z.H., Ma, W., Appels, R., Peña, R.J., Xia, X.C., 2009a. Characterization of low-molecular-weight glutenin subunit Glu-B3 genes and development of STS markers in common wheat (*Triticum aestivum* L.). Theor. Appl. Genet. 118, 525–539.

Wang, A., Liu, L., Peng, Y., Islam, S., Applebee, M., Appels, R., Yan, Y., Ma, W., 2015. Identification of low molecular weight glutenin alleles by matrix-assisted laser desorption/ionization time-of-flight mass spectrometry MALDI-TOF-MS in common wheat *Triticum aestivum* L. PloS One 109, e0138981.

Wang, L., Li, G., Peña, R.J., Xia, X., He, Z., 2010. Development of STS markers and establishment of multiplex PCR for Glu-A3 alleles in common wheat *Triticum aestivum* L. J. Cereal Sci. 51, 305–312.

Wang, X.B., Xia, C.X., Si, H.Q., Qiao, Y.Q., Chang, C., He, X.F., Xia, Y.X., 2009b. Gene markers for grain polyphenol oxidase activity in common wheat. Mol. Breed. 23, 163–170.

Wenzl, P., Li, H.B., Carling, J., Zhou, M.X., Raman, H., Paul, E., Hearnden, P., Maier, C., Xia, L., Caig, V., Ovesna, J., Cakir, M., Poulsen, D., Wang, J.P., Raman, R., Smith, K.P., Muehlbauer, G.J., Chalmers, K.J., Kleinhofs, A., Huttner, E., Kilian, A., 2006. A high-density consensus map of barley linking DArT markers to SSR, RFLP, and STS loci and agricultural traits. BMC Genomics 7, 206.

Wong, J.C., Lambert, R.J., Wurtzel, E.T., Rocheford, T.R., 2004. QTL and candidate genes phytoene synthase and zeta-carotene desaturase associated with the accumulation of carotenoids in maize. Theor. Appl. Genet. 108, 349–359.

Wrigley, C.W., 1995. Identification of Food-Grain Varieties. American Association of Cereal Chemists, St. Paul, MN.

Wusirika, R., Bohn, M., Lai, J., Kole, C., 2014. Genetics, Genomics and Breeding of Maize. CRC Press, FL, USA.

Xu, Y., Lu, Y., Xie, C., Gao, S., Wan, J., Prasanna, B.M., 2012. Whole-genome strategies for marker-assisted plant breeding. Mol Breed. 29, 833–854.

Xu, Y., Xie, C., Wan, J., He, Z., Prasanna, B.M., 2013. Marker-assisted selection in cereals: platforms, strategies and examples. In: Gupta, P.K., Varshney, R.V. (Eds.), Cereal Genomics II. Springer, Netherlands, (Chapter 14)..

Yan, J., Kandianis, C.B., Harjes, C.E., Bai, L., Kim, E.H., Yang, X., Skinner, D.J., Fu, Z., Mitchell, Q.L., Li, Q., Salas Fernandez, M.G., Zaharieva, M., Babu, R., Fu, Yu., Palacios, N., Li, J., DellaPenna, D., Brutnell, T., Buckler, E.S., Warburton, M.L., Rocheford, T., 2010. Rare genetic variation at *Zea mays* crtRB1 increases β-carotene in maize grain. Nat. Genet. 424, 322–327.

Yang, R.C., Ham, B.J., 2012. Stability of genome-wide QTL effects on malt α-amylase activity in a barley doubled haploid population. Euphytica 1881, 131–139

Ye, L., Huang, Y., Dai, F., Ning, H., Li, C., Zhou, M., Zhang, G., 2015. Identification of two key genes controlling chill haze stability of beer in barley *Hordeum vulgare* L. BMC Genomics 161, 449, Available from: http://www.biomedcentral.com/1471-2164/16/449.

Zemetra, R.S., Morris, R., Mattern, P.J., Seip, L., 1987. Gene locations for flour quality in winter-wheat using reciprocal chromosome substitutions. Crop Sci. 27, 677–681.

Zhang, Q., Dong, Y., An, X., Wang, A., Zhang, Y., Li, X., Gao, L., Xia, X., He, Z., Yan, Y., 2008. Characterization of HMW glutenin subunits in common wheat and related species by matrix-assisted laser desorption/ionization time-of-flight mass spectrometry MALDI-TOF-MS. J. Cereal Sci. 47, 252–261.

Zhang, W., Dubcovsky, J., 2008. Association between allelic variation at the Phytoene synthase 1 gene and yellow pigment content in the wheat grain. Theor. Appl. Genet. 116, 635–645.

Zhang, W., Gianibelli, M.C., Rampling, L.R., Gale, K.R., 2004. Characterisation and marker development for low molecular weight glutenin genes from Glu-A3 alleles of bread wheat *Triticum aestivum* L. Theor. Appl. Genet. 108, 1409–1419.

Zhang, W., Gianibelli, M.C., Xia, W., Rampling, L., Gale, K.R., 2003. Identification of SNPs and development of allele-specific PCR markers for gamma-gliadin alleles in *Triticum aestivum*. Theor. Appl. Genet. 107, 130–138.

Zhang, Y., Wu, Y., Xiao, Y., He, Z., Zhang, Y., Yan, J., Zhang, Y., Xia, X., Ma, C., 2009. QTL mapping for flour and noodle colour components and yellow pigment content in common wheat. Euphytica 165, 435–444.

Zhao, X.L., Xia, W., Gale, K.R., Lei, Z.S., He, Z.H., Sun, Q.X., Xia, X.C., 2007a. Identification of SNPs and development of functional markers for LMW-GS genes at Glu-D3 and Glu-B3 loci in bread wheat *Triticum aestivum* L. Mol. Breed. 20, 223–231.

Zhao, X.L., Xia, X.C., He, Z.H., Lei, Z.S., Appels, R., Yang, Y., Sun, Q.X., Xia, W., 2007b. Novel DNA variations to characterize low molecular weight glutenin Glu-D3 genes and develop STS markers in common wheat. Theor. Appl. Genet. 114, 451–460.

Zhou, G., Zhang, Q., Tan, C., Zhang, X.Q., Li, C., 2015. Development of genome-wide InDel markers and their integration with SSR, DArT and SNP markers in single barley map. BMC Genomics 16, 804.

Chapter 18

The Effects of Growth Environment and Agronomy on Grain Quality

Michael Gooding
Formerly University of Reading, Berkshire, United Kingdom

This chapter is reprinted from the first edition of this book

18.1 INTRODUCTION

The attainment of minimum grain quality standards is often an economic imperative for the farmer to secure a market, or achieve an adequate price. Grain that are: well-filled and plump for high flour extraction, energy content or alcohol production; free from impurities, infection and pests; and at a moisture content that allows safe transport and storage; are preferred by virtually all users and buyers of grain. Other criteria, such as protein concentration, protein type, enzyme activity, nutritional composition, taste and texture, milling properties and seed viability are important for specific markets. Additional contributions to marketability are: continuity of supply of grain meeting threshold levels of quality; the proximity of the supply chain; and the methods and locality of production (e.g. conventional vs. organic; traceability; regional affiliation). The extent to which farmers modify their agronomy to meet specific quality thresholds will depend partly on the financial reward for doing so (or penalty for not meeting the minimum standards). This is not a straightforward calculation. Quality is subject to the weather and soil conditions. Farmer interventions to grow for a specific market, which could involve expense such as choosing a lower yielding variety and/or increasing fertiliser or crop protection inputs, can be thwarted or rendered unnecessary by the environment. Furthermore, price premiums for grain of particular quality cannot often be predicted precisely before harvest. Growing for specific markets, therefore, depends upon the attitude to, and management of risk; and also on the experience, preference and skill of the farmer. In this chapter it is sought to identify the key environmental contributions to grain quality and to describe key interventions by which farmers can modify quality.

Cereal Grains. http://dx.doi.org/10.1016/B978-0-08-100719-8.00018-8

18.2 GROWING CONDITIONS AND ENVIRONMENT

The world-wide production of cereals exemplifies their range of adaptation. Cereal species and varieties vary in their tolerance of cold, heat, water-logging, drought and salinity. Furthermore cereal life-cycles can be controlled by major genes influencing the response to day length (photoperiod sensitivity) and cold periods (vernalisation), as well as many more genes affecting developmental rate. Varieties and species therefore vary in their response to temperature, vernalisation and photoperiod so that they can be broadly tailored to the prevailing photoperiod and climatic conditions of particular regions and sowing dates. Ideally, this adaptation ensures that: particularly sensitive developmental stages such as grain site formation and fertilisation do not coincide with severe frost or excessive heat; rapid canopy growth and grain filling occurs during periods of ample moisture and light; ripening and harvest is facilitated by drying conditions; and the life-cycle is completed within the structures of a rotation.

Growing varieties beyond their proposed geographic range or sowing window risks failure to meet yield and quality targets. Even for well-adapted varieties, the vagaries of the weather mean that abnormal conditions can account for large annual variations in grain yield and quality within a locality. It is, however, difficult to generalise about the effects of environment on quality, particularly on the physical attributes of grain. Wheat breeders have been tasked with achieving well-filled grain with high specific weights whatever the target environment. Additionally, cereals have significant plasticity in the numbers of grains they set in response to the environment. Grains may be well filled even in stressed conditions if, for example, the stress occurs early in grain site formation, rather than only in grain filling. However, when stresses are applied mostly during grain filling, they can have marked effects. Increasing drought severity during this latter stage reduces grain yield and mean grain weight, such that grains become shrivelled, have poor specific weight and contribute to screenings (Fig. 18.1). Drought often occurs simultaneously with warmer weather, which itself has similar effects to that of drought (Table 18.1). Figure 18.2 illustrates the combined effect of increasing temperature and restricting water supply to close to the wilting point (full details in Wan et al., 2009) during grain filling. Grain yield is reduced because grain filling is curtailed (Fig. 18.3). Although rate of grain filling can sometimes be increased by higher temperatures up to about 20°C, this is insufficient to overcome the negative effect of stress on duration. Even between closely related genotypes variations in ability to cope with the stress, or 'escape' it through altered periodicity of development are evident. The specific line labelled *x* in Fig. 18.2(A–D) is a high yielding line in cool wet conditions, equivalent to both its parents, but yield, mean grain weight, specific weight and screenings are much poorer in a hotter, drier environment.

Although drought and temperature stress can reduce yield and physical quality, additional confounding effects are evident in the field. In long-term experiments, specific weight has been positively associated with hours of sunshine during the grain filling period presumably due to positive effects on

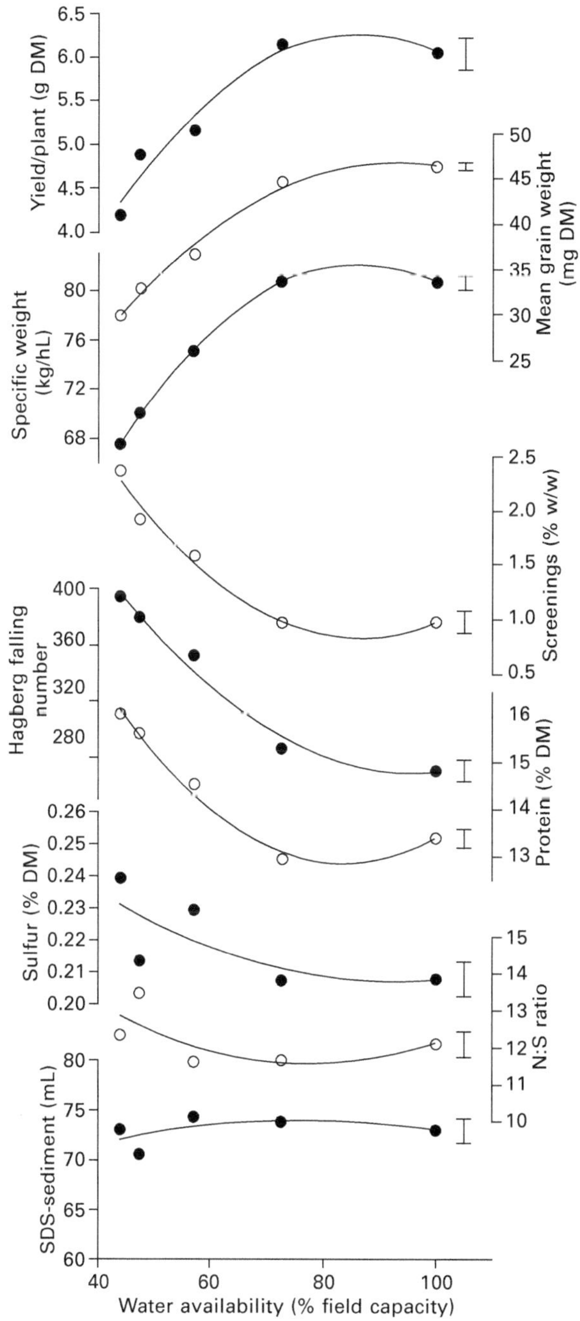

FIGURE 18.1 The effect of water availability during grain filling (from 15 to 28 days after anthesis) on the yield and quality of winter wheat (mean of daily max. and min. temperature during this period is 18.5°C). Points are the means of 32 plants grown as plots in each of four blocks in each of 2 years in enclosed poly-tunnels. *Open symbols refer to right-hand vertical axes.* Error bars are one SED (28 d.f.). *(Gooding, M.J., Ellis, R.H., Shewry, P.R., Schofield, J.D., 2003. Effects of restricted water availability and increased temperature on the grain filling, drying and quality of winter wheat. J. Cereal Sci., 37, 295–309.)*

TABLE 18.1 Effect of water availability and day/night temperature regime during grain filling (from 15 to 28 days after anthesis) on grain yield and quality of Hereward winter wheat

	100% field capacity		44% field capacity		
	23/15°C	28/20°C	23/15°C	28/20°C	SED
Grain yield (g DM/plant)	6.26	5.02	4.94	4.12	0.256
Mean grain weight (mg)	48.5	42.9	40.7	35.4	1.46
Specific weight (kg/hi)	78.4	75.5	74.0	69.0	1.36
Screenings (% w/w)	0.3	0.5	0.2	0.5	0.14
Protein (% DM)	12.0	13.4	13.5	13.7	0.433
Sulfur (% DM)	0.166	0.198	0.219	0.216	0.0232
N:S ratio	14.4	12.4	11.7	12.7	1.28
SDS-sediment (mL)	78.0	82.7	72 2	74.0	4.03
HFN	187	378	372	392	28.9

Means are from 84 plants grown as plots in each of two growth cabinets in each of 3 years.
HFN, Hagberg falling number.
Gooding, M.J., Ellis, R.H., Shewry, P.R., Schofield, J.D., 2003. Effects of restricted water availability and increased temperature on the grain filling, drying and quality of winter wheat, J. Cereal Sci., 37, 295–309.

photosynthesis and therefore improvements in grain filling, despite a positive association between temperature and sunshine hours. Similarly, frequency of wet days during grain ripening (after maximum grain dry matter has been attained) has been correlated with poorer specific weights, possibly because of the effects of weathering on grain surface texture (Atkinson et al., 2008). Increased temperatures during this late phase can thus be positively related to specific weight if it leads to more timely harvest and reduced deterioration in the field (Smith and Gooding, 1999).

Effects of climate and weather on grain protein concentrations are more predictable than those on grain specific weight. Increased temperature and reduced water availability have a less damaging effect on nitrogen accumulation compared to that on dry matter (Fig. 18.3). Increasing temperature and/or reducing water availability is, therefore, associated with increased grain protein concentration (Table 18.1, Figs 18.1 and 18.2E), an effect long-recognised in the field (Benzian and Lane, 1986). High protein concentration in wheat is often achieved in areas of the world where warmer and drier conditions are associated with shorter grain-filling periods. In areas where wheat matures in cooler and wetter conditions, associated with extended grain filling periods, meeting grain

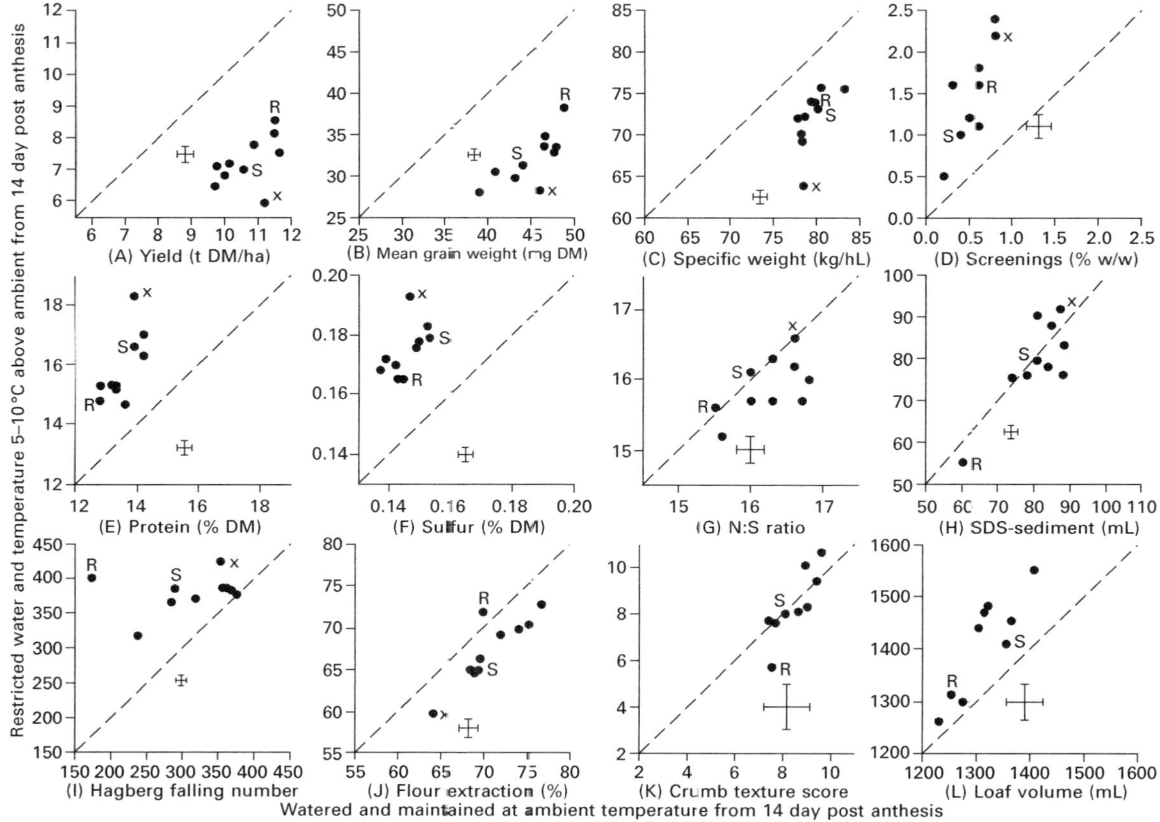

FIGURE 18.2 The effect of restricting water and increasing temperature (to a maximum of 28°C) on the yield and quality of wheat grown under plastic (mean ambient temperature is 18.5°C). Points are means from two seasons for eight doubled haploid progeny and the parents of Spark (S) times Rialto (R). *Dashed line*, 1:1; Error bars are one SED. *(Wan, Y., Underwood, C., Toole, G., Skeggs, P., Zhu, T., Leverington, M., Griffiths, S., Wheeler, T., Gooding, M.J., Poole, R., Edwards, K.J., Gezan, S., Welham, S., Snape, J., Mills, C., Mitchell, R., Shewry, P.R., 2009. A novel transcriptomic approach to identify candidate genes for grain quality traits in wheat, Plant Biotechnol. J., 7, 401–410.)*

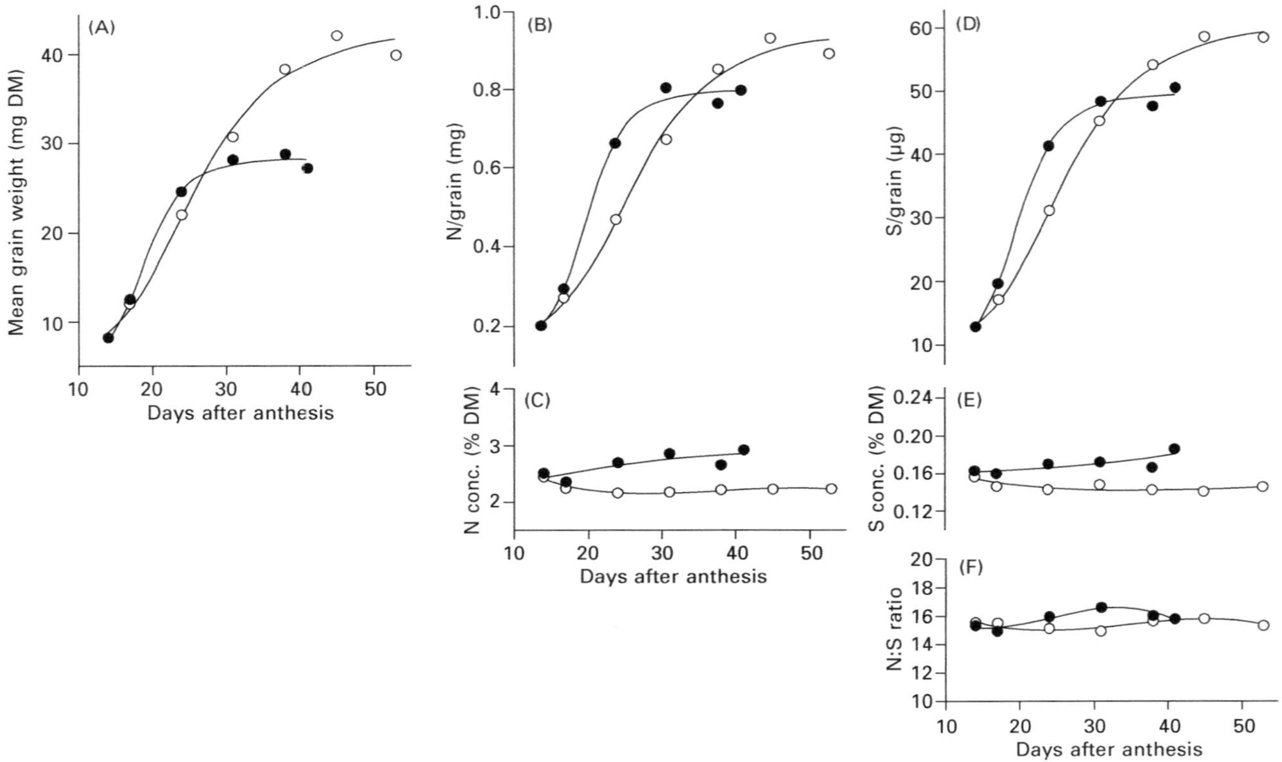

FIGURE 18.3 The effect of increasing temperature and reducing water availability on the grain filling of winter wheat with dry matter, nitrogen and sulfur. *Open symbols* are plants maintained at ambient temperature (mean of daily maximum and minimum temperature during this period is 18.5°C) and field capacity; *closed symbols* are at 5–10°C above ambient unless 28°C would be exceeded, and at close to wilting point from 14 days after anthesis. Points are means of eight lines in four blocks. *(Wan, Y., Underwood, C., Toole, G., Skeggs, P., Zhu, T., Leverington, M., Griffiths, S., Wheeler, T., Gooding, M.J., Poole, R., Edwards, K.J., Gezan, S., Welham, S., Snape, J., Mills, C., Mitchell, R., Shewry, P.R., 2009. A novel transcriptomic approach to identify candidate genes for grain quality traits in wheat, Plant Biotechnol. J., 7, 401–410.)*

protein thresholds for bread-making quality can be problematic, even when large quantities of nitrogen fertiliser are applied. Variety × environment interactions on grain protein concentration can often be explained by the equivalent reverse effects on grain yield, that is, varieties yielding comparatively poorly in hot and dry conditions often have particularly high protein concentrations in the same environment (Fig. 18.2E), in spite of much reduced total yields of protein (kg/ha).

Grain protein concentration is not only influenced by weather during grain filling. In the United Kingdom, North America and Australia rainfall during stem extension has been negatively associated with subsequent grain protein concentrations. In the United Kingdom, this appears to be accounted for by rainfall increasing losses of nitrogen following large top-dressings in early spring. In other areas, early rainfall may be associated with vegetative proliferation and dilution of protein, and/or by augmenting soil moisture such that succeeding drought during grain filling is delayed (Smith and Gooding, 1999).

Environmental conditions that influence grain protein concentrations are often related to effects on cereal-based products for which protein concentration is a recognised quality criterion. Although there are exceptions, environment does not usually have a large impact on the functional quality of the protein. For example, restricting moisture and increasing temperature during grain filling can increase loaf volumes (e.g. Fig. 18.2E vs. L). Grain sulfur and nitrogen accumulation are often influenced in similar ways by environment (Fig. 18.3) such that N:S ratio is not greatly affected (Figs 18.1 and 18.2). Wheat varieties with protein and other characteristics that are particularly suited to bread making have therefore been grown traditionally in environment zones conducive to achieve high protein concentrations, whereas cooler, wetter regions have been associated with biscuit (cookie) wheats, or with varieties destined for animal feed or alcohol production.

Despite the earlier assertion, reductions in SDS-sedimentation volume of wheat, implying changes in protein composition, have been associated with hours of temperatures above 32°C and hours of relative humidity below 40% during grain filling (Graybosch et al., 1995). More specifically, heat shocks (e.g. > 35°C during grain filling) have led to increases in the proportion of monomeric (gliadin) to polymeric (glutenin) fractions of storage protein (Blumenthal et al., 1993). This can lead to reduced dough strength and have a negative impact on the loaf quality (Blumenthal et al., 1991). There is, therefore, some effort to identify varieties with more stable protein functionality when exposed to hot weather, especially for use in regions where temperatures exceeding 35°C during grain filling are not uncommon (Castro et al., 2007).

One quality criterion for which there is a large interaction between variety and environment is α-amylase activity, and therefore Hagberg falling number (HFN). There are four sources of α-amylase that can lead to low HFN: pre-maturity sprouting (PrMS); post-maturity sprouting (PoMS); pre-maturity α-amylase activity (PMAA; or Late Maturity Amylase, LMA; or Late Maturity

Endosperm Amylase, LMEA); and retained pericarp α-amylase activity (RPAA). Of these, the most frequent to give problems are PoMS and PMAA. PoMS results from the alleviation of dormancy followed by germination in wet summers when harvest is delayed and grains start to sprout in the ear. PMAA often occurs in the crease region of the grain, emanating from the aleurone layer around the endosperm cavity in the absence of sprouting. The causes of PMAA are poorly understood but have been linked to low temperatures during grain development, slow grain drying rate, large grain size, temperature shocks (short periods of either high or low temperatures during grain filling) and grain cavity characteristics (Kindred et al., 2005). Sustained warmth and restricted water availability are generally associated with reduced α-amylase activity and increased HFN (Figs 18.1 and 18.2I; Table 18.1). Varieties grown in areas with cool and wet summers require a higher degree of sprouting resistance than types destined for areas with warm and dry conditions during grain filling. There is also a strong genetic component to PMAA (Mares and Mrva, 2008). For example, cv. Rialto in Fig. 18.2I is pre-disposed to PMAA (Joe et al., 2005). In warm and dry conditions Rialto achieves a high HFN, but unlike a number of its progeny when crossed with cv. Spark, it produces significant quantities of α-amylase and has low HFN when water availability is maintained at ambient temperatures. There is no correlation between the HFN achieved by a genotype in dry, warm conditions and its performance in cool, wet conditions.

18.3 SOWING

As already indicated, appropriate sowing dates are largely governed by the climate, rotational requirements, and the choice of species and variety. Rotational position has an impact on quality. One of the reasons is the impact of soil or trash borne diseases that build-up under successive cereal crops. Take-all (*Gaeumannomyces graminis*) attacks the roots and stem base of cereals and a number of grasses. In wheat (*G. g.* var. *tritici*) take-all, the disease builds up rapidly as two to four susceptible crops are grown in succession, particularly on neutral or alkaline soils. Infected tissue becomes blackened and water and nutrient uptake is restricted. Patches of infected plants are stunted and poorly competitive. Ears senesce pre-maturely and initially appear as bleached 'whiteheads' before becoming blackened with the invasion of sooty moulds. Take-all reduces mean grain weight, specific weight and sometimes HFN (Bateman et al., 1990). The sooty-moulds may contaminate and therefore discolour, flour.

Other than for effects related to disease, weed and pest control, effects of rotation on cereal quality are difficult to interpret. Rotations that involve breaks from cereal growing often benefit both the yield and nitrogen uptake of following cereal crops but this does not necessarily produce crops with high protein concentrations, indeed rotational cropping has a much larger effect on yield than on grain quality. When increases in grain protein concentration have occurred it has often been when a cereal has been preceded by a legume-rich crop or pasture, which has been growing for more than one year, and/or when a legume crop has been ploughed in as a green manure (Gooding and Davies, 1997).

Within sowing dates for which target yields can be approached, time of sowing often has little impact on grain quality. It is important that grain matures in appropriate conditions such that excessive heat, drought, or large rain falls during grain maturation are avoided. Timing of grain filling and maturation can be influenced by sowing date but variations in developmental stage often become smaller during development, particularly if there are strong photoperiod and vernalisation requirements. Delayed sowing, to the extent that yield will be reduced, can often increase grain protein concentrations, but this will rarely be a factor in determining sowing date, which is more often delayed due to limitations of logistics, soil conditions and weather.

Cereals can be established by a range of techniques ranging from drilling into a fully-prepared seedbed following complete inversion tillage with ploughing or discs, to zero- or direct-drilling into the soil and residue of the previous crop. Choice of system will depend on soil type and climate; the availability of machinery, fuel, time and labour; and weed pressures and the availability of effective herbicides. Cultivation technique will influence nutrient and moisture availability and distribution within the soil, particularly through impacts upon organic matter, soil water content at sowing and rooting ability. There are also implications for pest and disease control. As with sowing date, however, impacts of practices around those chosen to achieve target yields, have little or inconsistent effects on grain quality. Thin crops, either due to poor establishment, excessively low seed rates being used, or subsequent predation can be associated with poor yields, but these are mostly due to effects on grain populations rather than physical quality of the grain. In Fig. 18.4 reducing sowing density such that yield was reduced by 25% and was only associated with a 4% reduction in mean grain weight. There were even smaller effects on screenings, specific weight and HFN. Protein concentration does decline as yield increases, unless N supply is matched to target yield, and this effect can change the economically optimal seed rate for markets, where protein concentration is an important criterion (Gooding et al., 2002). However, this situation can reverse if weeds are not controlled adequately. Thin crops are less competitive against weeds and weeds can have important impacts upon quality.

18.4 WEED CONTROL

Weed control is particularly important for seed certification purposes. Closely related grass species predominate in this regard and quality standards will often define maximum numbers of named weed species per weight of seed. For example, one wild oat seed per kilogram of wheat, equating to a weed:crop seed purity ratio of 1 in about 20,000. At this level, hand roguing of seed crops can be economic, particularly if it safeguards the field for future use in growing seed crops. Grass weeds are also important alternative hosts for cereal pests and

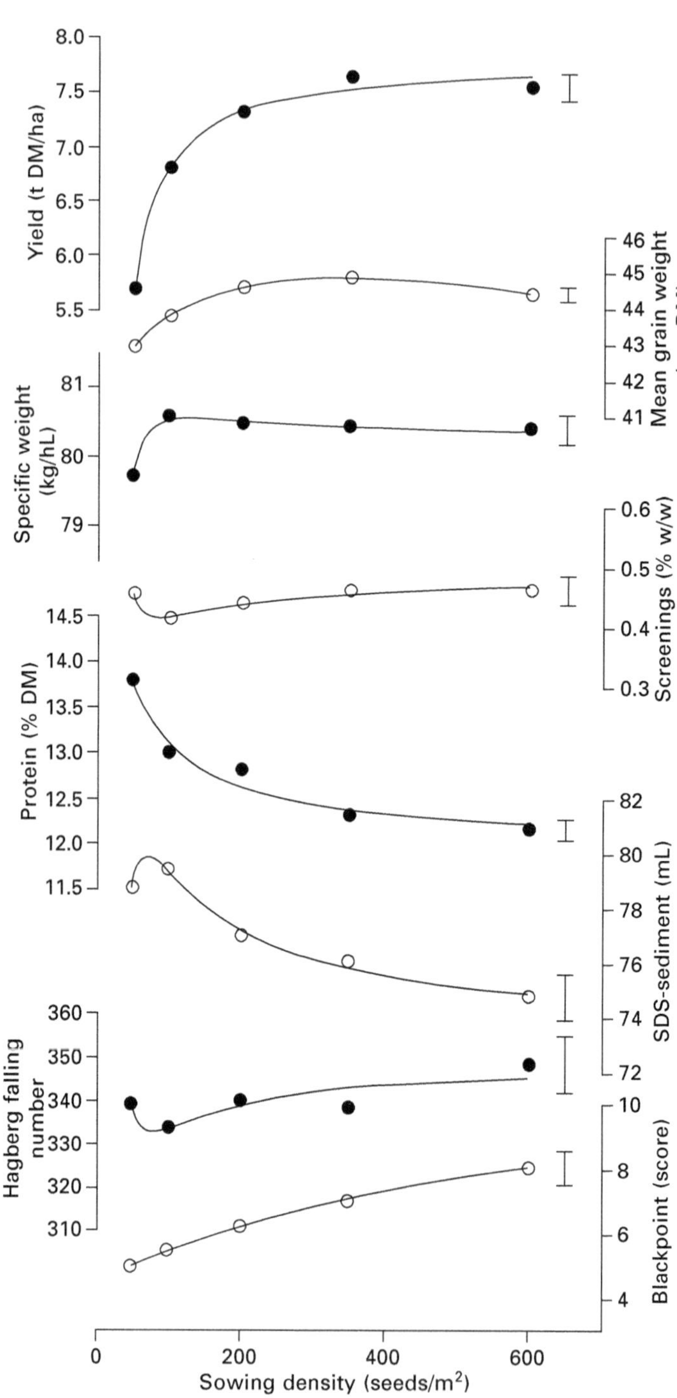

FIGURE 18.4 The effects of sowing density on the yield and quality of Hereward winter wheat. Points are the means from three field experiments (Gooding et al., 2002). *Open symbols refer to right-hand vertical axes. Vertical bars are one SED.*

diseases which impact quality. Of particular note here is ergot (*Claviceps* spp.). The fungus produces a sclerotium, a hard-purple mass which develops in the inflorescence in place of a grain. Grain samples containing more than 1% ergot can cause serious illness, and sometimes death, due to the presence of poisonous alkaloids. Grain batches containing ergot are, therefore, rejected by buyers. Ergot severity can be closely correlated with ingress of alternative grass hosts in cereal fields (Mantle et al., 1977). With regards to physical characteristics, weed seeds can contribute to screenings. Severe late-season competition from weeds can reduce mean grain weights, specific weights and protein concentrations (Cosser et al., 1996; Awan, 2002).

18.5 DISEASE CONTROL

Foliar diseases of cereals commonly reduce grain yield, particularly associated with reduced photosynthetic area, duration and function. Where infection pressures from fungal pathogens are high, for example in humid and fertile conditions, where rotations are dominated by cereals, and where varietal resistance is inadequate, fungicide-use to protect yield is often justified economically. In such circumstances, maintaining grain filling by controlling disease that would otherwise lead to premature death (senescence) of the canopy, increases yield (Fig. 18.5A). The relationship in Fig. 18.5A is curved, because in some cases it is possible to extend canopy life beyond the end of grain filling (Pepler et al., 2005). Nonetheless, around average canopy durations, maintaining canopy life by 10 days with fungicide is associated with a 16% increase in yield and a 13% increase in mean grain weight (Fig. 18.5B). Maintaining grain filling with disease control measures (Pepler et al., 2006) also reduces grain shrivelling, and improves specific weights (Fig. 18.5C). Here the relationship with canopy longevity is less marked, presumably because other factors such as surface texture, grain morphology unrelated to shrivelling and mean density of individual grains influence specific weight.

Despite the complexity of factors influencing specific weight, when severe disease has been controlled, fungicide-use is often associated with increased flour extraction rates.

Fungicide-use often has an impact on grain protein concentration. Disease control is associated with improved uptake and re-mobilisation of nitrogen (Gooding et al., 2005), so it is not inevitable that grain protein concentration is diluted when fungicide-use is associated with increased yield. Indeed, in wheat it appears that when rusts (*Puccinia* spp.) and/or powdery mildew are controlled by fungicides, yield and protein concentration can increase concurrently (Dimmock and Gooding, 2002). When *Septoria* spp. are controlled there can be a dilution of grain protein. Even in the United Kingdom, however, where *Septoria tritici* is usually the prevalent late-season foliar pathogen, reductions in protein concentration following fungicide-use is slight: a 0.16-fold increase in yield only being associated with a 0.02-fold reduction in grain protein

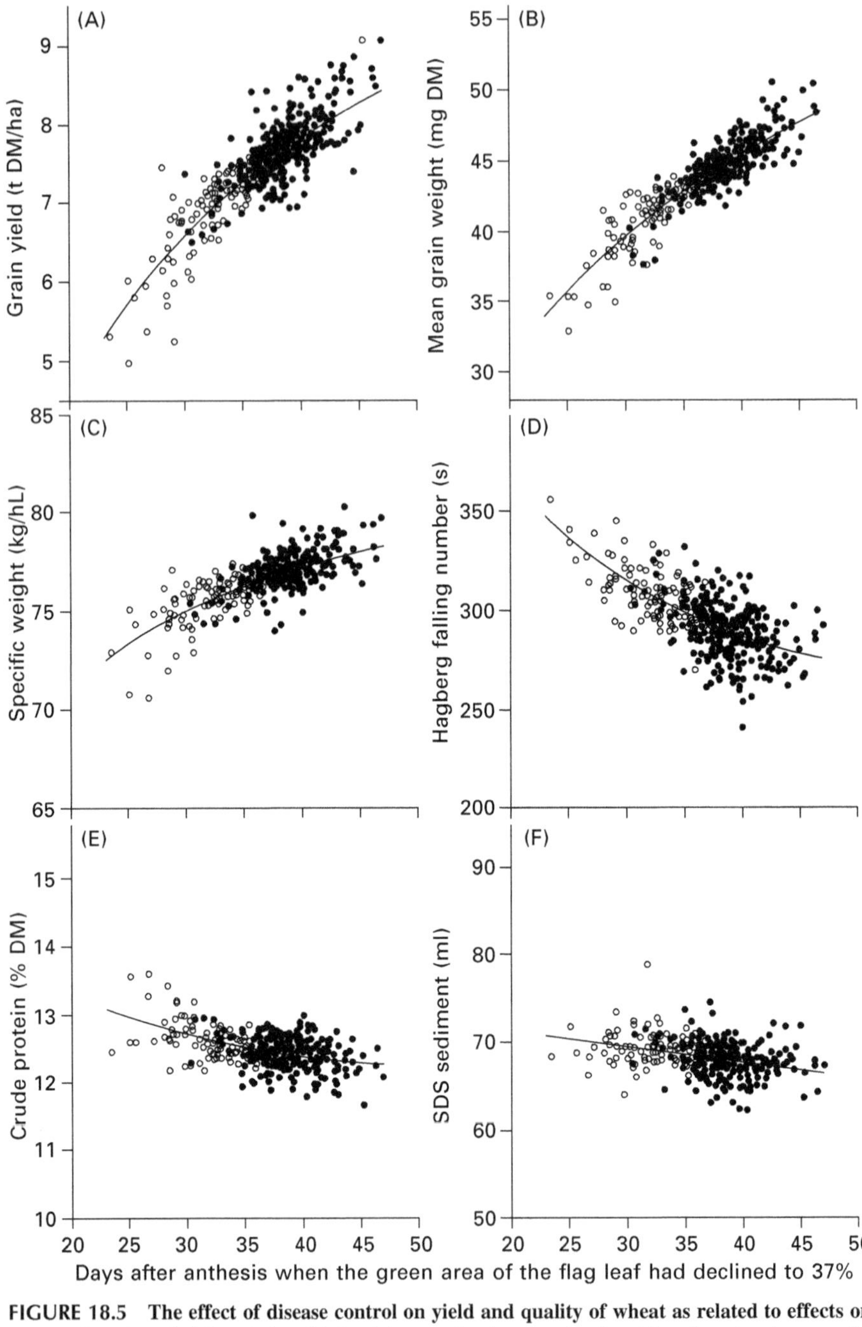

FIGURE 18.5 The effect of disease control on yield and quality of wheat as related to effects on canopy duration. *Solid symbols*, with fungicide; predominantly triazoles either alone in mixtures with strobilurins; applied at flag leaf emergence with usually second sprays either at the start of stem extension or at ear emergence; most prevalent disease controlled was *Septoria tritici,* followed by yellow and brown rusts *(Puccinia* spp.), and then powdery mildew *(Erysiphe graminis).* Points are means of at least three replicate plots (at least 2 m × 5 m) from 85 UK field site × season × cultivar combinations. Main and interacting effects of site × season × cultivar have been subtracted from the data to isolate the effect of fungicide.

concentration when canopy life is extended by 10 days (Fig. 18.5E). Fig. 18.5F shows SDS-sediment to decline in line with reduction in protein concentration, suggesting that protein quality for bread making has not been greatly influenced by foliar wheat pathogens. In isolated experiments, however, severe infection has been associated with: negative effects on SDS-sediment unrelated to protein concentration; reduced proportions of HMW-glutenin sub-units; and weaker doughs. Fungicide-use has sometimes increased loaf quality despite reducing protein concentration (Gooding et al., 1994). Effects of disease and its control on protein quality is not necessarily a direct influence of pathogens and fungicides. Controlling severe disease will alter the periodicity of, and hence mean temperature and rainfall during grain maturation.

Maintaining canopy-life in wheat with fungicides can increase α-amylase activity and/or reduce HFN (Fig. 18.5D). The basis for this effect has yet to be clearly defined but is associated with increased grain size and weight. The effect does not appear to depend on the specific pathogen controlled nor mode of action of the fungicide.

Disease control is particularly important for quality where pathogens discolour, taint, or release toxic compounds into the grain. *Fusarium* spp. infection, for example head blight or scab in small grained cereals, can cause grain shrivelling, altered protein composition and reduced seed viability, and is also a source of a number of mycotoxins that can cause severe illness in mammals. One such mycotoxin, deoxynivalenol (DON), is subject to maximum tolerable concentrations for grain and selected wheat products. *Fusarium* infection and mycotoxin concentration in wheat can be associated with: maize being the previous crop; wheat being established by minimum cultivation techniques rather than following plough-based, inversion tillage; and the growing of susceptible varieties. Suitable fungicides (e.g. most triazoles) applied at, or shortly after anthesis, can reduce *Fusarium* infection and mycotoxin concentrations (Beyer et al., 2006). However, choice of fungicide is critical as the ear is colonised by many saprophytic or only mildly pathogenic fungi and eradicating some or all of these competitors can actually increase *Fusarium* infection and/or mycotoxin production.

Small-grained cereals can become discoloured if they experience physical damage. Black point is a dark discolouration of the embryo region that can extend over the shoulders, and along the crease of grain. The discolouration derives from damaged tissue releasing phenols with peroxidases, and is often, but not always, associated with the presence of pathogenic fungi (e.g. *Alternaria* spp.; *Cochliobolus sativus*). Depending on fungal species, seed viability may also be reduced in severely black-pointed grain. The degree of black-pointing is related to the activity of the peroxidase isoenzymes released (Williamson, 1997) and hence black point severity depends strongly on variety. Control of black point with fungicides is not straightforward, as damage may occur when pericarp cells are crushed during normal grain filling, particularly in larger grain, unrelated to the quantity of any potential causal fungus. As fungicides controlling foliar diseases can increase grain size, their use can also actually increase black point severity when applied at flag leaf emergence. Reductions in black point severity appear more likely by fungicide applications at and after ear emergence, although even here control is far from total or reliable (Ruske et al., 2003).

18.6 CROP NUTRITION

In most growing conditions, nitrogen availability has the greatest impact on cereal quality compared with that of other nutrients. This is the case whether the nitrogen is released from the soil and crop residues; or spread as manures, composts or slurries; or applied as artificial fertiliser. Fertiliser nitrogen is commonly available straight as urea ($CO(NH_2)2$; 46% N) and ammonium nitrate (NH_4NO_3; 34.5% N); or in compounds containing other nutrients. Applying nitrogen to deficient crops can have marked effects on grain yield through increasing green area of the crop, light interception and particularly grain numbers per unit area. In regions of the world with high yield potential, where crop yield is more limited by light interception than by water availability, yield responses to nitrogen can be related to the amounts of nitrogen required in a canopy to produce sufficient green area to intercept near-asymptotic levels of light by anthesis. These large canopies often contain between 150 and 200 kg N/ha. Supplying significant quantities of nitrogen to achieve this level of uptake has a large impact on a range of quality criteria, economic margin and environmental impacts of cereal growing. Economic responses to high levels of nitrogen application are also often reliant on the use of short and stiff strawed varieties that can resist lodging, and/or the use of plant growth regulators; disease control measures such as resistant varieties and robust fungicide programmes; and pest management.

Increasing nitrogen availability at times to specifically improve grain yield, that is before and during stem extension, has minor effects on mean grain weight; the positive effects on yield nearly always being mostly due to effects on grain numbers. In some cases increases in yield can occur while mean grain weights decline in response to nitrogen. Reductions in mean grain weight appear particularly likely when large numbers of grain have been set in response to nitrogen application, but then are poorly filled if drought or lodging follows anthesis. In these instances nitrogen application is also often associated with increases in grain shrivelling, and sometimes reduced specific weights. However, nitrogen application can have effects on mean grain shape (Clarke et al., 2004), and can increase the density of individual grains such that there are small increases in specific weight when grain shrivelling is not excessive.

The largest and most reliable effect of nitrogen on grain quality is the effect on grain protein concentration. Where crops are highly deficient, a small increase in nitrogen availability can reduce grain protein concentration as it is diluted with a large increase in yield. At levels of availability seen in most commercial situations,

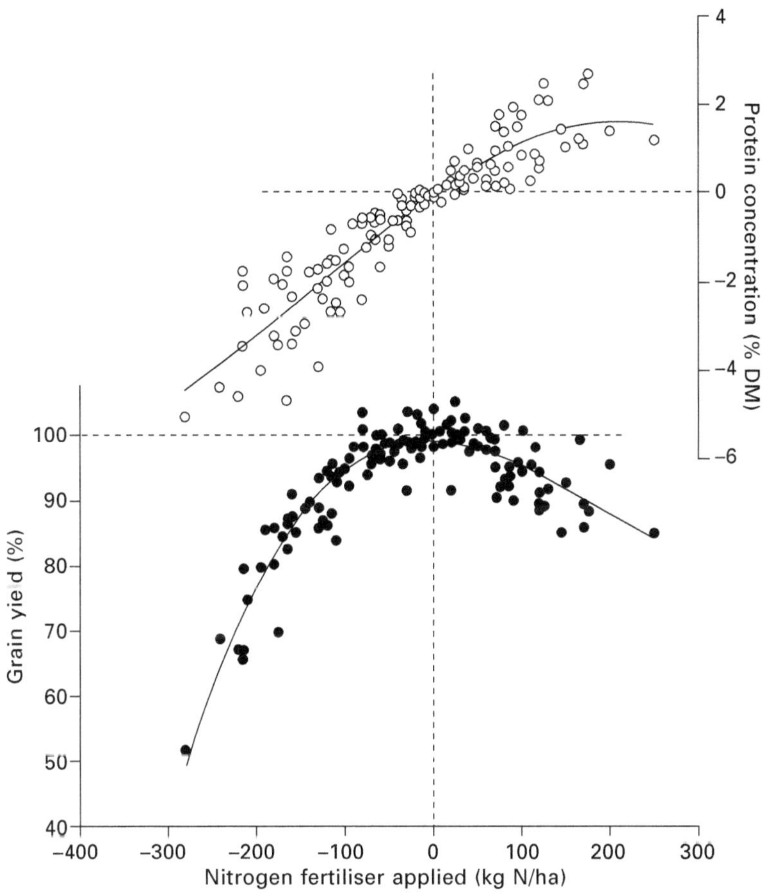

FIGURE 18.6 The response of grain protein concentration and grain yield to fertilizer nitrogen applied during stem extension to winter wheat in the UK. Data are from 23 variety × site × year combinations. The nitrogen applied has been rebased to the difference from the fitted rate that gave maximum yield (N_{max}) for each variety × site × year combination. Similarly yield is expressed as a percentage of the yield achieved at N_{max}, and protein concentration is the difference from that achieved at N_{max}.

however, increasing nitrogen application increases protein concentration. Indeed increasing nitrogen application beyond that required to maximise yield is associated with continued increases in protein concentration towards a significantly higher asymptote (Fig. 18.6). For markets where high protein concentration has an economic reward, applying more nitrogen than is required for maximal margin based solely on yield is sometimes justified. This is more likely in temperate, humid areas, such as NW Europe where large yielding crops often have marginal protein concentrations (and also low HFN) which limits the local supply of wheat for bread making. Applying nitrogen for purely a protein concentration response in wheat is usually most effective at or shortly after anthesis. At this time, better nitrogen recoveries are sometimes achieved by applications to the foliage in a spray of urea solution. Although nitrogen applications that are supra optimal for yield can give an economically-justified increase in protein concentration, the recoveries of this extra nitrogen are often poor (Gooding et al., 2007).

As protein becomes more concentrated with nitrogen fertiliser application, it is the storage proteins, for example the gliadins and glutenins in wheat, that are usually proportionately improved, at the expense of the more soluble proteins. The compositional response of the storage proteins to nitrogen application will depend on the availability of other key elements, particularly sulfur. Increasing nitrogen applications can increase the N:S ratio in the grain and, in S deficient situations, application of N can increase the proportion of sulfur-poor proteins. Sulfur deficiency has been increasing in many areas of the world as: industrial emissions are decreased and deposition on agricultural soils thereby reduced; use of sulfur-containing fertilisers has declined; reserves in soil organic matter have diminished; and crop demand has increased with increasing yields. Sulfur deficiency can be associated with reduced yields, reduced mean grain weights and increased grain shrivelling, all of which can be prevented by application of sulfur containing fertilisers such as ammonium- or magnesium sulfate. Sulfur has a significant impact on the functionality of storage protein for several end-uses because: S-containing amino-acids provide the inter-chain disulfide bonds that help maintain the polymer network of storage proteins; S availability influences the relative proportions of different storage proteins that are synthesised; and/or because glutathione, an S-containing tripeptide, can modify rheology. Consequently, for example pan bread quality made from wheat flour can be more closely associated with S concentration than with N concentration. Wheat has been classed sulfur-deficient if the grain has a higher N:S ratio than 17:1 and a grain-sulfur concentration of less than 0.12%. Sulfur application to deficient crops can sometimes increase grain protein concentrations, and in wheat, increase dough extensibility associated with an increased proportion of sulfur-containing storage proteins.

Increasing nitrogen application has variable effects on α-amylase activity and HFN. If application does not cause lodging, then increasing nitrogen availability often reduces α-amylase activity or increases HFN. This beneficial effect of nitrogen interacts with variety and its basis is not clearly understood; being variously attributed to delayed crop maturity, grain morphology and increased dormancy (Kindred et al., 2005).

18.7 LODGING AND GROWTH REGULATION

Lodging is the permanent displacement of cereal stems from the vertical (Berry et al., 2004); in the most serious cases, the crop is laid flat. Lodging occurs in all cereals, most often during the 2–3 months before harvesting. Lodging results from the stem breaking or buckling (stem lodging), or by the stem base anchorage being displaced (root lodging). Stem and root lodging can occur as wind increases in force above failure speeds (Berry et al. 2004). Root lodging becomes more likely as soil strength is reduced with rainfall, particularly on certain soil types. Stem lodging can also be the result of stem base diseases, such as eyespot (*Tapesia yallundae*) in wheat, weakening the lower internodes. Lodging can significantly affect yield, notwithstanding the reduced efficiency of harvesting laid crops. Lodging at or shortly after anthesis of small-grained cereals has reduced grain yields by 30–80%, declining to about 12% by the hard-dough stage of wheat (Berry et al., 2004). Severely lodged crops have grain with reduced mean weights, lower specific weights and increased α-amylase activity and protein concentration. The damp conditions around the ears of lodged crops favour fungal infections and mycotoxin production.

Lodging risk is increased at high nitrogen fertiliser application rates and by high seed rates. Lodging, and hence some reported effects of nitrogen and seed rate on yield and quality is strongly influenced by variety. Lodging susceptibility of a variety is a complex character but a major risk factor is crop height. Hence tall varieties have poor yields partly because lodging risk prevents the economic use of sufficient quantities of nitrogen to produce canopies large enough to maximise yield, or achieve sufficient protein concentrations.

In addition to using short, stiff-strawed varieties with strong anchorage, height and hence lodging risk can be reduced by the application of growth regulators, such as chlormequat, at the start of stem extension. When growth regulators have controlled lodging they have also often increased yield, mean grain weight, specific weight and HFN and reduced grain protein concentration.

18.8 FUTURE TRENDS

Climate change scenarios predict warmer and drier grain filling periods in many of the world's major cereal growing regions. Breeders and farmers will respond by selecting adapted material. Breeding for high yield is likely to remain the principal goal but there may well be shifts in the areas that are particularly suited to certain end-uses, for example it may be easier to attain bread-making quality thresholds for wheat in regions that are currently well suited to biscuit (cookie) wheats because of climatic effects on protein concentration. However, higher yields need to be achieved with improved resource-use efficiencies, particularly for nitrogen (and hence energy) and water. Increased grain production, at similar or reduced levels of fertiliser nitrogen input are likely to produce a downward pressure on grain protein concentrations.

In areas already with significant risk of excessive temperatures during grain filling there may be a need to alter the period of the growing cycle to escape predicted stresses, and/or to adopt more stress-tolerant varieties in terms of both yield and quality.

In the past, yields of cereals with specific quality characteristics (e.g. for bread making) have often been lower than those for markets with less stringent requirements (e.g. for animal feed). With more modern breeding techniques, such as with the use of molecular markers and doubled haploid populations, and with a better understanding of the genetic basis of quality, it is possible that high yields and quality will be combined more readily.

Variability in weather increases the need for more stable or resilient varieties, both from yield and quality perspectives. Annual fluctuations make it more difficult to predict when agronomic interventions to modify quality will be worthwhile. There have been a number of attempts to predict quality before harvest and to inform agronomic advice accordingly. These can be based on somewhat crude statistical associations (Smith and Gooding, 1999), understanding of medium to longer term weather cycles (Atkinson et al., 2008), or a more mechanistic modelling approach (Armour et al., 2004), with or without physical assessment of the crop prior to potential interventions (Bhandari et al., 2004). It is likely that improvements in quality prediction will continue, not only to inform agronomic variations from one crop to the next, but also spatially variable interventions within fields (Link et al., 2008).

SOURCES OF FURTHER INFORMATION

Gooding, M.J., Davies, W.P., 1997. Wheat Production and Utilization: Systems, Quality and the Environment. CAB International, Wallingford.

Kettlewell, P.S., 1996. Agronomy and cereal quality. In: Henry, R.J., Kettlewell, P.S. (Eds.), Cereal Grain Quality. Chapman and Hall, London, pp. 407–437.

REFERENCES

Armour, T., Jamieson, p., Zyskowski, R., 2004. Using the sirius wheat calculator to manage wheat quality—the canterbury experience. Proc. Agron. Soc. New Zealand 34, 171–176.

Atkinson, M.D., Kettlewell, P.S., Poulton, p.R., Hollins, p D, 2008. Grain quality in the broadbalk wheat experiment and the winter North Atlantic oscillation. J. Agric. Sci. 146, 541–549.

Awan N. A., 2002. Effect of herbicides, herbicidal weed control and nitrogen on the yield and bread-making quality of wheat *(Triticum aestivum),* PhD Thesis, University of Reading, UK.

Bateman, G.L., Hornby, D., Gutteridge, R.J., 1990. Effects of take-all on some aspects of grain quality of winter wheat. In: Milford, G.F., Kettlewell, P.S., Orson, J.H., Thomas, W.T.B., Pritchard, P.E., Myram, C. (Eds.), Aspects of Applied Biology 25, Cereal Quality II,. Association of Applied Biologists, Warwick, pp. 339–348.

Benzian, B., Lane, P.W., 1986. Protein concentration of grain in relation to some weather and soil factors during 17 years of English winter-wheat experiments. J. Sci. Food Agric. 37, 435–444.

Berry, P.M., Sterling, M., Spink, J.H., Baker, C.J., Sylvester-Bradley, R., Mooney, S.J., Tams, A.R., Ennos, A.R., 2004. Understanding and reducing lodging in cereals. Adv. Agron. 84, 217–271.

Beyer, M., Klix, M.B., Klink, H., Verreet, J.A., 2006. Quantifying the effects of previous crop, tillage, cultivar and triazole fungicides on the deoxynivalenol content of wheat grain—a review. J. Plant Diseases Prot. 113, 241–246.

Bhandari, D.G., Millar, S.J., Richmond, J.C., 2004. Prediction of grain protein content through portable FT-NIR measurement of developing wheat. In: Lafiandra, D., Masci, S., D'Ovidio,

R. (Eds.), Gluten Proteins. Royal Society of Chemistry Special Publications, London, pp. 312–315, 295.

Blumenthal, C.S., Bekes, F., Batey, I.L., Wrigley, C.W., Moss, H.J., Mares, D.J., Barlow, E.W.R., 1991. Interpretation of grain quality results from wheat variety trials with reference to high temperature stress. Aust. J. Agric. Res. 42, 325–334.

Blumenthal, C.S., Barlow, E.W.R., Wrigley, C.W., 1993. Growth environment and wheat quality: the effect of heat stress on dough properties and gluten proteins. J. Cereal Sci. 18, 3–21.

Castro, M., Peterson, C.J., Dalla Rizza, M., Diaz Dellavalle, P., Vazquez, D., Ibanez, V., Ross, A., 2007. Influence of heat stress on wheat grain characteristics and protein molecular weight distribution. In: Buck, H.T., Nisi, J.E., Salomon, N. (Eds.), Wheat Production in Stressed Environments. Springer, Dordrecht, pp. 365–371.

Clarke, M.P, Gooding, M.J., Jones, S.A., 2004. The effects of irrigation, nitrogen fertilizer and grain size on Hagberg falling number, specific weight and blackpoint of winter wheat. J. Sci. Food Agric. 84, 227–236.

Cosser, N.D., Gooding, M.J., Davies, W.P., Froud-Williams, R.J., 1996. Effects of wheat dwarfing genes on grain yield and quality of wheat in competition with *Alopecurus myosuroides,* Second International Weed Control Congress, Copenhagen, pp. 1089–1094.

Dimmock, J.P.R.E., Gooding, M.J., 2002. The influence of foliar diseases, and their control by fungicides, on the protein concentration in wheat gra. A. review. J. Agric. Sci. 138, 349–366.

Gooding, M.J., Davies, W.P., 1997. Crop nutrition and fertilizer use. In: Gooding, M.J., Davies, W.P. (Eds.), Wheat Production and Utilization. CAB International, Wallingford, pp. 146–183.

Gooding, M.J., Smith, S.P., Davies, W.P., Kettlewell, P.S., 1994. Effect of late-season applications of propiconazole and tridemorph on disease levels, senescence, grain development and breadmaking quality of winter wheat. Crop Prot. 13, 362–370.

Gooding, M.J., Pinyosinwat, A., Ellis, R.H., 2002. Responses of wheat grain yield and quality to seed rate. J. Agric. Sci. 138, 317–331.

Gooding, M.J., Gregory, P.J., Ford, K.E., Pepler, S., 2005. Fungicide and cultivar affect post-anthesis patterns of nitrogen uptake, remobilization and utilization efficiency in wheat. J. Agric. Sci. 143, 503–518.

Gooding, M.J., Gregory, P.J., Ford, K.E., Ruske, R.E., 2007. Recovery of nitrogen from different sources following applications to winter wheat at and after anthesis. Field Crops Res. 100, 143–154.

Graybosch, R.A., Peterson, C.J., Baenziger, P.S., Shelton, D.R., 1995. Environmental modification of hard red winter wheat flour protein composition. J. Cereal Sci. 22, 45–51.

Joe, A.F.T.W., Summers, R.W., Lunn, G.D., Atkinson, M.D., Kettlewell, P.S., 2005. Pre-maturity alpha-amylase and incipient sprouting in UK winter wheat, with special reference to the variety Rialto. Euphytica 143, 265–269.

Kindred, D.R., Gooding, M.J., Ellis, R.H., 2005. Nitrogen fertilizer and seed rate effects on Hagberg falling number of hybrid wheats and their parents are associated with α-amylase activity, grain cavity size and dormancy. J. Sci. Food Agric. 85, 727–742.

Link, J., Batchelor, W.D., Graeff, S., Claupein, W., 2008. Evaluation of current and model-based site-specific nitrogen applications on wheat *(Triticum aestivum* L.) yield and environmental quality. Precision Agric. 9, 251–267.

Mantle, P.G., Shaw, S., Doling, D.A., 1977. Role of weed grasses in etiology of ergot disease in wheat. Ann. Appl. Biol. 86, 339–343.

Mares, D., Mrva, K., 2008. Late-maturity alpha-amylase: low falling number in wheat in the absence of preharvest sprouting. J. Cereal Sci. 47, 6–17.

Pepler, S., Gooding, M.J., Ford, K.E., Ellis, R.H., 2005. A temporal limit to the association between flag leaf life extension by fungicides and wheat yields. Eur. J. Agron. 22, 365–376.

Pepler, S., Gooding, M.J., Ellis, R.H., 2006. Modelling simultaneously water content and dry matter dynamics of wheat grains. Field Crops Res. 95, 49–63.

Ruske, R.E., Gooding, M.J., Jones, S.A., 2003. The effects of adding picoxystrobin, azoxystrobin and nitrogen to a triazole programme on disease control, flag leaf senescence, yield and grain quality of winter wheat. Crop Prot. 22, 975–987.

Smith, G.P., Gooding, M.J., 1999. Models of wheat grain quality considering climate, cultivar and nitrogen effects. Agric. Forest Meteorol. 94, 159–170.

Wan, Y., Underwood, C., Toole, G., Skeggs, P., Zhu, T., Leverington, M., Griffiths, S., Wheeler, T., Gooding, M.J., Poole, R., Edwards, K.J., Gezan, S., Welham, S., Snape, J., Mills, C., Mitchell, R., Shewry, P.R., 2009. A novel transcriptomic approach to identify candidate genes for grain quality traits in wheat. Plant Biotechnol. J. 7, 401–410.

Williamson, P.M., 1997. Black point of wheat: *in vitro* production of symptoms, enzymes involved, and association with *Alternaria alternata*. Austr. J. Agric. Res. 48, 13–19.

Chapter 19

Analysis of Grain Quality at Receival

Stephen Delwiche* **and Diane Miskelly****
*USDA-ARS, Beltsville Agricultural Research Center, Food Quality Laboratory, Beltsville, MD, United States; **Westcott Consultants, Goulburn, NSW, Australia*

19.1 INTRODUCTION

The point of receival for grain marks the end of the line for the grower, but it marks the first point in the chain that takes grain from raw commodity to the first stage in processing and beyond—a chain that often includes the country elevator, an inland terminal elevator, a mill or refinery and an export terminal. At each transfer point, a grain lot typically has documentation that asserts its quality. For the dominant grains such as wheat, maize and rice, each major exporting country (or group of countries, as in Europe) has developed standards for grade and class. Administration of these standards is performed by government, government licensee (as in the United States and Canada) or grain receival agent (as in Australia). Because more than half of a country's production of a grain may be exported, it is the export potential that drives the need for its grain standards. And despite the fact that not all grain is required to undergo a country's official inspection (e.g. domestic sales in the United States do not require it), it is these standards that become the yardstick by which grain quality is considered, even at the first point of sale.

This chapter starts by describing the official practices and standards for the some of the World's major grain exporting countries. As with the corresponding chapter in the first edition of this book, focus is given to Australia, Canada, the European Union and the United States. However, in contrast to the earlier edition and in keeping with the trends in export share, the Black Sea Countries, Russia, Ukraine and Kazakhstan are now included. Because of the openly accessible and abundant information on US grain standards and procedures, the most extensive description of that country's system is provided. The remaining countries follow suit, with space given to features that make each country's system unique. It should be noted that there are many common factors among the

Cereal Grains. http://dx.doi.org/10.1016/B978-0-08-100719-8.00019-X

standards of the major exporting countries. For each country or region, a table is provided that summarises the official standard. The chapter emphasises the grain type wheat, because it is a common crop of export for these countries. Barley is described to a lesser extent, with essentially no coverage, for sake of brevity, given to the other major cereals.

Following the description of the country systems, this chapter describes the procedures and equipment that are commonly used at receival points. It should be noted, however, that the practices are widely varying, especially because official inspection at receival points is not generally needed. The remainder of the chapter is devoted to the description of procedures and methods that either are in use at stages further along the market chain or are in the midst of development that may have application at receival. It concludes with an update on the authors' thoughts on future trends in quality assessment at receival.

19.2 THE US GRAIN INSPECTION SYSTEM

Wheat, along with other grains and oilseeds must undergo official inspection when exported. Official standards exist for barley, canola, corn, flaxseed, mixed grain, oats, rye, sorghum, soybeans, sunflower seed, triticale and wheat. The official certificate must show grade designation and weight. For domestic sales, official inspection is not required; instead, these are performed upon request by state or private entities authorised by USDA GIPSA. If the results of an inspection are in dispute, the customer has three avenues for reconsideration. First, a re-inspection may be requested of the same laboratory. Next, an appeal inspection may be requested in which the inspection is performed at an FGIS field office. Last, the disputed sample may be inspected by the final arbitrator, the Board of Appeals and Review.

Wheat is grown throughout the contiguous region of the United States, with the preponderance for export grown in the central plains, ranging from Texas north and westward to the state of Washington. Maize has a similar wide region, but to the east of the wheat region. Grain exports are handled primarily through the ports at New Orleans, Louisiana and Portland, Oregon, with a small fraction shipped through the Great Lakes and St. Lawrence River. Inland facilities consist of country elevators and large terminals. Generally, grain is delivered by truck to the country elevator (Fig. 19.1), by train car to inland terminals and mills and by train car or barge to terminal ports.

The licensing of inland facilities is administered by the USDA Farm Service Agency. The United States Warehouse Act of 2002 authorised the USDA to license operators of agricultural warehouses, such as grain elevators, with an updating of regulations for storage of grain and rice in the November 2011 release of WA-402 (USDA-FSA, 2011). This program is voluntary, with 655 grain facilities (47% of US commercial space) being licensed under this Act in 2014. Licensees agree to pay annual fees ranging from (January 1, 2006 values, unchanged through 2014) \$155 [<150 000 bu (52 900 hL) capacity]

FIGURE 19.1 **Delivery of wheat from field to first point of sale in eastern Washington State.**

to $3680 [10,000,000 bu (3.52 million hL) capacity] and agree to comply with the Act's regulations. Among the requirements for obtaining a license is that the warehouse, depending on the commodity, has personnel who are qualified to weigh, sample, inspect and grade the commodity. Although the program is voluntary (hence, giving it the designation as a 'permissive regulatory act', USDA-FSA, 2008), a warehouse may become de-listed if it fails to comply. The purpose of the Act is sixfold: (1) providing protection for depositors of agricultural products, (2) establishing a uniform regulatory system for storing such products, (3) providing confidence to financial markets on the integrity and acceptability of warehouse receipts, (4) facilitating interstate and global commerce for the marketing of these products, (5) providing safe storage of the products at reasonable rates and (6) requiring operators to accept the products without discrimination.

The regulations for licensed warehouses state that US Government standards for kind, class and grade are used when such standards exist or, in the absence thereof, applicable state or municipal standards are used. Grain that is traded for domestic use does not have to receive 'official inspection'. For wheat, maize and rice, the official US standards prevail. The large number of warehouses prevents federal inspectors from being stationed at each. Rather, inspection is performed by trained local personnel. Either due to an appeal brought on by a dispute between the grower and the elevator operator or by interest of the elevator operator, a lot may be examined by the nearest federal field office or federally licensed office for official inspection at a pre-determined cost. In such cases, a split sample (2000 g minimum) is drawn and delivered to the inspection office. The USDA Grain Inspection, Packers and Stockyards Administration (GIPSA) and specifically the branch known as the Federal Grain Inspection Service (FGIS), is the government entity charged with official inspection. FGIS maintains a national service centre in Kansas City, MO and six field offices throughout the contiguous states. Complementing these offices are delegated or designated state offices and designated private agencies (approximately 50 altogether in 2015, hereafter termed 'official agency' offices) that have been approved by GIPSA to perform official inspection.

In the United States, the primary receival stations for grain sale are the country elevators. The overwhelming majority of grain lots delivered to the country elevator undergo unofficial inspection by elevator personnel for pricing and binning decisions. Visual assessment is made of loads for cleanliness and insects, while measurements may be made for test weight, moisture content, protein content and in weather-affected regions, falling number and deoxynivalenol level. Only grain at export terminals that is destined to be sold overseas is required to undergo official inspection.

A Quality Assurance/Quality Control (QA/QC) program is administered by GIPSA to keep inspection operations and equipment in alignment, from official agency through field office and on to a national technical services facility. Written agreements are established between each official agency and field office for quality monitoring. The agreement outlines quality factors specific to the locality that should be checked. For example scab damage may be problematic in a particular region, such that the agreement will specify that at least one sample containing this condition be submitted to the field office every month for quality assurance. Additionally, samples from randomly selected lots are submitted on a daily basis for a check of some or all grading factors of an official inspection. A similar mechanism of written agreements between field offices and the national office is utilised, also with conditions for daily, monthly and bi-monthly submission of check samples that are inspected by the national office's ultimate arbitrator, the Board of Appeals and Review. Time charts and graphs are developed and the data are statistically analysed to determine whether official agencies are in alignment with their respective field office and likewise between field offices and the BAR. Details of the USDA's quality assurance—quality control program are contained in *GIPSA's Quality Handbook* (USDA-GIPSA, 1996). Specific details of GIPSA's QA/QC program for NIR calibrations are found in Section 19.8.10.

19.2.1 Official Inspection Example: Wheat

This section summarizes the procedure for official inspection and standards for wheat in the United States. The standards undergo periodic review by GIPSA, with the last full review in the early 1990s and a solicitation through an Advance Notice for Proposed Rulemaking in 2010 for stakeholder comments on whether wheat standards and grading procedures needed any changes (USDA-GIPSA, 2011) and subsequent amendments to the standards, primarily in defining contrasting classes and reducing the allowable levels for shrunken and broken kernels of the top two grades (USDA-GIPSA, 2013a), which became effective in May 2014. Barley standards (USDA-GIPSA, 1997) are currently under review, with amendments expected to become finalised in 2015 (USDA-GIPSA, 2014a). US wheat is categorised into six 'pure' classes: Durum Wheat (with three subclasses—Hard Amber Durum, Amber Durum and Durum), Hard Red Spring (HRS, with three subclasses Dark Northern Spring, Northern Spring and Red Spring) Wheat, Hard Red Winter (HRW) Wheat, Soft Red Winter (SRW) Wheat, Hard White Wheat (HWW), Soft White Wheat (SWW, with three subclasses—Soft White, White Club and Western White). There are also two hybrid classes, unclassed wheat and mixed wheat, which arise from wheat that is intentionally not identified by class or wheat that is not at least 90% pure with respect to its most dominant class. In each of the first six classes, five grades exist, starting with No. 1 as the highest in quality and progressing downward to No. 5 (USDA-GIPSA, 2014b). One additional grade exists, US Sample grade, to catch the lots that do not meet the requirements for Grade Numbers 1 through 5.

The following procedures are applicable to grain lots that undergo official inspection (USDA-GIPSA, 1995), however, these procedures are also applicable as guidelines for unofficial operations. From each load (i.e. lot) of grain delivered by truck to the country elevator, a composite sample is collected by means of sub-sampling in order to produce a quantity (2000–2500 g) that is representative of the lot. Sampling is usually accomplished by either (1) probing the truck, (2) a diverter-type mechanical sampler that intermittently catches grain from the exit stream of the truck, or (3) a cup-type or pelican sampling device that is manually used to catch grain in sets (two per truck) from the elevator's load-in system. These devices are described later in the chapter.

Before the inspection sample is divided or has dockage removed, the inspector will check for any unusual odours. This can even occur while the grain is still in the truck. Odours are grouped into three broad categories. Sour odours can arise from fermentation of the grain, insects, heated grain during respiration or an off-odour picked up from storage. Musty odours can similarly arise from heated grain and insect activity, but may also arise from moulds and the dirt picked up during harvest operations. The last category, Commercially Objectionable Foreign Odours, consists of odours that are imparted to the stored grain from external sources. These include smoke, oil products, fumigant residue, insecticide, decaying animal and vegetable matter, skunk and weed seed. Presence of a distinct odour from any of these three categories, with exception of garlic or smut odour, results in the assignment to the lowest grade, 'US Sample grade'. Before relegating a lot to this status, the work portion is exposed to air in an open container for 4 h and then re-evaluated. Other conditions that may be assessed on the wheat lot as a whole are heat, caused from excessive respiration (often accompanied by a sour or musty odour) and insect infestation. For the latter condition, stored grain pests of primary importance include the rice weevil, granary weevil, maize weevil, cowpea weevil, lesser grain borer and the grain moth. With either condition, the elevator operator may choose to reject the lot.

If the results of an inspection are in dispute, the customer has three avenues for reconsideration. First, a re-inspection may be requested of the same laboratory. Next, an appeal inspection may be requested in which the inspection is performed at an FGIS field office. Last, the disputed sample may be inspected by the final arbitrator, the Board of Appeals and Review.

It should be noted that all the 'tests' prior to dockage removal are performed by visual or olfactory analysis, with the exception of moisture content. Moisture content, reported on a wet matter basis, is measured using either a Dickey-john Grain Analysis Computer (GAC 2500-UGMA, Auburn, IL) or a Perten AM 5200-A, which are capacitive (149 MHz) devices that have undergone the National Type Evaluation Program (NTEP) accreditation process by GIPSA. Approximately 350 g of grain is used by these instruments, which possess separate internal calibrations for all of the major grain species of the United States, including wheat, corn, oats, barley, sorghum, sunflower, rice and soybean.

Wheat inspection prior to dockage removal is performed on the original sample that has been divided using a Boerner or equivalent device to produce a work sample of 1000–1050 g (Fig. 19.2). Based on visual inspection of the work sample, the sample is checked for counts in excess of various threshold levels (in parentheses), including animal filth (2), castor beans (2), crotalaria seeds (3), glass (1), insect-damaged kernels (32), stones (4) and other foreign substances (5), such that an excessive count relegates the lot to the lowest grade (US Sample grade). GIPSA defines dockage as 'all matter other than wheat that can be removed from the original sample by use of an approved device' as well as 'underdeveloped, shriveled and small pieces of wheat kernels removed in

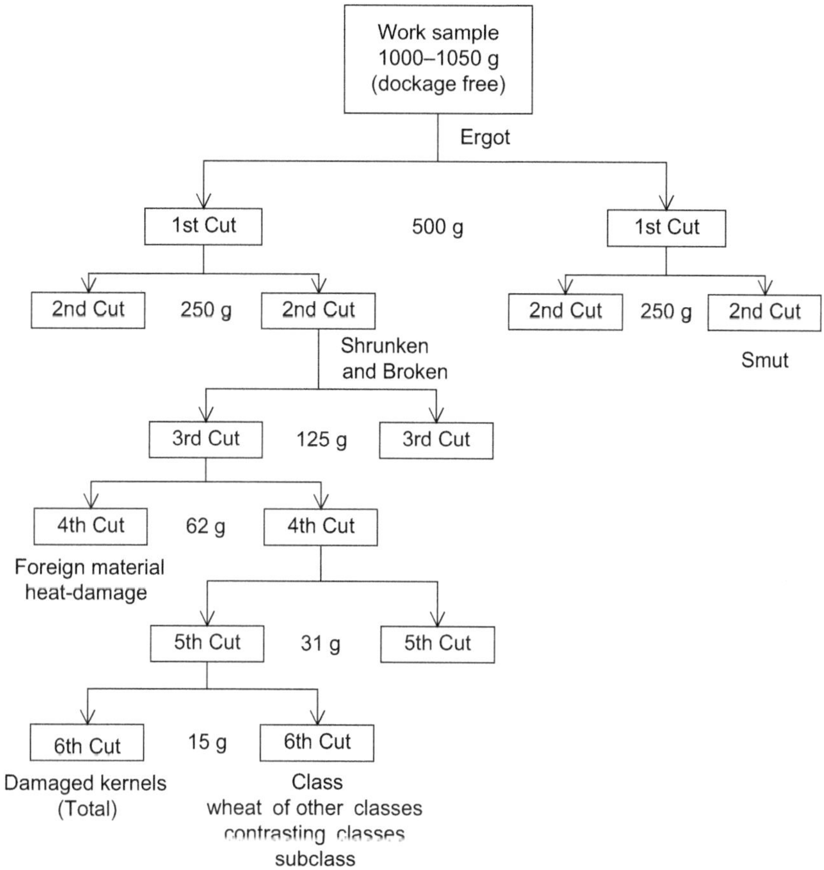

FIGURE 19.2 US Official Inspection of a work sample of wheat using a Boerner divider to reduce the material by one half with each pass. *(Adapted from USDA-GIPSA (2014c).)*

properly separating the material other than wheat and that cannot be recovered by properly re-screening or re-cleaning' (USDA-GIPSA, 2014b). In the United States, the Carter Dockage Tester is used for dockage determination. The device consists of a an inclined grate, or 'riddle' and two sieves, such that when combined with aspiration and motor-driven shaking, a clean work sample, consisting of chiefly intact millable kernels, is separated from the 'dockage' material. Dockage is determined on a 1000–1050 g portion on the tester configured with a No. 2 riddle (No. 25 for durum wheat), a No. 2 sieve in the middle and bottom sieve carriages. Dockage is the collection of material diverted by an air blower, over the riddle, except threshed and sprouted kernels, over the bottom sieve and in the bottom collection pan. The cleaned wheat is the portion that stays over the middle sieve.

The dockage-free work sample (1000–1050 g) undergoes additional steps in the inspection procedure, starting with the determination of test weight (bulk density). Test weight is measured using a longstanding procedure of pouring the grain from a fixed distance into a one-quart (0.95 L) cylindrical kettle of standard dimensions until overflowing, striking the kettle rim clean of excess grain and determining the weight of the kettle contents. With test weight determination completed, the entire amount of the work portion is recombined, visually examined for ergoty kernels, then repeatedly split by a Boerner divider to produce sub-samples for additional visual assessments. Ergot is evaluated on the undivided portion, shrunken and broken kernels [pieces that pass through a 0.064 in (1.6 mm) by 0.375 in (9.5 mm) oblong-hole sieve] or smut is evaluated on 250 g, insect-damaged kernels on 100 g and foreign material or heat-damaged kernels on 50 g. Finally, on a 15 g portion, the wheat is handpicked by moving kernels from a main pile to auxiliary piles defined by damage type, foreign material, wheat of other class, or other condition. Kernel counts or weight percentages are then determined for each pile, whereupon a grade for the lot or sub-lot is issued. Details of the maximum allowable levels of the various conditions for each grade are shown in Table 19.1. Considering that a 15 g portion of wheat may typically consist of 300–500 kernels, with each kernel separately examined, this stage of the inspection procedure is time consuming and can last 10–15 min for samples of borderline grades.

Barley in the United States falls under two classes, malting barley and barley, with each class divided into three classes. For malting barley class, the subclasses are six-rowed, six-rowed blue and two-rowed. The aleurone layer is the visual determinant for distinguishing between the six-rowed classes, with the blue colour present in the six-rowed blue subclass. For the barley class, the subclasses are six-rowed, two-rowed and barley. The third subclass catches any barley that does not meet the requirements for the other two subclasses. During official inspection (see USDA-GIPSA, 2013b), the work sample for barley is a dockage-free sample weighing between 850 and 950 g. The official standard for barley is similar to that for wheat, with specifications on test weight, damaged kernels, contamination and, in the case of malting barley, non-malting percentage (Table 19.2). Factors of garlic bulblets and Smutty are assessed on 500 g; ergot, plump barley and thin barley on 250 g; heat-damaged, heat-injured and kernel texture on 50 g; blighted, broken kernels, class/subclass, foreign material, frost-injured, mould-injured, other grains, skinned and broken kernels, sound barley and wild oats are all examined on 25 g. However, determinations of rodent pellets, bird droppings, other animal filth, broken glass, castor beans, cockleburs, crotalaria seeds, dockage, garlic, live insect infestation, large stones, moisture, temperature and unknown foreign substances are made on the whole sample. If a condition exists that is not caught in the representative sample, then the lot as a whole at the time of sampling becomes the basis for inspection. Unlike wheat, barley does not have sprout as a grading factor that would be caught in total damaged kernels. Rather, 'injured-by-sprout' is an 'official criteria' factor that is performed upon request and does not affect grade.

TABLE 19.1 Official Standard for US Wheat

	Grades US numbers				
Grading factors	1	2	3	4	5
Minimum pound limits of:					
Test weight per bushel					
Hard red spring wheat or white club wheat	58.0	57.0	55.0	53.0	50.0
All other classes and subclasses	60.0	58.0	56.0	54.0	51.0
Maximum percent limits of:					
Defects:					
Damaged kernels					
Heat (part of total)	0.2	0.2	0.5	1.0	3.0
Total	2.0	4.0	7.0	10.0	15.0
Foreign material	0.4	0.7	1.3	3.0	5.0
Shrunken and broken kernels	3.0	5.0	8.0	12.0	20.0
Total[a]	3.0	5.0	8.0	12.0	20.0
Wheat of other classes:[b]					
Contrasting classes	1.0	2.0	3.0	10.0	10.0
Total[c]	3.0	5.0	10.0	10.0	10.0
Stones	0.1	0.1	0.1	0.1	0.1
Maximum count (per 1,000 g wheat) limits of:					
Other material in 1 kg:					
Animal filth	1	1	1	1	1
Castor beans	1	1	1	1	1
Crotalaria seeds	2	2	2	2	2
Glass	0	0	0	0	0
Stones	3	3	3	3	3
Unknown foreign substances	3	3	3	3	3
Total[d]	4	4	4	4	4
Insect-damaged kernels in 100 g	31	31	31	31	31

US Sample grade is wheat that:
a. Does not meet the requirements for US Numbers 1, 2, 3, 4 or 5; or
b. Has a musty, sour, or commercially objectionable foreign odour (except smut or garlic odour) or
c. Is heating or of distinctly low quality.

[a]*Includes damaged kernels (total), foreign material, shrunken and broken kernels.*
[b]*Unclassed wheat of any grade may contain not more than 10.0% of wheat of other classes.*
[c]*Includes contrasting classes.*
[d]*Includes any combination of animal filth, castor beans, crotalaria seeds, glass, stones, or unknown foreign substance.*

Source: USDA-GIPSA, 2014b. Subpart M—United States Standards for Wheat. USDA-GIPSA-FGIS, Washington, D.C. Available from: http://www.gipsa.usda.gov/fgis/standards/810wheat.pdf

TABLE 19.2 Official Standard for Six-Rowed Malting Barley and Six-Rowed Blue Malting Barley

Grading factors	Grades US numbers			
	1	2	3	4
Minimum limits of:				
Test weight (lb/bu)	47.0	45.0	43.0	43.0
Suitable malting types (%)	95.0	95.0	95.0	95.0
Sound barley (%)[a]	97.0	94.0	90.0	87.0
Maximum limits of:				
Damaged kernels (%)[a]	2.0	3.0	4.0	5.0
Foreign material (%)	0.5	1.0	2.0	3.0
Other grains (%)	2.0	3.0	5.0	5.0
Skinned and broken kernels (%)	4.0	6.0	8.0	8.0
Thin barley (%)	7.0	10.0	15.0	15.0

[a]*Injured-by-frost kernels and injured-by-mould kernels are not considered damaged kernels or considered against sound barley.*
Source: USDA-GIPSA, 1997. Subpart B—United States Standards for Barley. USDA- GIPSA-FGIS, Washington, D.C. Available from: http://www.gipsa.usda.gov/fgis/standards/810barley97.pdf

19.3 THE CANADIAN GRAIN INSPECTION SYSTEM

Wheat (common and durum combined) is the most abundant export crop produced in Canada, capturing approximately 55% of the total tonnage of the cereals and oilseeds exported from 2013–14 crop year (CGC, 2014). For this reason, this section is primarily devoted to the issues of wheat quality. Until 2008, Canada relied on their longstanding method of wheat variety release and regulation, known as kernel visual distinguishability (KVD). This system required newly developed varieties to resemble one another within the same wheat class. By law, all varieties of wheat grown and sold in Canada were required to be registered by the Canadian Food Inspection Agency (CFIA). This led to problems with varieties being classed into the same class but having vastly different functionalities. It was also hampering wheat breeders' efforts in developing new varieties. A line with special traits such as resistance to disease would not be certified for release if it did not meet the visual appearance requirement. Conversely, an un-classed variety could sneak into the Canadian market system, just by having the visual characteristics of one of the classes. An example of this was a train shipment in 2002 that contained an excessive level of non-registered US wheat that resulted in the devaluation of the lot to feed status and the grain company's consequential loss of hundreds of thousands of dollars. Lastly, there has been the desire of growers to have high yielding non-milling varieties that can be used for feed or ethanol production (CGC, 2015a).

In 2008, Canada's replacement of the KVD system became effective for all eight western wheat classes [Canada Western Red Spring (CWRS), Canada Western Amber Durum (CWAD), Canada Western Red Winter (CWRW), Canada Western Extra Strong (CWES), Canada Prairie Spring Red (CPSR), Canada Prairie Spring White (CPSW), Canada Western Soft White Spring (CWSWS) and Canada Western Hard White Spring (CWHWS)]. The new system carries the name, 'Variety Eligibility Declaration' or VED. With the successor system, sellers are required to certify that a wheat lot consists of varieties that are on the approved list of varieties for that particular class. An additional class has been added to cover varieties for non-milling purposes, such as for ethanol production or for feed. While these varieties still require registration by the CFIA, the requirements for registration are limited to disease and agronomic variety considerations.

Up until 2012, the western provinces of Canada were required to sell their wheat and barley to the Canadian Wheat Board, which ran as a single desk system that established price schedules for these commodities and was the marketing agency for downstream sales. The single desk system terminated in August 2012, though the Board, which became simply CWB, continued to operate as a government-sponsored organisation until April 2015, when private ownership occurred for a 50.1% majority stake.

Legislation known as the Canadian Grain Act requires that elevators and grain dealers be licensed by the Canadian Grain Commission (CGC). Each licensee must provide security to the CGC so that in the case of a dispute in price, the producer will get fair payment for his grain after adjudication. The exception to the rule of licensing may occur if (1) the grain is not purchased directly from western grain producers, or (2) licensing a company is not required to maintain the quality, safe-keeping and orderly, efficient handling of grain, in which case an exemption is granted by either the Governor in Council (as defined by the Canadian Grain Act, CGC, 2015b), or the CGC itself. Exempted companies include elevators that ship only produce cars, agents acting on behalf of CGC-licensed companies, seed cleaning plants that do not purchase grain from producers, seed dealers that purchase seed from producers, elevators operated by non-profit organisations, feed mills, bulk handling facilities at port locations with no storage and container loading facilities located at ports. Other grain companies are outside the jurisdiction of the Canada Grain Act. These include feed lots, hog barns, standard farming operations who exclusively market their own grain products, eastern grain elevators and dealers who exclusively buy and sell grain produced in eastern Canada.

The licensed primary elevator bases the price of the wheat lot paid to the producer on the class, grade, moisture and dockage of the load. A document called a 'cash purchase ticket' is presented by the operator to the producer that is an entitlement of the holder of the document to payment, with purchase price stated on the document. The producer may dispute the price by having the elevator operator draw a sample representative of the load that is agreed upon by both parties. The 1 kg sample is sent to the CGC for an official inspection, with cost of the inspection, by agreement, assigned to either the producer or the operator. The difference in cost is then rectified upon return of the official inspection results. The inspection is completed within 5 days of receiving the sample. For the 2015 harvest the cost of an inspection is 48.51 CD. Appeals may be made within 15 days of the official inspection, whereupon the chief grain inspector for Canada performs the analysis and, if the grading result is different, a revised certificate is issued. The decision is final and not subject to further appeal or review by any court.

In addition to the nine western wheat classes, the CGC regulates and grades wheat grown in the eastern region of the country, though this wheat is primarily used for domestic consumption, with exports of eastern wheat accounting for about 2% of the tonnage of all Canadian wheat exported in the 2013–14 crop year (CGC, 2014). Divided into eight classes [Canada Eastern Red (CES), Canada Eastern Red Spring (CERS), Canada Eastern Hard Red Winter (CEHRW), Canada Eastern Soft Red Winter (CESRW), Canada Eastern Amber Durum (CEAD), Canada Eastern White Winter (CEWW), Canada Eastern Soft White Spring (CESWS) and Canada Eastern Hard White Spring (CEHWS)], these wheats also have grading standards.

Although the grading of wheat is generally similar among the classes in terms of the procedures and equipment used, the allowable levels of contaminants and defects for each grade are class dependent (CGC, 2015c), as seen in Table 19.3. The grading procedure is heavily reliant on human visual analysis by trained CGC inspectors, in which percentages by count or weight are assessed on specific conditions of contaminants (e.g. ergot, excreta, matter other than cereal grains, Sclerotinia and stones), wheat of other classes and damage (e.g. stained, immature, degermed, fireburnt, green or pink from immaturity, sawfly, midge, shrunken and broken, smudge and blackpoint and sprouted). In western wheat, insect damage (usually sawfly or midge) is generally less common than heat- and frost-damage. Two to five unique grades exist for the western classes, with grade No. 1 in each class being of highest quality. Only three grades exist for each of the eight eastern classes. Both western and eastern classes generally have a bottom class designated as ‘Feed’ for lots that do not meet the lowest grade within each class. In addition to the visual analyses, the equipment-based devices for grading include test weight apparatus and for some of the western classes, an instrument for protein content determination (usually performed by NIR). Protein content is reported on a 13.5% moisture basis.

Grade standards are further defined for wheat that has reached a port terminal in anticipation of overseas shipment. Essentially, the grade criteria remain the same with additional requirements on the tolerable levels of broken grain and slightly more stringent criteria on the total level of foreign material. The reason for the added stringency is to safeguard the cleanliness reputation of Canadian wheat.

TABLE 19.3 Official Standard for Canada Western Red Spring (CWRS) wheat

	Standard of quality							Foreign material			
Grade name	Minimum test weight kg/hL (g/0.5 L)	Variety	Minimum hard vitreous kernels %	Minimum protein %	Degree of soundness	Ergot %	Excreta %	Matter other than cereal grains %	Sclerotinia %	Stones %	Total %
No. 1 CWRS	75 (365)	Any variety of the class CWRS designated as such by order of the Commission	65	10	Reasonable well matured, reasonably free from damaged kernels	0.04	0.010	0.2	0.04	0.03	0.6
No. 2 CWRS	72 (350)	Any variety of the class CWRS designated as such by order of the Commission	No minimum	No minimum	Fairly well matured, may be moderately bleached or frost-damaged, reasonably free from severely damaged kernels	0.04	0.010	0.3	0.04	0.03	1.2
No. 3 CWRS	69 (335)	Any variety of the class CWRS designated as such by order of the Commission	No minimum	No minimum	May be frost-damaged, immature or weather-damaged, moderately free from severely damages kernels	0.04	0.015	0.5	0.04	0.06	2.4

(*Continued*)

TABLE 19.3 Official Standard for Canada Western Red Spring (CWRS) wheat (*cont.*)

	Standard of quality							Foreign material			
Grade name	Minimum test weight kg/hL (g/0.5 L)	Variety	Minimum hard vitreous kernels %	Minimum protein %	Degree of soundness	Ergot %	Excreta %	Matter other than cereal grains %	Sclerotinia %	Stones %	Total %
CW Feed	65 (315)	Any class or variety of wheat excluding amber durum and General Purpose	No minimum	No minimum	Reasonably sweet, excluded from other grades of wheat on account of damaged kernels	0.1	0.030	1	0.1	0.1	10
Grade, if specs for CW feed not met	Wheat, Sample CW Account Light Weight					Wheat, Sample CW Account Ergot	Wheat, Sample CW Account Excreta	Wheat, Sample CW Account Admixture	Wheat, Sample CW Account Admixture	2.5% or less—Wheat, Rjected grade, Account Stones Over 2.5% - Wheat, Sample Salvage	See Mixed grain

Grade name	Wheats of other classes or varieties[a]		Artificial stain, no residue %	Dark, immature %	Degermed %	Fireburnt %	Fusarium damage %	Grass green %	Grasshopper, army worm %	Heated	
	Contrasting classes %	Total %								Binburnt severely mildewed rotted, mouldy %	Total %
No. 1 CWRS	0.75	2.3	Nil	1	4	Nil	0.25	0.75	1	0.005	0.05
No. 2 CWRS	2.3	4.5	0.05	2.5	7	Nil	0.8	2	3	0.020	0.4
No. 3 CWRS	3.8	7.5	0.10	10	13	Nil	1.50	10	8	0.030	1.0
CW Feed			2	No limit	No limit	2	4	No limit	No limit	2.5	2.5
Grade, if specs for CW Feed not met	Over 10% amber durum—and/or General Purpose Wheat, Sample CW Account Admixture		Wheat, Sample CW Account Stained Kernels			Wheat, Sample CW Account Fireburnt	Wheat, Sample CW Account *Fusarium* damaged over 10%—Wheat, Commercial Salvage			Wheat, Sample CW Account Heated	

(Continued)

TABLE 19.3 Official Standard for Canada Western Red Spring (CWRS) wheat (*cont.*)

				Shrunken and Broken[b]			Smudge and blackpoint		Sprouted	
Grade name	**Natural stain %**	**Pink %**	**Sawfly, midge %**	**Shrunken %**	**Broken %**	**Total %**	**Smudge %**	**Total %**	**Severely sprouted %**	**Total %**
No. 1 CWRS	0.5	1.5	2.0	4	5	7	0.30	10	0.10	0.5
No. 2 CWRS	2	5	5	4	6	8	1	20	0.20	1.0
No. 3 CWRS	5	10	10	4	7	9	5	35	0.30	3.0
CW Feed	No limit	No limit	No limit	No limit	13	No limit within broken tolerances	No limit	No limit	No limit	No limit
Grade, if specs for CW Feed not met					Sample Broken Grain					

[a] *See working tolerance for 'Wheats of Other Classes or Varieties'*
[b] *See truncation rule for 'Shrunken and Broken'*
Source: CGC (2015c)

Determination of dockage consists of starting with a representative portion (≥900 g) of an official sample, in which the portion is generated from a Boerner-type divider. Canada defines dockage as 'any material intermixed with a parcel of grain, other than kernels of grain of a standard of quality fixed by or under this Act for a grade of that grain, that must and can be separated from the parcel of grain before that grade can be assigned to the grain' (CGC, 2015a). It includes wheat with long rootlets, un-threshed wheat heads and material other than wheat removed by a No. 25 riddle. Also, dockage includes material removed by a No. 5 buckwheat sieve, material removed by aspiration and soft earth pellets handpicked from a 'clean' sample upon operation of a Carter dockage tester. Unofficial samples (≥750 g) can also be assessed for dockage using the same procedure. After dockage removal, the 'cleaned' sample is analysed for additional contaminants and damage in order to determine grade. In this case the percentages by weight are based on the cleaned sample. The amount of grain that undergoes visual inspection is dependent on the contaminant or damage condition under scrutiny. Official inspection of wheat by the Canadian Grain Commission is mandatory for wheat delivered to terminal elevators but is optional for wheat delivered to the primary elevator. Under Section 30 of The Canada Grain Act (CGC, 2015b) official inspection for grain may be made at any place where provisions for inspection have been made. The CGC furnishes staff members to collect *official samples* and defines these separately from unofficial samples, in which the latter group may be collected by non-CGC members. For official samples, the CGC specifies that a mechanical sampler be used unless the physical structure of the sampling location prohibits the use of a mechanical sampler. An official sample for grade and dockage assessment must be at least 1 kg in size and depending on the location of collection in the market stream and the lot's next intended recipient, the sample is retained for a minimum period lasting from 20 days to 6 months. For unofficial sampling, the actual method of sampling varies with the elevator. For each load delivered, about 1 to 2 kg of wheat is drawn by one of three main techniques: a pneumatic probe driven into the truck bin, a manual scoop used during truck unloading and in rare circumstances, a diverter sampler.

Although there are no requirements on the models of quality assessment instruments at the primary elevator, most elevators possess a Carter Dockage tester, sets of hand sieves and a capacitive moisture meter. Further, most will also have an NIR instrument for protein content determination. Less abundant are elevators with mycotoxin test kits, such as those used in DON determination. The CGC provides ELISA-based DON testing on a mail-in basis, in which farmers or elevator operators pay $50 Canadian for each one-kilogram sample sent to a service centre or regional office.

At some elevators falling number testing may be performed for evidence of sprouting, though falling number is not a grading criterion. Rather, sprouting for grade determination is assessed by visual examination. Recently, the CGC has been pilot testing the Rapid ViscoAnalyser as an indicator of pre-harvest sprouting, with the reasons of a shorter analysis time and use of less water compared to the falling number method.

19.4 THE AUSTRALIAN GRAIN INSPECTION SYSTEM

19.4.1 Wheat

Wheat is the largest crop grown in Australia, with 5 year average production to 2014–15 of 25.8 million tonnes (ABARES, 2015). Production can be quite variable, depending on the seasonal conditions. Approximately 75% of the crop is available for export, with the remainder for domestic use as food, feed, industrial and seed. Major export markets for Australian wheat are South-east Asia, North Asia and Africa/Middle East. Exports are mainly in bulk.

Australian wheat is grown in the eastern, southern and western regions where the rainfall is approximately 300–600 mm. This crescent-shaped area is known as the wheat-belt. Most wheat varieties are spring wheat types, although sown in autumn and early winter months. Harvesting commences in the northern areas in September and progresses south with harvesting generally completed by late December/early January.

From 1939, the Australian Wheat Board (later privatised to AWB International) was the owner of all wheat produced in Australia. AWB was the single desk and the various bulk handling authorities in each State classified the wheat according to AWB receival standards and stored it until ready for delivery to the domestic market or loaded onto a vessel at the shipping terminal. This arrangement remained unchanged until the domestic market was deregulated in 1989, followed by the export market in 2008.

Since export market deregulation, the former bulk handlers have established themselves as major wheat marketers. There are six large bulk exporters plus about 12 other smaller traders and exporters. The market has seen considerable change with the entry of overseas marketers and major investment in port loading facilities. There has been an increase in container export over the past 5 years, with container trade accounting for 10–15% of total wheat exports.

The various wheat classes in Australia have evolved as marketing categories or grades of commercial wheat. The classes differ in wheat quality parameters, inherent processing characteristics and suitability for intended end-use. The hard wheat milling classes are Australian Prime Hard (APH), Australian Hard (AH), Australian Premium White (APW), Australian Standard White (ASW) and Australian General Purpose (AGP). Currently APH is available only in the traditional growing areas of NSW and Queensland. The largest classes are APW (about 35% of the crop) followed by AH. Australian Durum (ADR) is mainly grown in the high protein areas of northern NSW and Queensland.

There are two soft wheat classes. Australian Noodle Wheat (ANW) is mainly grown in Western Australia, where it is blended before export. Australian soft (ASFT) on the east coast is mainly for domestic consumption, although a small quantity is also grown in WA and available for export.

Wheat must be white grained to be eligible for the milling wheat classes and red wheats are automatically categorised as Feed wheat along with downgraded milling wheat which has been affected by seasonal conditions and has quality defects or high levels of contaminants.

New wheat varieties developed by wheat breeding companies must undergo classification before release. Classification is the term used to describe the process by which new wheat varieties are categorised into one of the declared classes of Australian milling wheat, which is a surrogate for attributing potential market value. This was formerly an 'industry good' function performed by AWB in the interests of the overall industry and was one of many functions that were an integral part of being the single desk marketer. After deregulation, this responsibility was taken on by a new entity, Wheat Quality Australia (WQA). WQA is an independent not-for-profit company owned by the Grains Research and Development Corporation and Grain Trade Australia and is responsible for wheat variety classification and maintaining the Wheat Variety Master List of variety x maximum classification for each classification zone. An expert group, the WQA Variety Classification Panel, compares the quality of a new line with established control varieties. Considerations include milling extraction, flour quality, rheological properties and end-product quality (bread types, noodles, biscuits, pasta). Breeders submit a minimum of 3 years multisite quality results from an accredited laboratory. A maximum classification is awarded in a classification zone which relates to a defined production area. In practice, breeders will usually target the APH, AH, APW, ANW, ADR and ASFT milling classes.

A class may contain several different binning grades, depending on the season. When a grower delivers wheat to a receival point, a combination of variety and environment determines the grade that is allocated. Grain standards are set by Grain Trade Australia (GTA), an industry trade organisation which reviews grain standards on an annual basis. The standards describe the qualities of various grades of commodities including wheat, barley, maize, oats and sorghum. They are primarily intended to be trade standards, but are generally adopted as receival or outturn standards by storage providers. That is, a grade is awarded based on variety plus assessed physical grain quality of a grower's load. Table 19.4 shows the major classes and grades for Australian wheat.

On delivery, a representative sample is drawn using a manual or vacuum probe in a vertical position with at least one probe sample being taken from the front, middle and rear of each bulk unit or trailer. A minimum sample size of 3 L is drawn for deliveries of up to 10 tonnes, with an additional litre taken for each additional 10 tonnes in the load. The wheat is composited and sub-sampled for subsequent testing for moisture, protein, test weight, screenings and un-millable material, defective grains, foreign seed contaminants and other contaminants. Moisture and protein content are determined using NIR and the instruments must be approved for trade purposes as payment is made to growers on the basis of protein content. Although the maximum allowable moisture content

TABLE 19.4 Main Physical Properties of Classes and Grades of Australian Wheats

Class	Australian Prime Hard (APH)	Australian Hard (AH)	Australian Premium White (APW)		Australian Standard White (ASW)	Australian Noodle Wheat (ANW)	Australian Soft (ASFT)	Australian durum (ADR)	Australian General Purpose	Australian Feed
Grade	APH2	H2	APW1	APW2	ASW1	ANW1	SFT1	ADR1	AGP1	FED1
Moisture %, maximum	12.5	12.5	12.5	12.5	12.5	12.5	12.5	12.5	12.5	12.5
Protein[a] (N × 5.7) %	13.0 min	11.5 min	10.5 min	10 min	—	9.5–11.5	9.5 max.	13.0 min	—	—
Test weight kg/hL, minimum	76	76	76	76	76	76	76	76	68	62
Screenings[b] %, maximum	5.0	5.0	5.0	5.0	5.0	5.0	5.0	5.0	10.0	15.0
Unmillable material[b] %, maximum	0.6	0.6	0.6	0.6	0.6	0.6	0.6	0.6	1.2	2.6
Falling number seconds, minimum	350	300	300	300	300	300	300	300	200	—
Vitreous kernels %, minimum	—	—	—	—	—	—	—	80	—	—

[a] *11% moisture basis.*
[b] *2.0 mm slotted screen.*
Source: Grain Trade Australia. Wheat Trading Standards. Available from: www.graintrade.org.au

is 12.5%, most deliveries are within the range of 10–11%. Protein content in Australia varies from 8% to 16% and is expressed on an 11% moisture basis. A minimum test weight of 76 kg/hL is required for classification into the major milling grades, although most deliveries are well above the minimum. Screenings are defined as the un-millable material passing through a 2.0 mm slotted screen after 40 shakes using an approved shaking device. Screenings include small, broken and cracked grains and small foreign seeds. Screenings must be below 5% by weight to be eligible for the main milling grades. Un-millable material above the screen is material other than wheat kernels and includes whiteheads, backbone and some large seedpods. Screenings (as measured in Australia) differs from dockage, which is the term often used in other countries. Dockage includes all levels of impurities including foreign matter, weed seeds, stones and small, shriveled and broken grain.

If grain defects are detected on visual inspection, 300 grains are assessed from the top of the 2.0 mm screen. Defects include stained, green, frost, heat, over-dried and insect damaged grain, field fungi and take-all affected grain and smut. There is a nil tolerance for sprouted grain and falling number will be determined on a load by load basis where sprouted grain is detected or in adverse harvest weather conditions where there has been pre-harvest rainfall. Also in this category, a minimum vitreous kernel count is required for ADR, but is not required for other grades.

Foreign seed contaminants fall into 11 categories and the standards refer to the maximum number allowable per half litre. Foreign seed contamination in Australia is generally at relatively low levels. Other contaminants is the last grading category and includes ryegrass ergot, insects, snails, loose smut, sand and earth. There is a nil tolerance for live insects and pickled grain.

Reference methods for the above tests are provided in the Wheat Trading Standards and visual standards for common defects, impurities and pests are provided online (www.graintrade.org.au). The sequence for sampling, testing and delivery is shown in Fig. 19.3 (A–D).

Since 1996, all new wheat varieties released in Australia carry an end point royalty (EPR). The EPR is set by the wheat breeder or variety owner and is paid by the grower on a tonnage basis at the first point of sale. The money collected is returned to the relevant wheat breeder to enable research and the development of new varieties suited to the Australian environment. EPR varieties now account for more than 85% of the total wheat grown.

Prior to export market deregulation, AWB was responsible for market promotion and market research. The Australian Export Grains Innovation Centre (AEGIC) was established in 2012 by the Grains Research and Development Corporation (GRDC) and the Department of Agriculture and Food, Western Australia. AEGIC activities include research and development, quality testing, economic and supply chain analysis, yield and production forecasting and promotional activities. Research is undertaken on improving quality and functional attributes of exported grains.

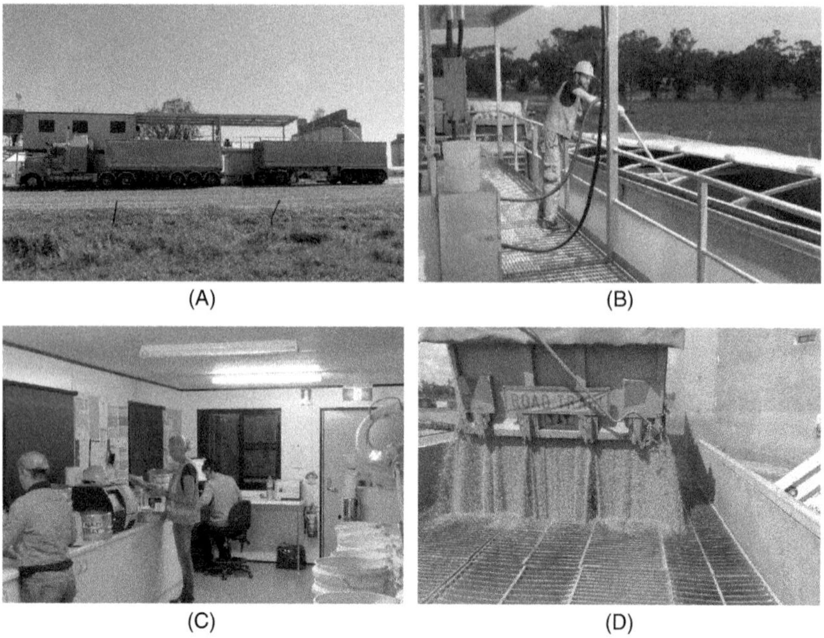

FIGURE 19.3 Harvest testing and receival in Australia. (A) The grain truck stops at the silo sampling stand, (B) grain is sampled using a vacuum probe, (C) inside the silo sampling stand showing testing underway and (D) grain is discharged for conveying to the appropriate bin. *(Photos by E. Williamson.)*

Much has changed since deregulation of the export single desk, but this has brought opportunities for many, as well as the need for the new marketers to understand classification, quality testing at receival and quality requirements of markets.

19.4.2 Barley

Barley is the second major cereal crop after wheat and is grown in the same geographic area as wheat. Average production for 5 years to 2014–15 was 8 million tonnes (ABARES, 2015). Australia is a net barley exporter, with about 65% exported and the remainder used on the domestic market for malting, feed and seed. The major export markets for Australian barley are China and the Middle East.

Barley is classified into Malt, Feed or Food grades. Depending on the season, about 30% of receivals are Malting grade. New malting barley varieties are accredited by Barley Australia and evaluated for domestic and export suitability through an expert panel comprised of technical malting and brewing experts. Malting barley varieties are tested for β-glucan, free alpha amino nitrogen and fermentability, as well as micromalting characteristics like extract, enzymes and wort viscosity.

When barley is delivered after harvest, it is tested according to that season's GTA Barley Trading Standards and a grade will be awarded after full assessment and varietal declaration. A representative sample is drawn using a manual sampling probe or vacuum probe and the method of compositing and sub-sampling is the same as already described for wheat. Moisture and protein content are determined by NIR and the protein content is reported on a dry basis. A minimum test weight of 65 kg/hL is required for classification as Malt 1. Retention is the grain retained above a 2.5 mm slotted screen after 40 shakes on the same approved device as for wheat. The minimum retention for Malt 1 is 70% w/w. Screenings is the percent by weight of material passing through a 2.2 mm slotted screen and includes small seeds, broken and small grains. A sliding scale for retention and screenings determines which grade is awarded (Table 19.5).

The sample is assessed visually and if defects are detected, they are determined as maximum tolerance per 100 grain sample (compared with 300 grain sample for wheat). Defective grains include sprouted and shot grain, for which there is a nil tolerance. If these are detected, or in the case of pre-harvest rainfall, falling number is tested and the minimum is 300 s for Malting and Food grades. Other defects include dark tipped grain, field fungi affected, skinnings, cleaved, broken, frost damaged, dry green, heat damaged, bin burnt or storage

TABLE 19.5 Main Physical Properties of Australian Malting and Feed Barleys

	Malting		Feed	
	Malt 1	Malt 2	Feed 1	Feed 2
Varietal Purity % by count, minimum	95	95		
Moisture %, maximum	12.5	12.5	12.5	12.5
Protein[a] % (N × 6.25), minimum	9.0	9.0	—	—
Protein[a] % (N × 6.25), maximum	12.0	12.0	—	—
Test weight kg/hL, minimum	65.0	65.0	62.5	60.0
Retention[b] %, minimum	70.0	62.0	—	—
Screenings[c] %, maximum	7.0	10.0	15.0	25.0
Germinative energy %, minimum	95	95	—	—
Falling number seconds, minimum	300	300	—	—

[a]*Dry basis.*
[b]*Remainder above 2.5 mm slotted screen.*
[c]*Matter passing through 2.2 mm slotted screen.*
Source: Grain Trade Australia. Barley Trading Standards. Available from: www.graintrade.org.au

mould affected grain which have similar tolerances for all Malting grades, but relaxed or no limit for Feed grades (Table 19.5).

Foreign seed contaminants fall into similar categories as wheat, although there are some differences in tolerances. Foreign cereal grain contamination is allowable in Malt 1 barley up to 85 grains per half litre.

In the other contaminants category, there is nil tolerance for smut, ergot, pickled grain, live stored grain insects and objectionable material. Full details and tolerances are given in the GTA Barley Trading Standards.

The varietal purity for Malting barley must be at least 95% and this is determined on the basis of grain morphology. Malting barley is segregated on the basis of a single variety, offering advantages for some buyers. Details of the eligible varieties can be found on the Barley Australia website (www.barleyaustralia.com.au). Food barley has the same grade standards as Malting barley and is a varietal-specific grade. All malting, food and feed varieties are two-row barleys.

Malting barley is required to maintain a minimum of 95% germinative energy and bulk handlers ensure that this is maintained during storage.

19.4.3 Future Trends for Australian Grain

The replacement of subjective visual inspection of grain at receival by objective assessment is high on the priority list for bulk handlers. This is a complex process, as multiple grain defects and weed seeds can be present in individual loads. Development of image analysis instruments such as the EyeFoss (see Section 19.9.1) have now reached the stage where they have been calibrated to predict visual grain inspection parameters such as black point and common weed seeds. A large scale project is underway and new applications are being developed in partnership with Australian bulk handlers and Curtin University.

19.5 THE EUROPEAN UNION (EU) GRAIN INSPECTION SYSTEM

Although it is not apparent that national rules are imposed on cereals sales at first point of sale in the EU, there does exist a price support or ‘intervention’ system for some of the major grain species (EC, 2008). Hence, these cereals that are taken over by intervention agencies must meet standards ratified by the Commission of the European Communities (EC). As with Canada and Australia, there has been a movement in the EU during the past several years towards less government regulation on the marketing of grain. The longstanding (since the 1960s) Common Market Organisation (CMO) for cereals, in its newest form, is described by EU Regulation 1308/2013, a document released in 2013 that replaced four regulations (EEC No 922/72, EEC No. 234/79, EC No 1037/2001 and EC No 1234/2007) dating back to 1972 (EU, 2013). The effect of this legislation has been to allow the trade of cereals to occur on more of a free market basis. This has been met by a lessening of a reliance on intervention. Initially available at fixed prices and no limits for wheat (common or durum), corn, barley, rye and sorghum, the intervention system for these commodities has been removed for sorghum or reduced to effectively zero for all other cereals except common wheat. A producer may offer up to 80 tonnes of EU-grown common wheat to the intervention agency. If the offer is admissible, the producer is instructed on when and where to deliver the wheat. Minimum standards for grain traded under the intervention program, including common wheat, durum wheat, barley, maize and sorghum, as defined in EU's regulation 742/2010 are presented in Table 19.6.

The standards address many of the same concerns of Australia, Canada and the United States, namely damaged grain (broken, shriveled, discoloured, overheated, sprouted, decayed and insect-damaged), contaminants (extraneous seeds, husks, ergot, dead insects and insect fragments, other cereals and extraneous matter), maximum moisture content, minimum protein content (durum and common wheat) and test weight. As with the standards of the other exporting countries, the EU standards for contaminants and defects are reliant on human visual analysis. Additional criteria include a minimum falling number (durum and common wheat) and Zeleny sedimentation volume (durum only).

19.5.1 United Kingdom

In the United Kingdom, the Home-Grown Cereals Authority (HGCA) is a non-profit cooperative that works in conjunction with the national government to establish industry policies for production, distribution and processing of cereals. The HGCA released a white paper in 2004 that provided outlines for establishing standards for grain testing laboratories in the U.K. (HGCA, 2004). Although such proposed standards are not necessarily targeted for the first point of sale, but rather for analytical laboratories servicing the cereals industry, the listed items are indicative of the perceived importance of quality factors in the United Kingdom. From this paper, the two main commodities of interest are wheat and barley. For wheat, the following tests are recommended: moisture content, test weight, protein content, falling number and hardness by SKCS (single kernel characterisation system—see Martin et al., 1993). For barley, the tests include nitrogen content, moisture content, test weight, germinative capacity and a sieving profile.

In 2004, HGCA Exports released a branding system for United Kingdom wheats to be sold on the export market, which represented almost 20% of production. Two export classes were established: ukp and uks. The ukp class has requirements for slightly higher quality indices than uks, specifically (for typical export specifications) in protein (11–13% vs. 10.5–11.5%), minimum test weight (76 vs. 75 kg/hL) and minimum falling number (250 vs. 220 s). Growers having the intention to supply the export market can select varieties from a recommended list. Placement on the list is dependent on the

TABLE 19.6 Minimum Requirements for Grain Traded Under Public Intervention in the European Union

	Durum wheat	Common wheat	Barley	Maize	Sorghum
A. Maximum moisture content (%)	14.5	14.5	14.5	13.5	13.5
B. Maximum percentage of matter which is not basic cereal of unimpaired quality (%)	12	12	12	12	12
1. Broken grains (%)	6	5	5	5	5
2. Grain impurities (%)	8.5	7	12	5	5
2.1 Impurities other than mottled grains (%)	5	7	12	5	5
(a) shriveled grains (%)	X	X	X	n.a.	n.a.
(b) other cereals (%)	3	X	5	X	X
(c) grains damaged by pests (%)	X	X	X	X	X
(d) grains in which the germ is discoloured (%)	X	X	n.a.	n.a.	n.a.
(e) grains overheated during drying (%)	0.50	0.50	3	0.50	0.50
2.2 Mottled grains (%)	3.5	n.a.	n.a.	n.a.	n.a.
3. Sprouted grains (%)	4	4	6	6	6
4. Miscellaneous impurities (%)	4.5[a]	3	3	3	3
of which:					
(a) extraneous seeds					
— noxious (%)	0.10	0.10	0.10	0.10	0.10
— other	X	X	X	X	X

(b) damaged grains					
— grains damaged by spontaneous heating or too extreme heating during drying (%)	0.05	0.05	X	X	X
— grains affected with fusariosis (%)	1.5	X	X	X	X
— other	X	X	X	X	X
(c) extraneous matter	X	X	X	X	X
(d) husks (cob fragments in the case of maize)	X	X	X	X	X
(e) ergot (%)	0.05	0.05	n.a.	n.a.	n.a.
(f) decayed grains	X	X	n.a.	n.a.	n.a.
(g) impurities of animal origin	X	X	X	X	X
C. Maximum percentage of wholly or partially non vitreous grains (%)	27	n.a.	n.a.	n.a.	n.a.
D. Maximum tannin content[b] (%)	n.a.	n.a.	n.a.	n.a.	1
E. Minimum specific weight (kg/hL)	78	73	62	n.a.	n.a.
F. Minimum protein content (%, dry basis)	11.5	10.5	n.a.	n.a.	n.a.
G. Hagberg falling number (seconds)	220	220	n.a.	n.a.	n.a.
H. Minimum Zeleny index (mL)	n.a.	22	n.a.	n.a.	n.a.

n.a., Not applicable, not requiring analysis; X, analysis required without specific limit but content to be taken into account for maximum limits set in points 2 and 4 of the table.
[a]*of which maximum 3% for impurities other than grains affected by fusariosis.*
[b]*as a percentage of dry matter.*
Source: EU, 2010. Commission Regulation (EU) No. 742/2010 of 17 August 2010 amending Regulation (EU) No 1272/2009 laying down common detailed rules of the implementation of Council Regulation (EC) No. 1234/2007 as regards buying-in and selling of agricultural products under public intervention. Off. J. Eur. Union, L217, pp. 4–11.

Alveograph measurements, particularly W and P/L. Exported wheat, including durum, averaged slightly more than 2 million tonnes per year over a 10 year period (2005–14).

In June 2015, the Home Grown Cereals Authority (HGCA) became part of the Agriculture and Horticulture Development Board (AHDB). The AHDB is a board that is mandated by the national government but funded by growers and others in the supply chain of agricultural commodities and is independent of both government and private industry.

19.5.2 France

The leading producer and exporter of cereals in Europe is France. In fact, France's wheat export tonnage in the 2014 crop year was third in line, behind the United States and Canada but ahead of Australia, with the north African countries being the largest buyers of French wheat. The French have traditionally divided their wheat into four groups, with the strongest one labelled as corrective, followed by superior, standard and lastly one for other uses. Factors such as protein content, Alveograph W and falling number determine what group a wheat lot falls under.

19.5.3 Germany

Germany is the second largest producer of cereals in Europe, behind France, with wheat being the most abundant, followed by barley, corn and rye at approximately one half, one quarter and one fifth the production of wheat. The quality groupings for wheat are similar to the French system, with a revamping in 1995 by the introduction of an elite group (E) for upward correction of lesser quality lots and for export followed by the existing groups A (formerly the best), B (bread wheat of all types), K (for cookies) and C (catch all). Formerly decided mainly by the volume of the baked product, the revamped system also includes subjective assessments of dough elasticity and surface quality, falling number, protein content, sedimentation value, water absorption and milling yield (Seibel, 2005). New variety releases are assigned to their respective groups by the Bundessortenampt (The Federal Plant Variety Office), the governmental agency that manages the Beschreibende Sortenliste, a comprehensive description of all German agricultural varieties (FPVO, 2015). The Bundesanstalt für Landwirtschaft und Ernährung (BLE), the regulatory body of Germany's federal agency for agriculture and food regulates and administers issues of storage, market intervention and monitors market trends.

19.6 THE ARGENTINE GRAIN INSPECTION SYSTEM

The development and implementation of quality standards for wheat and maize in Argentina is less advanced than other major exporting countries, such as Australia, Canada and the United States (Table 19.7). Complicating factors for

TABLE 19.7 Official Standard for Argentine Bread Wheat

Grade	Minimum test weight per hectolitre (kg/hL)	Percent maximum limits of								Sweet clover seeds/ 100 g
		Foreign material (%)	Damaged kernels: Heat damaged kernels (%)	Damaged kernels: Total (%)	Smutty kernels (%)	Yellow berry kernels (%)	Shrunken and broken kernels (%)	Insect bored kernels (%)	Moisture (%)	
1	79.0	0.20	0.50	1.00	0.10	15.0	0.50	0.50	14.0	8
2	76.0	0.80	1.00	2.00	0.20	25.0	1.20			
3	73.0	1.50	1.50	3.00	0.30	40.0	2.00			

Living insects and/or arachnids free for all grades. Protein content: basis 11% (moisture basis of 13.5%)
Source: Argentine Wheat, 2011. Argentine standard for wheat: New resolution Ex-SAGPyA 1262/04. Available from: <http://www.trigoargentino.com.ar/2011/ENG/texts/TRIGOPAN_CALIDAD_2011ENG.HTML>

wheat and maize especially include the strong intervention of the national government that regulates and in some years, eliminates the export of these two major cereal grains for supposed stabilisation of domestic prices of foods derived from them. Consequently, growers have been putting less land in wheat and maize, while increasing production of sorghum, barley and soybean, which do not fall under government control.

In recent years Argentina has dropped from being a big player in the wheat export market. By USDA statistics, Argentina slid from fourth place in 2006 to tenth place in 2013. Restrictions by the national government in the form of export taxes, which are 23% for wheat and export quotas, put in place to ostensibly protect the domestic consumer from runaway food prices, resulted in deincentivising growers to plant wheat, with exports dropping from 13 million tonnes in 2011 to 3 million tonnes in 2013 (USDA-NASS, 2013).

Beginning in 1998 there was a coordinated effort to introduce a variety classification system in Argentina, with a view to segregating the better quality varieties from the bulk, and establishing quality premiums in the domestic market. With enactment of a national resolution (No. 334) establishing a wheat quality program known by the acronym PRONACATRI (Programa Nacional de Calidad del Trigo), release of wheat germplasm by quality grouping has been practiced since 2004. This was driven by the Instituto Nacional de Tecnología Agropecuaria (INTA, the government's agricultural research organisation), grower groups and some private storage operators. As a result, all new varieties are required to pass through a coordinated trial program to enable them to be classified, they must be registered with the National Registry of Varieties and the commercialisation of new and certified seed must be authorised by Argentina's Secretariat of Agriculture, Livestock, Fishing and Food (SAGPyA).

Varieties are classified into three groups using seven test variables (from most to least important, based on their impact on processing quality and value): Alveograph strength (W), loaf volume, protein and gluten content, flour yield-to-ash content ratio, Farinograph dough stability, wet gluten content, protein content and test weight. The weighted sum of these variables, with each defined using a 0–9 scale, produces what is known as a Quality Index for bread wheat for each variety. Group 1 cultivars are extra strong wheats suitable for blending and baking. Group 2 cultivars are wheats adapted to traditional methods of baking (artisan breads of more than 8 h of fermentation time). The last division, Group 3, consists of cultivars suitable for direct methods of baking (artisan breads of less than 8 h of fermentation time).

The Argentine governmental agency, El Servicia Nacional de Sanidad y Calidad Agroalimentaria (SENASA), has developed quality standards for bread wheat and maize for the export market. Both wheat and maize are divided into three grades. Upon receival, elevator personnel perform initial assessment of grain. Moisture content is typically measured by capacitive instruments and protein content by NIR instruments. Lots that are determined to be of the highest grade (Grade 1) require no additional quality testing at receival. Those determined to be in either of the two lower grades (Grades 2 and 3) are required to be evaluated by a non-profit inspection company (Camara Arbitral de Cereales, or CAC) at the expense of both buyer and seller. Only in rare circumstances, when disagreement of the CAC evaluation persists between buyer and seller, does SENASA step in for a final evaluation.

SENASA's inspection standards for grade are largely based on visual inspection of a representative sample. In addition to stated limits for test weight, the other grade-determining factors fall into the broad categories of foreign material, shrunken and broken kernels and damaged kernels (Argentine Wheat, 2011). The latter category for wheat is divided into heated kernels, smut, yellow-berry, shrunken and broken and insect-bored. Shrunken and broken kernels are defined as whole or pieces of kernels that pass through a sieve having 1.6 mm by 9.5 mm slots. Although protein content is not a grade-determining component, it is included in the standard to indicate whether the price of the wheat is increased (> 11.0% protein = 2% bonus) or discounted (< 11.0% protein = 2% discount; < 10% protein = an additional 1% discount; < 9% protein = an additional 1% discount). Similarly, the standard includes the economic bonus or discount for wheat that is graded above or below the middle grade level (Grade 2). For the most recent year available (2011), Grades 1 and 3 were 1.5% higher and 1.0% lower in respective value than the middle grade.

19.7 THE BLACK SEA COUNTRIES' GRAIN INSPECTION SYSTEM

Ukraine, along with Russia and Kazakhstan, has increased its world market share in wheat exports since the 1990s. USDA Economic Resource statistics for 2013 (most recent year available of USDA's annual Agricultural Statistics) indicate that Ukraine exported 10 billion tonnes of wheat, corresponding to 6.2% of the World's trade. This is contrasted with the USA's share of 19.7% for the same year. If the two other Black Sea countries are included, the share rises to 21.0%. Based on late season projections for the 2015 harvest, world market share of wheat exports from these three countries has risen to 26.8% (USDA-FAS, 2015).

19.7.1 Ukraine

Ukraine releases their wheat standard translated into English. Known as DSTU 3768:2010 (Ukraine, 2010), the specifications for wheat arose from the former USSR's GOST standards, which are now standards adopted in whole or part by Russia and many of the independent countries (including Ukraine). As with the EU, Ukrainian wheat is divided into common and durum types, with separate grade specifications for each. It is not clear whether grading is required in domestic sales or just in exported wheat. Generally, wheat for feed purposes is exported in a larger amount than milling quality wheat. For common wheat, there exist two groups, A and B, of which the first group is used primarily in flour milling and baking, while the second group is used to some extent in foods but more typically in non-food and feed uses. The lowest quality grade of Group B, Grade 6, is used to catch the wheat falling out of specification of the other five grades, similar to Canada's General Purpose class. A similar specification structure, with five grades, is in place for durum wheat. Although allowances for damaged grains are generally higher for Ukrainian wheat, the specifications contain criteria that are absent in the other countries, such as wet gluten content (absent in Canada and US), protein content (absent in the US) and falling number (absent in Canada and US). The specifications for common wheat's six groups are contained in Table 19.8. Much of the inspection is performed by visual analysis, starting with sieving a test portion of 250 g with smaller openings first (1.0 × 20.0 mm), whereupon the overs are received, this time with larger openings (2.0 × 20.0 mm) to pass small and broken grains. Although not stated in the DSTU standard, it is assumed that the overs on the second sieve are then visually inspected and picked for unsound wheat kernels or non-wheat.

19.7.2 Russia

Russia's wheat exports accounted for 10.8% of the world's exports in 2013 (USDA-NASS, 2013), with projections in late 2015 translating into a 14.7% share. The proximity of the wheat growing areas of south western Russia to the Black Sea and to a lesser extent the Caspian Sea allow for the transport of wheat and other grain by ship to countries along the Middle East and Northern Africa. Availability of standards for Russian wheat in a language other than Russian is very limited. The applicable wheat standard is GOST 9353-90. The standard predates the breakup of the Soviet Union and with its close resemblance to the Ukranian standard, the GOST standard is assumed to be the basis for both Ukranian and Kazakhstan standards.

19.7.3 Kazakhstan

Kazakhstan, with 4% of the world's wheat export market, is the smallest of the Black Sea exporters despite a very large portion of the country's cropland, reported at 60% (USDA-FAS, 2014), dedicated to this commodity. However, because of the country's landlocked status which makes export of grain expensive and the government's recent policy of encouraging the production of other grains and oilseeds for animal feed with the aim of becoming an exporter of meat products, wheat has seen a 17% decrease in acreage over a 5 year period from 2009 (USDA-FAS, 2015). Of the 2014 year wheat export, more than 85% was traded to neighboring countries, including Afghanistan, Azerbaijan, Iran, Kyrgyzstan, Tajikistan and Uzbekistan. Exporting grain to more distant places requires the movement by boat from the Caspian Sea to Black Sea and is limited by the capacity of the Volga-Don shipping canal. Grain standards for Kazakhstan are not readily available via the Web, though are presumed to be similar to the former Soviet GOST standard.

TABLE 19.8 Ukrainian Common Wheat Grade Specifications.

	Description and limits for common wheat groups and grades[a]					
	A			B		
Parameters	**1**	**2**	**3**	**4**	**5**	**6**
Test weight minimum (g/L)[b]	760	740	730	710	690	NL
Vitreous kernels minimum (%)	50	40	NL	NL	NL	NL
Moisture content maximum (%)	14.0	14.0	14.0	14.0	14.0	14.0
Grain impurities maximum (%)	5.0	8.0	8.0	10.0	12.0	15.0
In particular						
Broken grains	5.0	5.0	5.0	GI	GI	GI
Grains of cereals	4.0	4.0	4.0	4.0	4.0	GI
Sprouted grains	2.0	3.0	3.0	4.0	4.0	GI
Waste impurities maximum (%)	1.0	2.0	2.0	2.0	2.0	5.0
In particular						
Mineral impurities, out of which:	0.3	0.5	0.5	0.5	0.5	1.0
Pebble, slag, ore	0.15	0.15	0.2	0.15	0.2	MI
Spoiled grains, out of which:	0.3	0.3	0.5	0.3	0.5	1.0
Fusarium affected grains	SG	SG	SG	SG	SG	SG
Harmful impurities, out of which:	0.2	0.2	0.2	0.2	0.2	0.5
Smut and ergot	0.05	0.05	0.05	0.05	0.05	0.1

(Continued)

TABLE 19.8 Ukrainian Common Wheat Grade Specifications. (*cont.*)

Parameters	Description and limits for common wheat groups and grades[a]					
	A			B		
	1	2	3	4	5	6
Trichodesma incanum	nil	nil	nil	nil	nil	nil
Corn-cockle	HI	HI	HI	HI	HI	HI
Each of any toxic seeds	0.05	0.05	0.05	0.05	0.05	0.1
Smutty grains maximum	5.0	5.0	8.0	5.0	8.0	10.0
Protein content minimum (% db)	14.0	12.5	11.0	12.5	10.5	NL
Wet gluten content minimum	28.0	23.0	18.0	NL	NL	NL
Gluten quality group	I–II	I–II	I–II	NL	NL	NL
Gluten quality range (IDK units)	45–100	45–100	20–100	NL	NL	NL
Falling number minimum (s)	220	180	150	150	130	NL

[a]*Abbreviations: GI, In limits for grain impurities (see DSTU 3768:2010 §3.5.2); HI, in limits for harmful impurities (see DSTU 3768:2010 §3.2.3); MI, in limits for mineral impurities (see DSTU 3768:2010 §3.2.1); NL, not limited; SG, in limits for spoiled grains (see DSTU 3768:2010 §3.2.4).*

[b]*1 kg/hL = 10 g/L*

Source: Ukraine, 2010. National Standard of Ukraine, Wheat: Technical specifications. DSTU 3768:2010.(Trans. from Ukrainian).

19.8 TRADITIONAL EQUIPMENT USED IN WHEAT INSPECTION AT RECEIVAL

Getting a representative sample of the grain lot is essential. Depending on the size of the grain receival facility, equipment can range from hand-held probes to programmable pneumatic samplers. Detailed discussion of sampling devices is beyond the scope of this chapter. However, common to all sampling plans is the need to ensure that the sample is representative of the lot or sub-lot from which it was drawn. For wheat in the United States, this typically entails gathering either a separate 2000 g work sample for every 500 bu (176 hL) sub-lot or, in the usual case of country elevator deliveries, a work sample that represents each truckload. The USDA-GIPSA sampling procedures have been described in Section 19.2.1. Alternatively, the International Standards Organisation (ISO) has developed recommendations for static sampling from a wide range of storage and conveyance devices, including bags, wagons and trucks (ISO 13690, 1999). For truckloads containing less than 15 tonnes of grain, five probe sites in a pattern are recommended. Similarly, ISO has developed a standard for automatic sampling by mechanical devices (ISO 6644, 2002), such as diverters, tubular samplers and cup samplers. With these devices, the recommended collected amount is broadly specified as 5 kg for lots up to 500 tonnes.

19.8.1 Sampling Devices

Probes are made of brass or aluminum consisting of two tubes, one inside the other. Eleven to twelve regularly spaced compartment windows are placed along the 1.5–1.8 m length of the probe assembly (Fig. 19.4). After insertion of the probe vertically in the truck, grain flows into the probe when the outer and inner tubes are twisted to align the windows. In the United States, recommendations exist for collecting a sample by probe. For a typical flat-bottom truck or trailer, the probe is inserted at three regularly spaced distances along the length of the truck and approximately 0.5 m from the inner wall of one side, with an emptying of the probe between insertions. Another three probings are made along the opposite side, as well as one in the centre of the truck. Combined, the grain sample should total at least 2000 g. Additional probing may have to be made for loads that are suspected to be of inferior quality, in which case the new sites are staggered with respect to the original ones.

Diverters are mechanical samplers that are mounted near the stream of the grain exiting the truck, at the end of a conveyor belt, or at the head of an elevator leg. Periodically, the diverter, or 'pelican' (typically 1.9 cm × 2.2 cm opening) moves across the stream of grain at a typical speed of 0.5 m/s, capturing a small portion of grain. The cycle time or period of the diverter depends on the flow rate of the grain (for specifications, see USDA-GIPSA, 2003), but usually ranges between 12 and 72 s. A secondary stage within the sampler may be used to return a portion of the diverted grain back to the main stream. Sampling

FIGURE 19.4 **Manual sampling of a flat-bottom truck in eastern Washington State.**

occurs throughout the duration of movement of the load, with a targeted sample mass of at least 2000 g.

Cups or 'pelicans' are manual sampling devices, known as an Ellis cup or a leather pelican pouch, that are placed in the stream of the moving grain. The sampler is required to periodically draw a set (equals three Ellis cupfuls or one 'pelican' cut) of grain from the stream every 500 bushels. For trucks, this translates into at least two sets, with a set typically representing 500 bushels. Similar to the other sampling devices, the amount of grain for inspection should be at least 2000 g.

19.8.2 Human Visual Analysis

Visual inspection of the grain lot as a whole, as well as careful inspection of a representative sample, be it officially or unofficially, has been, and for the near future continue to be, the overarching method of grain inspection at receival. Instruments such as digital imaging devices with pattern recognition algorithms have not been able to replace manual visual analysis, though field tests in Australia in 2016 with the EyeFoss instrument (see Sections 19.4.3 and 19.9.1) may demonstrate that imaging technology will soon be ready to supplant visual analysis. The reason why the challenge is so great lies in the diversity of conditions to be examined, which include broken and shriveled kernels, immature kernels, heat-damaged and frost-damaged kernels, staining, insect-damaged kernels, stored grain insects, ergot, mould-damaged kernels, foreign seeds, foreign matter, non-threshed grains, sprouted kernels and kernels of other classes.

The ability to discern the particular damage condition or defect for official analyses requires an extensive training program, as well as a check system to ensure alignment of field offices. In the United States, as explained in the Section 19.2.1, formal agreements exist between designated or delegated inspection offices and GIPSA field offices and likewise between the field offices and the national laboratory, that specify the types and number of check samples exchanged. Similar mechanisms exist for the other grain-exporting countries. Additionally, Australian, Canadian and American programs are reliant on a collection of digital images of normal, diseased and defective grains that are used as references.

19.8.3 Bulk Density (Test Weight)

These devices consist of a cylinder of standard dimension that receives grain when filled and levelled in a manner prescribed by the standard and weighed, whereupon the bulk density is determined by direct reading from a dedicated weighing scale or by calculation. The precision of this procedure is highly dependent on the level of adherence to the filling and levelling steps. In the United States, test weight is reported in English units as the number of pounds per Winchester bushel, with values ranging from ≥ 60.0 for No. 1 HRW to ≥ 50.0 for No. 5 grade of HRS. For Canada and elsewhere, test weight is reported on a kilogram per hectolitre basis and uses a cylinder of 0.5 L capacity typically (versus a one-quart kettle for US official analysis). Direct conversion between English and metric test weight values are to be made with caution due to the effects that container geometry and filling procedure have on the packing density of the kernels. For example the USDA-GIPSA specifies the following conversion formula when changing non-durum type wheat from pounds per bushel to kilograms per hectolitre: kg/hL = lb/bu × 1.292 + 1.419 (USDA-GIPSA, 2013d).

19.8.4 Moisture Content

Knowledge of grain moisture content is extremely important for reasons of avoiding spoilage from microbial activity (too much moisture), avoiding unsafe dust levels within the elevator (too little moisture) and having accurate knowledge of the concentrations of nutritional components of grain. Although moisture content is not generally a direct factor in determining wheat grade, there are requirements on the maximum moisture content allowed to avoid spoilage. Further, the concentrations of constituents such as protein content are adjusted to a fixed moisture level (e.g. for wheat: 12.0% in the United States,

13.5% in Canada, 0.0% in the EU). Because of this, moisture level can be a factor in calculating the sale price of the grain at receival. For grain that is traded on an 'as-is' weight basis, moisture content directly affects the sale price of the lot. Although moisture content can be easily measured by air-oven procedures, such as AACC International Approved Method 44-15A (130°C, 1 h, AACC 2015), primary receival points generally use faster methods that are based on either dielectric properties or the NIR response at water absorption bands at 950 nm or 1940 nm. Original dielectric devices such as the Motomco 919 m required separate algorithms for each cereal species. With a recent design by USDA FGIS, the frequency is substantially higher than the older design (149 MHz vs. 18 MHz), which, along with specifications provided on capacitive plate design, permits the use of a universal algorithm (relating the dielectric constant of the material between flat parallel plates to the moisture within the material and applying a correction factor for bulk density) that can be applied to all cereals. The two instruments described in Section 19.2.1 have gone through the US National Type Evaluation Program (NTEP), as described in Publication 14 of the National Conference on Weights and Measures (NCWM). USDA GIPSA has developed calibration accuracy criteria that are more restrictive than NTEP requirements, in which the maximum deviation on average between the dielectric meters and the air oven reference method is ±0.15 percentage units of moisture (USDA-GIPSA, 2015a).

19.8.5 Sieves and Dockage Tester

Essential to the determination of grain value is an accurate assessment of the amount of useable material within the lot compared to the total mass of material delivered. For wheat, this can be thought of as the amount of sound, mature, disease-free, undamaged millable kernels. By use of sieves, screens and aspiration, either through manual or automated mechanical operations, such as the Carter dockage tester, the percentage of useable material is determined. For United States wheat, dockage becomes a factor in determining whether a lot falls into the Sample grade category. For the higher US grades, the percentage of dockage is reported on the grade certificate. Screen settings for wheat are described in Section 19.2.1. For Canadian wheat, the amount of dockage serves as an estimation of the amount of material that must be removed by cleaning, because Canadian wheat for export must be 'commercially clean' of this type of material. Otherwise, Canadian wheat that is not commercially clean cannot be exported without special permission of the Canadian Grain Commission.

19.8.6 Protein Content

Most elevators possess a near-infrared (NIR) instrument that is configured in one of two formats: reflectance measurement of ground meal (original method) or transmittance measurement of whole kernels in bulk (newer method).

The latter method is often preferred because it does not require a sample to be ground and permits several hundred grams of grain to be analysed, as opposed to ground meal reflectance instruments that necessitate the use of a divider to obtain a representative sample (20–50 g) for grinding, packing an optical cell and cleanup of the cell afterward. Calibrations for protein content and other constituents of interest (e.g. moisture, oil) are often furnished by the instrument manufacturer or are developed and distributed by cooperatives or government offices. Nearly all NIR procedures are considered as secondary methods. Thus, a primary method, such as Kjeldahl or combustion for protein content, must be available at a central facility to keep a calibration in alignment through periodic checks. Because of the ease of use of NIR instruments, their accuracy (standard errors for protein content typically less than 0.20 percentage units when checked against the reference method) and rapidness (1–2 min per sample), this methodology for grain protein analysis has been adopted worldwide.

19.8.7 Wheat Hardness

The texture or hardness of wheat endosperm is critical in determining the suitability of wheat for various end products. Historically, wheat hardness was evaluated by particle size index (PSI), defined as the weight portion of a sample of flour passing through a very fine (i.e. No. 200) sieve. Starting in the 1980s, NIR reflectance spectroscopy has been used to measure hardness, first as a method that correlated to the PSI method (Williams and Sobering, 1986) and later as a primary method (Norris et al., 1989) approved by AACC International (Approved Method 39.70.02, AACC, 2015). In the latter method, the NIR reflectance instrument must be calibrated using a set of ten reference material samples, RM8441A, available from the US National Institute of Standards and Technology. The basis for the NIR method lies in dependency of the intensity of diffusely reflected light on the flour particle size distribution. This indirect method of light scatter, calculated from a measurement of the responses at two wavelengths (1680 nm and 2230 nm) of non-specific absorption, is effective in defining wheat hardness on a continuum scale, with hardness typically ranging from about 20 for the very soft wheats to 100 or slightly higher for durum wheats.

Starting in the early 1990s, an alternative method for wheat hardness was developed by the USDA-ARS in Manhattan, KS. An electrical/mechanical device, known as the single kernel characterisation system (SKCS) was developed that measured (1) the forces involved in crushing an individual kernel, (2) the size and mass of the kernel and (3) its D.C. conductance (an indicator of moisture content), whereupon a hardness coefficient, based on a scale similar to that of NIR hardness, was calculated (Martin et al., 1993). A minimum of 300 kernels are evaluated in order to characterise the distribution of kernel hardness of a sample. The technology for the SKCS was licensed to Perten instruments (Huddinge, Sweden) and the improved instrument (Model 4100) has been commercially available for several years. To date, the instrument has been acquired mainly by central government inspection laboratories and wheat breeding and genetics improvement laboratories. Cost and instrument complexity have precluded its wide adoption by wheat receival sites. SKCS performance may be checked with the NIST RM8441A samples and in fact as of 2011, the NIST documentation on these samples is now officially applicable to just the SKCS procedure. However, because the SKCS procedure uses the hardness scale that was originally developed for the NIR procedure involving ground meal (AACC International Method 39.70A), the NIST samples may also be used unofficially with the NIR procedure described in Norris et al. (1989).

19.8.8 ELISA Test Kits for Mycotoxins

The development of enzyme-linked immuno-sorbent assay (ELISA) test kits for cereal analysis has been underway for more than 20 years. The two most common applications are kits for mycotoxins (aflatoxin, deoxynivalenol, fumonisin, ochratoxin, T-2 toxin and zearalenone) and transgenic maize and soybean varieties. Despite the widespread occurrence of *Fusarium* head blight in all regions of the world that produce wheat and barley, elevators generally do not routinely test for mycotoxin levels, such as deoxynivalenol (DON), unless there is known problem in a particular crop season. Instead, operators will rely on the general appearance of the kernels in the work sample during visual inspection. In special circumstances, operators may use commercially available test kits for DON analysis in-house or may deliver samples to outside analytical laboratories on a fee basis. In the United States, the USDA-GIPSA's national centre in Kansas City, MO, will also conduct DON testing, either on a separate fee basis or as a non-grade determining criterion (similar to protein content) in an official inspection of wheat, barley, oats and maize. GIPSA has an approval program for rapid test kits (USDA-GIPSA, 2015b), for which manufacturers may receive approval. As of summer 2015, GIPSA's list of active approvals includes 24 aflatoxin kits, 12 DON kits, 5 fumonisin kits, 2 ochratoxin kits and 2 zearalenone kits. Similarly, there are 10 kits for biotechnology traits such as glyphosate-resistant (aka Roundup Ready) corn and soybeans, Cry2Ab (*Bacillus thuringiensis* or Bt) maize and LibertyLink (alternative to Roundup Ready) soybeans. Aflatoxin testing is mandatory in the United States for all exported maize.

19.8.9 Falling Number

The processing quality of wheat and barley declines sharply with increased levels of enzymatic activity brought on by pre-harvest sprouting. All visual inspection procedures include cues for detection of sprouted kernels; however, the level of α-amylase activity may be very high even though the kernel has no outward appearance of sprouting. This condition almost always occurs during rainy weather conditions occurring between the middle stages of grain development and harvest time, with susceptibility being a combination of genetic and environmental factors that are still the subject of research (Mares and Mrva, 2014). In rarer circumstances and particularly with white wheat, enzymatic activity may also be heightened by an effect known as late maturity α-amylase (LMA), which is a genetic defect that causes bread and durum wheats to synthesize the enzyme in the aleurone 20–30 days after anthesis (Mares and Mrva, 2008). Although not as prevalent as pre-harvest sprouting, the LMA condition is troublesome because visual recognition of affected kernels is not possible. In circumstances when the elevator operator knows that such conditions have occurred, an order may be placed for numerically characterising the effect. The two instruments available for indirectly testing α-amylase activity are the Falling Number (Hagberg) instrument and the Rapid ViscoAnalyser (RVA). In both methods, the starch within a meal-water mixture undergoes gelatinisation during heating. The α-amylase catalyses the hydrolysis of starch, which results in a reduction in starch paste viscosity. The older of the two methods, the Falling Number (Hagberg, 1960), simply reports the time needed for a machined stirring rod, including a 60 s period of up and down movement, to descend a fixed distance through a starch paste slurry contained in a precision test tube immersed in water at boiling temperature. Generally, two tubes are run simultaneously on portions from the same sample and the mean time is recorded as the falling number. The RVA, developed in Australia in the 1980s, is a specialised rheometer that tracks the torsional forces imparted on a paddle as it stirs a meal-water mixture at a prescribed rotational speed and heating regime. Although the Falling Number procedure is more universally used for sprout damage assessment, the RVA may be better adapted to conditions where the abundance of water (as needed for the first instrument) is limited, or when processing characteristics of starch and water mixtures are desired.

Falling number tests at primary receival points, such as the elevator, are usually not performed unless there is reason to believe that wheat or barley from a particular region was subjected to wet weather prior to harvest. Though frequently required on export sales contracts, falling number is not a requirement for official inspection in some countries (e.g. Argentina, Canada and the United States), or it may supersede a count by visual assessment for percentage of sprouted kernels in other countries (e.g. Australia), with required minimum values ranging from 200 to 350 s, depending on the grade. The EU regulations on qualifying grain lots for intervention agencies specifies a minimum falling number of 220 s for both durum and common wheat. In the United States and Canada, primary receival points will often set a cutoff of 300 s, below which a discount is levied on the wheat delivery. US federal and state laboratories officially perform FN testing according to FGIS Directive 9180.38 (USDA-GIPSA, 2013c), which provides criteria for equipment and operation beyond that specified in AAAC International Approved Method 56-81.03.

19.8.10 NIR Spectroscopy

Without question, NIR spectroscopy has been the most important 'breakthrough' in the last 50 years for rapid assessment of grain quality. Originating with the work of Karl Norris and co-workers of the USDA Agricultural Marketing Service (later with the USDA-ARS) in the 1950s, in which a rapid method was desired to monitor moisture gain and loss in five pound sacks of flour on the grocery shelf, this technology has evolved with the development of multivariate regression analysis methodologies. With thousands of applications published in journal articles, NIR technology is presently found in industries well beyond the cereals industry (e.g. petroleum, pharmaceutical, biomedical), though the food and feed industries are probably the ones that have most readily embraced it.

An explanation of the principles of NIR spectroscopy is the subject of numerous texts (Osborne et al., 1993; Williams and Norris, 2001; Burns and Ciurczak, 2001; Siesler et al., 2002) and is therefore avoided in this chapter. NIR research on cereals was reviewed by Delwiche (2004). Therefore, this section is devoted to how NIR spectroscopy is utilised in grain quality programs, using the official program of the United States as an example. Details of this program are provided in USDA-GIPSA's *Near-Infrared Transmittance Handbook* (USDA-GIPSA, 2006), which is available at the agency's website (http://www.gipsa.usda.gov/fgis/handbook/nirt_inspec.aspx). Through its field offices and approved agency offices, GIPSA maintains a network of more than 100 NIR instruments. Surrounding each of these offices are the country elevators and private analytical laboratories that unofficially (but legally) tie into the GIPSA network through the open access to the agency's calibration equations. According to the United States Grain Standards Act (Sections 800.125 and 800.135), the constituents of wheat, barley, maize and soybean that are typically requested in sales contracts, namely contents of protein, oil (maize and soybean) and starch (maize only), are not considered as grade-determining factors, but rather as 'official criteria'. Nevertheless, these factors are commonly used in establishing price, thus making the need for their accurate measurement critical. Fortunately, NIR methodology is well suited for such measurements. Largely through consolidation of the NIR grain equipment manufacturers, one instrument series, the Foss Infratec Grain Analyzer (Hillerød, Denmark), became the de-facto standard whole grain instrument in the United States and elsewhere throughout the world. However, in the past several years, other manufacturers, notably Perten Instruments and Unity Scientific (Brookfield, CT, USA), have successfully marketed comparable whole grain NIR instruments for use in networked operations.

Whole grain NIR analysis typically consists of collecting multiple spectral measurements of successive columns of grain passing through an inspection chamber. Essential to the accuracy and precision of the NIR instrument are (1) the diversity of the set of samples used in development of the calibration, (2) the accuracy of the primary reference method, (3) the stability of the NIR instrument and (4) the consistency of the operator in loading and operating the instrument. Generally, the first two requirements are the domain of a central laboratory, such as GIPSA's national centre in Kansas City. Because of the 20+ years of sampling and calibration development and the fact that primary methods such as combustion protein analysis are highly precise [RSD <1% (Bicsak, 1993)], central control of the first two requirements is well in order. For the remaining requirements, GIPSA monitors the instruments in its field offices and designated agencies by requiring that samples be submitted on a weekly basis to its central laboratory in Kansas City, MO. Typically, this consists of five samples for each grain species that is being inspected during that period, with two of the samples being the lowest and highest values encountered. These samples are then evaluated on a master instrument at the central office. Control charts are made that track the five-sample average difference between field and master instruments, as well as one that tracks the range difference (sample with greatest positive difference minus sample with most negative difference). Tolerance limits are prescribed for three conditions: (1) the average or range difference for the given week exceeding an 'absolute limit', (2) the differences for the given week and the preceding week exceeding a 'tolerance limit' and (3) the differences for four of five consecutive weeks exceeding a 'run limit'. When any of these limits are exceeded, corrective action must be taken immediately.

Additionally, each remote office performs a daily check of the NIR calibration for each commodity and constituent that it intends to analyse. This consists of running a set (i.e. standard reference set) of four to six check samples distributed by the central laboratory. Tolerance levels must be met for the constituent of interest, either on the basis of a single run's deviation from the established (baseline) value, the difference between duplicate runs, or the average deviation when all runs and duplicates are combined. Taking wheat protein content as an example, these levels are ±0.40%, ±0.20% and ±0.10% for the single run deviation, duplicate difference and combination deviation, respectively. Additional restrictions are placed on the deviations when compared to those of daily checks of the preceding two weeks (see *GIPSA NIRT handbook*). When a tolerance level is exceeded, a 'bias' correction is made to the constant term in the calibration equation by subtraction of the combination deviation.

19.9 NEW TECHNOLOGIES FOR USE AT GRAIN RECEIVAL

Current practices of grain inspection at receival continue to be heavily reliant on human visual analysis. To alleviate the time and labour demands that inspection requires, research has been active in the development of objective techniques that assist or replace traditional inspection operations. Underlying the successful adoption of these techniques at the country elevator will be the need for ruggedness, reliability, simplicity in use, speed and accuracy. NIR spectroscopy, first introduced to the grain industry in the 1970s, is probably the best example of a technology that transformed the grain trade. NIR whole-kernel transmittance is now used extensively in the measurement of protein content in wheat in national inspection programs, as well as in commerce. New technologies are sought with the same potential for adoption at receival sites as NIR spectroscopy. The technologies of active research are described in the following paragraphs.

19.9.1 Digital Imaging

Research on image analysis of grains dates back more than 25 years. Early research by the USDA-ARS in Manhattan, Kansas in the 1980s dealt with the identification of kernel morphological (size and shape) features that could be used to discriminate between the US wheat classes (Zayas et al., 1986), particularly during a time when a soft red winter wheat variety, 'Arkan', which had the appearance of a hard red winter wheat, was released and caused confusion in the wheat trade industry (Zayas et al., 1985). Similar work on classification was underway in Canada (Sapirstein et al., 1987; Symons and Fulcher, 1988), with the use of colour analysis soon following (Neuman et al., 1989a,b). The early work on grain imaging is reviewed in Sapirstein (1995). Since then, research has advanced into the detection of moulds (Ruan et al., 1998; Luo et al., 1999), image processing of touching kernels (Shatadal et al., 1995), kernel texture (Majumdar and Jayas, 2000) and kernel vitreosity (Xie et al., 2004). Inspection of brown rice has benefited from the combination of morphological feature extraction, colour reflectance and attenuation of transmitted light through single kernels in a dual camera system that is capable of sorting rice into thirteen quality states (Wan et al., 2002).

Commercial imaging instruments dedicated to grain inspection were tried since the early 2000s, starting with the Foss Cervitec and DuPont Canada's Acurum, but with limited success. Instrument cost and speed have been hindrances for their placement in receival points. The latest image based inspection system is the EyeFoss by Foss. Released in 2013, the EyeFoss has a 3-D laser and colour camera designed to identify defects (stained, frosted, sprouted and moulded) and non-wheat or non-barley seeds, particularly for the Australian market. The manufacturer claims that the system is capable of assessing 10,000 kernels in 4 min. The system was field-tested on almost 9,000 deliveries during the 2014–15 harvest at 7 CBH (Australia) receival sites in parallel with current defect testing and monitored using an external audit process. It was found that 99.4% of loads were correctly assessed by the EyeFoss, that is required no change to the grade as assessed by the sampler. The number of trial sites was increased to 13 during the 2015–16 harvest using improved calibrations and analysis of results is expected to be available during 2016. When tested on lentils for seed size characteristics (length, mass, bulk density and seed size index, an index that characterises the distribution of sizes, traditionally performed by sieving), this system produced close correlation with physical testing methods (LeMasurier et al., 2014).

19.9.2 Hyperspectral Imaging

The combination of NIR spectroscopy and digital imaging, called hyperspectral imaging, is presently a very active area of research in food and agriculture, as summarised in an early review by Gowen et al. (2007) and a more recent review by the same principal author focused on bacterial contamination (Gowen et al., 2015). Spectral readings, typically in the visible to the short wavelength region of the near-IR (400–1700 nm) are collected at every pixel within the image. Although this places a much larger demand on computer storage and processing, the hyperspectral image holds the promise of advantages in the detection of subtle spectral changes along the kernel surface undetectable by visual analysis through the simultaneous processing of images at more than one wavelength. Applications of this technology on wheat include use in classification (Choudhary et al., 2009; Mahesh et al., 2011), determining purity or cleanliness (Wallays et al., 2009; Fernández Pierna et al., 2012) and detecting *Fusarium* (Delwiche et al., 2011; Shahin and Symons 2011, 2012; Serranti et al., 2013; Barbedo et al., 2015), insect damage (Singh et al., 2010a,b), sprout damage (Singh et al., 2009; Xing et al., 2010), kernel discolouration from black point, field fungi, or pink stain (Berman et al., 2007) and hard vitreous kernels in durum wheat (Shahin and Symons 2008; Serranti et al., 2013). Because of enormity of hyperspectral image files, or 'hypercubes' and the accompanying intensive computer processing demands, hyperspectral imaging traditionally has been reserved as an analytical tool for scientific exploration. However, the research gleaned from hyperspectral imaging may result in the development of simpler one-to-three wavelength multispectral systems such in the application of detecting *Fusarium* damaged wheat kernels (Jallais et al. 2015).

19.9.3 ELISA Test Kits for Insect Activity

In addition to mycotoxin and transgenic variety testing by ELISA (see Section 19.8.8), researchers have explored the assessment of stored insect contamination. As an alternative to manual counting of insect fragments in the filter residue of acid-hydrolysed flour [with a US Food and Drug Administration (FDA) longstanding action level limit set at 75 fragments per 50 g flour (US Food and Drug Administration, 2014)], an immunoassay technique was developed for quantifying the level of stored grain insects, based on its sensitivity to myosin (Kitto, 1991). A collaborative laboratory study was conducted that compared this technique with other insect detection methods (X-ray analysis, cracking and flotation and insect fragment count), with results indicating that the ELISA technique had the highest precision (Brader et al., 2002). Follow-up research indicated that the concentration of detectable myosin declined with storage (for instance, after fumigation), such that the ELISA procedure could underestimate the level of insect infestation (Atui et al., 2007). Other research on the red flour beetle (*Tribolium castaneum*) demonstrated the ability of immunoassay techniques to detect stored grain insect contamination even after removal of the insects and larvae due to a sensitivity to faecal matter (Krizkova-Kudlikova and Hubert, 2008). Despite the commercial release of an ELISA test kit (Biotect, Austin, TX) for insect contamination, there has been reluctance on the part of the grain industry to adopt this method.

19.9.4 PCR-Based Detection Methods

Methods that rely on DNA-based polymerase chain reaction (PCR) procedures are becoming more commonplace in the cereals industry, particularly for the detection or quantification of GMOs. The common example for this lies with maize that has been genetically modified to contain a gene from the soil microbe *Bacillus thuringiensis* (*Bt*) which codes for the production of a toxin that is effective in control of the European corn borer (*Ostrinia nubilalis*) (Pan et al., 2005; Zhu et al., 2008). For many grain-importing countries, especially those of the European Union, strict limits (0.9% per ingredient in the EU) were imposed on the level and traceability of GM contamination in final food and feed products (EC, 2003a,b). Similar restrictions were placed on food products containing GMO soymeal produced from 'Roundup Ready' soybean varieties. However, in recent years adverse reaction in Europe to GMO plant-based food has softened, as noted with ongoing legislation designed to permit GMO plant-based food on a national or regional basis provided GMO foods are labelled and are traceable (EU, 2015).

PCR methodologies are also used in non-GMO situations, such as in Italy to ensure that pasta destined for export contain no more than 3% common wheat (Sonnante et al., 2009; Casazza et al., 2012) and in India to authenticate Basmati rice (Vemireddy et al., 2007; Ganapoulos et al., 2011; Kaur et al., 2015); however, these tests are generally performed at the national level. Hamada et al. (2012) developed a microsatellite PCR method for quantifying levels of *F. graminearum* in wheat seeds, however, this procedure was aimed at epidemiological investigations rather than commercial usage at sales points. Equipment and operational expenses of PCR methodologies are high and the required level of operator training is much higher than protein-based methods such as ELISA (Holst-Jensen et al., 2003). Use of PCR methods, therefore, is unlikely at the country elevator.

19.9.5 Electronic Noses

One component of the inspection process is the evaluation of the headspace above the grain for presence of off odours caused by moulds, insects, straw, herbs and other non-grain species, insects, decay and odours from other contamination sources. Reliance on the human nose for inspection has the drawback of disagreement between individual inspectors' perceptions of odours, as well as a health risk associated with the inhalation of the spores and vapours arising from the grain sample (Olsson et al., 2002). The electronic nose typically consists of several metal oxide semiconductor field effect transistors (MOSFETs) that are sensitive to the volatiles in the headspace during a regiment of warming (~50°C) the sample. Recent advances of combining metal oxide sensors with mass spectrometry have resulted in hybrid systems that reduce the traditional shortcomings of e-noses, such as drift, sensitivity and sensor poisoning (Loutfi et al., 2015). Multivariate calibrations, such as principal components analysis and partial least squares regression (Olsson et al., 2002) or artificial neural networks (Börjesson et al., 1996; Jonsson et al., 1997; Campagnoli et al., 2011; Lippolis et al., 2014) are used to relate the time response data from the detectors to a class of odour or a concentration of a compound, such as deoxynivalenol. There are several manufacturers of electronic noses, with operations based in Germany, France, Sweden, Italy, United Kingdom and the United States; however the emphasis for these devices in foods is in the higher valued commodities such as milk, wine, coffee, tea, fish and meat (Loutfi et al., 2015). Although the application of the electronic nose for grain inspection was investigated by the USDA-GIPSA in the 1990s, this technology was not implemented in field offices or country elevators, presumably for the reasons of complexity of operation and instrument cost. However, recent successes by Lippolis et al. (2014) in detecting DON contamination in durum wheat at the EU permitted maximum level of 1750 μg/kg, especially on ground meal, may pave the way for the use of electronic nose technology in high throughput screening in the future.

19.9.6 X-ray Imaging for Internal Insects

Research on the use of x-rays to detect insect eggs and larvae within the kernel has been ongoing for more than 50 years (Milner et al., 1950; Stermer, 1972; Schatzki and Fine, 1988; Keagy and Schatzki, 1993). Though x-ray analysis is useful in detection of the pre-adult stages of stored grain insects, the procedure of producing images on photographic film has been a detriment to rapid analysis. Advances in digital imaging technology now make it possible to capture and store the images without the need for film, whereupon the images can be viewed by the inspector (Haff and Slaughter, 2004), or image processing classification algorithms may be applied for unattended determination of internal infestation (Karunakaran et al., 2003). The use of X-ray imaging as well as vis/NIR hyperspectral imaging for detection of stored product insects and fungi was recently reviewed by Shah and Khan (2014). Even with the potential to shorten analysis time through real-time X-ray imaging for stored product insects, complexity, expense and worker safety issues of this technique stand in the way of it being used at receival points.

19.10 FUTURE TRENDS

Assessment of grain quality, particularly at the first point of sale, continues to be heavily reliant on manual inspection. The reason for this is largely because of the complexity of the inspection itself. In the case of wheat, this involves the need to determine various conditions of kernel damage and contaminants, which surprisingly have a high degree of commonality among the major exporting countries (Table 19.9). Today, instrumentation exists that can accurately

TABLE 19.9 Summary of Wheat Grade Factors of the Wheat Exporting Countries

Quality parameter	Argentina	Australia	Canada	European Union	France	Ukraine	United Kingdom	United States
Test variety	Grade No. 1	APH2	No. 1 CWRS	Common wheat	Class E	Common	UKP Bread Wheat	US No. 1 HRS
Protein Content	√	√	√	√	√	√	√	
Test Weight	√	√	√	√	√	√	√	√
Falling Number	√	√		√	√	√	√	
Moisture	√	√		√	√	√	√	√
Zeleny Index	√			√	√			
Wet Gluten	√				√	√	√	
Alveograph	√				√		√	
Farinograph	√				√			
Mixograph	√				√			
Ash	√						√	
Weight of 1000 kernels	√							
Odour		√	√			√	√	√
Hardness			√		√	√	√	
Baking Test	√	√						
Milling	√	√						
Sampling								
Size before splitting	4 kg	>3 L	900 g			250 g		

Size after splitting			>250 g					15 g
First cleaned by: (1) Dockage tester			√					√
(2) Sieve	√					√	√	√
(3) Other		√						
Defects/Damage								
Heat damage	√	√	√	√			√	√
Shrunken and broken kernels	√		√	√		√	√	√
Sprouted	√	√	√	√	√	√	√	√
Stained		√	√		√		√	
Fusarium damage		√	√		√		√	√
Smut	√	√	√	√		√		
Insect-damage	√						√	√
Chaff							√	√
Frost	√		√					√
Germ	√		√				√	√
Mould			√				√	√
Green or Sappy	√		√					
Non-vitreous		√			√			
Darkened kernels			√		√			

(Continued)

TABLE 19.9 Summary of Wheat Grade Factors of the Wheat Exporting Countries (*cont.*)

Quality parameter	Argentina	Australia	Canada	European Union	France	Ukraine	United Kingdom	United States
Test variety	**Grade No. 1**	**APH2**	**No. 1 CWRS**	**Common wheat**	**Class E**	**Common**	**UKP Bread Wheat**	**US No. 1 HRS**
Spotted					√			
Total Damage	√	√	√	√	√	√	√	√
Contaminants								
Dead Insects	√	√	√				√	√
Live Insects	√	√	√			√	√	√
Ergot		√	√	√	√	√	√	√
Excreta		√	√				√	√
Stones		√	√			√	√	√
Other grain seeds		√	√			√	√	√
Other seeds	√	√	√	√		√	√	√
Other foreign material	√	√	√	√	√	√	√	√

measure concentrations of mycotoxins and contaminants, biochemically related quality properties (protein, starch and oil), physical properties, odours, insect infestation and defects. Unfortunately, there is no single instrument capable of performing all quality monitoring tasks. Thus, a suite of instruments would be needed to replace the human inspector. This is unlikely due to the excessive cost of all equipment.

In lieu of relying on an arsenal of equipment, country elevators will continue to use the combination of manual inspection and, depending on the growing region, instruments specific to anticipated conditions of that region. For example falling number analysis may be conducted in areas known to have had rainy conditions prior to harvest. Likewise, ELISA tests for DON are conducted in years and regions of outbreaks of Fusarium Head Blight. Digital imaging may eventually become an integral part of the cereals inspection process, depending in part on the outcome of pilot trials in Australia using an off-the-shelf imaging system. With the ubiquity of mobile phones with built in cameras, macro photography may be accomplished with the addition of an external lens, whereupon, image processing routines for size, shape, colour and texture of cereal grains will be developed most likely. Such information will serve as an objective analysis of the lot's soundness to complement the human visual component of grain assessment at first point of sale. Smartphone technology has been making its way into agriculture, particularly in areas of weather and disease forecasting, planting decisions, pest monitoring, spray application rates and commodity pricing, with all but the last of these being pre-harvest operations. For the cereals in post harvest, some of the marketing companies (e.g. Emerald Grain Melbourne, Australia) provide applications that provide information in real time on prices, locations and operating hours of receival points and types of grain being accepted. Third party companies are also active in releasing grain management applications, for example an app called Grain Storage Manager (iNet Solutions Group, Omaha, NE, USA) can track the locations of grain in various bins and the ensuing sales transactions. Continued development of smartphone software for grain producers, handlers and processors will undoubtedly occur.

ACKNOWLEDGEMENT

The authors wish to thank Steven Rausch, USDA-ARS (Beltsville, MD) for providing helpful comments and reviewing early drafts.

SOURCES OF FURTHER INFORMATION AND ADVICE

The following websites provide extensive information on how cereal quality is assessed in the World's major grain exporting countries. Additionally, many provide crop-quality survey reports that are updated annually.

http://www.fsa.usda.gov/FSA/webapp?area=home&subject=coop&topic=was—USDA Farm Service Agency, Commodity Operations, Warehouse Services—information about licensing of grain elevators and their locations.

http://www.gipsa.usda.gov/fgis/fgis.aspx—USDA Grain Inspection, Packers and Stockyards Administration, Federal Grain Inspection Service - information on US grain standards and standard operating procedures.

www.aegic.org.au—Australian Export Grains Innovation Centre—supports trade and use of Australian grains, undertakes market research, quality and process research and testing, publishes Australian Wheat Quality Report.

www.barleyaustralia.com.au—Barley Australia—peak industry body which accredits Australian malting varieties, manages end-use market R&D projects and provides industry leadership.

www.cbh.com.au—CBH group—grower owned co-operative and largest western Australian bulk handler and exporter of Australian grain.

www.graincorp.com.au—GrainCorp—largest eastern Australian bulk handler and exporter of Australian grain, publishes eastern Australian Crop Report.

www.graintrade.org.au—Grain Trade Australia—sets wheat and coarse grain standards and distributes oilseeds and pulse standards; develops generic contracts and storage and handling agreements.

www.wheatquality.com.au—Wheat Quality Australia—ensures that the quality of classes of Australian wheat meet market processing and end-product requirements.

http://www.minagri.gob.ar/sitio/—Argentina's Ministry of Agriculture, Livestock and Fisheries and Food (in Spanish).

http://www.inta.gov.ar—Argentina's National Institute of Agricultural Technology, the government's agricultural research and extension organisation (in Spanish).

http://www.trigoargentino.com.ar—Commercial clearing house that releases a yearly quality report of Argentine wheat.

http://www.uswheat.org—US Wheat Associates - a private organisation that promotes the export of US wheat. Maintains down-loadable yearly crop-quality reports for several wheat classes.

http://cereals.ahdb.org.uk—a private organisation that provides market information, funds research and promotes the export of U.K. cereals and oilseeds.

http://www.ble.de/EN/01_Market/MarketOrganisation.html—The Federeal Office for Agriculture and Food in Germany.

http://eur-lex.europa.eu/en/index.htm—Portal to European Union law.

http://www.cwb.ca—Canadian Wheat Board—a private marketing agency that promotes domestic and export sales of Canadian wheat, renamed as G3 Canada Limited in summer 2015.

http://www.grainscanada.gc.ca—Canadian Grain Commission—a government agency that establishes grain standards and conducts inspections and research.

REFERENCES

AACC, 2015. Approved Methods Analysis, eleventh ed. St. Paul, MN, AACC International. Available from: <http://methods.aaccnet.org/toc.aspx>

ABARES, 2015. Australian crop report. Australian Bureau of Agricultural and Resource Economics and Sciences, Canberra, Australia. No. 174. Available from: <www.agriculture.gov.au/abares/publications>

Argentine Wheat, 2011. Argentine standard for wheat: New resolution Ex-SAGPyA 1262/04. Available from: <http://www.trigoargentino.com.ar/2011/ENG/texts/TRIGOPAN_CALIDAD_2011ENG.HTML>

Atui, M., Flinn, P.W., Lazzari, S.M.N., Lazzari, F.A., 2007. Detection of *Rhyzopertha dominica* larvae in stored wheat using ELISA: the impact of myosin degradation following fumigation. J. Stored Prod. Res. 43, 156–159.

Barbedo, J.G., Tibola, C.S., Fernandes, J.M., 2015. Detecting *Fusarium* head blight in wheat kernels using hyperspectral imaging. Biosyst. Eng. 131, 65–76.

Berman, M., Connor, P.M., Whitbourn, L.B., Coward, D.A., Osborne, B.G., Southan, M.D., 2007. Classification of sound and stained wheat grains using visible and near infrared hyperspectral image analysis. J. Near Infrared Spectrosc. 15, 351–358, doi: 10.1255/jnirs.754.

Bicsak, R.C., 1993. Comparison of Kjeldahl method for determination of crude protein in cereal grains and oilseeds with generic combustion method: collaborative study. J. AOAC 76, 780–786.

Börjesson, T., Eklöv, T., Jonsson, A., Sundgren, H., Schnürer, J., 1996. Electronic nose for odor classification of grains. Cereal Chem. 73, 457–461.

Brader, B., Lee, R.C., Plarre, R., Burkholder, W., Kitto, G.B., Kao, C., Polston, L., Dorneanu, E., Szabo, I., Mead, B., Rouse, B., Sullins, D., Denning, R., 2002. A comparison of screening methods for insect contamination in wheat. J. Stored Prod. Res. 38, 75–86.

Burns, D.A., Ciurczak, E.W., 2001. Handbook of Near-Infrared Analysis, second ed. Marcel Dekker, New York, pp. 814.

Campagnoli, A., Cheli, F., Polidori, C., Zaninelli, M., Zecca, O., Savoini, G., Dell'Orto, V., 2011. Use of the electronic nose as a screening tool for the recognition of durum wheat naturally contaminated by deoxynivalenol: a preliminary approach. Sensors 11 (5), 4899–4916.

Casazza, A.P., Morcia, C., Ponzoni, E., Gavazzi, F., Benedettelli, S., Breviario, D., 2012. A reliable assay for the detection of soft wheat adulteration in Italian pasta is based on the use of new DNA molecular markers capable of discriminating between *Triticum aestivum* and *Triticum durum*. J. Cereal Sci. 56, 733–740.

CGC, 2014. Canadian Grain Exports (annual): Crop year 2013–2014. Canadian Grain Commission, Winnipeg. Available from: https://www.grainscanada.gc.ca/statistics-statistiques/cge-ecg/cgem-mecg-eng.htm

CGC, 2015a. Consultation document—February 20, 2015. Canadian Grain Commission, Winnipeg. Available from: https://www.grainscanada.gc.ca/consultations/2015/classes-01-en.htm

CGC, 2015b. Consolidation: Canada Grain Act. R.S.C., 1985, cG-10, Minister of Justice, Canada. Available from: http://laws-lois.justice.gc.ca/eng/acts/g-10/index.html

CGC, 2015c. Wheat—Chapter 4 Official Grain Grading Guide. Canadian Grain Commission, Winnipeg. Available from: http://www.grainscanada.gc.ca/oggg-gocg/04/oggg-gocg-4-eng.htm

Choudhary, R., Mahesh, S., Paliwal, J., Jayas, D.S., 2009. Identification of wheat classes using wavelet features from near infrared hyperspectral images of bulk samples. Biosyst. Eng. 102 (2), 115–127.

Delwiche, S.R., 2004. Analysis of small grain crops. In: Roberts, C.A., Workman, J., Reeves, J.B. (Eds.), Near-Infrared Spectroscopy in Agriculture. American Society of Agronomy, Madison, pp. 269–320.

Delwiche, S.R., Kim, M.S., Dong, Y., 2011. *Fusarium* damage assessment in wheat kernels by Vis/NIR hyperspectral imaging. Sens. Instrum. Food Qual. Saf. 5 (2), 63–71.

EC, 2008. Commission Regulation No 687/2008 of 18 July 2008 establishing procedures for the taking-over of cereals by intervention agencies of paying agencies and laying down methods of analysis for determining the quality of cereals. Off. J. Eur. Union L192, 20–48.

EC, 2003a. Regulation (EC) No 1829/2003 of the European Parliament and of the Council of 22 September 2003 on genetically modified food and feed. Off. J. Eur. Union L268, 1–23.

EC, 2003b. Regulation (EC) No 1830/2003 of the European Parliament and of the Council of 22 September 2003 concerning the traceability and labeling of genetically modified organisms and the traceability of food and feed products produced from genetically modified organisms and amending Directive 2001/18/EC. Off. J. Eur. Union L268, 24–28.

EU, 2013. Regulation (EU) No 1308/2013 of the European Parliament and of the Council of 17 December 2013 establishing a common organisation of the markets in agricultural products and repealing Council Regulations (EEC) No 922/72, (EEC) No 234/79, (EC) No 1037/2001 and (EC) No 1234/2007. Off. J. Eur. Union 56, 671–854.

EU, 2015. Directive (EU) 2015/412 of the European Parliament and the Council of 11 March 2015 amending Directive 2001/18/EC as regards the possibility for the Member States to restrict or prohibit the cultivation of genetically modified organisms (GMOs) in their territory. Off. J. Eur. Union 68, 1–8.

Fernández Pierna, J.A., Vermeulen, P., Amand, O., Tossens, A., Dardenne, P., Baeten, V., 2012. NIR hyperspectral imaging spectroscopy and chemometrics for the detection of undesirable substances in food and feed. Chemometr. Intell. Lab. Syst. 117, 233–239.

FPVO, 2015. Bundessortenamt. Federal Plant Variety Office. Available from: http://www.bundessortenamt.de/internet30/index.php?id=3&L=1

Ganapoulos, I., Argiriou, A., Tsaftaris, A., 2011. Adulterations in basmati rice detected quantitatively by combined use of microsatellite and fragrance typing with high resolution melting (HRM) analysis. Food Chem. 129 (2), 652–659.

Gowen, A.A., O'Donnell, C., Cullen, P.J., Downey, G., Frias, J.M., 2007. Hyperspectral imaging–an emerging process analytical tool for food quality and safety control. Trends Food Sci. Technol. 18 (12), 590–598.

Gowen, A., Fen, Y., Gaston, E., Valdramidis, V., 2015. Recent applications of hyperspectral imaging in microbiology. Talanta 137, 43–54.

Hagberg, S., 1960. A rapid method for determining alpha-amylase activity. Cereal Chem. 37, 218–222.

Haff, R.P., Slaughter, D.C., 2004. Real-time x-ray inspection of wheat for infestation by the granary weevil, *Sitophilus granarius* (L.). Trans. ASAE 47, 531–537.

Hamada, M., Yin, Y., Ma, Z., 2012. Rapid detection of *Fusarium graminearum* complex in wheat seeds using species-specific PCR primer designed based on a microsatellite region. Cereal Res. Commun. 40 (1), 85–94.

HGCA, 2004. Grain Testing Standards for testing. Home-Grown Cereals Authority, London, pp. 6. Available from: http://www.hgca.com/document.aspx?fn=load&media_id=940&publicationId=1362

Holst-Jensen, A., Rønning, S.B., Løvseth, A., Berdal, K.G., 2003. PCR technology for screening and quantification of genetically modified organisms (GMOs). Anal. Bional. Chem. 375, 985–993.

International Standards Organisation, 1999. Cereals pulses and milled products—sampling of static batches. ISO 13690, pp. 18.

International Standards Organisation, 2002. Flowing cereals and milled cereal products—Automatic sampling by mechanical means. ISO 6644, pp. 9.

Jallais, B., Roumet, P., Pinson-Gadais, L., Bertrand, D., 2015. Detection of Fusarium head blight contamination in wheat kernels by multivariate imaging. Food Control 43, 250–258.

Jonsson, A., Winquist, F., Schnürer, J., Sundgren, H., Lundström, I., 1997. Electronic nose for microbial quality classification of grains. Int. J. Food Microbiol. 35, 187–193.

Karunakaran, C., Jayas, D.S., White, N.D.G., 2003. X-ray image analysis to detect infestations caused by insects in grain. Cereal Chem. 80, 553–557.

Kaur, S., Panesar, P.S., Bera, M.B., Kaur, V., 2015. Simple sequence repeat markers in genetic divergence and marker-assisted selection of rice cultivars: a review. Crit. Rev. Food Sci. Nutr. 55, 41–49.

Keagy, P.M., Schatzki, T.F., 1993. Machine recognition of weevil damage in wheat radiographs. Cereal Chem. 70, 696–700.

Kitto, G.B., 1991. A new rapid biochemical technique for quantitating insect contamination in grain. Association of Operative Millers Technical Bulletin, March, 5835–5838.

Krizkova-Kudlikova, I., Hubert, J., 2008. Development of polyclonal antibodies for the detection of *Tribolium casteneum* contamination in wheat grain. J. Agric. Food Chem. 56, 8035–8040.

LeMasurier, L.S., Panozzo, J.F., Walker, 2014. A digital image analysis method for assessment of lentil size traits. J. Food Eng. 128, 72–78.

Lippolis, V., Pascale, M., Cervellieri, S., Damascelli, A., Visconti, A., 2014. Screening of deoxynivalenol contamination in durum wheat by MOS-based electronic nose and identification of the relevant pattern of volatile compounds. Food Control 37, 263–271.

Loutfi, A., Coradeschi, S., Mani, G.K., Shankar, P., Rayappan, J.B.B., 2015. Electronic noses for food quality: a review. J. Food Eng. 144, 103–111.

Luo, X., Jayas, D.S., Symons, S.J., 1999. Identification of damaged kernels in wheat using a colour machine vision system. J. Cereal Sci. 30, 49–59.

Mahesh, S., Jayas, D.S., Paliwal, J., White, N.D.G., 2011. Identification of wheat classes at different moisture levels using near-infrared hyperspectral images of bulk samples. Sens Instrum Food Qual Saf. 5 (1), 1–9.

Majumdar, S., Jayas, D.S., 2000. Classification of cereal grains using machine vision: III. Texture models. Trans. ASAE 43, 1681–1687.

Mares, D.J., Mrva, K., 2008. Late-maturity α-amylase: low falling number in wheat in the absence of preharvest sprouting. J. Cereal Sci. 47, 6–17.

Mares, D.J., Mrva, K., 2014. Wheat grain preharvest sprouting and late maturity alpha-amylase. Planta 240, 1167–1178.

Martin, C.R., Rousser, R., Brabec, D.L., 1993. Development of a single-kernel wheat characterization system. Trans. ASAE 36, 1399–1404.

Milner, M., Lee, M.R., Katz, R., 1950. Application of x-ray technique to the detection of internal insect infestation of grain. J. Econ. Entomol. 43, 933.

Neuman, M.R., Sapirstein, H.D., Shwedyk, E., Bushuk, W., 1989a. Wheat grain color analysis by digital image processing. I. Methodology. J. Cereal Sci. 10, 175–182.

Neuman, M.R., Sapirstein, H.D., Shwedyk, E., Bushuk, W., 1989b. Wheat grain color analysis by digital image processing. II. Wheat class discrimination. J. Cereal Sci. 10, 183–188.

Norris, K.H., Hruschka, W.R., Bean, M.M., Slaughter, D.C., 1989. A definition of wheat hardness using near infrared reflectance spectroscopy. Cereal Food World 34, 396–705.

Olsson, J., Börjesson, T., Lundstedt, T., Schnürer, J., 2002. Detection and quantification of ochratoxin A and deoxynivalenol in barley grains by GC-MS and electronic nose. Int. J. Food Microbiol. 72, 203–214.

Osborne, B.G., Fearn, T., Hindle, P.H., 1993. Practical NIR Spectroscopy with Applications in Food and Beverage Analysis, second ed. Pearson Education, Essex, pp. 227.

Pan, A., Yang, L., Xu, S., Yin, C., Zhang, K., Wang, Z., Zhang, D., 2005. Event-specific qualitative and quantitative PCR detection of MON863 maize based upon the 3'-transgene integration sequence. J. Cereal Sci. 43, 250–257.

Ruan, R., Ning, S., Song, A., Ning, A., Jones, R., Chen, P., 1998. Estimation of *Fusarium* scab in wheat using machine vision and a neural network. Cereal Chem. 75, 455–459.

Sapirstein, H.D., 1995. Variety identification by digital image analysis. In: Wrigley, C.W., Identification of Food-Grain Varieties. American Association of Cereal Chemists, St. Paul, pp. 91–130.

Sapirstein, H.D., Neuman, M.R., Wright, E.H., Shwedyk, E., Bushuk, W., 1987. An instrumental system for cereal grain classification using digital image analysis. J. Cereal Sci. 6, 3–14.

Schatzki, T.F., Fine, T.A., 1988. Analysis of radiograms of wheat kernels for quality control. Cereal Chem. 65, 233–239.

Seibel, W., 2005. Warenkunde Getreide: Inhaltsstoffe, Analytik, Reinigung, Trocknung, Lagerung, Vermarktung, Verarbeitung.

Serranti, S., Cesare, D., Bonifazi, G., 2013. The development of a hyperspectral imaging method for the detection of *Fusarium*-damaged, yellow berry and vitreous Italian durum wheat kernels. Biosyst. Eng. 115 (1), 20–30.

Shah, M.A., Khan, A.A., 2014. Imaging techniques for the detection of stored product pests. Appl. Entomol. Zool. 49, 201–212.

Shahin, M.A., Symons, S.J., 2008. Detection of hard vitreous and starchy kernels in amber durum wheat samples using hyperspectral imaging. NIR News 19 (5), 16–18.

Shahin, M.A., Symons, S.J., 2011. Detection of Fusarium damaged kernels in Canada Western Red Spring wheat using visible/near-infrared hyperspectral imaging and principal component analysis. Comput. Electron. Agric. 75 (1), 107–112.

Shahin, M.A., Symons, S.J., 2012. Detection of fusarium damage in Canadian wheat using visible/near-infrared hyperspectral imaging. J. Food Meas. Charact. 6 (1–4), 3–11.

Shatadal, P., Jayas, D.S., Bulley, N.R., 1995. Digital image analysis for software separation and classification of touching grains. II. Classification. Trans. ASAE 38, 645–649.

Siesler, H.W., Ozaki Y., Kawata, S., Heise, H.M., 2002. Near-Infrared Spectroscopy: Principles, Instruments, Applications, Wiley VCH, Weinheim, pp. 348.

Singh, C.B., Jayas, D.S., Paliwal, J., White, N.D., 2009. Detection of sprouted and midge-damaged wheat kernels using near-infrared hyperspectral imaging. Cereal Chem. 86 (3), 256–260.

Singh, C.B., Jayas, D.S., Paliwal, J., White, N.D., 2010a. Detection of midge-damaged wheat kernels using short-wave near-infrared hyperspectral and digital colour imaging. Biosyst. Eng. 105 (3), 380–387.

Singh, C.B., Jayas, D.S., Paliwal, J., White, N.D., 2010b. Identification of insect-damaged wheat kernels using short-wave near-infrared hyperspectral and digital colour imaging. Comput. Electron. Agric. 73 (2), 118–125.

Sonnante, G., Montemurro, C., Morgese, A., Sabetta, W., Blanco, A., Pasqualone, A., 2009. DNA microsatellite region for a reliable quantification of soft wheat adulteration in Durum wheat-based foodstuffs by real-time PCR. J. Agric. Food Chem. 57, 10199–10204.

Stermer, R.A., 1972. Automated x-ray inspection of grain for insect infestation. Trans. ASAE 15, 1081–1085.

Symons, S.J., Fulcher, R.G., 1988. Determination of wheat kernel morphological variation by digital image analysis: I. Variations in eastern Canadian milling quality wheats. J. Cereal Sci. 8, 211–218.

Ukraine, 2010. National Standard of Ukraine, Wheat: Technical specifications. DSTU 3768:2010. (Translated from Ukrainian).

USDA-FAS, 2014. Kazakhstan—Republic of: Grain and Feed Annual 2014. GAIN Report 4/1/2014.

USDA-FAS, 2015. Grain: World Markets and Trade: Black Sea wheat exports projected at a record, 8/12/2015. Available from: http://www.fas.usda.gov/data/grain-world-markets-and-trade

USDA-FSA, 2008. Warehouses Licensed Under the U.S. Warehouse Act: As of December 31, 2014. USDA-FSA, Washington, D.C. Available from: http://www.fsa.usda.gov/Internet/FSA_File/whselst2014.pdf

USDA-FSA, 2011. WA-402: Licensing Agreement for Grain and Rice Warehouse Operators. USDA-FSA, Washington, D.C. Available from: http://forms.sc.egov.usda.gov/efcommon/eFileServices/eForms/WA402.pdf

USDA-GIPSA, 1995. Grain Inspection Handbook: Book I: Grain Sampling. USDA-GIPSA-FGIS, Washington, D.C. Available from: http://www.gipsa.usda.gov/fgis/handbook/grain-insp/grbook1/bk1ch1.pdf

USDA-GIPSA, 1996. Quality Handbook. USDA-GIPSA-FGIS, Washington, D.C. Available from: http://archive.gipsa.usda.gov/reference-library/handbooks/qualbook/qualhb.pdf

USDA-GIPSA, 1997. Subpart B—United States Standards for Barley. USDA- GIPSA-FGIS, Washington, D.C. Available from: http://www.gipsa.usda.gov/fgis/standards/810barley97.pdf

USDA-GIPSA, 2003. Mechanical Systems Handbook. USDA-GIPSA-FGIS, Washington, D.C. Available from: http://www.gipsa.usda.gov/fgis/handbook/mechanical_inspec.aspx

USDA-GIPSA, 2006. Near-Infrared Transmittance Handbook (NIRT). USDA-GIPSA-FGIS, Washington, D.C. (most recent modification on 24-03-2008). Available from: http://www.gipsa.usda.gov/fgis/handbook/nirt_inspec.aspx

USDA-GIPSA, 2011. 2011 Annual Report. USDA-GIPSA-FGIS, Washington, D.C. Available from: http://www.gipsa.usda.gov/fgis/publication/ar/2011_fgis_AR.pdf

USDA-GIPSA, 2013a. 2013 Annual Report. USDA-GIPSA-FGIS, Washington, D.C. Available from: http://www.gipsa.usda.gov/fgis/publication/ar/2013_fgis_AR.pdf

USDA-GIPSA, 2013b. Grain Inspection Handbook: Book II, Chapter 2 Barley. USDA-GIPSA-FGIS, Washington, D.C. Available from: http://www.gipsa.usda.gov/fgis/handbook/grain-insp/grbook2/barley.pdf

USDA-GIPSA, 2013c. Falling number determination for wheat. Directive 9180.38, USDA-GIPSA-FGIS, Washington, D.C. Available from: http://www.gipsa.usda.gov/laws/fgisprogramdirectives.aspx

USDA-GIPSA, 2013d. Grain Inspection Handbook: Book II, Chapter 1 General Information. USDA-GIPSA-FGIS, Washington, D.C. Available from: http://www.gipsa.usda.gov/fgis/handbook/gihbk2_inspec.aspx

USDA-GIPSA, 2014a. 2014 Annual Report. USDA-GIPSA-FGIS, Washington, D.C. Available from: http://www.gipsa.usda.gov/fgis/publication/ar/2014_fgis_AR.pdf

USDA-GIPSA, 2014b. Subpart M—United States Standards for Wheat. USDA-GIPSA-FGIS, Washington, D.C. Available from: http://www.gipsa.usda.gov/fgis/standards/810wheat.pdf

USDA-GIPSA, 2014c. Grain Inspection Handbook: Book II, Wheat, USDA-GIPSA-FGIS, Washington, D.C (Chapter 13). Available from: http://www.gipsa.usda.gov/fgis/handbook/grain-insp/grbook2/wheat.pdf.

USDA-GIPSA, 2015a. Official moisture calibrations for unified grain moisture algorithm (UGMA) compatible meters. Directive 9180.61, 8/1/2015. Available from: http://www.gipsa.usda.gov/fgis/moisture.aspx

USDA-GIPSA, 2015b. Rapid test kit evaluation program (TKE). USDA-GIPSA-FGIS, Washington, D.C. Available from: http://www.gipsa.usda.gov/fgis/rapidtestkit.aspx

USDA-NASS, 2013. 2013 Agricultural Statistics Annual: Grain and feed, USDA-NASS, Washington, D.C. Available from: http://www.nass.usda.gov/Publications/Ag_Statistics/2013/index.asp

US Food and Drug Administration (FDA), 2014. Defect Levels Handbook: The Food Defect Action Levels: Levels of Natural or Unavoidable Defects in Foods That Present no Health Hazards for Humans. FDA, College Park, MD. Available from: http://www.fda.gov/Food/GuidanceRegulation/GuidanceDocumentsRegulatoryInformation/SanitationTransportation/ucm056174.htm#CHPTR

Vemireddy, L.R., Abchak, S., Nagaraju, J., 2007. Capillary electrophoresis is essential for microsatellite marker based detection and quantification of adulteration of Basmati rice (*Oryza sativa*). J. Agric. Food Chem. 55, 8112–8117.

Wallays, C., Missotten, B., De Baerdemaeker, J., Saeys, W., 2009. Hyperspectral waveband selection for on-line measurement of grain cleanliness. Biosyst. Eng. 104, 1–7.

Wan, Y.N., Lin, C.M., Chiou, J.F., 2002. Rice quality classification using an automatic grain quality inspection system. Trans. ASAE 45, 379–387.

Williams, P.S., Norris, K.H., 2001. Near-Infrared Technology in the Agricultural and Food Industries, second ed. American Association of Cereal Chemists, St. Paul, 296.

Williams, P.C., Sobering, D.C., 1986. Attempts at standardization of hardness testing of wheat. I. The grinding/sieving (particle size index) method. Cereal Foods World 31 359, 362–364.

Xie, F., Pearson, T., Dowell, F.E., Zhang, N., 2004. Detecting vitreous wheat kernels using reflectance and transmittance image analysis. Cereal Chem. 81, 594–597.

Xing, J., Symons, S., Shahin, M., Hatcher, D., 2010. Detection of sprout damage in Canada Western Red Spring wheat with multiple wavebands using visible/near-infrared hyperspectral imaging. Biosyst. Eng. 106 (2), 188–194.

Zayas, I., Lai, F.S., Pomeranz, Y., 1986. Discrimination between wheat classes and varieties by image analysis. Cereal Chem. 63, 52–56.

Zayas, I., Pomeranz, Y., Lai, F.S., 1985. Note: discrimination between arthur and arkan wheats by image analysis. Cereal Chem. 62, 478–480.

Zhu, H., Zhao, X., Jia, J., Sun, J., Zhao, K., 2008. A specific qualitative and real-time PCR detection of MON863 maize based on the 5'-transgene integration sequence. J Cereal Sci. 48, 592–597.

Chapter 20

Maintaining Grain Quality During Storage and Transport

Ian Batey
Sunset Cereal Services, Meadowbank, NSW, Australia

20.1 INTRODUCTION

Chapter 1 of this book presents the 'grain chain' as a continuum from breeding, through growing and management of the crop, to harvesting, segregation, storage and transport, to processing and final consumption. Most of these stages are covered in detail in earlier or later chapters. This chapter is concerned with the storage and transport of essentially unprocessed grain, the latter term usually referring to products that retain the original structural form of the harvested grain kernel. This is the portion of 'grain chain' from receival to processing.

Grain is rarely processed or consumed immediately after harvest. It is almost always involved in a period of storage, and usually also with transport from the field where it is grown to one or more intermediate places of storage. It was this storability of grain that allowed the change from a nomadic society to an agrarian one. Indeed, it was the ability to store and control grain supplies that gave rise to the most important kingdoms in the ancient Middle East (Paulette, 2016). Control over grain became control over people. As one writer expressed at the time:

> *Whoever has silver, whoever has jewels, whoever has cattle, whoever has sheep shall take a seat at the gate of he who has grain, and pass his time there. (Quoted in Paulette, 2016).*

Fig. 20.1 shows grain storage pits in an Iron Age settlement at Buoux, France. From this early form of storage, there have developed far more complex facilities, sometimes involving high-tech equipment and computer control.

The overall function of the supply chain is to move grain efficiently from the paddock to the consumer at a rate, and with quality characteristics, that meet the users' requirements. Most cereal grains are not consumed directly after harvesting. Even in those places where there is a significant proportion of

Cereal Grains. http://dx.doi.org/10.1016/B978-0-08-100719-8.00020-6

FIGURE 20.1 **Grain storage pits dating from the Iron Age, Buoux, France.**

the crop retained for the farmer's own use, it is necessary for it to be stored for up to a year until the next harvest is ready. Where the grain is traded and exported, large-scale storage facilities are needed and it is not uncommon for grain to be stored for more than a year to provide a continuous supply to end users and to iron out the year-to-year fluctuations in production. Approximately 20% of world cereal production is carried over from one year to the next (see Chapter 3). Methods for the assessment of grain quality during storage and transport are mostly the same as those used at receival from the farmer (Chapter 19) or at the next stage of processing (Chapter 22). Aspects of storage related to food safety are covered in Chapter 15. This chapter is concerned with how grain is stored and moved, from segregation to processing, to maintain the quality requirements of the consumer.

20.2 GENERIC CHARACTERISTICS OF STORAGE

For most cereals, the major issue of storing grain is to maintain a safe product for consumption as food or feed, and to prevent physical and economic loss from infestation and consumption by animals and insects. This requires facilities suitable for the volume and type of grain, and the nature of the distribution process from the store, and, if necessary, drying and fumigation. Figs. 20.2–20.5 contrast the storage of sorghum and rice in local and village stores with bulk wheat storage. For wheat, rice and barley in particular, storage conditions play an important role in maintaining the quality which the end user requires and for which the maximum price may be paid. Particular aspects of storage of these grains are described in detail in other sections of this chapter.

Storage begins with grain segregation, the process initiated at receival, the aspect covered in Chapter 19, in which the quality basis of segregation is described, together with the testing needed to assess quality. Many aspects of grain

FIGURE 20.2 Sorghum granary, near Lake Victoria, Kenya. *(Photo courtesy of Prof J R N Taylor.)*

FIGURE 20.3 Village rice store, Laos. *(Photo courtesy of Dr Melissa Fitzgerald.)*

storage are common to all grains. Grain at harvest is a living organism and is therefore subject to biological ageing and deterioration. Some of that deterioration in storage is a result, directly or indirectly, of respiration of the grain. In storage, grain is also a large source of biologically available energy making it susceptible to consumption and damage by 'pests', both animal and microbial species. These biological factors mean that, apart from merely supplying grain in a timely fashion, the storage part of the supply chain needs to meet the additional objectives of preventing quality and material loss. Thus, it is necessary

FIGURE 20.4 Bulk wheat storage facility at Bellata, NSW, Australia showing a variety of storage types. *(Photo courtesy of GrainCorp, Australia.)*

to hold grain under conditions that prevent spoilage from microbes, insects and rodents, and that maintain (or at least minimise) any loss of the quality present at harvest. Grain moisture content (water activity) must be kept below critical levels to prevent microbial growth, but at the same time, grain should not be very dry because extreme dryness can create problems of dust generation with the potential for dust explosions.

Key issues with grain storage are moisture levels, temperature, mechanical damage due to handling, and infestations with microbes, insects, birds and rodents. Any form of storage must allow these factors to influence the quality of the grain minimally. Damage from some of these factors is inter-related. For example, moisture levels influence both the growth of micro-organisms and breakage of grains during handling. Moisture levels can also affect the suitability of the storage environment for different insects. Temperature also affects the growth of insects and micro-organisms.

During storage, there is a natural exchange between the gaseous environment and the grain itself. This is due to respiration and it results in a depletion of the oxygen content and an increase in carbon dioxide. At the same time, water is freed and heat is given off. As with most organisms, the rate of respiration increases with higher moisture and increased temperature. Control of temperature and moisture is also a means of controlling the respiration that will inevitably occur when whole grain is stored.

(A)

(B)

FIGURE 20.5 Bunker storage (A) being filled and (B) being sealed in rural Australia. *(Photo A courtesy of Di Miskelly. Photo B courtesy of Eric Williamson.)*

Grain temperature during storage is critical to maintaining quality as it will determine the relative humidity and the rate of respiration. All cereals are poor conductors of heat, and unless there is some intervention, the temperature at harvest will be maintained for some time, despite ambient conditions outside. This is not a major problem where the harvest occurs in autumn or early winter, but in tropical areas the temperature of grain at harvest will be quite high and above the temperature that is optimal for maintaining quality. In countries like Australia, where the main grain harvest occurs in late spring to mid-summer, the grain is often above a desirable temperature when it is delivered to storage facilities. If the temperature of the grain remains too high, there is a loss of quality, particularly evident in wheat and barley. Aeration can be applied using ambient air at the coolest time of the day. It is standard practice for wheat in Australia to be stored in bins that may be aerated with external air. Cool air is drawn in and blown upwards through the grain, while warmer air is exhausted out of the top of the bin. The whole process may be under computer control to optimise the times of aeration so that the coolest air available is drawn through the stored grain at an optimal speed.

Aeration of stored grain can also achieve a degree of moisture control, thus helping to control the risk of insect infestation. In experiments in Australia, wheat grain was placed in aerated storage bins with moisture approximately 1% above the then acceptable level of 12% (Desmarchelier and Ghaly, 1993). Grain was reduced to levels below 12% within a month of being placed in storage by aerating with fans. There was no evidence of any microbial or fungal growth. Despite the lack of fumigation, there was no increase in insect population during the trial. In many developing countries, grain is stored in sacks rather than in bulk storage. Sealed sacks can be used to control the moisture and composition of the included atmosphere inside the sack, for maintaining the grain quality (Gras et al., 1990).

20.2.1 Control of Grain Moisture Levels

The safe level of moisture is considered to be equivalent to the equilibrium moisture of the grain at 75% relative humidity and 25°C. This moisture content varies slightly for the different grains, but is generally between 14% and 15% moisture (Table 20.1). However, the moisture content alone is not enough to assure safe storage conditions. Despite a safe moisture level, averaged across all of the grain stored in a particular container, localised variations can occur as a result of external temperature changes. These localised places may be the start of 'hot spots' of spoilage.

Grain drying is one approach to moisture control. Where the crop is harvested with moisture above levels acceptable for receival into storage, it is necessary for some process to be applied to reduce it. For rice, where it is routine for the harvest moisture to be well above the acceptable limits, grain drying is a

TABLE 20.1 Equilibrium Moisture Content of Cereals at 75% Relative Humidity (Based on Bushuk and Lee, 1978)

Cereal	Moisture (%)
Barley	14.3
Maize	14.3
Oats	13.4
Rice	14.0
Sorghum	15.3
Wheat	15.0

standard procedure. For other crops, it is usually employed only when necessary. In Australia, the moisture level of wheat at harvest is usually well below the maximum acceptable level when it is delivered for storage. While reducing the risk of spoilage by micro-organisms, drying introduces risks of other quality damage occurring. Drying rice too fast will cause cracking to occur in the grain. When it is milled, the resulting rice grain may break, and the presence of too many broken kernels will cause the batch to be downgraded. Application of excessive heat to wheat and barley will also cause a reduction in grain quality. This deterioration can be observed in dried grain by measuring the germinability, an essential quality in barley for malting and a useful indicator of damage to the proteins responsible for quality in wheat.

The simplest form of drying grain is to lay it out in the sun. This can only be done in a suitable climate, and when the required reduction in water content can be achieved quickly. Prolonged exposure in this way exposes the grain to other risks of loss or damage. Aeration of storages is a more sophisticated way of achieving air-drying without having to spread it out in the sun, but it requires that the storage facility be fitted with suitable equipment for moving the air around. More complicated dryers use heated air driven through the grain mechanically. Available designs range from rotating drum dryers to fluidised beds.

20.2.2 Control of Insects

The 'pesticide revolution', following the end of World War II, promised a reduction in the loss of food during growth and storage due to insects and mites. Despite initial successes, the promise was not delivered, as it was subsequently discovered that many of the compounds used were damaging to the environment and/or had serious toxicity issues with humans. In addition, insects soon developed resistance to the various types of insecticides being used, requiring higher concentrations to maintain the level of protection. The consequence of this approach has been the desire of consumers to avoid foods with any traces of pesticides, and many countries have a zero tolerance of some pesticides in all food imports. Nevertheless, it is necessary to employ some means of control to ensure that damage due to insects is prevented. For stored grain, fumigation with gaseous toxins is the preferred way in which this is done, although in the past liquids have also been applied. Liquids may comprise solids in a carrier solvent, or substances such as carbon disulfide which can be applied directly or in a diluted form with another solvent. Some fumigants and insecticides that have been or are being used in grain storage are listed in Table 20.2. This list is not exhaustive but covers many chemicals that are being or have been used.

A wide variety of substances have been used—organo-chlorines, organophosphates, carbamates and pyrethroids being the main classes (Annis, 2016a). Many of these are not volatile so that residues may remain in the grain after treatment. The organo-phosphates have relatively short half-lives, particularly in humid conditions while organo-chlorines tend to be persistent as they are

TABLE 20.2 Common Pesticides (Fumigants and Insecticides) in Common Use on Grain, Past and Present

Aldrin	Fenitrothion
Carbaryl	Malathion
Carbon disulfide	Methyl bromide
Chlordane	Phosphine
DDT	Pirimiphos-methyl
Dichlorvos	Pyrethrin
Endosulfan	Sulfuryl fluoride

not broken down quickly. Maximum residue levels (MRL) are set for each of the substances that are permitted for use in cereals. The US Department of Agriculture maintains an online database of MRL for a wide range of commodities and pesticides (http://www.globalmrl.com). It includes levels for overseas markets as well as for the US domestic market. A table of typical MRL from this database is given in Table 20.3 for some chemicals commonly used with grain. Maximum recommended levels for cereals from Codex Alimentarius Commission are also given.

The main gaseous fumigants are methyl bromide, carbon dioxide, phosphine and very volatile liquids such as carbon-disulfide and ethyl formate. The latter are usually applied as a solution in another solvent. Phosphine may be introduced directly as a gas, or it may be applied to the grain as aluminium phosphide. This latter material reacts with ambient moisture in the atmosphere to give phosphine and a residue of aluminium oxide. It can be applied as tablets distributed throughout the grain store, or in a plastic blanket situated at the top of a storage bin. Tablets react with the moisture in the grain and release phosphine gas killing insects in the vicinity. When used in a plastic blanket at the top of the store, the gas produced will circulate throughout the grain by convection currents. Provided that the store is sealed, this achieves the necessary kill rate of insects in the store (Williams et al., 1996).

As well as toxicity to workers and consumers, the effect of the fumigant on the environment must be considered. For example, methyl bromide was phased out in 2005 because of its depleting effect on the ozone layer, except for some critical-use exemptions. None of these exemptions apply to cereals (https://www.epa.gov/ods-phaseout/methyl-bromide). New fumigants are being developed, and chemicals such as sulfuryl fluoride and ethyl formate are being considered (Annis, 2016b). The testing and approval process is lengthy and they are not yet in widespread use. Sulfuryl fluoride has been introduced into Australia for fumigation, but its use requires trained and accredited operators. Its introduction has been necessitated by the development of resistance to phosphine by

TABLE 20.3 Maximum Residue Levels in Milligram per Kilogram for Some Pesticides and Fumigants Used on Cereals[a]

	Barley	Maize	Millet	Oats	Rice	Sorghum	Wheat
Phosphine							
US	0.1	0.1	0.1	0.1	0.1	0.1	0.1
EU	0.1	0.1	0.1	0.1	0.1	0.1	0.1
Japan	0.1	0.1			0.1		0.1
Codex							0.1
Malathion							
US	8	8	8	8	8	8	8
EU	8	8	8	8	8	8	8
Japan	2	2			0.1		8
Codex		0.05					10
Pyrethrins							
US	3	3		1	3	1	3
EU	3	3		3	3	3	3
Japan	3	3			3		3
Codex	0.03	0.03		0.03	0.03	0.03	0.03
Chlorpyriphos-methyl							
US	6				6	6	6
EU	3				3	3	3
Japan	6				0.1		6
Codex	10	0.05			0.1	10	10
Sulfuryl fluoride							
US	0.1	0.1	0.1	0.1	0.04	0.1	0.1
EU	0.05	0.05	0.05	0.05	0.05	0.05	0.05
Japan	0.1	0.1			0.04		0.1
Codex	0.05	0.05	0.05	0.05	0.05	0.05	0.05

[a]*There is no value provided for those uses and markets left blank.*
Sources: USDA MRL database http://www.globalmrl.com and http://www.fao.org/fao-who-codexalimentarius/standards/pestres/en/

some insects of stored grain. Sorption is more rapid than phosphine, but slower than other fumigants. Desorption occurs very rapidly when the grain is aerated. (Hwaidi et al., 2015), but residues are cumulative with repeated treatments. Durum grain fumigated with sulfuryl fluoride shows no loss of end-product quality, but there is a significant reduction in germinability (Hwaidi et al., 2016). Hence this not a suitable fumigant for grain that will be used for seed purposes.

A novel approach to fumigation has been the use of bio-gas from cow dung (Mohan and Gopalan, 1992). This gas comprises about 40% methane and 60% carbon dioxide. Reports claim good success in ridding the grain of insects, and propose it as a suitable means of fumigation in developing countries but there seems to have been little uptake of the research. This may be related to the potential fire risks in using this mixture, together with a desire to reduce greenhouse gas emissions.

Modified atmospheres (MA), including the use of carbon dioxide, are commonly used in grain storage. Nitrogen is also used with the primary goal of reducing the oxygen content of the gases in the storage facility, and thus either kill or inactivate any insects present. This approach is widely used in many countries on many different cereals, but it requires good containment. Unless the storage is sealed well, the level of oxygen is not reduced sufficiently, or maintained at the required level (<1% v/v). There also has to be a cheap source of nitrogen readily available or the cost of the process may make it uneconomic. Low-oxygen MA can also be achieved with the exhaust of gas burners. Although these have been available in several countries, there has not been widespread application. Low-oxygen MA is more widely used by holding grain in airtight bins under a low-oxygen atmosphere. Relief valves may be fitted to overcome differences in pressure as the temperature changes.

Under MA conditions, carbon dioxide is effectively being used as a fumigant. It has a lethal effect on insects when the concentration is above 40% (v/v). At 60% concentration, the action is rapid and all insect stages are killed in a four-day exposure at 26°C.

Finally, one form of insect control is simple hygiene on the farm. Many pieces of equipment, such as harvesters and augers, and storage facilities have various corners, crevices, and other dead spaces where grain can collect and remain when the facility is not in use. Grain left undisturbed becomes a perfect breeding ground for insects and, if the space is accessible, rodents and birds. Removal of all grain or fumigation of the equipment or store before use in following seasons will ensure that the next season's harvest does not become contaminated almost immediately. The recommended procedure is to clean and treat the equipment or storage bin immediately after use, and then again before use the following season. Some insects of stored wheat are shown in Fig. 20.6.

20.2.3 Storage of Barley

For malting purposes, it is required that the viability of the grain is maintained during storage of barley. For other cereals, this is only an issue when the grain is to be used for seed purposes. A supply of barley with germination >95% is essential for malting, and this must be available throughout the year. Thus storage for at least 12 months after harvest may be required and the germination must be maintained at this minimum acceptable level for the whole time. Moisture and temperature are important for preservation of malting quality. For safe

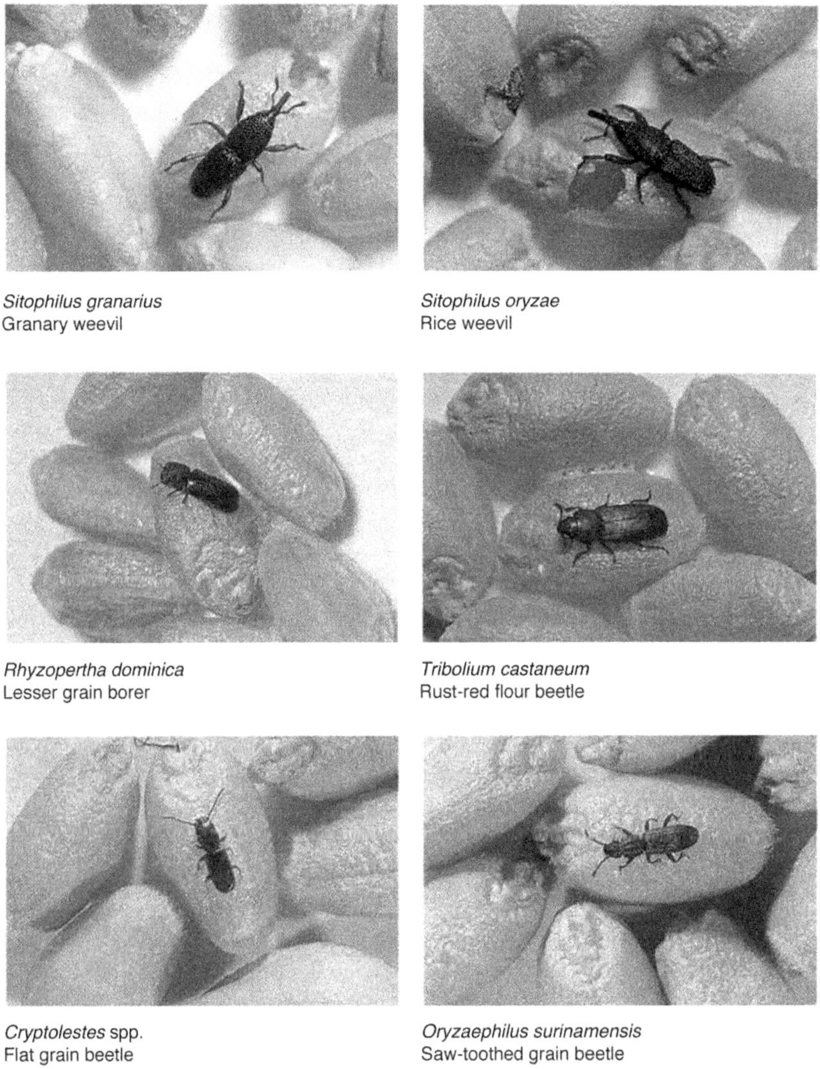

FIGURE 20.6 Some common insect pests of stored wheat, from Ferns et al. (1978). *(Reproduced, by permission of CSIRO. © CSIRO.)*

storage up to 1 year, moisture must be less than 13%; grain of <11% moisture may be stored up to 5 years (Henry, 2016). Temperature and moisture also interact in determining the safe storage period. While grain at 13% moisture retains its viability for over 12 months at 15°C, at 20°C the period is reduced to 6–9 months and at 30°C it may only be stored for about 2 months. Thus reduction of both the moisture and the temperature is needed to maximise storage life. Pre-harvest sprouting is another factor that reduces the 'safe' storage life of barley (Bason et al., 1993a). A mathematical model has been developed to predict 'safe' storage life, based on the degree of sprout damage, plus storage conditions and barley variety (Bason et al., 1993b). For quality assurance purposes, it is necessary to monitor germination at regular interval during storage.

Some barley varieties have a period of dormancy immediately after harvest. In these cases, the malting of new season's barley may not be commenced for several weeks or months, thus requiring that some of the previous season's grain be stored for over 12 months to provide a regular supply of raw grain over this period. Dormancy is affected by genotype (variety), environment (growth conditions), and storage. Some varieties with a strong dormancy may be subjected to a short period of warm storage before malting to remove the dormancy. If this procedure is applied, the grain must have been dried to a suitably low moisture level.

As in other grains, modified atmosphere conditions have been employed for barley storage to kill insects or control their growth and development. Reduction in oxygen levels may inhibit insect growth and reproduction. Storage under carbon dioxide will kill active insects and prevent their progression through the various larval stages of growth. However, very high concentrations of CO_2 may also reduce O_2 to dangerously low levels. In malting barley, this will cause a reduction in germinative capacity.

For maximum germination, grain must be harvested at maturity, and at many receival stations the moisture level needs to be below 13%. Models have been developed to predict the storage life of barley at different temperatures and moisture levels. These models may be used to optimise the value of harvested grain by predicting those batches with an expected short storage life, thus allowing them to be processed before other batches with longer predicted storage life, due to better combinations of temperature and moisture conditions.

Barley varieties may differ in their rates of germination, thus providing a further reason for the segregation of barley according to variety. Maltsters require that each batch of barley must germinate at the same rate; therefore the mixing of incompatible grain lots must be avoided. The specifications required by an end user may not be achievable in a single malt sample and blending must be used to achieve a particular value for a parameter, for example protein content. In malting barley, it is usually not feasible to blend the grain prior to malting. Batches of a single barley variety with different protein contents and from different growth environments will usually have differing germination rates, just as different varieties with the same protein content will also vary in their germination rate. Thus, it is necessary for the maltster to achieve a particular set of specifications by blending the malts. It is equally necessary for the grain handler to ensure that barley binned together has the same germination time.

20.2.4 Storage of Rice

With increasing urbanisation and movement of people from small farms to cities, there has been a concomitant increase in the need for storage. Just after

World War II, almost 75% of rice was either consumed by the farmer or used for bartering with other local producers. This is still the case in some developing countries, but the trend is away from this local consumption to a central storage (Barber, 1972). These changes have been encouraged by the practice of many governments to stabilise food prices, allowing rice producers to sell their crop immediately after harvest and use the money thus gained to purchase food later as required. In many Asian countries, the return to the farmer ranges from 50% to 80% of the retail cost, much more than in developed countries where the farmer may only receive 20–30% of the retail cost. Local storage can lead to high losses (10–50%) due to spoilage, so there is an economic incentive for the farmer to sell his produce soon after harvest and reduce his risk of economic loss. In most Asian countries, rice is usually stored already bagged (Fig. 20.7).

Rice is usually harvested at a moisture level far too high for safe storage (>20%), so it is necessary to first dry the grain. Dryer designs include cross-flow dryers, concurrent-flow dryers, mixed-flow dryers, rotary dryers and fluidised-bed dryers (Brooker et al., 1992; Jaiboon et al., 2009). During the drying process with such equipment, a moisture gradient is created from the centre to the surface of the grain, leading to fissuring within the grain. Unacceptable drying conditions may result in cracking of the kernel, thus leading to more broken grains during milling and reducing the quality and acceptability of the rice. Rapid drying tends to increase the proportion of broken grains. The design of some dryers is such that, although the grain may be dried so that the average moisture is at the optimum level (12–14%), there may be large pockets of grain where the moisture is well above this level, leading to the growth of spoilage organisms. It is also necessary for grain to be dried soon after harvest if the moisture is >21%. When paddy rice is held at these levels of moisture for 48 h, significant amounts of volatile microbial metabolites have been identified

FIGURE 20.7 Bulk store of bagged rice. *(Photo courtesy of IRRI, Philippines.)*

in the milled rice obtained from the stored paddy. These metabolites produce unacceptable odours and flavours. At moisture levels below 21%, there was no impact on the flavour and odour of the milled rice after holding paddy rice for this time (Champagne et al., 2004).

Following drying, rice may be stored as either paddy or milled rice. Under conditions that do not cause microbial growth, there is still an effect of ageing. This is greater in paddy rice, presumably due to the presence of the metabolic enzymes present in the germ, so there is a preference for storage of grain after milling. However, even milled rice may develop undesirable characteristics if stored under poor conditions. Volatile components arise from the oxidation of lipids and affect the odour and flavour of the rice, and may also cause an increase in yellowness. Storage at low temperatures reduces this effect, but in a tropical environment it is not easy to achieve low-temperature storage. There is also a yellowing of the grain during storage. As consumers expect rice to be white, this change in colour leads to a perceived loss of quality. Finally, there is a change in textural properties during storage. Rice tends to become harder after cooking when it has been stored for lengthy periods.

20.2.5 Storage of Wheat

Unlike rice, a high proportion of the world wheat crop is placed into bulk storage on a large scale (Grundas and Wrigley, 2016). Wheat is usually processed before consumption and almost all is milled to flour prior to the next step in the chain. In some countries, significant amounts of grain are stored at the farm, either sold or to be sold under contract to the miller. This is an economic decision of the farmer—whether or not to pay the cost of storage in anticipation of achieving a higher price some time after harvest. The alternative is for the farmer to sell the grain to a trader where it will be held with grain of the same protein content, grade and supposed quality after segregation as described in Chapter 19. Most of the grain destined for export from the major exporting countries is stored in this way. There is sometimes the option for the farmer to hedge on selling his grain, by placing it into storage at a local or regional elevator, and paying for the storage. There is the anticipation once again for the farmer to receive a higher price at a time after harvest when supply and demand factors may allow this to occur. There is usually an agreed price, less storage and freight charges, which the farmer will receive from the grain trader.

As stated earlier, storage conditions play an important role in maintaining the quality of wheat. Under Australian conditions, wheat is commonly received at temperatures in excess of 25°C, sometimes in excess of 30°C. Unless the storage temperature is reduced to about 23°C, major changes in dough properties occur (Wrigley and Bekes, 1999), namely, an increase in dough strength (as R_{max}), a decrease in extensibility (measured in the Extensigraph) and an increase in dough development time (measured in the Farinograph). A decrease in loaf volume on baking is also observed. When the temperature of the wheat was reduced to 23°C or below, there was little change in quality over a 12-month period provided that the moisture was also controlled at 12% or below.

20.3 TRANSPORT

Transport of grain occurs in many ways, from being hand-carried by individuals at one extreme through to the loading of large bulk ocean-going carriers. The mode of transport employed will be appropriate for the origin and size of a 'shipment' and the distance and terrain over which it needs to be moved (Figs. 20.8–20.10). The cost and availability are the key factors determining what will be used in a particular situation. This choice will not have a great impact on the quality of the cereal being transported. The key factor, however, will be the traceability, or identity preservation, of the grain. It is essential that the customer receives the type and quality of grain needed for the process into which it will be fed. The implementation of traceability begins when the grain is first delivered into storage. Grain being moved directly from a farm store to a processor will be tested while in store, and again on arrival at the mill or other processing plant. In this situation, traceability is not such an issue and is in control of the farmer and processor. Where it passes through a number of transport and storage modes, identity preservation becomes necessary at all points where grain can be mixed with other grain, or incorrectly shipped to the next destination. This is particularly important for wheat and barley where admixture may result in severe quality loss from the end-use perspective.

Grain in a rural storage elevator is assigned a quality grade on receival based on variety and protein content. Sampling of all grain placed in a single bin or cell allows testing to provide average quality attributes for the batch. In larger stores, there may be individual smaller cells of the same grade, but which may vary in properties. If the quality properties sought by a customer are present in

FIGURE 20.8 Rice being carried by hand after harvest. *(Photo courtesy of IRRI, Philippines.)*

FIGURE 20.9 **River transport of rice.** *(Photo courtesy of IRRI, Philippines.)*

FIGURE 20.10 **Grain samples being taken from a truck-load at the 'sampling stand' before being delivered at the nearby 'silo' storage in the Australian wheat belt.**

one cell, but not all, it becomes necessary to identify and keep separate those cells of inferior quality. Where there are some cells of superior quality, it could be possible to blend these with cells of lower quality to achieve an average of the desired quality. This blending may involve grain storage at different local elevators, and will require movement of the grain to a regional store where the grain can be blended and stored. From here, it may be shipped directly to a mill, or to a seaboard terminal for international transport. International bulk purchases usually involve grain of a particular grade and protein content. Average quality for the grade will have been determined by the export agent and the entire shipment will be expected to meet this quality as a minimum standard. Sampling and testing of all grain blended should ensure this standard.

For wheat export, grain will also be tested at stages during transport, particularly at the export terminal. While the grain is being loaded, it will ideally be measured for protein content by online sampling and NIT measurement. Prediction of some dough properties may also be done at this point (see Chapter 5). If the protein falls below the minimum required, loading can be stopped or higher protein grain included to increase the protein. However, these are steps of last resort and maintaining good control at all previous points in the chain should avoid this necessity.

The key to arriving at the final point in the grain chain with the correct grain is good testing and documentation along the way. A sample and records of each batch are retained, and when the sample is aggregated with other samples in a bin or car-lot, a representative sample is also taken and kept. It is usually not possible to test samples while being transported, except when they are moved to large storages and held for some time, perhaps even months, pending sale or export. In the case of dispute, the samples taken along the line may be tested.

Most grain-handling agents have software and databases to help them achieve effective identity preservation and traceability systems. Many of them are proprietary systems, and may be linked to the control of equipment moving the grain for loading into trucks, trains or ships. In some cases, a warning is given before samples are mixed as a means of minimising the risk of blending unsuitable samples.

20.4 FUTURE TRENDS

In the area of storage and transport, the expected trends mostly follow past trends. There is an increasing desire for consumers to have their food without pesticide residues, so this will encourage less insecticide use and more modified atmosphere storage. Storage under controlled conditions of atmosphere and humidity should allow better preservation of the grain and reductions in losses due to spoilage. With demand for cereals starting to outstrip supply, improved storage conditions provide an opportunity to allow better utilisation of the grain that is grown. This is a very desirable outcome in those areas where food is sometimes in short supply and losses due to spoilage are high.

There is much research being directed to improving storage in developing countries. In these countries, the cost of complex storage facilities is prohibitive, so local solutions are being investigated. Some are the application of low-technology approaches such as hermetic sealing (Likhayo et al., 2016) or the use of termite mound clay to construct storage bins (Omobowale et al., 2016). The use of insecticide-treated netting and bags has also been trialled. In this case, it was essential for the grain to be free of insects before being placed in storage (Dowell and Dowell, 2016). Any change in storage conditions that reduces the amount of wasted grain in these area is desirable.

One approach has been the application of silo bags. These are large plastic bags, up to 150 m long, which are filled with grain, sealed and stored in the open, often in the fields where the grain is harvested (Fig. 20.11). It is possible

FIGURE 20.11 Silo bag storage in South Africa. *(Photo provided by Silo Bags International, Nagambie, Vic. Australia.)*

to dry the grain in the bags where necessary. The concept of silo bags originated in Argentina and allowed farmers to maintain the value of their grain at a time when the monetary system was in total disarray. By storing it themselves in silo bags, they were able to trade it at the current value, usually by barter, whenever they accessed it. Its use has since expanded to other countries, where its main purpose is to provide safe storage in environments where other forms of storage are unavailable. In one location, waste has been reduced from as much as 40 down to only 0.02% over 1.5 million tonnes of grain. There is one report that under certain conditions, storage for more than 3 months increases the risk of holes developing in the plastic membrane with a concomitant decrease in grain quality (GRDC, 2011). In some countries, however, grain has been stored for up to 3 years with no loss of quality. Proper preparation and management of the storage site, combined with the use of good quality bags, should overcome the risk of quality loss anywhere.

Although there is a demand for pesticide-free grain, there will always be a need for the use of chemical control under certain circumstances. Some storage facilities are unsuitable for controlled atmosphere storage and to avoid losses, insecticides must be used. The trend for insects to become resistant to chemicals after they have been in use for some time will also continue. This will require the development of new insecticides to maintain the effectiveness of treatments.

The possibility of transport of grain by pipeline has been canvassed many times in the past. The use of liquids such as vegetable oil or super-critical liquefied carbon dioxide have been suggested and used in experiments but neither has been satisfactory. It is difficult to remove the oil from the grain, and liquid CO_2 extracts lipid from the grain (Grundas and Wrigley, 2016). This latter effect may not necessarily be undesirable, as certain lipids when present may cause a deterioration in dough and baking quality of wheat, and if they were to be removed preferentially during transport, the increase in value of the grain in having a higher quality may partially offset the cost of pumping the slurry.

20.5 SOURCES OF FURTHER INFORMATION AND ADVICE

Randolph Barker, R., Herdt, R.W. and Rose, B., 1985. The Rice Economy of Asia. Washington, DC, Resources for the Future, Inc.

Web sites of different grain-handling organisations (list not comprehensive)

In Australia:

Cooperative Bulk Handling Ltd—www.cbh.com.au
GrainCorp—www.graincorp.com.au
Grain Producers—grainsguide.grainproducers.com.au

In Canada:

AGT Food and Ingredients—www.agtfoods.com
Canadian Grains Commission—www.grainscanada.gc.ca

In Europe:

Wessex Grain—www.wessexgrain.co.uk
Gleadell Agriculture—www.gleadell.co.uk
International Grains Council—www.igc.org.uk

In USA:

Archer Daniels Midland—www.adm.com
Bunge North America—www.bunge.com
US Grains Council—www.grains.org

REFERENCES

Annis, P.C., 2016a. Chemicals for grain production and protection. Wrigley, C., Corke, H., Seetharaman, K., Faubion, J. (Eds.), Encyclopedia of Food Grains, vol. 4, Elsevier, Oxford, UK, pp. 99–104.

Annis, P.C., 2016b. Stored Grain: Invertebrate Pests. In: Wrigley, C., Corke, H., Seetharaman, K., Faubion, J. (Eds.), Encyclopedia of Food Grains, vol. 4. Elsevier, Oxford, UK, pp. 110–116.

Barber, S., 1972. Milled rice and changes during ageing. In: Houston, D.F. Rice (Ed.), Chemistry and Technology. AACC International, St Paul, MN, pp. 215–263.

Bason, M.L., Ronalds, J.A., Wrigley, C.W., Hubbard, L.J., 1993a. Testing for sprouting damage in malting barley using the Rapid Visco Analyser. Cereal Chem. 70, 247–251.

Bason, M.L., Ronalds, J.A., Wrigley, C.W., 1993b. Prediction of "safe" storage life for sound and weather-damaged malting barley. Cereal Foods World 38, 361–363.

Brooker, B.B., Bakka-Arkema, F.W., Hall, C.W., 1992. Drying and Storage of Grains and Oilseeds. Van Nostrand Reinhold, New York.

Bushuk, W., Lee, J.W., 1978. Biochemical and functional changes in cereals: maturation, storage and germination. In: Hutton, H.O., Milner, M. (Eds.), Post-Harvest Biology and Biotechnology. Food and Nutrition Press, Westport, CT, pp. 1–33.

Champagne, E.T., Thompson, J.F., Bett-Garber, K.L., Mutters, R., Miller, J.A., Tan, E., 2004. Impact of storage of freshly harvested paddy rice on milled white rice flavour. Cereal Chem. 81, 444–449.

Desmarchelier, J.M., Ghaly, T., 1993. Effects of raising the receival moisture content on the storability of Australian wheat. Aust. J. Exp. Agric. 33, 909–914.

Dowell, F.E., Dowell, C.N., 2016. Reducing grain storage losses in developing countries. Qual. Assur. Saf. Crop. Foods, 1–8.

Ferns, G.K., Fitzsimmons, R.W., Martin, R.H., Wrigley, C.W., 1978. Australian wheat varieties. Supplement No. 1. Eleven recent wheats, grain defects and appendixes on foreign seeds and insects. CSIRO, Melbourne, Australia.

Gras, P.W., Bason, M.L., Esteves, L.A., Sabio, G.C., Annis, P.C., van Graver, J.E., 1990. Quality changes in white maize stored in sealed bag sacks. J. Stored Prod. Res. 26, 199–206.

GRDC, 2011. Silo bags handy, but limit to short-term use. Ground Cover #95. www.grdc.com.au/Media-Centre/Ground-Cover/Ground-Cover-Issue-95-November-December-2011/Silo-bags-handy-but-limit-to-shortterm-use.

Grundas, S., Wrigley, C.W., 2016. Wheat: harvesting, transport and storage. In: Wrigley, C., Corke, H., Seetharaman, K., Faubion, J. (Eds.), Encyclopedia of Food Grains, vol. 4. Elsevier, Oxford, UK, pp. 42–49.

Henry, R.J., 2016. Barley. Harvesting, transport and storage. In: Wrigley, C., Corke, H., Seetharaman, K., Faubion, J. (Eds.), Encyclopedia of Food Grains, vol. 4. Elsevier, Oxford, UK, pp. 50–53.

Hwaidi, M., Collins, P.J., Sissons, M., Pavic, H., Nayak, M.K., 2015. Sorption and desorption of sulfuryl fluoride by wheat, flour and semolina. J. Stored Prod. Res. 62, 65–73.

Hwaidi, M.I., Sissons, M., Sissons, M., Pleming, D., Collins, P.J., 2016. Does fumigation of durum wheat and semolina with sulfuryl fluoride affect quality of the grain, semolina and derived spaghetti and bread? Cereal Chem. 93, 482–486.

Jaiboon, P., Prachayawarakorn, S., Devahastin, S., Soponronnarit, S., 2009. Effects of fluidized bed drying temperature and tempering time on quality of waxy rice. J. Food Eng. 95, 517–524.

Likhayo, P., Bruce, A.Y., Mutambuki, K., Tefera, T., Mueke, J., 2016. On-farm evaluation of hermetic technology against maize storage pests in Kenya. J. Econ. Entomol. 109 (4), 1943–1950.

Mohan, S., Gopalan, M., 1992. A study on the use of biogas from cow dung for storage insect control. Bioresour. Technol. 39, 229–232.

Omobowale, M.O., Armstrong, P.R., Mijinyawa, Y., Igbeka, J.C., Maghirang, E.B., 2016. Maize storage in termite mound clay, concrete, and steel silos in the humid tropics: comparison and effect on bacterial and fungal counts. Transactions of the ASABE 59, 1039–1048.

Paulette, T., 2016. Grain, storage, and state making in Mesopotamia (3200–2000 BC). In: Manzanilla, L.R., Rothman, M. (Eds.), Storage in Ancient Complex Societies: Administration, Organization, and Control. Routledge, New York, pp. 85–109.

Williams, P., Nickson, P.J., Braby, M.F., Henderson, A.P., 1996. Phosphine fumigations of wheat in 2500 m^3 ateel bins without recirculation facilities. J. Stored Prod. Res. 32, 153–162.

Wrigley, C.W., Bekes, F., 1999. Glutenin-protein formation during the continuum from anthesis to processing. Cereal Foods World 44, 562–565.

Chapter 21

Biosecurity: Safe-Guarding Quality at All Stages of the Grain Chain

David R. Eagling*, Bill Magee*, Pat J. Collins**, Yong Lin Ren†, Yang Cao‡ and Fujun Li‡

**Plant Biosecurity Cooperative Research Centre, Canberra, ACT, Australia; **Department of Agriculture and Fisheries, Agri-Science Queensland, Brisbane, QLD, Australia; †School of Biological Science and Biotechnology, Murdoch University, Perth, WA, Australia; ‡Academy of State Administration of Grain, Beijing, China*

21.1 BIOSECURITY DEFINITIONS AND THE CONCEPT OF THE BIOSECURITY CONTINUUM

21.1.1 Biosecurity Definitions

Biosecurity is a relatively modern term with multiple meanings and definitions that vary according to various disciplines.

Our interest is in grains and this connects the use of the term biosecurity to its first appearance, which was in agricultural and environmental communities. In this context the term was established to convey a set of preventive measures designed to reduce the risk of transmission of infectious diseases in crops and livestock, quarantined pests, invasive alien species and living modified organisms (Koblentz, 2010).

In some countries, such as Australia, the definitional parameters of the term have remained largely unchanged by the government. Plant Health Australia, the body responsible for the national coordination of government and industry partnership in plant biosecurity, recently defined the term in Australia's Emergency Plant Pest Response Deed as '*a set of measures designed to protect a crop, crops or a sub-group of crops from plant pests at national, regional and individual farm levels*'. This definition will be used as the basis for the chapter.

However, the authors note that through the 1990s the term's parameters were extended by some countries, notably the USA, to encompass the concept of

Cereal Grains. http://dx.doi.org/10.1016/B978-0-08-100719-8.00021-8

biological terrorism. This extension of the term's scope was largely in response to the view that civilian research projects in medicine had the potential (or risk) to be used in military applications. In this context the National Academies of Science define biosecurity as 'security against the inadvertent, inappropriate, or intentional malicious or malevolent use of potentially dangerous biological agents or biotechnology, including the development, production, stockpiling, or use of biological weapons as well as outbreaks of newly emergent and epidemic disease'.

21.1.2 Biosecurity as a Continuum

Biosecurity threats are considered to have increased in the last 2 decades, essentially as a reflection of increased global movements of people and goods, and to some extent climate change.

This has led governments to place a greater emphasis on managing the whole biosecurity continuum—onshore, at the border and offshore—rather than focusing primarily on interventions at a country's border.

The global community relies on the exchange of food and many countries have agricultural sectors and rural communities that are dependent on continued access to overseas markets. Countries and their agricultural sectors benefit from the World Trade Organization's (WTO's) system of rules-based trade.

The WTO is the global international organisation that deals with the rules of trade between nations. At its heart are the WTO agreements, negotiated and signed by the bulk of the world's trading nations and ratified in their parliaments. The goal is to help producers of goods and services, exporters and importers conduct their business.

Under this structure, the 161 countries that comprise the WTO Members, must adhere to a science-based process for assessing import risks, including access requests from trading partners for new products.

A country's biosecurity policy provides for a standard of quarantine that manages risks to a very low level. All countries trade commodities such as grain knowing that attaining zero risk on imported goods is unattainable due to the shifting nature of global trade, climate change and the impossibility of examining each and every shipment of goods imported.

In practical terms, adoption of a zero-risk threshold would imply the cessation of international travel, trade and tourism. Even then, risk would remain from the natural migration of birds and unpredictable climatic events that could bring windborne pests and diseases to a country.

21.1.3 Biosecurity—Offshore Activities

Offshore activities aim to reduce the risks of exotic pests and diseases reaching a country with the government of an individual country seeking to understand global risks (including emerging threats), working with international trading partners, the private sector and engaging with travellers about biosecurity requirements.

Commodities such as grain are covered under the term 'plant biosecurity' and involve regional plant protection organisations and adherence to the principles established through the International Plant Protection Convention to develop and implement harmonised standards for their safe movement.

The government also works directly with a country's individual trading partners to minimise the risks associated with imported goods, conveyances and associated packaging materials. Specific offshore activities include cooperation in multi-lateral forums, import risk analyses, intelligence gathering and audit activities.

Offshore activities to reduce risk rely upon better intelligence and monitoring of emerging threats through the scientific community, analysis using geospatial data, strengthened international agreements, standards and partnerships (including capacity building) and effective treatments in the country of origin. Offshore activities also focus on ongoing monitoring and understanding of biosecurity within other countries, including through system and supply chain audits.

21.1.4 Biosecurity—Border Activities

Border activities seek to intercept biosecurity risks that present at airports, seaports, mail centres and along a country's geographical borders. Border activities include import permit decisions, inspection of passengers, goods, vessels and mail, audit activities and post-entry quarantine.

For grains, as for many commodities, the movement of imported grain involves many entities in the overall supply chain, including owners of goods, service providers, such as brokers and freight forwarders and goods carriers such as vessels and shipping companies.

A government commonly uses inspection, audit and surveillance activities, certification and targeted operations to ensure the compliance of goods with its requirements. This involves the capture and analysis of data to assess compliance risks to enable a government to adopt appropriate performance-based monitoring strategies and set intervention levels suited to the client's compliance behaviour.

21.1.5 Biosecurity—Onshore (Post-Border) Activities

In the event that there is an incursion of a pest or disease of biosecurity risk, onshore arrangements are designed to reduce the likelihood that the pest or disease will become established.

Examples of onshore activities include monitoring and surveillance activities for exotic animals and plant pests and diseases, and emergency preparedness and response plans. Formal national arrangements exist for managing responses to emergency animal and plant pests and diseases and food-safety issues in aquatic and terrestrial environments.

21.2 GRAINS IN THE CONTEXT OF PLANT-BIOSECURITY SYSTEMS TO SAFEGUARD QUALITY

There are several elements of a plant-biosecurity system that are involved in safeguarding the quality of commodities such as grains and that are enacted by a country.

21.2.1 Risk Assessment and Development of Import Conditions

Countries conduct various forms of risk assessments in accordance with the principles set out in the International Standards for Phytosanitary Measures. The output of such assessments is a set of risk-management measures, usually referred to as import conditions. The conversion of the output from a risk assessment into a set of practical, commercially viable and clearly understood import conditions is a very important step in ensuring the safe importation of commodities such as grain. Clarity in import conditions is also very important to ensure that certifying bodies and all commercial parties in the supply chain are able to comply with the conditions of import.

21.2.2 International Standard Setting

The 161 countries under the WTO are active participants in the international and regional standard-setting processes under the International Plant Protection Convention (IPPC). Such standards seek to minimise the spread of pests in international trade and promote, to the greatest extent possible, harmonised measures for addressing pest threats.

21.2.3 Offshore Audits and Verification

Biosecurity specialists are also engaged in a range of offshore audits, verifications and inspections especially prior to the commencement of new trade pathways. The purpose of these activities is to ensure that exporting countries can comply with a destination country's biosecurity requirements and to prevent the arrival of non-compliant consignments at the border.

21.2.4 Border Activities

Border activities aim to detect biosecurity threats at the border and in the case of grains include inspection of international transport vessels, cargo, conveyances, mail (e.g. to detect the presence of exotic weed seeds that would affect grain-production systems), plants and plant products arriving in Australia. An important component of this work is raising awareness of biosecurity amongst travellers, importers and transport operators. Techniques employed at the border include use of detector dogs, X-ray machines, and surveillance and physical inspection activities.

21.2.5 Post-Entry Quarantine

Import of live plants and some species of seeds considered to pose a risk of seed-borne diseases are subject to a period of growth in post-entry quarantine. For high-risk plants requiring specialist diagnostic techniques, many countries operate post-entry quarantine facilities which may also be complemented by scientific and research institutions and by privately owned facilities that have government approvals.

21.2.6 Domestic Quarantine

A number of countries operate domestic quarantine systems to oversight internal movement and trade for the purpose of minimising the potential spread of pests via the movement of plants, plant products, soil or contaminated equipment. This process also operates to restrict the spread of pests that are highly regionalised to particular areas within a region.

21.2.7 Pest Management

Grain production, along with other plant-production industries, often employs a range of pest-management techniques as a multi-faceted approach to controlling the most economically threatening pests.

Chemical control of pests is highly regulated in many countries through a national registration system, which requires that chemicals used are registered, effective and used in accordance with label requirements to avoid residue problems.

Other pest-management practices used in the grain industry to manage quality include cultural and mechanical control as well as stored-product pest-management practices, including fumigation, controlled atmosphere and other technologies.

21.2.8 Surveillance

Plant-biosecurity systems employ a combination of targeted and general surveillance programs carried out by governments, plant industries and the wider community. Surveillance has several objectives including: early detection of new pest incursions; to support area freedom statements for international trade purposes; for delimiting surveys following an incursion and to improve pest management.

21.2.9 Diagnostics

Expertise in diagnostics across the wide spectrum of pests and pathogens affecting plants and plant products is a fundamental component of plant biosecurity. Imported plants and seeds which are grown in post-entry quarantine are subjected to a range of diagnostic techniques ranging from visual inspection by plant pathologists, use of herbaceous indicators, virus indexing and molecular techniques.

Diagnostics is also an essential component of effective targeted surveillance, eradication programs and to support continued market access for plants and plant products.

21.2.10 Emergency Response

When new pests enter a country and are detected post-border, this often initiates formal mechanisms developed by the government to rapidly and effectively respond to plant pests to minimise negative impacts.

Pests that are damaging to grain quality can enter a country via natural means, such as wind currents (e.g. rusts) or through the high volume of movement of cargo and people into the country.

Some countries have highly formalised systems. For example Australia has an Emergency Plant Pest Response Deed between the government and the plant industries with both a policy and an operational framework for responding to pests covered under the Deed. The objective of the Deed is to ensure a rapid and effective response and provide clarity around the roles of all affected parties and cost-sharing arrangements.

21.2.11 Export Inspection and Certification

Countries that export grain do so under internal requirements (e.g. an *Export Control Act)* and in accordance with the phytosanitary requirements of the importing country.

The destination country provides export inspection services and certification for grains according to their phytosanitary requirements. An exporting country often maintains strong linkages with grain supply chain members to ensure that export produce is accurately described and complies with all legislative and importing country requirements.

21.2.12 Implementing and Resourcing Approaches to Biosecurity

The FAO has published a policy and regulatory framework to improve biosecurity coordination (http://www.fao.org/food/food-safety-quality/a-z-index/biosecurity/en/). Published as a 'toolkit' the document sets out guidance to develop and implement national biosecurity frameworks. The document comprises:

Biosecurity Principles and Components (Part 1): provides a broad introduction to biosecurity, and discusses the characteristics, requirements and benefits of a more harmonised approach.

Biosecurity Capacity Assessment Tool (Part 2): outlines a guided process to assess capacity needs across the entire biosecurity arena.

Biosecurity Risk Analysis Manual (Part 3): presents a framework to structure and guide the application of risk analysis (encompassing risk assessment, risk management and risk communication) in biosecurity.

All countries face an ongoing challenge in quantifying the value of biosecurity in the supply chain, relative to other priority areas of funding for agriculture. In many cases governments are generous in offering in-principle support for biosecurity initiatives but cautious in the allocation of funding—at least until there has been a biosecurity incident and the consequences of an incursion become all too clear.

Case studies which describe and quantify the financial and social impacts of pests on production systems and market access are often used as powerful tools to assist governments in the recognition of biosecurity as an essential component of safe and enduring trade in agricultural commodities.

Given that the development of the institutional and operational systems to support biosecurity represent a considerable and long-term investment, many developing countries are initially reliant on the adoption of international phytosanitary standards as a cost-effective way to implement biosecurity measures for imports and exports.

In addition to the adoption of international phytosanitary standards by National Plant Protection Organisations, there are significant advantages in educating exporters, transport operators and owners of storage facilities in the practical aspects of plant biosecurity. There are clear commercial benefits when the private sector embraces phytosanitary requirements and particularly if they become a pre-requisite for contractual arrangements.

The creation of a phytosanitary standard for the international movement of grain is under consideration by the IPPC and the involvement of the private sector in this process has been recognised as vital for the development and adoption of this standard.

21.3 THE CHANGING NATURE OF BIOSECURITY RISK IN GRAINS

21.3.1 Production

The Worldwatch Institute has published data for global grain production showing a record production of 2.316 billion tons in 2007 (Anon, 2016) representing almost a tripling of the global harvest since 1961.

Biosecurity threats to global production of grains are many but the Russian Wheat aphid (RWA), a pest that can cause significant losses in cereal crops, provides a useful example of the key issues.

Biosecurity threats move along transport and natural entry pathways, which are constantly changing due to factors as diverse as climate change, demographics and trade. In the case of RWA, data published by the Centre for Agriculture and Bioscience International (CABI) shows it is considered native to south-western parts of Asia where it is widespread (CABI, 2016) although it is still spreading to the major wheat-growing areas of central China (Zhang, 1991).

As a biosecurity threat pathway overwintering eggs are considered more likely to be transported alive on harvested grains or fodder rather than nymphs or adult RWA. However, adult RWA can spread by actively flying throughout the crop or using wind currents. Long-distance dispersal also occurs by 'hitch-hiking' on machinery, clothes or plant material and climate-matching models have been developed to predict infestations (Hughes and Maywald, 1990).

The CABI datasheet (CABI, 2016) references that RWA has spread throughout Africa, to Chile and Argentina, to western states in the USA and western provinces in Canada. RWA remains within a limited distribution in central and northern Europe and has not been detected in Australia or New Zealand.

A key determinant of biosecurity threat is the capacity of the threat to establish in a new area. In this context plant growth stage and temperature are important factors in estimating RWA survival (Girma et al., 1990) as is commodity (Behle and Michels, 1990). The biosecurity incursion of RWA in the United States has shown that overwintering mortality can be 100% if temperatures are low enough for long enough (Butts, 1992).

As Australia is one of the last major grain-producing countries to remain free of the RWA preparation has focussed on development of a contingency plan. The plan, developed in consultation with industry, provides background information on the pest biology and available control measures to assist with preparedness for an incursion into Australia as well as the guidelines and options for steps to be undertaken and considered when developing a Response Plan for a possible incursion.

21.3.2 Harvest and Storage

The timing of the harvest of cereal grains varies across the world. In some countries, such as Australia, harvest occurs throughout the summer periods resulting in grains typically coming into storage at about 30°C.

Grains are then stored over the hottest part of the year and unless actively cooled, provide conditions that favour insect pest development and infestation.

From the perspective of biosecurity, there is a complex of many species of beetles, moths and other insects that can attack stored grain. The literature identifies 242 species of insects that have been recorded in stored grains of which 130 are considered to be associated with damage or contamination of grain and oil seeds (Jin, 2007).

There are five main beetle species that are cosmopolitan in distribution throughout warm temperate and tropical regions of the world: the red flour beetle, *Tribolium castaneum* (Herbst); the lesser grain borer, *Rhyzopertha dominica* (F.); the rice weevil, *Sitophilus oryzae* (L.); the sawtoothed grain beetle, *Oryzaephilus surinamensis* (L.); and the rusty grain beetle, *Cryptolestes ferrugineus* Stephens.

Other species that are tropical grain pests include the Khapra beetle, *Trogoderma granarium* Everts, and the larger grain borer, *Prostephanus truncatus* (Horn). In cooler climates the list of grain pests expands to include the granary weevil, *Sitophilus granarius* (L.) and the confused flour beetle, *Tribolium confusum* (Jaquelin du Val).

Several species of psocids, most importantly *Liposcelis entomophila* (Enderlein), *L. decolor* (Pearman), *L. paeta* Pearman and *L. bostrychophila* (Badonnel) (Nayak et al., 2014) can also be quite troublesome, particularly in moist environments such as maritime ports and in humid, tropical regions of the world.

Insect infestation of grain was a key issue that was mitigated in the mid-1960s when grain protectant chemicals were introduced and assisted countries such as Australia facilitate quality standards for grain trade (Murray, 1981).

By the mid 1990's the development of resistance to protectants and market preference for grain with nil or very low chemical residues had driven the uptake and use of the fumigant phosphine. The only feasible alternative fumigant available at that time, methyl bromide, was being restricted in its use under international agreement because of its ozone-depleting characteristics (Anon, 2015). Although protectants are still in use, most grain for domestic or international markets is now fumigated with phosphine. In addition to its low residue status, phosphine is a relatively cheap, multi-commodity and easy to apply compared with both chemical and non-chemical alternatives.

Increasing reliance on phosphine, however, was soon followed by a steady increase in the frequency of resistance in target species. Initially, resistance was moderate or 'weak' and was not a major concern for the industry (Herron, 1990).

However, in 1997, a quantitative change in resistance levels was detected in a major pest, *R. dominica* (Collins, 1998). This occurred when the frequency of weak resistance phenotypes in populations of this species reached about 80% (Collins, 2006). Genetic analysis revealed that weak resistance was controlled by one major factor and that it was this gene plus the selection of a second resistance gene that produced the new 'Strong resistance' phenotype (Collins et al., 2002; Schlipalius et al., 2002).

It was not long before strong resistance phenotypes were also detected in two other species, *T. castaneum* (Jagadeesan et al., 2012). Meanwhile very high levels of resistance to phosphine had already been detected in *R. dominica*, *C. ferrugineus*, *S. oryzae* and *O. surinamensis* from Dhaka, Bangladesh (Tyler et al., 1983).

Following this initial detection, a survey revealed high resistance levels in populations of these species, and in *T. castaneum*, from Bhutan, Brazil, Mali, Nepal, Pakistan and Sri Lanka. In some instances (e.g. Bhutan, Nepal) resistant insects had apparently been imported with grain shipments (Taylor and Halliday, 1986). Similarly, high resistance was detected in *C. ferrugineus* populations in England on imported produce (Dyte and Halliday, 1985). Widespread high-level resistance was also reported in many strains of *T. castaneum, R. dominica* and *S. oryzae*, *O. surinamensis* in India (Rajendran, 1998).

Surveys in India (Kaur et al., 2015) indicate frequencies of greater than 95% incidence of strong resistance in *R. dominica* in some regions and high frequencies in other species. Widespread high-level resistance has also been reported in *R. dominica* (Lorini et al., 2007) in Brazil and Philippines (Sayaboc and Gibe, 1997) and in *S. oryzae*, *T. castaneum* and *R. dominica* in Morocco (Benhalima et al., 2004) and China (Zeng, 1998; Cao et al., 2008), and in *T. castaneum* in Turkey (Koçak et al., 2015) and in *T. castaneum* and *R. dominica* populations in the USA (Opit et al., 2012).

Evolution of high-level resistance to phosphine in Australia was a major challenge to the grain industry. Research showed, however, that strong resistance levels known in Australia could be controlled with phosphine by increasing either gas concentration or fumigation period (Collins et al., 2005). This increased costs significantly but it allowed farmers and central storage to regain control.

Effective fumigations, however, now required high standards of gas-tightness in grain storages to maintain fumigant concentrations at required levels. Monitoring indicated that frequency of strong resistance was highest in *R. dominica*. This species also showed the highest level of resistance followed by *S. oryzae* and *L. bostrychophila* and new fumigation protocols were based on controlling these insects (Collins et al., 2005; Nayak et al., 2003).

Despite these changes, results from a national resistance-monitoring program (Collins et al., 2003; Emery et al., 2011) indicated that frequencies of both weak and strong resistance were increasing. This, together with the possibility that resistance could also increase in strength, as appeared to have occurred in other regions (Taylor and Halliday, 1986; Mills, 1983), motivated the various industry sectors to collaborate to develop and adopt a national phosphine-resistance management strategy (GTA, 2015).

This strategy has four major components:

1. A *national monitoring program* to provide early detection of resistance and strategic and tactical information.
2. A *research capability* to estimate the impact of resistance and to develop effective strategies to combat resistance.
3. An *extension network* to promote fumigation best-management practice.
4. A *resistance management implementation plan.*

The strategy is based on the best scientific information on insect pest biology and resistance selection available. In addition, because the grain industry operates within regulatory, operational and market-driven boundaries, these constraints have been recognised and incorporated into the plan.

Key underpinning elements of the strategy are (1) reduce selection pressure and (2) destroy resistant insects. Selection pressure is reduced by limiting the number of fumigations applied to any one parcel of grain ideally to three per year and by broader and improved application of non-chemical options such as cooling and hygiene.

Resistant populations are destroyed by ‘making every fumigation count’, that is, by only using approved rates of phosphine that are researched and known to control resistant insects, and by using alternative fumigants or protectants where available.

Farm extension emphasises best practice fumigation in sealable silos, coupled with the use of aeration cooling, assiduous hygiene and monitoring (GRDC, 2015). A national standard for sealing farm silos was also developed (AS 2628-2010, 2016), and registrations for aluminium phosphide and gaseous phosphine were updated in collaboration with registrants (APVMA, 2015). By adhering to specified concentrations and exposure periods, control failures in central storages and on-farm, due to the presence of strongly phosphine-resistant *R. dominica* and other species, decreased significantly (Collins, 2006).

In 2007, several central storages in eastern Australia experienced severe infestations of *C. ferrugineus* in bulk wheat (Nayak et al., 2013). Although resistance had been noted in this species previously in Australia (Collins, 1998) and was reported from overseas (Mills, 1983; Tyler et al., 1983), it had generally been regarded as a minor pest in the Australian system.

Investigation of these outbreaks revealed that this species had now developed a very high level of resistance to phosphine (resistance factor $>1000\times$), much stronger than that shown by other species, with these insects able to survive several of the registered phosphine application rates (Kaur and Nayak, 2015). In addition, subsequent resistance surveys across Australia established that weak resistant *C. ferrugineus* populations were prevalent throughout all grain-production regions indicating the potential of strong resistance to develop more extensively (Nayak et al., 2013). In some circumstances, effective control with phosphine required much longer fumigation times and higher concentrations than could be achieved in many storages (Kaur and Nayak, 2015).

The development of resistance has prompted central storage managers and farmers to seek alternatives, in some cases returning to use of protectants, or opting to apply another fumigant, sulfuryl fluoride (SF), which had been registered for disinfestation of grain in Australia only a few years earlier. Research indicated that insects resistant to phosphine were susceptible to SF, that is, there is no cross-resistance (Jagadeesan et al., 2015).

Research is continually aimed at maximising the usefulness of SF to the grain industry (Nayak et al., 2015), particularly as an alternative to phosphine as part of the phosphine resistance management strategy. SF, however, is more expensive to apply than phosphine and may potentially result in fluoride residue levels above MRLs if over-used (Hwaidi et al., 2015). It is also potentially susceptible to insect resistance and must be used judiciously if we are to gain optimum and long-term use of it.

Despite success at managing the problem, the development of resistance to phosphine in all major beetle pests infesting stored grain calls into question the long-term sustainability of heavy reliance on fumigant- or chemical-based biosecurity. There is a need for cost-effective, non-chemical solutions to either replace or support chemical tactics.

In several countries the development of resistance has re-awakened an interest in the role of hygiene practices in storage, emphasising regular inspections of equipment, the quality of sealing of storage doors and windows and the regular maintenance of operation areas to remove residual grain, dust, dirt and debris, caulking holes, holes and cracks.

Other technologies that are 're-emerging' include controlled atmosphere and biological control. In the future, grain-protection systems will rely on combinations of non-chemical treatments such as aeration cooling, controlled atmosphere, enhanced formulations of diatomaceous earths, heating and pro-active management based on advanced insect-detection methods to achieve insect-free grain. The role of fumigants is likely to focus on the provision of strategic intervention in these systems.

REFERENCES

Anon, 2015. The Montreal protocol on substances that deplete the ozone layer. Available from: http://ozone.unep.org/en/treaties-and-decisions/montreal-protocol-substances-deplete-ozone-layer

Anon, 2016. Grain harvest sets record, but supplies still tight. Available from: https://www.worldwatch.org/node/5539

APVMA, 2015. Australian Pesticides and Veterinary Medicine Authority. Phosphine fumigant ECO$_2$FUME, Product No: 50177.

AS 2628-2010. SAI Global, 2016. Available from: https://www.saiglobal.com/

Behle, R.W., Michels, Jr., G.J., 1990. Russian wheat aphid development, reproduction and survival on wheat and rye grown in four host-plant media. Southwestern Entomol. 15 (2), 109–121.

Benhalima, H., Chaudhry, M.Q., Mills, K.A., Price, N.R., 2004. Phosphine resistance in stored-product insects collected from various grain facilities in Morocco. J. Stored Prod. Res. 40, 241–249.

Butts, R.A., 1992. Cold hardiness and its relationship to overwintering of the Russian wheat aphid (Homoptera: Aphididae) in southern Alberta. J. Econ. Entomol. 84 (4), 1140–1145.

CABI, 2016. Diuraphis noxia datasheet. Available from: http://www.cabi.org/isc/datasheet/9887

Cao, Y., Li, G., Zhou, J., Li, Y., Qu, G., 2008. The Inhibition Effect of Low Oxygen on Four Species of Stored Grain Insect Pests, 8th International Conference on Controlled Atmosphere and Fumigation in Stored Product (CAF2008). 21–25 September, 2008. pp. 39–44.

Collins, P.J., 1998. Resistance to grain protectants and fumigants in insect pests of stored products in Australia. In: Banks, H.J., Wright, E.J., Damcevski, KA., (Eds.), Stored grain in Australia. Proceedings of the Australian postharvest technical conference, Canberra, 26–29 May, 1998.

Collins, P.J., 2006. Resistance to chemical treatments in insect pests of stored grain and its management (keynote presentation). In: Lorini, I., Bacaltchuk, B., Beckel, H., Deckers, D., Sundfeld, E., Santos, J.P., Biagi, J.D., Celaro, J.C., Faroni, L.R., Bortolini, L.O.F., Sartori, M.R., Elias, M.C., Guedes, R.N.C., Fonseca, R.G., Scussel, V.M., (Eds.), Proceedings of the 9th International

Working Conference on Stored Product Protection, 15–18 October 2006, Campinas, Brazil, Brazilian postharvest Association, pp. 277–282.

Collins, P.J., Daglish, G.J., Bengston, M., Lambkin, T.M., Pavic, H., 2002. Genetics of resistance to phosphine in *Rhyzopertha dominica* (Coleoptera: Bostrichidae). J. Econ. Entomol. 95, 862–869.

Collins, P.J., Daglish, G.J., Pavic, H., Kopittke, R.A., 2005. Response of mixed-age cultures of phosphine-resistant and susceptible strains of lesser grain borer, *Rhyzopertha dominica*, to phosphine at a range of concentrations and exposure periods. J. Stored Prod. Res. 41, 373–385.

Collins, P.J., Emery, R.N., Wallbank, B.E., 2003. Two decades of monitoring and managing phosphine resistance in Australia. In: Credland, P.F., Armitage, D.M., Bell, C.H., Cogan, P.M., Highley, E., (Eds.), 8th International Working Conference on Stored Product Protection—Technology into Action. 22–26 July, 2002, CAB International, York, UK, pp. 570–575.

Dyte, C.E., Halliday, D., 1985. Problems of the development of resistance to phosphine by insect pests of stored grains. EPPO Bull. 15, 51–57.

Emery, R.N., Nayak, M.K., Holloway, J., 2011. Lessons learned from phosphine resistance monitoring in Australia. Stewart Postharvest Rev. 7 (3), 1–8.

Girma, M., Wilde, G., Reese, J.C., 1990. Influence of temperature and plant growth stage on development, reproduction, life span, and intrinsic rate of increase of the Russian wheat aphid (Homoptera: Aphididae). Environ. Entomol. 19 (5), 1438–1442.

GRDC, 2015. Stored grain information hub. Available from: http://storedgrain.com.au/

GTA, 2015. Phosphine strategy. Available from: http://www.graintrade.org.au/sites/default/files/file/NWPGP/Phosphine%20Resistance%20Strategy.pdf

Herron, G.A., 1990. Resistance to grain protectants and phosphine in coleopterous pests of grain stored on farms in New South Wales. J. Aust. Entomol. Soc. 29, 183–189.

Hughes, R.D., Maywald, G.F., 1990. Forecasting the favourableness of the Australian environment for the Russian wheat aphid, Diuraphis noxia (Homoptera: Aphididae), and its potential impact on Australian wheat yields. Bull. Entomol. Res. 80 (2), 165–175.

Hwaidi, M., Collins, P.J., Sissons, M., Pavic, H., Nayak, M.K., 2015. Sorption and desorption of sulfuryl fluoride by wheat, flour and semolina. J. Stored Prod. Res. 62, 65–73.

Jagadeesan, R., Collins, P.J., Daglish, G.J., Ebert, P.R., Schlipalius, D.I., 2012. Phosphine resistance in the rust red flour beetle, *Tribolium castaneum* (Coleoptera: Tenebrionidae): Inheritance, gene interactions and fitness costs. PLoS One 7, e31582.

Jagadeesan, R., Nayak, M.K., Pavic, H., Chandra, K., Collins, P.J., 2015. Susceptibility to sulfuryl fluoride and lack of cross-resistance to phosphine in developmental stages of the red flour beetle, *Tribolium castaneum* (Coleoptera: Tenebrionidae). Pest Manag. Sci. 71, 1379–1386.

Jin, Z., 2007. Development of Grain Storage Science and Technology. Sichuan Publishing House, Chengdu, China.

Kaur, R., Nayak, M.K., 2015. Developing effective fumigation protocols to manage strongly phosphine-resistant *Cryptolestes ferrugineus* (Stephens) (Coleoptera: Laemophloeidae). Pest Manag. Sci. 71, 1297–1302.

Kaur, R., Subbarayalu, M., Jagadeesan, R., Daglish, D.J., Nayak, M.K., Naik, H.R., Ramasamy, S., Subramanian, C., Ebert, P.R., Schlipalius, D.I., 2015. Phosphine resistance in India is characterised by a dihydrolipoamide dehydrogenase variant that is otherwise unobserved in eukaryotes. Heredity 115, 188–194.

Koblentz, G.D., 2010. Biosecurity reconsidered: calibrating biological threats and responses. Int. Secur. 34 (4), 96–132.

Koçak, E., Schlipalius, D., Kaur, R., Tuck, A., Ebert, P., Collins, P., Yılmaz, A., 2015. Determining phosphine resistance in rust red flour beetle, *Tribolium castaneum* (Herbst.) (Coleoptera: Tenebrionidae) populations from Turkey. Turkish J. Entomol. 39, 129–136.

Lorini, I., Collins, P.J., Daglish, G.J., Nayak, M.K., Pavic, H., 2007. Detection and characterisation of strong resistance to phosphine in Brazilian *Rhyzopertha dominica* (F.) (Coleoptera: Bostrychidae). Pest Manag. Sci. 63, 358–364.

Mills, K.A., 1983. Resistance to the fumigant hydrogen phosphide in some stored product species associated with repeated inadequate treatments. Mitt. Dtsch. Ges. Allg. Angew. Entomol. 4, 98–101.

Murray, W.J., 1981. The storage of grain in Australia. In: Williams, P., Amos, T.G., Proceedings of First Australian Stored Grain Pest Control, Conference, Melbourne, May, 1981. CSIRO, Melbourne, pp. 4–11.

Nayak, M.K., Collins, P.J., Pavic, H., Kopittke, R.A., 2003. Inhibition of egg development by phosphine in the cosmopolitan pest of stored products *Liposcelis bostrychophila* (Psocoptera: Liposcelididae). Pest Manag. Sci. 59, 1191–1196.

Nayak, M.K., Collins, P.J., Throne, J.E., Wang, J.-J., 2014. Biology and management of psocids infesting stored products. Annu. Rev. Entomol. 59, 279–297.

Nayak, M.K., Holloway, J.C., Emery, R.N., Pavic, H., Bartlet, J., Collins, P.J., 2013. Strong resistance to phosphine in the rusty grain beetle, *Cryptolestes ferrugineus* (Stephens) (Coleoptera: Laemophloeidae): its characterisation, a rapid assay for diagnosis and its distribution in Australia. Pest Manag. Sci. 69, 48–53.

Nayak, M.K., Jagadeesan, R., Kaur, R., Daglish, G.J., Reid, R., Pavic, H., Smith, L., Collins, P.J., 2015. Developing sulfuryl fluoride as a 'phosphine resistance breaker'—the Australian experience. In: Arthur, F.H., Kengkanpanich, R., Chayaprasert, W., Suthisut, D., (Eds.) Proceedings of the 11th International Working Conference on Stored Product Protection, 24–28 November, 2014, Chiang Mai, Thailand, pp. 475–479.

Opit, G.P., Phillips, T.W., Aikins, M.J., Hasan, M.M., 2012. Phosphine resistance in *Tribolium castaneum* and *Rhyzopertha dominica* from stored wheat in Oklahoma. J. Econ. Entomol. 105, 1107–1114.

Rajendran, S., 1998. Phosphine resistance in stored grain insect pests in India. In: Jin, Z., Liang, Q., Liang, Y., Tan, X., Guan, L., (Eds.). Stored Product Protection. Proceedings of the 7th International Working Conference on Stored Product Protection, 14–19 October, 1998, Beijing, China, Sichuan Publishing House of Science and Technology, pp. 635–641.

Sayaboc, P.D., Gibe, A.J.G., 1997. Resistance of Rhyzopertha dominica (Coleoptera: Bostrychidae) to phosphine in the Philippines. In: Donahaye, E.J., Navarro, S., Varnava, A., (Eds.), Proceedings International Conference on Controlled Atmosphere and Fumigation in Stored Products. 21–26 April 1996, Printco Ltd, Nicosia, Cyprus, pp. 513–517.

Schlipalius, D.I., Qiang Cheng, Reilly, P.E.B., Collins, P.J., Ebert, P.R., 2002. Genetic linkage analysis of the lesser grain borer *Rhyzopertha dominica* identifies two loci that confer high-level resistance to the fumigant phosphine. Genetics 161, 773 782.

Taylor, R.W.D., Halliday, D., 1986. The geographical spread of resistance to phosphine by coleopterous pests of stored products. Proceedings of the Crop Protection Conference, Pests and Diseases, Brighton, England, pp. 607–613.

Tyler, P.S., Taylor, R.W., Rees, D.P., 1983. Insect resistance to phosphine fumigation in food warehouses in Bangladesh. International Pest Control 25, 10–21.

Zeng, L., 1998. Development and countermeasures of phosphine resistance in stored grain insects in Guangdong of China. In: Jin, Z., Liang, Q., Liang, Y., Tan, X., Guan, L., (Eds.). Stored Product Protection. Proceedings of the 7th International Working Conference on Stored product Protection, 14–19 October, 1998, Beijing, China, Sichuan Publishing House of Science and Technology, pp. 642–647.

Zhang, G., 1991. Russian Wheat Aphid (RWA) in ChinaMisc Publ Agric Exp Stn Okla State Univ, Stillwater, USA, pp. 327-328.

Part IV

Processing Cereal Grains to Optimize Product Quality

Chapter 22

Assessing and Managing Wheat-Flour Quality Before, During and After Milling

Diane Miskelly* and Dai Suter**
*Westcott Consultants, Goulburn, NSW, Australia; **Pymble, NSW, Australia

22.1 INTRODUCTION

For almost all purposes, wheat is milled before further processing. Some milling processes are very minimal, for example in the production of 'bulgur', which is steamed or par-boiled and then cracked and flaked. However, wheat is mostly milled extensively through a series of roller mills to produce a fine flour plus the by-products of bran and germ (see Section 22.5.3). Wheat milling is mostly a dry milling process, although there have been some wet milling processes developed (see Chapter 23). Semolina is a coarse fraction of the endosperm, a product of the initial breaking of the wheat grain before the particle size is reduced further to give a fine patent flour. Its main use is for the manufacture of the various forms of pasta, for which purpose durum semolina is sought—and legally required—in many countries. The coarse fraction called 'semolina' may also be obtained from bread wheats. Couscous is also made from semolina, traditionally durum semolina, although forms made from high-protein hard bread wheats are popular in some parts of the world.

22.2 PROCUREMENT OF WHEAT FOR PROCESSING

This stage of wheat processing depends on the use to which the wheat will be put. The end-users (baker, noodle maker, starch producer) will purchase flour with fairly rigid specifications for their particular purpose. These specifications are based on the experience of the user and the needs of their particular process, but almost always do not place any direct requirements on wheat quality other than those relating to safety (chemical or microbial contamination). For two processors making the same or similar products, the flour specifications are often quite different and are set by the requirements of the process being used in the manufacture of

Cereal Grains. http://dx.doi.org/10.1016/B978-0-08-100719-8.00022-X

the product and thus, various baking processes have different flour specifications. For pasta manufacture, the processor will usually specify 100% durum wheat, and will also require a clean semolina (no specks) with good yellow colour. It is usually the role of the miller to purchase wheat to give flour with the properties required by the customer. This is achieved by the blending the lowest cost grist to consistently meet the customer's needs by the most efficient milling processes.

22.3 CHOICE OF WHEATS FOR BLENDING

Except where it has been purchased directly from a grower under contract, wheat is rarely available as a single variety. Grain of a suitable range of genotypes (varieties) from several growers will generally be mixed and stored as a single grade with a narrow (<1%) protein range, for example 11–12%. A range of tests are performed at the time of grain receival, as described in Chapter 19, to determine the relevant grade for binning purposes. Millers will purchase wheat based on established grades and reported grain quality from Crop Reports of major exporting countries and domestic harvest testing, or samples will be made available to prospective purchasers for their own testing. Such testing will involve grain quality parameters and test-milling on a small scale, followed by rheological testing of the flour. Most millers do not specify the varietal mix, as this is already known when selecting the particular grade being purchased. On occasions, millers will seek or avoid grades which have specific varieties, but this is not always practised, or even possible.

Wheat of exactly the desired protein content and quality is frequently not available, and it is necessary for millers to purchase grades of different protein content and/or quality for blending. Often this will include wheat from different origins with the aim of producing flour of consistent quality. A high-protein wheat, with dough properties too strong for most purposes, will be blended with a wheat with a lower protein and weaker dough-quality potential, to give flour with an intermediate protein content and appropriate dough strength. While the protein content is usually determined with a linear relationship based on the proportions of the wheats being blended, dough properties are usually not linear (Bekes et al., 2001) and careful selection of grades must be made. Another quality factor to consider when blending wheats is falling number, a measure of pre-harvest rain damage. Blends of wheats composed of wheat with different individual falling number will have a much lower falling number than would be predicted on the basis of the arithmetic mean. Falling number must be converted to liquefaction number before averaging (see Method 107/1 of the ICC Methods for an explanation of how this is calculated).

22.4 HISTORY OF MILLING

Grains and seeds have been ground or crushed to flour using primitive mortars and stones for many thousands of years. The Egyptians developed saddle-stone grinders which allowed grinding of larger amounts of grain. Scenes depicted on the walls of Egyptian tombs from around 2600 BC such as those in the tomb of King Ramses III show that milling and baking activities were well established at that time. The next major improvement in around 400 BC was the development of the rotary action hour-glass mill or conical quern mill. Larger quantities of ground wheat could be produced using oxen or slave labour to turn the mill stones. Smaller domestic versions were also available. The quern is probably the forerunner of the two large horizontal mill-stones used for centuries in water, wind and steam-powered mills, and stone ground flour is still produced to-day. By the 19th century, millstones weighed up to 1 tonne and had a diameter of up to 137 cm (Jones and Jones, 1990). Sifting was employed to separate coarser endosperm (flour) particles from the outer branny layers, and the separated endosperm was re-ground. Purification on the basis of particle size was also introduced.

Although the concept of roller milling can be traced to 16th century Spain, the first attempt to commercialise a roller mill was not until 1820 (Walker and Eustace, 2016). However, complete re-design of mill layout was required, together with the introduction of pairs of rolls with automated adjustable clearances between the rolls to manage changing feed pressures. Modern roller flour milling was developed in Hungary in the 1870s, where it was found that wheat produced by the roller milling process gave a higher yield of flour of superior quality. Other improvements followed in the development of a gradual reduction system, and improved equipment and mechanisation. Despite some initial hesitation because of the investment involved, millers worldwide adopted the new process, and those that did not went out of business.

22.5 WHEAT MILLING

The wheat grain is composed of about 82% endosperm, 15% bran and 3% germ. The aim of milling is the efficient separation of bran and germ from the endosperm (Fig. 22.1) and to reduce the particle size of the endosperm to flour.

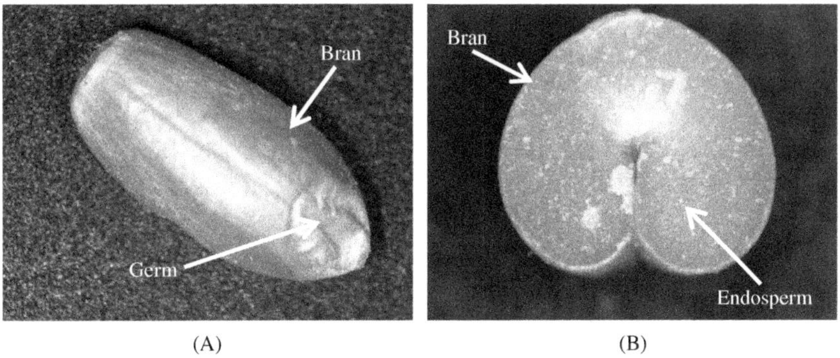

FIGURE 22.1 Light micrographs of (A) wheat grain and (B) transverse section of wheat grain (not to scale). *(Provided by K. McGee)*

Prior to milling, there are a number of essential steps required to ensure a quality product.

Wheat is cleaned to remove impurities which can include weed seeds, seeds of other cereals crops, and stones, sand, nails, wire and wood. These need to be removed as they can damage milling machinery and/or contaminate the final mill products. In the screen room, cleaning machines, such as separators, combinators, magnets and aspirators separate impurities on the basis of size, shape, specific gravity, magnetic properties and air resistance (aspiration). Small and shrivelled grains which will have low milling extraction can also be removed in the cleaning process.

Wheat is conditioned or tempered before milling by addition of sufficient water to bring the grain to the optimum moisture for milling and then allowing sufficient time (lying time) for the moisture to equilibrate. Conditioning toughens the bran, making it less friable and easier to separate from the endosperm. It also modifies endosperm hardness making it easier for the milling process to reduce the particle size and to optimise starch damage. The amount of water added in the conditioning process will depend on the initial moisture content of the wheat, grain hardness, climatic conditions and on the target moisture content of the milled products. Generally hard wheat is conditioned to 15.5–17% moisture, durum wheat to 16–17% and soft wheat to 14–15.5% moisture. Conditioning may be a single or multistage operation. Lying time varies from 6 to 24 h, with soft and durum wheats having a lower lying time requirement. Conditioning regimes will vary according to the mill, initial wheat moisture content and origin or type of wheat or wheats milled. In all cases, optimum conditioning will ensure efficient separation of bran from endosperm and correct moisture content of products.

22.5.1 The Milling Process

While the milling process as a whole consists of many operations, it can be divided into three main stages: breaking, separation on particle size and density, and reduction. A simplified mill flow system is shown in Fig. 22.2.

22.5.1.1 Break System

The first grinding operation is through a series of pairs of fluted or corrugated break rolls, which open the grain and scrape the bran to release endosperm particles. There are generally 4–5 break roll passages (1BK, 2BK, 3BK, etc.) which have successively finer spiral flutes. Between each of the break operations, the stock is sieved in a plansifter separating flour from bran and larger endosperm particles. The sizings system grades the endosperm, producing a narrower particle-size range, thus optimising the performance of the reduction system. Bran finishers and quiver sifters remove any remaining flour from the coarse bran after the break system.

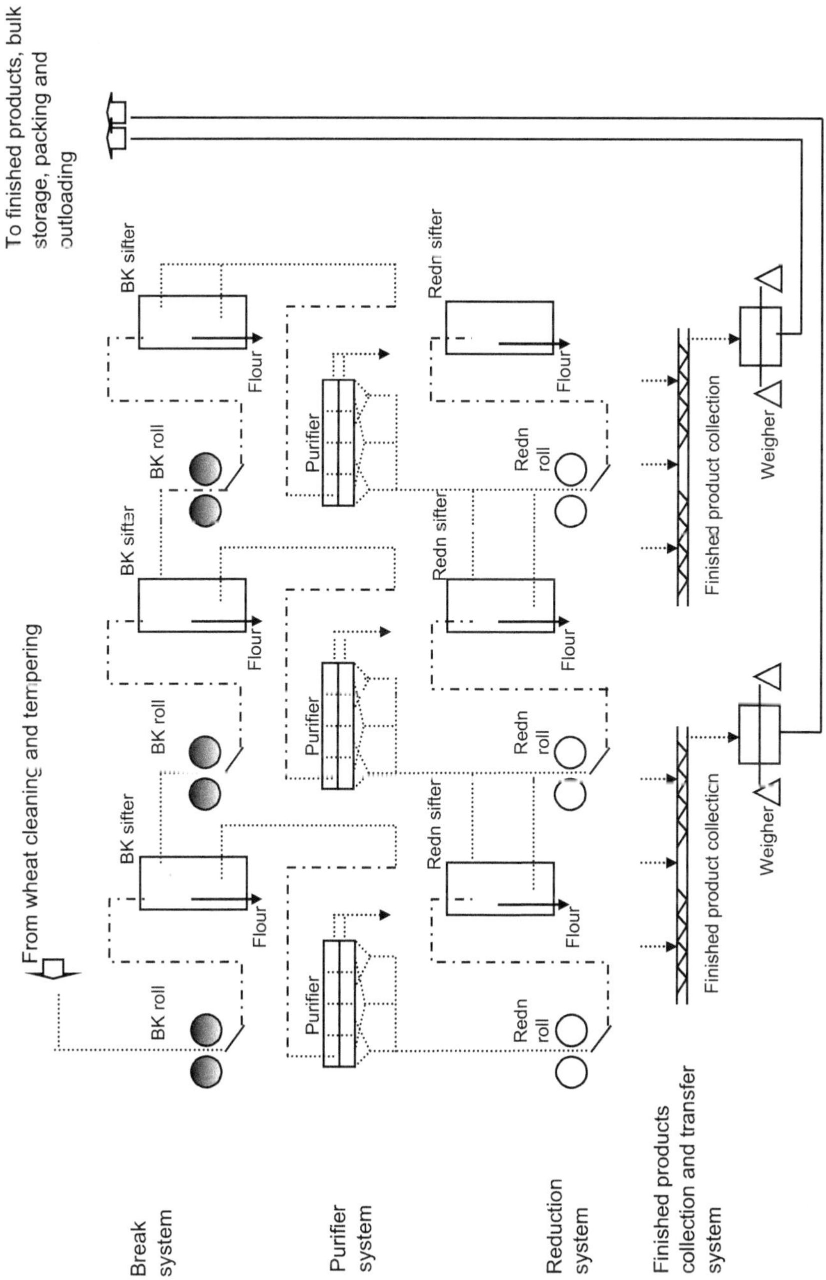

FIGURE 22.2 General mill diagram for a commercial flour mill. *(Provided by E. McGraffin)*

22.5.1.2 Separation on Particle Size and Density

Plansifters separate on the basis of particle size and density. They are composed of a number of tiered sieve boxes, which are clothed with nylon, polyester or wire screening material. Sieve sizes are selected to allow gradual sifting or grading of bran, semolinas and flour.

Purifiers separate the fine bran from the semolina (coarse endosperm, also known as farina) on the basis of size, density and specific gravity, using sieves and air currents. Coarse and fine endosperm particles are separated in the purifiers before being sent to the reduction system.

22.5.1.3 Reduction System

The reduction system aims to reduce the size of the endosperm chunks and produces the required starch damage levels. There are from 8 to 12 reduction-roll passages; these are between pairs of smooth, frosted rollers and are known as A, B, C or 1M, 2M, 3M, etc. The rolls operate at slightly different speed differentials of 1:1.25. Flour is removed in multisection plansifters between reduction roll passages. Germ is removed in the reduction system and flaked. Remaining fine bran, endosperm and germ materials are removed as pollard ('red dog' in the United States).

22.5.1.4 Differences Between Wheat Milling and Durum Milling

Generally mills will be dedicated to wheat or durum milling, as the two have different designs and numbers of milling machines. Compared with bread wheat, durum wheat has a shorter conditioning time, and mills have finer fluting of break rolls, few reduction rolls and many more purifiers. The durum semolina produced is usually between 250 and 500 μm, which is much coarser than the durum flour by-product or flour milled from bread wheat.

22.5.1.5 Blending and Packing

At the end of the milling process, flour streams will be combined either in total—as a 'straight run' or 'straight grade' flour; alternatively, selected streams may be combined to give a number of divides or flours of different quality which are used for different purposes. Flour is stored in bulk bins prior to bulk tanker delivery or automated packaging into bags of various sizes.

22.5.1.6 Debranning

A recent development in milling technology has been the introduction of debranning using the PeriTec system. The outer bran and layers are sequentially removed prior to milling (www.satake-europe.com). This permits the use of a simpler mill flow, and produces flour that has lower levels of bran contamination but higher ash. Debranning has particular applications in durum milling, giving improved semolina yield and colour (Dexter and Wood, 1996).

22.5.2 Extraction Rate Versus Grist Rate

Flour extraction rate (the amount of flour produced from a known quantity of grain) is a measure of efficiency of milling separation in real time. Flour extraction can be expressed on the basis of the original wheat, as dirty, clean or conditioned grain. For example for the straight run flour in Table 22.1, and expressed on a conditioned wheat weight,

Extraction rate = 100 × Weight of straight run flour/Weight of wheat to 1 BK

Flour extraction rates have increased in recent decades, due to selection of good milling lines in breeding programs and improvements in milling equipment. The extraction rate of a straight run flour is generally 74–76%, but it is not uncommon for higher extraction rates to be achieved, especially when high quality wheats are available. An improvement of even 1% in extraction rate represents considerable value to the flour miller.

Another method of measuring extraction is grist rate determined by the number of tonnes of dirty wheat required to make one tonne of straight-run flour. In this instance, the level of screenings, un-millable material and moisture can have a large impact on the result. These variables are often beyond the control of the miller and are dependent on the grades and types of wheat purchased.

22.5.3 Mill Products and Their Uses

The major mill products are:

- Flour—straight run and divide. Used for all types of breads, starch and gluten production, noodles, cakes, pastry, household, industrial.
- Semolina (farina, middlings)—coarse and fine. Used for pasta, continental breads, baby food, dusting.
- Bran—coarse and fine. Used for feed, human consumption and breakfast cereal.
- Germ—Used for human consumption and pharmaceuticals.
- Pollard (red dog)—Used for feed.
- Wholemeal or wholewheat flour—increasingly used for breads and all types of end products

The names and compositions of the various products will vary according to country and customer requirements. For example, the term 'semolina' may be reserved in some countries for durum, rather than for common bread wheat. Milling by-products, such as bran and pollard, may be combined to make products called millrun or offal which are used in animal-feed formulations. Flour types are discussed later in this chapter.

TABLE 22.1 Mill Stream Analysis Commercial Flour Mill

Stream	Extraction (wheat to 1BK) %	Protein (%, as is)	Moisture (%)	Ash (%, as is)	Starch damage (AACC %)	Water absorption (%)	Development time (min.)	Maximum resistance (BU)	Extensibility (cm)
Break									
1BK	5.5	13	14.6	0.63	6.1	61.6	8.5	395	28.3
2BK	3.8	13.8	14.7	0.54	5.3	61.1	10.5	485	28.8
3BK	4.8	16.4	14.3	1.01	6.1	65.3	5.7	250	28
3BKBC	1.6	16.3	13.8	0.91	6.6	66.8	6.9	235	27.8
4BK	1.9	16.4	13.6	1.43	12.6	68.6	5.5	165	27.8
4RE	0.8	14.6	13.2	1.39	10.9	68.4	5		
SZ1	1.2	12.4	14.7	0.67	9.5	63.5	7.4	425	24.8
SZ2	0.6	13.1	14.5	0.83	7.9	62.7	6	270	28.3
Reduction									
A1	5.7	10.7	14.3	0.36	8.9	63.1	9.5	450	21.6
A2	6.1	11.4	14.2	0.38	7.4	62	9.6	495	22.7
B1	15.7	10.7	14.2	0.38	7.2	62.2	8.1	470	19.7
B2	14.2	12.6	14	0.52	7.2	64.7	5.2	310	25.6
C	8.4	11	13.1	0.54	10.1	68.9	3	335	19.1
B-2	1.5	12.3	13.8	0.73	9.9	67.6	4.7	225	23.7
D	2.4	11.3	12.8	0.61	9.3	69.7	4	270	19.5
E	1.5	11.2	11.7	0.71	12.1	74.2	3.5	175	17.8
F	0.6	14.5	12.7	1.25	10.6	73.8	4.7		
G	2.3	12.4	11.3	1.15	13.1	80.4	4		
H	1	15.5	10.2	2.2	14.8	89.5	4.7		
Straight run	78.7	12.3	14.1	0.6	8.3	64.8	5	320	23.5

Source: Data provided by Flour Millers Council of Australia.

22.6 QUALITY ASSESSMENT DURING AND AFTER MILLING

Some quality assessment can be carried out during milling. Rapid tests that are applicable include: protein and moisture content determined by near infrared spectroscopy (NIR), degree of starch damage, sieving to measure granularity of mill products, the Pekar test for flour colour, and determination of flour-water absorption. These will be discussed later in this chapter. Online NIR systems can also be used to estimate moisture, protein content and ash.

A range of tests must be carried out post-milling to ensure conformity to specification, thus to meet customer requirements. These can be divided into chemical, physical, rheological and end-product tests.

22.6.1 Moisture Content

The moisture content of flour is defined as the loss in weight on drying at a standard temperature for a set time (usually at 130°C for 60 min). Standard methods are detailed in Table 22.2. While moisture can be determined by weight loss, it is routinely determined by NIR. Moisture is an important characteristic, serving as a quality and economic indicator of milled flour to ensure conditioning regimes have been correct and that it is not below specification. As many other flour quality tests are corrected to a constant moisture basis, it is often a pre-requisite for further testing. Flour moisture content should be checked during production and on out-loading, as flour moved by blow lines within the plant tends to lose a small amount of moisture.

22.6.2 Protein Content

Protein content is one of the main determinants of suitability for end use and is on almost every flour specification. Protein content is measured as nitrogen by the Kjeldahl or Dumas methods (Table 22.2). However, routinely NIR is used. For flour, protein content is usually expressed on the basis of nitrogen content multiplied by 5.7 (N × 5.7), and it may also be corrected to a constant moisture basis, such as 14% or a dry basis (% db). The protein content of flour is dependent on the protein content of the original wheat, and on extraction rate. Typically there is a drop in protein content from wheat to flour of 0.6–2.0%.

22.6.3 Measurement of Flour Colour and Flour Purity

Assuming no foreign matter contamination, flour colour is dependent on bran contamination and the presence of xanthophylls, which are flour pigments that impart a creamy or yellowish colour to the flour. There are a number of methods to assess flour colour and flour purity.

The traditional method of measuring the efficiency of milling, the bran contamination and the purity of the flour produced is the ash test (Table 22.2). Ash is the mineral residue that remains after combustion. Thermogravimetric

TABLE 22.2 Methods for Analysis of Flour

Method	Procedure	Method references
Moisture	Oven drying	AACC 44-15.02, ICC 110/1, CCAT 08, AGSA 01-01
Protein	Kjeldahl	AACC 46-12.01, ICC 105/2, AGSA 02-01
	Dumas	AACC 4630.01, ICC 167, CCAT 19, AGSA 02-03
	NIR	AACC 39-11.01, calibrated against standard chemical methods
Wet gluten	Gluten washing	AACC 38-12.02, ICC 137/1, CCAT 13
Ash	Incineration at high temperature	AACC 08-01.01, ICC 104/1, CCAT 12
	NIR	AACC 08-21.01
Flour colour	Tristimulus colour	CCAT 25, AGSA 09-02
	Agtron	AACC 40-30.01
	Pekar	AACC 14-10.01
Starch damage	AACC	AACC 76-30.02
	Megazyme	AACC 76-31.01, ICC 164, CCAT 21, AGSA 04-02
	Chopin SDmatic	AACC 76-30.01, CCAT 24
	Farrand	CCAT 05
	NIR	Calibrated against standard chemical methods

AACC-AACC International. Approved methods of Analysis, 11th edn., St Paul, MN, USA. Available online www.aaccnet.org.
AGSA-AGSA (formerly RACI), 2003. Official Testing Methods of the Australasian Grain Science Association, 4th edition. Australasian Grain Science Association, Narrabri, NSW, Australia.
CCAT-Manual of Methods of the Cereals and Cereal Applications Testing Working Group, Campden BRI. Gloucestershire, UK. Available from www.campdenbri.co.uk.
ICC ICC, 2011. Standard Methods of the International Association for Cereal Science and Technology (ICC), Vienna, Austria.

analysis (TGA) is a new instrumental method allowing both ash and moisture analysis to be available in 4–5 h, with a higher throughput than the conventional methods. Use of ash content as a measure of flour purity is based on the assumption that the endosperm has lower ash content than the bran, aleurone and germ. A high ash flour will have a darker colour than a low ash flour. Ash content can be expressed on an 'as analysed', constant moisture or dry basis, so it is important to understand which of these has been applied in a flour specification or when comparing flour analytical data.

The ash content of wheat grain can vary significantly, depending on the wheat variety and on environment factors, such as soil type and climatic conditions. Although there is a positive correlation between wheat ash and flour ash (www.satake.co.uk), it is really bran contamination that affects flour quality, product colour and end-product performance. Despite these reservations, ash content is still used in many flour specifications and as a measure of milling performance.

Tristimulus colour meters have gained acceptance due to their ease of use on dry flours and end products. Colour meters operate according to the CIE (Commission Internationale de l'Eclairage) colour space parameters. The most commonly used parameters are:

L* (brightness/whiteness): 0–100 (a value of 0 is black)
a* (red/green): −60 to +60 (a value of −60 is green)
b* (yellowness): −60 to +60 (a value of −60 is blue)

Most flours have an L* value of >90 and a b* value of 5–12. Higher L* values indicate more cleanly milled flours, and higher b* values indicate creamy yellow flours. Dry flour measurements have the disadvantage that finer particle size appears lighter and less yellow, resulting in an increase in L* and a decrease in b* (Oliver et al., 1992).

Digital image analysis techniques depend on the contrast between light (endosperm) and dark (coloured components of bran layers) materials. Bran content is estimated as % light to dark and the number of individual specks can be accounted above a minimum speck sizes, based on the number of pixels. Branscan is an example of this type of instrument (www.branscan.com).

The Fluoroscan (www.branscan.com) uses image analysis to measure aleurone content by ultraviolet autofluorescence and pericarp content by visible light absorption. Data processed from these measurements are used to predict flour ash content. The sample is presented as a dry powder and results are available for all three parameters within two minutes with real-time image display. The equipment is available in laboratory and online configurations.

Flour-grade colour, or colour grade, was originally developed by Kent-Jones and Martin in 1950 to give a measure of bran contamination in a flour/water slurry at 540 nm. It has largely been replaced by other methods as outlined above.

The Agtron (www.agtron.net) also measures the reflectance of a flour/water slurry. Results are independent of flour bleaching. The same instrument can be used to estimate flour yellowness at a wavelength of 436 nm.

The Pekar test is a simple qualitative visual assessment of flour colour traditionally favored by flour millers. Flour samples are packed onto a glass plate using a flour slick, and then immersed in cold water (Table 22.2). With time, dark specking associated with the bran enzyme polyphenol oxidase (PPO) appears, and the background colour due to xanthophylls become visible, varying from white to creamy yellow.

Flour colour is important for quality prediction of products made from white flour. Flour colour affects bread crumb colour, the colour of frozen pastry and many other end products. It is particularly important for Asian products, such as steamed bread and noodles. Many noodles are sold uncooked and any bran contamination will be apparent as darkening and specking with time. Consumers often wrongly interpret this as a microbiological-spoilage issue instead of darkening as a result of PPO activity.

22.6.4 Starch Damage and Water Absorption

During the milling process, some of the starch is mechanically damaged. Damaged starch absorbs up to 3 times more water than undamaged starch (Greer and Stewart, 1959) and in doughs this form of starch can be readily attacked by flour amylases yielding fermentable sugars. Hard wheat has a compact endosperm structure in which starch granules are tightly bound within the protein matrix. Flours from hard wheat have a high level of starch damage compared with soft wheat which has a less compact endosperm structure and is readily milled to flour with low starch damage. There are optimum starch damage levels suited to different end products, with bread having a requirement for high (but not excessive) starch damage, and biscuits and cakes having a requirement for low starch damage. Morgan and Williams (1995) divided starch damage into high (7–10.25 AACC units) and low (3.25–5.5 AACC units), corresponding to soft and hard wheat populations. Pasta manufacture also has a requirement for low starch damage, and for this reason it is produced from semolina which does not undergo the heavy milling needed to produce fine flour.

Starch damage is increased during the reduction stage of the milling process, with each successive pass having a higher degree of starch damage (Table 22.1). Commercial millers have control over starch damage levels and a number of options are available for increasing starch damage levels. These include increased reduction roll pressure, increased wheat hardness and decreased conditioning moisture levels.

Methods for determining starch damage are listed in Table 22.2. The once widely used Farrand method has now been largely replaced by the Megazyme method, which uses enzyme digestion followed by spectrophotometric measurement. This handy kit can be used in a batch type process with a high throughput (www.megazyme.com). The Chopin SDmatic is an amperomeric method that is rapid and convenient for mill use. Suitably calibrated NIR can also be utilised. Equations for conversion from one chemical method to another are available (Gibson et al., 1993; Morgan and Williams, 1995).

22.6.5 Water Absorption

Water absorption is the amount of water, expressed on a flour-weight basis, that is required to produce a dough of appropriate consistency. Water absorption is of prime importance in baked products, affecting every stage of the process, as well as the yield and quality of the finished product. Protein content, damaged starch level and non-starch polysaccharide content (mainly pentosans) all influence water absorption. Protein absorbs approximately twice its weight in water as the dough is mixed; some polysaccharides are even more effective in absorbing water.

In the flour mill laboratory, water absorption is routinely measured for process control and quality assurance using the Farinograph (www.brabender.com) or similar dough-mixing instruments (Table 22.3). The standard dough consistency for the Farinograph is at 500 Brabender Units (BU) (or 600 BU in the UK). Flour water absorptions vary from 50 to about 70%, depending on wheat hardness, milling conditions and the targeted flour specifications.

TABLE 22.3 Methods for Measurement of Dough Rheology

Instrument	Measures	Method references
Farinograph	Water absorption, development (peak) time, arrival time, departure time, stability mixing, tolerance index	AACC 54-21.02, ICC115/1, CCAT 04, AGSA 06-02
DoughLab	Water absorption, peak consistency, development time, stability, mixing energy, mixing tolerance index, torque	AACC 54-21.02, 54-70.01 (high speed), AGSA 06-02, ICC 115/1
Mixolab	Water absorption, mixing time, stability starch gelatinisation enzymatic activity starch retrogradation	AACC 54-60.01, ICC 173
Alveograph	Maximum pressure (P), length of curve to rupture (L), area under the curve (W), ratio between dough strength and extensibility (P/L)	AACC 54-30.02, ICC 121
Consistograph	Water absorption, mixing time	AACC 55-50.01, ICC 171
Extensograph	Maximum resistance, extensibility, area under the curve	AACC 54-10.01, ICC 114/1, CCAT 03, CCAT 16, AGSA 06-01
Viscograph	Pasting temperature, peak viscosity, peak temperature, final viscosity, setback	ICC 126/1, AGSA 06-03
RVA	Peak time, peak viscosity, holding strength, final viscosity	AACC 76-21.01, ICC 162

22.6.6 Physical Dough Properties

When flour is mixed with water to form a dough, the storage proteins are hydrated to form a gluten matrix. The resultant dough properties depend on protein content and protein quality which are determined by the type, variety and quality of the wheat. Most wheat-based foods have a specific requirement for dough properties which can be classified in terms of dough strength (Table 5.5), extensibility and the balance between these two attributes.

Recording dough mixers (Table 22.3) measure the resistance of a dough to mixing at controlled temperature and mixing speed. As water addition influences dough consistency, several attempts may be required to achieve correct consistency. Development time gives an indication of optimum mixing time in bakery applications. Both development time and stability are indicators of dough strength, with higher values indicating greater strength in both cases. Together with water absorption, these are usually found in flour specifications and certificates of analysis. Fig. 22.3 shows the measurements that can be derived from a mixing curve using the DoughLab instrument (www.perten.com).

The Extensograph ('Extensigraph' in the United States) records the force required to stretch a dough piece to breaking point with a hook which moves at constant speed. The dough has been previously mixed in the Farinograph and includes salt. Dough strength is measured as maximum resistance and extensibility as the distance which the dough piece stretches before rupture. Interpretation of Extensograph curves permits the classification of dough properties as

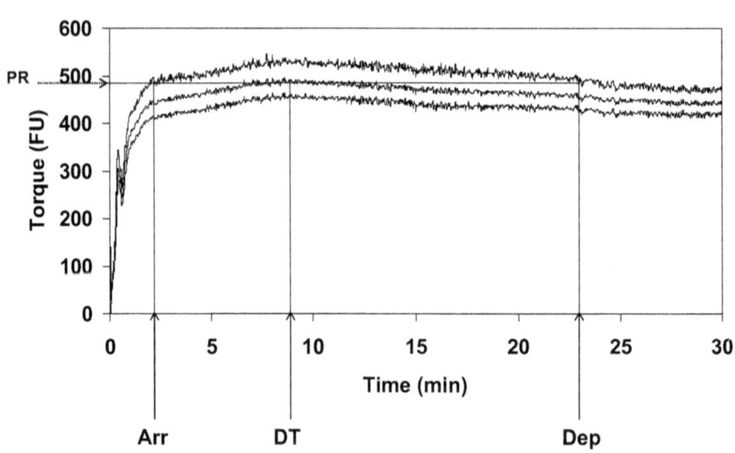

FIGURE 22.3 DoughLab mixing curve showing some of the parameters measured. *PR*, Peak resistance, the highest value of the mean mixing curve; *DT*, development time, the time of peak resistance; *Arr*, arrival time, the time when the value of the maximum mixing curve reaches PR; *Dep*; departure time, the time when the maximum mixing curve falls below PR. Stability is the difference between the departure time and the arrival time. For a full range of DoughLab parameters see Cauvain and Young (2009).

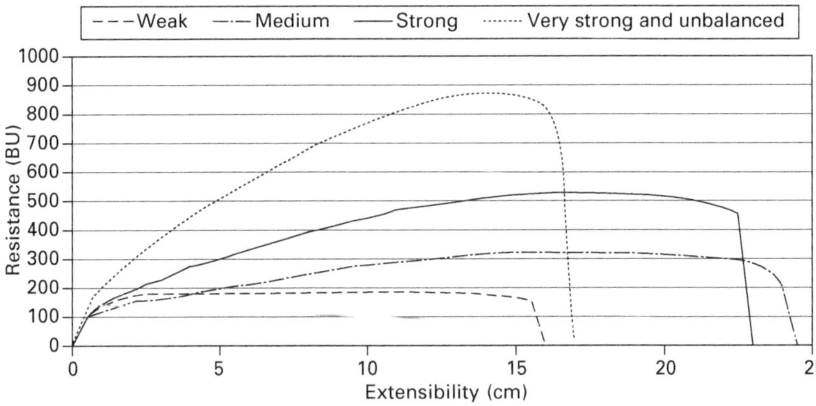

FIGURE 22.4 Extensograph curves of weak, medium, strong and very strong but unbalanced flours.

weak, medium and strong (Fig. 22.4). Doughs must have appropriate extensibility for their strength, otherwise they will be regarded as 'unbalanced'.

The Alveograph (www.chopin.fr/en) measures the air pressure and time to inflate an extruded dough piece to form a bubble and burst. It was originally developed for weak French flours and traditionally operates on a fixed flour:water ratio known as constant hydration (Table 22.3). The Alveograph is used in Europe, South America and parts of the Middle East and Asia. Flour specifications often stipulate the area under the curve—W, which is an indication of baking strength. Flours with high water absorption may produce doughs which are too dry with a consequent overestimation of dough strength. The AlveoConsistograph enables alveographs to be performed at optimum water addition, which is better suited to flours with higher starch damage levels.

Whereas most recording dough mixers measure the behaviour of hydrated proteins, the Mixolab not only measures a mixing curve, but also measures the behaviour of starch in the dough after the mixing stage into the final stage of heating. As temperature increases, starch swelling, gelatinisation and breakdown occur. Fig. 22.5 shows the differences in mixing and starch properties of a strong baker's flour and a weak biscuit flour dough using the Mixolab.

22.6.7 Evaluation of Starch Pasting Properties

Starch is the major component of wheat flour and is often overlooked in terms of its functional properties and contribution to end-product quality. Cakes, Japanese Udon noodles, some instant noodles, hot-plate products, batters, soups and wafers all rely on starch gelatinisation to confer the correct textural characteristics. Flours that are suitable for these uses require high paste viscosities. In baking, starch gelatinisation assists in the formation and stabilisation of the bread crumb, and is a factor in bread staling.

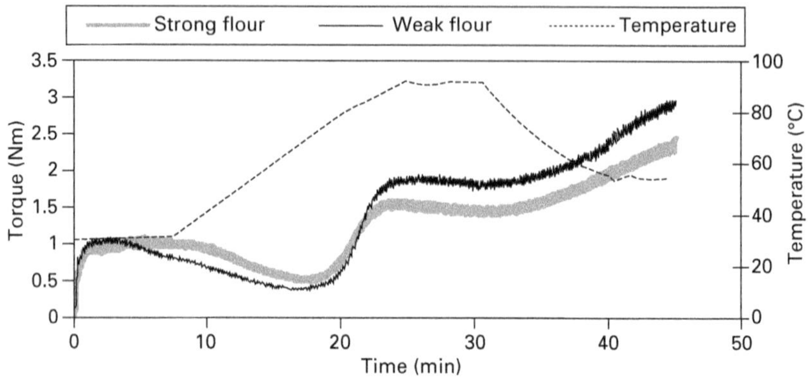

FIGURE 22.5 Mixolab profiles of weak and strong flours showing mixing curve and gelatinisation profile. *(Provided by H. Salman)*

The cooking and pasting properties of flour/water and starch/water mixtures during heating and cooling are measured as resistance to the stirring action of paddles or pins, recorded as changes in viscosity with time. The Viscograph and the Rapid Visco Analyser (RVA) are two such instruments (Table 22.3). The RVA, using only a small sample, offers flexibility in terms of programmable temperature profiles and shear rates. The STD 1 profile for the RVA is completed after 13 minutes. A typical RVA curve is shown in Fig. 22.6.

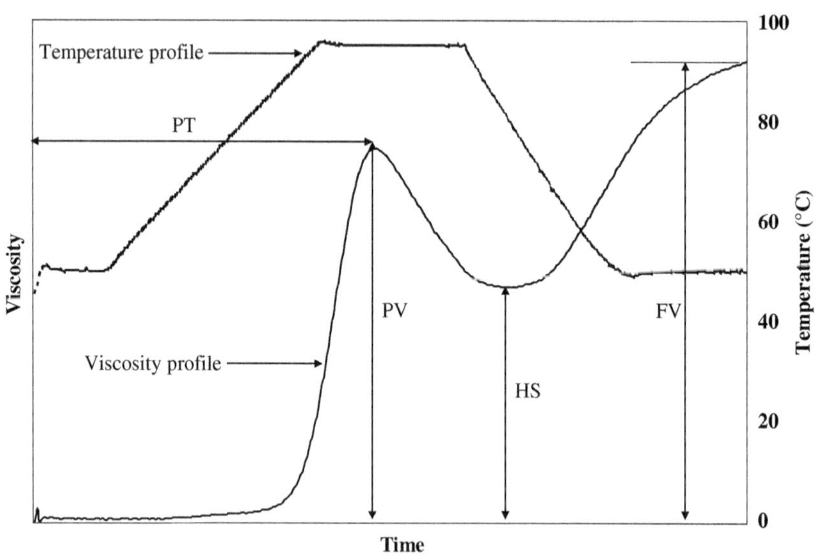

FIGURE 22.6 RVA curve showing main parameters measured. *PT*, peak time; *PV*, peak viscosity; *HS*, holding strength (or trough viscosity); *FV*, final viscosity. Other parameters often considered are breakdown (PV minus HS) and setback (FV minus HS).

Low peak viscosities, indicating thin viscous pastes, may be an indication of sprout damage to the grain due to rain at harvest. There is a strong correlation between the RVA stirring number method (AACC Method 22-08.01) and Falling Number (Bason and Blakeney, 2007). Additionally, both instruments may be used to measure the effect of cereal or malt flour α-amylase added to flour.

22.6.8 Wet Chemistry Versus Near Infrared Spectroscopy (NIR)

22.6.8.1 Calibration of NIR Instruments

NIR, introduced in the grains industry in the 1970s, gained rapid acceptance not only in food and agriculture, but in many other application such as pharmaceuticals and medicine. NIR (Near Infrared Reflectance) and NIT (Near Infrared Transmittance) are recognised as providing easy-to-use, rapid, and non-destructive methods of analysis. The technique is based on absorption in the infrared region of the electromagnetic spectrum in the wavelength region 780–2500 nm. Instrumentation utilises grating monochromators, diode array spectrometers or fixed filter instruments, the latter being the cheapest and generally limited to protein, moisture, grain-hardness and fat analysis. There are many instruments available for flour-mill wheat intake and quality control; selection of the most appropriate model depends on the application.

NIR/NIT is a secondary analysis method. This necessitates calibration of the instrument using a large sample population and standard methods of analysis. It may be a challenge for a flour-mill laboratory to achieve appropriate calibration if the mill has only a limited range of analysed samples for the analytical property of interest. Calibration involves an understanding of the application of chemometrics plus a high degree of analytical accuracy and precision in manual testing methods. Some instruments may have calibrations included, but these will need validation using a range of local samples (Delwiche and Millar, 2010).

22.6.8.2 Applications of NIR in the Mill

While the original and most successful usage of NIR/NIT was for protein and moisture content, other applications of interest to the miller such as wheat hardness (Osborne, 1991) and starch damage (Osborne et al., 1982) have been developed. The potential application of the technique to other traditional flour and wheat quality parameters was initially hindered by high standard errors of cross validation (SECV) and the influence of cross-correlations with protein content (Dowell et al., 2006). Commercial applications, which offer a suite of calibrations based on large sample sets, coupled with a data-management package and technical support, are now available and in use by flour millers (Dominy, 2003). NIR/NIT methods have higher standard errors than the wet chemistry methods on which they are based and their accuracy can be influenced by seasonal conditions. However, this may be outweighed by the convenience and repeatability of the NIR technique.

Online NIR instruments are available integrated with milling systems. Originally used for determining flour protein content, they could be integrated into a feedback control system to control gluten addition to flour. The newer, more reliable systems can be used in flour quality assurance and process control of flour protein, moisture, starch damage, ash and colour, with real-time results fed back to the mill control room every 10 s.

22.6.9 Stream Selection—Straight Run and Divide Flours

Flours from each of the different milling stages have quite different properties as shown by mill-stream analysis in Table 22.1. These differences are dependent on their origin within the endosperm, on contamination with non-endosperm components, and on the degree of processing. Knowledge of the properties of the various streams allows decisions to be made regarding which, if any, streams are to be blended or removed in the final flour.

22.6.9.1 Mill Stream Analysis

The first parameter considered is usually ash. Ash content is lower in reduction streams such as A1, A2 and B1, which are derived from the central endosperm (Table 22.1). Ash content increases as the streams contain more non-endosperm material. To construct a cumulative ash curve, the cumulative ash is calculated for each stream and plotted against extraction rate, starting with the stream with the lowest ash and proceeding with the next lowest in turn (Fig. 22.7). The objective in designing a good milling process is to produce a curve which has a low gradient at the base, rising to a steep gradient at the end where the dirty flour streams are included. Ash content can be used to predict commercial-mill yield.

As can be seen from Table 22.1, the protein range across the streams is over 5%, and may be as high as 7%. This is because there is a protein gradient across

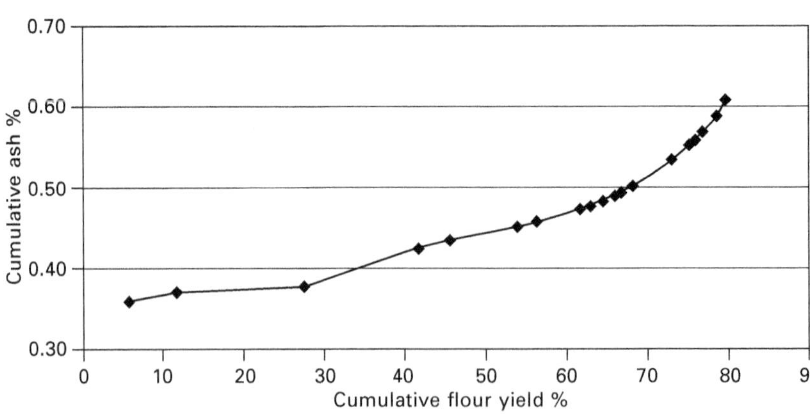

FIGURE 22.7 Cumulative ash curve for a commercial flour mill.

the grain and the bran layers are higher in protein content than endosperm; consequently the protein content is higher in dirty streams. Protein content is related not only to cost, but also to functionality. A clean stream such as A1, which is at the beginning of the reduction system, has strong dough properties as indicated by a high Farinograph development time, high Extensograph resistance and high extensibility, but a moderate protein content, namely 10.7%. The influence of bran and germ on strength can be seen by the weak dough properties in the dirty reduction streams B-2, D and E, with E also exhibiting a decrease in extensibility.

The starch damage in flours from the early break rolls is quite low, as there is not a lot of grinding early in the milling process and the central endosperm is more friable than the outer grain layers. Starch damage is additive with successive passes through the reduction rolls. This can be seen from a comparison of A2, with a starch damage of 7.4, and C with a starch damage of 10.1.

Water absorption in the streamed flours is impacted by starch damage, protein content and the moisture content of the flours. Reduction streams such as G and H, which have very high water absorptions, are quite sticky and contain finely divided bran and germ as well as high levels of pentosans.

22.6.9.2 Divide Flours

If all the streams are combined, a straight run or straight grade flour is produced. This is the simplest milling system. The extraction rate depends on milling efficiency, wheat-bran colour (white or red) and whether ash content is an issue in meeting flour specifications. It is possible to produce a wider range of flours by blending or removing various streams. The following scenarios illustrate some of the divides that can be achieved:

1. Remove dirty or tail-end streams. In the example of the cumulative ash curve in Fig. 22.7, a patent flour with an extraction rate of 70% would correspond to an ash content of about 0.5%. This flour would have better colour, and show an increase in dough strength and baking quality. The residual or clear flour would be used for feed, plywood or alcohol production.
2. Remove one or two clean streams. This accounts for about 10% of the cleanest flour, with an ash content <0.45%. The remainder of the flour can be used as a straight-run or main-grade flour with a small loss in quality and end-product performance. The low ash flours are considered as high quality in the market and attract a premium. Depending on the grist, these flours may be used for pastry, noodles and cake.
3. In some Asian countries, up to three or four divides or grades of flour can be produced from a single milling. For example, the lowest ash flour (<0.40%) could be used for noodles, second grade flour (ash <0.70%) for confectionery and household use and third grade flour for industrial applications.

In each case, the aim is to maximise the extraction that can be achieved, while maintaining quality specifications, end-product performance and consistency.

22.6.10 Flour Treatments and Additives

22.6.10.1 Chlorination

Chlorination treatment bleaches flour by destroying xanthophyll pigments; it modifies lipid, protein and starch properties and decreases the pH to about 4.5. Flour pH can be determined using AACC Method 02-52.01. Chlorinated flour is used in the production of high-ratio cakes, which contain more sugar than flour. These cakes have improved volume and crumb strength and white crumb colour. Due to environmental and health concerns, chlorination has been banned in a number of European and South American countries, the UK and Japan.

22.6.10.2 Heat Treatment

Where chlorination has been withdrawn, flour heat treatment is used. In general, heat treated flours produce cakes of inferior quality to those made from chlorinated flours and the flour cost is considerably greater. Heat treatment of flour is also used to reduce flour moisture content for use in packet cake mixes, and infant foods. The RVA may be used to monitor the heat treatment process (www.perten.com).

22.6.10.3 Bleaching

Although flour bleaching agents were commonly used in flour milling in the past, their use has decreased or has been discontinued in many countries. A typical flour bleaching agent is benzoyl peroxide, which bleaches the xanthophyll pigments responsible for the creamy yellow colour of flour. Reaction time is 24–48 h and the effect of bleaching can be readily monitored by tristimulus b* measurements.

22.6.10.4 Flour Modifying Agents

Flour improving agents for bread making may be added at the mill. Typical additives, where use is permitted, are ascorbic acid, malt flour and fungal and bacterial amylases. Addition of additives at the mill can improve flour performance and consistency (www.muehlenchemie.de).

22.6.10.5 Vitamin and Mineral Enrichment

Wheat is an important source of B group vitamins and minerals such as calcium, iron and magnesium. In countries such as the United States and United Kingdom, flour has long been used as a source for replacement of vitamins and minerals lost in processing, by enrichment to higher levels. Thiamin, niacin, riboflavin and more recently folic acid are added, along with calcium and iron. The global Food Fortification Initiative (FFI) aims to prevent nutritional anaemia and neural tube birth defects worldwide. To date, it is estimated that 85 countries have legislated to fortify flour; thus about 30% of the world's flour is fortified (www.ffinetwork.org/).

22.6.11 End-Product Performance for Specific End-Product Functions

According to Table 1.1, the food manufacturer is the next 'customer' after flour milling, and his expectation is that the flour supplied will be of consistent quality and suited to his process. While chemical and physical properties of flour can be useful, properly conducted test baking or small-scale end-product evaluation is the best guide to predict the processing performance for any particular flour. It would be impossible for a flour mill to carry out end-product evaluation to accommodate every customer's formulation and process, but fortunately there is a degree of commonality in the flour required for different end-products. As an example, high protein flour suitable for the production of hamburger buns is also suitable for artisan bread. Many mills prefer to carry out their own in-house tests; others send samples for external evaluation; a further alternative is to rely on protein and dough testing results and the miller's knowledge of market-quality requirements. Testing the end-product performance and suitability for different end products is discussed in Chapter 24.

22.6.12 Predictive Quality Tests

Rapid prediction of protein quantity and quality and of suitability for end-use allows millers to receive, segregate and blend wheat prior to milling. Protein content is readily measured within a few minutes by NIR/NIT, but other predictive tests take a longer time. Among the various predictive tests, flour Zeleny (AACC Method 56-61.02) and wheatmeal SDS sedimentation (CCAT Method 10) have been widely used in breeding programs. Sedimentation tests rely on the swelling and flocculation properties of glutenin protein in dilute lactic acid, and results are correlated with gluten strength and baking quality. Of the above two methods, the SDS sedimentation test, which is suitable for testing wholemeal, gives better discrimination between samples. The application of a modified swelling index of gluten (SIG) test has been applied to receival samples for the prediction of soft wheat quality (Uthayakumaran et al., 2007).

Wet gluten can be washed from wheat or flour using automated gluten washers (Table 22.2). The results are closely related to protein content as shown in Fig. 22.8. Nevertheless, many wheat and flour specifications, particularly in Asia, require flour wet-gluten content.

The gluten-index method provides an indication of gluten quality as well as quantity, with higher values (>85) indicating strong gluten properties. It is defined as:

$$\text{Gluten index} = 100 \times \text{wet gluten remaining on an 88 } \mu\text{m sieve } (g)/\text{total wet gluten (g)}.$$

The Solvent Retention Capacity (SRC) test was originally developed to assess soft-wheat quality (AACC Method 56-11.02). Four flour solvents—water,

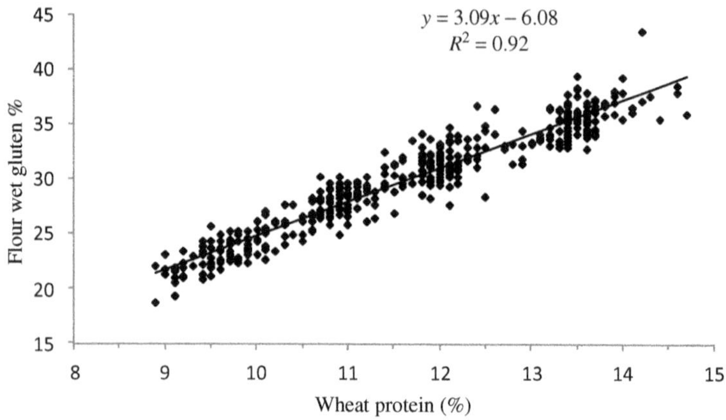

FIGURE 22.8 Relationship between flour wet gluten and wheat protein content for Australian wheat. *(Source: data from GrainCorp Australian Crop Reports, www.graincorp.com.au)*

5% lactic acid, 5% sodium carbonate and 50% sucrose—are used as alternatives. SRC values for lactic acid are associated with glutenin quality, sodium carbonate with starch damage, sucrose with pentosans, and the water-based result is affected by all water-absorbing components.

22.6.13 Food Safety

Food safety needs to be maintained at all points during the food chain to minimise or eliminate the risks to the consumer. HACCP (Hazard Analysis Critical Control Points) originated as part of the space program to ensure the health of astronauts. HACCP principles are applied to review the process and products (http://fsrio.nal.usda.gov).

Hazards are classified as chemical, physical or biological. After a hazard analysis has been carried out, Critical Control Points (CCP) and critical limits are established. These vary according to the specific country and mill needs. Critical control points in a flour mill include wheat intake (where there must be checks for moisture content, insect infestation, the presence of foreign matter and mycotoxins), rollers and bran finishers (checking for ferrous-metal contamination) and sifters (checking for non-ferrous metal contamination). Hence the maintenance of sieves, metal detectors, magnets and security screens is vital. An important part of the process is documentation and the establishment of corrective actions.

22.6.14 Quality Systems

Quality-management systems provide a systematic framework to meeting customer requirements, defining processes, process control, measuring performance and facilitating continual improvement. Management systems have the best outcomes when they are fully integrated into the business, as this creates an environment where all people are fully committed. ISO 9001 (www.iso.org) is one of the management systems used in flour mills, but it is becoming more common to use ISO2200 which integrates food safety and provides international recognition of third party certification. Quality-management and food safety systems are often the responsibility of those involved in mill quality assurance. Further information about these systems is available in Chapter 15.

22.6.14.1 Microbiology

Flours destined for sensitive products such as baby food, frozen and refrigerated pastries and pasta, shelf-stable noodles and batters have a requirement for microbiological testing.

Fungal and bacterial species are concentrated on the outer bran and germ layers of the grain. Low levels of spoilage microorganisms and of bacteria such as *Salmonella* species, *Escherichia coli*, *Bacillus cereus* and other coliforms have been reported in wheat and flour. High levels of these bacteria would generally result from poor hygiene or poor pest control during transport and handling. A wet harvest period and low harvest temperatures may be indicative of higher microbial loads with higher specific plate count and yeast and mould count. Generally, wheat and flour are not regarded as high-risk food ingredients, and very few incidents of food poisoning have been documented with wheaten flour. This is due to their low moisture content and water activity, and the need for subsequent cooking before consumption. However, wet harvests and high grain moisture in Europe may result in high levels of mycotoxins which will need to be managed throughout the supply chain (see Chapter 15). Microbiological test results for flour are lower than for the wheat from which it was milled. Patent (low-extraction) flours will be lower than straight-run flours and any second- or third-grade flour divides which contain higher bran contamination. Bran and germ have the highest microbial loads. During wheat conditioning, as moisture content is raised there can be an increase in aerobic mesophilic and *E coli* counts (Berghofer et al., 2003). Poor mill sanitation, particularly in conditioning bins and equipment, hot and humid environments and build-up of damp residual material in flour storage bins can result in localised contamination.

22.7 WHOLE-GRAIN FLOUR TESTING

Wholemeal and whole-grain flours are usually manufactured by reconstituting white flour together with the bran and germ components in the proportions found in the whole grain.

It is usual to determine the quality parameters of the white flour, as described in Section 22.6, to ensure that it has the capability to meet end-product performance requirements when recombined with the whole-grain components. One of the key quality parameters of importance includes protein content to ensure sufficient quality gluten to support the grain or meal in the flour. Another critical quality parameter is to ensure sufficient shelf life of the flour by excluding insect infestation and rancidity development (increase in free fatty acids and peroxide value) which can cause odour or off taste and poor baking performance.

Unless the added whole-grain premix has been heat-treated, shelf life of wholegrain flours will be limited unless stored under refrigeration. For wholegrain flours, the particle size of the bran is important as excessive quantities of fine particles can interfere with the gluten structure and cause loaf collapse.

22.8 HEALTHY GRAIN

There is a current trend for increased interest and concern about health issues with wheat-based products. On the one hand, there is an emphasis on the increased consumption of whole-grain foods in western countries as one way to improve the diet; also as a way of overcoming the increasing incidence of obesity and the related illnesses such as cardio-vascular disease and diabetes. The fibre content of cereals is also believed to reduce the risk of colo-rectal cancer. More foods are becoming available with a cereal base, but with higher fibre content. Whole grain foods are also higher in vitamins and essential minerals as well as an important range of phytochemicals, such as phytic acid, antioxidants, flavonoids, phytoestrogens, etc. Novel wheats have recently been developed that have significantly increased wheat endosperm fibre content by increasing the starch amylose content.

In the last 20 years there has been a huge increase in the prevalence of obesity; it is now estimated that there are 400 million people in the world living with type II diabetes with a further 50% undiagnosed. Providing foods with low glycaemic index has been found to be very beneficial for those with type II diabetes but also in preventing the onset of diabetes. In Australia and New Zealand low glycaemic index wheat-based products have been on the market for several years. This concept is now gaining interest in North America and Europe. A key challenge for researchers is to produce low glycaemic index wheat-based products without the addition of wholegrain. With significant numbers of the population being diagnosed with macular degeneration, there are opportunities to reduce this risk through increasing the lutein content of the diet. Lutein is found in durum and other highly pigmented wheats.

Such advances in the nutritional quality of wheats are likely to be enhanced through the use of genetic manipulation (GM) technology in the breeding of wheats, though in some instances researchers have used GM technology to explore the feasibility and then used non-GM routes for production of commercial products. Several international consumer surveys have shown that consumers will readily accept GM products when there is a significant health benefit attached. While GM technology has been used commercially for a number of crops including maize, soybean, cotton, etc., there are currently no commercial GM wheat varieties in the market. This is mainly due to its unacceptability in large markets worldwide but also because the use of GM technology in wheat is significantly more complicated than for other cereals and grains due to the hexaploid nature of wheat.

22.9 WHEAT INTOLERANCE

There is a growing concern that there are increased numbers of people displaying sensitivity to cereals. This sensitivity may take the form of an allergy (2.5–3.0% of western populations), celiac disease (1–2%) or what has become known as 'non-celiac gluten sensitivity' (suggested to be 10% by Dr Rodney Ford). Several popular books, for example, Wheat Belly (Davis, 2011) have been written about the potentially adverse health effects of wheat and gluten in recent years. Unfortunately much of the material quoted has little scientific basis. However books and articles written by Dr Rodney Ford in New Zealand provide compelling scientific evidence for the existence of non-celiac gluten intolerance (www.drrodneyford.com). Market surveys in Australia (CSIRO) and the UK indicate that at least 10% of the population is looking for gluten-free products. Current processing methods now allow many baked products, including bread, pasta, biscuits and cakes, to be made with non-gluten containing grains with quite acceptable quality for consumers. Recent nutritionist claims that only cereal-based baked products provide adequate nutrition are not true. Most gluten-free grains contain good quantities of fibre, vitamins, minerals, etc., equivalent to those found in cereals and many other grains (legumes) may be lower in glycaemic index than cereals.

During the last few years there is good evidence to show the link between the FODMAP (fermentable oligosaccharides disaccharides monosaccharides and polyols) content of foods and irritable bowel syndrome (IBS) which afflicts 15% of Western populations (www.med.monash.edu/cecs/gastro/fodmap). It has also been shown that wheat-based products have significant FODMAP content (fructans) which may stimulate IBS symptoms. Many gluten-free products are low in FODMAP content and therefore useful for consumers with IBS. It has also been shown recently that spelt contains significantly less fructan than conventional bread making wheats and is useful for those with IBS. While it is clear that there is a good scientific basis for the effect of FODMAPs for those suffering IBS there are recent reports (Biesiekierski et al., 2011; Biesiekierski et al., 2013a,b) suggesting that non-celiac gluten intolerance does not exist and is caused by FODMAPs. This may have some basis for consumer problems associated with gut problems but has no basis whatsoever for the large range of other symptoms associated with gluten consumption.

As this market grows, there are likely to be some reductions in conventional wheat milling and baked product volumes, but so far there is no convincing data to support this prediction. However, in Asian countries there has been increasing inclusion of wheat-based products in place of rice. This includes noodles, breads, cakes and pasta and for many Asian countries suitable quality wheat is imported from North America and Australia.

22.10 SUSTAINABLE FARMING AND EFFECTS OF CLIMATE CHANGE

Sustainable farming techniques for wheat production are now common practice in many of the wheat producing countries of the world. These include precision agriculture techniques using global positioning systems (GPS) to map the variation in crop yields across farms, apply fertilisers, nutrients and other techniques such as minimum tillage to minimise soil compaction and moisture loss.

The predictive effects of global climate change on wheat production include significant changes to rainfall in some traditional wheat growing areas, thus reducing or eliminating wheat production. With increasing CO_2 concentrations, wheat yields are likely to increase significantly but at the expense of reduced protein content requiring increased fertiliser application rates which in turn further increase greenhouse gas emissions. With higher temperatures during grain filling, heat shock will become more common which can affect the quality of the gluten in the grain.

22.11 FUTURE TRENDS

The aim of the flour miller is to produce the highest yield of flour of the right quality for the customer at the lowest possible cost and to ensure consistency of that quality. Milling technology has continued to advance with subsequent improvements in processing performance. Collection of accurate and reliable processing and analytical data are integral to this process and wireless technology is one of the new tools being used. Correctly calibrated online NIR systems can enable remote monitoring of important flour parameters and the opportunity for real-time process control. Online particle size measurement is now also possible for monitoring particle size during milling and of the end products including flour and semolina. Colour sorting is another new technology based on image analysis that can separate foreign seeds and grains, discoloured and damaged grains and foreign materials prior to milling.

Customer demand for wholemeal and whole-grain products has presented challenges in terms of the short wholemeal flour shelf life and lower flour functionality which need to be addressed by development of low-cost effective solutions, whether wheat-based or ingredient technology. Grain mixes, grains with health benefits, organic grain and low glycaemic index products will be increasingly part of the global food market and so flour functionality (particularly protein content and protein quality) will need to be sufficient to give products with similar quality to those made from white flour, but with additional flavour and texture.

SOURCES OF FURTHER INFORMATION AND ADVICE

Cauvain, S., Young, L., 2009. The ICC Handbook of Cereals, Flour. DEStech Publications, Lancaster, PN, USA.

www.healthgrain.org—HEALTHGRAIN Forum—outcomes of EU project which identified pathways for new technologies and promotion of the health potential of wholegrains.

Khan, K., Shewry, P.R., 2009. Wheat: Chemistry and Technology, Fourth Edition AACC International, St Paul, MN, USA.

Posner, E.S., Hibbs, A.N., 2004. Wheat Flour Milling, Second Edition AACC International, St Paul, MN, USA.

WQOG, 2011. Understanding Australian Wheat Quality, GRDC, Canberra, ACT, Australia. Available from: www.grdc.com.au

REFERENCES

Bason, M.L., Blakeney, A.B., 2007. Enzyme methods. In: Crosbie, G.B., Ross, A.S. (Eds.), The RVA Handbook. AACC International, St Paul, MN, USA, Chapter 10.

Bekes, F., Larroque, O., Hart, P., O'Riordan, B., Miskelly, D., Wrigley, C.W., 2001. Achieving consistency of wheat quality by the predictive use of multi-stream blending. In: Wootton, M., Batey, I.L., Wrigley, C.W. (Eds.). Cereals 2000 Proceedings of the Eleventh International ICC Cereals and Bread Congress. Cereal Chemistry Division, Royal Australian Chemical Institute, Melbourne, Australia, pp. 464–469.

Berghofer, L.K., Hocking, A.D., Miskelly, D., Jansson, E.J., 2003. Microbiology of wheat and flour milling in Australia. Int. J. Food Microbiol. 85, 137–149.

Biesiekierski, J.R., Newnham, E.D., Irving, P.M., Barrett, J.S., Haines, M., Doecke, J.D., Shepherd, S.J., Muir, J.G., Gibson, P.R., 2011. Gluten causes gastrointestinal symptoms in subjects without celiac disease: a double-blind randomized placebo-controlled trial. Am. J. Gastroenterol. 106, 508–514.

Biesiekierski, J.R., Muir, J.G., Gibson, P.R., 2013a. Is gluten a cause of gastrointestinal symptoms in people without celiac disease? Curr. Allergy Asthma Rep. 13, 631–638.

Biesiekierski, J.R., Peters, S.L., Newnham, E.D., Rosella, O., Muir, J.G., Gibson, P.R., 2013b. No effects of gluten in patients with self-reported non-celiac gluten sensitivity after dietary reduction of fermentable, poorly absorbed, short-chain carbohydrates. Gastroenterology 145, 320–328.

Davis, W., 2011. Wheat Belly. Rodale Inc, New York.

Delwiche, S.R., Millar, S., 2010. A new AACC International approved method for assessment of a calibration from a near infrared spectrometer. Cereal Food. World 55, 293–296.

Dexter, J.E., Wood, P.J., 1996. Recent applications of debranning of wheat before milling. Trends Food Sci. Tech. 7, 35–41.

Dominy, S.F., 2003. Leveraging the power of NIR. World Grain Dec. 2003, 18–20.

Dowell, F.E., Maghirang, E.B., Xie, F., Lookhart, G.L., Pierce, R.O., Seabourn, B.W., Bean, S.R., Wilson, J.D., Chung, O.K., 2006. Predicting wheat quality characteristics and functionality using near-infrared spectroscopy. Cereal Chem. 83, 529–536.

Gibson, T.S., Kaldor, C.J., McCleary, B.V., 1993. Collaborative evaluation of an enzymatic starch damage assay kit and comparison with other methods. Cereal Chem. 70, 47–51.

Greer, E.N., Stewart, B.A., 1959. The water absorption of wheat flour: relative effects of protein and starch. J. Sci. Food Agric. 10, 248–252.

Jones, L., Jones, P., 1990. The Flour Mills of Victoria 1840-1990. The Flour Millers' Council of Victoria, Melbourne, Australia, p. 3.

Morgan, J.E., Williams, P.C., 1995. Starch damage in wheat flours: a comparison of enzymatic, iodometric and near-infrared techniques. Cereal Chem. 72, 209–212.

Oliver, J.R., Blakeney, A.B., Allen, H.M., 1992. Measurement of flour color in space parameters. Cereal Chem. 69, 546–551.

Osborne, B.G., 1991. Measurement of the hardness of wheat endosperm by near infrared spectroscopy. Postharvest News Inf. 2, 331–334.

Osborne, B.G., Douglas, S., Fearn, T., 1982. The application of near infrared reflectance analysis to rapid flour testing. J. Food Technol. 17, 355–363.

Uthayakumaran, S., Barker, N., Batey, I.L., Dines, J., Miskelly, D., Wrigley, C.W., 2007. Rapid methods to predict soft-wheat dough quality for specific food products. Cereal Chem. 84, 522–526.

Walker, C.E., Eustace, W.D., 2016. Milling and Baking History. Wrigley, C., Corke, H., Seetharaman, K., Faubion, J. (Eds.), Encyclopedia of Food Grains, vol. 3, Elsevier, Oxford, United Kingdom, pp. 299–306.

Chapter 23

Industrial Processing of Grains Into Co-Products of Protein, Starch, Oils and Fibre

Ian Batey*, Colin Wrigley** and Diane Miskelly†
*Sunset Cereal Services, Meadowbank, NSW, Australia; **QAAFI, University of Queensland, St. Lucia, QLD, Australia; †Westcott Consultants, Goulburn, NSW, Australia

23.1 INTRODUCTION

The cereal grain as harvested, mature, is an integrated whole system, ready to perform its natural task of producing another plant when the moisture and temperature conditions are right to trigger germination. Man intervenes in this natural cycle to claim the cereal grain for his own needs. The successful propagation of the cereal grains worldwide is due to the grain being a rich source of nutrients for man and other predators (animals, insects, fungi, etc.). Primarily, we value the cereal grains for their ability to provide the wide range of foods described in this book.

In addition, the cereal grain is a source of valuable chemical components, whose value is enhanced if they can be derived in relatively pure form. The most obvious chemical component is starch, which takes up the majority of the mass of cereal grains. Purified starch offers great potential as a diverse food additive and a chemical feedstock with many uses. Its agricultural source provides the environmentally attractive advantage of being renewable, as distinct from petroleum-based feedstocks.

Grain storage protein, the second most available component, offers itself as a valuable food additive, with functional properties that differ, depending on the grain-species source and the purification process. Additional components of value include oils, extracted from cereal-grain germ, and non-starch polysaccharides, such as glucans and cellulose.

Given the value of these several components as near-pure chemical entities, processing methods have been developed on an industrial scale for their isolation in forms suitable for further processing or incorporation in a range of

Cereal Grains. http://dx.doi.org/10.1016/B978-0-08-100719-8.00023-1

food products. This general approach has been taken to the logical extreme with "whole-plant" fractionation and utilisation (see Chapter 25, Fig. 25.2).

This branch of applied chemistry was termed 'chemurgy' when proposed in the 1930s by the American William J. Hale in a book entitled 'The Farm Chemurgic'. Initial products included the use of plant-based cellulose to make motion-picture film, thereby creating links between farmers and the chemical industry (Holmes, 1935). At about the same time, the chemurgy concept attracted automobile manufacturer Henry Ford into making car components (e.g. gear-shift knobs) from soybean components, culminating in the concept of the 'Soybean car' (Ford, Henry; Gregorie, Eugene T. (January 13, 1942). 'Patent 2,269,452, Automobile Chassis Construction'. United States Patent Office. p. 1. Retrieved October 20, 2012. (https://www.thehenryford.org/research/soybeancar.aspx). Ford is also reported to have experimented with the production of diesel fuel from plant sources.

Several of the aspects of grain processing are listed under the stage of Processing in Table 1.1 (Chapter 1). The major targets for isolating chemical components in recent years have been the starch and protein of maize and wheat, and these two are described in the following sections as being the only major industrial cereals for starch-protein processing. In addition, the germ oil of maize, wheat and rice are significant industrial activities as described in this chapter. Non-starch polysaccharides are prepared from several cereal grains. Additional industries operate on the non-grain parts of cereal plants (Fig. 25.2 in Chapter 25).

23.2 STARCH

Starch is the predominant component in the grains of all cereal species, so it is the most obvious target for 'chemurgy'. The abundance of starch in the cereal grain is seen in Figs. 4.19 and 5.1 (Chapters 4 and 5), occurring in the form of granules embedded in a matrix of storage protein within the endosperm. Despite the ubiquity of starch in the endosperm of the cereals, the sizes and shapes of the granules differ from one species to another, even permitting species identification possible to a limited extent. See examples in Fig. 4.18 (Chapter 4). The internal structure of the granule is illustrated in Fig. 4.20 (Chapter 4), presumably indicating day–night variations in the deposition of starch in different crystalline forms.

Isolation of the starch in its native granular form is generally desirable, and this involves operating below the gelatinisation temperature of the granules. Starch isolation is simplified by the removal of the outer bran layers of the grain, and also by dry-crushing the endosperm, but some of the industrial procedures involve soaking the whole grain before proceeding into crushing the grain.

As well as being the major source of energy for the growing plant, starch is also a major source of energy in our diet, given our reliance on cereal grains for food. In addition, starch offers important functional properties in food technology and in a wide range of industrial applications. Starch has the attractive properties of forming a paste or gel when its suspension in water is heated and subsequently cooled (BeMiller and Whistler, 2009; Stoddard, 2016).

Stated most simply, starch is a polymer of glucose units (α-1,4 linked) in helices, with some α-1,6 branches. It comprises two types of polymer of glucose—amylose and amylopectin—with amylose being a linear molecule with few branches and amylopectin being a highly branched polymer. There are also differences in their molecular weights. Amylose has a molecular weight of several hundred thousand Daltons while amylopectin has a molecular weight measured in millions. Nevertheless, starches from the diversity of cereal grains exhibit a wide range of functional properties, depending on the source species; this diversity is increasing with the introduction of mutant genotypes of various species (Eliasson, 2004). Due to the considerable differences in properties of starch and the various end-product requirements, the correct analysis of starch quality is critical, as described by Gunaratne and Corke (2016). The main chemical and physical properties of starch are amylose content, starch granule size, swelling power, hot and cold paste viscosity and gelatinisation temperature. Some of these characteristics are related, for example high swelling power is associated with high hot paste viscosity and lower amylose content.

Quality characteristics of starch are important. Viscosity and amylose content are the main quality parameters. The former is measured in either a Viscograph or a Rapid Visco Analyser. Amylose is usually measured by the intensity of a solution with iodine, reading the amylose content from a standard curve prepared from mixtures of amylose and amylopectin. Methods using high performance liquid chromatography, differential scanning calorimetry or binding with concanavilin A have also been proposed. A kit is commercially available to measure amylose using this last method (Megazyme kit K AMYL). Both amylose and viscosity values will usually be specified by the customer. Gelatinisation temperature is also important. Where the starch will be heat processed, the temperature of the process needs to exceed the gelatinisation temperature. Some high amylose starches gelatinise above 100°C. Gelatinisation temperature is usually measured by differential scanning calorimetry. It may also be measured on a temperature-controlled hot stage with a microsope having a polarising filter. Starch granules exhibit a Maltese cross-effect when viewed in a microscope under polarised light. The cross disappears on gelatinisation. Starch granule size is measured by a particle size analyser of various types.

23.2.1 Uses of Cereal Starches in General

Starch is an essential raw material for food and industrial use. The source of the starch is generally the largest starch-containing crop grown in the region. In North America, it is maize (corn) whereas in Australia it is wheat. In some parts of the world, starch is produced industrially from non-cereal sources, but this is not relevant to this book. However, it should be noted that for many purposes, for example fermentation or hydrolysis to glucose syrups, the source of the starch does not matter and a processor will purchase the most economic material. In both food and non-food uses, starch can provide adhesion, viscosity, film-forming, binding and moisture retention. In food uses, starches serve as thickeners, gelling aids, texturisers, emulsifiers, bulking agents, adhesives for coating agents, water binders, and fat substitutes (Lawton, 2016). We are likely to be more familiar with the food uses of corn starch (e.g. in cakes, pies, soups, sweets, desserts, mayonnaise, puddings, ice cream, nougats) than in non-food uses (e.g. in explosives, diapers, wallpaper, adhesives, printing, tyres, plastics, bookbinding, mining and fireworks!).

An important non-food use is paper sizing. One of the earliest detected non-food uses of starch was for coating the surface of papyrus reeds to provide a smooth surface for writing. Today, corn-based starches account for most of the starch used in paper made in the United States. Much of that starch has been physically or chemically modified.

Much of the starch used in processed food has been modified in some way, thus to enhance the desirable properties of the starch used, to minimise undesirable properties and often to introduce novel characteristics (BeMiller and Whistler, 2009; Lawton, 2016). This approach of modification greatly expands the potential value of starch utilisation, given the already enormous range of properties for native starches from the many species and genotypes of the cereals. See BeMiller (2016) for amplification of the range of modifications and the ways in which these changes provide better functional properties.

Starch is used as an 'adjunct' in the brewing of beer, to complement the endogenous starch of the malt and wort, either barley or some other grain (MacLeod and Evans, 2016). Further fermentation applications relate to the production of distilled spirits (Palmer, 2016) and of fuel ethanol (Guragain et al., 2016).

23.2.2 Corn Starch

Industrial production of starch is greatest from maize (corn), with an estimated 85% or more of industrially produced starch worldwide coming from maize (Rausch and Eckhoff, 2016b). Starch yield from the dry grain is about 66%. Thus maize is known as 'a starch crop'. Maize itself is also the cereal species grown to the greatest extent worldwide (over a billion tonnes of grain annually; Chapter 2). See also Chapters 1, 4, and 11.

Maize is a staple food for large groups of people in Africa, Latin America, North America, and Asia. Given its widespread consumption, there is also great diversity of foods based on maize (Serna-Saldivar, 2016), and in general their preparation and properties rely on the starch characteristics of this grain.

There are two major approaches to the industrial fractionation of the maize kernel to separate the starch from the remaining components, namely, dry milling (Rausch and Eckhoff, 2016a) and wet milling (Rausch and Eckhoff, 2016b). The latter is the more popular (responsible for >85% of starch produced from corn). Despite its popularity and recent efficiency, the wet-milling process is a large consumer of energy and of water, involving five stages:

- steeping (soaking for a day or two, generally with lactic acid or sulfur dioxide);
- germ recovery (the embryo/germ contains negligible starch but valuable oils);
- fibre recovery (to release purified starch);
- protein recovery (a useful by-product, inappropriately known as 'corn gluten');
- and finally starch washing and drying.

See Rausch and Eckhoff (2016b) for full schematics of the process, plus discussions of alternatives and efficiency factors.

The maize dry-milling process is designed to fractionate the kernel into germ, pericarp (husk) and endosperm components by the crushing and separation of these components using air streams to take advantage of their distinctive aerodynamic properties. Although termed 'dry' milling, the success of the process is dependent on the addition of water to the dry grain, thus to condition it to about 22% moisture (Rausch and Eckhoff, 2016a).

A related 'dry' process involves the production of fuel ethanol from maize by grinding and jet cooking, followed by the stages of liquefaction, saccharification and fermentation (Rausch and Eckhoff, 2016a). See Guragain et al. (2016) for schematics of ethanol production from corn using the 'dry-grind process'.

There are different types of maize starch, usually described as 'normal', 'waxy', and 'high amylose'. Normal starch has an amylose content in the range of 20–30% and comes from wild-type plants. Waxy starch, so named because of the appearance of the seeds of its original discovery, contains no amylose as it has a recessive gene for the enzyme that catalyses amylose synthesis (granule bound starch synthase—GBSS). High amylose starch is available with amylose contents ranging from 40 to 80+%. Each type of starch has its own uses and they are definitely not interchangeable. Waxy starch or grain may be identified rapidly by the colour it stains with iodine. The red colour of waxy starch is very different from the blue colour of starch containing amylose. The term 'waxy starch' is, by convention, used to describe starch of any origin containing no amylose.

23.2.3 Wheat Starch

The preparation of starch from wheat is also big business internationally, especially in Europe, North America and Australia. Like maize, wheat starch may be isolated by wet or dry processing, but the wet process is by far the more popular one. In the latter, wheat is dry milled (see Chapter 22) to a flour before separation of the starch and protein. The bran and germ are thus separated from the endosperm in the dry state and it is only the flour that is used in the wet-separation process (Batey and Huang, 2016). The processes, however, differ from those for maize because the storage protein of the wheat grain forms into the visco-elastic gluten complex on wetting. Compared to maize, wheat gluten is much stronger in its retention of the starch granules, but it is also more coherent; this coherence holds the gluten together even permitting rather rough handling.

Another distinction between wheat and maize is the bipolar size distribution of wheat starch granules, with A granules ranging in size from about 15 μm to over 25 μm as distinct from the B granules (about 3–10 μm). The B granules greatly outnumber the A granules, but the mass proportion of B granules is low (about 20%). The size differences mean that the isolation of A granules is easier and cleaner than for the B granules. As a result the term 'prime starch' is used for a fraction that is mainly A-granules but still contains significant numbers of small (B-) granules; B starch is usually regarded as second class due to the higher content of protein that is isolated with these small starch granules, but it also includes damaged starch. The roller milling process to provide the flour for starch extraction inevitably results in some starch damage. Damaged starch granules absorb water rapidly, which reduces their density and causes them to sediment less easily than the undamaged large granules. Intact B-granules also sediment less easily than the A-granules.

There have been several approaches to wet-milling whole wheat, and at least two of them have been translated into commercial processes; these are described later in relation to wet-gluten production (Section 23.3.2). Nevertheless, most production of wheat starch and gluten occurs from flour, the product of an earlier dry-milling step.

Ideally, wheat grain for starch-gluten production should form a strong dough and have a high milling yield. A soft-grained texture is desirable to minimise starch damage during flour milling, and thus maximise the yield of starch and minimise losses to the waste stream.

In practice, soft wheats of suitable protein content are rarely available, and blends of hard wheats need to be used. Milling needs to be performed to avoid as much starch damage as possible. The complex economics of the wheat price, the value of products and the protein content of the wheat will often determine the actual grade of wheat used. The desired protein content of the grain selected may vary depending on the preferred mix of products planned: a higher grain-protein content means a greater yield of gluten and less starch, and vice versa. Methods for selection of flour of suitable quality are those described in the aforementioned milling section.

Until recently, there has not been the variation in wheat starch properties that has existed in maize and rice. As wheat is a hexaploid species, there is a degree of redundancy in many of the enzymes of starch synthesis with genes being present in all three genomes. The starch of most bread wheats has an amylose content of 22–28% and waxy and high amylose variants were unknown until about 20 years ago. Alleles for GBSS have been identified on chromosomes 7A, 7D and 4A, the last being the result of a natural translocation from chromosome 7B. Null alleles have been identified for all alleles and lines with a single null allele have been found to contain starch with an amylose content reduced by a small amount (2–5%). Lines with two null alleles have a further reduction of amylose by a similar amount. Selective crossing of lines with null alleles has allowed the production of lines with no GBSS containing little or no amylose (Zhao and Sharp, 1998). Such lines are referred to as 'waxy', and although there have been several reports of successful breeding of waxy wheats, none have been commercially successful to date. Incorporation of waxy wheats into blends of regular wheat flours results in bread of poor quality (Garimella Purna et al., 2011). However, lines which have the null-4A allele have been shown to be associated with good quality for Udon noodle production (Zhao et al., 1998). Varieties or lines with one or two alleles for GBSS present are referred to as 'partial waxy' (Graybosch, 1998).

There have been reports of varieties with starch of higher than normal amylose content (Hung et al., 2008), but these have only amylose content in the 35–40% range. For the amylose content to be significantly higher, amylopectin biosynthesis needs to be suppressed. Starch branching enzymes (SBE) are associated with the synthesis of amylopectin. In diploid species, there are natural mutants with low levels of SBE and conventional breeding has given rise to high amylose strains such as high amylose maize. The hexaploid nature of the wheat genome makes this more complicated. Any deficiency in SBE from one genome will be compensated for by the other two genomes. Lines have been developed with much higher levels of amylose (>70%) by suppressing the SBEII enzyme (Regina et al., 2016) and there are potential health benefits from wheat cultivars with high levels of amylose (Regina et al., 2006).

23.2.4 Starch From Other Grains

Although starch may be extracted from all cereals, industries have not developed for the specific isolation of starch from other species. Factories have been built, with the ability to process grains other than wheat or maize. The process for separation of starch from maize may be applied with minor modifications to other grains. In practice, factories of this nature have only been used for other grains when the economics of grain price have justified it. The starch produced has been used for purposes for which any starch would be suitable.

23.3 PROTEIN

The process of isolating the starch from cereal endosperm inevitably leads to the isolation of storage protein, the second largest component of the grain. However, the resulting protein fraction is likely to be contaminated with small starch granules, lipids and fibrous material (largely non-starch polysaccharides).

23.3.1 Maize Protein

In the case of maize, this protein fraction is known as 'corn gluten', although it is unrelated to the gluten of wheat (or of wheat's close relatives) which are harmful to people with coeliac disease (see Chapter 14). In addition, corn gluten has none of the useful rheological properties of wheat gluten. Accordingly, there are no rheological tests for corn-gluten quality, apart from its protein content and food-safety evaluation. Its use is mainly as a food additive, to enhance protein content, and for incorporation into animal feed (Black, 2016).

23.3.2 Wheat Gluten

There are two main processes used for the production of gluten—the Martin process and the Batter process—both of which have been used for many years but have been modified extensively by individual processors to incorporate new technology such as hydrocyclones (Batey and Huang, 2016). The main difference is the amount of water mixed with the flour at the start of the process. The Martin process forms a dough which is mechanically 'worked' in a stream of water which removes the starch. The Batter process uses a thick suspension of flour in water which is stirred slowly to separate the starch and protein. A variety of sieves, centrifuges and hydrocyclones are then used to isolate and purify each fraction.

The Alfa-Laval Raisio process, strictly speaking, does not wet-mill whole wheat. The dry milling of wheat is a modified milling process before passing directly to the separation process (Dahlberg, 1978). The first stage is a high-shear centrifugation step, followed by more conventional steps. Original publications claimed good yields of high-quality gluten, but the process in its entirety has not been adopted widely. The great advantage of the Raisio process is the significant reduction in water usage which has a particularly important environmental benefit.

Other processes, not dissimilar to the Alfa-Laval Raisio method, have been developed and practised commercially. These start with roller-milled flour, and utilise high-shear conditions in centrifuges and hydrocyclones for the mixing of flour and water. These are variously known as the Hydrocyclone process, High-Pressure Disintegration process, Barr and Murphy process and the Westfalia process.

The Farmarco process also used a dry pin-milling step, before applying a conventional Martin process to the flour obtained (Grace, 1989). However, this was a crude flour, compared to the product of roller milling, and gluten quality was compromised as a result.

The only true wet-milling process for whole grain was one devised by Pillsbury in the 1970s, in which the wheat was steeped in dilute hydrochloric acid. This process became the basis of a commercial plant, but this ceased in 1979 after only a short period of operation.

The quality requirements of the various starch-gluten manufacturing processes have been discussed by Sayaslan (2004).

The assessment of dry gluten quality is important because the separation and drying stages may cause loss of rheological properties, thereby reducing the value of dry gluten for many uses in food technology. Commercial gluten contains some residual fibre and starch, as well as much of the free lipid in the original flour and some non-prolamin proteins (globulins and albumins).

Wet gluten is susceptible to degradation with excessive heat and to obtain dry gluten the drying process needs to be carefully controlled as the dough properties are easily destroyed by careless practice. Ring drying was applied to gluten in the first half of the 20th century and has since been the basis of dry gluten production (Knight, 1965). The principle is simple: wet gluten with a moisture content of ~70% is mixed with sufficient dry, vital gluten to reduce the moisture to ~20%. This lowered-moisture material is ground and subjected to flash drying. Some of the dried gluten is removed and the rest is returned to the drying cycle to reduce the moisture content of more wet gluten. The material is still very sensitive to excessive heat in the drier, but careful control of the temperature produces a vital wheat gluten that retains the visco-elastic properties that are characteristic of the original gluten.

Protein content is measured by the Dumas or Kjeldahl methods, to meet the minimum specifications of the customer. Starch is required to have a maximum protein content of 0.3%, also measured in either of these ways. In addition, the dough strength and extensibility of the gluten are measured using the Extensograph, using a dough prepared in a Farinograph. For this test, a standard mixture of 15% gluten and 85% starch is mixed with water to form a dough. In France, the strength of the gluten is usually measured with an Alveograph.

Wheat gluten is used in many ways to enhance the protein content and textural properties of foods, as described by Day et al. (2006). There are also many non-food applications (Bietz and Lookhart, 1996; Day et al., 2006; Batey and Huang, 2016).

23.4 OTHER COMPONENTS

Following the fragmentation of the cereal grain, and the isolation of the starch and protein from the endosperm, there remain the germ (rich in valuable oils), bran tissues and endosperm cell walls (rich in valuable non-starch polysaccharides). These 'other components' also attract industrial processing to contribute materials for various food and non-food applications.

23.4.1 Oils

Maize, rice and wheat are the main cereals whose germ tissues are used for oil extraction industrially (Godber, 2016). However, the success with which these efforts are pursued depends on the extent to which the preliminary processing procedures have been designed to maximise the quality of the oil-bearing tissues.

Rice oil is obtained from the germ and bran, removed during the polishing of brown rice to produce white rice (the main form in which rice is consumed). Heat treatments may be required to prevent the action of lipolytic enzymes. The oil is mainly found in the germ, which represents about 2% of the rice-kernel mass. However, rice oil is an underexploited resource, with rice-oil production representing less than 10% of its potential, based on world rice production, almost all of which is milled to white rice.

On the other hand, the germ of the maize grain comprises 10–14% of the seed weight, so the extraction of germ oil is an attractive option. However, the utilisation of maize is so diverse that only about 10% of maize production may go into the milling industry and thus contribute potentially to oil production (Godber, 2016).

Solvent extraction is the usual method of oil removal, using a solvent such as hexane. Subsequent stabilisation, filtration and refining are complex aspects of oil production. Both types of oil are mixtures of several fatty acids: oleic (C18:1), linoleic (C18:2), palmitic (C16:0), smaller amounts of stereic (C18:0), and low levels of myristic (C14:0) and linolenic (C18:3) (Godber, 2016). Generally corn oil has a higher content of linoleic acid than rice oil, and rice oil has more oleic and palmitic acids. Both oils also contain beneficial compounds, particularly vitamin E vitamers, tocotrienols, tocophenols and (for rice) oryzanols.

Both oil types are used as cooking oils and in the manufacture of margarine and mayonnaise. These cereal-grain oils are obviously in competition with oils from oilseed species, such as canola, soy, cottonseed, linseed, flax, safflower and sunflower - all coming from dicotyledonous plants (not monocotyledonous cereals).

23.4.2 Non-Starch Polysaccharides

After the removal of starch, oils and protein from cereal grains, the residue largely consists of various carbohydrates. This wide range that excludes starch, includes structural materials such as cellulose, pectic polysaccharides, β-glucans, fructans, glucomannans, and more—all polymers of various monosaccharides (Fincher, 2016). Most form the outer bran layers and the inner cell walls of the grains. For wheat, for example, these cell-wall materials come from the endosperm, aleurone, pericarp and seed coat (see Chapter 4) In total, the non-starch polysaccharides may constitute about 10% of the whole grain weight.

As integral (endogenous) parts of all cereal grains, the non-starch polysaccharides play significant roles in the processing functions, such as brewing, baking, breakfast-cereal manufacture, stock feed pelleting and in the manufacture of pharmaceuticals for treating various disease issues. The β-glucans of oats and barley have attracted particular attention as being functional dietary-fibre components, thus helping to reduce the risks of cardiovascular disease and diabetes

(Izydorczyk, 2016). In addition to these health benefits, the β-glucans are used in the food industry as fat replacers, stabilisers and thickening agents.

Industrial isolation of β-glucans, mainly from oats and barley, involves extraction of aleurone or endosperm preparations with water or alkaline solutions, removal of starch and protein using hydrolytic enzymes and precipitation of β-glucans with alcohol (Izydorczyk, 2016).These semi-purified isolates can be used in pharmacological products and as food additives in various baked products, thus to enhance dietary-fibre ingestion (Malcolmson et al., 2014).

23.5 BIOFUELS

Concern about the environment, climate change and crude oil resources has meant that there is renewed interest in plant-based products such as fuel and plastics (see Section 23.6). The principal biofuels have been ethanol and biodiesel, the latter from oilseed crops (Yan, 2016). This concern goes back several decades and in 2013, worldwide production of fuel ethanol reached 85,000 ML (Guragain et al., 2016). The main sources of the feedstock for this production are the starch-rich cereal grains and sucrose-rich crops such as sugar cane, sugar beets and sweet sorghum. In parts of Europe, starch from potato is also used for ethanol production.

Maize (corn) is the main cereal used for ethanol production, although in Canada and Australia it is wheat. Production from maize may involve a separation of the components of the grain as described previously, but often whole grain is ground and subjected to fermentation. In this case, there is a by-product known as dry distillers grains (DDG) which is used as animal feed. The by-products of ethanol production from corn are an important component of the economics of the industry, whether they are DDG or the more valuable protein and oils coming from the starch separation step. Isolating the starch first is more capital and energy intensive but gives by-products of greater purity and value.

Wheat may also be fermented before or after the separation of the components. If whole grain is used, a product similar to the maize DDG is co-produced. Separation of the starch and gluten also gives gluten as a valuable by-product. As wheat is roller milled before going into a starch-gluten separation process, the bran is separated in a dry state and there is no need to dry it before using it for animal feed. There is at least one starch/gluten/ethanol plant that uses the B-starch as a feedstock for ethanol production. This reduces the requirement for both water and energy in the starch separation process as there is less need for washing and centrifuging the starch components to maximise the yield of prime (A-) starch.

The use of food crops for this purpose has raised concern about global food supplies, and there is increasing study of the conversion of ligncellulosic material to ethanol. The structural carbohydrates of all crops are rich in cellulose which is also a polymer of glucose and conversion of this material to ethanol would have little effect on food supplies. The conversion of cellulose to ethanol is more difficult and less efficient than using starch as the raw material. At present, cellulose-based ethanol is not economically viable and the only plants in operation are heavily subsidised (Guragain et al., 2016).

Biorefineries are factories using plant crops as a feedstock and producing a range of products. The concept provides for utilisation of the whole plant—grain/seed and stalks. Products would be starch, protein, biofuels and bioplastics. The concept has not really been applied: some factories have been built utilising crops in novel ways but not as whole plants. One such ethanol plant was built in a major sorghum-growing area of Australia, with the concept of processing crops locally to reduce transport costs and provide additional benefits to growers. It was not hugely successful and was under threat of closure.

23.6 BIOPLASTICS

Polymers abound in nature and plant- and animal-based polymers have been used for food, furniture and clothing for thousands of years. The first artificial thermoplastic polymer, 'celluloid', was invented in the 1860s. Since then, many new polymeric compounds derived from renewable resources have been developed. Most remained at the laboratory stage due to the large-scale application of oil-based synthetic polymers. In the past 20 years or so, there has been a renaissance of interest in plastics from renewable feedstocks. Starch has been re-discovered as a plastic material and other polymers such as poly-lactic acid (PLA) and poly-hydroxy-alkanoates (PHA) can be manufactured from fermentable sugars and vegetable oils, respectively.

Starch is the main material used in the manufacture of bioplastics, and is the basis of many biodegradable plastic products. It may be made into films by the inclusion of plasticisers or it may be part of a copolymer with oil-based polymers such as polyethylene. Materials made out of starch alone are fully biodegradable, but the use of normal starch can lead to brittle materials. Films made from high amylose starch are much stronger.

Worldwide, the amount of starch-based polymers has been increasing by 40–50% per year, and is expected to reach 3.45 million tonnes by 2020. This is only a small proportion of the 270 million tonnes of polymers made from petrochemicals, but it is estimated that potentially 90% of this could be replaced by plant-based polymers.

23.7 RESISTANT STARCH

The implications for human health of increasing the amount of dietary fibre have been recognised for some time (Fuentes-Zaragoza et al., 2010; Topping et al., 2010). One important component of dietary fibre is resistant starch (RS). RS is defined as the total amount of starch and the products of starch degradation that resists digestion in the small intestine of healthy people (Asp, 1992).

It is generally accepted that there are four types of RS, designated RS1, RS2, RS3 and RS4. In RS1, the starch is physically inaccessible to digestion due to intact cell walls, as in whole grains. RS2 comprises native starch granules protected from digestion by the structure of the starch granule. RS3 is retrograded starch formed when starch-containing food is cooked then cooled. RS3 forms, principally from the amylose component, as the product cools. RS4 are chemically modified starches that do not occur naturally but are resistant to digestion (Topping and Clifton, 2001).

RS1 and RS2 are present in unprocessed and uncooked foods, but RS3 and RS4 can be formed and included in food by selection of appropriate ingredients. Grains with high amylose content are particularly suitable for the preparation of RS3 as the long amylose chains become entwined as the product cools, creating retrograded starch which is resistant to enzymic attack. Resistant starch is fermented in the colon to give short chain fatty acids (SCFA), which have been shown to be beneficial to colonic health. The most common form of high amylose starch has come from maize, in which there are several varieties with this characteristic. High amylose lines of barley (Morell et al., 2003, 2011) and wheat (Regina et al., 2006, 2016) have also been developed. There are a number of products which incorporate high amylose starch, ranging from breads through to breakfast cereals (Bird et al., 2008; King et al., 2008).

One potential use of resistant starch is as a carrier of nutrients or bioactives through the digestive tract (Augustin and Hemar, 2009). This may be by encapsulation or by binding of the desired component to protect it from digestion or destruction as it passes through the upper parts of the digestive tract. Dehydration as a result of gastroenteric illnesses such as cholera is a major cause of death in many parts of the world. Inclusion of resistant starch in oral rehydration therapy has been shown to be useful. SCFAs generated in the colon allow water and sodium ions to be absorbed (Binder et al., 2014). In this role, it has the potential of saving many lives every year. Zinc deficiency has been shown to be a cause of low immunity in certain populations. Zinc absorption from the colon is also enhanced when supplied with resistant starch (Gopalsamy et al., 2015).

23.8 FUTURE TRENDS

There is a current trend for increased interest and concern about health issues with cereal-based products. On the one hand, there is an emphasis on the increased consumption of whole-grain foods in western countries as one way to improve the diet as a way of overcoming the increasing incidence of obesity and the related illnesses of cardio-vascular disease and diabetes. The fibre content of cereals is also believed to reduce the risk of colo-rectal cancer. More foods are becoming available with a cereal base, but with higher fibre content. It is likely that resistant starch will become a major ingredient in foods of many types, not only those based on cereals. Whole grain foods are also higher in vitamins and essential minerals as well as an important range of phytochemicals, such as phytic acid, antioxidants, flavonoids, phytoestrogens, etc. Research is currently exploring the opportunities to increase wheat endosperm fibre content by significantly increasing the starch amylose content, while other research programmes are looking at ways to reduce the Glycaemic Index of wheat-based products. With significant numbers of the population being diagnosed with macular degeneration, there are opportunities to reduce this risk through increasing the lutein content of the diet. This is found in durum and other highly pigmented wheats. Resistant starch will also become the basis of pharmaceutical treatments for some illnesses.

Many advances in the nutritional quality of wheats are only likely to be possible through the use of GM wheats. Several international consumer surveys have shown that consumers will readily accept GM products when there is a significant health benefit attached.

On the other hand, there is a growing concern that there are increased numbers of people displaying sensitivity to cereals. This sensitivity may take the form of an allergy, coeliac disease or what has become known as 'non-coeliac gluten sensitivity'. A market survey in 2008 by the LA Times indicated that 8% of US consumers were looking for gluten-free (GF) products. A Rural Industry Research and Development Corporation report in Australia in 2006 suggested that 5% of Australians were looking for GF grain products. As this market grows, there are likely to be some significant reductions in conventional wheat milling and baking volumes. To cater for the needs of these individuals, it is likely that more gluten-free foods based on cereals other than wheat will become available. Many gluten-free foods will require replacement of gluten with a range of starches, both native and modified.

In terms of the use of wheat gluten as an ingredient, there is a trend opposing the demand for gluten-free foods. This is the desire for consumers in Western countries to move away from what has become the standard white sliced bread to specialty breads. Many of these require the addition of gluten to increase the protein content of the flour and produce a loaf of the necessary quality. There have been many predictions of the over-supply of wheat gluten as wheat starch production increased in the past but none have been fulfilled to any large extent. New uses have been found for any increased production, including modified gluten products. Many of these are soluble, and could be used to replace soluble protein derivatives from other sources. However, it is unlikely that regulatory authorities will allow the replacement of soluble proteins with a solubilised wheat gluten in foods that would otherwise not contain wheat.

Ethanol production will continue to grow. World oil prices have put a brake on the increase in ethanol production to some extent, but many countries have implemented (or have plans to implement) a requirement for a proportion of their automotive to come from renewable sources. The major promise of biofuels from lignocellulosic biomass, such as cereal straw and husks, is yet to achieve economic and large-scale production, as so-called 'second-generation bioethanol' (Guragain et al., 2016). The rise of such technology would further justify the concept of 'biorefineries' and the use of the whole cereal plant, as described in Chapter 25 (Fig. 25.2).

The use of bioplastics will also increase and will put pressure on starch production. Cellulose-based plastics would reduce this pressure but currently most bioplastics are based on starch. The rate of growth of bioplastics has reduced with drops in the price of crude oil (Shen et al., 2009) but will become more economic as and when the price increases again. One aspect of the industrial manufacture of ethanol is to convert it by dehydration to ethylene, a precursor of many industrial chemicals and polymers. Currently, this process is not competitive with ethylene from crude oil but this could change in the future (Shen et al., 2009).

New breeding technologies will continue to provide genetic sources of innovative types of starches and proteins (see Chapter 16). As a result, there will be new opportunities for applications in the downstream processing of native starch products with novel properties. The need for these new native starches may become more pressing if there is consumer reaction against the use of chemically modified starch additives.

REFERENCES

Asp, N.G., 1992. Resistant starch—Proceedings from the Second Plenary Meeting of EURESTA-European FLAIR Concerted Action, 11 on Physiological Implications of the Consumption of Resistant Starch in Man. Preface. Eur. J Clin. Nutr. 46 (suppl. 2), S1.

Augustin, M.A., Hemar, Y., 2009. Nano-and micro-structured assemblies for encapsulation of food ingredients. Chem. Soc. Rev. 38 (4), 902–912.

Batey, I.L., Huang, W., 2016. Gluten and modified gluten. In: Wrigley, C., Corke, H., Seetharaman, K., Faubion, J. (Eds.), Encyclopedia of Food Grains. second ed. Academic Press, Oxford, pp. 408–413.

BeMiller, J.N., 2016. Starch: modification. Wrigley, C., Corke, H., Seetharaman, K., Faubion, J. (Eds.), Encyclopedia of Food Grains, vol. 3, Elsevier Ltd, Oxford, UK, pp. 282–286.

BeMiller, J.N., Whistler, R.L., 2009. Starch Chemistry and Technology, third ed. Academic Press, New York.

Bietz, J.A., Lookhart, G.L., 1996. Properties and non-food potential of gluten. Cereal Foods World 41, 376–382.

Binder, H.J., Brown, I., Ramakrishna, B.S., Young, G.P., 2014. Oral rehydration therapy in the second decade of the twenty-first century. Curr. Gastroenterol. Rep. 16 (3), 1–8.

Bird, A.R., Vuaran, M.S., King, R.A., Noakes, M., Keogh, J., Morell, M.K., Topping, D.L., 2008. Wholegrain foods made from a novel high-amylose barley variety (Himalaya 292) improve indices of bowel health in human subjects. Br. J. Nutr. 99 (05), 1032–1040.

Black, J.L., 2016. Animal feed. Wrigley, C., Corke, H., Seetharaman, K., Faubion, J. (Eds.), Encyclopedia of Food Grains, vol. 3, Elsevier Ltd, Oxford, UK, pp. 215–223.

Dahlberg, B.I., 1978. A new process for the industrial production of wheat starch and wheat gluten. Starch/Starke 30, 8–12.

Day, L., Batey, I.L., Wrigley, C.W., Augustin, M.A., 2006. Gluten uses and food industry needs. Trends Food Sci. Technol. 17, 82–90.

Eliasson, A.-C. (Ed.), 2004. Starch in Food. Structure, Function and Applications. Woodhead Publishing, Cambridge, UK.

Fincher, G.B., 2016. Cereals: chemistry and physicochemistry of non-starchy polysaccharides. Wrigley, C., Corke, H., Seetharaman, K., Faubion, J. (Eds.), Encyclopedia of Food Grains, vol. 2, Elsevier Ltd, Oxford, UK, pp. 208–223.

Fuentes-Zaragoza, E., Riquelme-Navarrete, M.J., Sánchez-Zapata, E., Pérez-Álvarez, J.A., 2010. Resistant starch as functional ingredient: a review. Food Res. Int. 43 (4), 931–942.

Garimella Purna, S.K., Miller, R.A., Seib, P.A., Graybosch, R.A., Shi, Y.-C., 2011. Volume, texture, and molecular mechanism behind the collapse of bread made with different levels of hard waxy wheat flours. J. Cereal Sci. 54, 37–43.

Godber, J.S., 2016. Oil from rice and maize. Wrigley, C., Corke, H., Seetharaman, K., Faubion, J. (Eds.), Encyclopedia of Food Grains, vol. 3, Elsevier Ltd, Oxford, UK, pp. 453–457.

Gopalsamy, G.L., Alpers, D.H., Binder, H.J., Tran, C.D., Ramakrishna, B.S., Brown, I., Manary, M., Mortimer, E., Young, G.P., 2015. The relevance of the colon to zinc nutrition. Nutrients 7 (1), 572–583.

Grace, G., 1989. Preparation of vital wheat gluten. In: Applewhite, T.H. (Ed.), Vegetable Protein Utilization in Human Foods and Animal Feedstuffs. American Oil Chemists' Society, Chicago, IL.

Graybosch, R.A., 1998. Waxy wheats: origin, properties and prospects. Trends Food Sci. Technol. 9, 135–142.

Gunaratne, A., Corke, H., 2016. Starch, analysis of quality. Wrigley, C., Corke, H., Seetharaman, K., Faubion, J. (Eds.), Encyclopedia of Food Grains, vol. 2, Elsevier Ltd, Oxford, UK, pp. 198–207.

Guragain, K.N., Probst, K.V., Vadlani, P.V., 2016. Fuel alcohol production. Wrigley, C., Corke, H., Seetharaman, K., Faubion, J. (Eds.), Encyclopedia of Food Grains, vol. 3, Elsevier Ltd, Oxford, UK, pp. 235–244.

Holmes, H.N., 1935. The farm chemurgic (Hale, William J.). J. Chem. Educ. 12, 99.

Hung, P.V., Maeda, T., Miskelly, D., Tsumori, R., Morita, N., 2008. Physicochemical characteristics and fine structure of high-amylose wheat starches isolated from Australian wheat cultivars. Carbohydr. Polym. 71, 656–663.

Izydorczyk, M.S., 2016. β-Glucans: measurement and processing. Wrigley, C., Corke, H., Seetharaman, K., Faubion, J. (Eds.), Encyclopedia of Food Grains, vol. 3, Elsevier Ltd, Oxford, UK, pp. 287–298.

King, R.A., Noakes, M., Bird, A.R., Morell, M.K., Topping, D.L., 2008. An extruded breakfast cereal made from a high amylose barley cultivar has a low glycemic index and lower plasma insulin response than one made from a standard barley. J. Cereal Sci. 48 (2), 526–530.

Knight, J., 1965. The Chemistry of Wheat Starch and Gluten and Their Conversion Products. Leonard Hill, London, pp 10–11.

Lawton, J.W., 2016. Starch: uses of native starch. Wrigley, C., Corke, H., Seetharaman, K., Faubion, J. (Eds.), Encyclopedia of Food Grains, vol. 3, Elsevier Ltd, Oxford, UK, pp. 282–286.

MacLeod, L., Evans, E., 2016. Barley: malting. Wrigley, C., Corke, H., Seetharaman, K., Faubion, J. (Eds.), Encyclopedia of Food Grains, vol. 3, Elsevier Ltd, Oxford, UK, pp. 423–433.

Malcolmson, L., Lukie, C., Swallow, K., Sturzenegger, T., Han, J., 2014. Using barley flour to formulate foods to meet health claims. Cereal Foods World 59, 235–242.

Morell, M.K., Kosar-Hashemi, B., Cmiel, M., Samuel, M.S., Chandler, P., Rahman, S., Buleon, A., Batey, I.L., Li, Z., 2003. Barley sex6 mutants lack starch synthase IIa activity and contain a starch with novel properties. Plant J. 34 (2), 173–185.

Morell, M.K., Topping, D., Batey, I.L., 2011. Barley with reduced SSII activity and starch containing products with a reduced amylopectin content. U.S. Patent No. 7,888,499.

Palmer, G.H., 2016. Beverages: distilled. Wrigley, C., Corke, H., Seetharaman, K., Faubion, J. (Eds.), Encyclopedia of Food Grains, vol. 3, Elsevier Ltd, Oxford, UK, pp. 193–205.

Rausch, K.D., Eckhoff, S.R., 2016a. Maize: dry milling. Wrigley, C., Corke, H., Seetharaman, K., Faubion, J. (Eds.), Encyclopedia of Food Grains, vol. 3, Elsevier Ltd, Oxford, UK, pp. 458–466.

Rausch, K.D., Eckhoff, S.R., 2016b. Maize: wet milling. Wrigley, C., Corke, H., Seetharaman, K., Faubion, J. (Eds.), Encyclopedia of Food Grains, vol. 3, Elsevier Ltd, Oxford, UK, pp. 467–481.

Regina, A., Berbezy, P.G.L., Chanliaud, E.M.A.I., Duperrier, B., Morell, M.K., 2016. High amylose wheat-II. U.S. Patent No. 9,357,722. Washington, DC: U.S. Patent and Trademark Office.

Regina, A., Bird, A., Topping, D., Bowden, S., Freeman, J., Barsby, T., et al., 2006. High-amylose wheat generated by RNA interference improves indices of large-bowel health in rats. Proc. Nat. Acad. Sci. USA 103 (10), 3546–3551.

Sayaslan, A., 2004. Wet-milling of wheat flour: industrial processes and small-scale test methods. Leben. Wissen. Technol. 37, 499–515.

Serna-Saldivar, S.O., 2016. Maize: foods from maize. Wrigley, C., Corke, H., Seetharaman, K., Faubion, J. (Eds.), Encyclopedia of Food Grains, vol. 3, Elsevier Ltd, Oxford, UK, pp. 97–109.

Shen, L., Haufe, J., Patel, M.K., 2009. Product Overview and Market Projection of Emerging Bio-Based Plastics. University of Utrecht, The Netherlands.

Stoddard, F.L., 2016. Starch chemistry. Wrigley, C., Corke, H., Seetharaman, K., Faubion, J. (Eds.), Encyclopedia of Food Grains, vol. 2, Elsevier Ltd, Oxford, UK, pp. 174–180.

Topping, D., Clifton, P., 2001. Short-chain fatty acids and human colonic function: roles of resistant starch and non-starch polysaccharides. Physiol. Rev. 81, 1031–1064.

Topping, D.L., Segal, I., Regina, A., Conlon, M.A., Bajka, B.H., Toden, S., Clarke, J.M., Morell, M.K., Bird, A.R., 2010. Resistant starch and human health. In : van der Kamp, J.W., Jones, J., McCleary, B., Topping, D., (Eds.), Dietary Fibre: New Frontiers in Food and Health, 311–321.

Yan, Y., 2016. Biodiesel. Wrigley, C., Corke, H., Seetharaman, K., Faubion, J. (Eds.), Encyclopedia of Food Grains, vol. 3, Elsevier Ltd, Oxford, UK, pp. 245–250.

Zhao, X.C., Batey, I.L., Sharp, P.J., Crosbie, G., Barclay, I., Wilson, R., Morell, M.K., Appels, R., 1998. A single genetic locus associated with starch granule properties and noodle quality in wheat. J. Cereal Sci. 27, 7–13.

Zhao, X.C., Sharp, P.J., 1998. Production of all eight genotypes of null alleles at "waxy" loci in bread wheat, *Triticum aestivum* L. Plant Breeding 117, 488–490.

Chapter 24

Optimisation of End-Product Quality for the Consumer

Diane Miskelly
Westcott Consultants, Goulburn, NSW, Australia

24.1 INTRODUCTION

Cereals are the basis for many plant-based foods throughout the world, providing a source of carbohydrate, protein, fibre and micronutrients in the human diet. They are also utilised in beverages, animal feeds or in industrial uses such as starch–gluten manufacture and biofuels.

Although production and consumption of individual cereals were previously regionally based according to the area of production, this is becoming less common through world trade and globalisation of diets. For example rising wheat consumption in the traditionally rice-consuming countries of south-east Asia has seen the increased production of both traditional wheat-based foods, such as noodles, as well as western baked products adapted to the local market.

Cereals are usually processed in one or more stages to produce a wealth of different end-products. This can vary from polishing brown rice, puffing for breakfast cereals or fermenting to alcohol. Milling is the most common initial processing step. In the case of rice, 'milling' means the removal of the bran layer (see Chapter 12). On the other hand, 'milling' for wheat and related cereals, involves the provision of flour milled from the endosperm or from the wholegrain (see Chapter 22 for a description of wheat flour milling). The milled flour is then used in the production of a range of breads, noodles, cakes, cookies and numerous other products. Some products are consumed as purchased and others require further processing at home.

The raw materials and manufacturing processes used govern the type and quality of the end-product produced. It is not easy or even possible to substitute one cereal for another and maintain the expected product characteristics. Even within a cereal type, certain quality characteristics are essential for individual end-uses. Wheat-based products have the most sophisticated range of raw

Cereal Grains. http://dx.doi.org/10.1016/B978-0-08-100719-8.00024-3

material specifications and expected processing performances. Laboratory and pilot-scale end-product tests are useful to evaluate the potential performance of raw materials in processing and measure the quality of the end product.

24.2 MEETING CONSUMER-PERCEIVED QUALITY NEEDS

Today's consumers are more sophisticated than ever before and the internet brings instant, often conflicting information on a range of food and diet related issues. Consumers' perception of food quality is complex and modern consumers regard taste, texture, colour, health, convenience and price as important when making purchasing decisions, but may also consider other factors such as organic, GM-free and gluten-free. There is a trend for consumers to be differentiated, and marketers will develop products targeting certain age or demographic groups.

Manufacturers must not only provide for consumers' perceived quality needs, but work within a framework of quality and food-safety management programs. Technical product specifications are necessary for ensuring the quality of incoming raw materials and the quality of the finished product. Quality-management systems monitor the manufacturing process and the product during and after manufacture, measuring appropriate parameters such as volume, colour and weight. Food-safety programs are integrated to identify and evaluate potential food-safety hazards linked to incoming raw materials and manufacture. These procedures work to ensure there is no food-safety hazard such as pathogenic bacteria, or accidental contamination with a potential allergen such as nuts, or gluten in gluten-free products. Product recall procedures are in place to enable tracing of manufactured product distribution and enable prompt withdrawal from the market place.

Fig. 24.1 shows some examples of consumer perceptions of product quality and how these are accommodated within quality-management systems in the manufacturing process.

24.3 THE BREAD FAMILY

The ancient Egyptians are thought to be the first to make leavened bread. Knowledge of bread-making was passed to the Greeks and then to the Romans before spreading throughout Europe (Qarooni et al., 1992). At first bread was made from barley and emmer wheat, evolving to bread wheat as cultivation developed, then changing to the better-adapted rye in the cold climates of northern and eastern Europe. Unique styles have developed ranging from the flat breads of the Middle East, Africa, India and South America to the pan and hearth breads of Europe, which are now spread worldwide. Steamed breads and stuffed buns originated in China more than 2000 years ago and became a staple part of the diet in the wheat-growing areas of northern China before spreading to south-east Asia.

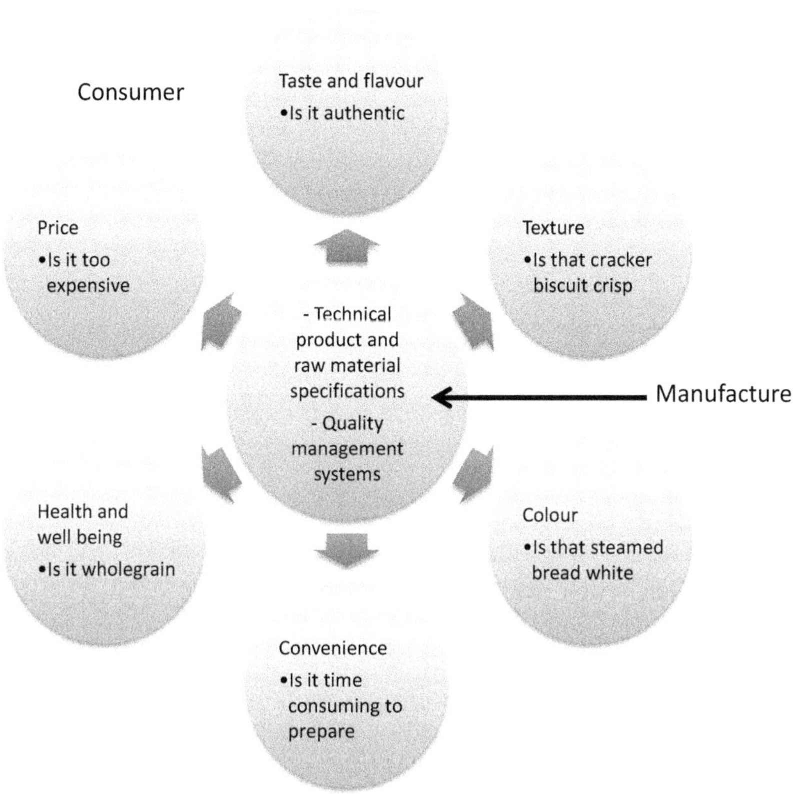

FIGURE 24.1 **Examples of consumer-perceived needs and how they are met by integration into manufacturing process and quality systems.**

24.3.1 Leavened Pan and Hearth Bread

The basic ingredients for making bread are flour, water, yeast and/or sourdough and salt. Bakery water addition is generally 55–65% on a flour weight basis (baker's percentage). Salt is added to provide flavour, as bread without salt is bland; salt also toughens the gluten. Other ingredients are used in the different bread-making processes and styles; they include oxidants (to aid in developing the dough), fats and emulsifiers (to aid dough stability and crumb softness), soy flour, malt flour and enzymes (to aid fermentation or retard staling) and sugar.

Bread-making consists of a number of steps—mixing, fermentation, proofing and baking. Baking can be carried out in a tin or pan (hence pan bread) or on the hearth or stone of an oven (hence hearth bread). A lid is used on top of the tin to create the square shape popular for sandwiches. The different types are shown in Fig. 24.2. Modern bread-making methods differ in terms of minor ingredients used, the mixing methods to develop the gluten and the fermentation

(A) (B)

FIGURE 24.2 (A) Pan bread baked without and with a lid. (B) Bread baked on the hearth of the oven.

times. These variations have resulted in the evolution of four major dough-processing methods detailed as follows.

24.3.1.1 The Straight-Dough Method

In this traditional method, ingredients are mixed to a dough which is then allowed to ferment. During bulk fermentation, the dough matures and becomes more elastic and the gluten is fully developed. Dough is briefly remixed or 'knocked back' to release some of the carbon dioxide before dividing and moulding, after which it is deposited into baking trays for a short proof followed by baking. The total process generally takes 4–8 h from mixing to when the bread emerges from the oven.

24.3.1.2 The Sponge-and-Dough Method

The sponge-and-dough process is used in the United States and Asia, and is often used elsewhere for hamburger buns in multi-national fast-food franchises. In this two-step procedure, a portion of the flour is mixed with a portion of the water plus the yeast; the resulting soft dough is allowed to ferment for about 2 h. The rest of the ingredients are then added, the dough is remixed and allowed a secondary fermentation. The remainder of the process is carried out as in the straight-dough method. Total process time is 6–8 h. It is common to add high levels of fat and sugar, so that the resulting bread has a fine, even texture and sweet taste. Due to the long fermentation time and the fat and sugar additions, a high-protein high-strength flour is usually specified.

24.3.1.3 The Rapid-Dough Method

This method is popular in Australian bakeries. The main difference from other dough-making methods is that gluten-modifying agents such as ascorbic acid are added and the doughs are developed fully in the mixer. Without the need for bulk fermentation, dough is divided, moulded, proofed and baked as before. The entire process can be completed in 2–2.5h. Medium-protein, medium-strength flour is suitable for use in the rapid-dough method.

24.3.1.4 Mechanical Dough Development (Chorleywood Bread Process)

Doughs are mixed for 2–4 min in high-intensity mixers (often under vacuum) to a specified maximum energy input. Gluten-modifying agents are incorporated in the formulation. The remainder of the process is similar to the rapid-dough method and the time taken is similar. Pan bread produced by this process has a fine, uniform structure, but the method can be adjusted for any type of fermented product (Brown, 1993). The Chorleywood bread process is used in the United Kingdom, New Zealand and many Australian plant bakeries.

24.3.1.5 Assessing Leavened-Bread Quality

Bread products vary in their size, shape, texture and appearance according to their style, and comparisons should be made within rather than between styles. A sandwich loaf will have a symmetrical external appearance and soft, uniform crumb texture, whereas artisan-style bread should have a crisp crust and a more irregular, open crumb structure with a firmer, resilient texture (Fig. 24.3).

No matter which style of bread is considered, it should have a good volume, which is normally measured by displacement or by laser-scanning instrumentation (www.stablemicrosystems.com). The crust should be evenly baked to a golden-brown colour with the external appearance typical of the style, whether sandwich, baguette, artisan or other. The loaf is cut for an evaluation of internal characteristics. For white bread, a white or creamy white colour is usually preferred and this can be measured subjectively or by using a tristimulus colour meter. For sandwich bread, cell structure should be fine and uniform. Such bread will have a soft texture, and will not crumble or tear when cut. Image analysis can be used to quantify appearance and cell structure with images retained for

FIGURE 24.3 Sandwich bread showing a fine texture and artisan bread with a more open irregular crumb.

later comparative use (www.c-cell.info). Texture can be measured subjectively, or objectively using a texture analyser (www.stablemicrosystems.com). Bread becomes firmer as it stales and shelf life is an important consideration for consumers. The rate of staling can be measured by testing firmness with a texture analyser daily over 3–5 days.

Commonly, bread is allocated a score for each of its component characteristics using a scoring system which assigns relative importance to each characteristic. The highest scoring characteristic is usually allocated to volume. Examples of these scoring systems can be found in the methods given in Table 24.1. The flour-quality attributes for different bread types are documented in Table 5.5 and are probably the best known of all the different end products.

24.3.2 Flat Breads

Flat breads are the oldest of all bread products and are consumed in North Africa, Southern Europe, the Middle-East, Indian subcontinent, Turkey and Central America (Qarooni, 1996). They are usually made from wheat flour, but mixtures with corn, rye, millet and barley are common. High-extraction flours (up to 85% flour yield) are used depending on the product; the bakery water addition varies from 50–70%. After the fermentation stage, doughs are sheeted, die-cut or pressed before proofing and baking. High baking temperatures (up to 550°C) are used.

Flat breads form one or two layers during baking depending on whether the sheeted dough has been docked (pricked). The pocket is a distinctive characteristic of the two-layered flat bread and is formed as steam is generated during baking, forcing the dough layers apart.

Many different types of flat bread are consumed and some examples are shown in Fig. 24.4. They have become popular in many western countries, finding a new range of uses as sandwich wraps and pizza bases.

Chapattis, parathas (Fig. 24.5), roti and puri are common single-layered flat breads of the Indian subcontinent which are either chemically leavened or unleavened. Chapattis are made from 100% extraction atta flour and the unleavened doughs are divided, rounded and sheeted by hand. Sheeted dough pieces are cooked on both sides on a hot plate. Tandoori bread (Fig. 24.5), including naan, is a leavened dough made by a similar process, but baked in a clay oven or tandoor at temperatures in excess of 330°C (Hashmi and Wootton, 2000).

24.3.2.1 Assessing Flat-Bread Quality

Depending on the type, flat breads should have a round, uniform shape and a smooth crust free from blisters and cracks (Qarooni, 1990; Quail et al., 1990; Hashmi and Wootton, 2000). For two-layered flat breads, a clean separation should be achieved between top and bottom layers. Crumb appearance should be fine and uniform, and crumb should be creamy-white. Bread should be able to be rolled and folded without cracking or breaking as it restricts the ability to add filling. Methods and scoring systems for flat-bread are given in Table 24.1.

TABLE 24.1 Methods and Quality Parameters for Assessment of Wheat-Based End-Product Characteristics

	Product	Product evaluation parameters	Method references
Breads	Straight dough	Volume, external appearance, crumb colour, texture, structure, softness	AACC 10-09.01 (long fermentation), AACC 10-10.03 (optimised), AGSA 07-02; Wheat Marketing Center (2008)
	Rapid dough	Volume, external appearance, crumb colour, texture structure and softness	AGSA 07-03; Preston et al. (1982) (Canadian short process)
	Sponge and dough	Volume, external appearance, crumb colour, texture, structure and softness	AACC 10-11.1; Lever et al. (2005)
	Flat breads	Shape, crust appearance, flexibility, crust and crumb colour, crumb appearance	Hashmi and Wootton (2000); Qarooni et al. (1987); Quail et al. (1990); Wheat Marketing Center (2008)
	Steamed breads	Specific volume, spread ratio, external appearance and colour, internal appearance, crumb colour, crumb firmness, elasticity, stickiness	Huang et al. (1993); Huang et al. (1998); Wheat Marketing Center (2008)
Noodles	White salted	Colour, colour stability, luster, cooked noodle smoothness, softness and elasticity (springiness), flavour	AGSA 07-05; Ross and Hatcher (2005); Solah and Crosbie (2016); Wheat Marketing Center (2008)
	Yellow alkaline	Colour, colour stability, cooked noodle smoothness, softness and elasticity (springiness), flavour	AGSA 07-06; Ross and Hatcher (2005); Miskelly (1996)
Biscuits	Semi-sweet	Length, width, height, checking, hardness	AGSA 07-04; Chow et al. (2003)
	Cookies	Width, thickness, spread factor	AACC 10-50.05; Wheat Marketing Center (2008)
Cakes	Sponge; high-ratio	Volume, shape, external appearance and colour, internal appearance and colour, crumb softness, flavour	AACC 10-90.01; Wheat Marketing Center (2008)
Durum products	Pasta	Appearance and colour, cooking characteristics, cooked pasta firmness and stickiness, cooked pasta colour	AACC 66-41.01; Marti et al. (2016); Wheat Marketing Center (2008)

AACC International. Approved Methods of Analysis, eleventh ed. AACC International, St. Paul, MN, USA. Available online from: www.aaccnet.org
AGSA (formerly RACI). Supplement 4th Edition Official Testing Methods of the Australasian Grain Science Association, 2010, Australasian Grain Science Association; Narrabri, NSW, Australia.

FIGURE 24.4 Examples of flat breads (clockwise from left)—Lebanese *khubz*, Greek pita, Turkish *pide*.

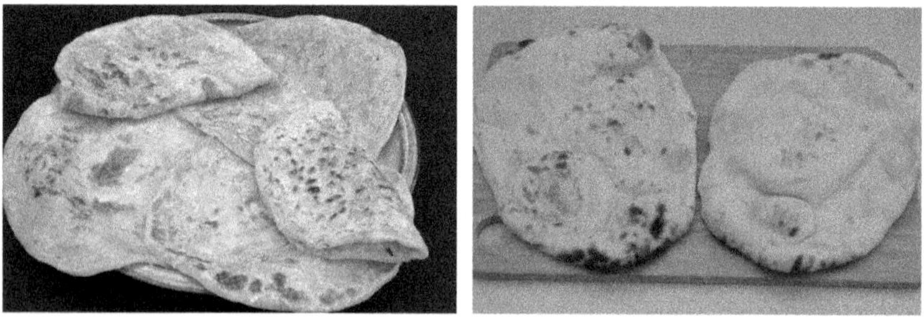

(A) (B)

FIGURE 24.5 Indian flat breads. (A) Paratha. (B) Tandoori bread.

Protein content and quality, grain hardness and starch damage are all important quality criteria for Arabic bread quality. As detailed in Table 5.5, optimum wheat protein content is 10–12% and flours should have intermediate dough strength (Qarooni, 1990).

24.3.2.2 Pizzas

Pizza can be traced back to its origin in the Persian Empire around 500 BC when flat bread was topped with cheese and dates. Tomato was added by the Romans (Qarooni et al., 1992). Pizza coated with spices, oils and sesame seeds is still popular in the Middle-East (*maneesh*). Modern pizza is a firmly established food in the west, with the crust produced by a sheeting and die-cut procedure, hot or cold pressing, or shaping by hand. Dough retardation (where dough is held at refrigerator temperatures) is often used in fast-food franchises to enable timely meeting of customer orders. There are two types of crusts: thick and thin. Pizza should have a uniform golden-colour top and bottom, a shiny surface and a creamy white crumb colour. Texture and crispness depend on the style. The dough must have sufficient strength to support the toppings of cheese, tomato paste, meat etc. Generally a flour with a minimum protein content of 10–12% with medium-strong but extensible dough properties is used in pizza production.

24.3.3 Tortillas (Corn)

Although tortillas can be made from either wheat or corn, only those made from corn will be considered here. Corn tortillas originated in South America and are now popular worldwide, but particularly in the United States (Fig. 24.6). Traditionally, the corn was soaked in water containing 1% lime to soften and hydrate the kernels by a process known as nixtamalisation. The soaked kernels were then ground on a stone-mill to produce wet masa, which was sheeted into thin disks before baking on a griddle (Rooney and Suhendro, 1999). Today, dry masa flour is produced by several methods including steam injection of lime-steeped kernels prior to hammer-milling. A typical method of production is to produce a dough using dry masa flour before sheeting and cutting using a dye or rotary cutter prior to baking.

Like other flat breads, tortillas stale quickly, becoming brittle and unable to be rolled smoothly around fillings. Rollability can be measured objectively using texture analysis. Addition of gums to formulations can help retain flexibility, as can the use of modified atmosphere packaging. Colour and flavour are also important quality characteristics. Colour and cooking characteristics depend on the correct selection of maize hybrids during the breeding process. The standard corn-quality tests of test weight, grain size and vitreousness are detailed in Chapter 11. In addition, corn-kernel density is measured as floating kernels in

FIGURE 24.6 Corn tortillas.

sodium nitrate solutions of different density and is regarded as a measure of grain hardness (Wu and Bergquist, 1991).

24.3.4 Steamed Breads

Steamed breads made without fillings are a staple food in the wheat-growing areas of northern China, but are also popular in southern China. There are three major styles of steamed bread—northern, southern and Guandong. Steamed buns (as distinct from steamed breads) are made with savoury fillings (such as meat, vegetables and seafood) or sweet fillings (such as bean pastes or custard). They are consumed as a snack food. Ingredients for dough are very simple: flour, yeast and water with little or no added salt. Water is added at lower levels than for flat and pan bread (45–50% bakery absorption), giving a product with a more chewy texture. In the southern provinces of China and other parts of Asia, fat and/or sugar are added. As with pan bread, there are a number of different but unique dough-processing methods for making steamed bread, but the same basic operations of mixing, fermentation and proofing are involved. Steamed bread is cooked in a steamer rather than baked in an oven, resulting in a thin white skin rather than the baked crust of western bread. Examples of steamed bread and buns are shown in Fig. 24.7.

24.3.4.1 Assessing Steamed Bread Quality

A white, shiny, smooth skin free from blemishes is preferred by consumers. The skin colour can be measured using a tristimulus colour meter. Consumers also prefer a fine crumb texture, which is elastic, cohesive and does not stick to the teeth when chewing. These characteristics can be evaluated subjectively using a trained sensory panel, or using a texture analyser (see Table 24.1 for methods).

FIGURE 24.7 **Steamed breads and buns.**

Steamed bread quality is influenced by protein content and strength as well as flour extraction rate. Flour colour, including yellowness due to flour xanthophylls, plays an important role in determining steamed bread colour. Optimum flour protein content should be within the range of 8–12%, depending on the style (Huang and Miskelly, 2016). Steamed bread is perhaps the most sensitive of all end products to the damaging effects of the starch-degrading enzymes of rain-damaged wheat, indicated by flour falling numbers below 200 s. When such flour is used, the resulting steamed bread has small volume and poor appearance, rough texture and sticky eating quality (Zhou et al., 1994). The importance of falling number as a wheat receival test is discussed in Chapter 19.

24.3.5 Whole-Grain, Multigrain and Composite Breads

Bread consumption patterns are changing in some western countries with consumer preferences shifting away from white bread towards more wholegrain, multigrain, seeded, specialty and ethnic breads. This is driven by a consumer desire for healthier and more nutritious products, freshness and more complex flavour profiles. Some of the health benefits including higher fibre, vitamins, nutrients and anti-oxidants have been covered in the chapters on individual grains. Many options are available for a diverse range of grain and wholegrain breads with functional properties [see Rahaie et al. (2014) for a review].

Breads containing other wholegrain, grains or non-wheat flour may have a smaller volume, firmer, denser texture and darker colour than white bread, yet there are many high-quality commercial products available (Fig. 24.8). The addition of bran, wholemeal, kibbled grains, seeds such as linseed and sunflower and other cereal flours such as rye, barley and oats cause physical and chemical changes in doughs and weaken the gas cell structure. The protein

FIGURE 24.8 Wholegrain, grain and wholemeal breads.

is diluted, so a higher protein flour or added gluten is required. Doughs are tighter, so water addition needs to be increased, and formulations including fats, emulsifiers, bread improvers and enzymes will need to be optimised. The impact on quality will depend on the level of addition of the non-white flour component and its particle size. For example adding kibbled (cracked) grains or wholegrain flour will have fewer deleterious product quality effects than entire wholegrains. Finer bran particle size and pre-treatments such as heat treatment may offer potential for counteracting the detrimental effects of bran addition (Hemdane et al., 2016).

The quality of wholegrain, multi-grain and composite breads is evaluated using a modified scoring system based on those used for white breads (Table 24.1). In addition, the non-flour components contribute flavour, aroma and mouthfeel, which need to be taken into consideration in the assessment process.

24.3.6 Sourdough

The use of sourdough for making bread is one of the oldest forms of fermentation and is also one of the most modern of popular food trends. Sourdough can be used for all types of bread, including wheat and rye, as well as traditional cakes and crackers. There are several ways of producing sourdoughs. Traditional methods are often used in artisan bread applications. The sour is generated from one day to the next and there is generally a long fermentation time. Another method is to use dried sourdough starters (containing lactic acid bacteria) together with yeast. Sourdough is based on lactic acid fermentation by *Lactobacillus* spp., which create an optimum pH and chemical products which contribute to the characteristic aroma and flavour as well as affecting dough properties and taste.

24.3.6.1 Assessing Sourdough Bread Quality

Sourdough is a complex system. Measurement of pH and titratable acidity can assist in monitoring the maturity of the dough during the fermentation phase and identifying any needed ingredient or process adjustments.

Sourdough bread, in particular, artisan and hearth bread styles should have an attractive shape and external appearance (Fig. 24.9). Flavour and aroma are key sensory attributes preferred by consumers. The internal crumb depends on the style, with some types having a more open and irregular crumb and resilient texture. Volume, specific density and shelf life are also important quality characteristics.

24.4 NOODLES

Noodles are a traditional food in Asian diets. They originated in China before spreading to other Asian countries and can now be found worldwide. Today, most noodles are produced in automated or semi-automated production lines

FIGURE 24.9 A wide variety of bread products can be made using the sourdough process.

which have replaced domestic or manual labour. There are a number of different types based on ingredients used singly or in blends—wheat, buckwheat and rice flours; and rice, wheat, corn, mung bean, tapioca, potato and tapioca starches. Regional preferences exist as to the product type, size, shape and texture.

Wheat-based noodles are the most common noodles and are classified as white-salted noodles (WSN) and yellow alkaline noodles (YAN) dependent on the type of salt used in the formulation (Fig. 24.10). Noodles can be sold fresh, boiled, refrigerated, frozen, pasteurised, semi-dried, dried, steamed, steamed and dried or steamed and fried.

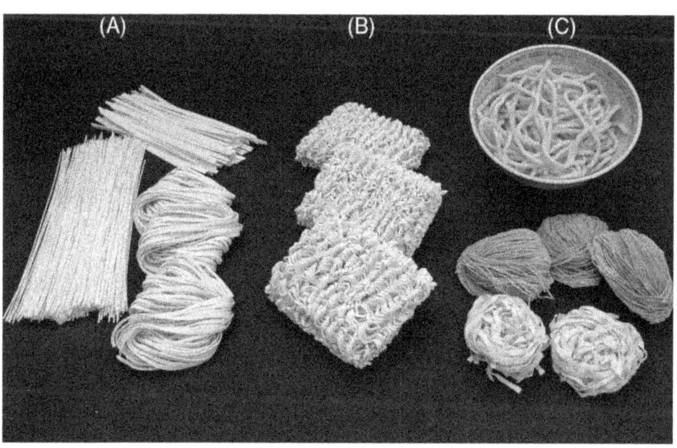

FIGURE 24.10 White-salted noodles (A), instant noodles (B) and yellow alkaline noodles (C).

24.4.1 White-Salted Noodles

WSN are made by mixing flour, 1–3% salt (sodium chloride) with approximately 34% water. The uniform, crumbly dough is passed through a series of sheeting rolls during which gluten development occurs. Further sheeting is required to reduce dough to the required thickness. A resting stage may be incorporated after mixing, or before dough is reduced in thickness. This assists in even dough hydration and dough relaxation. The sheeted dough is cut to the required width through standard size cutting rolls according to the type of noodle and ranging from 0.6 to 6.0 mm slit width. Noodles maybe dried and packaged or sold fresh.

Dried noodles are the most common type of WSN produced in China and many other parts of Asia. They are produced from flour with 9–11% protein content, and when cooked should have a firm and elastic texture. A white colour is preferred by consumers.

Udon noodles are boiled noodles popular in Japan. The noodle quality requirements are perhaps the most demanding for all noodle categories and include texture, colour, surface appearance and flavour (Table 24.1). Flour for Japanese *udon* noodles is milled from soft-grained wheat or blends of soft and hard wheat to give a protein content in the range 8–10%. High starch paste viscosity and high swelling starch characteristics are important to obtain the required soft and elastic texture (Crosbie, 1991). In order to satisfy this requirement, breeders select new varieties with a null allele for granule-bound starch synthase protein on chromosome 4A (null 4A). Noodles should have a creamy-yellow colour and be free from discolouration and bran specks. Low extraction flour with ash content less than 0.40% is required for *udon* noodles. The level of natural yellow pigments, mainly xanthophylls, in flour is a good indicator of the degree of creamy-yellowness of WSN. This is dependent on wheat variety with some environmental interaction.

24.4.2 Yellow-Alkaline Noodles

YAN are made using a similar process to that used for WSN. YAN can be sold fresh, boiled (e.g. Hokkien), dried, steamed, steamed and dried or steamed and fried (instant). The formula includes flour, water, salt and alkaline salts at about 1% on a flour weight basis. Alkaline salts commonly used include potassium or sodium carbonate; or a mixture of the two which are dissolved in part of the water before addition during the mixing stage. The alkaline salts modify dough rheological properties by toughening the dough and increasing paste viscosity. Doughs have a high pH (9–11), which results in the expression of a yellow colour from the naturally occurring flour flavonoid compounds. These compounds are colourless at acidic or neutral pH. The yellow colour of noodles is enhanced due to the additive effect of flour yellow pigments.

24.4.2.1 Assessing YAN Quality

YAN should have a firm and elastic texture and smooth surface. This is assessed using sensory and instrumental techniques (Table 24.1). Noodles should be light yellow in appearance, and free from specks and brown or black discolouration. Darkening of noodles with time is a serious problem for all fresh noodle manufacturers, whether WSN or YAN. Noodle colour is measured using a Minolta (or similar) colour meter by measuring L* and b* on a portion of sheeted dough at 0 and 24 h after sheeting. The colour stability (L*0-L*24) is used as measure of the darkening. Darkening is due to a combination of the enzyme polyphenol oxidase (PPO) and non-PPO darkening. PPO is concentrated in the bran layer and accounts for about 50% of the total darkening, appearing as dark specks concentrated on bran particles. Flour millers control bran contamination by reducing the extraction rate of the flour, but as PPO is under genetic control, breeders can select low PPO varieties for noodle applications. The mechanism of non-PPO darkening is still not fully described, but appears to be due to a combination of (1) physical changes in the noodle surface properties and is correlated with protein content and (2) a PPO-like enzyme which oxidises protein tyrosine groups (Asenstorfer et al., 2014). This presents a dilemma for noodle manufacturers, who must balance the requirement for the firmer texture that higher protein flour confers, but at the price of poorer noodle colour.

24.4.3 Other Noodles

24.4.3.1 Instant Noodles

Instant noodles were developed in 1958 as 'chicken ramen' and were a means of providing fast-cooked noodles in post-war Japan. They have continued their role in emergency food aid in the wake of many natural disasters to this present day, having a shelf life of up to 6 months. Found in supermarkets worldwide, they are mostly consumed as a snack. Preparation is simple, the bagged type requiring about 2 min cooking in boiling water, and cup-type (in heat-resistant foam cups) needing only the addition of boiling water; soup and seasoning sachets are included in the packet.

Instant noodles are similar to alkaline noodles, but contain only about 0.2% alkali, plus a range of stabilisers such as CMC, gums and alginates which function as water binders and improve re-hydration and textural properties. Some instant noodles contain modified starches which function in a similar manner. The processing steps are the same as for other noodles up until the cutting stage, where a wave former is used to produce the characteristic waved shape. Instant noodles can be steamed and dried, or more commonly steamed and fried.

24.4.3.2 Assessing the Quality of Instant Noodles

There is a wide variation in preferences for colour, eating quality and texture according to region. The protein content and protein quality positively impact the firmness of the noodles, and protein is inversely related to oil uptake during frying. Instant noodles are normally made with a flour protein content greater than 9%. Flour should have extensible dough properties and moderately strong dough strength to enable sheeting without shrinkage. Starch properties are also important, with some manufacturers preferring flours with high starch paste viscosity as the lower gelatinisation temperature assists with more rapid hydration as noodles are prepared for consumption. Colour is also an important characteristic, with bright noodles preferred, and the yellow colour is dependent on consumer and regional preference. Instant noodles are usually of moderate firmness with a slightly chewy texture (Azudin, 1998). Both firmness and colour can be measured instrumentally.

24.4.3.3 Rice Noodles

There are many types of rice noodles which are widely consumed in China and many parts of south-east Asia. These include fresh sheets or noodles cut from fresh sheets (e.g. *chanfen, kueh teow*), fresh extruded round rice noodles (*zhafen*, *laksa*) and dried rice sticks or rice vermicelli (Fig. 24.11). The traditional process for rice vermicelli involves soaking the rice, which can vary from a few hours to days in the case of fermented noodles. The rice is ground using a wet process, or dry rice flour is sometimes substituted. The rice slurry is partially steamed to gelatinise the starch, and the slurry/dough is kneaded before it is extruded and cut. This is followed by steaming or boiling. As rice does not contain gluten, pre-gelatinisation of the starch is necessary to create a uniform structure. Rice noodles are made from long-grain rice with intermediate to high amylose content and starch pasting properties, as measured with the RVA, play an important role in the quality of the product (Bhattacharya et al., 1999). Additives such as emulsifiers and modified and unmodified starches are sometimes used to improve the textural properties of the noodles.

FIGURE 24.11 Dried and fresh rice noodles.

Assessment of rice noodle quality includes cooking quality (cooking loss) and sensory evaluation (taste and aroma, lustre, stickiness, hardness and toughness) (Jianfen, 2009). Textural properties can also be measured using instrumental methods.

24.5 BISCUITS, COOKIES, WAFERS AND SNACK FOODS

The term 'biscuit' has two common meanings. In North America it often refers to a soft bread-like product elsewhere known as a scone. Elsewhere, biscuits are recognised as thin low-moisture baked products of diverse size, shape and composition. It is postulated that biscuits were first produced as non-perishable rations for long journeys such as on early sea voyages. Biscuit formulations may contain many ingredients for flavour, colour, texture or as leavening. There are many types of biscuits including crackers, hard-sweet, short-sweet, cookies and wafers. Filled (creamed) and chocolate-coated (enrobed) biscuits are popular.

24.5.1 Cookies

Cookies are a universally recognised product made principally from sugar, fat, flour and water (Fig. 24.12). Cookie doughs are very soft and contain only minimal additions of water. Cookie doughs are deposited as clumps and during baking they spread to the specified manufacturing diameter and thickness. Control of cookie spread is crucial to automated production and packaging. While flour properties can affect the spread, as assessed by the cookie-spread test (Table 24.1), it is easier, more convenient and more effective to control the spread during production by manipulation of the major dough ingredients. Fat content (and solid fat content as measured by NMR) will affect the spread of the

FIGURE 24.12 Cookies.

FIGURE 24.13 Marie (top left) and other hard-sweet biscuits.

dough piece as will sugar granulation and water addition. Many cookie formulations include leavening agents which can also have an effect on cookie spread.

24.5.2 Hard-Sweet and Semi Hard-sweet Biscuits

An example of a hard-sweet biscuit is a Marie biscuit (Fig. 24.13). Hard-sweet biscuits are made from flour, sugar, shortening, leavening agents and water and may contain egg or dairy products. Semi hard-sweet doughs as used to manufacture sandwiched or creamed biscuits are very stiff as only minimal water additions are necessary. Doughs are sheeted to the required thickness before being cut with a reciprocating or rotary cutter. These biscuits typically have a firm, crisp bite. They are susceptible to checking (cracking which may occur sometime after packaging) which is due to uneven moisture distribution within the biscuit. This can be avoided by baking to the specified moisture and cooling to enable moisture equilibration.

24.5.3 Crackers

Crackers (Fig. 24.14) are manufactured mainly from flour and water with relatively small quantities of shortening. Some cracker doughs can be prepared with yeast as the leavening for either long or short fermentation times. The sponge-and-dough method can be used, although the long fermentation time does not fit well into modern manufacturing methods. Some cracker doughs are chemically leavened using either reducing agents (sodium metabisulfite) and/or proprietary enzymic mixtures of amylases and proteases.

Cracker doughs may require low-protein soft flours where the machining of the dough needs soft dough properties and a tender finished biscuit texture is important. Crackers that undergo a fermentation stage often require a

FIGURE 24.14 Examples of crackers.

higher-protein hard-wheat flour with strong dough properties. Regardless of hardness or protein content of the wheat, a biscuit dough that is sheeted and cut to form the biscuit will require good extensibility to cope with the sheeting process.

24.5.4 Wafers

Wafers differ from other biscuit products in that they are made by a batter process. The best-known products are ice cream cones and wafer biscuits which are sandwiched with cream filling (Fig. 24.15). Wafers are made from soft,

FIGURE 24.15 Wafer biscuits are often sandwiched with cream and can be chocolate-coated.

low-protein flour which must be free from pre-harvest sprouting due to the detrimental effects of α-amylase on the starch in the batter and hence the product. Other ingredients may include starch, sugar (optional), vegetable oil, emulsifiers (such as lecithin), leavening agents (sodium and/or ammonium bicarbonate), salt and optional milk powder, flavourings or colours. Water is added at about 130–150% of flour weight to achieve the desired batter consistency. It is important to minimise gluten aggregation and control batter viscosity during the mixing phase. The batter is then pumped to a supply tank before it is spread accurately onto the wafer plates which close before being conveyed through the oven. After baking, the wafer sheets are released from the plates and are sent to a temperature and humidity-controlled cooling tunnel where the moisture equilibrates and increases slightly.

A common problem in wafers is cracking due to a weak wafer structure. This can be caused by use of the wrong wheat, rain-damaged wheat and/or failure to control moisture during or after production.

24.5.5 Specifications and Assessment of Quality

It is commonly thought that all biscuits, cookies and crackers are made from 100% soft wheat, but in practice blends of soft and hard wheat flour are often used in different varieties. The ratio of blending is dependent on the product and required product characteristics. Flour specifications for biscuit, cookie and cracker production will generally contain ranges for protein, moisture and ash. Also important are the rheological properties and specifications that will include Farinograph water absorption and development time, dough sheeting properties as measured by the Extensograph and finally a flour-paste viscosity (Viscograph or RVA) to ensure the flour is made from sound wheat.

Assessment of quality includes the biscuit dimensions (thickness, width, shape and spread), colour and texture. Some references to methods are given in Table 24.1. Biscuit dimensions (height × width or diameter) and the biscuit thickness (packet length) as measured by the stack height of a nominated number of pieces are critical where products are machine wrapped into printed wrappers of specified dimensions. This ensures that the packet meets both the minimum weight but also does not exceed the manufacturer's allowable maximum, yet the contents still fit within the sealed wrapping. Spread ratio (width/thickness) is an important physical parameter for some types as it provides information about the biscuit shape. Colour can be measured subjectively, or objectively as L*, a*, b* by tristimulus colour meters or using machine vision. Required texture—hardness, crispness or brittleness—will depend on the type, and is assessed by sensory analysis or using texture analysis. Flavour is important to ensure minimum variation batch to batch and also to ensure no off-flavours have been introduced from raw materials or the process. Finally, moisture content of the finished products should be determined to maximise shelf life and minimise defects.

24.5.6 Snack Foods

The broad definition of snack foods is food eaten between meals. A recent development has been the inclusion of wholegrain snacks to promote a healthier image. The product category includes ready-to-eat (RTE) cereal snack foods such as muesli bars, corn chips, rice crackers, popcorn and extruded snacks. The extrusion process involves mixing, heating, shearing, forming, cooking, and extrusion or co-extrusion. The products can go through a 'half-product' or pellet stage, or directly extruded and expanded in a one-step process. After extrusion, the shaped products are dried, fried or baked. A wide range of raw materials are used including rice flour, soft wheat flour and/or corn grits, together with other starches, water, gluten and oil. Colourings and flavourings are usually added, the latter as a dust or spray at the end of the process. Some examples of extruded snack foods are shown in Fig. 24.16.

A high expansion ratio is an important characteristic and can be measured as radial expansion (diameter of product cross-section/die). Specific volume is also important as a measure of expansion and is necessary to enable the product to fill the packaging at the required packet weight. Colour can be assessed by sensory (external and internal appearance) or instrumental methods. Flavour and aroma should be characteristic for the type of product. Final product moisture is an important process-control parameter.

Extruded snack foods are crisp and crunchy and the texture is affected both by the ingredients and the process (feed rate, temperature, time, pressure, shear and other extrusion variables). The texture can be evaluated by sensory analysis such as crispness, hardness, sticking to the teeth and tooth packing. There are a number of applicable instrumental techniques [see Anton and Luciano (2007) for a review].

FIGURE 24.16 Snack foods based on (front) wheat, (rear left and centre) corn, (rear right) wholegrain corn and wheat.

24.6 CAKES

Cakes and confectionary items are popular worldwide and there are many varieties including distinct types which originated in different regions. Special cakes are often eaten on festive occasions such as birthdays and weddings. Two examples, chemically leavened cakes and doughnuts will be considered here.

24.6.1 Chemically Leavened Cakes

The basic ingredients in cakes are flour, sugar, eggs, fats and emulsifiers and aerating agents. Flavourings, colours, fruits and vegetables and chocolate are a few of the optional extras.

Cakes generally fall into two primary categories: batter type (e.g. pound cake, chocolate cake) and a lighter, fluffier product known as foam type which is made by whipping the eggs and sugar prior to adding the flour (e.g. sponge cake, Swiss roll). A third type of cake combines batter and foam and the resultant cake is intermediate between the two (e.g. chiffon cake). Some examples of traditional cake types are shown in Fig. 24.17.

All the ingredients have a functional role, contributing to product quality. Flour provides structure and texture through starch gelatinisation. Unlike breadmaking, formation of gluten structure is detrimental to cakes, and consequently low-protein soft wheats (8–9% protein) or blends are used to mill cake flour. Flour for sponge cakes is sometimes chlorinated to a pH less than 5.0 or heat treated to denature the protein and increase flour viscosity. A fine particle size is preferred, and this can be achieved by air classification of the flour after milling.

Sugar adds sweetness and bulk, raises the gelatinisation temperature of the starch, extends shelf life and improves browning of the crust. Eggs bind the ingredients and form the foam structure in sponge cakes and provide colour and

FIGURE 24.17 **Variety of cakes.**

flavour. Fats provide softness, moistness and shelf life. Aerating agents function to improve the volume, tenderise the crumb and give a uniform crumb fine structure. Milk or milk powder is an optional ingredient which has nutritional benefits as well as promoting crust browning and shelf life.

24.6.1.1 Assessing Cake Quality

Cakes should be assessed for shape, dimensions, volume, texture, flavour and external appearance. The layer cake-measuring template in AACCI method 10-91.01 provides indices for volume, symmetry and uniformity of a layer cake. The top of the cake should show no signs of collapse or cracks, and the crust should have a uniform well-baked colour. The internal texture should be fine and even, but texture will vary depending on the type of cake. Crumb firmness and colour can be assessed subjectively or using instrumental methods. Cakes should have good flavour and aroma, and a sweet taste. The flavour will depend largely on the optional ingredients added to the cake. Methods for the baking quality of cake flour are given in Table 24.1.

Formulation balance is of utmost importance in cake manufacture and unbalanced formulations or processing problems can result in a myriad of manufacturing faults from partial or complete collapse to tough texture (see http://www.nzbakels.co.nz for further information).

24.6.2 Doughnuts

There are many fried confectioneries and doughnuts are arguably the best known. Machines for making doughnuts were developed in the United States in 1920 and today can range from a simple mixer, depositor and deep fryer suitable for cake shops to fully-automated plants. Ring-shaped doughnuts are the most recognisable, and there are other shapes including shells, cylinders and twists. Doughnuts fall into two categories—yeast-raised or cake-type, depending on whether they are yeast or chemically-leavened (Fig. 24.18). Ingredients include flour, sugar, shortening, eggs, milk, salt, dough conditioners and emulsifiers. The process consists of mixing a dough in the case of yeast-raised doughnuts, or

(A) (B)

FIGURE 24.18 (A) Cake doughnuts. (B) Yeast-raised doughnuts.

batter for cake doughnuts. The dough is allowed a short rest period or floor time before depositing or extruding, or for yeast-raised, the dough is often sheeted and cut before proofing and frying at about 190°C. Fried doughnuts are often dipped in sugar, glazed, iced or filled with cream and fruit or jam.

24.6.2.1 Assessing Doughnut Quality

Important product quality characteristics include: appearance which should be smooth, free from blisters, pitting, cracks and collapse; colour which is neither too pale nor too dark; flavour and oil uptake. Doughnuts should maintain their shape after frying (measured as spread and height). They should have a tender texture which can be measured subjectively or using texture analysis. Cake doughnuts have a more dense texture and are more cake-like than yeast-raised doughnuts.

Flour used in doughnut production is generally a blend of medium to high protein flour (10.5—12%) with low-protein soft flour (8.5—9.5% protein). The ratio will vary around 50:50. The proportion of high protein flour may need to be increased to support additional ingredients such as fruit; but if the protein is too high, the texture may be too firm with product faults such as blisters or collapse. Batter viscosity is an important parameter for cake doughnuts and is affected by flour cold-batter viscosity.

Due to the need to balance the interaction of ingredients, it is common for flour millers to produce premixes containing all ingredients, thus reducing ingredient inventories for manufacturers, franchises and in-store bakeries and ensuring finished-product consistency.

24.7 PASTA AND COUSCOUS

24.7.1 Pasta

Pasta has been a traditional food of Mediterranean countries for centuries and is now eaten worldwide because of its ease of manufacture and preparation. Traditional domestic production involved rolling dough into a sheet which was then cut and/or shaped. A wooden press was used in Italy in the early 19th Century in the first mechanised production of extruded pasta.

Durum semolina is mixed with water (about 28–30%) and if required, additional ingredients such as egg, spinach or spinach powder, herbs etc. This first stage of the process distributes and hydrates the ingredients. The crumbly dough is then extruded through a rough or smooth die, the shape of which determines the final product shape. Pasta is classified as long (e.g. spaghetti, fettuccine) or short (e.g. shells, elbow). Pasta can be sold fresh, frozen or pasteurised and chilled, but the most common form is dried. Drying to about 12% moisture must be carried out under carefully controlled temperature and humidity regimes to prevent checking (cracking) which is a serious defect. Some examples of pasta products are shown in Fig. 24.19.

FIGURE 24.19 **Varieties of long and short pasta.**

24.7.2 Assessing Pasta Quality

Appearance, cooking quality and texture are important quality attributes of pasta (Table 24.1). Both uncooked and cooked pasta should have a translucent, bright yellow colour and be free from white, black and brown specks or discolouration. The yellow colour of pasta is due to high levels of endosperm yellow pigments, which is a priority in durum breeding programs (see Chapter 6). During cooking, pasta should absorb water and maintain its shape, and the strands should not stick together. Pasta is cooked to 'al dente' firmness, after which the texture–firmness and resilience—are evaluated. Pasta firmness can be evaluated using a sensory panel or by instrumental methods. Cooking loss and water uptake are other characteristics that can be measured. High protein content (>13% in the wheat) and high protein quality are essential for premium pasta. Durum wheat should be free from pre-harvest sprouting (as measured by high falling number) as α-amylase can cause stickiness and higher cooking loss. In some instances, hard-wheat semolina or even bread-wheat flour are substituted or added to durum semolina for a lower cost product.

24.7.3 Couscous

Couscous is a popular North African staple food. At home, it is made by rubbing moistened durum semolina between the hands to form small granules and removing larger particles by sieving. Uniformity of granule size is important for cooking quality, as smaller granules preferentially absorb water during cooking. Couscous is steamed before drying (often in the sun), after which it can be stored. In commercial production, semolina is blended with water to agglomerate the particles, after which excessively large and small granules are removed before steaming. The couscous is dried to 10–12% moisture before grading by sieving into coarse, medium and fine particles (see www.buhlergroup.com for a description of the process).

Couscous is prepared for consumption by moistening the grains, steaming in the top of a couscoussier and serving with a sauce of meat and/vegetables cooked at the same time. Couscous should rapidly absorb the sauce and maintain its firmness without becoming sticky. The time of rehydration, cooking time, cooked weight, stickiness and mouth feel are important quality characteristics (Debbouz and Donnelly, 1996). Colour, size and uniform shape are also important. Optimum wheat protein content is 10–12%.

24.8 BREAKFAST CEREALS

24.8.1 RTE Cereals

RTE cereals became available in the United States in the latter part of the 19th century. They are consumed with milk or water and do not need cooking before consumption. RTE cereals are mainly made from wheat, corn, oats, rice and barley. These can be wholegrains or grain products such as corn grits (milled from corn endosperm), flakes or bran. Other ingredients include malt, salt, sweeteners (sugars), flavourings, colours and often vitamins, minerals and fibre. Batch processing usually involves cooking using direct steam injection followed by flaking and toasting. In continuous extrusion cooking, ingredients are mixed under heat and shear before pressing through a die to form an intermediate pellet or finished product.

Popular breakfast cereals include flakes, puffed cereals, breakfast biscuits and combinations with seeds, nuts and dried fruit (Fig. 24.20).

FIGURE 24.20 **Breakfast cereals (clockwise from front) based on corn, wheat, mixture wheat and oats, and rice.**

Different cereal types have different raw material requirements. Those processed from unmilled wholegrain require grains of large size and high bulk density (test weight). Milled grits for cereals have a specification and range for particle size.

Finished-product assessment may include size, shape, bulk density and weight (important for packaging), colour, hydration and water-holding capacity (for bowl life) and pasting properties (measured by RVA). Texture is considered an important attribute and includes breaking strength and deformation resistance. These attributes can be measured by sensory or instrumental methods. Texture will vary according to the product from crunchy, hard-and-dense to crispy-and-expanded.

24.8.2 Hot Cereals, Porridge

Porridge is made by cooking whole or milled cereals in water or milk. Typically served at breakfast, porridge is sweetened with sugar or served as a savoury dish with spices and other ingredients. There are many kinds of porridge. *Kasha* in Russia and the Ukraine is made from intact grains such as wheat, barley, oats and rye. Barley meal porridge is common in the Middle East and North Africa and maize grits are boiled to make polenta in Italy. Polenta is more than just a breakfast food as it can be further cooked, cooled, sliced and re-cooked by frying, baking or grilling. *Congee* and *juk* are rice porridges and many variants are consumed in Asian countries with accompaniments including onions, vegetables and eggs. Porridge can also be made from sorghum and millets (see Chapter 13).

24.8.2.1 Oat Porridge

Oat porridge is favoured in many western countries and has a number of health benefits, including high fibre and β-glucan. Oats are milled and processed before cooking (see Chapter 10). Coarse oatmeal requires long cooking, whereas fine oatmeal cooks more quickly but has a smoother texture. Rolled oats are partially cooked during manufacture to inactivate lipase enzymes before flaking through rollers. Ready-cooked oats are flaked more thinly and the starch is completely gelatinised, thus the only cooking required is the addition of boiling water.

24.8.2.2 Assessing Quality of Oats

Oats used for rolled oats should have high test weight, large grain size, high percentage of groats (dehulled oats) and low screenings. Important quality tests for rolled oats include moisture, fibre, free fatty acids (to test for lipase inactivation). Colour is important in many markets, and can be measured using a tristimulus colour meter.

Texture and flavour of the cooked oats are assessed by a sensory panel. Pasting properties are important to predict the cooking properties and the RVA presents a convenient test method (AACCI 76-22.01).

24.9 CEREALS EATEN AS THE ENTIRE GRAIN

Most grains require a processing step before consumption, but in these instances processing is relatively minimal, and some qualify as wholegrain.

24.9.1 Rice

Worldwide, the most widely eaten cereal is rice, after processing (milling) to remove the outer husks, bran and germ.

Sushi is a good example of a popular rice-based food. Sushi originated in Japan, and has become a part of a more global food culture. Firstly, special rice vinegar is mixed with cooked sushi rice. There are many types of sushi. For one popular type, the vinegared rice is hand-moulded and topped with raw fish or other seafood. In the rolled variant, a special bamboo mat is used and nori (a thin sheet of prepared seaweed) is wrapped around the rice and fillings such as omelette or cucumber. 'Fusion' styles can be found in the United States and elsewhere. Examples of sushi are shown in Fig. 24.21.

As rice is eaten as a wholegrain physical qualities such as appearance, uniformity, size and shape are important determinants of acceptance (see Chapter 12). For sushi, a glossy appearance and pearly-white colour are important, as is the taste and flavour. The preferred texture is slightly sticky, and this is assessed by sensory or instrumental methods. Premium rice for sushi making is short grain, with very low amylose content. Medium grain rice is used in some global markets.

24.9.2 Popcorn

Popcorn is made from a special type of corn, which was known to the Aztecs and indigenous people of north and south America. Later it was adopted in mainstream America as part of festivals or celebrations such as Christmas, but

FIGURE 24.21 **Sushi is a popular global food.**

FIGURE 24.22 **Popcorn before (A) and after (B) popping.**

popcorn is now a popular snack food and has a 'healthy' image as it is a wholegrain and low in fat and sugar. Popcorn is sold as pre-popped or the more popular un-popped which is ready to pop at home in a microwave or small popping appliance. It is usually coated with oil, butter and salt to add flavour (Fig. 24.22).

White or yellow corn can be used in popcorn production. Firstly, kernels are separated from the husk, cleaned and graded to separate small kernels. Unpopped corn needs to be packaged to maintain a moisture content of about 13%, which gives expansion volume after popping. During the popping process, the steam is released inside the kernel as it swells and then pops at about 220°C.

24.9.2.1 Assessing Popcorn Quality

Popcorn expansion is measured as the volume of popped corn compared to the original weight, or the volume of popped corn compared to the weight of popped kernels (Sweley et al., 2012). The popping rate is calculated as the number of popped grains (www.popcorn.org). Shape, colour, aroma, taste and texture are other important parameters for consumer acceptance.

24.9.3 Pearled Barley

Pearled barley is used in meat and vegetable soups, casseroles, porridge, soy sauce and as a rice extender (Baik and Ullrich, 2008). Pearling is achieved through a number of abrasion stages using carborundum or emery grinding stones. This process gradually removes the outer grain layers leaving approximately 40–50% of the grain. Another product, pot barley, is minimally processed by dehulling and removal of a small proportion of the outer grain. Pearled barley can be further processed by milling, flaking or grinding.

Barley to be used for pearling should have large and uniform grain size, high test weight and uniform colour. For use as a rice extender, waxy types are preferred, hardness is important and the whiteness of the endosperm should complement the colour of the rice (Edney et al., 2002).

24.9.4 Bulgur

Bulgur or burghul is a pre-cooked wheat product popular in Central Asia, Turkey, the Middle East and north African cuisines, where it is used in salads such as tabouleh, soups and baked products such as kibbeh, soups or boiled and consumed like rice. As enzymes are inactivated during the cooking process, bulgur has a long shelf life. Traditional artisan preparation utilises sun drying and debranning by hand before crushing with a stone or coarse mill. Modern production consists of cleaning, washing, tempering to 40% moisture, cooking to 95°C, drying, de-branning, cracking and sifting. Starch is gelatinised during the cooking process (Belibagli et al., 2009)

24.9.4.1 Assessing Bulgur Quality

Important quality characteristics are moisture content, particle size, shape, colour, texture and flavour. Maximum moisture content of 10–12% is necessary for keeping quality. Bulgur is sold in both coarse and fine particle sizes for different food applications. In both instances particles should have a smooth oval shape. Bulgur must maintain its integrity during the cooking process. A light-yellow or amber-yellow colour is usually preferred, and durum wheat is favoured (Fig. 24.23). Hard wheat is sometimes used.

FIGURE 24.23 Bulgur.

24.10 BEVERAGES FROM GRAINS

Fermented beverages from grains originated in China, Mesopotamia and Egypt about 7000 years ago (Poelmans and Swinnen, 2011). Alcoholic beverages made from grains are mostly fermented (beers) or distilled (whisky, shochu, white spirit such as moutai). During fermentation, cereal starch is converted to alcohol. Distillation concentrates the alcohol to produce a much higher alcohol content.

24.10.1 Fermented Beverages

Beer can be made from rice, wheat, corn, millet, sorghum and rye, but barley is most commonly used to produce malt and will be considered here. Specific malting barley varieties are utilised in brewing and they must satisfy quality standards such as maximum protein content (see Chapter 19 for examples of barley grain inspection standards in the United States and Australia and Chapter 9 for assessing malting barley quality).

Beer is generally produced from malted barley, water, hops, yeast and various carbohydrate adjuncts such as brewer's liquid glucose, liquid sucrose, unmalted barley and cereal starches. There are many variations of the process.

Barley is malted by steeping the grains in water to initiate germination. Germination lasts for several days to promote the formation of enzymes to be used during brewing. Germination is terminated by kilning the grains to storage-stable moisture content before delivery to the brewery. The malt is then milled and the grist is transferred with hot water to the mash tun where the mash is heated to about 65°C. This allows conversion of the starches to simple sugars and beta-limit dextrins. The mash tun is drained and the spent grains sparged or rinsed. The resulting filtrate, known as wort, is then boiled with the hops and cooled before transferring to the fermenter where the remaining ingredients are added. At the end of fermentation, the beer is centrifuged to remove the yeast before being transferred to storage tanks for maturation. It is then filtered before packaging.

24.10.1.1 Assessing Beer Quality

Important parameters for assessing beer quality are taste and flavour, colour, specific gravity, oxygen and carbon dioxide content, total nitrogen content, alcohol content, haze stability, bitterness and pH.

24.10.2 Distilled Beverages

Although any fermentable carbohydrate source can be used for the production of distilled beverages, the most common sources are cereals such as barley, corn, wheat, rye, rice and sorghum or a mixture of several cereals. Malt whiskey is produced from malted barley. Other ingredients are water and brewer's yeast. The process for producing distilled beverages is somewhat similar to that for beer, although there are many variations. After the fermentation process, the alcohol content is increased by continuous or batch distillation, with the temperature and other processing conditions determining the percentage of alcohol in the final product. Water can be added to reduce the alcohol content. Filtration, aging in oak barrels or earthenware jars and blending are some of the techniques used to produce unique, high-value products.

Distilled beverages are evaluated according to their style, which is very characteristic. Parameters include colour, alcohol content, specific gravity and sensory characteristics including aroma, taste and mouth feel.

24.11 FEED USES

Cereal grains are major sources of energy for pets, monogastric animals (poultry, pigs and horses), ruminants (sheep, goats and cattle) and in aquaculture (fish, shrimp etc). In 2015/16, about 36% of world cereal production was utilised for animal feed, and this proportion is increasing at the rate of 1% annually. Feeds can be consumed as grain; by-products from food or industrial production (e.g. fibre, distillers grains, corn gluten); or processed to pellets as compound feeds by extrusion cooking. Feeds are formulated and supplemented on a least-cost basis and according to the age and specific nutritional requirements by animal type.

Important quality characteristics of feeds include protein content (expressed as $N \times 6.25$ on a dry-matter basis), moisture content, fat, dietary fibre, ash, amino-acid profile, digestibility and energy value. Many of these characteristics can be determined using NIR. Physical properties such as size and density are important in extruded feeds used in aquaculture. For bottom feeders such as shrimp, the pellets need to maintain their integrity and not disperse.

Mycotoxins are a potential toxic hazard in many cereal crops and raw materials should be monitored where grain is sourced from geographic areas prone to mycotoxin occurrence.

24.12 INDUSTRIAL USES

Cereals find uses beyond food and feed in industrial sectors. Separation into starch and gluten provides a source of further ingredients for the food and manufacturing industries and a renewable resource for production of biofuels.

24.12.1 Products Derived From Starch

Starch is a major component of tubers, roots of plants and cereals. Commonly used cereal starches are derived from corn (normal and waxy), wheat, sorghum and rice. Different starches have different physical and chemical properties. Commercial production of starch depends on the availability and price of the raw materials in a particular country. In the United States, corn starch is a major industry, whereas in Australia it is primarily wheat starch. The industry has developed into a worldwide trade. Starch and starch derivatives are used in a diverse range of industries, including food, paper, mining, pharmaceuticals, packaging and adhesives.

The separation of starch and gluten from wheat flour is described in Chapter 23. Corn starch is separated by a wet milling process in which the corn kernels are first steeped in warm water containing sulfur dioxide before germ is separated and the remainder is milled before separation of the starch and gluten, after which the starch is dried (see www.corn.org for a description of the process).

Starch is used in an unmodified (native) form or can be modified by chemical, physical or enzymic processes. Chemically modified starches differ in gelatinisation properties and reactions to temperature, pH, stirring and freezing. Starch can be pre-cooked (pre-gelatinised) so that it forms a paste when mixed in cold or warm water, finding applications in 'instant' foods. Further, starch can be hydrolysed to dextrins and further broken down to solid and liquid glucose, dextrose and caramels. These conversion products function as sweeteners and colourings, and are used in fast-setting adhesives.

Starch functions as a thickener and stabiliser in many foods such as puddings and fillings, serves to disperse other ingredients, and functions as a binder, humectant and texturiser. In corrugated board manufacture, it forms a strong adhesive film, functioning as an adhesive.

24.12.1.1 Assessing Starch Quality

Depending on the type of starch, specifications include moisture, maximum protein, maximum ash, pH, particle size and maximum sulfur dioxide. These mainly act as indicators of process control. Functional attributes include the colour when made into a paste (clear or opaque) and texture (varying from long and cohesive as in waxy corn to short and heavy as in normal corn and wheat). Cold-batter viscosity is important in some applications and can be measured using a Ford cup or similar. Gelatinisation properties can be measured using a Viscograph or RVA and critical parameters, which dictate usage include gelatinisation temperature, peak viscosity, final viscosity and setback (Fig. 22.6).

24.12.2 Biofuels

Interest in biofuels has been increasing as a means of reducing reliance on fossil fuels, reducing greenhouse gases and as a source for providing 'clean energy'. Some countries already mandate the use of biofuels as a blend with gasoline.

Biofuels can be either biodiesel (derived from vegetable or animal fats) or bioethanol (derived from organic material that can be converted to sugar). Feedstocks for bioethanol production use cereals that include corn, wheat and sorghum, depending on the country of manufacture as the price of the raw material is a major determinant of the cost.

Cereals are firstly wet or dry milled and the starch is converted to simple sugars by hydrolysis before fermentation and distillation to alcohol. Inferior-grade grains such as sprout-damaged, insect-damaged or mould-affected grain can be used in manufacture and there are no standard grain specifications, but some guidelines are available for farmers (http://cereals.ahdb.org.uk).

Potential ethanol yield depends on the starch content, conversion rate of the cereal starch to sugar and fermentation efficiency (Smith et al., 2006). The ethanol yield can be expressed as kg EtOH per kg cereal feedstock.

The use of cereal grains in bioethanol production has engendered ongoing debate over the potential competition between biofuels and food, and concerns over food security.

24.13 FUTURE TRENDS

Consumer trends will be the most important driver of new products, which will determine the variety and types of end products available. While there will always be a demand for staple cereal foods such as bread, the growth areas are expected to be in niche and specialty products. New product test and evaluation methods will need to be developed for these new products. More globalisation has seen the manufacture and acceptance of non-traditional foods. Manufacturers need to source suppliers of raw materials of the right quality type as well as produce quality end products which are outside their traditional manufacturing base. In addition, manufacturers need to satisfy consumer demands for varied, convenient and healthy food.

SOURCES OF FURTHER INFORMATION

Bao, I.S., 2016. Rice, eating quality. In: Wrigley, C., Corke, H., Seetharaman, K., Faubion, J., (Eds.), Encyclopedia of Food Grains, vol. 3. Elsevier, Oxford, UK, pp. 166–175.

Chavan, R.S., Chavan, S.R., 2011. Sourdough technology—a traditional way for wholesome foods: a review. Compr Rev. Food Sci. Food Saf. 10, 169–182.

Fast, R.B., 2000. In: Fast, R.B., Caldwell, E.F. (Eds.), Manufacturing Technology of Ready-to-Eat Cereal. second ed. AACCI, St. Paul, MN, USA, (Chapter 2).

Gou, G.G. (Ed.), 2010. Asian Noodles: Science, Technology, and Processing. John Wiley & Sons, Inc, Hoboken, NJ, USA.

Guragain, Y.N., Probst, K.V., Vadlani, P.V., 2016. Fuel alcohol production. Wrigley, C., Corke, H., Seetharaman, K., Faubion, J. (Eds.), Encyclopedia of Food Grains, vol. 3, Elsevier, Oxford, UK, pp. 235–244.

Izydorczyk, M.S., Dexter, J.E., Barley, 2016. Milling and processing. In: Wrigley, C., Corke, H., Seetharaman, K., Faubion, J. (Eds.), Encyclopedia of Food Grains, vol. 4. Elsevier Ltd, Oxford, UK, pp. 434–445.

Li, Y., Liang, J.F., Yang, M.Y., Chen, J.Y., Han, B.Z., 2015. Traditional Chinese rice noodles: history, classification, and processing methods. Cereal Food World 60, 123–127.

Nebraska Corn Board has information on corn research projects as well as end uses of corn including biofuels. Available online : http://www.nebraskacorn.org

Sissons, M.J., Abecassis, J., Marchylo, B., Carcea, M., 2012. Durum Wheat: Chemistry and Technology. AACCI, St. Paul, MN, USA.

Smith, T.C., Kindred, D.R., Brosnan, J.M., Weightman, R.M., Shepherd, M., Sylvester-Bradley, R., 2006. Wheat as a feedstock for alcohol production. HGCA Res. Rev., 61, Available online: https://cereals.ahdb.org.uk.

Solah, V.A., Crosbie, G.B., 2016. Noodles: testing for quality. Wrigley, C., Corke, H., Seetharaman, K., Faubion, J. (Eds.), Encyclopedia of Food Grains, vol. 3, Elsevier, Oxford, UK, pp. 154–160.

Tiefenbacher, K., Wrigley, C., 2016. Wafers: methods of manufacture. In: Wrigley, C., Corke, H., Seetharaman, K., Faubion, J., (Eds.), Encyclopedia of Food Grains, vol. 3. Elsevier, Oxford, UK, pp. 51–58.

Zhou, W., Hui, Y.H. (Eds.), 2014. Bakery Products Science and Technology. second ed. Wiley Blackwell, Chichester, West Sussex, UK.

REFERENCES

Anton, A.A., Luciano, F.B., 2007. Instrumental texture evaluation of extruded snack foods: a review. Cienc. Tecnol. Aliment. 5, 245–251.

Asenstorfer, R.E., Appelbee, M.J., Kusznir, C.A., Mares, D.J., 2014. Toward an understanding of mechanisms involved in non-Polyphenol Oxidase (non-PPO) darkening in Yellow Alkaline Noodles (YAN). J. Agric. Food Chem. 62, 4725–4730.

Azudin, M.N., 1998. Screening of Australian wheat for the production of instant noodles. In: Blakeney, A.B., O'Brien, L. (Eds.), Pacific People and Their Food. AACCI, St. Paul, MN, USA, pp. 101–121.

Baik, B.K., Ullrich, S.E., 2008. Barley for food: characteristics, improvement, and renewed interest. J. Cereal Sci. 48, 233–242.

Belibagli, K., Vardin, H., Dalgic, A., 2009. Application of food safety and quality management systems to bulgur processing. Ital. J. Food Sci. 21, 499–515.

Bhattacharya, M., Zee, S.Y., Corke, H., 1999. Physicochemical properties related to quality of rice noodles. Cereal Chem. 76, 861–867.

Brown, J., 1993. Advances in breadmaking technology. In: Kamel, B.S., Stauffer, C.E. (Eds.), Advances in Baking Technology. Blackie Academic & Professional, Glasgow, UK, pp. 135–151.

Chow, C., Kennett, A., Tonkin, R., Cornish, G., Miskelly, D., Dines, J., 2003. Non-linear interactions in soft wheat flour blending: II: Marie style biscuits. In: Black, C.K., Panozzo, J.F (Eds.), Cereals 2003 Proceedings of the 53rd Australian Cereal Chemistry Conference, Royal Australian Chemical Institute, Melbourne, pp.183–186.

Crosbie, G.B., 1991. The relationship between starch swelling properties, paste viscosity and boiled noodle quality in wheat flours. J. Cereal Sci. 13, 145–150.

Debbouz, A., Donnelly, B.J., 1996. Process effect on couscous quality. Cereal Chem. 73, 668–671.

Edney, M.J., Rossnagel, B.G., Endo, Y., Ozawa, S., Brophy, M., 2002. Pearling quality of Canadian barley varieties and their potential use as rice extenders. J. Cereal Sci. 36, 295–305.

Hashmi, I., Wootton, M., 2000. Test baking and evaluation of tandoori bread. American Institute of Baking. Tech. Bull. XXII 11, 1–6.

Hemdane, S., Jacobs, P.J., Dornez, E., Verspreet, J., Delcour, J.A., Courtin, C.M., 2016. Wheat (*Triticum aestivum* L.) Bran in bread making: a critical review. Compr. Rev. Food Sci. Food Saf. 15, 28–42.

Huang, S., Betker, S., Quail, K., Moss, R., 1993. An optimised processing procedure by response surface methodology (RSM) for northern-style Chinese steamed bread. J. Cereal Sci. 18, 89–102.

Huang, S., Miskelly, D.M., 2016. Steamed Bread: Ingredients, Processing and Quality. Woodhead Publishing, Cambridge, UK, Chapter 3.

Huang, S., Quail, K., Moss, R., 1998. The optimisation of a laboratory processing procedure for southern style Chinese steamed bread. Int. J. Food Sci. Technol. 33, 345–357.

Jianfen, L., 2009. Chinese rice noodles. In: Zaigui, L., Hongzhuo, T. (Eds.), Traditional Chinese Foods: Production and Research Progress. Nova Science Publishers, New York, (Chapter 3).

Lever, T., Kelly, A., De Faveri, J., Martin, D., Sheppard, J., Quail, K., Miskelly, D., 2005. Australian wheat for the sponge and dough bread making process. Aust. J. Agric. Res. 56, 1049–1057.

Marti, A., D'Egidio, M.G., Pagani, M.A., 2016. Pasta: quality testing methods. Wrigley, C., Corke, H., Seetharaman, K., Faubion, J. (Eds.), Encyclopedia of Food Grains, vol. 3, Elsevier, Oxford, UK, pp. 161–165.

Miskelly, D.M., 1996. The use of alkali for noodle processing. In: Matsuo, R.R., Kruger, J., Dick, J. (Eds.), Pasta and Noodle Technology. AACC International, St. Paul, MN, USA, pp. 227–273.

Poelmans, E., Swinnen, J.F., 2011. From monasteries to multinationals (and back): a historical review of the beer economy. LICOS Discussion Paper 294. Available from: http://www.econstor.eu/handle/10419/74857

Preston, K.R., Kilborn, R.H., Black, H.C., 1982. The GRL Pilot Mill. II. Physical dough and baking properties of flour streams milled from Canadian Red Spring Wheats. Can. Inst. Food Sci. Technol. J. 15, 29–36.

Qarooni, J., 1990. Flat breads. American Institute of Baking. Tech. Bull. XII 12, 1–6.

Qarooni, J., 1996. Flat Bread Technology. Chapman and Hall, New York, USA.

Qarooni, J., Orth, R.A., Wootton, M., 1987. A test baking technique for Arabic bread quality. J. Cereal Sci. 6, 69–80.

Qarooni, J., Ponte, J.G., Posner, E.S., 1992. Flat breads of the world. Cereal Food World 37, 863–865.

Quail, K.J., McMaster, G.J., Tomlinson, J.D., Wootton, M., 1990. Effect of baking temperature/time conditions and dough thickness on Arabic bread quality. J. Sci. Food Agric. 53, 527–540.

Rahaie, S., Gharibzahedi, S.M.T., Razavi, S.H., Jafari, S.M., 2014. Recent developments on new formulations based on nutrient-dense ingredients for the production of healthy-functional bread: a review. J. Food Sci. Technol. 51, 2896–2906.

Rooney, L.W., Suhendro, E.L., 1999. Perspectives on nixtamalization (alkaline cooking) of maize for tortillas and snacks. Cereal Foods World 44, 466–470.

Ross, A.S., Hatcher, D.W., 2005. Guidelines for the laboratory manufacture of Asian wheat flour noodles. Cereal Foods World 50, 296–304.

Sweley, J.C., Rose, D.J., Jackson, D.S., 2012. Hybrid and environment effects on popcorn kernel physiochemical properties and their relationship to microwave popping performance. J. Cereal Sci. 55, 188–194.

Wheat Marketing Center, 2008. Wheat and Flour Testing Methods—A Guide to Understanding Wheat and Flour Quality, Version 2. Kansas State University, KS, USA. Available from: www.wheatflourbook.org

Wu, Y.V., Bergquist, R.R., 1991. Relation of corn grain density to yields of dry-milling products. Cereal Chem. 68, 524–544.

Zhou, W., Lin, H., Liu, H., Chao, F., Lu, H., 1994. The relationships between Chinese steam bread and flour quality factor. In: Xie, G.F., Ma, Z.D., (Eds.), Proceedings 94, International Symposium & Exhibition on New Approaches in the Production of Food Stuffs and Intermediate Products from Cereal Grain and Oil Seeds. Beijing, China, pp. 701–709.

Part V

Summary and Appendices

Chapter 25

Grain Quality: The Future is With the Consumer, the Scientist and the Technologist

Colin Wrigley*, Ian Batey** and Diane Miskelly†
*QAAFI, University of Queensland, St. Lucia, QLD, Australia; **Sunset Cereal Services, Meadowbank, NSW, Australia; †Westcott Consultants Goulburn, NSW, Australia

25.1 INTRODUCTION

The future of cereal grains is largely in the hands of consumers, who vote with their purchasing power and who influence others with their gossip. These contributions to general trends are complemented by the innovations of scientists and technologists who contribute possibilities and realities that can be transformed into new products for consumers to adopt or reject.

The quality and acceptability of the grain-based products that reach the consumer are determined by quality assessment and management throughout the grain chain, as detailed in Fig. 1.1. Improvements in these quality-determining processes involve all of the people at all stages—breeders, growers, grain handlers, processers and retailers. In fact, each successive person in the chain is a consumer of the ones above. Thus innovation at each stage has great potential to produce benefits elsewhere in the sequence of events. Likewise, problems at any stage have implications for all others involved.

Future innovations can be expected at all these stages of the grain chain. For example, novel starch properties in the form of high amylose or resistant starch offer processors the opportunity to develop products rich in dietary fibre. Such starches are considered to improve bowel health and to reduce the risk of type-2 diabetes. Novel starch properties also offer the processor opportunities to enhance crumb texture and staling rate, as well as producing novel snack foods using extrusion technologies. Nutritional advantages are envisaged as the concept of 'nutraceuticals' moves into providing enhanced levels of antioxidants, reputed to reduce the risk of cancer and other diseases due to strong free-radical scavenging activities.

Cereal Grains. http://dx.doi.org/10.1016/B978-0-08-100719-8.00025-5

This chapter complements comments about the future, made in each of the earlier chapters, in the sections entitled 'Future trends'.

25.2 THE BIG PICTURE

This book emphasises the advantages of improving the analysis and management of grain quality. This must be seen in the context of the overall 'grain-chain' concept, described initially in Fig. 1.1 and expanded in chapters on specific cereal grains. As explained at the end of Chapter 1, the industrial grain chain of Fig. 1.1 is complemented by the biochemical version illustrated in Fig. 25.1. It is a sequence that also ends with the consumer, and also starts with the breeder (who designs the genome [set of genes] of a new variety) in partnership with the grower (responsible for the way in which the grain is grown). These growth conditions (E) interact with the variety (the genotype, G) to determine grain composition. Complementary to the climatic part of the growth environment (E) (which cannot be controlled) is farm management (M) in Fig. 25.1. The quality of the final product also depends (much later) on processing conditions (P) that lead to the final product to be assessed by the consumer.

The future is the result of a process of grain-quality innovation, involving three major groups of people:

- the scientist (who develops notional feasibilities), represented especially by the 'how-it-works' sequence of Fig. 25.1;
- the technologist (who needs to dwell on realities), represented especially by the 'doing-it' sequence of Fig. 1.1;
- the consumer (who ultimately determines the success or otherwise of specific innovations), represented by you and me.

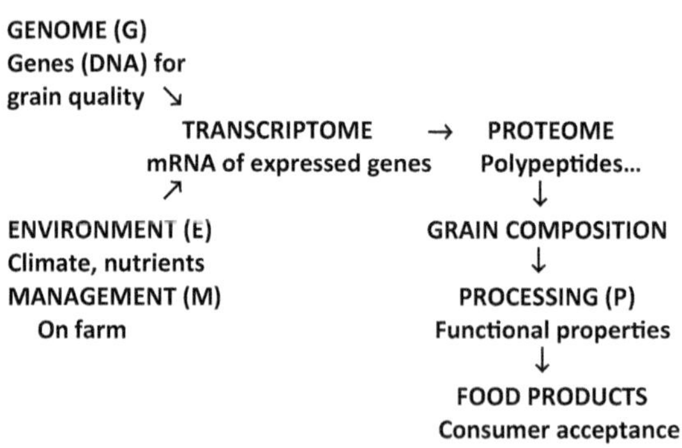

FIGURE 25.1 The biochemical 'grain chain' sequence, starting with the interaction of genotype (G, genome) with growth conditions/environment (E) and grower input as management (M), thereby determining grain composition via expressed genes (transcriptome) and the consequent polypeptides (proteome) which assemble as functional proteins, to become the biological catalysts responsible for synthesizing all the other chemical components of the grain—starch, non-starch polysaccharides, lipids and other nutrients. Thereafter, the figure emphasises the role that processing (P) plays, leading to the final product for the consumer. The figure shows graphically the G x E x M x P interactions.

25.2.1 The Scientist

The grain chain of Fig. 25.1 emphasises the role of the scientist in advancing future developments. As fundamental research moves towards elucidating the genetic control of grain-quality attributes, for example, these findings facilitate the breeding and selection of lines that will more accurately suit the utilisation needs further down the grain chain. Thus scientific advances become the source of new concepts and possibilities.

As an example at the fundamental research level, Shewry et al. (2009a,b, 2012) have laid a foundation for re-examined developing grains of wheat, detailing the accumulation of the grain's major storage biopolymers—starch and the gluten-forming storage proteins. They used the combination of biochemical and transcriptome analyses. Protein accumulation continued until the start of grain desiccation when there was 'a dramatic increase in the proportion of large glutenin polymers', marking the formation of polymerised storage protein that would become functional glutenin during dough making. Starch accumulation was roughly parallel to that of the gluten-forming proteins.

In parallel, associated transcriptome studies are providing a reference dataset for studying the possible manipulation of proteins and starch in the cereal grain generally. The advantage of research at the transcriptome level is that the genes being studied are those that are actually functioning in protein synthesis (and subsequently starch synthesis) under the existing environmental conditions and development stage, in contrast to studies at the genome level where most of the genes are not functional at the stage being examined (Barrero et al., 2015).

Thus, a detailed picture can be formulated to show how specific parts of the genome (the genetic constitution of the plant) interacts with the growth environment via the transcriptome, indicating how specific genes are expressed in that environment to give the proteome (all the polypeptides of a tissue, e.g. endosperm) and on down the metabolic/processing sequence (Fig. 25.1) through processing to the consumer.

Genetic manipulation of grain proteins at the research level has already assisted greatly in elucidating, at the molecular level, which changes in protein structure would be most effective in creating useful improvements in the functional properties of the glutenin subunits as they function in the glutenin polymers of wheat. Such studies have involved the addition and removal of cysteine residues to polypeptides of wheat and barley, plus the insertion of amino-acid chains of modest length, to extend polymer length, and consequently how, in theory, proteins of non-wheat cereals might be given dough-like properties (Tamas et al., 2002).

Some of these studies have generated novel gluten-like proteins that contribute to dough properties so strongly that they cannot even be mixed, without dilution with other protein (Rooke et al., 1999; Butow et al., 2003). However, this type of experiment shows the way to providing more effective gluten that would give satisfactory dough strength at lower grain-protein content. Such studies relate strongly to the design of varieties needed into a future of an atmosphere richer on carbon dioxide (Guzman et al., 2016) and a grain crop consequently deficient in grain protein.

Related studies have shown that there are naturally occurring over-expressions of glutenin-coding genes that deliver greater dough strength (Vawser and Cornish, 2004). Yet we still have much to elucidate about the genetic and environmental controls on the two main variables that determine the functional properties of gluten proteins—the monomeric-to-polymeric ratio and the molecular-weight-distribution of the glutenin polymeric structure.

Scientists are also active further down the grain chain to improve biosecurity, grain storage, nutrition, milling, processing, packaging and food safety.

25.2.2 The Technologist

Breakthroughs in basic science may remain as interesting 'marvels' unless they can be translated into reality at the technological level. For example, Day et al. (2009) tell the story of carrying an idea through from laboratory experimentation to pilot-scale processing; the idea involved the replacement of sodium chloride with ammonium chloride as a dough-strengthening agent in the water-washing of gluten from dough, thereby providing wet gluten of better quality on an industrial scale. Despite the success at the pilot scale, the procedure was not implemented at the technological level (at least at the time of pilot-scale proving).

Development of new products is one of the most obvious roles of the technologist. New products created can be extensions of existing products or a completely new and innovative product. The latter are designed to keep up with consumers' needs or even drive market trends and can open entirely new market segments. Examples are the inclusion of ancient grains, pseudocereals and seeds in existing products such as bread and gluten-free formulations in bakery products and snack foods (see Chapter 14). Gluten-free formulations require the addition of many ingredients to mimic the function of gluten and produce an acceptable product (Engleson and Atwell, 2008; Casper and Atwell, 2014).

25.2.3 The Consumer

To become even more practical, the attitudes of the consumer must be considered. How well has grain and product quality been assessed and managed through the grain chain? The consumer votes on this question by selecting this product or not; by endorsing it in conversation or not; ultimately by using the 'power of the purse', choosing the most appealing food products (breads, noodles, pasta, snack foods, beverages, rice, sweetening agents and so much more) on the basis of a wide range of attributes—appearance, taste, value, perceived nutritional and health qualities (see Fig. 24.1). The consumer's vote determines whether a particular brand or type of food continues to prosper financially or whether low sales mean that it is discontinued. Increasingly, internet and social media availability means food trends can become quickly apparent, together with the demographic differences underlying consumer choices.

Badaracco (2016) has used the term 'Pandora's Box' to describe these food trends and fads; once opened and generally accepted, these popular food trends are unlikely to be closed—like Pandora's mythical Box. It thus seems inevitable that food manufacturers must pander to the consumer's demands to concentrate buying preferences on labels depicting 'natural', 'free-from artificial ...' and 'pure', even though what these terms mean may not be very clear. A 2016 US Food and Drug Administration (FDA) public enquiry into the use of the term 'natural' in labelling showed a spectrum of responses as to what the definition should be. These ranged from excluding foods 'containing chemicals' to allowing the use of foods derived from GM sources (www.confectionerynews.com), leaving the FDA the task of navigating what has become in some cases a legal minefield. The current trend for 'superfoods' or 'supergrains' is being driven by a desire for healthy and less refined alternatives to wheat. This category increased by 200% globally in the period 2011–15 (www.foodbusinessnews.net) and includes ancient grains and pseudocereals such as quinoa.

At face value, these consumer trends may be labelled as fads, but for some consumers, the fads lead to the dismal consequence of the distinct eating disorder of *Orthorexia nervosa*, characterised by extreme and excessive preoccupation with the exclusive eating of food believed to be 'healthy', illogically to the exclusion of a wide range of nutritious foods. Included in these exclusions are various grain-related foods such as gluten-free products.

Nevertheless, because of the consumer's buying power, popular attitudes and preferences must be followed closely by everyone involved along the grain chain (Table 1.1). The breeder must provide the genetic potential for grain quality that meets the needs of everyone further down the chain. The grain grower must target quality attributes that suit the needs of the grain marketer and the processor who, in turn, send pricing signals that reward the grower's efforts. The grain processor must be close to the consumer, and thus be well aware of retail and sociological trends.

For example, in recent decades, consumers in many western cultures have moved from a tradition of eating three meals a day around the family meal table to 'grazing' (Asp, 2004)—helping themselves to snatched food snacks as they pass through the kitchen *en route* elsewhere. At the same time, there are many more women in the workforce, with little time (or inclination) for the traditional forms of food preparation.

Consequently, food processors have moved towards value-added convenience foods, even for single-serving size, permitting a family member to heat a single meal in the microwave oven at any time, independent of others in the family. Thus the traditional skill of home cooking has moved up the processing chain. The mass production of specific food servings requires more attention to analysis (the correct quality specifications for raw materials) and management (the correct handling in transport, storage, processing, packaging and retailing).

Greater communication worldwide, travel and immigration internationally have led to diversification of grain-based foods, so that the consumption of noodles, chapattis and Chinese-steamed breads is no longer confined to the countries and cultures of their origins. Foods such as sushi and Vietnamese rice noodle soup 'pho', once considered 'exotic' in western countries, are now enjoyed every day everywhere. Thus, flour millers and food manufacturers in western countries have needed to understand the flour specifications and processing requirements that are specific to a wider range of products. Conversely, regions where rice or corn have long been the traditional foods have moved towards wheat-based foods, such as western-style breads or locally adapted foods that rely on the dough properties of wheat flour (Morris, 2016a,b).

Again, flour millers and manufacturers in making these new products need to choose the right raw materials and processes as well as setting up systems for evaluation of finished product quality (see Chapter 24). Because of the commonality of quality requirements of many wheat-based foods, this task is not as difficult as it may appear and requires a working knowledge of the manufacturing processes of the various products. A common trend is the production of specific-purpose bakery and cake 'premixes' which contain not only the correct flour, but most of the dry ingredients. This reduces the need for separate storage and weighing of raw materials, as well as ensuring a finished product of consistent quality provided the make-up instructions are followed.

25.2.4 Designing the Future in Terms of G X E X M X P

What do we see, peering into the crystal ball of future cereal-grain products? We can see a diversity of possibilities, but many of these potential products may not persist into the hard, cold realities in the market place and the consumer, who is the culmination of the grain chain (Fig. 25.1). Designing and producing novel products involves cereal varieties (genotype, G) appropriate for a specific product; also involved are the expected growth environments (E), the grower contribution via farm management (M) and the nature of the processing (P) for which the grain will be used. The 'big picture' of all stages of the grain chain must be considered by everyone involved in grain science as the future is viewed.

25.3 BREEDING FOR GRAIN QUALITY

The traditional goals of plant breeding appear to be moving beyond yield-associated breeding towards adaptation to a changing growth environment and towards selection for consumer-targeted concerns about grain quality and composition, including (in the words of Chapter 16) 'various types of starches, dietary fibre, vitamins, micronutrients, etc., which exert positive physiological effects on human health.'

25.3.1 New Genes for Novel Genotypes

Many of these traits for grain quality and composition may be met using existing germplasm and available parent lines. However, novel sources are needed because the selection of appropriate gene sources, and thus parent lines, is basic to all aspects of breeding. Related to this basic strategy is the search for novel genes, thereby expanding the gene pool and the range of parent lines for breeding, so that a wider range of phenotypes may be considered. The extension of the gene pool still involves expeditions to conserve existing grass plants. For example, Klindworth et al. (2009) reported the use of chromosome substitution lines from wild emmer wheats to expand the pasta-making potential of durum wheat.

Basic to these endeavors is the race to conserve genetic reserves so that future research can take advantage of previously undiscovered genes. Accordingly grain-growing nations have so-called gene banks—repositories of seeds. In addition, there is the Svalbard Global Seed Vault on the Norwegian island of Spitsbergen about 1,300 km from the North Pole. In this icy vault is preserved a wide variety of plant seeds that are duplicate samples of seeds held in gene banks worldwide, thereby 'future-proofing' the world's germplasm.

Adding a new generation of opportunities is the recent production of a genome 'assembly' for bread wheat—essentially a first draft of the entire sequence of bread wheat DNA—by the International Wheat Genome Sequencing Consortium (www.wheatgenome.org/). To quote: 'if the wheat genome [is imagined] as a book, the assembly is the first draft. The second draft is akin to breaking a single block of text into chapters—the individual chromosomes. The final stage is the completed, checked and edited genome ready for use by breeders.'

25.3.2 Genetic Engineering

Expansion of the gene pool is also possible now by the introduction of genes from other species using new-generation breeding methods, including 'genetic engineering' (Privalle, 2016), so that genes can be used that would not previously have been considered with conventional cross breeding (Jones, 2005; Jones et al., 2009; Henry, 2014).

Whereas genetically engineered (GM) crops in commercial production do not yet include most of the cereals, GM maize has become widely accepted, now accounting for about 30% of world corn production, but a lower proportion than estimates of 82% of world production for soybeans and 68% for cotton, and similar to the 20% take-up of GM canola. See website: http://www.isaaa.org/resources/publications. Nevertheless, genetic engineering technology is likely to extend the ability of breeders to improve the quality and agronomic performance of all cereal grains.

Already GM crops in general are claimed to have increased grain yields by 22% and reduced pesticide use by 37% (Chapter 11; James, 2014) and GM technologies can be expected to continue transforming the grains produced and the foods that we eat (Blair et al., 2015). Yet public pressure will also continue to demand interventions such as labelling to identify products containing GMO grains (Mathiowetz and Jones, 2016). Methodologies for GM detection have thus been necessary parts of the quality management of relevant cereal grains (Shillito, 2016).

Despite the controversy created by GM seeds, there is no proof of harm to humans or stock as a result of ingesting grain of approved GM grain (Goodman, 2014). Yet there is no global mechanism by which such facts may be acknowledged and by which official 'approval' may be provided as general assurance because different countries have different regulatory systems for GM crops.

Nevertheless, it seems inevitable that GM plants of the remaining cereal species will become acceptable in the future, if clear advantages are thus provided, especially if these advantages relate to aspects of grain quality that would be attractive to consumers. For example, genetic engineering has been used as a direct means of improving specific aspects of rice nutrition, such as increased beta-carotene levels (Beyer et al., 2002) in 'golden rice' and higher folate content (Storozhenko1 et al., 2007). Risk factors to be considered in genetically modified crops for nutrition and health have been reviewed by Magaña-Gómez and Calderón de la Barca (2009) and by Mathiowetz and Jones (2016).

A National Academies of Sciences, Engineering, and Medicine review concluded that new molecular tools are blurring the formerly clear distinctions between conventional and GM breeding, and that many gains and greater progress can be achieved using both methods rather than one or the other alone (National Academies of Sciences, 2016).

25.3.3 Traditional Breeding

On the other hand, various nutritional advantages are also being provided by traditional breeding methods to enhance the nutrient deficiencies in cereal species. A prime example is maize, for which the opaque hybrids (e.g. quality protein maize) offer increases in essential amino acids, especially lysine (see Chapter 11).

Tanner et al. (2016) describe the breeding of an ultra-low gluten barley suitable for use in barley beer. The ultra-low gluten barley has hordein levels less than 5 mg/kg, which is below the CODEX limit of 20 mg/kg for description of gluten-free foods. This was achieved using conventional breeding methods in which hordein single null lines were crossed to produce double null progeny which were then backcrossed with the variety Sloop to produce a hordein triple null variety Kebari. Although grain size and some malting characteristics were reduced, the quality of the beer was acceptable in micro-malting trials. The gluten-free barley beer has been released commercially in Europe.

There is also research directed towards the further possibility of developing gluten-free wheat varieties that would not be toxic to people with coeliac disease. This involves genetic engineering to reduce the immunotoxic components of gluten by gene-silencing technology to delete the genes coding for gliadin proteins (Rosell et al., 2014; Shewry and Tatham, 2015).

Research uses of genetic modification technologies have brought significant benefits for traditional breeding via the identification of gene markers for marker-assisted selection (MAS) of desirable breeding lines in the early stages of breeding. Although MAS technology may not substitute for the actual testing of phenotype in later generations, MAS does reduce the likelihood of unnecessary propagation of poor-quality progeny (DePauw and O'Brien, 2016). The novel 'GMO' technologies, successful for improving maize and soy, are expected to provide quality improvements of the wider range of cereal species.

25.3.4 Genotype Versus Phenotype

Despite our increasing knowledge of genotype (the genome, the genes), it will always be phenotype (the actual grain) that matters most in breeding, namely, the actual performance of grain of a new variety, irrespective of what is expected based on the genes present or absent. As Fig. 25.1 shows, manipulation of genotype (G) alone is not the only influence affecting grain composition and processing quality; growth environment (E) also contributes, plus the contribution of the grain grower's management (M) of the growth conditions. For this reason, selection for traits with strong G X E interactions, such as grain quality, has been difficult and usually delayed until late generation, when sister lines can be grown at a range of sites. In recent times, the use of doubled-haploid lines has been used so that homozygosity might be achieved more quickly than by conventional crossing methods.

Grain-quality traits involve many genes (they are called 'quantitative traits') thus being difficult to genetically manipulate. Molecular mapping of quantitative-trait loci (QTLs) is proving to be a valuable approach to this problem as reported, for example, for gluten strength in durum wheat by Patil et al. (2009), for noodle-quality traits (Zhang et al., 2009), flour colour (reviewed by Schulthess and Schwember, 2013), β-glucan levels in barley (Li et al., 2008) and quality traits in rice (Lou et al., 2009; Bao and Corke, 2016). In this way, breeders are gaining access to valuable genetic markers that can help in assessing quality attributes via MAS, even though the nature of the specific coding genes is not fully elucidated (Varshney et al., 2005) (see Chapter 16).

With the progressive elucidation of the genomes of cereal species (Akhunov, 2016; Tiwari et al., 2016), an explosion is anticipated with the continuing study of functional genomics, transcriptomics and proteomics (e.g. Shewry et al., 2009a; Shewry and Tatham, 2015), permitting geneticists to build up a picture of the chromosomes gene-by-gene and marker-by-marker (Devos et al., 2009), leading to the use of 'perfect markers' for grain-quality traits (Varshney et al., 2005). However, as the picture builds up, there is the need to be aware of gene–gene interactions, and probably further complications that we are hardly aware of at present. Nevertheless, we look forward to a simplification of the processes of selecting desirable progeny in practical breeding, whereby genome-wide analyses of specific genes and markers can be performed efficiently on a large scale (Akbari et al., 2006; Xu et al., 2012), thus increasing the speed of breeding and selection through to releasing a variety that demonstrates the phenotype that was targeted in the breeding and selection program.

25.3.5 Greater Precision in Selecting for Specific Quality Traits

Efficiency is improving to pin-point quality traits using new selection methods that offer greater precision in the results obtained, more convenience in testing, and less labour (Ross and Bettge, 2009). Selection for specific glutenin subunits in wheat is an example of helping the breeder to defer the need for processing and product testing by going further back up the sequence in Fig. 25.1 to the proteome, where the identity and amounts of specific polypeptides can be assessed. The traditionally tedious assessment procedure of gel electrophoresis for this purpose has been replaced by capillary electrophoresis, by mass spectrometry used alone, or more likely in combination with DNA-based markers (see Section 17.3.1)

Advances are also being made in our understanding of how processing quality relates to grain composition, thereby helping to fill the gap between processing and the genome-environment interaction in Fig. 25.1 (Wrigley et al., 2009). Falling in this 'gap' are first the fully formed proteins, especially enzymes (a step beyond the component polypeptides of Fig. 25.1) and below these protein-based enzymes and hormones are starch, the non-starch polysaccharides, lipids and a range of other aspects of grain composition.

Whereas the analysis of grain composition at the level of all these components is tedious at present, it is encouraging to speculate on the possibility that non-destructive methods, suited to whole-grain analysis, for example NIR (Chapter 19), might be expanded to analyse more aspects of grain composition than the present range (moisture, protein, oil, as well as grain hardness). For example, NIR is reported to provide good assessment of β-glucan in barley, facilitating the selection of progeny rich in β-glucan for human food use or β-glucan-low lines for malting and feed uses (Schmidt et al., 2009).

Beyond this speculation about easier grain analysis is the dream that our knowledge of composition-processing relationships would be so advanced that we could use composition analyses to predict product-quality potential directly. This advance might thus permit us to dispense with intermediate testing such as dough quality for wheat or malt extract for barley. Such achievements in selection for quality in breeding obviously translate directly to improved quality assessment and management all the way down the grain chain.

25.4 GRAIN GROWING

The 'E' of G x E x M x P has received considerable media publicity (and also research attention). The media paint a picture like this: 'Environmental 'stresses', such as climate change, rainfall and loss of pollinators, are predicted to have major impacts on agriculture and food production over the coming decades' (https://wellcome.ac.uk). Sweeping statements such as this relate particularly to food production broadly, but it is also true that progressive changes in the growth environment for cereal grains have the potential to alter grain quality. While most aspects of the natural growth environment cannot be manipulated (nor even predicted) farm management (M) is subject to improvement by ongoing agronomic research.

25.4.1 Grain Quality, Harvest and Financial Viability

Grain yield has long been the ultimate financial consideration for grain growers, but many are realising that the volume of grain is not the only issue; other factors impinge on profitability (and ultimately on the viability of the farm). In most grain-growing regions, growers have the option of obtaining premium returns by delivering grain of superior quality and (often) of specified varietal content. As a result, the overall return is the combination of yield (grain volume) and premium (grain quality), less the many input costs. Increasingly therefore, growers continue to target grain that will attract a premium at harvest time, whilst assessing the extra costs involved in this strategy.

Breeders and agronomists also realise these considerations and they continue their research with the aims of providing the required tools, especially improved varieties and predictive means of assessing strategies, e.g. software to determine best sowing time to counter the combined threats of late frost (GRDC, 2016) and early heat stress (see Section 25.4.5 and Chapter 18). In addition, progressive growers combine their efforts as grower associations to lobby for research funds, for government assistance and for viable means of quality-based grain segregation and identity preservation throughout storage and transport, through to marketing.

Decisions about the balance of aiming for yield and/or grain quality are made long before sowing, in the vacuum of predictions for the future, long before actual cost/price structures are definite and before actual climate can be known. Given the importance of all these factors, grain growing of the future will see greater research accent on assisting grain growers in making more intelligent decisions, and in providing better predictions of the expected climate—both the economic climate and the physical climate.

Contract growing is another aspect of these considerations, similar to other agricultural pursuits, such as poultry and livestock production, for which farmers take the risky parts of the production chain in cooperation with corporate interests. Already, some innovative grain growers have ventured into the contract growing of specialty crops, such as naked barley with high β-glucan content and wheat varieties with distinctive properties, such as purple grain colour or extremes of starch composition.

25.4.2 The Entrepreneurial Grain Growers of the Future

More entrepreneurial is the move down the grain chain by other growers, adding value to their crop by further processing (see Fig. 1.1). Thus, the grower becomes the processor and the retailer of specialty products, such as noodles, pasta, beer, snack foods and milled and packaged rice, based on growing their own grain genotypes with special quality attributes; then proceeding to maximise returns by processing, packaging and wholesaling (e.g. see http://www.bellatagold.com.au/).

Given the many uses of cereal plants—food, feed, forage, fodder and fuel (see Chapter 1)—there are also ongoing trends for enterprising grain growers to capture opportunities for profit from more than just the grain. The obvious first opportunity is the use of the young emerging cereal plants as forage for ruminant stock. Varieties of wheat (dual purpose) and related cereals for this approach are being bred to be late maturing, so that the stock can be put onto early sown cereals and subsequently taken off allowing the young plants to grow, come into head and mature as usual for harvest. Thereafter the opportunity is given for feeding stock with the hay (fodder), discarded when the grain is harvested. Thirdly there is the stubble remaining in the ground for further stock feed. This approach requires the appropriate choice of varieties and of agronomic management. Long-established is the whole-plant use of immature corn plants as silage, continuing the theme of using so much more of the cereal plant than just the grain. Rice farmers are also extremely enterprising in their use of the rice plant remaining after grain harvesting.

A further extension of this approach is the whole-plant 'biorefinery' at the local level, as advocated by Munck (2016) whereby a local factory is equipped to receive and process the whole above-ground cereal plant. In this strategy, cereal plants are cut off at ground level and the whole plant is transported to a local 'Center' (Fig. 25.2) where the plants are dried (a necessity in some northern regions), chopped and fractionated into grain, straw and leaf components. Some of the resulting components are returned to the farm, other parts to further processing into particle board, starch for food and paper, oil and pelleted stock feed.

Is it further possible that the grower can promote his grains and the resulting products through tourism? The farm stay and Bed-and-Breakfast approaches to tourism are well established and thriving in places where tourism is possible. A new approach is the 'virtual farm visit' whereby the source of the food grain would be seen by the customer. This scenario is described by Western Australian wheat grower Brad Jones: 'Noodle packets would have a QR code so while someone in, say, Beijing, is cooking they can scan the code and connect with an interactive video showing wheat being grown on a farm in Western Australia … taking the cook in Beijing into the header cab [of the harvester] and build a relationship that makes them feel a part of the process' (Collis, 2016). His aim is to see his business in a broader context and 'facilitate the development of closer links between his production and the end customer', thus taking in the whole production-processing chain portrayed in Fig. 1.1.

25.4.3 Organic or Intensive Management?

Farm management is a constant need to balance input costs and predicted returns. A significant cost pressure is the increasing scarcity and rising cost of the

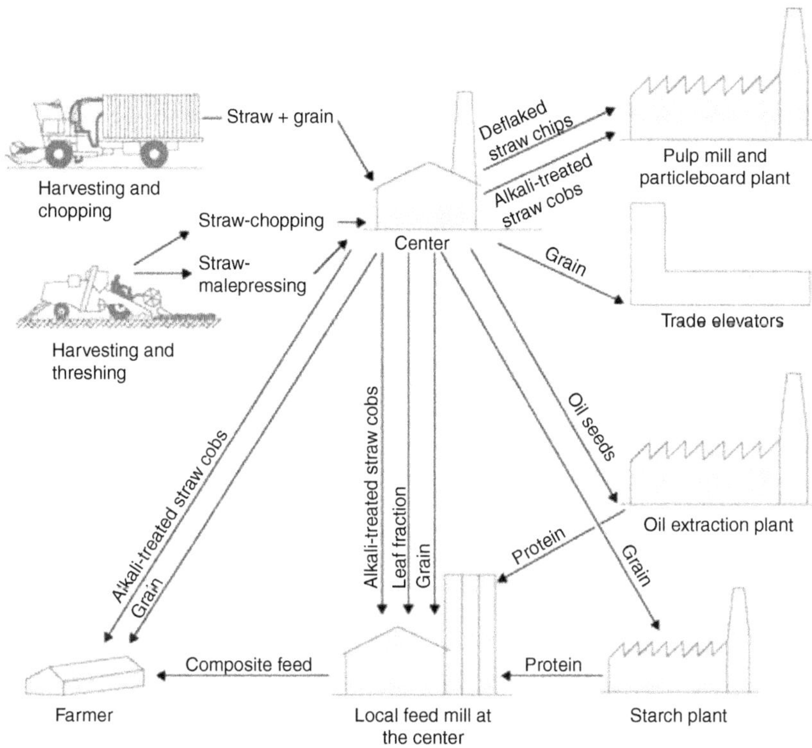

FIGURE 25.2 The agricultural biorefinery concept involving the fractionation of the whole cereal plant into value-added components for ongoing processing. *(Reproduced with permission from Petersen, P., Munck, L., 1993. Whole crop utilisation of barley including new potential uses. In: MacGregor, A.W., Batty, R.S. (Eds.), Barley: Chemistry and Technology. AACC International, St. Paul, MN, USA, pp. 437–474.)*

fertilisers that are an essential part of intensive cropping. A partial alternative, as a substitute for nitrogenous fertilisers, is the rotation of cereals with legumes. An example (Garofalo et al., 2009) is the long-term rotation of durum wheat with faba beans, especially in low-input situations with low rainfall. The increasing popularity of organically farmed crops is a reaction against the use of 'chemicals', and the perception of nutritional benefits from food grown under more natural conditions (Gelinas and David, 2016) and the expectation of premium prices for organic grain. However, there is an ongoing need for the consumer to be assured that the regulations for growing crops organically are adhered to (Siderer et al., 2005). A study of organic cropping of rice failed to demonstrate any resulting advantage with respect to antioxidants in whole grain rice (Chen and McClung, 2015). Nevertheless, the production of organic grains is reported to be a growth opportunity for the grain industry (Haumann, 2009). This promise is being fulfilled with recorded growth in the Australian organic grain sector rising by 20% between 2011 and 2014 (www.http://austorganic.com).

Nitrogen-based fertilisers are needed to provide levels of grain protein in wheat to meet premium-grade specifications, but in addition, sulfur fertiliser has become a requirement (see Chapter 18). This aspect of soil nutrition has become critical where wheat-growing no longer profits from the adventitious provision of sulfur from the atmosphere due to strict pollution controls. Sulfur deficiency in wheat grain shows itself as altered dough properties (e.g. poorer extensibility) despite adequate protein content, due to restricted synthesis of sulfur-rich gluten proteins (see Chapter 18). Test methods are available for sulfur-deficient grain (Uthayakumaran et al., 2007), thus to identify grain deliveries that may give unexpected dough quality and to identify soils that require sulfur fertilisation before the next crop.

In many cereal-growing countries, the greatest scarcity is water, both as rain (preferably at the appropriate time) and as irrigation. As a result, conflicts are likely to increase between regions and between countries that share common rivers and water-sheds. On the other hand, rising sea levels are predicted to flood some cereal-growing regions. Simple solutions to these problems are not readily available. It is to be hoped that the prophecies of 'water wars' will not become realities.

25.4.4 The Data-Rich Approach to Farm Management

Traditionally, the older generation of farmers carried a wealth of experience from which to make decisions about management. The future farmer must combine this traditional wisdom with the 'new data'—the wealth of advice available at the fingertips at a range of web sites—plus the data that can be obtained on site on-farm; information from these sources and more are being termed 'big data'. These many sources of data concern climate, soil analyses, frost risk, potential heat stresses, biosecurity hazards, variety performance results, grain prices and 'cloud-based technologies' for decision making about the whole range of management issues (Jennings, 2016). While the data sources are now large, the equipment can be small—hand-held mobile phones and tablets.

Precision agriculture (PA) is one of the means of producing on-site data to guide decision making, thus to manage the growing of crops for better yield and quality (Leonard, 2016). PA is a means of obtaining detailed information about the performance of a growth area down to a small spatial basis using Global Positioning Systems (GPS) so that the same area can be scanned repeatedly to superimpose different types of information (Fig. 25.3) (Bramley, 2009; Anon, 2009; Leonard, 2016).

A new phase, related to PA, is the use of unmanned aerial vehicles (UAVs, most simply drones) (Fig. 25.4) to scan the growing crop to detect areas of water stress, weed infestations or disease problems, plus the elementary knowledge of asset mapping of the property (Leonard, 2015).

The main accent of PA so far has been on optimising grain yields, but in addition, precision agriculture can also help to improve grain quality (Wrigley, 2005).

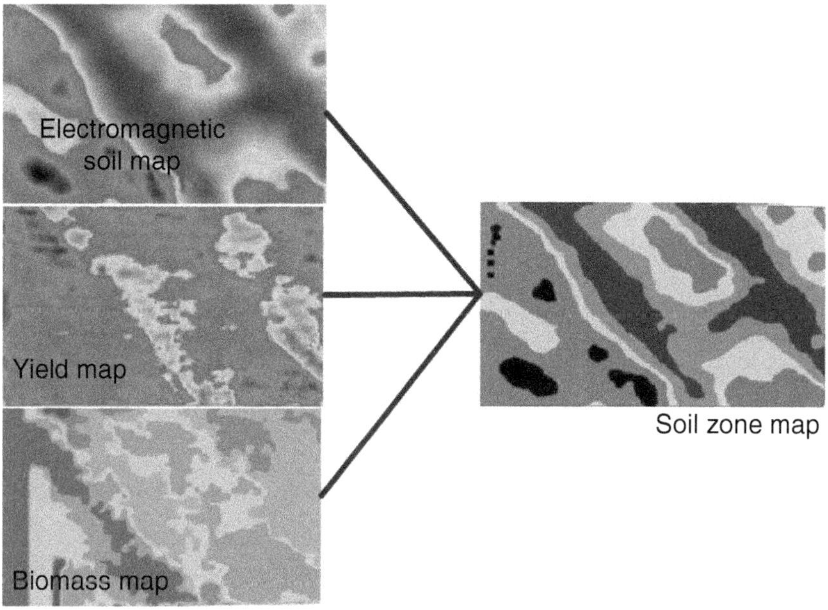

FIGURE 25.3 Analysis of a region of a cereal-growing field using precision agriculture. The grain yield map (left middle) is combined with the same area that has been scanned for plant biomass (bottom left) and for soil type using electromagnet sensing (upper left). These data are combined into the map at right for use in analysing local parts that may perform poorly, with a view to remedial action prior to the next season's sowing. *(Reproduced with permission from Leonard, E.C., 2016. Precision agriculture. In: Wrigley, C., Corke, H., Seetharaman, K., Faubion, J. (Eds.). Encyclopedia of Food Grains, vol. 4, Elsevier, Oxford, pp. 162–167)*

FIGURE 25.4 Unmanned aerial vehicles fitted with infra-red cameras can measure crop growth over the crop, providing data to compare with measurements taken via satellite and on-the-spot observations using traditional approaches. *(Reproduced with permission from Birchip Cropping Group, Vic., Australia.)*

Building up a map of the harvest area with respect to protein content is possible by fitting 'on-the-go' NIR equipment to the combine harvester beside the yield monitor (Whelan et al., 2009). In this way, a continuous indication of protein content is provided, assisting the grower to harvest selectively so as to meet the protein content of a specific premium grade. In addition, pre-harvest sampling and quality analysis can be used as an aid to intelligent harvesting, for example to meet a protein content target.

Avoidance of down-grading due to sprout damage is another possible quality target associated with rain at harvest. A common scenario is that sprout damage is more severe in some parts of the harvest area than others. Pre-harvest sampling and testing (Wrigley, 2005) offer the opportunity of selective harvesting, avoiding the parts of the field known to have sprouted grain that would otherwise cause the whole crop to be down-graded.

Leonard (2016) has contrasted the decision-based systems of the past and future as follows: 'It is only about 100 years since tractors started to replace horses as the power source for grain production. Few farmers in the developed world would now know how to harness a horse team. Within the next 100 years, it is likely that farmers will lose the skills of driving a tractor but will require and gain skills in computer science and electronics.'

25.4.5 Climates of the Future

Future decades hold uncertainties for grain growing, including changed patterns of rainfall, droughts and floods, plus decreased frost frequency and rising sea levels, leading to changes in growth temperature (+2°C predicted in the next 50 years) and heat stress (more frequent and more severe episodes). The main basis for these predictions is the increasing atmospheric levels of CO_2 (http://www.ipcc.ch/). Irrespective of the causes, the level of CO_2 in the atmosphere is increasing, as shown by long-term monitoring at the Mauna Loa site in Hawaii (Fig. 25.4). The level of CO_2 has increased from about 315 parts per million 50 years ago to over 400 ppm now (http://www.esrl.noaa.gov/gmd/ccgg/trends/).

Rising CO_2 levels will act as a 'carbon fertiliser' raising grain yields for C3 cereals such as wheat, barley and rice. In parallel, progressive lowering of grain-protein levels is expected, with implications for the uses of all grain species, especially for the processing of wheat flour. Some research reports indicate that higher CO_2 levels will produce smaller grains (possibly reducing flour yield in milling) and a higher proportion of large starch granules (good for starch-gluten manufacturers) (Blumenthal et al., 1996; Pleijel and Danielsson, 2009; Högy et al., 2009).

On the other hand, the scenario is different for C4 cereals, such as maize and sorghum, which may not show the yield increases predicted for the C3 grasses. There are two different photosynthetic pathways plants can use. Plants using the C3 pathway rely on the enzyme rubisco to fix carbon atoms from carbon dioxide. The first stable product in this process is based on three carbon atoms, hence the 'C3 pathway'. C4 plants use the enzyme, PEP carboxylase, leading to a four-carbon molecule and the 'C4 pathway' and different implications for rising levels of CO_2.

Associated climate changes point to increasing temperatures, less frost and more frequent heat-stress episodes. To the extent that such changes may occur during the grain-filling period, stronger dough may be the result for wheat in regions that currently experience the range 15–30°C (Moss and Randall, 1990; Gibson et al., 1998), especially for 'cooler' wheat-growing regions in the Northern Hemisphere. In contrast, dough weakening may be the result for regions subject to heat stress (at least a few days with maxima over 35°C) (Blumenthal et al., 1995; Gibson et al., 1998). This latter scenario applies to wheat-growing regions of southern Europe, parts of South America and Africa, Australia and the mid-west of the USA.

Rice quality is also affected detrimentally by high-temperature stress during grain filling, due to the formation of chalky grains as a result of loose packing of amyloplasts. Ishimaru et al. (2009) used magnetic resonance images to examine rice caryopses that had been affected by high-temperature. Scanning electron microscopy showed that around the centre of the endosperm, there was disorganised development of amyloplasts associated with lower water content. Grain chalkiness caused by high-temperature stress may become a major problem for rice growers if and as global warming continues.

The effects of temperature on grain quality (not only for wheat) have been shown to be genotype dependent, with some varieties showing less alteration in grain quality than others (Blumenthal et al., 1995; Panozzo and Eagles, 1998; Gibson et al., 1998). For example, varieties with the 5 + 10 combination of glutenin subunits in wheat have been shown to be more tolerant to the dough-weakening effects of heat-stress than those with 2 + 12 glutenin subunits (Blumenthal et al., 1995). The genotypic aspect of the altered grain quality offers the opportunity for varieties to be identified, or developed, that are tolerant to the effects of heat stress on dough quality. In addition, the prospect of fewer and earlier frost episodes offers the strategy that crops may be planted earlier than at present, thereby to mature before the onset of heat stress.

Another consequence of climate change may be altered patterns of rainfall, with consequent droughts more often in traditional grain-growing regions. As a result, there may be the need to move grain-growing to new areas where higher rainfalls can be assured. The implications for grain quality may be that fresh breeding and selection is needed for 'new' situations. Furthermore, irrespective of the grain-growing region, new models of grain performance (for yield and quality) will be required as the agricultural world moves into an era of more variable climate (Hochman et al., 2009). Crop simulation models which are currently restricted to grain quality parameters could be extended in the future to include functionality, milling characteristics and end-use value (Nuttall et al., 2016) (Fig. 25.5).

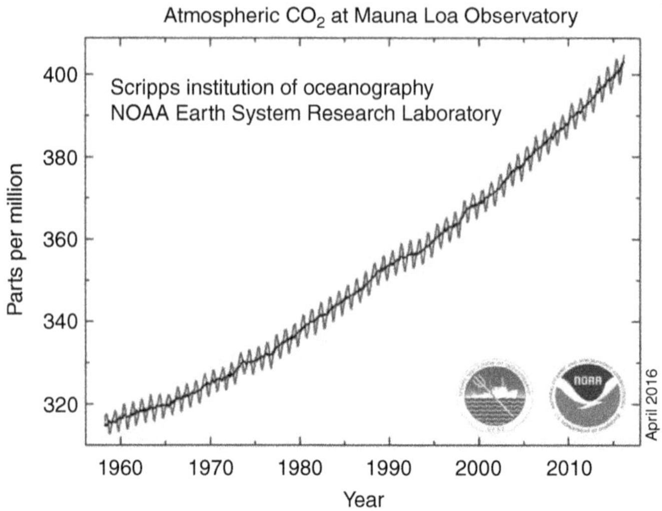

FIGURE 25.5 Monthly mean atmospheric carbon dioxide at Mauna Loa Observatory, Hawaii. The *smooth curve* represents the seasonally corrected data. The jagged trace is CO_2 data measured continuously, showing the annual up-and-down cycle due to the enhanced uptake of CO_2 by the extensive forests of the northern hemisphere during northern summers. *(Reproduced with permission from the National Oceanic and Atmospheric Administration.)*

25.5 GRAIN RECEIVAL AND SEGREGATION

25.5.1 Improved Quality-Screening Methods

Improved methods of screening for specific aspects of grain quality are needed both by the breeder (see Section 25.3.3) and at the site of grain receival, be it the grain elevator (silo) or the flour or feed mill. However, these two stages of the grain chain differ markedly with respect to the conditions of quality assessment. The breeder requires large numbers of small samples (even half seeds) to be tested, with low labour costs but little need for speed. In contrast, at grain receival, testing should ideally be fast, often with the limited resources of the country elevator, as detailed in Table 1.4.

While objective testing methods for grain receival standards such as test weight, protein, moisture, test weight and falling number are well established, subjective visual inspection is still the basis of much grain testing at receival. Grain defects, identification of foreign seeds and other contaminants are currently assessed by visual inspection. Digital imaging analysis has the potential to recognise and quantify at least some of these defects and has been shown to be capable of predicting black tip and some common weed seeds (see Sections 19.4.3 and 19.9.1).

Whereas NIR (ground grain) and NIT (whole grain) technologies have revolutionised the analysis of all grain types for moisture, protein content and oil content, there is the prospect that these methods will in future be used to assess a wider range of quality attributes, based on specific aspects of grain composition.

For example NIR/NIT analysis of β-glucan in barley (Schmidt et al., 2009) might provide a general distinction between barley varieties suited for human food (rich in β-glucans) versus barleys for malting and animal feed (low in β-glucans).

NIR technology has also been adapted to assess quality in grains one-at-a-time for the full range of conventional assays, as well as the distinction between normal and waxy wheat kernels, that is those with no amylose (Dowell et al., 2009). Grain-by-grain analysis would also be important for malting barley, for which uniform germination is a critical quality attribute (Chapter 9). Going beyond the strict assessment of grain composition, NIR and NIT are already used reliably to assess grain hardness in wheat. An additional application is the detection by NIR of the mycotoxin Vomitoxin (deoxynivalenol) due to Fusarium infection (Peiris et al., 2015). NIR is also reported to be effective in determining the degree to which grain samples are defective with respect to sprouting and midge damage (Singh et al., 2009). Might it someday be possible for NIT to assess dough strength potential, based for example on chemical information about the length of glutenin-polymer chains?

An additional requirement at grain receival, compared to breeding, is the need for varietal identity to be checked against variety declaration on delivery, because only a few specified varieties of similar quality are acceptable for a specific grade or class. Present and future approaches to verification of the variety declaration are described in Chapters 17 and 19.

On the other hand, testing at grain receival has hardly begun to use any of the growing number of marker-assisted tests that are now entering into the selection process of breeding. However, there is the possibility that tests for specific aspects of grain quality may become so simple and rapid that their use will make it unnecessary to specify variety or to use methods of varietal identification. Such methods may employ quality-specific markers, possibly based on DNA analysis, or more likely on the presence of marker proteins, because the extraction of proteins from the grain is easier than that of DNA. Test assessment based on protein markers might take advantage of the specificity of antibodies and the rapid immuno-assay systems that have been developed for medical testing. Currently, ELISA testing is used for the detection of mycotoxins and genetically modified grain and these methods are fast, cost effective and can be used on-site at grain receival or at a regional laboratory.

Nevertheless, verification of variety declaration will still be needed for countries where royalties for Plant Breeders' Rights are based on the delivery of harvested grain (end-point royalties), rather than on the seed bought for sowing (Cooke and Wrigley, 2004). In addition, maltsters will probably continue to prefer to buy barley grain of a single variety (and uniform grain size), to ensure uniform germination and consistent malt quality.

25.5.2 Identity Preservation

The segregation of grain of one quality type from grain of different quality is pointless unless the distinction can be preserved throughout storage and transport, so that the grain lots suited to distinct processing are delivered, identity preserved, to the target mill or factory. The process of identity preservation, from farm gate to miller or food processor, is costly. Thus, the financial advantage must be real, involving as it does quality testing at receival and probably at various stages of storage and transport, as well as requiring segregated storage. Nevertheless, the cost should be reflected in more efficient processing, so that grain is provided that will better meet the specifications of the next stage of the grain chain.

25.6 GRAIN STORAGE AND TRANSPORT

25.6.1 Whole Grain

The harvest of any specific crop occurs once a year in most places. The harvested grain may be stored and transported during the following year, or even longer. Safeguarding this valuable resource, so as to maintain its quality, is an essential part of the management of grains. The suitability of the cereal grains for storage is a major reason why they are as attractive as a food source, but nevertheless, stored grain is also attractive to many predators—insects, animals, birds and microorganisms. Thorough assessment of grain soundness and cleanliness is thus essential to storage management. Recent research has been directed towards technology for detecting the presence of insects in grain en route to storage. This assessment must include the presence of eggs and larvae, as well as live insects. The management of microbiological safety also starts when the grain is first assessed before going into storage; monitoring this aspect of grain health, though more difficult, is gaining research attention.

Processors are obviously sensitive to the presence of any form of infestation in grain received for processing, but increasingly they are also sensitive to the presence of residues from any form of grain protectant that may have been used in storage. For this reason, the use of fumigants and inert atmospheres are proving to be more popular, as they are expected to leave no residues (e.g. Rajendran and Muralidharan, 2001). In this category, the use of carbon dioxide-rich atmospheres is still attractive (Jayas and Jeyamkondan, 2002), because carbon dioxide can kill insects as well as just preventing proliferation, as is claimed for other inert atmospheres, and the development of insect resistance to carbon dioxide appears to be less likely than for 'chemical' pesticides. Nevertheless, chemical protectants and fumigants are likely to continue in use, partly because of their fast action and because they do not depend greatly on the need for a complete seal of the storage container.

The future may see the realisation of dreams and diverse approaches to grain transportation that have been suggested, such as carrying grain by pipeline as a slurry in an inert liquid. This technology has proved feasible on a commercial scale for the efficient transport of inert granulated materials such as iron-ore pellets, but difficulties are foreseen for biological materials such as grains.

Canola oil and super-critical carbon dioxide have been suggested and trialled as suitable carriers, but the removal of the oil residue is a problem with the former, and the latter has been found to extract lipid from whole wheat grain. As fuel costs rise, older ideas such as these may warrant re-investigation.

25.6.2 Milled Products

The storage and transport of whole grain is safer with respect to quality stability than after milling, due to the consequent disruption of cellular structure and the likelihood that degradative enzymes are provided with access to their substrates. Lipases are especially of concern in this respect, and heat treatment has been the traditional approach to their inactivation. However, in the heat-treatment process there has also been loss of antioxidant activity. Therefore, Rose et al. (2008) set about devising treatments that would stabilise milled grain against lipase action and rancidity without decreasing antioxidant activity. This objective proved to be achievable with careful application of dry heat, microwave or steam treatment.

25.7 GRAIN PROCESSING

25.7.1 Milling

Milling operations for cereal grains in general have become highly automated for large mills in western countries, relying more and more on computerised controls and sensors to detect problems, such as the 'light-out' wheat mill (Posner, 2009). Flour millers need to be flexible to utilise wheat blends of different origins and quality in the quest for least-cost gristing while maintaining flour quality. New technology is being implemented in wheat milling and de-branning provides for the removal of much of the bran before the crushing stage of flour milling. This approach offers the advantage of also removing contaminants associated with the bran, such as pesticides and microorganisms, and de-branning is followed by a shorter break system, since the remaining task of the rollers is to reduce the particle size of the mill streams (Posner, 2009). Particle size of flour and semolina can now be measured inline and adjustments made to first break rolls to aid process control in real-time. Other new technology is the use of optical colour sorters prior to milling which can be used to separate diseased and discoloured grains or mycotoxin-affected grain.

Cryomilling of rice starch has been proposed as a means of modifying the structure and properties of rice starch by non-chemical means (Devi et al., 2009). The temperature of milling trialled was very low (in liquid nitrogen). The paste viscosity of the cryomilled starch decreased significantly as a result of this treatment.

The quality and value of milled rice are largely determined by the efficiency of milling. Control of the removal of the outer layers in the process is critical. NIR methodology has been proposed for this purpose, using continuous NIR measurement of surface lipid content (Saleh et al., 2008).

The accurate monitoring of quality attributes is still a significant factor in milling the various types of grain. NIR is a major means of monitoring the moisture and protein content of incoming grain and product streams on a continuous basis, most commonly for wheat milling, but also for other cereals. This information is vital to the blending of wheat-flour streams for the purposes of blending to achieve the quality specifications of flour customers (Posner and Hibbs, 2005). Prediction of the outcome of blending is a relatively simple task when formulating grain lots or flour streams to achieve composition attributes (e.g. protein content), because relationships are linear.

However, a linear relationship cannot be assumed when blending wheat-flour streams to achieve specific targets for dough properties, water absorption, starch quality, product quality or enzyme-related quality (e.g. falling number) or because of the non-linearity of these attributes. For the blending of wheat varieties with respect to dough properties, the greatest departure from linearity on blending was for combinations of varieties that are most different with respect to the composition of their glutenin subunits (Larroque et al., 2000). On the other hand, dough-quality relationships approach linearity for combinations of varieties with similar glutenin subunits. The composition of glutenin subunits of the high-molecular-weight group (*Glu-1*) is more important in this respect than those of the low-molecular-weight subunits (*Glu-3*).

25.7.2 Ingredients to Improve Nutrition, Processing and Product Quality

Although food production from cereal grains is an ancient art, there is still more for us to discover about the interactions of the various endogenous constituents during processing. A recent review attempts to summarise the current knowledge and describe what else we need to know (Goesaert et al., 2005). In addition, the food industry has long profited from the use of a wide range of additives to enhance processing efficiency, flavour, nutrition, shelf life and consumer acceptance. These additives include vitamins and minerals, especially added to cereal-based breakfast foods and industrially-milled wheat, corn and rice flours. The latter is promoted through the Food Fortification Initiative (http://www.ffinetwork.org/) as a means of preventing neural tube birth defects and anaemia.

Salt (sodium chloride) is one of the oldest additives, used for bread production, as well as for porridge, ready-to-eat breakfast cereals and many other foods. Salt is considered the fourth essential ingredient of leavened bread, after flour, water and yeast. Salt is acknowledged to strengthen dough structure, stabilise yeast fermentation rate and enhance bread flavour beyond the actual salty taste (Miller and Hoseney, 2008). However, cereal-based foods have been targeted as sources of excessive salt in the diet. In fact, sodium reduction in the diet is cited as a major nutritional goal (Heidolph, 2008). The cereal-processing industry is therefore seeking alternatives to sodium chloride as means of substituting for its several low-cost advantages (Day, 2016).

Other additives, especially for bread production, serve to provide better product quality. One of these is the use of potassium bromate, a mild oxidising agent that has recently become controversial. Gelroth et al. (2009) have reviewed its use, pointing out its value as a processing aid, and that it is still a permitted additive in the USA, although restricted in Europe, Canada, China, and some other countries. With ongoing compliance with baking guidelines its continuing use is likely to be permitted, at least in the USA, but this depends on continued monitoring by the US Food and Drug Administration. Where potassium bromate is no longer used, bread of good quality can be made using ascorbic acid as the oxidising agent together with amylases, xylanases and lipases.

The production of frozen doughs appears to extend the requirement for additives to enhance bread quality following frozen storage and transport prior to baking. A range of traditional and novel additives, reviewed by Selomulyo and Zhou (2007), includes various emulsifiers, hydrocolloids and anti-staling agents.

Additives also include various enzymes, and these have generally been well accepted. The most commonly used enzymes in commercial bread baking are amylases, xylanases, proteases and lipases. The enzyme transglutaminase (of microbial origin) is a newer candidate for consideration as a bread additive, offering the advantage of increasing dough strength by covalently cross-linking between the lysine and glutamine residues of proteins (Medina-Rodríguez et al., 2009). The potential value of this type of additive may increase into the future as protein levels fall in the wheat crop due to the added expense of nitrogen-based fertilisers and increasing levels of atmospheric carbon dioxide.

25.7.3 No More Colour Problems with Noodles?

Studies of the colour and darkening (colour stability) of Asian noodles with time has involved gaining an understanding of the various physical and chemical mechanisms that play a role. The colour of yellow alkaline noodles is due to a combination of flour flavonoids at high pH plus naturally occurring xanthophylls. The rate and extent of darkening is governed by the enzyme polyphenol oxidase (PPO) plus non-PPO darkening (see Section 24.4.2).

Null-PPO alleles have been identified and included in crosses which when milled and made into noodles produced product with improved colour stability (Hystad et al., 2016). This provides noodle-wheat breeders with an opportunity to develop varieties with good colour stability. Non-PPO darkening accounts for about 50% of all noodle darkening, so in the future there is the potential for further gains to be made by investigating the mechanisms involved and identifying potential non-PPO darkening markers for use in breeding programs.

25.7.4 Gluten-Free Foods

Whereas wheat is uniquely suited to bread making because of its gluten proteins, there is an increasing requirement for gluten-free bread and other products. This need relates partly to the requirement that a life-long gluten-free diet must be pursued by people with coeliac disease—an 'inflammatory disorder of the upper small intestine triggered by the ingestion of wheat, rye, barley and possibly oat products' (Wieser and Koehler, 2008). In addition to coeliacs, there is an increasing number of people who claim to be gluten intolerant. In many cases, this intolerance is real, even though the condition is not so well defined or diagnosable as coeliac disease is. It has been termed 'non-coeliac gluten sensitivity' (NCGS) (Suter and Fleming, 2008). In other cases, there may be an element of 'food fad'. For the coeliac gluten-free diet, the need for gluten-free foods extends beyond wheat to include rye, barley and possibly also oats and beyond solid foods to include beverages (Arendt and Dal Bello, 2009). Gluten-free food is occupying increasingly more shelf space in supermarkets and in 2015 1 in 10 new products launched globally were gluten-free, with the largest segments in bakery (including biscuits) and snack foods. Many basic snack food ingredients such as corn are already gluten-free and gluten-containing cereals can be replaced by a wide variety of other ingredients.

Suitable gluten-free ingredients include corn, millet, sorghum, rice, pseudocereals, coconut flour, inulin and legumes together with various starches, gums, enzymes and hydrocolloids. Formulations and processing conditions have been well described by Casper and Atwell (2014). There are numerous recipes for gluten-free home baking on various internet web sites. The challenge in gluten-free baking is to produce a quality bread without the visco-elastic properties that gluten confers. While many commercial gluten-free breads are available that have quite good quality, the texture is more like cake than that of bread. Many commercial products are fortified with vitamins and minerals.

Some close relatives of wheat are claimed to suit people with some forms of wheat intolerance, though not safe for coelics. 'Kamut' is the registered name of korasan, a wheat-related cereal that is claimed to be unchanged from its ancient origins (see Section 14.3.3). It is probably derived from a natural hybrid between *T. durum* and *T. polonicum*, which occurred in the Fertile Crescent (Khlestkina et al., 2006). Spelt is another wheat-related species in use for possibly alleviation of some forms of wheat intolerance (see Section 14.3.4). Spelt has a unique 'nutty' flavour and can be used in bread, pastry noodles and confectionery. In addition, sorghum has come to new prominence in the western world as a gluten-free alternative grain that has been shown to be adaptable to many novel food products (Lemlioglu-Austin, 2014).

25.7.5 A Greater Diversity of Uses for Grains and Their by-Products

The development of gluten-free foods is one example of the trends towards broadening the use of cereal grains other than wheat in food manufacture. Thus there have recently been publications advocating the wider food use of barley (Baik and Ullrich, 2008), oats (Mishra and Monro, 2009), sorghum (Taylor et al., 2006), maize and rice (Sabanis and Tzia, 2009; Matalanis et al., 2009).

The extended use of by-products of other processing is also advocated, for example brewers' spent grain (Mussatto et al., 2006) as well as sources of novel starch properties (e.g. Wongsagonsup et al., 2008).

Blending milled or cracked non-wheat grain with wheat flour can improve the nutritional value of composite breads (minerals, fibre, antioxidants), although there are a number of technological challenges to overcome in formulation and processing (Collar, 2014).

25.7.6 Biofuels

Greater diversity in grain use for the production of fuel ethanol is also being advocated. Until recently the most popular species for biofuels production has been maize and wheat to a lesser extent (see Chapter 23). Already, the increase in cereal-grain use for biofuels production has caused a distortion of grain production and processing, due to the diversion of significant volumes of grain for fermentation (Balat and Balat, 2009). In particular, sorghum has been suggested as an alternative to corn (Zhao et al., 2008a,b; Chuck-Hernandez et al., 2009).

Into the future, biorefineries are envisaged as factories that will go beyond the present practice of fermenting cereal starch into ethanol, to the production of 'biohydrogen' and 'biogas' (Kaparaju et al., 2009). An important change with respect to the processing of cereal grains will be the use of lignocellulose for the production of bioethanol, if that proves to be economically practicable on a commercial scale. Cereal straw may be the first substrate for this type of process, but diversification to a wider range of plant sources is likely (Sanchez and Cardona, 2008).

25.8 NUTRITIONAL BENEFITS OF GRAIN—REAL AND PERCEIVED

The cereal grains have long had a good image of being 'healthy'. In general, this image is well deserved, but some aspects of a 'healthy image' border on nutritional mythology, so it will in future be important for medical professionals to concentrate on advocating the aspects of grain nutrition that can be justified (Slavin, 2003; Wrigley and Asp, 2004). Yet the healthy image of grain-based foods becomes confused by the apparent need for manufacturers to go to 'Clean Label'—apparent necessity to show that packaged foods are free of artificial ingredients (flavours, colours, preservatives, etc.).

25.8.1 Whole-Grain Foods

The nutritional benefits of cereal grains in the diet have long been recognised, especially when consumed as whole-grain foods (Solah et al., 2016). These advantages are enshrined in the nutritional guidelines of many countries (Wrigley and Asp, 2004; Jones et al., 2015), with grains often indicated to be the basis of a good diet, for example, as the base of a pyramid diagram. The ingestion of dietary fibre has traditionally been a major reason for promoting grains, and it has been the outer (bran) layers of the cereal grains that are the basis for whole-grain promotion (Black and Lewis, 2009; Anderson et al., 2009).

There has been an increase in the number of new whole-grain products available, but the standard definition of whole-grain—that all parts of the grain endosperm, bran and germ should be present in their original proportions—is yet to be universally adopted by the food industry. Yet there is the fact of life that a majority of grain-based foods, especially wheat-based, involves the inclusion of refined grains (mainly endosperm), and the case of this fact being a valid contribution to a balanced diet has been made by Miller (2014).

There has also been the realisation that the starch fraction of the cereal grain can make a valuable contribution to dietary health, especially if its rate of hydrolysis and absorption is slowed, due to the starch being 'resistant'. This has been part of the concept of the glycaemic index (GI), and the labelling of some foods as 'low-GI' or 'high-GI'. As a result, geneticists and food technologists have combined to produce grains and foods, respectively that provide increased levels of resistant starch (e.g. Wongsagonsup et al., 2008; Woo et al., 2009).

25.8.2 Functional Foods

From these roots has grown the 'nutraceutical' concept that foods can be more than merely sources of energy and nutrients, so that some food sources are imbued with medicinal properties. 'Functional ingredients' are thus promoted for a range of foods, especially those that are grain-based (Gallagher and Barmann, 2009). Relevant to grains are the families of phytochemicals, anthocyanins and antioxidants (Castaneda-Ovando et al., 2009; Barba de la Rosa et al., 2009), which are reputed to help in reducing the risk of many diseases, including heart disease, high blood pressure, strokes, various forms of cancer and even diabetes and osteoporosis (Fardet et al., 2008; Barba de la Rosa et al., 2009; Shen et al., 2009; Gallagher and Barmann, 2009; Castaneda-Ovando et al., 2009). Whereas these types of compound have long been recognised in fruits and vegetables, it is only in recent years that their presence in cereal grains is being promoted, especially for pigmented grains.

'Black rice' is an example of this type of grain. It is strongly pigmented, mainly in the bran layers, due to the presence of anthocyanins (especially cyaniding-3-glucoside and peonidin-3-glucoside), polyphenolics, flavonoids, tocopherols, tocotrienols and gamma-oryzanol. These compounds have strong free-radical scavenging and antioxidant activities, thus reportedly reducing DNA damage and lipid peroxidation. Such compounds are also present in the bran of normal (white) rice, but the world-wide practice of eating milled rice, after removal of the bran layers, means that even these low levels of antioxidants are largely lost to the diet.

Because the bran and germ tissues of rice are largely regarded as by-products of rice milling, this material offers new opportunities for further processing to recover important bioactive compounds. For example rice-bran oil (about 25% of rice bran) is being commercially exploited for gamma-oryzanol and tocols (vitamin E compounds) (Lerma-Garcia et al., 2009).

Antioxidants are found in all the cereal grains, particularly in pigmented genotypes, and the range of their occurrence has been reviewed by Fardet et al. (2008). Wheat and barley are no exceptions; blue and purple varieties of both have long been recognised. The genetic control of such pigmentation is relatively simple, involving only one or two genes in wheat, so the breeding of high-anthocyanin wheats is feasible (Knievel et al., 2009).

A Europe-based survey was conducted to create an extensive database of bioactive components in cereal grains, providing data on dietary fibre components and on phytochemicals considered to have potential health benefits - the 'HEALTHGRAIN Cereal Diversity Screen' (Ward et al., 2008). For example, bioactives were studied in rye grain (Nyström et al., 2009) and folate in wheat (Piironen, 2008). This information now facilitates the selection of specific lines of cereal species in which high levels of desirable bioactive components would be present. In addition, information is available concerning their presence in grain tissues, not only in the bran layers. Addition information relates to yield and processing quality. However, the lines were grown at only one site, so information is lacking about the extent to which the production of bioactive compounds might be influenced by growth conditions.

25.9 GOVERNMENT CONCERNS, POPULATION PRESSURES AND WORLD-TRADE POLICIES RELATING TO CEREAL GRAINS

Cereal grains are at the forefront of government concerns about food security for their people, as grain is seen as being basic to providing sustenance to the population of each country, especially those with significant numbers of people on the poverty line. In these countries especially, populations are expected to continue to grow into the future, making the ongoing task of providing adequate food supplies even more difficult. The statistic has been presented (Section 1.2.1) that at present, sufficient cereal grain is produced to supply each person on earth with one kilogram of grain per day.

There is thus the challenge to make a more equitable distribution of this apparently adequate production of grain. World-trade policies form a significant part of the movement of grain from sites of (over-)production to places of need. Changes in such policies should be more readily altered for the betterment of world hunger than problems associated with water and fertiliser supply, for example, but changing these policies requires the good will and compassion of the affluent nations.

The problems of food supply have been discussed by Gilland (2002), who pointed out the potential for inadequacies of water and fertiliser supplies to limit future grain supplies, whilst also concluding that the most critical factors are the need to limit population growth. Dixon (2007) has written more optimistically by stating that the population growth of 90% during the second half of the twentieth century was accompanied by a 115% increase in world food supplies, representing a theoretical food-supply increase of 25% per person.

The inevitable increase in world population is accompanied by migration of people from rural to urban areas (Dixon, 2007). This migration means that there are fewer workers involved in agricultural practice and more food consumers. In addition, the expansion of urban areas means that agricultural land is shrinking as cities expand. Global warming adds a further threat to the available area of arable land as low-lying land, some of the most productive for grains, is threatened by inundation and salting by rising sea levels.

In addition to volume of supply, grain quality is of concern into the future. The maintenance of grain quality, and thus grain value, requires ongoing attention to the development of adequate and efficient methods of assessment of quality at all stages of the grain chain, plus the implementation of HACCP policies to ensure continued food safety.

25.10 CONCLUSIONS

The production, processing and consumption of cereal grains faces a 'changing landscape' due to climate uncertainties, food insecurities, changing treats from pests and diseases, potential advances such as GMO varieties with nutritional benefits (e.g. high zinc and bioactives) and adaptations to harsh growth environments, such as water-limited environments.

On the other hand, there are now signs that venture capital investment is being made in innovative ideas that could reshape the agriculture and food industry in a similar way to the rise of the internet economy. 'Ventures feature accelerators, angel investors, incubators and venture capitalists willing to support ideas that a decade ago may have been perceived as impractical.' (www.foodbusinessnews.net).

Finally, in answer to the question: 'Will yield increases continue so that we can continue to feed the world?' Robert Saik (Agri-Trend agronomy advisor) stated: 'The biggest threat to agriculture being able to feed the growing global population is the anti-GM non-science movement' (Anon, 2016).

SOURCES OF FURTHER INFORMATION

Literature

Anon, 2016. GRDC Booklet, Crop Yields and Global Food Security. Will Yield Increases Continue to Feed the World? Grains Research & Development, Canberra, Australia.

Graybosch, R.A., Liu, R.H., Mandl, R.L., Shi, Y.-C., Wang, D., Wu, X., 2009. New uses for wheat and modified wheat products. In: Carver, B.F. (Ed.), Wheat: Science and Trade. Wiley-Blackwell, Ames, Iowa, pp. 521–550.

Wrigley, C., Corke, H., Seetharaman, K., Faubion, J., 2016. Encyclopedia of Food Grains, vols. 1–4. Elsevier, Oxford, UK.

Website

http://www.agritrend.com/services/agrology-consulting/service-overview.aspx (Canadian agronomy service)

REFERENCES

Akbari, M., Wenzl, P., Caig, V., Carling, J., Xia, L., Yand, S., Uszynski, G., Mohler, V., Lehmensiek, A., Kuchel, H., Hayden, M.J., Howes, N., Sharp, P., Vaughan, P., Rathmell, B., Hutter, E., Kilian, A., 2006. 'Diversity arrays technology (DArT) for high-throughput profiling of the hexaploid wheat genome'. Theor. Appl. Genet. 113, 1409–1421.

Akhunov, E., 2016. Wheat genetics and genomics. Wrigley, C., Corke, H., Seetharaman, K., Faubion, J. (Eds.), Encyclopedia of Food Grains, vol. 4, Elsevier, Oxford, UK, pp. 382–387.

Anderson, J.W., Baird, P., Davis, Jr., R.H., Ferreri, S., Knudtson, M., Koraym, A., Waters, V., Williams, C.L., 2009. Health benefits of dietary fiber. Nutr. Rev. 67, 188–205.

Anon, 2009. Use of precision agriculture by the Australian grains industry. Crop Pasture Sci. 60, 795–798.

Anon., 2016. Crop yields and global food security. Will yield increases continue to feed the world? GroundCover. Grains Research & Development, Canberra, Australia, Issue 122, p. 40.

Arendt, E.K., Dal Bello, F., 2009. The Science of Gluten-free Foods and Beverages. AACC International Press, St Paul, MN.

Asp, E., 2004. Consumer trends in consumption. Wrigley, C., Corke, H., Walker, C. (Eds.), Encyclopedia of Grain Science, vol. 1, Elsevier, Oxford, pp. 322–330.

Badaracco, S., 2016. Strategies for successfully entering popular food trends. Cereal Food World. 61, 81.

Baik, B.-K., Ullrich, S.E., 2008. Barley for food: characteristics, improvement, and renewed interest. J. Cereal Sci. 48, 233–242.

Balat, M., Balat, H., 2009. Recent trends in global production and utilization of bio-ethanol fuel. Appl. Energy 86, 2273–2282.

Bao, J.S., H Corke, H., 2016. Rice: genetics. Wrigley, C., Corke, H., Seetharaman, K., Faubion, J. (Eds.), Encyclopedia of Food Grains, vol. 4, Elsevier, Oxford, UK, pp. 398–406.

Barba de la Rosa, A.P., Fomsgaard, I.S., Laursen, B., Mortensen, A.G., Olvera-Martınez, L., Silva-Sáchez, C., Mendoza-Herrera, A., González-Castaneda, J., De Leon-Rodrıguez, A., 2009. Amaranth (*Amaranthus hypochondriacus*) as an alternative crop for sustainable food production: phenolic acids and flavonoids with potential impact on its nutraceutical quality. J. Cereal Sci. 49, 117–121.

Barrero, J.M., Cavanagh, C., Verbyla, K.L., Tibbits, J.F.G., Verbyla, A.P., Huang, B.E., Rosewarne, G.M., Stephen, S., Wang, P., Whan, A., Rigault, P., Hayden, M.J., Gubler, F., 2015. Transcriptomic analysis of wheat near-isogenic lines identifies PM19-A1 and A2 as candidates for a major dormancy QTL. Genome Biol. 16, 93.

Beyer, P., Al-Babili, S., Ye, X., Lucca, P., Schaub, P., Welsch, R., Potrykus, I., 2002. Golden rice: Introducing the ß-carotene biosynthesis pathway into rice endosperm by genetic engineering to defeat vitamin A deficiency. J. Nutr. 132, 506S–510S.

Black, P.D., Lewis, D., 2009. Dietary fiber: consumer attitudes and preferences. Cereal Food World 54, 213–215.

Blair, R., Joe, M., Regenstein, J.M., 2015. Genetic Modification and Food Quality: A Down-to-Earth Analysis. Wiley-Blackwell Publishing, UK.

Blumenthal, C., Bekes, F., Gras, P.W., Barlow, E.W.R., Wrigley, C.W., 1995. Identification of wheat genotypes tolerant to the effects of heat stress on grain quality. Cereal Chem. 72, 539–544.

Blumenthal, C., Rawson, H.M., McKenzie, E., Gras, P.W., Barlow, E.W.R., Wrigley, C.W., 1996. Changes in the grain quality of field-grown wheat due to doubling the level of atmospheric carbon dioxide. Cereal Chem. 73, 762–776.

Bramley, R.G.V., 2009. Lessons from nearly 20 years of precision agriculture research, development, and adoption as a guide to its appropriate application. Crop Pasture Sci. 60, 197–217.

Butow, B.J., Ma, W., Gale, K.R., Cornish, G.B., Rampling, L., Larroque, O., Morell, M.K., Bekes, F., 2003. 'Molecular discrimination of Bx7 alleles demonstrates that a highly expressed high-molecular-weight glutenin allele has major impact on wheat flour dough strength'. Theor. Appl. Genet. 107, 1524–1532.

Casper, J.L., Atwell, W.A., 2014. Gluten-Free Baked Products. AACC International, St Paul, MN, USA.

Castaneda-Ovando A, de Lourdes Pacheco-Hernandez, Ma. Elena Paez-Hernandez, 2009. Cereal diversity screen: concept, results and prospects. J. Agric. Food Chem. 56, 9699–9709.

Chen, M.-H., McClung, A.M., 2015. Effects of cultivars, organic cropping management, and environment on antioxidants in whole grain rice. Cereal Chem. 92, 364–369.

Chuck-Hernandez, C., Perez-Carrillo, E., Serna-Saldivar SO, 2009. Production of bioethanol from steam-flaked sorghum and maize. J. Cereal Sci. 50, 131–137.

Collar, C., 2014. Barley, maize, sorghum, millet and other cereal grains. In: Zhou, W., Hui, Y.H. (Eds.), Bakery Products Science and Technology. second ed. Wiley, pp. 107–126.

Collis, B., 2016. Opening the gate to virtual visitors. Grains Research & Development, Ground-Cover, Canberra, Australia, Issue 122, pp. 14–15.

Cooke, R.J., Wrigley, C.W, 2004. Variety registration and breeders' rights. Wrigley, C., Corke, H., Walker, C. (Eds.), Encyclopedia of Grain Science, vol. 3, Elsevier, Oxford, pp. 314–321.

Day, L., 2016. Cereal food production with low salt. Wrigley, C., Corke, H., Seetharaman, K., Faubion, J. (Eds.), Encyclopedia of Food Grains, vol 3, Elsevier, Oxford, UK, p. 396.

Day, L., Augustin, M., Pearce, D.J., Batey, I.L., Wrigley, C.W., 2009. Enhancement of gluten quality combined with reduced lipid content through a new salt-washing process. J. Food Eng. 95, 365–372.

DePauw, R., O'Brien, L., 2016. Wheat breeding: exploiting and fixing genetic variation by selection and evaluation. Wrigley, C., Corke, H., Seetharaman, K., Faubion, J. (Eds.), Encyclopedia of Food Grains, vol. 4, Elsevier, Oxford, pp. 279–286.

Devi, A.F., Fibrianto, K., Torley, P.J., Bhandari, B., 2009. Physical properties of cryomilled rice starch. J. Cereal Sci. 49, 278–284.

Devos, K.M., Dolezel, J., Feuillet, C., 2009. Genome organization and comparative genomics. In: Carver, B.F. (Ed.), Wheat: Science and Trade. Wiley-Blackwell, Ames, Iowa, pp. 327–367.

Dixon, J., 2007. The economics of wheat: value chains from research to field to fork. In: Buck, H.T., Nisi, J.E., Salomón, N. (Eds.), In: Proceedings of the Seventh International Wheat Conference. Springer, The Netherlands, pp. 9–22.

Dowell, F.E., Maghirang, E.B., Graybosch, R.A., Berzonsky, W.A., Delwiche, S.R., 2009. Selecting selecting and sorting waxy wheat kernels using near-infrared spectroscopy. Cereal Chem. 86, 251–255.

Engleson, J., Atwell, B., 2008. Gluten-free product development. Cereal Food World 53, 180–184.

Fardet, A., Rock, E., Rém, C., 2008. Is the in vitro antioxidant potential of whole-grain cereals and cereal products well reflected in vivo? J. Cereal Sci. 48, 258–276.

Gallagher, J., Barmann, K., 2009. Working with functional foods. Cereal Food World 54, 210–212.

Garofalo, P., Di Paolo, E., Rinaldi, M., 2009. Durum wheat (*Triticum durum* Desf.) in rotation with faba bean (*Vicia faba* var. minor L.): long-term simulation case study. Crop Pasture Sci. 60, 240–250.

Gelinas, P., David, C., 2016. Organic grain production and food processing. Wrigley, C., Corke, H., Seetharaman, K., Faubion, J. (Eds.), Encyclopedia of Food Grains, vol. 4, Elsevier, Oxford, pp. 154–161.

Gelroth, J., Sanders, L., Cogswell, T., Zvaners, 2009. The use of potassium bromate by the commercial baking industry. Cereal Food World 54, 205–209.

Gibson, L.R., McCluskey, P.J., Tilley, K.A., Paulsen, G.M., 1998. Quality of hard red winter wheat grown under high temperature conditions during maturation and ripening. Cereal Chem. 75, 421–427.

Gilland, B., 2002. World population and food supply. Can food production keep pace with population growth in the next half-century? Food Policy 27, 47–63.

Goesaert, H., Brijs, K., Veraverbeke, W.S., Courtin, C.M., Gebruers, K., Delcour, J.A., 2005. Wheat flour constituents: how they impact bread quality, and how to impact their functionality. Trends Food Sci. Technol. 16, 12–30.

Goodman, R.E., 2014. GMOs: are they a regulatory or food safety issue? Cereal Food World 59, 164–166.

Graybosch, R.A., Liu, R.H., Mandl, R.L., Shi, Y.-C., Wang, D., Wu, X., 2009. New uses for wheat and modified wheat products. In: Carver, B.F. (Ed.), Wheat: Science and Trade. Wiley-Blackwell, Ames, Iowa, pp. 521–550.

GRDC, 2016. Managing Frost Risk. Grains Research & Development Corporation, Australia, www.grdc.com.au.

Guzman, C., Xiao, Y., Crossa, J., Gonzalez-Santoyo, H., Huerta, J., Singh, R., Dreisigacker, S., 2016. Sources of the highly expressed wheat bread making (wbm) gene in CIMMYT spring wheat germplasm and its effect on processing and bread-making quality. Euphytica 209, 689–692.

Haumann, B., 2009. U.S. organic sales continue to grow. Cereal Food World 54, 164–165.

Heidolph, B.B., 2008. Is sodium reduction the next target in the "better-for-you" formulation trend? Cereal Food World 53, 21–27.

Henry, R.J., 2014. Wheat genomics for grain quality improvement. Cereal Food World 59, 22–24.

Hochman, Z., van Rees, F.H., Carberry, P.S., Hunt, J.R., McCown, R.L., Gartmann, A., Holzworth, D., van Rees, S., Dalgliesh, N.P., Long, W., Peake, A.S., Poulton, P.L., McClelland, T., 2009. Re-inventing model-based decision support with Australian dryland farmers. 4. Yield Prophet® helps farmers monitor and manage crops in a variable climate. Crop Pasture Sci. 60, 1057–1070.

Högy, P., Zörb, C., Langenkämper, G., Betsche, T., Fangmeier, A., 2009. Atmospheric CO_2 enrichment changes the wheat grain proteome. J. Cereal Sci. 50, 248–254.

Hystad, S.M., Giroux, M.J., Martin, J.M., 2016. Impact of null polyphenol oxidase alleles on white salted noodles. Cereal Chem. 93, 255–262.

Ishimaru, T., Horigane, A., Ida, M., Iwasawa, San-oh, Y., Nakazono, M., Nishizawa, N.K., Masumura, T., Kondo, M., Yoshida, M., 2009. Formation of grain chalkiness and changes in water distribution in developing rice caryopses grown under high-temperature stress. J. Cereal Sci. 50, 166–174.

James, C., 2014. Global status of commercialized biotech/GM crops: 2014. 2014. ISAAA Brief No. 49. ISAAA: Ithaca, NY.

Jayas, D.S., Jeyamkondan, S., 2002. Postharvest technology: modified atmosphere storage of grains, meats, fruits and vegetables. Biosyst. Eng. 82, 235–251.

Jennings, R., 2016. The silver lining in data-filled clouds. GroundCover. Grains Research & Development, Canberra, Australia, Issue 122, p. 31.

Jones, H.D., 2005. Wheat transformation: current technology and applications to grain development and composition. J. Cereal Sci. 41, 137–147.

Jones, H.D., Sparks, C.A., Shewry, P.R., 2009. Transgenic manipulation of wheat quality. In: Khan, K., Shewry, P.R. (Eds.), Wheat Chem. Technol. fourth ed. MN, AACC International, St. Paul, pp. 437–451.

Jones, J.M., Peña, R.J., Korczak, R., Braun, H.J., 2015. Carbohydrates, grains, wheat, in nutrition and health: An Overview. Part I. Role of carbohydrates in health. Cereal Food World 60, 224–233.

Kaparaju, P., Serrano, M., Thomsen, A.B., Kongjan, P., Angelidaki, I., 2009. Bioethanol, biohydrogen and biogas production from wheat straw in a biorefinery concept. Bioresour. Technol. 100, 2562–2568.

Khlestkina, E.K., Röder, M.S., Grausgruber, H., Andreas Börner, A., 2006. A DNA fingerprinting-based taxonomic allocation of Kamut wheat. Plant Genet. Resour. 4, 172–180.

Klindworth, D.L., Hareland, G.A., Elias, E.M., Faris, J.D., Chao, S., Xu, S.S., 2009. Agronomic and quality characteristics of two new sets of Langdon durum–wild emmer wheat chromosome substitution lines. J. Cereal Sci. 50, 29–35.

Knievel, D.C., Abdel-Aal, E-S.M., Rabalski, I., Nakamura, T., Hucl, P., 2009. Grain color development and the inheritance of high anthocyanin blue aleurone and purple pericarp in spring wheat (*Triticum aestivum* L.). J. Cereal Sci. 50, 113–120.

Larroque, O.R., Bekes, F., Wrigley, C.W., Rathmell, W.G., 2000. Analysis of gluten proteins in grain and flour blends by RP-HPLC. In: Shewry, P.R., Tatham, A.S. (Eds.), Wheat Gluten. Royal Society of Chemistry, Cambridge, UK, pp. 136–139.

Lemlioglu-Austin, D., 2014. Sorghum: obliging alternative and ancient grain. Cereal Food World 59, 12.

Leonard, E.C., 2015. Farming's new eyes in the sky. GroundCover. Grains Research & Development, Canberra, Australia, Issue 118, pp. 32–33.

Leonard, E.C., 2016. Precision agriculture. Wrigley, C., Corke, H., Seetharaman, K., Faubion, J. (Eds.), Encyclopedia of Food Grains, vol. 4, Elsevier, Oxford, pp. 162–167.

Lerma-Garcia, M.J., Herrero-Martinez, J.M., Simo-Alfonso, E.F., Mendonça, C.R.B., Ramis-Ramos, G., 2009. Composition, industrial processing and applications of rice brad γ-oryzanol. Food Chem. 115, 389–404.

Li, J., Båga, M., Rossnagel, B.G., Legge, W.G., Chippar, R.N., 2008. Identification of quantitative trait loci for β-glucan concentration in barley grain. J. Cereal Sci. 48, 647–655.

Lou, J., Chen, Y., Yue, G., Lou, Q., Mei, H., Xiong, L., Luo, L., 2009. QTL mapping of grain quality traits in rice. J. Cereal Sci. 50, 145–151.

Magaña-Gómez, J.A., Calderón de la Barca, A.M., 2009. Risk assessment of genetically modified crops for nutrition and health. Nutr. Rev. 67, 1–16.

Matalanis, A.M., Campanella, O.H., Hamaker, B.R., 2009. Storage retrogradation behavior of sorghum, maize and rice starch pastes related to amylopectin fine structure. J. Cereal Sci. 50, 74–81.

Mathiowetz, A., Jones, J.M., 2016. Genetically modified grains and the consumer. Wrigley, C., Corke, H., Seetharaman, K., Faubion, J. (Eds.), Encyclopedia of Food Grains, vol. 2, Elsevier, Oxford, pp. 50–56.

Medina-Rodríguez, C.L., Torres, P., Martínez-Bustos, F., Ramirez-Wong, B., Paredes-López, O., Castaño-Tostado, E., 2009. Effect of microbial transglutaminase on dough proteins of hard and soft (*Triticum aestivium*) and durum (*Triticum durum*) wheat cultivars. Cereal Chem. 86, 127–132.

Miller, R.A., 2014. The case for refined grains in a balanced diet. Cereal Food World 59, 137–140.

Miller, R.A., Hoseney, R.C., 2008. Role of salt in baking. Cereal Food World 53, 4–6.

Mishra, S., Monro, J.A., 2009. Digestibility of starch fractions in wholegrain rolled oats. J. Cereal Sci. 50, 61–66.

Morris, 2016a. Cereals: overview of uses. Accent on wheat grain. Wrigley, C., Corke, H., Seetharaman, K., Faubion, J. (Eds.), Encyclopedia of Food Grains, vol. 3, Elsevier, Oxford, pp. 1–7.

Morris, 2016b. Evaluation of wheat-grain quality attributes. Wrigley, C., Corke, H., Seetharaman, K., Faubion, J. (Eds.), Encyclopedia of Food Grains, vol. 3, Elsevier, Oxford, pp. 251–256.

Moss, H.J., Randall, P.J., 1990. Some effects of temperature regime during grain filling in wheat. Austr. J. Agric. Res. 41, 603–617.

Munck, L., 2016. Utilization of the whole cereal plant to maximize its value. Wrigley, C., Corke, H., Seetharaman, K., Faubion, J. (Eds.), Encyclopedia of Food Grains, vol. 3, Elsevier, Oxford, pp. 228–234.

Mussatto, S.I., Dragone, G., Roberto, I.C., 2006. Brewers' spent grain: generation, characteristics and potential applications'. J. Cereal Sci. 43, 1–14.

National Academies of Sciences, Engineering, and Medicine, 2016, Genetically.

Nuttall, J.G., O'Leary, G.J., Panozzo, J.F., Walker, C.K., Barlow, K.M., Fitzgerald, G.J., 2016. Models of grain quality in wheat—A review. Field Crop. Res.

Nyström, L., Lampi, A.-M., Andersson, A.A.M., Kamal-Eldin, A., Gebruers, K., Courtin, C.M., Patil, R.M., Oak, M.D., Tamhankar, S.A., Rao, V.S., 2009. Molecular mapping of QTLs for gluten strength as measured by sedimentation volume and mixograph in durum wheat (*Triticum turgidum* L. ssp durum). J. Cereal Sci. 49, 378–386.

Panozzo, J.F., Eagles, H.A., 1998. Cultivar and environmental effects on quality characters in wheat. I. Starch, Aust. J. Agr. Res. 49, 757–766.

Patil, R.M., Oak, M.D., Tamhankar, S.A., Rao, V.S., 2009. Molecular mapping of QTLs for gluten strength as measured by sedimentation volume and mixograph in durum wheat (*Triticum turgidum* L. ssp durum). J. Cereal Sci. 49, 378–386.

Peiris, K.H.S., Bockus, W.W., Dowell, F.E., 2015. Near-infrared spectroscopic evaluation of single-kernel deoxynivalenol accumulation and *Fusarium* head blight resistance components in wheat. Cereal Chem. 93, 25–31.

Piironen, V., 2008. Phytochemicals and dietary fiber components in rye varieties in the HEALTHGRAIN diversity screen. J. Agric. Food Chem. 56, 9758–9766.

Pleijel, H., Danielsson, H., 2009. Yield dilution of grain Zn in wheat grown in open-top chamber experiments with elevated CO_2 and O_3 exposure. J. Cereal Sci. 50, 278–282.

Posner, E.S., 2009. Wheat flour milling. In: Khan, K., Shewry, P.R. (Eds.), Wheat Chemistry and Technology. fourth ed. AACC International, St. Paul, MN, pp. 119–152.

Posner, E.S., Hibbs, A.N., 2005. Wheat Flour Milling, second ed. American Association of Cereal Chemists, Inc, St Paul, MN.

Privalle, L., 2016. Development of genetically modified grains. Wrigley, C., Corke, H., Seetharaman, K., Faubion, J. (Eds.), Encyclopedia of Food Grains, vol. 4, Elsevier, Oxford, pp. 343–349.

Rajendran, S., Muralidharan, N., 2001. Performance of phosphine in fumigation of bagged paddy rice in indoor and outdoor stores. J. Stored Prod. Res. 37, 351–358.

Rooke, L., Békés, F., Fido, R., Barro, F., Gras, P., Tatham, A.S., Barcelo, P., Lazzeri, P., Shewry, P.R., 1999. Overexpression of a gluten protein in transgenic wheat results in greatly increased dough strength. J. Cereal Sci. 30, 115–120.

Rose, D.J., Ogden, L.V., Dunn, M.L., Pike, O.A., 2008. Enhanced lipid stability in whole wheat flour by lipase inactivation and antioxidant retention. Cereal Chem. 85, 218–223.

Rosell, C.M., Barro, F., Sousa, C., Mena, C., 2014. Cereals for developing gluten-free products and analytical tools for gluten detection. J. Cereal Sci. 59, 354–364.

Ross, A.S., Bettge, A.D., 2009. Passing the test on wheat end-use quality. In: Carver, B.F. (Ed.), Wheat: Science and Trade. Wiley-Blackwell, Ames, Iowa, pp. 455–493.

Sabanis, D., Tzia, C., 2009. Effect of rice, corn and soy flour addition on characteristics of bread produced from different wheat cultivars. Food Bioprocess Technol. 2, 68–79.

Saleh, M.I., Meullenet, J.F., Siebenmorgen, T.J., 2008. Development and validation of prediction models for rice surface lipid content and color parameters using near-infrared spectroscopy: a basis for predicting rice degree of milling. Cereal Chem. 85, 787–791.

Sanchez, O.J., Cardona, C.A., 2008. Trends in biotechnological production of fuel ethanol from different feedstocks. Bioresour. Technol. 99, 5270–5295.

Schmidt, J., Gergely, S., Schönlechner, R., Grausgruber, H., Tömösközi, S., Salgó, A., Berghofer, E., 2009. Comparison of different types of NIR instruments in ability to measure β-glucan content in naked barley. Cereal Chem. 86, 398–404.

Schulthess, A., Schwember, A.R., 2013. Improving durum wheat (*Triticum turgidum* L. var durum) grain yellow pigment content through plant breeding. Ciencia e Investigación Agraria 40 (3), 475–490.

Selomulyo, V.A., Zhou, W., 2007. Frozen bread dough: Effects of freezing storage and dough improvers. J. Cereal Sci. 45, 1–17.

Shen, Y., Jin, L., Ziao, P., Lu, Y., Bao, J., 2009. Total phenolics, flavonoids, antioxidant capacity in rice grain and their relations to grain color, size and weight. J. Food Sci. 49, 106–111.

Shewry, P.R., Tatham, A.S., 2015. Improving wheat to remove coeliac epitopes but retain functionality. J. Cereal Sci. 67, 12–21.

Shewry, P.R., Underwood, C., Wan, Y., Lovegrove, A., Bhandari, D., Toole, G., Mills, E.N.C., Denyer, K., Mitchell, R.A.C., 2009a. Storage product synthesis and accumulation in developing grains of wheat. J. Cereal Sci. 50, 106–112.

Shewry, P.R., D'Ovidio, R., Lafiandra, D., Jenkins, J.A., Mills, E., Bekes, F N.C., 2009b. Wheat grain proteins. In: Khan, K., Shewry, P.R. (Eds.), Wheat Chemistry and Technology. fourth ed. AACC International, St. Paul, MN, pp. 223–298.

Shewry, P.R., Mitchell, R.A.C., Tosia, P., Wan, Y., Underwood, C., Lovegrovea, A., Freeman, J., Toole, G.A., Mills, E.N.C., Ward, J.L., 2012. An integrated study of grain development of wheat (cv. Hereward). J. Cereal Sci. 56, 21–30.

Shillito, R.D., 2016. Detection of genetically modified grains. Wrigley, C., Corke, H., Seetharaman, K., Faubion, J. (Eds.), Encyclopedia of Food Grains, vol. 4, Elsevier, Oxford, pp. 350–359.

Siderer, Y., Maquet, A., Anklam E, 2005. Need for research to support consumer confidence in the growing organic food market. Trends Food Sci. Technol. 16, 332–343.

Singh, C.B., Jayas, D.S., Paliwal, J., White, N.D.G., 2009. Detection of sprouted and midge-damaged wheat kernels using near-infrared hyperspectral imaging. Cereal Chem. 86, 256–260.

Slavin, J., 2003. Why whole grains are protective: biological mechanisms. Proc. Nutr. Soc. 62, 129–134.

Solah, V.A., Menq, X., Kerr, D.A., Zhu, K., Devine, A., Prince, R., Binns, C.W., 2016. Effects of whole grain food consumption in older Australian women. Cereal Food World 61, 51–58.

Storozhenkol, S., Veerle De Brouwer, V., Volckaert, M., Navarretel, O., Blancquaert, D., Zhang, G.-F., Lambert, W., Van Der Straeten, D., 2007. Folate fortification of rice by metabolic engineering. Nat. Biotechnol. 25, 1279.

Suter, D.A.I., Fleming, F., 2008. Cereal food allergy issues, opportunities and the way forward for industry. Food Austr. 16, 523–525.

Tamas, L., Gras, P.W., Solomon, R.G., Morell, M.K., Appels, R., Bekes, F., 2002. Chain extension and termination as a function of cycteine content and length of central repetitive domains in storage proteins. J. Cereal Sci. 36, 313–325.

Tanner, G.J., Blundell, M.J., Colgrave, M.L., Howitt, C.A., 2016. Creation of the first ultra-low gluten barley (*Hordeum vulgare* L.) for coeliac and gluten-intolerant populations. Plant Biotechnol. J. 14, 1139–1150.

Taylor, J.R.N., Schober, T.J., Bean, S.R., 2006. Novel food and non-food uses for sorghum and millets. J. Cereal Sci. 44, 252–271.

Tiwari, V.K., Faris, J.D., Friebe, B., Gill, B.S., 2016. Genome mapping. Wrigley, C., Corke, H., Seetharaman, K., Faubion, J. (Eds.), Encyclopedia of Food Grains, vol. 4, Elsevier, Oxford, pp. 365–375.

Uthayakumaran, S., Zhao, F.J., Sivri, D., Roohani, M., Batey, I.L., Wrigley, C.W., 2007. Defect identification in wheat grain by micro-fluidic electrophoresis: sulfur deficiency and bug damage. Cereal Chem. 84, 301–303.

Varshney, R.K., Graner, A., Sorrells, M.E., 2005. Genomics-assisted breeding for crop improvement. Trends Plant Sci. 10, 621–630.

Vawser, M., Cornish, G.B., 2004. Over-expression of HMW glutenin subunit Glu-B1 7x in hexaploid wheat varieties (*Triticum aestivum*). Austr. J. Agric. Res. 55, 577–588.

Ward, J.L., Poutanen, K., Gebruers, K., Piironen, V., Lampi, A.M., Nyström, L., Andersson, A.A.M., Åman, P., Boros, D., Rakszegi, M., Bedö, Z., Shewry, P.R., 2008. The HEALTHGRAIN.

Whelan, B.M., Taylor, J.A., Hassall, J.A., 2009. Site-specific variation in wheat grain protein concentration and wheat grain yield measured on an Australian farm using harvester-mounted on-the-go sensors. Crop Pasture Sci. 60, 808–817.

Wieser, H., Koehler, P., 2008. The biochemical basis of celiac disease. Cereal Chem. 85, 1–13.

Wongsagonsup, R., Varavinit, S., BeMiller, J.N., 2008. Increasing slowly digestible starch content of normal and waxy maize starches and properties of starch products. Cereal Chem. 85, 738–745.

Woo, K.S., Maningat, C.C., Seib, P.A., 2009. Increasing dietary fiber in foods: the case for phosphorylated cross linked resistant starch, a highly concentrated form of dietary fiber. Cereal Food World 54, 217–223.

Wrigley, C.W., 2005. Precision agriculture—a means of improving grain quality? Cereal Food World 50, 143–144.

Wrigley, C.W., Asp, E.A., 2004. Nutrition: guidelines for grain-based foods. Wrigley, C., Corke, H., Walker, C. (Eds.), Encyclopedia of Grain Science, vol. 2, Elsevier, Oxford, pp. 325–328.

Wrigley, C.W., Asenstorfer, R., Batey, I.L., Cornish, G.B., Day, L., Mares, D., Mrva, K., 2009. The biochemical and molecular basis of wheat quality. In: Carver, B.F. (Ed.), Wheat: Science and Trade. Wiley-Blackwell, Ames, Iowa, pp. 495–520.

Wrigley, C., Corke, H., Seetharaman, K., Faubion, J., 2016. Encyclopedia of Food GrainsElsevier, Oxford, UK.

Xu, Y., Lu, Y., Xie, C., Gao, S., Wan, J., Prasanna, B.M., 2012. Whole-genome strategies for marker-assisted plant breeding. Mol. Breed. 29, 833–854.

Zhang, Y., Wu, Y., Xiao, Y., Yan, J., Zhang, Y., Zhang, Y., Ma, C., Xia, X., He, Z., 2009. QTL mapping for milling, gluten quality, and flour pasting properties in a recombinant inbred line population derived from a Chinese soft × hard wheat cross. Crop Pasture Sci. 60, 587–597.

Zhao, R., Bean, S.R., Wang, D., 2008a. Sorghum protein extraction by sonication and its relationship to ethanol fermentation. Cereal Chem. 85, 837–842.

Zhao, R., Bean, S.R., Wu, X., Wang, D., 2008b. Assessing fermentation quality of grain sorghum for fuel ethanol production using rapid visco-analyzer. Cereal Chem. 85, 830–836.

Appendix 1

Composition of Grains and Grain Products

Denise L. King, Bhaskarani Jasthi and Janet Pettit
Nutrition Coordinating Center, Epidemiology and Community Health, University of Minnesota, Minneapolis, MN, United States

A1.1 INTRODUCTION

The cereal grains are basic to the diet for most people in the world. However, there are major differences from one region to another and from one culture to another with respect to the specific cereal species that are processed and eaten. Many grains are under-utilised and offer unexplored new product opportunities. There is thus the need for nutritionists, food scientists and individuals in general to be aware of the advantages and deficiencies of specific cereals in our diet. Therefore, this appendix provides information (Tables A1.1–A1.6) about the nutrients in wholegrains of the cereal species described in this book, including major grains such as wheat, triticale, rye, barley, oats, corn, rice, sorghum and millet, as well as some less-common cereal grains such as bulgur, teff, couscous, kamut and spelt and 'pseudo-grains' amaranth, buckwheat and quinoa. Pseudo-grains are false grains because they have different botanical origins (non-grasses), but are included here because they are similar to true cereal grains in their nutritional profile and use. Seed grains are composed of bran, germ and starchy endosperm (flour) and when a grain is refined most of the bran and some of the germ are removed, resulting in nutrient losses. Nutrient composition is provided for milled products when available.

Data provided for products processed from grains go no further than the result of milling in these tables. Further processing may cause significant changes in the availability of nutrients, such as the partial destruction of some vitamins, or the increased availability of vitamins, as is the case of alkali treatment of maize meal. For example see Schakel et al. (2004) for the nutrient content of foods processed from grain. See also the individual chapters on each cereal species in this book.

Cereal Grains. http://dx.doi.org/10.1016/B978-0-08-100719-8.00026-7

TABLE A1.1 Nutrient Contents of Grains and Grain-Related Products From Wheat

Nutrients (per 100 g)	Wheat bran, dry	Wheat germ	Wheat, cracked whole wheat	Wheat, hard red spring, dry	Wheat, hard red winter, dry	Wheat, soft red winter, dry	Wheat, sprouted	White all-purpose flour, unenriched	Whole wheat flour
Energy (kJ)	904	1506	1368	1377	1368	1385	828	1523	1423
Total protein (g)	15.6	23.2	12.6	15.4	12.6	10.4	7.5	10.3	13.2
Tryptophan (g)	0.28	0.32	0.16	0.20	0.16	0.14	0.12	0.13	0.17
Threonine (g)	0.50	0.97	0.37	0.43	0.37	0.34	0.25	0.28	0.37
Isoleucine (g)	0.49	0.85	0.46	0.54	0.46	0.40	0.29	0.36	0.44
Leucine (g)	0.93	1.57	0.85	1.04	0.85	0.76	0.51	0.71	0.90
Lysine (g)	0.60	1.47	0.34	0.40	0.34	0.32	0.25	0.23	0.36
Methionine (g)	0.23	0.46	0.20	0.23	0.20	0.17	0.12	0.18	0.23
Cystine (g)	0.37	0.46	0.32	0.40	0.32	0.30	0.13	0.22	0.28
Phenylalanine (g)	0.60	0.93	0.59	0.72	0.59	0.51	0.35	0.52	0.68
Tyrosine (g)	0.44	0.70	0.39	0.44	0.39	0.33	0.28	0.31	0.28
Valine (g)	0.73	1.20	0.56	0.68	0.56	0.50	0.36	0.42	0.56
Arginine (g)	1.09	1.87	0.60	0.70	0.60	0.52	0.43	0.42	0.65
Histidine (g)	0.43	0.64	0.29	0.33	0.29	0.26	0.20	0.23	0.36
Alanine (g)	0.77	1.48	0.45	0.56	0.45	0.41	0.30	0.33	0.49
Aspartic acid (g)	1.13	2.07	0.64	0.81	0.64	0.56	0.45	0.44	0.72

Glutamic acid (g)	2.87	4.00	4.00	4.95	4.00	3.66	1.87	3.48	4.33
Glycine (g)	0.90	1.42	0.53	0.62	0.53	0.46	0.31	0.37	0.57
Proline (g)	0.88	1.23	1.29	1.68	1.29	1.17	0.67	1.20	2.08
Serine (g)	0.68	1.10	0.59	0.66	0.59	0.54	0.34	0.52	0.62
Total fat (g)	4.25	9.72	1.54	1.92	1.54	1.56	1.27	0.98	2.50
Total saturated fatty acids (g)	0.63	1.67	0.27	0.31	0.27	0.29	0.21	0.16	0.43
Total monounsaturated fatty acids (g)	0.64	1.37	0.20	0.30	0.20	0.18	0.15	0.09	0.28
Total polyunsaturated fatty acids (g)	2.21	6.01	0.63	0.77	0.63	0.66	0.56	0.41	1.17
Total carbohydrate (g)	64.5	51.8	71.2	68.0	71.2	74.2	42.5	76.3	72.0
Total dietary fibre (g)	42.8	13.2	12.2	12.2	12.2	12.5	1.1	3.0	10.7
Insoluble dietary fibre (g)	38.4	12.2	10.2	10.2	10.2	10.4	0.9	1.5	8.9
Soluble dietary fibre (g)	4.4	1.0	2.0	2.0	2.0	2.1	0.2	1.5	1.8
Pectins (g)	0.2	0.7	0					0	
Total sugars (g)	0.4	8.7	0.4	0.4	0.4	0.4	5.5	0.3	0.4
Glucose (g)	0.1	0.7	0	0	0	0	0.7	0	0
Fructose (g)	0.1	0.5	0.1	0.1	0.1	0.1	0.7	0	0.1
Sucrose (g)	0.3	7.4	0.4	0.4	0.4	0.4	4.1	0.2	0.4

(Continued)

TABLE A1.1 Nutrient Contents of Grains and Grain-Related Products From Wheat (*cont.*)

Nutrients (per 100 g)	Wheat bran, dry	Wheat germ	Wheat, cracked whole wheat	Wheat, hard red spring, dry	Wheat, hard red winter, dry	Wheat, soft red winter, dry	Wheat, sprouted	White all-purpose flour, unenriched	Whole wheat flour
Maltose (g)	0	0.1	0	0	0	0		0.1	0
Starch (g)	16.1	20.4	58.6	55.4	58.6	61.3	35.9	73.0	57.8
Ash (g)	5.8	4.2	1.6	1.9	1.6	1.7	1.0	0.5	1.6
Minerals									
Calcium (mg)	73	39	29	25	29	27	28	15	34
Iron (mg)	10.6	6.3	3.2	3.6	3.2	3.2	2.1	1.2	3.6
Magnesium (mg)	611	239	126	124	126	126	82	22	137
Phosphorus (mg)	1013	842	288	332	288	493	200	108	357
Potassium (mg)	1182	892	363	340	363	397	169	107	363
Zinc (mg)	7.3	12.3	2.7	2.8	2.7	2.6	1.7	0.7	2.6
Selenium (mcg)	77.6	79.2	70.7	70.7	70.7	60.8	42.5	33.9	61.8
Vitamins									
Thiamin (mg)	0.52	1.88	0.38	0.50	0.38	0.39	0.23	0.12	0.50
Riboflavin (mg)	0.58	0.50	0.12	0.11	0.12	0.10	0.16	0.04	0.17
Niacin (mg)	13.58	6.81	5.46	5.71	5.46	4.80	3.09	1.25	4.96
Pantothenic acid (mg)	2.18	2.26	0.95	0.94	0.95	0.85	0.95	0.44	0.60

Vitamin B-6 (mg)	1.30	1.30	0.30	0.34	0.30	0.27	0.27	0.04	0.41
Total folate (mcg)	79	281	38	43	38	41	38	20	44
Total vitamin A activity (IU)	9	103	9	9	9	0	21	0	9
α-Carotene (mcg)	0	0	0	0	0	0	6	0	0
β-Carotene (mcg)	6	62	5	5	5	0	6	0	5
Lutein + zeaxanthin (mcg)	240	790	220	220	220	0	0	18	220
Vitamin E (IU)	2.2	32.5	1.5	1.5	1.5	1.5	0.1	0.1	1.1
Vitamin K (mcg)	1.9	9.9	1.9	1.9	1.9	1.9	33.0	0.3	1.9
Others									
Phytic acid (mg)	3658	4088	721	724	721	729	348	83	740
Oxalic acid (mg)	457	84	62	63	62	63	67	40	64

TABLE A1.2 Nutrient Contents of Grains and Grain-Related Products From Rye, Teff, Triticale, Bulgur, Couscous, Kamut and Spelt

Nutrients (per 100 g)	Rye flour	Rye, whole grain, dry	Teff, dry	Triticale flour, wholegrain	Bulgur, dry	Couscous, dry	Kamut, flour	Spelt, dry
Energy (kJ)	1460	1414	1536	1414	1431	1573	1410	1414
Total protein (g)	10.9	10.3	13.3	13.2	12.3	12.8	14.5	14.6
Tryptophan (g)	0.12	0.11	0.14	0.16	0.19	0.16	0.13	0.13
Threonine (g)	0.34	0.29	0.51	0.41	0.35	0.34	0.44	0.44
Isoleucine (g)	0.25	0.21	0.50	0.48	0.46	0.49	0.57	0.55
Leucine (g)	0.61	0.56	1.07	0.92	0.83	0.87	1.11	1.07
Lysine (g)	0.21	0.29	0.38	0.37	0.34	0.25	0.42	0.41
Methionine (g)	0.14	0.15	0.43	0.21	0.19	0.20	0.25	0.26
Cystine (g)	0.19	0.33	0.24	0.28	0.29	0.36	0.31	0.33
Phenylalanine (g)	0.49	0.44	0.70	0.64	0.58	0.62	0.77	0.74
Tyrosine (g)	0.18	0.20	0.46	0.39	0.36	0.34	0.35	0.38
Valine (g)	0.35	0.32	0.69	0.62	0.55	0.54	0.69	0.68
Arginine (g)	0.47	0.45	0.52	0.68	0.58	0.47	0.70	0.69
Histidine (g)	0.16	0.19	0.30	0.31	0.29	0.26	0.38	0.36
Alanine (g)	0.41	0.41	0.75	0.49	0.44	0.37	0.54	0.53
Aspartic acid (g)	0.54	0.56	0.82	0.79	0.63	0.52	0.80	0.76

Glutamic acid (g)	2.53	2.29	3.35	4.05	3.88	4.60	4.84	4.65
Glycine (g)	0.43	0.42	0.48	0.57	0.50	0.40	0.58	0.60
Proline (g)	0.85	0.80	0.66	1.20	1.28	1.40	1.59	1.63
Serine (g)	0.48	0.46	0.62	0.60	0.58	0.60	0.76	0.73
Total fat (g)	1.5	1.6	2.4	1.8	1.3	0.6	2.1	2.4
Total saturated fatty acids (g)	0.18	0.20	0.45	0.32	0.23	0.12	0.20	0.41
Total monounsaturated fatty acids (g)	0.19	0.21	0.59	0.18	0.17	0.09	0.21	0.45
Total polyunsaturated fatty acids (g)	0.70	0.77	1.07	0.79	0.54	0.25	0.62	1.26
Total carbohydrate (g)	75.4	75.9	73.1	73.1	75.9	77.4	70.6	70.2
Total dietary fibre (g)	11.8	15.1	8.0	14.6	18.3	5.0	11.1	10.7
Insoluble dietary fibre (g)	7.9	12.0	7.2	13.0	15.2	3.9	9.3	9.4
Soluble dietary fibre (g)	3.9	3.1	0.8	1.6	3.1	1.1	1.8	1.3
Pectins (g)								
Total sugars (g)	1.1	1.0	1.8	1.0	0.4	2.1	7.8	6.8
Glucose (g)	0.1	0.2	0.7	0.1	0	0.3	0.6	0.8
Fructose (g)	0.1	0.1	0.5	0.1	0	0.3	0.1	0.2
Sucrose (g)	0.9	0.7	0.6	0.4	0.3	1.5	0.6	0.6
Maltose (g)	0	0	0	0.4	0	0	6.6	5.2

(Continued)

TABLE A1.2 Nutrient Contents of Grains and Grain-Related Products From Rye, Teff, Triticale, Bulgur, Couscous, Kamut and Spelt (*cont.*)

Nutrients (per 100 g)	Rye flour	Rye, whole grain, dry	Teff, dry	Triticale flour, wholegrain	Bulgur, dry	Couscous, dry	Kamut, flour	Spelt, dry
Starch (g)	50.6	49.6	36.6	58.0	56.7	69.3	52.4	53.9
Ash (g)	1.2	1.6	2.4	1.9	1.5	0.6	1.7	1.8
Minerals								
Calcium (mg)	24	24	180	35	35	24	22	27
Iron (mg)	2.5	2.6	7.6	2.6	2.5	1.1	3.8	4.4
Magnesium (mg)	63	110	184	153	164	44	130	136
Phosphorus (mg)	225	332	429	321	300	170	364	401
Potassium (mg)	374	510	427	466	410	166	403	388
Zinc (mg)	2.2	2.7	3.6	2.7	1.9	0.8	3.7	3.3
Selenium (mcg)	14.4	13.9	4.4	67.2	2.3	11.0	81.5	11.7
Vitamins								
Thiamin (mg)	0.29	0.32	0.39	0.38	0.23	0.16	0.57	0.36
Riboflavin (mg)	0.11	0.25	0.27	0.13	0.12	0.08	0.18	0.11
Niacin (mg)	1.73	4.27	3.36	2.86	5.11	3.49	6.38	6.84
Pantothenic acid (mg)	0.49	1.46	0.94	2.17	1.05	1.24	0.95	1.07
Vitamin B-6 (mg)	0.27	0.29	0.48	0.40	0.34	0.11	0.26	0.23
Total folate (mcg)	34	38	85	74	27	20	43	45

Total vitamin A activity (IU)	0	11	9	10	9	3	10	10
α-Carotene (mcg)	0	0	0	0	0	0	2	0
β-Carotene (mcg)	0	7	5	6	5	2	5	5
Lutein + zeaxanthin (mcg)	210	210	66	215	220	85	301	169
Vitamin E (IU)	2.1	1.3	0.1	1.3	0.1	0.2	0.9	1.2
Vitamin K (mcg)	5.9	5.9	1.9	3.9	1.9	1.0	1.8	3.6
Others								
Phytic acid (mg)	890	703	528	603	1512	325	739	745
Oxalic acid (mg)	51	32	45	65	59	13	67	36

TABLE A1.3 Nutrient Contents of Grains and Grain-Related Products From Oats and Rice

Nutrients (per 100 g)	Oat bran, dry	Oat flour, wholegrain	Oatmeal, dry	Brown rice flour	Brown rice, raw	Rice bran, dry	White rice flour	White rice, raw
Energy (kJ)	1029	1586	1586	1519	1548	1322	1531	1527
Total protein (g)	17.3	13.2	13.2	7.2	7.9	13.4	6.00	7.1
Tryptophan (g)	0.34	0.18	0.18	0.09	0.10	0.11	0.07	0.08
Threonine (g)	0.50	0.38	0.38	0.27	0.29	0.56	0.21	0.26
Isoleucine (g)	0.67	0.50	0.50	0.31	0.34	0.57	0.24	0.31
Leucine (g)	1.37	0.98	0.98	0.60	0.66	1.02	0.49	0.59
Lysine (g)	0.76	0.64	0.64	0.28	0.30	0.65	0.21	0.26
Methionine (g)	0.34	0.21	0.21	0.16	0.18	0.31	0.14	0.17
Cystine (g)	0.58	0.46	0.46	0.09	0.10	0.32	0.11	0.15
Phenylalanine (g)	0.91	0.67	0.67	0.37	0.41	0.64	0.32	0.38
Tyrosine (g)	0.67	0.40	0.40	0.27	0.30	0.41	0.31	0.24
Valine (g)	0.96	0.69	0.69	0.42	0.47	0.88	0.35	0.44
Arginine (g)	1.28	0.85	0.85	0.55	0.60	1.06	0.52	0.59
Histidine (g)	0.41	0.28	0.28	0.18	0.20	0.36	0.15	0.17
Alanine (g)	0.87	0.56	0.56	0.42	0.46	0.97	0.33	0.41
Aspartic acid (g)	1.58	1.22	1.22	0.68	0.74	1.31	0.55	0.67
Glutamic acid (g)	3.75	2.83	2.83	1.47	1.62	1.85	1.10	1.39
Glycine (g)	0.95	0.64	0.64	0.36	0.39	0.88	0.27	0.33
Proline (g)	0.98	0.45	0.45	0.34	0.37	0.67	0.28	0.34

Serine (g)	0.89	0.71	0.71	0.37	0.41	0.66	0.31	0.38
Total fat (g)	7.0	6.5	6.5	2.8	2.9	20.9	1.4	0.7
Total saturated fatty acids (g)	1.33	1.11	1.11	0.56	0.58	4.17	0.39	0.18
Total monounsaturated fatty acids (g)	2.38	1.98	1.98	1.01	1.06	7.55	0.44	0.21
Total polyunsaturated fatty acids (g)	2.77	2.30	2.30	1.00	1.04	7.46	0.38	0.18
Total carbohydrate (g)	66.2	67.7	67.7	76.5	77.2	49.7	80.1	80.0
Total dietary fibre (g)	15.4	10.1	10.1	4.6	9.5	21.0	2.4	1.3
Insoluble dietary fibre (g)	9.6	5.2	5.2	4.0	8.3	19.1	2.4	1.3
Soluble dietary fibre (g)	5.8	4.9	4.9	0.6	1.3	1.9	0	0
Pectins (g)	0.3	0.5	0.5					0
Total sugars (g)	1.5	1.0	1.0	0.9	0.9	0.9	0.1	0.1
Glucose (g)	0.1	0	0	0	0	0.2	0	0
Fructose (g)	0.1	0	0	0	0	0.2	0	0
Sucrose (g)	1.1	1.0	1.0	0.8	0.9	0.5	0	0.1
Maltose (g)	0	0	0	0	0	0	0	0
Starch (g)	45.9	57.9	57.9	65.4	64.9	27.0	77.6	0
Ash (g)	2.9	1.8	1.8	1.5	1.5	10.0	0.6	0.6
Minerals								
Calcium (mg)	58	52	52	11	23	57	10	28
Iron (mg)	5.4	4.3	4.3	2.0	1.5	18.5	0.4	4.3
Magnesium (mg)	235	138	138	112	143	781	35	25

(Continued)

TABLE A1.3 Nutrient Contents of Grains and Grain-Related Products From Oats and Rice (*cont.*)

Nutrients (per 100 g)	Oat bran, dry	Oat flour, wholegrain	Oatmeal, dry	Brown rice flour	Brown rice, raw	Rice bran, dry	White rice flour	White rice, raw
Phosphorus (mg)	734	410	410	337	333	1677	98	115
Potassium (mg)	566	362	362	289	223	1485	76	115
Zinc (mg)	3.1	3.6	3.6	2.5	2.0	6.0	0.8	1.1
Selenium (mcg)	45.2	28.9	28.9	8.1	23.4	15.6	15.1	15.1
Vitamins								
Thiamin (mg)	1.17	0.46	0.46	0.44	0.40	2.75	0.14	0.58
Riboflavin (mg)	0.22	0.16	0.16	0.08	0.09	0.28	0.02	0.05
Niacin (mg)	0.93	1.13	1.13	6.34	5.09	34.00	2.59	4.19
Pantothenic acid (mg)	1.49	1.12	1.12	1.59	1.49	7.39	0.82	1.01
Vitamin B-6 (mg)	0.17	0.10	0.10	0.74	0.51	4.07	0.44	0.16
Total folate (mcg)	52	32	32	16	20	63	4	231
Total vitamin A activity (IU)	0	0	0	0	0	0	0	0
α-Carotene (mcg)	0	0	0	0	0	0	0	0
β-Carotene (mcg)	0	0	0	0	0	0	0	0
Lutein + zeaxanthin (mcg)	180	180	180	0	0	220	0	0
Vitamin E (IU)	1.5	0.6	0.6	1.8	1.8	7.3	0.2	0.2
Vitamin K (mcg)	3.2	2.0	2.0	1.9	1.9	1.9	0	0.1
Others								
Phytic acid (mg)	1399	680	680	920	521	6550	486	145
Oxalic acid (mg)	55	18	18	37	22	123	5	5

TABLE A1.4 Nutrient Contents of Grains and Grain-Related Products From Corn

Nutrients (per 100 g)	Corn bran, dry	Corn flour, wholegrain, blue	Corn flour, wholegrain, white	Corn flour, wholegrain, yellow	Corn grain, white, dry	Corn grain, yellow, dry	Cornmeal, wholegrain, white, dry	Cornmeal, wholegrain, yellow, dry
Energy (kJ)	937	1523	1510	1510	1527	1527	1515	1515
Total protein (g)	8.4	8.8	6.9	6.9	9.4	9.4	8.1	8.1
Tryptophan (g)	0.06	0.05	0.05	0.05	0.07	0.07	0.06	0.06
Threonine (g)	0.32	0.22	0.25	0.26	0.35	0.35	0.31	0.31
Isoleucine (g)	0.30	0.30	0.25	0.25	0.34	0.34	0.29	0.29
Leucine (g)	1.03	1.23	0.85	0.85	1.16	1.16	1.00	1.00
Lysine (g)	0.24	0.14	0.20	0.20	0.27	0.27	0.23	0.23
Methionine (g)	0.18	0.20	0.15	0.15	0.20	0.20	0.17	0.17
Cystine (g)	0.15	0.19	0.13	0.13	0.17	0.17	0.15	0.15
Phenylalanine (g)	0.41	0.45	0.34	0.34	0.46	0.46	0.40	0.40
Tyrosine (g)	0.34	0.24	0.23	0.28	0.38	0.38	0.33	0.33
Valine (g)	0.42	0.42	0.35	0.35	0.48	0.48	0.41	0.41
Arginine (g)	0.41	0.31	0.35	0.35	0.47	0.47	0.41	0.41
Histidine (g)	0.26	0.22	0.21	0.21	0.29	0.29	0.25	0.25
Alanine (g)	0.62	0.69	0.52	0.52	0.71	0.71	0.61	0.61
Aspartic acid (g)	0.58	0.57	0.43	0.48	0.66	0.66	0.57	0.57
Glutamic acid (g)	1.57	1.78	1.30	1.30	1.77	1.77	1.53	1.53

(*Continued*)

TABLE A1.4 Nutrient Contents of Grains and Grain-Related Products From Corn (*cont.*)

Nutrients (per 100 g)	Corn bran, dry	Corn flour, wholegrain, blue	Corn flour, wholegrain, white	Corn flour, wholegrain, yellow	Corn grain, white, dry	Corn grain, yellow, dry	Cornmeal, wholegrain, white, dry	Cornmeal, wholegrain, yellow, dry
Glycine (g)	0.35	0.27	0.28	0.28	0.39	0.39	0.33	0.33
Proline (g)	0.73	0.91	0.61	0.61	0.82	0.82	0.71	0.71
Serine (g)	0.39	0.42	0.33	0.33	0.45	0.45	0.39	0.39
Total fat (g)	0.9	5.1	3.9	3.9	4.7	4.7	3.6	3.6
Total saturated fatty acids (g)	0.13	0.63	0.54	0.54	0.67	0.67	0.51	0.51
Total monounsaturated fatty acids (g)	0.24	1.28	1.02	1.02	1.25	1.25	0.95	0.95
Total polyunsaturated fatty acids (g)	0.42	2.57	1.76	1.76	2.16	2.16	1.64	1.64
Total carbohydrate (g)	85.6	73.9	76.9	76.9	74.3	74.3	76.9	76.9
Total dietary fibre (g)	79.0	8.4	7.3	7.3	7.1	7.3	7.3	7.3
Insoluble dietary fibre (g)	75.3	7.4	6.4	6.4	6.2	6.4	6.4	6.4
Soluble dietary fibre (g)	2.0	1.0	0.9	0.9	0.8	0.9	0.9	0.9
Pectins (g)	2.1	0.1	0.1	0.1	0.1	0.2	0.2	0.2
Total sugars (g)	0.0	1.6	0.6	0.6	0.6	0.6	0.6	0.6

Glucose (g)	0	0.1	0.2	0.2	0.2	0.2	0.2	0.2
Fructose (g)	0	0.1	0.1	0.1	0.1	0.1	0.1	0.1
Sucrose (g)	0	1.2	0.3	0.3	0.3	0.3	0.3	0.3
Maltose (g)	0	0.2	0.1	0.1	0.1	0.1	0.1	0.1
Starch (g)	6.6	63.9	66.4	66.4	66.6	66.3	69.0	69.0
Ash (g)	0.4	1.4	1.5	1.5	1.2	1.2	1.1	1.1
Minerals								
Calcium (mg)	42	5	7	7	7	7	6	6
Iron (mg)	2.8	1.7	2.4	2.4	2.7	2.7	3.5	3.5
Magnesium (mg)	64	110	93	93	127	127	127	127
Phosphorus (mg)	72	263	272	272	210	210	241	241
Potassium (mg)	44	381	315	315	287	287	287	287
Zinc (mg)	1.6	2.2	1.7	1.7	2.2	2.2	1.8	1.8
Selenium (mcg)	16.5	2.2	15.4	15.4	15.5	15.5	15.5	15.5
Vitamins								
Thiamin (mg)	0.01	0.16	0.25	0.25	0.39	0.39	0.39	0.39
Riboflavin (mg)	0.10	0.23	0.08	0.08	0.20	0.20	0.20	0.20
Niacin (mg)	2.74	2.60	1.90	1.90	3.63	3.63	3.63	3.63
Pantothenic acid (mg)	0.64	0.55	0.66	0.66	0.42	0.42	0.43	0.43
Vitamin B-6 (mg)	0.15	0.47	0.37	0.37	0.62	0.62	0.30	0.30
Total folate (mcg)	4	18	18	18	17	19	18	25

(*Continued*)

TABLE A1.4 Nutrient Contents of Grains and Grain-Related Products From Corn (*cont.*)

Nutrients (per 100 g)	Corn bran, dry	Corn flour, wholegrain, blue	Corn flour, wholegrain, white	Corn flour, wholegrain, yellow	Corn grain, white, dry	Corn grain, yellow, dry	Cornmeal, wholegrain, white, dry	Cornmeal, wholegrain, yellow, dry
Total vitamin A activity (IU)	71	0	3	214	0	214	3	214
α-Carotene (mcg)	21	0	0	63	0	63	0	63
β-Carotene (mcg)	32	0	1	97	0	97	1	97
Lutein + zeaxanthin (mcg)	1355	90	5	1355	0	1355	5	1355
Vitamin E (IU)	0.6	1.0	0.6	0.6	0.8	0.7	0.6	0.6
Vitamin K (mcg)	0.3	0.1	0.3	0.3	0.3	0.3	0.3	0.3
Others								
Phytic acid (mg)	50	940	940	940	908	940	940	940
Oxalic acid (mg)	7	39	39	39	37	39	39	39

TABLE A1.5 Nutrient Contents of Grains and Grain-Related Products From Barley, Millet and Sorghum

Nutrients (per 100 g)	Barley (pearled), dry	Barley flour	Barley flour, wholegrain	Barley, hulled	Millet flour	Millet, dry	Sorghum flour, wholegrain	Sorghum, dry
Energy (kJ)	1473	1443	1439	1481	1561	1582	1502	1377
Total protein (g)	9.9	10.5	10.0	12.5	10.8	11.0	8.4	10.6
Tryptophan (g)	0.17	0.18	0.15	0.21	0.17	0.12	0.11	0.12
Threonine (g)	0.34	0.36	0.34	0.42	0.35	0.35	0.31	0.35
Isoleucine (g)	0.36	0.38	0.35	0.46	0.47	0.47	0.31	0.43
Leucine (g)	0.67	0.71	0.69	0.85	1.54	1.40	1.09	1.49
Lysine (g)	0.37	0.39	0.37	0.47	0.14	0.21	0.17	0.23
Methionine (g)	0.19	0.20	0.19	0.24	0.32	0.22	0.15	0.17
Cystine (g)	0.22	0.23	0.21	0.28	0.18	0.21	0.17	0.13
Phenylalanine (g)	0.56	0.59	0.55	0.70	0.68	0.58	0.44	0.55
Tyrosine (g)	0.28	0.30	0.29	0.36	0.33	0.34	0.23	0.32
Valine (g)	0.49	0.52	0.48	0.61	0.58	0.58	0.39	0.56
Arginine (g)	0.50	0.53	0.50	0.63	0.37	0.38	0.33	0.36
Histidine (g)	0.22	0.24	0.22	0.28	0.26	0.24	0.17	0.25
Alanine (g)	0.39	0.41	0.43	0.49	1.28	0.99	0.76	1.03
Aspartic acid (g)	0.62	0.66	0.65	0.78	0.71	0.73	0.56	0.74
Glutamic acid (g)	2.59	2.74	2.61	3.26	2.60	2.40	1.74	2.44
Glycine (g)	0.36	0.38	0.37	0.45	0.27	0.29	0.31	0.35
Proline (g)	1.18	1.25	1.16	1.48	0.91	0.88	0.65	0.85
Serine (g)	0.42	0.44	0.43	0.53	0.78	0.64	0.41	0.46

(Continued)

TABLE A1.5 Nutrient Contents of Grains and Grain-Related Products From Barley, Millet and Sorghum (*cont.*)

Nutrients (per 100 g)	Barley (pearled), dry	Barley flour	Barley flour, wholegrain	Barley, hulled	Millet flour	Millet, dry	Sorghum flour, wholegrain	Sorghum, dry
Total fat (g)	1.2	1.6	1.7	2.3	4.3	4.2	3.3	3.5
Total saturated fatty acids (g)	0.24	0.34	0.36	0.48	0.54	0.72	0.53	0.61
Total monounsaturated fatty acids (g)	0.15	0.21	0.22	0.30	0.92	0.77	0.94	1.13
Total polyunsaturated fatty acids (g)	0.56	0.77	0.83	1.11	2.62	2.13	1.40	1.56
Total carbohydrate (g)	77.7	74.5	77.0	73.5	73.1	72.9	76.6	72.1
Total dietary fibre (g)	15.6	10.1	16.7	17.3	3.5	8.5	6.6	6.7
Insoluble dietary fibre (g)	12.2	3.5	13.9	14.3	3.2	7.7	6.2	6.3
Soluble dietary fibre (g)	3.4	6.6	2.8	3.0	0.3	0.8	0.5	0.5
Pectins (g)			0.2	0.6				
Total sugars (g)	1.3	0.8	0.9	0.8	1.7	1.5	1.9	2.5
Glucose (g)	0.3	0	0.1	0.1	1.3	0.6	1.3	1.1
Fructose (g)	0.2	0	0.1	0.1	0	0.2	0.2	1.0
Sucrose (g)	0.6	0.4	0.4	0.5	0.3	0.6	0.5	0.3
Maltose (g)	0.2	0.2	0.2	0.2	0.1	0.1	0	0.2
Starch (g)	62.2	56.4	54.3	41.4	69.9	62.9	68.0	60.7
Ash (g)	1.1	1.3	1.6	2.3	1.2	3.3	1.3	1.4
Minerals								
Calcium (mg)	29	32	52	33	14	8	12	13

Iron (mg)	2.5	2.7	2.4	3.6	3.9	3.0	3.1	3.4
Magnesium (mg)	79	96	100	133	119	114	123	165
Phosphorus (mg)	221	296	235	264	285	285	278	289
Potassium (mg)	280	309	310	452	224	195	324	363
Zinc (mg)	2.1	2.0	2.4	2.8	2.6	1.7	1.6	1.7
Selenium (mcg)	37.7	37.7	32.6	37.7	32.7	2.7	12.2	12.2
Vitamins								
Thiamin (mg)	0.19	0.37	0.35	0.65	0.41	0.42	0.33	0.33
Riboflavin (mg)	0.11	0.11	0.17	0.29	0.07	0.29	0.06	0.10
Niacin (mg)	4.60	6.27	3.98	4.60	6.02	4.72	4.50	3.69
Pantothenic acid (mg)	0.28	0.15	0.25	0.28	1.27	0.85	0.54	0.37
Vitamin B-6 (mg)	0.26	0.40	0.24	0.32	0.37	0.38	0.33	0.44
Total folate (mcg)	23	8	18	19	42	85	25	20
Total vitamin A activity (IU)	22	0	19	22	43	43	0	33
α-Carotene (mcg)	0	0	0	0	0	0	0	0
β-Carotene (mcg)	13	0	11	13	26	26	0.5	20
Lutein + zeaxanthin (mcg)	160	160	138	160	197	220	12	66
Vitamin E (IU)	0	0.9	0.5	0.9	0.2	0.1	0.8	0.8
Vitamin K (mcg)	2.2	2.2	1.9	2.2	0.8	0.9	6.4	0.9
Others								
Phytic acid (mg)	480	555	467	656	950	511	513	810
Oxalic acid (mg)	12	56	18	25	24	45	54	45

TABLE A1.6 Nutrient Contents of Grains and Grain-Related Products From Amaranth, Buckwheat and Quinoa

Nutrients (per 100 g)	Amaranth, dry	Buckwheat whole groat, dry	Buckwheat whole groat flour	Quinoa flour	Quinoa, dry
Energy (kJ)	1552	1448	1402	1540	1540
Total protein (g)	13.6	11.7	12.6	14.1	14.1
Tryptophan (g)	0.17	0.17	0.18	0.17	0.17
Threonine (g)	0.53	0.45	0.48	0.42	0.42
Isoleucine (g)	0.56	0.44	0.47	0.50	0.50
Leucine (g)	0.84	0.74	0.79	0.84	0.84
Lysine (g)	0.72	0.60	0.64	0.77	0.77
Methionine (g)	0.22	0.15	0.16	0.31	0.31
Cystine (g)	0.18	0.20	0.22	0.20	0.20
Phenylalanine (g)	0.52	0.46	0.50	0.59	0.59
Tyrosine (g)	0.32	0.21	0.23	0.27	0.27
Valine (g)	0.65	0.60	0.65	0.59	0.59
Arginine (g)	1.02	0.87	0.94	1.09	1.09
Histidine (g)	0.37	0.27	0.29	0.41	0.41
Alanine (g)	0.77	0.66	0.71	0.59	0.59
Aspartic acid (g)	1.21	1.00	1.08	1.13	1.13
Glutamic acid (g)	2.16	1.81	1.95	1.87	1.87
Glycine (g)	1.57	0.91	0.98	0.69	0.69
Proline (g)	0.67	0.45	0.48	0.77	0.77

Serine (g)	1.10	0.61	0.65	0.57	0.57
Total fat (g)	7.0	2.7	3.1	6.1	6.1
Total saturated fatty acids (g)	1.46	0.59	0.68	0.71	0.71
Total monounsaturated fatty acids (g)	1.69	0.83	0.95	1.61	1.61
Total polyunsaturated fatty acids (g)	2.78	0.83	0.95	3.29	3.29
Total carbohydrate (g)	65.3	75.0	70.6	64.2	64.2
Total dietary fibre (g)	6.7	10.3	10.0	7.0	7.0
Insoluble dietary fibre (g)	3.8	3.2	5.1	6.1	6.1
Soluble dietary fibre (g)	2.9	7.1	4.9	0.9	0.9
Pectins (g)				0	
Total sugars (g)	1.7	1.9	2.6	6.1	6.1
Glucose (g)	0.3	0.2	0.2	2.8	2.8
Fructose (g)	0	0	0	0.6	0.6
Sucrose (g)	1.4	1.7	1.7	2.7	2.7
Maltose (g)	0	0	0	0	0
Starch (g)	57.3	62.8	58.0	52.2	52.2
Ash (g)	2.9	2.2	2.5	2.4	2.4
Minerals					
Calcium (mg)	159	17	41	47	47
Iron (mg)	7.6	2.5	4.1	4.6	4.6
Magnesium (mg)	248	221	251	197	197

(Continued)

TABLE A1.6 Nutrient Contents of Grains and Grain-Related Products From Amaranth, Buckwheat and Quinoa (*cont.*)

Nutrients (per 100 g)	Amaranth, dry	Buckwheat whole groat, dry	Buckwheat whole groat flour	Quinoa flour	Quinoa, dry
Phosphorus (mg)	557	319	337	457	457
Potassium (mg)	508	320	577	563	563
Zinc (mg)	2.9	2.4	3.1	3.1	3.1
Selenium (mcg)	18.7	8.4	5.7	8.5	8.5
Vitamins					
Thiamin (mg)	0.12	0.22	0.42	0.36	0.36
Riboflavin (mg)	0.20	0.27	0.19	0.32	0.32
Niacin (mg)	0.92	5.14	6.15	1.52	1.52
Pantothenic acid (mg)	1.46	1.23	0.44	0.77	0.77
Vitamin B-6 (mg)	0.59	0.35	0.58	0.49	0.49
Total folate (mcg)	82	42	54	184	184
Total vitamin A activity (IU)	2	0	0	14	14
α-Carotene (mcg)	0	0	0	0	0
β-Carotene (mcg)	1	0	0	8	8
Lutein + zeaxanthin (mcg)	28	220	220	163	163
Vitamin E (IU)	1.8	0.5	0.5	3.6	3.6
Vitamin K (mcg)	0	7.0	7.0	0	0
Others					
Phytic acid (mg)	410	1072	1957	1180	1180
Oxalic acid (mg)	229	143	269	184	184

A1.2 PREPARATION OF THE NUTRIENT TABLES

The grain nutrient-composition tables (Tables A1.1–A1.6) were developed by the Nutrition Coordinating Center (NCC) at the University of Minnesota, Minneapolis. They contain values compiled from the US Department of Agriculture (USDA) National Nutrient Database for Standard Reference, the scientific literature, manufacturers' information, other food databases and also imputed values. Wherever analytical values were not available, they were derived using the standardised procedures. They were derived from a different form of the same food (e.g. raw vs. cooked), from a similar food belonging to the same genus or family or by calculation from ingredients of a recipe (Schakel et al., 1997). Nutrient data are provided on an 'as is' basis (say 14% moisture content), unless indicated as 'dry basis'. Protein values were calculated from the amount of nitrogen in the food, using a conversion factor specific for each food group.

A1.3 NUTRIENT CONTENTS OF GRAIN AND GRAIN-RELATED PRODUCTS

Although a more realistic approach would be to provide a median value and a normal range for each combination of nutrient and grain material, given the large amount of information provided in these tables, it is possible to provide only one entry for each combination of nutrient and material. However, it must be realised that there is a large extent of biological diversity for the materials involved, and variation in growing conditions and cultivars that make a single numerical figure inadequate in describing any combination of nutrient and material. Therefore, it is important that these data are interpreted as being indicators of the general range for nutrient status.

For example the protein content of wheat grain can range from below 8% to over 16%, but in a set of tables like this, it has been necessary to provide only one estimate for each type of grain. While Table A1.1 gives the protein content as 10.4% for soft red winter wheat, 12.6% for hard red winter and 15.4% for hard red spring wheat, these values are approximations based on available data and cannot necessarily be applied to a specific wheat sample. As is indicated in Chapters 3 and 5, wheat is intentionally grown and segregated to provide consignments of grain with protein contents in specific and distinct ranges, thus to suit the various specifications of food manufacturers. On the other hand, the tables clearly indicate that rice and corn grain have lower protein contents in general than wheat and rye.

When utilising data from Tables A1.1–A1.6, please refer to the 2015 Nutrition Coordinating Center (NCC) Food and Nutrient Database, which is developed and maintained by NCC, located at the Division of Epidemiology and Community Health in Minneapolis, University of Minnesota.

SOURCES OF FURTHER INFORMATION

A wider range of nutrient information is available from the Nutrition Coordinating Center, Division of Epidemiology and Community Health, School of Public Health, 1300 South Second Street, Suite 300, Minneapolis, Minnesota 55454-1015, United States and from the website: http://www.ncc.umn.edu/.

The USDA National Nutrient Database for Standard Reference from the website: http://www.ars.usda.gov/ba/bhnrc/ndl.

REFERENCES

Schakel, S., Buzzard, I., Gebhardt, S., 1997. Procedures for estimating nutrient values for food composition databases. J. Food Comp. Anal. 10, 102–114.

Schakel, S.F., Van Heel, N., Harnack, J., 2004. Grain composition tables. Wrigley, C., Corke, H., Walker, C. (Eds.), Encyclopedia of Grain Science, vol. 3, Elsevier, Oxford, pp. 431–447.

Appendix 2

Equivalence Between Metric, US and Chinese Units for the Grain Industry

Colin Wrigley
QAAFI, University of Queensland, St. Lucia, QLD, Australia

A2.1 INTRODUCTION

Effective trading requires general agreement about units for weights and measures, whether for grain, processed foods or any other commodity. Although this necessary agreement may be established within regions, uniformity is still lacking internationally, specifically between nations using Imperial units (US or British versions) and those who have adopted the SI system of metric units (*Systeme Internationale d'Unites*). This appendix answers some questions about the strange diversity of units. It also provides tables for conversion between metric, US and Chinese units (Tables A2.1–A2.3).

Examples in the grain industry include questions such as the difference between a 'ton' and a 'tonne'. Is a bushel a measure of volume or of mass (weight); why and how does the size of the US bushel differ between grain species? How large is a 'section' in the USA? Why are there 100 pounds in the American hundredweight, but 112 pounds in the British hundredweight? The unit known as an 'ounce' differs in quantity depending on whether it is a measure of mass or of volume; if volume, whether it is a dry or liquid measure; if a weight, whether it is an ounce Troy, an Apothecaries' ounce or an ounce Avoirdupois! The old British unit of a 'barleycorn' appears to be particularly appropriate to an article on units for the grain industry. What is a 'barleycorn'? Furthermore, there are also strange and confusing terms in the many traditional Chinese units; for example the similar terms (e.g. the fen [分] and the li [厘]) are used for units of length, area and mass. See Table A2.3 for clarification.

Cereal Grains. http://dx.doi.org/10.1016/B978-0-08-100719-8.00027-9

TABLE A2.1 Conversion From Metric to US Units

Metric units		US units
Mass		
1 gram (g)	=	0.0353 ounce
1 kilogram (kg)	=	2.205 pounds (lb)
1 tonne (t) (metric ton)	=	1.102 US short ton (a short ton is 2,000 lb)
1 tonne (t)	=	0.9843 US long ton (a long ton is 2,240 lb)
50 kg	=	1 US bag of grain
1 quintal (q) (100 kg)	=	220.5 pounds
Volume		
1 cubic centimetre (cc)	=	0.0610 cubic inch
1 millilitre (mL)	=	0.3382 fluid ounce
1 litre (L)	=	0.2642 US liquid gallon
1 hectolitre (hL)	=	26.42 US liquid gallons
1 hectolitre	=	2.838 US bushels
1 litre	=	0.02838 US bushel
Bulk Density		
1 kilogram/hectolitre	=	0.7770 pound/US bushel
Grain Yield		
1 quintal/hectare	=	0.79 US cwt/acre
0.1235 tonne/hectare	=	1 bag/acre
6.73 tonnes/hectare	=	100 bushels/acre for wheat
1 tonne/hectare (t/ha)	=	7.92 bushels/acre for oats
1 tonne/hectare	=	11.89 bushels/acre for barley
1 tonne/hectare	=	13.87 bushels/acre for maize
1 tonne/hectare	=	13.87 bushels/acre for rye
1 tonne/hectare	=	14.86 bushels/acre for wheat
Temperature		
0°C (Celsius)	=	32°F (Fahrenheit)
20°C	=	68°F
100°C	=	212°F
Length		
1 centimetre (cm)	=	0.3937 inch (in)
1 metre (m)	=	3.2808 feet (ft)

TABLE A2.1 Conversion From Metric to US Units (*cont.*)

Metric units		US units
1 metre	=	1.094 yard (yd)
1 kilometre (km)	=	0.6213 mile
Area		
1 square centimetre (cm^2)	=	0.1550 square inch (in^2)
1 square metre (m^2)	=	10.76 square feet
1 square metre	=	1.196 square yard
1 square metre	=	0.0002471 acre (ac)
1 hectare (ha)	=	2.471 acres
1 hectare	=	0.00386 square mile
Power		
1 kilowatt (kW)	=	1.341 horsepower (hp)
Energy		
1 joule (J)	=	0.239 calorie (cal)
1 kilojoule (kJ)	=	0.945 British Thermal Unit (BTU)

Source: Adapted from Huang, W., Wrigley, C.W., 2016. Units of grain science and trade: Equivalence between US, Chinese and metric units. In: Wrigley, C., Corke, H., Seetharaman, K., Faubion, J. (Eds.), Encyclopedia of Food Grains, vol. 2, Elsevier Ltd, Oxford, UK, p. 308..

TABLE A2.2 Conversion From US to Metric Units

US units		Metric units
Mass		
1 ounce	=	28.35 grams
1 pound	=	0.4536 kg
1 US hundredweight (cwt) (100 lb)	=	45.36 kilograms
1 US short ton	=	0.9072 tonne
1 US long ton	=	1.016 tonne
1 bag of grain	=	50 kg
Volume		
1 cubic inch	=	16.38 cubic centimetres
1 fluid ounce	=	29.57 millilitres
1 US liquid gallon	=	3.785 litres

(*Continued*)

TABLE A2.2 Conversion From US to Metric Units (*cont.*)

US units		Metric units
1 US bushel	=	35.24 litres
1 quart (liquid)	=	0.9463 litre
Bulk Density		
1 pound/bushel	=	1.287 kg/hL
Grain Yield		
1 US cwt/acre	=	1.265 quintal/hectare
1 bushel/acre for wheat	=	6.73 kg/ha
1 bushel/acre for maize	=	7.21 kg/ha
1 bushel/acre for rye	=	7.21 kg/ha
1 bushel/acre for barley	=	8.41 kg/ha
1 bushel/acre for oats	=	12.62 kg/ha
Temperature		
32°F (Fahrenheit)	=	0°C (Celsius)
68°F	=	20°C
212°F	=	100°C
Length		
1 inch (in)	=	25.4 millimetre (mm)
1 foot (ft)	=	0.3048 metre
1 yard	=	0.9144 metre
1 mile	=	1.6093 kilometre
Area		
1 square inch	=	645.2 square millimetres
1 acre	=	0.4047 hectare
1 square mile (1 'section')	=	259.0 hectares
Power		
1 horsepower	=	0.7457 kilowatt
Energy		
1 calorie	=	4.187 joules
1 British Thermal Unit	=	1.055 kilojoule

Source: Adapted from Huang, W., Wrigley, C.W., 2016. Units of grain science and trade: Equivalence between US, Chinese and metric units. In: Wrigley, C., Corke, H., Seetharaman, K., Faubion, J. (Eds.), Encyclopedia of Food Grains, vol. 2, Elsevier Ltd, Oxford, UK, p. 308.

TABLE A2.3 Chinese Units Equivalent to Metric (SI) and US Units

Chinese units	Metric units	US units
Length		
1 hao (毫)	0.0333 millimetre (mm)	0.0013 inch (in)
1 li (厘)	0.3333 mm	0.0131 in
3.333 li (厘)	1 mm	0.0394 in
1 fen (分)	0.3333 centimetre (cm)	0.1312 in
3.3333 fen (分)	1 cm	0.3937 in
1 cun (寸)	3.3333 cm	1.3123 in
1 chi (尺)	0.3333 metre (m)	1.094 feet (ft)
3 chi (尺)	1 m	1.094 yard (yd)
1 zhang (丈)	3.3333 m	3.647 yd
1 li (里)	0.5 kilometre (km)	0.3107 mile
2 li (里)	1 km	0.6213 mile
Area		
1 square li (平方厘)	0.1111 square millimetre (mm^2)	0.0002 square inch (in^2)
9 square li (平方厘)	1 mm^2	0.0015 in^2
90 square hao (平方毫)	1 mm^2	0.0015 in^2
1 square fen (平方分)	0.1111 square centimetre (cm^2)	0.0172 in^2
9 square fen (平方分)	1 cm^2	0.1550 in^2
1 square cun (平方寸)	11.111 cm^2	1.722 in^2
1 square chi (平方尺)	0.1111 square metre (m^2)	1.195 square feet
9 square chi (平方尺)	1 m^2	1.196 square yard
1 square zhang (平方丈)	11.111 m^2	13.289 square yard
1 square li (平方里)	25 hectare (ha)	61.775 acre (ac)
0.04 square li (平方里)	1 ha	0.00386 square mile (1 "section")
1 li (厘)	6.666 m^2	7.973 square yard
1 fen (分)	66.666 m^2	79.733 square yard
1 mu (亩)	666.66 m^2	797.33 square yard
15 mu (亩)	1 ha	0.00386 square mile (1 "section")
1 qing (顷)	6.666 ha	0.0257 square mile (1 "section")

(Continued)

TABLE A2.3 Chinese Units Equivalent to Metric (SI) and US Units (*cont.*)

Chinese units	Metric units	US units
Volume		
1 cuo (撮)	1 cubic centimetre (cc)	0.0610 cubic inch
1 cuo (撮)	1 millilitre (mL)	0.3382 fluid ounce
1 shao (勺)	10 Ml	3.382 fluid ounce
1 he (合)	100 mL	33.82 fluid ounce
1 sheng (升)	1 liter (L)	0.2642 US gallon
1 dou (斗)	10 L	2.642 US gallon
1 shi (石)	1 hL	2.838 US bushels
Mass		
1 hao (毫)	5 milligram (mg)	0.0001765 ounce
0.2 hao (毫)	1 mg	0.0000353 ounce
1 li (厘)	50 mg	0.001765 ounce
1 fen (分)	500 mg	0.01765 ounce
1 qian (钱)	5 gram (g)	0.1765 ounce
0.2 qian (钱)	1 g	0.0353 ounce
1 liang (两)	50 g	1.765 ounce
1 jin (斤)	0.5 kilogram (kg)	1.1025 pounds (lb)
2 jin (斤)	1 kg	2.205 lb
1 dan (担)	50 kg	110.25 lb
1 dan (担)	0.5 quintal (q)	1.1025 US hundredweight (cwt)
2 dan (担)	1 q	2.205 cwt
1 dan (担)	0.05 tonne (t) (metric ton)	0.0551 US short ton
20 dan (担)	1 t	0.9843 US long ton

Source: Reproduced with permission from Huang, W., Wrigley, C.W., 2016. Units of grain science and trade: Equivalence between US, Chinese and metric units. In: Wrigley, C., Corke, H., Seetharaman, K., Faubion, J. (Eds.), Encyclopedia of Food Grains, vol. 2, Elsevier Ltd, Oxford, UK, p. 308.

A2.2 HISTORICAL CONFUSIONS WITH UNITS OF MEASUREMENT

The Imperial and metric systems have their origins in Britain and France of the 18th century. The familiar units of length in France, prior to the French Revolution of 1789, were the *pied-de-roi* (king's foot), *pouce* and *perche*. These corresponded to the British foot, inch and perch, respectively, but the lengths were different for the corresponding French and British units. Even within France, the *pied-de-roi* varied in length. The Paris *pied* was 11% longer than the Strasbourg *pied* and 10% shorter than the Bordeaux *pied*. In 1790, the French *Academie des Sciences* set about producing a unified system of weights and measures (Freeman, 1990). A pendulum system was initially proposed as the basis of the unit of length, as previously considered in collaboration with the Royal Society of Britain, whereby the *pendule* would be the length of a pendulum with a swing period of 1 s, leading to secondary units of *kilopendules*, *millipendules* and *cubic pendules* (Freeman, 1990).

Nevertheless, the *Academie des Sciences* decided upon the *metre* as one ten millionth part of one quarter of the earth's meridian. The original package of units also included a 10-h clock, a circle divided into 400 *grades* (still in limited use) and a Republican calendar. The Republican year started on September 22nd (the equinox) with 12 renamed months (e.g. *Pluviose* for January, *Floreal* for April, *Thermidor* for July), each month being divided into three 10-day *decades*, plus a 5-day annual holiday (*fête*) (6 days' holiday in a leap year) (Freeman, 1990). The package of changes met public resistance, plus some confusion, and in 1812 the government of the day proclaimed that the traditional system may be used in parallel with the new one. Nevertheless, many aspects of the new metric system survived and have gone on to acceptance in many countries, but not all.

The initiative of the French *Academie des Sciences* included an approach to the King of England for British collaboration in this process, even though France and England were officially at war at the time. The *Philosophical Transactions of the Royal Society* record the following (as quoted by Freeman, 1990, and unedited):

> *"Some curious Gentlemen both of the ROYAL SOCIETY of London and of the ROYAL ACADEMY of SCIENCES at Paris, thinking it might be of good Use, for better comparing together the Success of Experiments made in England and in France, proposed some time since, that accurate Standards of the Measures and Weights of both Nations, carefully examined, and made to agree with each other, might be laid up and presented in the Archives both of the ROYAL SOCIETY here, and of the ROYAL ACADEMY of SCIENCES at Paris. Which proposal having been received with the general Approbation of both those Bodies, they were therefore pleased to give the necessary Directions for the bringing the same into Effect."*

The wording suggests that there was cordial cooperation between these scientific bodies. However, the cooperation did not occur, and centuries of division on units of measurement followed. Perhaps this offer was not relayed to the English or perhaps it was rebutted at a high level. Nevertheless, French-based efforts continued to formulate a system of metric units.

A2.3 THE INTERNATIONAL SYSTEM OF UNITS (SI SYSTEM)

The SI system of metric units (*Systeme Internationale d'Unites*) has developed as a result of successive General Conferences on Weights and Measures, starting in 1889, and continuing about four-yearly to the present, with world-wide representation. At the 10th such conference (in 1954) seven base units were adopted: the metre, the kilogram, the second, the ampere, the Kelvin (for temperature), the mole and the candela (for luminous intensity).

All other metric units are derived from these. At the 11th such conference in 1960, the official symbol 'SI' was allocated for the *Système International d'Unités* as the 'modernised metric system' providing a practical system of units for unifying techniques for measurement used in industry, commerce, research and education (reviewed by Lentner, 1981). The 11th conference also confirmed prefixes (*pico-*, *nano-*, *micro-*, *mega-*, *giga-* and *tera-*) to indicate powers of ten as prefixes to the SI units. The expanded list of these is provided in Table A2.4.

A2.4 IMPERIAL UNITS IN TWO VERSIONS

If the Anglo-French cooperation of the 18th century had been successful a few centuries ago, it could have led to a more unified world today. The lack of uniformity of units was compounded by the spread of the earlier British Imperial System to North America where it remained in its original form, despite a subsequent British revision of Imperial units defined by the 'British Weights and Measures Act' of 1824 (the 'foot-pound-second' system).

However, continued use of the pre-1824 British system in the USA created the following differences between the British Imperial System of 1824 and the US Customary System (Adapted from Huang and Wrigley, 2016):

- One US liquid gallon is 3.785 litres, smaller than the British gallon, dry or liquid (4.546 litres), the volume of ten pounds (avoirdupois) of water at 62°F. British units of liquid capacity are thus about 20% larger than the corresponding American units.
- One US dry gallon is 4.405 litres, one-eighth of a US Winchester bushel.
- One US 'short ton' is 2,000 pounds (20 US hundredweight (cwt), 907.2 kilograms), a 'shorter' measure than the British 'long ton' of 2,240 pounds (1,016 kilograms), made up of 20 British hundredweight of 112 pounds each.
- The US hundredweight (45.36 kilograms) is one hundred pounds (as the name suggests), significantly less than the British hundredweight of 112 pounds (50.80 kilograms). This anomaly has been explained by the changes in mass incurred during the shipping of wool from Australia to Britain in the 19th century. Although bales of wool started as 100 pounds on the dry Australian sheep station, they gained moisture during the long ocean voyage in the wool Clippers to become 112 pounds on arrival in Britain. Evidently the British mill owners preferred to change the definition of the hundredweight than to pay for 12 pounds of water!

TABLE A2.4 SI Prefixes to Indicate the Sizes of SI Units, Based on Powers of Ten

Factor	Prefix	Symbol
10^{24}	yotta	Y
10^{21}	zetta	Z
10^{18}	exa	E
10^{15}	peta	P
10^{12}	tera	T
10^{9}	giga	G
10^{6}	mega	M
10^{3}	kilo	k
10^{2}	hecto	h
10^{1}	deca	da
10^{-1}	deci	d
10^{-2}	centi	c
10^{-3}	milli	m
10^{-6}	micro	μ
10^{-9}	nano	n
10^{-12}	pico	p
10^{-15}	femto	f
10^{-18}	atto	a
10^{-21}	zepto	z
10^{-24}	yocto	y

Source: Adapted from Huang, W., Wrigley, C.W., 2016. Units of grain science and trade: Equivalence between US, Chinese and metric units. In: Wrigley, C., Corke, H., Seetharaman, K., Faubion, J. (Eds.), Encyclopedia of Food Grains, vol. 2, Elsevier Ltd, Oxford, UK, p. 308.

- The word 'ton' (pronounced 'tun') used in the British and US systems is similar to the metric 'tonne' (pronounced 'ton', to rhyme with 'Don'), also known as a 'metric ton' (especially in the USA). However, the ton (in its various forms) and the tonne are slightly different quantitatively (see Tables A2.1 and A2.2). One US long ton is 1.016 tonne (metric ton) and one US short ton is 0.9072 tonne.
- A barrel (dry) is 105 US dry quarts (26.25 dry US gallons) or 36 British gallons (liquid and dry are the same). A barrel of flour, sometimes in US use, is 196 pounds. Otherwise, the mass of the US barrel obviously depends on the commodity being measured. One US liquid barrel is 31.5 US gallons.

A2.5 THE US BUSHEL—BOTH DRY VOLUME AND MASS

The bushel is a key measure for the grains industry, but it is also an outstanding example of ongoing confusion in units, especially in international trade. The US bushel (35.24 litres), a dry measure of volume, is 2,150 cubic inches, 8 US dry gallons, 4 pecks or 32 dry quarts. The US bushel differs from the British bushel (8 British gallons), which is 36.37 litres. The British bushel is thus about 3% larger in volume than the US bushel. The term 'Winchester bushel' originated in Britain as a result of the choice of the city of Winchester by King Alfred the Great as the capital of his kingdom of Wessex in the ninth century; the term 'Winchester measure' did not appear in British law until the 17th century. The 'Winchester bushel' crossed the Atlantic and stayed on after the term was abandoned by the British in 1824.

The US bushel is primarily a measure of volume, but it has also become a measure of mass (weight); however, its weight differs for each of several grain species (Tables A2.1 and A2.2). In the metric system, grain yield is measured in tonnes per hectare; that is, units of mass per unit of area. However, the US tradition has been to measure yield in bushels per acre; that is, units of volume (originally not mass) per unit of area. The use of volume measurement units for grain in the US makes it difficult to provide direct conversion between metric and US units for grain yield. To overcome this problem, the US Government has defined the bushel for the following grains in terms of mass, irrespective of the true test weight:

- 1 bushel of wheat = 60 pounds.
- 1 bushel of maize = 56 pounds.
- 1 bushel of oats = 32 pounds.
- 1 bushel of barley = 48 pounds.
- 1 bushel of rye = 56 pounds.

As a result, conversion factors for grain yields depend on the grain being considered. For wheat, 100 bushels/acre becomes 6,000 pounds per acre, equal to 2,723 kg per 0.4047 ha, or 6.73 tonnes per hectare (Tables A2.1 and A2.2). A rough 'rule-of-thumb' for wheat is to take the yield in lb/acre and increase it by 10% to obtain an approximation to kg/ha. Correspondingly, 1 tonne/hectare equates to 14.86 bushels/acre for wheat, 13.87 bushels/acre for maize, 7.92 bushels/acre for oats, 11.89 bushels/acre for barley and 13.87 bushels/acre for rye.

Farmers use yet another measure of grain yield, namely, 'bags to the acre'. However, this colloquial and informal estimate of grain yield is difficult to define. The British bag measures three British bushels (0.109 cubic metre), so a grain yield of one tonne per hectare is about 15 British bushels per acre and about 5 bags per acre. The US bag is 50 pounds.

- The British 'bag' is three bushels (0.109 cubic meters); it is not in general use in the USA, but it is used by farmers in former British colonies, such as Australia.

50 kg	=	1 'bag' of grain
0.1235 tonne/hectare	=	1 bag/acre

A2.6 CHINESE UNITS OF MEASUREMENT (ADAPTED WITH PERMISSION FROM HUANG AND WRIGLEY, 2016)

China has a very long history of grain growing, during which many different units of grain measurement have been used. In different dynasties and various places, the units of grain may be not uniform. But, with historical changes, most of the traditional units are no longer used. Now, in most cases, metric (SI) units are used in China. Nevertheless, some Chinese traditional units are still in use. It is thus valuable in trade to be aware of these units, which are listed in Table A2.3.

For length units, there are the hao (毫), li (厘), fen (分), cun (寸), chi (尺), zhang (丈) and li (里). For the area units, there are square hao (平方毫), square li (平方厘), square fen (平方分), square cun (平方寸), square chi (平方尺), square zhang (平方丈), and square li (平方里). For units of area, there are li (厘), fen (分), mu (亩) and qing (顷). For mass units, there are hao (毫), li (厘), fen (分), qian (钱), liang (两), jin (斤 and dan (担).

The unit liang (两) used now is not equal to what it was in the past. Now, one jin (斤) is equal to ten liang (两), but in the past one jin (斤) was equal to sixteen liang (两). For volume units, there are cuo (撮), shao (勺), he (合), sheng (升), dou (斗) and shi (石).

The relationships between Chinese units, Metric (SI) and US units are shown in Table A2.3.

A2.7 CONCLUSIONS

Despite the lack of general adoption of the metric system in the USA, its limited use there has also involved spellings changes that further confound the possibility of international standardisation. Specifically, the tonne, litre and metre of the SI Metric System have become the metric ton, the liter and the meter, respectively, in US spelling and terminology.

The US is officially on the metric system. All US units are defined in terms of SI metric units. The use of Customary US units continues to be permitted in trade, but they are not mandated. In many cases, including consumer foods, such as canned foods and packaged baking mixes, they are dual-labelled in US Customary Units as well as in their metric equivalent.

Until there is worldwide agreement on units and universal adoption of one system of weights and measures, we need to rely on conversion tables to permit reliable ‘translation’ from one system to another. Tables A2.1 and A2.2 provide for conversion between the two systems in most general use, namely, the SI metric system and US units.

Finally, to answer the trivia question in the Introduction section, a ‘barleycorn’ is an old British unit of length, equal to $^1/_3$ inch or 0.84667 cm. In the absence of a suitable measuring rule, grains of barley were placed end-to-end and were counted.

SOURCES OF FURTHER INFORMATION

- http://www.bipm.org/en/home/: Describes the International Bureau of Weights and Measures (BIPM) which has the mandate to provide the basis for a single, coherent system of measurements throughout the world, traceable to the International System of Units (SI). The headquarters of BIPM is near Paris, France.
- http://www.sizes.com/units/: Provides definitions of a wide diversity of units, graphical conversion charts and lists of symbols and acronyms for units.

REFERENCES

Freeman, J.H., 1990. The metre and the pendulum. Nature 348, 105–106.

Lentner, C. (Ed.), 1981. Geigy Scientific Tables, vol. 1, Ciba-Geigy Ltd, Basle, Switzerland.

Huang, W., Wrigley, C.W., 2016. Units of grain science and trade: equivalence between US, Chinese and metric units. Wrigley, C., Corke, H., Seetharaman, K., Faubion, J. (Eds.), Encyclopedia of Food Grains, vol. 2, Elsevier Ltd, Oxford, UK, p. 308.

Index

B

G

H

I

K

L

M

N

X

Y

Z